Computational Methods in Lanthanide and Actinide Chemistry

Computational Methods in Lanthanide and Actinide Chemistry

Edited by

MICHAEL DOLG

Institute for Theoretical Chemistry,
University of Cologne, Germany

WILEY

This edition first published 2015

Registered office
John Wiley & Sons Ltd, The Atrium, Southern Gate, Chichester, West Sussex, PO19 8SQ, United Kingdom

For details of our global editorial offices, for customer services and for information about how to apply for permission to reuse the copyright material in this book please see our website at www.wiley.com.

Library of Congress Cataloging-in-Publication Data
Dolg, Michael, 1958-
Computational methods in lanthanide and actinide chemistry / Michael Dolg.
pages cm
Includes bibliographical references and index.
ISBN 978-1-118-68831-1 (cloth)
1. Rare earth metals. 2. Actinide elements. 3. Chemistry, Inorganic. I. Title.
QD172.R2D65 2015
546′.41–dc23

2014040220

A catalogue record for this book is available from the British Library.

Set in 10/12pt Times LT Std by SPi Publisher Services, Pondicherry, India
Printed and bound in Malaysia by Vivar Printing Sdn Bhd

1 2015

Contents

Contributors

Raymond Atta-Fynn, Department of Physics, University of Texas, Arlington, USA

Jochen Autschbach, Department of Chemistry, University at Buffalo, State University of New York, USA

Zoila Barandiarán, Department of Chemistry, Autonomous University of Madrid, Spain.

Enrique R. Batista, Theoretical Division, Los Alamos National Laboratory, USA

Donald R. Beck, Department of Physics, Michigan Technological University, USA

Eric J. Bylaska, Pacific Northwest National Laboratory, USA

Xiaoyan Cao, Theoretical Chemistry, University of Cologne, Germany

Ludovic Castro, Laboratory of Physics and Chemistry of Nano-Objects, National Institute of Applied Sciences, France

Carine Clavaguéra, Laboratoire de chimie moléculaire, Départment de chimie, École Polytechnique CNRS, France

Wibe A. de Jong, Lawrence Berkeley National Laboratory, USA

Jean-Pierre Dognon, Laboratoire de Chimie Moléculaire et de Catalyse pour l'énergie, CEA/Saclay, France

Michael Dolg, Institute for Theoretical Chemistry, University of Cologne, Germany

Kenneth G. Dyall, Dirac Solutions, USA

Ephraim Eliav, School of Chemistry, Tel Aviv University, Israel

Niranjan Govind, Pacific Northwest National Laboratory, USA

Uzi Kaldor, School of Chemistry, Tel Aviv University, Israel

Christos E. Kefalidis, Laboratory of Physics and Chemistry of Nano-Objects, National Institute of Applied Sciences, France

Andrew Kerridge, Department of Chemistry, Lancaster University, UK

Laurent Maron, Laboratory of Physics and Chemistry of Nano-Objects, National Institute of Applied Sciences, Toulouse, France

Richard L. Martin, Theoretical Division, Los Alamos National Laboratory, USA

Hiroko Moriyama, Graduate School of Natural Sciences, Nagoya City University, Japan

Steven M. O'Malley, Atmospheric and Environmental Research, USA

Lin Pan, Physics Department, Cedarville University, USA

Lionel Perrin, Laboratory of Physics and Chemistry of Nano-Objects, National Institute of Applied Sciences, France

Kirk A. Peterson, Department of Chemistry, Washington State University, USA

Pekka Pyykkö, Department of Chemistry, University of Helsinki, Finland

Florent Réal, Université Lille CNRS, Laboratoire PhLAM, CNRS UMR 8523, France

Trond Saue, Laboratoire de Chimie et Physique Quantiques, Université Paul Sabatier (Toulouse III), France

Bernd Schimmelpfennig, Institute for Nuclear Waste Disposal (INE), Karlsruhe Institute of Technology (KIT), Germany

Luis Seijo, Department of Chemistry, Autonomous University of Madrid, Spain.

André Severo Pereira Gomes, Université Lille CNRS, Laboratoire PhLAM, CNRS UMR 8523, France

Lidia Smentek, Department of Chemistry, Vanderbilt University, USA

Hiroshi Tatewaki, Graduate School of Natural Sciences, Nagoya City University, Japan

Valérie Vallet, Université Lille CNRS, Laboratoire PhLAM, CNRS UMR 8523, France

Lucas Visscher, Theoretical Chemistry, Amsterdam Center for Multiscale Modeling, VU University Amsterdam, The Netherlands

Ulf Wahlgren, Department of Physics, Stockholm University, AlbaNova University Centre, Sweden

John W. Weare, Department of Chemistry and Biochemistry, University of California San Diego, USA

Anna Weigand, Theoretical Chemistry, University of Cologne, Germany

Florian Weigend, Karlsruhe Institute of Technology, Germany

Ahmed Yahia, Laboratory of Physics and Chemistry of Nano-Objects, National Institute of Applied Sciences, France

Shigeyoshi Yamamoto, School of International Liberal Studies, Chukyo University, Japan

Ping Yang, Environmental Molecular Sciences Laboratory, Pacific Northwest National Laboratory (PNNL), USA

Preface

Lanthanides and actinides comprise about one-quarter of the known chemical elements collected in the periodic table. Because of their complex electronic structure, the significant electron correlation effects, and the large relativistic contributions, the f-block elements are probably the most challenging group of elements for electronic structure theory. In 1987 Pyykkö reviewed the available relativistic electronic structure calculations for f-element molecules (Inorganica Chimica Acta 139, 243–245, 1987). Of the 59 listed studies, 53 dealt with actinides and only 10 with lanthanides. The applied computational methods comprised *ab initio* Dirac-Hartree-Fock one-center expansion and Dirac-Hartree-Fock-Slater calculations, quasirelativistic all-electron X_α-studies, and semiempirical valence-only approaches like relativistic extended Hückel theory. None of these studies took into account static electron correlation explicitly using a multi-configurational wavefunction or included at an *ab initio* level the effects of dynamic electron correlation. No applications of modern density functional theory to f-element molecules were reported either. The treatment of relativity included the Dirac one-particle relativity in a few cases explicitly, but mostly in some approximate form, whereas corrections due to the Breit two-particle interaction or arising from quantum electrodynamics were entirely neglected. Relativistic effective core potentials were only available for a few actinides, which certainly also hampered a routine exploration of lanthanide and actinide chemistry with quantum chemical approaches.

Tremendous progress was made in dealing with lanthanide and actinide systems since the 1987 review of Pyykkö appeared, and the field continues to develop quickly. The current book aims to provide the reader an overview of those state-of-the-art electronic structure theory approaches that have been successfully used for f-element systems so far and summarizes examples of their application. The 16 chapters were written by leading experts involved in the development of these methods as well as their application to various aspects of f-element chemistry. From the results of several studies discussed in these contributions it becomes apparent that quantum chemists successfully conquered the field of lanthanide and actinide chemistry and can provide very valuable contributions not merely supplementing experimental studies, but also frequently guiding their setup and explaining their outcome. Moreover, with largely improved theoretical methods and computational resources at hand, it also became possible to obtain new insights with respect to the interpretation of the electronic structure of f-element compounds.

Despite these many encouraging developments, it is appropriate to say that when it comes to lanthanides and actinides modern electronic structure theory currently can accomplish many things, but certainly not all. It is also clear that this book can only provide a snapshot of the current state of affairs. A number of promising computational approaches, e.g., local electron correlation schemes or F12-dependent wavefunctions, are currently developed and already successfully applied to non-f-element systems. They will during the next years most likely significantly extend the array of available methods for quantum chemical studies of lanthanides and actinides. Thus more exciting developments can be anticipated.

Finally I would like to thank all authors of the chapters for their excellent contributions. My thanks also go to the staff at Wiley, i.e., Sarah Higginbotham, Sarah Keegan, and Rebecca Ralf, for their guidance and support during this book project. Last but not least, I'm grateful to Mrs. Peggy Hazelwood for copy-editing and to Mr. Yassar Arafat at SPi for final handling of the proofs.

Michael Dolg
Cologne
May 2014

1

Relativistic Configuration Interaction Calculations for Lanthanide and Actinide Anions

Donald R. Beck[1,*], *Steven M. O'Malley*[2] *and Lin Pan*[3]

[1]*Department of Physics, Michigan Technological University*
[2]*Atmospheric and Environmental Research*
[3]*Physics Department, Cedarville University*

1.1 Introduction

Lanthanide and actinide atoms and ions are of considerable technological importance. In condensed matter, they may be centers of lasing activity, or act as high temperature superconductors. Because the f-electrons remain quite localized in going from the atomic to the condensed state, a lot of knowledge gained from atoms is transferable to the condensed state. As atoms, they are constituents of high intensity lamps, may provide good candidates for parity non-conservation studies, and provide possible anti-proton laser cooling using bound-to-bound transitions in anions such as La^- [1].

In this chapter we will concentrate on our anion work [2–4], which has identified 114 bound states in the lanthanides and 41 bound states in the actinides, over half of which are new predictions. In two anions, Ce^- and La^-, bound opposite parity states were found, making a total of 3 [Os^- was previously known]. Bound-to-bound transitions have been observed in Ce^- [5] and may have been observed in La^- [6]. We have also worked on many properties of lanthanide and actinide atoms and positive ions. A complete list of publications can be found elsewhere [7].

*Corresponding author. E-mail: donald@mtu.edu

Computational Methods in Lanthanide and Actinide Chemistry, First Edition. Edited by Michael Dolg.

1.2 Bound Rare Earth Anion States

In 1994, we began our first calculations on the electron affinities of the rare earths [8]. These are the most difficult atoms to treat, due to the open *f*-subshells, followed by the transition metal atoms with their open *d*-subshells. At that time, some accelerator mass spectrometry (AMS) measurements of the lanthanides existed [9, 10] which were rough. Larger values might be due to multiple bound states, states were uncharacterized as to dominant configuration, etc.

Local density calculations done in the 1980s had suggested anions were formed by *4f* attachments to the incomplete *4f* subshell. Pioneering computational work done by Vosko [11] in the early 1990s on the seemingly simple Lu^- and La^- anions using a combination of Dirac-Fock and local density results suggested instead that the attachment process in forming the anions involved *p*, not *f*, electrons.

Our 1994 calculation on a possible Tm anion was consistent with this, in that it showed *4f* attachment was not a viable attachment process. Our calculations are done using a Relativistic Configuration Interaction (RCI) methodology [12], which does a Dirac-Hartree-Fock (DHF) calculation [13] for the reference function (s) (dominant configurations). The important correlation configurations (e.g., single and pair valence excitations from the reference configuration[s]) are then added in, using the DHF radials and relativistic screened hydrogenic function (called virtuals), whose effective charge (Z*) is found by minimization of the energy matrix, to which the Breit contributions may be added, if desired.

Experience gained in the mid-1990s suggested that good candidates for bound anion states might be found by combining observed ground and excited state neutral spectra with the computational knowledge that closing an *s*-subshell might lower the energy ~1.0 eV or adding a *6p*-electron to a neutral atom state (*7p* in the actinides) might lower the energy ~0.25 eV. The variety of energetically low-lying configurations in the observed spectrum of La and Ce suggests a potential for a large number of bound anion states, which has now been computationally confirmed.

As an example of the process, a $Tm^-\,4f^{14}6s^2$ anion state might be bound if there were a $4f^{14}6s^1$ state observed in the neutral atom that was less than 1 eV above the ground state. The use of excited states with s/p attachment also has the computationally attractive feature that it avoids, to a good level of approximation, having to compute correlation effects for *d* and/or *f* electrons. An *s* attachment to an excited state can be equivalent to a *d* attachment to the ground state. The angular momentum expansions for such pair excitations converge slowly, and a lot of energy is associated with (nearly) closed *d* and/or *f* subshells. Clearly, it is best to reduce such problems if usable experimental results exist.

It has always been our position to use no more than moderately size *d* wavefunction expansions. Current limits are about 20,000 symmetry adapted wavefunctions built from fewer than 1 million Slater determinants, and use of two virtuals per *l*, per shell (n). This allows the "physics" (systematics) to be more visible and reduces the need for "large" computational resources that were frequently unavailable in the "old" days. Development of systematic "rules" is one of the main goals of our research. Some examples follow: (i) determining which correlation effects are most important for a specific property [14, 15], (ii) near conservation of f-value sums for nearly degenerate states [15, 16], (iii) similar conservation of g-value sums [16], (iv) similar conservation of magnetic dipole hyperfine constants [17, 18]. This approach does mean near maximal use of symmetry, creating extra auxiliary computer codes, and increases the necessity of automating data preparation and file manipulation. Much stricter development of this automation is one of the two factors that reduced calculation of the entire actinide row to less than the time it used to take to complete the calculation for one anion (>4 months for Nd^-). Use of moderately sized wavefunctions also requires careful selection of which property-specific configurations to include and careful optimization of the virtual radial functions.

1.3 Lanthanide and Actinide Anion Survey

In 2008 and 2009 our group presented a series of three papers [2–4] representing an unprecedented and complete survey of the bound lanthanide and actinide anion states predicted by valence level RCI calculations. The first of these [2] was a study of all *6p* attachments to $4f^n\, 6s^2$ ground and excited states of the lanthanide neutral spectra (then and throughout the discussion here we use n as an occupancy of N-2 where N is the total number of valence electrons in the neutral atom configuration, including the core-like *4f*/*5f* subshells). The second paper [3] completed the lanthanide survey with *6p* attachments to $4f^m\, 5d\, 6s^2$ thresholds and *6s* attachments to $4f^m\, 5d^2\, 6s$ thresholds ($m \equiv N-3$). The final paper in the series [4] included the equivalent *7s* and *7p* attachments to corresponding actinide neutral thresholds as well as additional states in Th^- and Pa^- representing *7p* attachments to $5f^q\, 6d^2\, 7s^2$ thresholds ($q \equiv N-4$). The approach used to handle the complexity of these calculations represented the culmination of over three decades of group experience in developing techniques and computational tools for RCI basis set construction. The path that led to this comprehensive lanthanide and actinide anion survey was somewhat circuitous and developed originally through adjustments to increasing difficulties with each step toward more complex systems. In the following subsections we discuss some milestones leading up to the survey, the computational issues and solutions, the improved analytical tools that were needed, and a summary of results of the survey.

1.3.1 Prior Results and Motivation for the Survey

Throughout the 1990s and early 2000s our group had been steadily pushing our methodology towards more and more complex atomic systems. The ability to do so was partly from techniques described in Section 1.3.2 but also largely due to ever-increasing computer power. Mid-row transition metal studies had become fairly routine, e.g., binding energies of Ru^- [19], Os^- [20], and Tc^- [21]. However, the added complexity of a near-half-full f subshell over that of a d had relegated us for the most part to the outer edges of the lanthanide and actinide rows, e.g., Ce^- [22, 23], Th^- [24], Pr^- [25], U^- [26], Pa^- [27], La^- [28], and Lu^- [29]. During the mid- to late-1990s, we were twice enticed by the unique case of Tb to attempt to skip to the center of the lanthanide row [30]. The Tb ground state is $4f^9\, 6s^2$, but the low-lying first excited state (~35 meV [31]) is of the opposite parity $4f^8\, 5d\, 6s^2$ configuration, and the possibility of opposite parity Tb^- bound anion states resulting from the same *6p* attachment mechanism was a tempting prize. Unfortunately, those initial attempts at this mid-row anion were premature and Tb^- would have to wait to use basis set construction techniques that we eventually developed in the mid-2000s.

As we were gradually working our way inwards from the ends of the lanthanide and actinide rows, our papers began to take on a back-and-forth dialog with the work of the experimental atomic physics group at University of Nevada, Reno (Thompson and co-workers) usually within our accessible range of elements, e.g., La^- [32], Lu^- [33], Tm^- [34], Ce^- [35], Pr^- [36], but occasionally outside our purview in the mid-row, e.g., Eu^- [37] (although this case is a much more approachable problem than it may appear at first glance since the $4f^7$ electron subgroup can be effectively restricted to ^8S and ^6P terms [3]). This observation coupled with a similar result in our work on Eu^- auto-detachment lifetimes in metastable states [38] helped lead to the realization that angular momentum couplings of the $4f^n$ subshell could be considerably restricted with little accuracy loss, with great gains in computational efficiency (see Section 1.3.2.2). The Reno group employed the laser photodetachment electron spectroscopy (LPES) technique, which relies upon identification of peaks in the kinetic energy spectrum of detached photoelectrons, and their electron affinities were often considerably higher than contemporary AMS values [39]. Of particular interest to us were the many features in their published Ce^- spectrum [35], and a brief overview of the history of the electron affinity measurements and computations for Ce^- will perhaps set the stage to better illustrate the wide gap in predicted/measured values between our two groups and the importance of their eventual reconciliation.

In the early 1990s, AMS experiments such as Garwan *et al.* [9] had detected Ce^-, and high yields seemed to indicate a large electron affinity (>0.6 eV) or several moderately bound states. At the time, density functional theory computations of Vosko *et al.* [11] were suggesting that Ce^-'s (and other lanthanides') bound states would be formed by *6p* and possibly *5d* attachments rather than *4f* as previously assumed. Our earliest Ce^- calculations in 1994 studied both of these attachments to the $4f\ 5d\ 6s^2$ ground state configuration and predicted a $4f\ 5d\ 6s^2\ 6p$ Ce^- anion ground state with electron affinity of 259 meV. These calculations also predicted four additional bound even states and an odd $4f\ 5d^2\ 6s^2$ state with a binding energy of 178 meV. In 1997 Berkovitz *et al.* used a combination of laser excitation and AMS techniques to obtain a measured electron affinity of 700(10) meV. In the midst of our renewed push towards the center of the lanthanide and actinide rows, we revisited our RCI Ce^- computations in 2000 [23]. A recent re-dimensioning of our main RCI code [12] allowed us to now include extensive second-order effects. These took the form of relaxed *j* restrictions on electron subgroups and configurations representing triple/quadruple replacements with respect to the configurations of interest; no direct impact on their energies, only indirect through lowering of nearby single/double replacement (first-order) configurations. These results flipped the predicted RCI ground state to $4f\ 5d^2\ 6s^2$ with an electron affinity increased to 428 meV with an additional six odd bound excited states and eight even $4f\ 5d\ 6s^2\ 6p$ bound states with largest binding energy of 349 meV. Just when the gap between experiment and theory seemed to be improving, Davis and Thompson (Reno group) reported their 2002 measurements [35] with an LPES electron affinity of 955(26) meV and at least two excited states at 921(25) and 819(27) meV. By 2004, Cao and Dolg [40] had produced their own computational results using the relativistic energy-consistent small core pseudo-potential methodology that predicted the same $4f\ 5d^2\ 6s^2$ ground state configuration with an electron affinity of 530 meV.

In 2006, we again revisited Ce^- with the RCI methodology [41]. This time we were able to include some limited core-valence correlation involving the *5s* and *5p* subshells and the careful use of a high J excited neutral threshold to define the relative positions of the anion manifolds with respect to the neutral ground state configuration. (We will see in Section 1.3.3 that the latter introduced its own difficulties.) The result was an increase in the RCI electron affinity to 511 meV in good agreement with the recent Cao and Dolg value, but this latest RCI study also resulted in an astounding 26 bound anion states for both parities. Despite these advancements, there seemed to be no missing elements in our bases that could possibly account for the near doubling of the two computational results to approach the experimental values of Davis and Thompson.

Discussions with Thompson [42] suggested that the remaining difference of ~450 meV might be a misinterpretation of the most prominent feature of their photoelectron kinetic energy spectrum when they used it as the anion ground state to neutral ground state transition. In fact, some crude initial photodetachment calculations suggested that one should expect the most prominent feature to be from the $4f\ 5d^2\ 6s^2$ ground state configuration detaching via $6s$ to εp leaving the neutral atom in a $4f\ 5d^2\ 6s$ excited state. Based on this idea, the experimental analysis would be off by ~350 meV, but the two sets of computations were also still too low by >100 meV. We performed an ambitious set of cross section computations for all our bound anion states to all neutral thresholds within the range of the 2.410 eV experimental incident photon energy [35] using a frozen core approximation. By using the experimental photon energy, we were able to simulate an expected spectrum by adding together Gaussian peaks at each transition energy scaled by the appropriate relative cross sections and a few different effective anion temperature distributions. The Gaussian width was arbitrarily adjusted until the scaled simulated spectrum produced a best qualitative fit to the experimental data (i.e., subjective "by eye" comparison), and alignment of the prominent peaks then produced an effective electron affinity of 660 meV. The re-analysis was met with enthusiasm from the experimenters [43], and additional experimental results from Walter *et al.* [44] using tunable laser photodetachment threshold spectroscopy with incident photon energies in the range of 610-750 meV produced an electron affinity of ~650 meV. We thus regarded this Ce^- case as an excellent example of theory and experiment working together toward a common goal of understanding a particularly complex atomic anion. Section 1.4 discusses more recent work on Ce^- photodetachment cross section.

Having made the case for the usefulness of simulated spectra in analysis of photodetachment studies, we endeavored to take the next step in other systems, the intention being to provide additional computational details beyond simple binding energy numbers. Our Nd^- work from early 2008 [45] was this next step. It was successful in the sense that much like Ce^-, detachment to higher excited state configurations was found to be much more likely, provided an experimenter attempted an LPES study with the same 2.410 eV incident photon energy that had been used by Davis and Thompson [35]. In this case $4f^4\ 6s^2\ 6p$ anion states were found to detach with much higher cross section via 6s to εp leaving the neutral atom in excited $4f^4\ 6s\ 6p$ states rather than a $6p$ detachment back to the neutral $4f^4\ 6s^2$ ground state configuration (primarily due to resonances arising from coincidence with the incident energy and the excited threshold positions). Of course, more practically the real effect of this result was to warn an experimenter to instead choose a lower photon energy to avoid these $4f^4\ 6s\ 6p$ thresholds altogether.

In the long run, the most important products of this Nd^- study were the tools that were developed to help deal with basis set construction and *jls* restrictions placed on the $4f^4$ electron subgroup. We had reached that limit where necessity had forced us to invent a new approach in order to move further into the row. Having gone through the pains of developing a new technique to piece together multi-electron basis members (see Section 1.3.2), it wasn't until we began to select the next project that we realized just how powerful was the approach that had been applied to Nd^-. The analysis of $4f^{n,m}$ subgroup composition could be left to fairly straightforward, manageable computations (few configurations) on each neutral atom, and the compositional analysis and basis selection for the remaining portion of the anion and neutral states could be performed on a system near the end of a row and systematically reapplied to all the elements of the row, including near the center where a full un-trimmed basis set would be orders of magnitude larger than our 20k limit. These techniques lent themselves particularly well to binding energy studies where analysis need only be done on basis set composition and energy contribution, as opposed to including additional criterion based on contributions to hyperfine structure, transition probabilities, or photodetachment cross sections. Because the cross section computations had by far taken up the bulk of CPU and human time in the Ce^- and Nd^- projects, it was not too difficult a decision to temporarily forgo that avenue of inquiry in favor of returning to pure binding energy studies, particularly now that doing so meant the opportunity to study whole rows of heavy elements in the 6-8 months we had typically spent on one anion.

1.3.2 Techniques for Basis Set Construction and Analysis

The robustness of the RCI methodology lies primarily in the flexibility of basis set construction. The process begins with the generation of 1-electron basis functions by solving the multi-configurational Dirac-Fock (MCDF) equations using Desclaux's program [13]. In order to do Dirac-Fock calculations on a reference manifold for rare earths (and transition metals), the number of R^k integrals allowed was increased to 500K, the integrals were stored on disk, and the number of configuration state vectors (called "parents") was increased to 1000. The structure input is produced by our RCI code [12]. To extrapolate from fractional Z (> N+1) down to N+1, so necessary for anions, Z was converted from an integer to a real variable. At each stage in the basis set construction we are free to select or omit individual functions as needed while considering the delicate balance of reduction in basis size *vs* shift in energy positions (typically we aim for losses of less than a few meV if possible). The following sections describe the basis set construction and analytical tools in more detail.

1.3.2.1 Simple Omission of "Small" Basis Members

At any particular stage in the development of an RCI basis set, one is typically adding correlation in one of several potential forms. It is important to saturate the 1-electron virtual basis typically up to two or three virtuals of each symmetry, so one could be adding, say vp'' when vp and vp' are already present. One could also be extending the virtuals to higher l, e.g., correlation of the form *5d 6p* to *vf vg* has been found to be significant, so the equivalent *5d 6p* to *vg vh* is tried. Often one includes second order effects after determining the importance of first order correlation, e.g., $6s^2$ to vp^2 replacements from a $5d^x\ 6s^2$ configuration is typically a large contributor to correlation, and adding the triple replacement $5d\ 6s^2$ to $vp^2\ vd$ will have no direct impact on the zeroth order configuration but may lower the double replacement and increase its contribution. Finally, if the system is not too complex (open valence electrons or close to full subshells), one may have room to add some core-valence effects, e.g., *5p 6s* to *vp vd* in a transition metal $5d^x\ 6s^2$ calculation.

Regardless of the "trial" that is being performed, one would typically perform an analysis of energy contributions using intermediate normalization [46] and RCI coefficients of individual basis members and make decisions on how much to trim the new basis set before moving on to the next step. For a particular J calculation this requires careful removal of basis members with little contribution in all levels of interest. When considering more than a few levels, the effectiveness of simple removal of basis functions can lose much of its usefulness as the set of functions small in all levels of interest is much smaller than the set of small contributors for any particular level. If one is concerned about using the wavefunction to compute some other property, impact on that calculation needs to be considered as well, e.g., transition probabilities to several potential initial or final states of other J-parity calculations.

Although useful, even on a simple system, this approach can only trim an RCI basis size by a factor of ~2–3 at most. For the lanthanide and actinide survey we are discussing here, we needed an order of magnitude or more for the mid-row anions to make the calculations feasible.

1.3.2.2 Applying jls Restrictions to $4f^{n,m}$ Subgroups

One of the insights into our early Tb^- calculations had been the realization that the $4f^9\ 6s^2\ 6p$ calculation had negligible mixing of any basis functions that did not restrict the $4f^9$ group of electrons to j = 15/2. With hindsight this is now obvious given that it was an attachment to the $4f^9\ 6s^2$ J = 15/2 neutral ground state. In our work on Nd^- we did LS analysis of the anion vs. neutral $4f^4$ composition. Although the RCI method is purely relativistic, we can approximate LS states or LS terms on subgroups by straightforward rotation of the *jj* basis set based on an approximate diagonalization of L^2+S^2 applied to the spinors' major components. The rotation does not affect the completeness of the basis. For Nd^- we developed an auxiliary code that allowed us to make such a rotation of the *jj* basis of the MCDF portion of an RCI calculation's wavefunction after the fact in order to obtain both *jj* and *ls* analyses. It was then that we began to realize the strong connection between $4f^4$ composition between anion and neutral states.

At the time our RCI program [12] had two optional approaches to generate basis functions within the program. The first was a full-configuration option that was typically applied to a single relativistic configuration (e.g., $4f^2_{5/2}\ 4f^2_{7/2}\ 6s^2\ 6p^2_{3/2}$) but could also include all relativistic configurations that made up a non-relativistic one (e.g., $4f^4\ 6s^2\ 6p$) to allow this pseudo-rotation to an LS basis on the full configuration or *ls* terms within a subgroup. To use the latter option one necessarily complicates the basis set because all basis functions now contain mixing of all the determinants instead of just the ones from individual relativistic configurations (which is why one would want to do the LS analysis after the fact as mentioned above).

The second approach split a non-relativistic configuration into two parts, created determinant information on the two groups, and then pasted them together. This necessarily requires a range of *js* for each piece, and a pasting together based on total J of the calculation and combinations using standard angular momentum vector coupling theory. For Nd^- the natural split is to separate $4f^4$ from the remaining 3-electron group of each correlation configuration.

The first option happened to be more limited in terms of the complexity of the size of the configuration it could handle. The second could handle a ten fold larger set of basis functions, but it didn't support the LS rotation option. Since our analysis was suggesting that an overall set of jls restrictions on $4f^4$ would give us the edge we needed to allow a reasonable amount of valence correlation within our 20k limit, we extracted the method with the LS-rotation option to an external program where it could be re-dimensioned without adversely affecting other aspects of the RCI program. This auxiliary code [47] allows output to be read seamlessly. Even with this improvement, some of the more complicated correlation functions were running into memory limits and taking a day or more to complete.

As we began to think about next steps towards more complex anions, we realized that this new approach needed further improvement. It made no sense to keep re-creating the same $4f^{n,m}$ *ls* terms for each configuration. The process could also benefit from automation since files were being generated by hand and were vulnerable to typos. Ultimately, we ended up creating a database of partial configuration files. Using the program just described, we generated $4f^{n,m}$ and remaining 3- and 4-electron groups for this survey separately as if they were full configurations.

A second auxiliary code was then developed that would take a list of files for each part and, given the range of *js* for each piece and desired total J, piece together complete configurations with *jls* restrictions on the $4f^{n,m}$ part and relativistic *jj* bases on the remaining part. Further scripting and analysis was developed to scan an RCI program input file, recognize which external files would need to be available, and auto-generate a script to piece them together. This hybrid approach was an order of magnitude faster in CPU time and the human time and saved likely several months of tedious file editing.

1.3.2.3 Improvements of 1-Electron Bases in p Attachments

The MCDF calculation of our 1-electron bases includes an energy minimization of a specified level within each J-parity combination [13]. In order to generate a solution for anion states, one typically needs to step the nuclear charge down from some fractional value slightly greater than the actual Z of the anion, reusing the solutions from each previous step as an initial estimate. In the case of a $p_{1/2}$ attachment the singly occupied $p_{1/2}$ subshell of the final iteration in the process (the one with the correct integer Z), will be quite diffuse, but not overly so compared to the other d and s valence subshells ($\sim$1 a.u. beyond the *s*). However, the $p_{3/2}$ subshell generated by this optimized calculation will typically have drifted to a much more diffuse $\langle r \rangle$ with a difference of $>$1 a.u. compared to $\sim$0.1 a.u. typical of pairs of relativistic subshells in neutral or multiply-occupied anion subshells. The opposite is true for an optimization to a $p_{3/2}$ attachment.

This disparity of the $\langle r \rangle$ of the outer DF *p* subshells became problematic when we began to add correlation involving *p* virtuals. If the MCDF radials were optimized to a $p_{1/2}$ level, we would see large correlation contributions from the single replacement *p* to $vp_{3/2}$ configuration in a $p_{3/2}$ attachment, indicating that energy minimization of our *vp* orbitals was selecting Z* to act as a correction to the poorly optimized DF $p_{3/2}$ radial. The disparity in DF $\langle r \rangle$ propagated through the virtual *ps* as the $j = 1/2$ radials would have much higher Z* than the corresponding $j = 3/2$ radial.

Our solution to this issue was to allow mixed 1-electron radial bases with outer DF *p* radials each taken from a computation optimized to a corresponding attachment, avoiding either of the diffuse orbitals that had "drifted away." This works because as the RCI program reads in the 1-electron radial functions it performs an additional Gram-Schmidt orthogonalization outwardly through all functions of each symmetry. Using these mixed radial bases allows the virtual *ps* to then properly optimize important double replacement configurations representing *s p* to *vs vp* + *vp vd*. The result was useful for basis set construction because we were then able to stop at two sets of virtuals, i.e., *vp* and *vp*′.

1.3.2.4 Notation Conventions for Analysis

In dealing with the large number of bound states reported in this survey [2–4] it was useful to develop some shorthand notations when describing composition of anion levels. In addition to the typical total LS breakdown within the dominant MCDF configurations (with total J indicated on the leading term only for simplicity), we included composition within an alternate *jj* basis selected to reveal the mixing of attachment mechanism within each state. For *p* attachments we denoted a $p_{1/2}$ attachment term as "(*j*)," where *j* is the value for the neutral $f^n\ s^2$ or $f^m\ ds^2$ portion of the anion wavefunction. The $p_{3/2}$ attachments were distinguished by use of braces instead of parentheses: "{*j*}." By rounding the contributions to integer percent we were able to convey a great deal of information on level composition within fairly concise tables. For example, the Pr^- ground state was simply described by two sets of compositions [2]: "$^5K_5$ 77, ^3I 21, ^5I 1, ^3H 1" and "(9/2) 99, [9/2] 1."

When it came to *s* attachments to excited $f^m\ d^2\ s$ excited neutral thresholds, it was not possible to split the s^2 subgroup in the anion to separate out the neutral portion of the wavefunction. Instead, we approached the *s*-attachment analysis by providing similar *jj* composition within our neutral wavefunctions, minus the *s* electron, i.e., on the $f^m\ d^2$ subgroup. To distinguish from the p attachments we here used angled brackets in our notation: "$\langle j \rangle$". For example, in Gd^- the two lowest $4f^7\ 5d^2\ 6s$ levels, $^{11}F_2$ and $^{11}F_3$, were described by their $4f^5\ d^2$ composition as "$\langle 3/2 \rangle$ 88, $\langle 5/2 \rangle$ 12" and "$\langle 5/2 \rangle$ 86, $\langle 7/2 \rangle$ 14," respectively. By investigating this composition, we thus showed that the two lowest $4f^7\ 5d^2\ 6s^2$ levels, $^{10}F_{3/2}$ and $^{10}F_{5/2}$, can be characterized primarily as *6s* attachments to the J = 2 and J = 3 thresholds, respectively.

1.3.3 Discussion of Results

In developing our basis sets for this survey, the leap to tackling entire rows of the periodic table at once was predicated by the profound realization that the composition of bound anion states' $f^{n,m}$ electron subgroups remained consistent with those of the corresponding low-lying neutral thresholds to which we were attaching. In transition metals we used to treat d attachments to $d^x s^2$ neutral ground states as *s* attachments to excited $d^{x+1} s$ levels to avoid needing to include the large disruption of perhaps multiple eV of correlation energy in d^2 pair replacements. (These should have small anion-to-neutral differences when comparing levels with the same d occupancy of x + 1 in this example.) This is an actual description of the anion states. The anion "knows" it is a weakly bound attachment to a particular neutral threshold. This consistency of $f^{n,m}$ composition is true not only within the MCDF manifolds of both anion and neutral, but also for all correlation configurations that do not attempt to change the *f* occupancy (within the wavefunctions of the low-lying levels of interest).

Our approach was to extract dominant *lsj* terms for each element on a moderately sized neutral calculation, freeze that smaller $f^{n,m}$ set of terms, and use this simpler basis set with the methodology discussed in Section 1.3.2.2 when pasting together all the correlation configurations for the J-parity calculation of an anion or neutral. For p attachment to $f^n s^2$ thresholds, this is very straightforward since the f^n lsj composition is identical to the full configuration LSJ due to the closed s^2 subshell. For attachments to $f^m(d+s)^3$ neutral configurations, it is a bit trickier since one needs to consider the ls terms within the f^m group for each j as they appear in the levels of interest in a range of total Js, but it is nonetheless a well-defined process. Once the methodology was well automated, this initial stage was the primary focus of the "human time" spent on each anion. Test calculations on the smaller neutral calculations with different levels of $f^{n,m}$ basis trimming could be compared to the moderately sized un-trimmed calculations to determine the trade-off between fewer retained jls terms and energy losses (again targeting a few meV or less if possible).

The second great realization was that the remaining 2- to 4-electron piece of the anion and neutral bases were also remarkably consistent from system to system despite changing Z and n/m occupancy. This allowed us to tailor that piece of the bases on a simpler system such as Nd^- and apply the same replacements with j restrictions when performing the more complex calculations such as Tb^-.

Finally, we also used the fact that the mathematics of piecing two groups of electrons together is essentially independent of the ls. That is, the file that was used for $4f^m\ 5d_{3/2}\ vd_{3/2}\ vf_{7/2}$ for a given total J would be identical to the same set of subshell js, such as $4f^m\ 6p_{3/2}\ vd_{3/2}\ vg_{7/2}$. Together with other automation mentioned in Section 1.3.2.2, we were able to set up file naming conventions so that moving from one system to the next in the row was simplified to making a few global search-and-replace edits to the thousands-of-lines long RCI program input files.

Our results of the 3-paper lanthanide and actinide survey [2–4] are summarized in Table 1.1. The wealth of bound anion states is attributed, particularly on the left sides of each row, to the many low-lying neutral thresholds over a wide range of Js due to the density of states arising from the $f^{n,m}$ group (compared to, say, a transition metal with no such group). Despite the fact that we have included few second-order effects, nor added any core-valence correlation, and frozen the occupancy of the core-like f subshell for each calculation, the results seem consistent with experiment where applicable. For example, in the Ce^- case, we re-evaluated our placement of anion *vs* neutral manifolds by using the actual J=4 neutral ground state rather than a simpler high-J excited state [41]. By using central Js we were comparing anion and neutral calculations with similar density of states (levels per eV) so any second-order type deficiencies in the basis sets that caused improper positioning of higher but near by excited levels tended to wash out between anion and neutral. The revised ab initio RCI electron affinity [3] was more consistent with the re-evaluation of the Davis and Thompson value [35] and that of Walter *et al.* [44]. A new measurement of the EA of Ce has yielded a value of 570(20) meV [48].

At the other end of the lanthanide row, our prediction of the Tm^- electron affinity of 22 meV is in reasonable agreement with the Nadeau *et al.* measurement of 32 meV [39]. In between, our results predict a linearly decreasing binding of p attachments moving left to right across each row mimicking a similar linear relation long known for transition metal s attachments to $d^x\ s$ thresholds [49]. This trend that arises from our ab initio computations improves our confidence that experimenters searching for anion states in the middle of the rows would be well recommended to begin with our predictions.

In addition to the linear relation of p binding energies, our analysis showed a trend in p attachments that produced an anion $p_{1/2}$ attachment on either side of each low-lying neutral

Table 1.1 *Binding energy (meV) relative to neutral ground state for the lowest state of each lanthanide and actinide anion configuration. The numbers in parentheses are the number of predicted bound states for that configuration*

Anion	Z	n	m	q	$f^n s^2 p$	$f^m ds^2 p$	$f^m d^2 s^2$	$f^m s^2 p^2$	$f^q d^2 s^2 p$	$f^q d^3 s^2$
La^-	57		0			434(8)	545(7)			
Ce^-	58		1			300(10)	660(22)			
Pr^-	59	3	2		177(6)		24(1)			
Nd^-	60	4			167(7)					
Pm^-	61	5			154(8)					
Sm^-	62	6			130(8)					
Eu^-	63	7			117(6)					
Gd^-	64		7			234(12)	70(3)			
Tb^-	65	9	8		85(2)	88(6)				
Dy^-	66	10			63(2)					
Ho^-	67	11			50(2)					
Er^-	68	12			38(2)					
Tm^-	69	13			22(2)					
Lu^-	71		14			353(3)	78(1)			
Ac^-	89		0			221(5)				
Th^-	90			0					368(6)	364(3)
Pa^-	91		2	1		384(2)	134(1)		62(1)	52(1)
U^-	92		3			373(2)	260(1)			
Np^-	93		4			313(2)	103(1)			
Pu^-	94	6			85(1)					
Am^-	95	7			76(2)					
Cm^-	96		7			321(5)				
Bk^-	97	9			31(2)					
Cf^-	98	10			18(2)					
Es^-	99	11			2(1)					
Lr^-	103		14			296(1)		465(2)		

threshold (J±1/2) with similar relative binding. Four additional $p_{3/2}$ levels were then also present for the range surrounding the neutral J of J-3/2 to J+3/2 where allowed by the triangle inequality, though these had significantly less relative binding than the two $p_{1/2}$ attachments. Often, many of these states ended up unbound relative to the neutral ground state, but the analysis described in Section 1.3.2.4 was invaluable in understanding the wealth of bound states and low-lying resonances. While the *s* attachments were fewer and did not seem to follow a linear trend, their relative bindings were relatively constant for each row: ~990 meV for lanthanides [3] and ~860 meV for actinides [4].

Because of the linear relationship of the *p* attachment for lowest binding energies across each row, a few precisely measured electron affinities should suffice to rescale our work to create higher quality estimates for the remaining anions. The survey also may be used to identify practical uses for these species. For example, we have suggested [1] that use of La^- opposite parity bound-to-bound transitions may provide a more efficient means of laser cooling of antiprotons than does Os^-, which has been examined by Kellerbauer and Walz [50].

1.4 Resonance and Photodetachment Cross Section of Anions

Like other negative ions, lanthanide and actinide anions have transient states embedded in their continuous spectra, which extends from and beyond the ground state of the neutral atom. These transient states are degenerate with the continuum states of the same energy. During photodetachment, where negative ions absorb energy from photons and eject the excess electron, these transient states may manifest themselves as resonant structure in the photodetachment spectrum. In the following discussion, they are called resonant states.

Photodetachment spectra are an important tool in studying properties and electronic structure of negative ions. In laser photodetachment threshold (LPT) measurement, the relative photodetachment cross section of the negative ion is measured as a function of photon energy, using a continuously tunable laser beam. With the laser photon's energy starting from below the EA, the measured cross section spectrum reveals the opening of the lowest photodetachment threshold, from which the EA can be inferred. However, the spectrum almost always consists of other features. If the negative ion has multiple bound states, which is true with many lanthanide and actinide anions [3, 4], the spectrum may consist of peaks due to transitions between the bound anion states. These bound states will lead to more photodetachment thresholds. Also, resonance structures in the form of peaks, bumps, or windows may arise due to interaction between the resonant states and the continuum states. Given the complexity, it can be understood that the analysis of the experimental spectrum could be greatly facilitated by computation of the photodetachment cross section spectrum so that the negative ion's EA, number of bound states, and their BEs can be more accurately determined. Next, we give a brief description of the RCI calculation of photodetachment cross sections.

Before calculating the cross section, it's important to conduct a survey of possible resonant states lying in the continua. In the independent-particle approximation, these states are represented by a single configuration, which can be predicted by "exciting" a valence electron in a negative ion bound state configuration into an unfilled valence subshell except for the $4f$ subshell even though it is partially filled. As an illustration, for the Ce^- bound states $4f\,5d^2\,6s^2$ ($4f\,5d\,6s^2\,6p$), one possible resonance configuration is $4f\,5d^3\,6s$ ($4f\,5d^2\,6s\,6p$).

Quite often one will find there are multiple resonant states lying in the continua under consideration. Some of them may interfere with each other through their mutual interaction with the same continua. The interference may dramatically alter the profile of the resonance structure from what they'd look like if no interference occurred. These resonant states are said to "overlap." On the other hand, there is usually more than one available channel within the range of photon energy under study. The formalism for photodetachment calculations should then be able to treat the case of multiple discrete states embedded in multiple continua. The configuration interaction in the continuum (CIC) theory is our theory of choice.

The groundwork of using CI in the continuum was laid by Fano when studying autoionization during electron scattering of He [51]. His theory requires a pre-diagonalized subset of discrete states for resonance configurations and a pre-diagonalized subset of continuum states. The total continuum state is a linear combination of the discrete states and continuum states. Complete solution was developed for three cases: one discrete state embedded in one continuum; multiple discrete states embedded in one continuum; one discrete state embedded in two continua. The profile formula for resonance features was also derived, enabling analysis of experimental plots. Later, Fano and Prats reformulated the problem [52] where the prediagonalized basis was avoided. Later, Altick and Moore carried out the numerical work [53] for multiple discrete states lying in one continuum in the photoionization of He. Their CI calculation gave a reliable description of the differential oscillator strength as well as the resonant states involved. In their formalism, the assumption of prediagonalized basis was removed.

Extension of Fano's formalism to the general case of multiple discrete states embedded in multiple continua was first conducted by Mies [54]. Like Fano, Mies assumed a pre-diagonalized basis. By imposing the asymptotic condition for the continuum state from the scattering theory, Mies derived the complete solution to the total continuum problem and gave formulas for energies and widths of resonances.

Ramaker and Schrader [55] also extended Fano's theory and derived the expressions for the resonance energy, width, and total cross section. Their formalism was developed in line with Mies's. However, the pre-diagonalization assumption was lifted and non-orthonormality of basis vectors was treated explicitly. This more general treatment has led to the following: first, the solution to the linear coefficients in the total continuum state now involves solving a set of coupled integral equations; second, integrations involving two continuum functions occur. The formalism therefore incorporates interaction between two continua.

A concurrent work with that of Ramaker and Schrader was done by Davis and Feldkamp [56]. Their work was similar to Mies's, and pre-diagonalization was assumed. Their approach divided the solution into two cases – when there are more discrete states than continuum states and when there are more continuum states than discrete states. For the first case, their approach was expected to reduce computation time. Unlike Mies [54], scattering theory was not employed in the derivation.

Although the abovementioned theories were formulated in the context of either photoionization of atoms or electron scattering with atoms, they lend themselves equally well to photodetachment of negative ions, where an essential question is again to construct the wavefunction for the total continuum state in the presence of transient states embedded in continua. The main results from Mies's work [54] for calculating cross sections are summarized below.

1.4.1 The Configuration Interaction in the Continuum Formalism

Suppose for an atomic system, some discrete states ϕ_n are embedded in a range of continuum states $\psi_{\beta E}$. The total continuum state $\Psi_{\alpha E}$ is then a linear combination of the two sets of states:

$$\Psi_{\alpha E} = \sum_n B_{\alpha n}\phi_n + \sum_\beta \int_0^\infty dE' C_{\alpha\beta}\psi_{\beta E'} \tag{1.1}$$

where $B_{\alpha n}$ and $C_{\alpha\beta}$ are mixing coefficients.

In solving for Bs and Cs, two matrices are introduced: The CI matrix V,

$$V_{n\beta} = \langle \phi_n | H - E | \psi_{\beta E} \rangle \tag{1.2}$$

and the interaction matrix F,

$$F_{nm}(E) = E_n \delta_{nm} + P \int_0^{\infty} dE' \sum_{\beta} \frac{V_{n\beta} V_{m\beta}}{E - E'} \tag{1.3}$$

where E_n is the position of discrete state ϕ_n before being perturbed by the continua, P stands for the Cauchy principle integral around the singularity in the denominator.

The discrete states can be transformed so that F_{nm} becomes diagonal:

$$F = G^{+} \varepsilon G \tag{1.4}$$

The transformed discrete states are

$$\Phi_n^{shift} = \sum_m G_{nm} \phi_m \tag{1.5}$$

The eigenvalues ε_n are the "positions" of these transformed discrete states. Accordingly, the CI matrix V is transformed into:

$$v_{n\beta}(E, E') = \sum_m G_{nm} V_{m\beta} \tag{1.6}$$

With the introduction of a "modified resonance state,"

$$\Phi_n = \Phi_n^{shift} + \sum_{\beta} P \int_0^{\infty} dE' \frac{v_{n\beta} v_{\beta E'}}{E - E'} \tag{1.7}$$

the total wavefunction is determined as:

$$\Psi_{\alpha E} = \sum_{\beta} Z_{\alpha\beta} (\psi_{\beta E} + \sum_n \frac{v_{n\beta}}{E - \varepsilon_n} \Phi_n) \tag{1.8}$$

where Z is the normalization matrix, obtained by finding the inverse of the matrix $I + iK$. (This is Mies's equation (25), with $\tilde{A}^{\circ}$ removed. The matrix $\tilde{A}^{\circ}$ is diagonal [55]. Since its inclusion makes no difference to transition rate, it thus can be dropped when calculating cross sections.) The reaction matrix K is defined in terms of the transformed CI matrix v:

$$K_{\alpha\beta}(E) = \sum_n \frac{\pi v_{n\alpha} v_{n\beta}}{E - \varepsilon_n} \tag{1.9}$$

The total differential oscillator strength for transitions from the initial bound state Ψ_i into all channels is:

$$df/dE = \tilde{t}^{-*} t^{-} \tag{1.10}$$

where t^- is the column vector:

$$t^- = \{t_\alpha^-\} = \{\langle \Psi_{\alpha E} |T| \Psi_i \rangle\} \quad (1.11)$$

The following expression [57] converts *df/dE* into cross section:

$$\sigma = 4\pi^2 \alpha a_o^2 \frac{df}{dE} = 8.067 \frac{df}{dE}(Mb) \quad (1.12)$$

where α is the fine-structure constant, a_0 is the Bohr radius.

The application of the theory shown above to computation of photodetachment cross sections of lanthanide negative ion Ce^- is summarized below. After that, the limitations and approximations in the calculation will be discussed. The work combined with experiment has identified bound-to-bound transitions in Ce^- for the first time [5].

1.4.2 Calculation of the Final State Wavefunctions

The wavefunctions for the initial bound state Ψ_i, the resonant states ϕ_n, and the neutral thresholds are generated in separate valence RCI calculations. The valence RCI calculation for the resonant state ϕ_n yields energy positions of unperturbed resonant states, i.e., the E_ns in equation 1.3. Since virtual orbitals are used for the correlation configurations that capture not only the bound orbitals but also a portion of the continuum orbitals, it's important to avoid in ϕ_n the correlation configurations that are equivalent to the continuum state. For example, in Ce^- [5], $4f\ 5d\ 6s^2\ vp$ and $4f\ 5d\ 6s^2\ vf$ were excluded from the basis set for resonant state $4f\ 5d\ 6s^2\ 6p$. Otherwise, the variational optimization for $4f\ 5d\ 6s^2\ 6p$ may collapse into the continuum $4f\ 5d\ 6s^2\ \varepsilon p\ (f)$ in which it lies.

Since the wavefunctions for the resonant states and the neutral thresholds are prepared by separate RCI calculations, non-orthonormality (NON) must be taken into account when evaluating the CI matrix *V* (equation 1.2). This is treated partially by using common DF radials and common first set of virtual orbitals for the two states. The second set of virtual orbitals, however, is optimized separately. Experience has shown the radial overlaps between the second virtual orbital and its corresponding orbitals of the same symmetry are very close to 1 (e.g., larger than 0.97). Orbitals of different symmetry are orthogonal. Usually two sets of virtual orbitals suffice.

The wavefunction for the unperturbed continuum state $\psi_{\beta E}$ is constructed by coupling the wavefunction of a free electron to that of a neutral state [45]. It is assumed that angular part of the wavefunction of a free electron takes the same form as that of a bound electron. The radial function is energy dependent and is generated in the frozen-core Dirac-Fock approximation, using a modified version [45] of the relativistic continuum wave solver code of Perger *et al.* [58, 59]. The εl thus generated is orthogonal to all the DF radials of the same symmetry in the neutral atom.

In addition to extensive shell scripts and codes that prepare data and manage runs, which have greatly improved our efficiency in carrying out computations, following are two important modifications made to continuum codes that have facilitated the accuracy of our calculations.

The continuum wave function solver of Perger *et al.* [58, 59] allows only one basis function from the DF configuration for the core when generating the radial function for the continuum electron εl_j. As a result, only the dominant basis function was used in our earlier calculations [60]. This however, may impact the orthogonality of εl_j to the DF radial functions *nl* since εl_j will be made orthogonal to only one of the relativistic DF radials due to the use of one basis function. We have updated this code [45] to allow up to 1000 basis functions in the core. Now the continuum radial function is properly orthogonalized to DF radials of the same symmetry regardless of the *j* of εl.

The CI matrix element, $V_{n\beta}$, which is needed when applying Mie's CIC theory, is evaluated by the continuum integral solver code by Perger *et al.* [61]. The code evaluates only R^K integrals and not the one-electron *I* integrals. This produced difficulty when evaluating the matrix element for Brillouin single excitations (e.g., $4f\,5d^2\,6s\,6p$ *vs* $4f\,5d^2\,6s\,\varepsilon p$) where the numerical value of the matrix element approaches a non-zero constant as ε increases to large values. We have identified the cause to be omission of I integrals and so implemented the ability to evaluate I integrals into the code. The asymptotic behavior has then been corrected and the matrix element approaches zero at high photon energies.

Due to the nature of the generated continuum electron εl, approximations have to be made when evaluating $V_{n\beta}$ (equation 1.2). Although the angular structure of all the R^k and *I* integrals can be generated by the RCI code [12], difficulty occurs when computing the radial integrals. Since the εl radial is generated orthogonal to the DF radials *nl* only and not orthogonal to the virtual orbital radials *vl*, the value of any radial integral that contains both εl and *vl* will be erroneous. Two approximations are made to avoid radial integrals involving both εl and *vl* of the same symmetry. The first is to use a truncated basis set when calculating the angular structure. With a careful choice of the DF radials, especially those of the same symmetry *l* as the continuum electron, and a choice of the first set of virtual orbitals, correlation configurations involving *vl* should have trivial weights in both ϕ_n and the neutral core. Leaving out configurations whose weight falls below a certain threshold (e.g., 0.05%) we can exclude all such configurations from the basis set for the neutral core and most such configurations from the basis set for ϕ_n. The summation of the weights of the left-out configurations is less than 0.1%. The truncation therefore makes a trivial impact on $V_{n\beta}$. The same strategy was employed by Beck [62]. After the truncation, there may still be configurations containing *vl* in the basis set for ϕ_n. Since their summed weight is not significant (around 2% or less), they are left out from the truncated basis set for ϕ_n when evaluating its CI matrix element with a continuum state that contains εl of the same symmetry. This treatment has shown to be acceptable in our calculation on Ce^- [5].

In our calculation for Ce^- [5], photodetachment cross sections of the three lowest bound states were calculated and superimposed. To match the measured spectrum, the features in the calculated spectrum had to be shifted. One cause to this shift is the inadequate treatment of core correlation in the negative ion and the neutral atom. Although correlation effects involving the closed core are similar among bound states of the negative ion, especially those of the same configuration, difference in correlation effect between the negative ion and the neutral atom are more significant. Thus, the failure to include this differing correlation effect will result in an almost uniform shift in the EA and BE of the negative ion.

Another cause for the not-matching peaks might be the improper positioning of the unperturbed resonant states. In the basis set for ϕ_n, there are often significant configurations other than the dominant one, which were not equally correlated. This missing correlation can be partially and economically compensated for by shifting the diagonal elements of these significant configurations. In the case of resonant state Ce^- $4f\ 5d\ 6s^2\ 6p$, shifting correlation configuration $4f\ 5d^2\ 6s\ 6p$ in the basis set lined up the computed peaks with those in the experimental plot. Depending on the J value of the resonant state, different shifts were needed. Specifically, shifts ranging from 0.05–0.20 eV resulted in the resonance peaks moved by 0.014–0.039 eV. Despite the different amount of shift, the sequence of the peaks remained the same.

The missing correlation effects mentioned above do not count for other small discrepancies, however. For example, while compensating for missing correlation effect places the peaks in better alignment with the experiment, the width and the relative amplitude of the peaks may deviate more from the experiment. We suspect several things contribute to this behavior. First, the orthogonality assumption between the subset of ϕ_n and the subset of $\psi_{\beta E}$ may play a role. By assuming ϕ_n and $\psi_{\beta E}$ are orthogonal, E was removed from the inner product in equation 1.2. However, any non-zero contribution from this term would have been carried over to the CI matrix V, then the interaction matrix F, and eventually to the position of the resonances and their widths. Secondly, the NON between ϕ_n and the neutral core of $\psi_{\beta E}$ was treated partially by using a common set of DF and first set of virtual orbitals, at the price of a preferably optimized radial set for the resonant state. Thirdly, the diagonalization of the unperturbed continuum states was simply assumed.

Some of the above approximations can be removed from future calculations. The full treatment of NON can be realized numerically. Actually, in our code [63] for differential oscillator strength, NON is fully treated following the formalism of King *et al.* [64, 65]. Also, the orthogonality between ϕ_n and $\psi_{\alpha E}$ can be checked and evaluated by modifying the current code [61].

Another thing to take into account might be the assumption of prediagonalization in Mies's theory [54], which excludes interaction between two continua. Ramaker and Schrader [55] removed this assumption from their formalism. Accordingly, the solution to the total continuum wavefunction becomes more complicated and involves solving a set of coupled integral equations. However, their formalism allows interaction between two continua with the evaluation of CI matrix element between two channels. Implementing this requires a code that evaluates the radial integral involving two relativistic continuum electrons. A non-relativistic code for this task has been written by Belling [66]. On the other hand, Komninos and Nicolaides have implemented inter-channel coupling in their non-relativistic CI formalism [67] using a multichannel quantum defect approach. In their relatively recent application [68] to the photodetachment of Be^- metastable state, it was shown that velocity form was more stable with the inclusion or omission of correlation effect and inter-channel coupling. This is consistent with what we have observed in Ce^-, where the relativistic equivalent, Coulomb gauge, is more stable (inter-channel coupling was absent in our calculation). They have also shown that inter-channel coupling was not significant and HF approximation without inter-channel coupling gives very reasonable results in the velocity form, that correlation effect together with inter-channel coupling does not change the shape of the velocity form of the spectrum, even though they do improve agreement between the velocity form and the length form of total cross section. The change in the absolute value of total cross section due to correlation, for velocity form, is larger at lower photon energies (about 25%). These seem to indicate that computation without interaction between the continua is still able to give a reasonable and qualitative description of features in the experimental plot, which is exactly what we have seen in Ce^-.

The CI formalism has been shown to be an effective way of describing resonances [53]. One has the freedom to leave out correlation configurations that have trivial impact on the property of interest, either based on experience or knowledge of the theory. This enables one to maintain the size of the basis at a minimal level, which makes a significant difference in computation time. As an example, the computation time for evaluating the CI matrix element V increases dramatically with photon energies due to more oscillations in the continuum electron's radial function. With the truncated basis sets mentioned earlier, in Ce^- [5], the computer time was halved to 1.5 h for the largest photon energy. CI also provides a natural way to identify resonant state configurations. Working on the bound states of the negative ion, those eigenstates lying in the higher spectrum and above the ground state of the neutral threshold can be identified and become candidates for resonant states. One can then isolate them and do a separate CI calculation to examine their significance on the cross section spectrum.

Acknowledgments

Work supported by the National Science Foundation, Grant # PHY-0652844. Thanks to Ms. M. H. Abdalmoneam for help in preparing this manuscript.

References

[1] L. Pan and D. R. Beck, Candidates for laser cooling of atomic anions: La^- versus Os^-, *Phys. Rev. A* **82**(1), 014501(3), (2010).
[2] S. M. O'Malley and D. R. Beck, Valence calculations of lanthanide anion binding energies: 6p attachments to $4f^n$ $6s^2$ thresholds, *Phys. Rev. A* **78** (1), 012510(7) (2008).
[3] S. M. O'Malley and D. R. Beck, Valence calculations of lanthanide anion binding energies: *6p* and *6s* attachments to $4f^m(5d + 6s + 6p)^3$ thresholds, *Phys. Rev. A* **79**(1), 012511(15) (2009).
[4] S. M. O'Malley and D. R. Beck, Valence Calculations of Actinide Anion Binding Energies: All Bound *7p* and *7s* Attachments *Phys. Rev. A* **80**(3), 032514(9), (2009).
[5] C. W. Walter, N. D. Gibson, Y.-G. Li, D. J. Matyas, R. M. Alton, S. E. Lou, R. L. Field, III, L. Pan, and D. R. Beck, Experimental and theoretical study of bound and quasibound states of Ce^-, *Phys. Rev. A,* **84**(3), 032514 (11) (2011).
[6] C. W. Walter, private communication.
[7] www.phy.mtu.edu/~donald/.
[8] D. Datta and D. R. Beck, Possibility of Formation of Rare-Earth Negative Ions byAttachment of f-electrons to the Atomic Ground State, *Phys. Rev. A* **47**(6), 5198-5201 (1993).
[9] M. A. Garwan, A. E. Litherland, M. J. Nadeau, and X. L. Zhao, Lanthanide negation ion detection using accelerator mass Spectrometry, *Nucl. Instr. and Meth. in Phys. Res. B* **79**(2), 631–634 (1993).
[10] D. Berkovits, S. Ghelberg, O. Heber, and M. Paul, Weakly-bound negative ions studied by laser excitation and AMS, *Nucl. Instr. and Meth. in Phys. Res. B* **123**(2) 515–520 (1997).
[11] S. H. Vosko, J. B. Lagowski, I. L. Mayer, and J. A. Chevary, Theoretical study of even- and odd-parity states in La^- and Ac^-: Evidence for the uniqueness of La^-, *Phys. Rev. A* **43**(11), 6389–6392, (1991).
[12] D. R. Beck, RCI program, unpublished, 1978- present (we update our program code frequently).
[13] J. P. Desclaux, A Multiconfiguration Relativistic Dirac-Fock Program, *Comput. Phys. Commun.* **9**, 31–45, (1975).
[14] D. R. Beck and C. A. Nicolaides in Excited States in Quantum Chemistry, C. A. Nicolaides and D. R. Beck, editors, D. Reidel Dordrecht (1979), p. 105–142.

[15] D. R. Beck, Ab Initio Electric Dipole f values for Fe II ($3d^6$ 4s+$3d^7$) J = 9/2 $3d^6$ 4p J=9/2 Transitions, *Phys. Scr.* **71(5)**, 447–452 (2005).
[16] D. R. Beck and M. H. Abdalmoneam, Relativistic Configuration Interaction Lifetimes and Transition Probabilities for W II, Bull.Am.Phys.Soc.57, DAMOP K1.5 (2012).
[17] D. R. Beck, Magnetic Hyperfine Structure Constants for the nd^2 Triplet States of Sc II and Y II, *Phys. Rev. A,* **45**(3), 1399–1402 (1992).
[18] D. R. Beck and M. Abdalmoneam, Role of $d^{n-1}s$ configurations in hyperfine structure of d^n levels of transition metal atoms and ions: Application to V II DAMOP 2014, submitted.
[19] P. L. Norquist, D. R. Beck, R. C. Bilodeau, M. Scheer, R. A. Srawley, and H. K. Haugen, Theoretical and Experimental Binding Energies for the d^7s^2 ^4F Levels in Ru^-, including hyperfine structure and M1 decay rates, *Phys. Rev. A* **59**(3), 1896–1902 (1999).
[20] P. L. Norquist and D. R. Beck, Binding Energies, hyperfine structure, and Magnetic dipole decay rates for Os^- $5d^76s^2$ ^4F levels, *Phys. Rev. A* **61**(1), 014501(3), (1999).
[21] S. M. O'Malley and D. R. Beck, Binding Energies of $4d^6$ $5s^2$ States in Tc^-, *Phys. Rev. A* **65**(6), 064502(4), (2002).
[22] K. D. Dinov, D. R. Beck, and D. Datta, Electron Affinities of Six Bound States of Ce^- Formed by Attachment of *6p* and 5*d* electrons to Ce, *Phys. Rev. A* **50**(2), 1144–1148 (1994).
[23] S. M. O'Malley and D. R. Beck, Electron Affinities and E1 f-values for 15 Bound States of Ce^- Formed by 6*p* and 5*d* Attachment, *Phys. Rev. A* **61**(3), 034501(4) (2000).
[24] D. Datta and D. R. Beck, Electron Affinities of Opposite Parity Bound States in Th^-: Relativistic Configuration Interaction Studies, *Phys. Rev. A* **50**(2), 1107–1111 (1994).
[25] K. D. Dinov and D. R. Beck, Electron Affinities of 6*p* Electrons in Pr^-, *Phys. Rev. A* **51**(2), 1680–1682, (1995).
[26] K. D. Dinov and D. R. Beck, Electron Affinities and Hyperfine Structure for U^- and U I Obtained from Relativistic Configuration Interaction Calculations *Phys. Rev. A* **52**(4), 2632–2637 (1995).
[27] K. D. Dinov and D. R. Beck, Electron Affinity of Pa by 7*p* attachment and Hyperfine Structure Constants for Pa^-, *Phys. Rev. A* **53**(6), 4031–4035 (1996).
[28] S. M. O'Malley and D. R. Beck, Electron Affinities and E1 f-values for 11 Bound States of La^- Formed by 6*p* and 5*d* Attachment, *Phys. Rev. A* **60**(3), 2558–2561 (1999).
[29] S. M. O'Malley and D. R. Beck, Relativistic Configuration Interaction of Electron Affinities for Three Bound States of Lu^- Formed by 6*p* Attachment, *J. Phys. B* **33**(20), 4337–4342, (2000).
[30] K. Dinov, unpublished (1995); S. M. O'Malley and D. R. Beck, Electron Affinities of Tb by Attachment of a 6p Electron, Bull. *Am. Phys. Soc.* **42**, 1087 (1997).
[31] W. C. Martin, R. Zalubas, and L. Hagan (Eds.), Atomic Energy Levels - The Rare-Earth Elements, Natl. Bur. Stand. Ref. Data Ser. Natl. Bur. Stand. (U.S.) Circ. No. 60 (U.S. GPO, Washington, D.C. (1978).
[32] A. M. Covington, D. Calabrese, J. S. Thompson, and T. J. Kvale, Measurement of the electron affinity of lanthanum *J. Phys. B* **31**(20), L855-L860, (1998).
[33] V. T. Davis and J. S. Thompson, Measurement of the electron affinity of lutetium, *J Phys. B* **34**(14), L433-L437 (2001).
[34] V. T. Davis and J. S. Thompson, Measurement of the electron affinity of thulium, *Phys. Rev. A* **65**(1), 010501(3), (2001).
[35] V. T. Davis and J. S. Thompson, Measurement of the electron affinity of Cerium, *Phys. Rev. Lett.* **88**(7), 073003(4), (2002).
[36] V. T. Davis and J. S. Thompson, Measurement of the electron affinity of praseodymium, *J. Phys. B* **35**(1), L11-L14, (2002).
[37] V. T. Davis and J. S. Thompson, An experimental investigation of the atomic europium anion, *J. Phys. B* **37**(9), 1961–1965, (2004).
[38] S. M. O'Malley and D. R. Beck, Autodetachment lifetime calculations of long lived metastable states of Ba^- and Eu^-, *J. Phys. B* **38** (15), 2645–2655 (2005).
[39] M. J. Nadeau, M. A. Garwan, X. L. Zhao, and A. E. Litherland, A Negative Ion Survey Toward the Completion of the Periodic Table of Negative Ions, *Nucl. Instr. and Meth. in Phys. Res. B* **123**(2), 521–526 (1997).

[40] X. Cao and M. Dolg, Electron Affinity of Ce and electronic states of Ce^-, *Phys. Rev. A* **69**(4), 042508(5) (2004).
[41] S. M. O'Malley and D. R. Beck, Calculation of Ce- binding energies by analysis of photodetachment partial cross sections, *Phys. Rev. A* **74**(4), 042509(10), (2006).
[42] J. S. Thompson, private communication to D. R. Beck.
[43] V. T. Davis and A. M. Covington, private communication.
[44] C. W. Walter, N. D. Gibson, C. M. Janczak, K. A. Starr, A. P. Snedden, R. L. Field III, and P. Andersson, Infrared photodetachment of Ce^-: Threshold spectroscopy and resonance structure, *Phys. Rev. A* **76**(5), 052702(8), (2007).
[45] S. M. O'Malley and D. R. Beck, Calculations of Nd^- binding energies and photodetachment partial cross sections, *Phys. Rev. A* **77**(1), 012505(6), (2008).
[46] D. R. Beck, Electron quadrupole and hexadecapole moment of Ni II ground state, *J. Phys. B* **45**(22), 225002 (4) (2012).
[47] S. M. O'Malley, program LSDIAG, unpublished.
[48] J. Felton, M. Ray, and C. Chick Jarrold, The Electron Affinity of Ce, Revisited, *Phys. Rev. A*, submitted for publication.
[49] C. S. Feigerle, R. R. Corderman, S. V. Bobashev, and W. C. Lineberger, Binding Energies and structure of transition metal negative Ions, *J. Chem. Phys.* **74**(3), 1580–1598 (1981).
[50] A. Kellerbauer and J. Walz, A novel cooling scheme for antiprotons, *New J. Phys.* **8**(3), 45(9) (2006).
[51] U. Fano, Effects of configuration interaction on intensities and phase shifts, *Phys. Rev.*, **124**(6), 1866–1878 (1961).
[52] U. Fano and F. Prats, On the Connection between the theories of collision and of atomic spectra, *J. Natl. Acad. Sci.* (India), **33**, Pt. IV, 553–562 (1963).
[53] P. L. Altick and E. Neal Moore, Configuration interaction in the Helium continuum, *Phys. Rev.*, **147**(1), 59–65 (1966).
[54] F. H. Mies, Configuration interaction theory: Effects of overlapping resonances, *Phys. Rev.*, **175**(1), 164–175 (1968).
[55] D. E. Ramaker and D. M. Schrader, Multichannel configuration-interaction theory: Application to some resonances in helium, *Phys. Rev. A*, **9**(5), 1980–1991 (1974).
[56] L. C. Davis and L. A. Feldkamp, Interaction of many discrete states with many continua, *Phys. Rev. B*, **15**(6), 2961–2969 (1977).
[57] R. D. Cowan. *The Theory of Atomic Structure and Spectra*, Univ. California Press, Berkeley, CA (1981).
[58] W. F. Perger, Z. Halabuka, and D. Trautmann, Continuum wavefunction solver for GRASP, *Comput. Phys. Commun.*, **76**(2), 250–262 (1993).
[59] M. G. Tews and W. F. Perger, Standalone relativistic continuum wavefunction solver, *Comput. Phys. Commun.*, **141**(1), 205–216 (2001).
[60] S. M. O'Malley and D. R. Beck, Three long lived excited states of Tm^-, *Phys. Rev.* A **70**, 022502(7) (2004).
[61] W. F. Perger and V. Karighattan, Relativistic continuum wavefunction solver, *Comput. Phys. Commun.*, **66**(2–3), 392–402 (1991).
[62] N. Miura, T. Noro, and F. Sasaki, Theoretical study of the $2s2p^4$ ^4P resonance state in the photodetachment of C^-, *J. Phys. B: Atomic, Molecular and Optical Physics*, **30**(23), 5419–5427 (1997).
[63] D. R. Beck, program rfv, unpublished.
[64] H. F. King, R. E. Stanton, H. Kim, R. E. Wyatt and R. G. Parr, Corresponding orbitals and the nonorthogonality problem in molecular quantum mechanics, *J. Chem. Phys.*, **47**(6), 1936–1941 (1967).
[65] D. R. Beck and Z. Cai, Relativistic many-body methodology of electric dipole oscillator strengths with application to Tl^+ $6s^2 \rightarrow 6s\,6p$, *Phys. Rev. A*, **41**(1), 301–311 (1990).
[66] J. A. Belling, The evaluation of slowly converging radial integrals occurring in atomic collision problems, *J. Phys. B*, **1**(1), 136 (1968).

[67] Y. Komninos and C. A. Nicolaides, Many-electron approach to atomic photoionization: Rydberg series of resonances and partial photoionization cross sections in helium, around the n=2 threshold, *Phys. Rev. A,* **34**(3), 1995–2000 (1986).

[68] C. Sinanis, Y. Komninos, and C. A. Nicolaides, Asymmetry parameter and total cross section for the photodetachment of the metastable Be^- $1s^2 2s2p^2$ ^4P state, *Phys. Rev. A,* **51**(4), R2672–R2675 (1995).

2

Study of Actinides by Relativistic Coupled Cluster Methods

Ephraim Eliav and Uzi Kaldor

School of Chemistry, Tel Aviv University

2.1 Introduction

This chapter will mostly address high-accuracy calculations of actinide systems. Lanthanide systems, where the broader availability of experimental data provides more extensive checks of computed properties, will also be discussed. Predictions for the eka-actinide atoms E121 (eka-Ac) and E122 (eka-Th) will also be presented.

Actinide elements and their compounds have become the subject of intensive research in recent years (see the reviews [1–3]). In technology, interest is stimulated by the role of these elements in nuclear fuel and waste products, as well as the development of the actinide-organometallic chemical industry. In basic physics, actinide ions are prime candidates for observing parity non-conservation effects, which may reveal possible inconsistencies of the Standard Model [4]. It is interesting to note that actinides may have been involved in the creation of life on our planet [5]. In spite of recent progress, many spectroscopic and other physical properties of actinides are still unknown, or known with very low accuracy, due in part to the relative scarcity, toxicity, and radioactivity of these elements.

Actinides (as well as their lanthanide homologues) present severe problems to theory and computations, caused by large, non-additive relativistic and correlation effects, further complicated by the multireference character of many electronic states, involving the $5f$, $6d$, $6p$, and $7s$ orbitals. This makes the chemical bonding of actinide compounds rather complex, requiring elaborate methodology and substantial computational effort. Still, theoretical investigation provides a powerful, safe, and cheap alternative to experimental research of actinides. Using state-of-the-art computational methods, it is possible to better understand

Computational Methods in Lanthanide and Actinide Chemistry, First Edition. Edited by Michael Dolg.

the actinide chemical and physical processes, supporting and complementing experimental studies. Advanced methods, such as those discussed below, can provide calculated properties with accuracy comparable or even superior to experiment. High-level first-principles approaches aimed at the actinide compounds should be based on size-extensive, size-consistent, and balanced treatment of the dynamic and nondynamic correlation effects, and include relativity from the outset, at a level consistent with the treatment of electron correlation. Here we give an overview of a state-of-the-art ab initio approach used in computational actinide chemistry and atomic physics: the relativistic Fock-space coupled cluster (RFSCC) method. RFSCC is an all-order multireference correlation approach, which satisfies the criteria of high-level ab initio methods enumerated above; it also has a unique potentiality to consistently include high-order relativistic and quantum electrodynamics (QED) effects in an "all order" fashion, similar to the treatment of correlation effects. This outstanding capability of the FSCC method is a consequence of its "valence universality," the applicability of the method to systems with different numbers of particles (a fundamental feature of QED systems). A pilot variant of RFSCC, called double FSCC (DFSCC), has been proposed recently [6, 7]. The DFSCC is derived directly from QED, using the Lagrangian formalism and covariant evolution operator (CEO) of Lindgren [8]. This approach includes the treatment of both electronic and photonic degrees of freedom on equal quantum footing. DFSCC yields a possible avenue for covariant treatment of heavy relativistic multielectronic systems with high precision (see [6, 7]). It is still under development, and will not be discussed here.

The standard quantum chemical description is based on approximate relativistic Hamiltonians, which are not covariant and use quantization procedure only for electronic degrees of freedom, treating the electromagnetic and other interactions classically. The most precise relativistic many-body Hamiltonian currently used in quantum chemistry is the four-component Dirac-Coulomb-Breit (DCB) no-virtual-pairs Hamiltonian. The DCB Hamiltonian includes leading relativistic effects up to order α^2, which is currently regarded as satisfactory with respect to chemical accuracy, but may not be sufficient for spectroscopic (and other, mainly atomic, properties) precision. Four-component methods, complemented by high-level treatment of correlation, provide the most accurate approach to studies of heavy atomic and molecular systems, including actinide compounds. The combination of infinite order many-body RFSCC with the DCB Hamiltonian yields perhaps the most precise computational method currently available in quantum chemistry. Further refinement of this computational approach demands inclusion of QED and nuclear corrections, currently achieved by low-order perturbation theory. The DCB-RFSCC approach may serve as a benchmark method for obtaining highly precise results and for the calibration of less accurate and less expensive approaches, which use a more approximate relativistic Hamiltonian and simpler correlation methods. The RFSCC approach is computationally demanding, limiting its scope to actinide atoms and small molecules. For recent, more general reviews of RFSCC and its applications to other heavy and superheavy systems, refer to Eliav and Kaldor [6, 7].

The methods used, including the relativistic equations, the FSCC scheme, and the intermediate Hamiltonian approach, are described in Section 2.2. The following section lists representative applications of the methodology to actinide atoms and molecules, demonstrating the level of accuracy achieved where comparison with experiment is available, as well as predictions that can be made for the properties still not experimentally available. The final section provides a summary.

2.2 Methodology

The methods used here have been presented earlier, in connection with the applications cited below. Comprehensive methodological discussions appear in our 2010 reviews [6, 7]. Only brief descriptions will be given in this section.

2.2.1 The Relativistic Hamiltonian

The Dirac-Coulomb-Breit Hamiltonian is

$$H = \sum_i \left[c\boldsymbol{\alpha}_i \cdot \boldsymbol{p}_i + c^2(\beta_i - 1) + V_{\text{nuc}}(i)\right] + \sum_{i<j} V(i,j), \tag{2.1}$$

where V_{nuc} is the attraction to the nucleus, modelled by a uniform or Gaussian charge distribution, and the Coulomb-Breit 2-electron potential is

$$V(1,2) = \frac{1}{r_{12}} - \frac{1}{2r_{12}}\left[\boldsymbol{\alpha}_1 \cdot \boldsymbol{\alpha}_2 + (\boldsymbol{\alpha}_1 \cdot \boldsymbol{r}_{12})(\boldsymbol{\alpha}_2 \cdot \boldsymbol{r}_{12})/r_{12}^2\right]. \tag{2.2}$$

The no-virtual-pair approximation (NVPA) is invoked so that negative-energy solutions of the SCF equations are discarded.

The DCB Hamiltonian (2.1) is correct to second order in the fine-structure constant α but is not covariant. Properly designed four-component many-electron NVPA methods are currently the most accurate approaches for neutral and weakly-ionized atoms and molecules [9], and are used for benchmark calculations. Most of the many-body approaches implemented were adapted from the non-relativistic realm by using relativistically invariant double point groups, as well as Kramers (time-reversal) symmetry when applicable. In the atomic case, the high symmetry allows separation of radial and angular degrees of freedom. The angular part can be solved analytically with the help of Racah algebra [10], whereas the radial equations may be solved by finite difference methods. In molecular calculations, one has to resort to the *algebraic approximation*, the use of finite basis set expansions. This approach is also used in most atomic applications.

2.2.2 Fock-Space Coupled Cluster Approach

The coupled cluster (CC) approach is the most powerful and accurate of generally applicable electron correlation methods. This has been shown in many benchmark applications of 4-component relativistic CC methods to atoms [11–18] and molecules [19–31]. The CC method is an all-order, size-extensive, and systematic many-body approach. Multireference variants of relativistic 4-component CC methods capable of handling quasidegeneracies, which are important for open-shell heavy atomic and molecular systems, have been developed in recent years [15, 17–19, 21, 31]. In particular, the multireference FSCC scheme [32, 33] is applicable to systems with a variable number of particles, and is an ideal candidate for merging with QED theory to create an infinite-order size-extensive covariant many-body method applicable to systems with variable numbers of fermions and bosons [6, 7].

The development of a general multiroot multireference scheme for treating electron correlation effects usually starts from consideration of the Schrödinger equation for a number (d) of target states,

$$H\Psi^{\alpha} = E^{\alpha}\Psi^{\alpha}, \alpha = 1, ..., d. \tag{2.3}$$

The physical Hamiltonian is divided into two parts, $H = H_0 + V$, so that V is a small perturbation to the zero-order Hamiltonian H_0, which has known eigenvalues and eigenvectors, $H_0|\mu\rangle = E_0^{\mu}|\mu\rangle$.

The case of exact or quasi-degeneracy, occurring in many open shell heavy compound systems, involves the equality or near equality of some energy values E_0^{α}. By adopting the NVPA approximation, a natural and straightforward extension of the nonrelativistic open-shell CC theory emerges. The multireference valence-universal Fock-space coupled-cluster approach is presented here briefly; a fuller description may be found in Refs. [32, 33]. FSCC defines and calculates an effective Hamiltonian in a d-dimensional model space $P = \sum |\mu\rangle\langle\mu|$, $\mu = 1, .., d$, comprising the most strongly interacting zero order many-electron wavefunctions. All other functions are in the complementary Q-space so that $P + Q = 1$. All d eigenvalues of $H_{\rm eff}$ coincide with the relevant eigenvalues of the physical Hamiltonian,

$$H_{\rm eff}\Psi_0^{\alpha} = E^{\alpha}\Psi_0^{\alpha}, \alpha = 1, ..., d. \tag{2.4}$$

The functions $\Psi_0^{\alpha} = \sum_{\mu} c_{\mu}^{a}|\mu\rangle$, with $\alpha = 1, ..., d$, describe the projections $P\Psi^{\alpha}$, which constitute the major part of Ψ^{α}. The effective Hamiltonian has the form [11, 35]

$$H_{\rm eff} = PH\Omega P, \; H_{\rm eff} = H_0 + V_{\rm eff}. \tag{2.5}$$

Ω is the normal-ordered wave operator, mapping the eigenfunctions of the effective Hamiltonian onto the exact ones, $\Omega\Psi_0^{\alpha} = \Psi^{\alpha}, \alpha = 1, ..., d$. It satisfies intermediate normalization, $P\Omega P = P$. The effective Hamiltonian and wave operator are connected by the generalized Bloch equation, which for a complete model space P may be written in the compact linked form [35]

$$Q[\Omega, H_0]P = Q(V\Omega - \Omega H_{\rm eff})_{\rm linked}P. \tag{2.6}$$

Ω is parametrized exponentially in the coupled cluster method. A particularly compact form is obtained with the normal ordered form $\Omega = \{\exp(S)\}$.

The Fock-space approach starts from a reference state (closed-shell in our applications, but other single-determinant functions may also be used), correlates it, then adds and/or removes electrons one at a time, recorrelating the whole system at each stage. The sector (m, n) of the Fock space includes all states obtained from the reference determinant by removing m electrons from designated occupied orbitals, called valence holes, and adding n electrons in designated virtual orbitals, called valence particles. The practical current limit is $m + n \leq 2$, although higher sectors have also been tried [36]. The excitation operator S, defined by the exponential parametrization of Ω, is partitioned into sector operators

$S = \sum_{m\geq 0}\sum_{n\geq 0} S^{(m,n)}$. This partitioning allows for partial decoupling of the open-shell CC equations according to the so-called subsystem embedding condition [32]. The equations for the (m,n) sector involve only S elements from sectors (k,l) with $k \leq m$ and $l \leq n$ so that the very large system of coupled nonlinear equations is separated into smaller subsystems, which are solved consecutively: first, the equations for $S^{(0,0)}$ are iterated to convergence; the $S^{(1,0)}$ (or $S^{(0,1)}$) equations are then solved using the known $S^{(0,0)}$, and so on. This separation, which is exact, reduces the computational effort significantly. The effective Hamiltonian (2.5) is also partitioned by sectors. An important advantage of the method is the simultaneous calculation of a large number of states.

The FSCC equations for a particular (m,n) sector of the Fock space are derived by inserting the normal-ordered wave operator into the Bloch equation (2.6). The final form of the FSCC equation for a complete model space includes only *connected* terms [11, 35],

$$Q[S_l^{(m,n)}, H_0]P = Q\{(V\Omega - \Omega H_{\text{eff}})_l^{(m,n)}\}_{\text{conn}}P, \tag{2.7}$$

$$H_{\text{eff}}^{(m,n)} = P(H\Omega)_{\text{conn}}^{(m,n)}P. \tag{2.8}$$

After converging the FSCC equation (2.7), the effective Hamiltonian (2.8) is diagonalized, yielding directly transition energies. The effective Hamiltonian in the FSCC approach has diagonal structure with respect to the different Fock-space sectors. From (2.8) it follows that two Fock-space sectors belonging to a common Hilbert space, with the same net number of particles, e.g. (0,1) and (1,2), do not mix even if they have strongly interacting states. This means that important nondynamic correlation effects are approximated. The mixed sector CC [37] can eliminate this problem.

The FSCC equation (2.7) is solved iteratively, usually by the Jacobi algorithm. As in other CC approaches, denominators of the form $(E_0^P - E_0^Q)$ appear, originating in the left-hand side of the equation. The well-known intruder state problem, appearing when some Q states are close to and strongly interacting with P states, may lead to divergence of the CC iterations. The intermediate Hamiltonian method avoids this problem in many cases and allows much larger and more flexible P spaces.

2.2.3 The Intermediate Hamiltonian CC method

The accuracy and convergence of the Fock-space coupled cluster method depend on an appropriate partitioning of the function space into P and Q subspaces. Ideally, the P space should include all functions important to the states considered, since the effective Hamiltonian is diagonalized in P, whereas Q-space contributions are included approximately. On the other hand, convergence of the coupled cluster iterations is enhanced by maximal separation and minimal interaction between P and Q. These requirements are not always easy to reconcile. Relatively high P functions have often strong interaction with or are energetically close to Q states, making convergence slow or impossible. The offending functions are usually included in P because of their significant contribution to the lower P states, and we may not be particularly interested in the correlated states generated from them by the wave operator; however, the FSCC is an all-or-nothing method, and lack of convergence for some states means that no states at all are obtained. The intermediate Hamiltonian coupled cluster method [38] addresses this problem, making possible larger and more flexible P spaces, thereby extending the scope of the coupled cluster method and increasing its accuracy.

An additional advantage of the ability to use extended model spaces may be reducing the need for including high excitation levels in the formalism. The need for high excitations (triple and higher) is usually limited to a small group of virtual orbitals. If such orbitals are brought into P, all excitations involving them are included to infinite order by diagonalizing the effective Hamiltonian, avoiding the need for the (usually expensive) treatment of their contribution to dynamical correlation.

The concept of the intermediate Hamiltonian has been proposed by Malrieu *et al.* [39] in the framework of degenerate perturbation theory. The P space is partitioned into the main P_m subspace, which includes all the states of interest, and the intermediate P_i subspace, serving as a buffer between P_m and the rest of the functional space Q. The corresponding operators satisfy the equations

$$P_m + P_i = P, \qquad P + Q = 1. \tag{2.9}$$

The rationale for this partitioning follows: the relatively high states in P contribute significantly to the states of interest, which evolve from the lower P states, but couple strongly with intruders from Q and spoil the convergence of the iterations; they should therefore be treated differently from the lower states. This goal is achieved by partitioning P and allowing more approximate treatment of P_i states. The intermediate Hamiltonian H_I is constructed in P according to the same rules as the effective Hamiltonian,

$$H_I = PH\Omega P, \tag{2.10}$$

but only $|\Psi_m\rangle$ states, with their largest part in P_m, are required to have energies E_m closely approximating those of the physical Hamiltonian,

$$H_I P|\Psi_m\rangle = E_m P|\Psi_m\rangle. \tag{2.11}$$

The other eigenvalues, which correspond to states $|\Psi_i\rangle$ with the largest components in P_i, may be more or less accurate. This leads to some freedom in defining the relevant eigenfunctions and eigenvalues, and, therefore, in the evaluation of problematic QSP_i matrix elements. To control this freedom and make the approach more general and flexible, we also use the partitioning

$$Q = Q_i + Q_m. \tag{2.12}$$

This additional partitioning narrows the overlap of the P and Q energies, which becomes limited to P_i and Q_i subspaces (see Figure 2.1), reducing the number of problematic amplitudes, now Q_iSP_i.

Partitioning the P and Q projectors of the FSCC equation (2.7) into the main and intermediate parts by formulas (2.9, 2.12) yields four coupled CC equations,

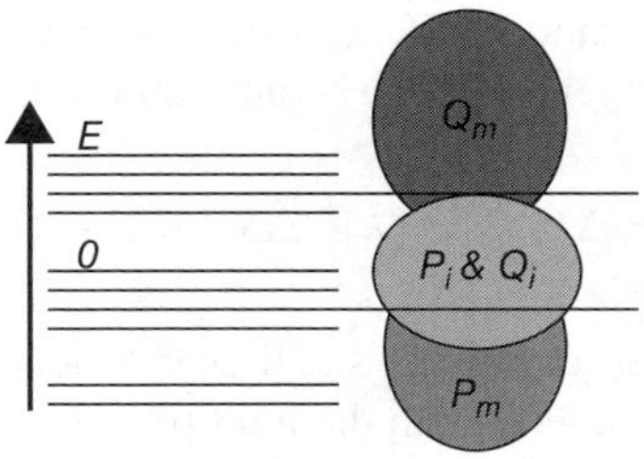

Figure 2.1 *Model spaces in the modified intermediate Hamiltonian method*

$$Q_m[S, H_0]P_m = Q_m\{V\Omega - \Omega H_{\text{eff}}\}_{\text{conn}}P_m \tag{2.13}$$
$$Q_i[S, H_0]P_m = Q_i\{V\Omega - \Omega H_{\text{eff}}\}_{\text{conn}}P_m \tag{2.14}$$
$$Q_m[S, H_0]P_i = Q_m\{V\Omega - \Omega H_{\text{eff}}\}_{\text{conn}}P_i \tag{2.15}$$
$$Q_i[S, H_0]P_i = Q_i\{V\Omega - \Omega H_{\text{eff}}\}_{\text{conn}}P_i. \tag{2.16}$$

Only the last of these can cause convergence problems. Successful replacement of this equation by another, based on physical considerations, is the central point of the IH method. The new equation to be used instead of equation 2.16 will be called the *IH condition* (IHC). Ideally, it should satisfy the following demands:

- be free of convergence problems;
- have minimal impact on the other coupled equations (2.13–2.15).

Subject to these demands, the IHC should be as close to equation 2.16 as possible.

Several IH FSCC methods have been developed and applied recently, based on different IH conditions. The first such approach [38], denoted IH1, uses the condition

$$Q_i\Omega P_m H\Omega P_i = Q_i H\Omega P_i, \tag{2.17}$$

which is similar to the equation proposed by Malrieu and applied up to the 3rd order of degenerate perturbation theory [39]. Although Malrieu's scheme could not go beyond 3rd order, because terms with small denominators appear, the IHCC variants developed in our group are all-order and may be used in the framework of any multireference CC formulation.

The next IH FSCC scheme (IH2) is based on the perturbation expansion of the problematic Q_iSP_i amplitudes. In the lowest order we simply take

$$Q_iSP_i = 0. \tag{2.18}$$

This type of IH condition has also been used for developing a new type of hybrid multireference coupled cluster schemes, including the mixed sector CC [37].

Another IH condition leads to the most flexible and useful scheme, the extrapolated IH (XIH) [40, 41], which can yield correct solutions both for P_m and P_i, thereby recovering the whole effective Hamiltonian spectrum in the extended model space P. This can be accomplished even when the standard FSCC approach using the same model space P has intruder states leading to divergence. The IH condition for the XIH approach has the form

$$Q_i[S, H_0 + P_i \Delta P_i]P_i = Q_i\{\beta \Delta S + V\Omega - \Omega H_{\text{eff}}\}_{\text{conn}} P_i. \tag{2.19}$$

Δ is an energy shift parameter, correcting small energy denominators for the problematic intruder states. A compensation term with the multiplicative parameter β ($\beta \leq 1$) is added on the right-hand side. For $\beta = 1$, the $P_i \Delta P_i$ term on the left-hand side is fully compensated so that equation 2.19 is equivalent to equation 2.16. Proper choice of the two parameters makes it possible to reach convergence in equation 2.19 and thus in the non-problematic equations (2.13–2.15). Several calculations with different values of the parameters allow extrapolation of both P_m and P_i level energies to the limit $\Delta \rightarrow 0$ or $\beta \rightarrow 1$. This extrapolation was found to be robust, in most cases linear for P_m states and quadratic for states in P_i. In the extrapolation limit the IH method transforms into the effective Hamiltonian approach. The XIH approach is asymptotically size extensive and in many cases size consistent, even for incomplete P_m, requiring only that the entire model space P is complete. A somewhat similar IH FSCC scheme has been proposed by Mukhopadhyay *et al.* in 1992 [42], but to the best of our knowledge has never been implemented.

The intermediate Hamiltonian approaches presented here may be applied within any multiroot multireference infinite order method. Recently [43] we implemented the XIH scheme to another all-order relativistic multiroot multireference approach, the Hilbert space or state universal CC, which is the main alternative to and competitor of Fock-space CC. This approach will not be discussed here. We only mention that it allows mixing P-space sectors, which can interact strongly, e.g., 1-particle with 2-particle 1-hole [37].

2.3 Applications to Actinides

Quantitative description of heavy systems requires high-level inclusion of both relativity and correlation. These two effects are non-additive, as demonstrated in our early work on the gold atom [15] and later on lanthanum [16]. The reasons are well understood: the spatial distribution of the relativistic orbitals differs significantly from that of nonrelativistic counterparts (*s* and *p* orbitals undergo contraction, whereas *d* and *f* orbitals expand), affecting the correlation energy. Additional complications and challenges involving these systems arise from the size of the systems, the number of electrons, and the close energetic proximity of many electronic states.

Representative applications of the NVPA Fock-space CC method to actinide systems are presented below. Many calculations have been carried out over the last 15 years, involving various heavy and superheavy atomic and molecular systems (not limited to actinides), with dozens of transition energies calculated per system. Most atomic results agreed with experiment within a few hundredths of an eV. Molecular applications of the RFSCC are less precise, due to the symmetry limitations on molecular basis sets. Still, our calculations of heavy molecular systems, including actinide compounds, yield state-of-art benchmark molecular parameters. A fuller description may be found in the original publications and in our recent reviews [6, 7].

The spherical symmetry of atoms, which leads to angular decomposition of the wave function and coupled cluster equations, is used at both the Dirac-Fock-Breit [34] and CC [15, 44] stages of the calculation. The energy integrals and CC amplitudes, which appear in the Goldstone-type diagrams defining the CC equations, are decomposed in terms of vector coupling coefficients, expressed by angular momentum diagrams, and reduced Coulomb-Breit or S matrix elements, respectively. The reduced equations for single and double excitation amplitudes are derived using the Jucys-Levinson-Vanagas theorem [11] and solved iteratively. This technique makes possible the use of large basis sets with high l values, as a basis orbital gives rise to two functions at most, with $j = l \pm 1/2$, whereas in Cartesian coordinates the number of functions increases rapidly with l. Typically we go up to h ($l = 5$) or i ($l = 6$) orbitals, but higher orbitals (up to $l = 8$) have also been used. To account for core-polarization effects, which may be important for many systems, we correlate at least the two outer shells, usually 20–50 electrons, but as many as 119 electrons were correlated for the anion of element 118 [45]. Finally, uncontracted Gaussians are used, since contraction leads to problems in satisfying kinetic balance and correctly representing the small components. On the other hand, it has been found that high-energy virtual orbitals have little effect on the transition energies we calculate, since these orbitals have nodes in the inner regions of the atom and correlate mostly the inner-shell electrons, which we do not correlate anyway. These virtual orbitals, with energies above 80 or 100 hartree, are therefore eliminated from the CC calculation, constituting in effect a post-SCF contraction.

The Fock-space coupled cluster and its intermediate Hamiltonian extension have been incorporated into the DIRAC package [46], opening the way to molecular applications. The heavy actinide species NpO_2^+, NpO_2^{2+}, and PuO_2^{2+} were calculated, giving access to the ground and many excited states and leading to reassignment of some of the observed spectroscopic peaks [23, 24]. Later applications addressed UO_2, UO_2^+ [25] and UO_2^{2+} [26], to resolve contradictions between several high-level molecular electronic spectra calculations. Visscher and coworkers used RFSCC and IH-RFSCC in recent benchmark calculations on uranyl-containing molecules [28–30].

The applications presented below involve actinide atomic and molecular systems, and show the power of the relativistic FSCC and IH approaches.

2.3.1 Actinium and Its Homologues: Interplay of Relativity and Correlation

Actinium is the first element of the actinide group. This element and its lighter homologue La have $nd_{3/2}(n+1)s^2\ {}^2D_{3/2}$ ground states, with $n = 5$ in La and 6 in Ac. The heavy eka-Ac (element 121) has been predicted [16] to have an $8s^28p\ {}^2P_{1/2}$ ground state, due to the sizable relativistic stabilization of the $8p_{1/2}$ electron. These three atoms were the subject of an early application of our RFSCC approach [16].

Two sets of calculations were carried out for each of the elements. The first set started from the closed-shell rare-gas configuration of M^{3+} (M = La, Ac, or E121). Two electrons were added to the reference states, one at a time, to get the sequence

$$M^{3+}(0,0) \rightarrow M^{2+}(0,1) \rightarrow M^{+}(0,2). \tag{2.20}$$

The 40 electrons in the $(n-2)f^{14}(n-1)s^2(n-1)p^6(n-1)d^{10}ns^2np^6$ shells were correlated in Ac^{3+} ($n = 6$) and $E121^{3+}$ ($n = 7$); 26 electrons were correlated in La^{3+} ($n = 5$), which has no occupied f shells. Electrons were then added in the $(n-1)f$, nd, $(n+1)s$, and $(n+1)p$ orbitals, to yield states of M^{2+} and M^{+} (because of convergence problems, the f and p orbitals were omitted in the Ac calculation). The second set of calculations started from the [RG]$(n+1)s^2$ configuration of the M^+ ions, where RG stands for the appropriate rare gas. Forty-two electrons were correlated (28 in the case of La), and electrons were added in the nd and $(n+1)p$ orbitals to give the states of the neutral atoms and the anions, according to the scheme

$$M^+(0,0) \rightarrow M(0,1) \rightarrow M^-(0,2). \tag{2.21}$$

The basis sets used and other technical details of the calculations may be found in the original paper [16].

The ionization potentials (IP) and excitation energies (EE) of La and its ions are shown and compared with experiment in Table 2.1. Note that most states appearing in Tables 2.1–2.3 are listed by their LS designations. Since L and S are not good quantum numbers in the relativistic Hamiltonian, these designations should be taken with caution. Still, they are useful for identification purposes, and have therefore been retained. Large relativistic effects are observed for the states of the relatively light La, up to 2.5 eV, rivaling even the heavier gold atom [15], which shows a local maximum in these effects [47]. There are major changes in the level structure of the cations, including altered ground states. The nonrelativistic ground state of La^{2+} is [Xe]$4f$, with [Xe]$5d$ higher by 1.42 eV; relativity reverses this order, predicting (in agreement with experiment) a $5d$ ground state. The net relativistic effect on the $5d$–$4f$ spacing is 2.4–2.6 eV. A similar reversal has been observed [48] for the isoelectronic Ba^+. There, the nonrelativistic ground state was [Xe]$5d$, changed relativistically to the correct [Xe]$6s$.

Large effects occur in La^+ too, with the $5d$ IP increasing relativistically by 1 eV. Again, the nonrelativistic ground state ($4f5d$) is different from the relativistic (and experimental) $5d^2$. As may be expected, the relativistic stabilization of s orbitals reduces strongly the energies of the $5d \rightarrow 6s$ excitations: the one-electron excitation goes from 1.1 to about 0.3 eV, and the two-electron excitation changes from 3.1 to 0.9 eV. The Dirac-Coulomb and Dirac-Coulomb-Breit energies are in good agreement with experiment, while the nonrelativistic results are far off. In neutral La, the IP is reduced from 7.135 to 5.582 eV (experimental 5.577 eV).

The transition energies of Ac are collected in Table 2.2 and compared with available experimental data [49, 50]. Nonrelativistic calculations are too far from reality to be worthwhile, and were dispensed with. Relativistic effects are even larger than in La, causing significant modifications in level ordering and spacing. The $7s$ orbital is stabilized relative to $6d$, changing the ground state configuration of the cations from [Xe]$5d$ and $5d^2$ of La^{2+} and La^+ to [Rn]$7s$ and $7s^2$ for the corresponding Ac species. The $7s^2$, $6d7s$, and $6d^2$ manifolds in Ac^+ are well separated, unlike the overlap and mixing occurring in La^+. Neutral La and Ac have similar $ns^2(n-1)d$ ground states, but the lowest excitation energies $(n-1)d \rightarrow np$, are much lower in the latter. The Ac anion is predicted to have a $7s^27p6d$ $^1D_2$ bound state with an EA of 0.35 eV, similar to the DFT result of Vosko *et al.* [51].

Transition energies in eka-actinium and its ions are collected in Table 2.3. The higher nuclear charge leads to larger relativistic effects than in actinium. The most striking

Table 2.1 *Ionization potentials (IP) and excitation energies (EE) of La and its cations (eV) by the nonrelativistic (NR), Dirac-Coulomb (DC), and Dirac-Coulomb-Breit (DCB) FSCCSD method*

	Final state		NR	DC	DCB	Expt[a]
		La^{2+}, ground state $5d\ ^2D_{3/2}$				
IP	[Xe]	$^1S_0$	19.647	19.159	19.166	19.18
EE	$5d$	$^2D_{5/2}$	0.	0.205	0.198	0.199
	$4f$	$^2F_{5/2}$	−1.420	1.013	0.932	0.892
		$^2F_{3/2}$	−1.420	1.210	1.114	1.078
	$6s$	$^2S_{1/2}$	2.669	1.628	1.637	1.685
	$6p$	$^2P_{1/2}$	5.997	5.177	5.191	5.209
		$^2P_{3/2}$	5.997	5.564	5.574	5.593
		La^+, ground state $5d^2\ ^3F_2$				
IP	$5d$	$^2D_{3/2}$	10.311	11.323	11.330	11.43[b] 11.06[a]
EE	$5d^2$	$^3F_3$	0.	0.132	0.127	0.126
		$^3F_4$	0.	0.254	0.246	0.244
	$5d^2$ [c]	$^1D_2$	0.293	0.171	0.172	0.173
	$5d6s$	$^3D_1$	1.067	0.229	0.237	0.235
		$^3D_2$	1.067	0.319	0.323	0.321
		$^3D_3$	1.067	0.401	0.404	0.403
	$5d^2$	$^3P_0$	0.586	0.671	0.669	0.651
		$^3P_1$	0.586	0.732	0.727	0.709
		$^3P_2$	0.586	0.798	0.791	0.772
	$6s^2$	$^1S_0$	3.112	0.899	0.911	0.917
	$5d^2$	$^1G_4$	0.737	0.927	0.920	0.927
	$5d6s$ [c]	$^1D_2$	1.605	1.311	1.309	1.252
		La, ground state $6s^25d_{3/2}\ ^2D_{3/2}$				
IP[d]	$5d^2$	$^3F_2$	7.135	5.587	5.582	5.577
EE	$6s^25d_{5/2}$	$^2D_{5/2}$	0.	0.157	0.153	0.131
	$6s^26p_{1/2}$	$^2P_{1/2}$	2.668	2.021	2.032	1.887
	$6s^26p_{3/2}$	$^2P_{3/2}$	2.668	2.240	2.249	2.018

[a]Martin *et al.* [52]. [b]Moore [49]. [c]Mixed state, $5d6s + 5d^2$.
[d]$La(6s^25d) \rightarrow La^+(5d^2)$, calculated as the difference of $La(6s^25d) \rightarrow La^+(6s^2)$ and $La^+(5d^2) \rightarrow La^+(6s^2)$.

expression of relativity is the ground state electron configuration. Both La and Ac have $ns^2(n-1)d\ ^2D_{3/2}$ ground states, as do the lighter group 3 elements Sc and Y. The excitation energies to the $ns^2np\ ^2P_{1/2}$ state are 1.31 eV in Y [49], 1.89 eV in La [52], and 0.98 eV in Ac (present work). The reversal of the trend in Ac is due to relativity; the lowering of this transition energy is enhanced in eka-actinium, to the extent that $8s^28p\ ^2P_{1/2}$ becomes the ground state, lower by 0.4 eV than the lowest $8s^27d$ energy. This is similar to what happens

Table 2.2 *Ionization potentials (IP) and excitation energies (EE) of Ac and its cations (eV) by the Dirac-Coulomb (DC) and Dirac-Coulomb-Breit (DCB) FSCCSD method*

		Final state	DC	DCB	Expt[a]
		Ac^{2+}, ground state $7s\ ^2S_{1/2}$			
IP	[Rn]	$^1S_0$	17.518	17.512	
EE	$6d$	$^2D_{3/2}$	0.157	0.145	0.099
		$^2D_{5/2}$	0.588	0.569	0.521
		Ac^{+}, ground state $7s^2\ ^1S_0$			
IP	$7s$	$^2S_{1/2}$	11.91	11.90	12.1
EE	$6d7s$	$^3D_1$	0.634	0.623	0.588
		$^3D_2$	0.702	0.690	0.653
		$^3D_3$	0.976	0.960	0.921
		$^1D_2$	1.194	1.176	1.127
		Ac, ground state $7s^26d_{3/2}\ ^2D_{3/2}$			
IP	$7s^2$	$^1S_0$	5.31	5.32	5.17[b]
EE	$7s^26d_{5/2}$	$^2D_{5/2}$	0.274	0.290	0.277
	$7s^27p_{1/2}$	$^2P_{1/2}$	0.969	0.984	
	$7s^27p_{3/2}$	$^2P_{3/2}$	1.573	1.583	

[a]Moore [49]. [b]Martin *et al.* [50].

in group 11 (the coinage metals), where the $(n-1)d^2ns \rightarrow (n-1)dns^2$ excitation energy goes down from Ag to Au and becomes negative for roentgenium, making $7d8s^2$ the ground state of element 111 [53]. An $8s^28p$ ground state for E121 has been predicted by previous work, using Dirac-Fock functions plus Breit correction and semiempirical correlation [54], Dirac-Fock-Slater [55–57], or density functional [58] methods.

Other states of E121 also show large relativistic effects. Compared with corresponding transitions in Ac, already affected by relativity, the $s \rightarrow d$ transitions go up by 2–3 eV in the mono- and dication. The lowest $s^2 \rightarrow d^2$ excitation in $E121^+$ is nearly 6 eV, compared with 1.7 eV for Ac^+. The lowest state of the eka-actinium anion is $8s^28p^2$, compared to $7s^27p6d$ for Ac^-, showing again the stabilization of p relative to d orbitals.

The Breit effect is rather small (up to 0.01 eV for La, 0.02 eV for E121) for most states. The only exceptions are levels with occupied f orbitals, where the Breit term contributes 0.06–0.1 eV; in all these cases it improves considerably the agreement with experiment. Another important contribution to the f states comes from the i orbitals in the basis. It changes these states by up to 0.1 eV, while the effect on other levels is much smaller.

All three atoms are predicted to have positive electron affinities, with 0.33 eV for La, 0.35 eV for Ac, and 0.57 eV for E121. Recently we started a new series of calculations of the same elements using more advanced IH-FSCC and a substantially larger basis set than possible in 1998. More accurate results are expected, particularly for the electron affinities, for which the basis used may not have been sufficient. Similar improvement was achieved for the alkali atoms [59].

Table 2.3 *Ionization potential (IP) and excitation energies (EE) of element 121 and its cations (eV) by the Dirac-Coulomb (DC) and Dirac-Coulomb-Breit (DCB) FSCCSD method*

	Final state		DC	DCB
	$E121^{2+}$, ground state $8s\ ^2S_{1/2}$			
IP	[E118]	$^1S_0$	18.67	18.65
EE	$7d$	$^2D_{3/2}$	2.883	2.859
		$^2D_{5/2}$	3.507	3.478
	$8p$	$^2P_{1/2}$	4.186	4.196
		$^2P_{3/2}$	6.831	6.819
	$6f$	$^2F_{5/2}$	4.557	4.471
		$^2F_{7/2}$	5.050	4.960
	$E121^{+}$, ground state $8s^2\ ^1S_0$			
IP			12.67	12.66
EE	$7d8s$	$^3D_1$	2.764	2.744
		$^3D_2$	2.850	2.829
		$^3D_3$	3.209	3.185
		$^1D_0$	3.800	3.775
	$8s8p$	$^3P_0$	3.001	3.010
		$^3P_1$	3.331	3.338
		$^3P_2$	5.293	5.284
	$5f8s$	$^3F_2$	5.149	5.082
		$^3F_3$	5.314	5.238
		$^3F_4$	5.635	5.558
	E121, ground state $8s^2 8p_{1/2}\ ^2P_{1/2}$			
IP	$8s^2$	$^1S_0$	4.458	4.447
EE	$8s^2 7d_{3/2}$	$^2D_{3/2}$	0.412	0.389
	$8s^2 7d_{5/2}$	$^2D_{5/2}$	0.738	0.714
	$8s^2 8p_{3/2}$	$^2P_{3/2}$	1.436	1.424

2.3.2 Thorium and Eka-thorium: Different Level Structure

The Fock-space and intermediate Hamiltonian coupled cluster methods were applied to the ground and excited levels of the second actinide element, thorium, and its heavy homologue eka-thorium (E122) [60]. Two Fock-space schemes are used. The first starts with the M^{4+} ion, with a closed-shell structure corresponding to a rare gas, and adds two electrons,

$$M^{4+}[(0)\ \text{sector}] \rightarrow M^{3+}[(1)\ \text{sector}] \rightarrow M^{2+}[(2)\ \text{sector}]. \tag{2.22}$$

Table 2.4 Valence orbitals in Fock-space and Intermediate Hamiltonian calculations

	Th	E122
	Scheme (2.22)	
FS[a]	$7s, 6d, 5f$	$8s, 7p, 7d, 6f$
IH: P_m	7-8s,6-7p,6-7d,5-6f	8-9s,7-8p,7-8d,6-7f
IH: P	7-12s,6-11p,6-10d,5-9f,5-7g, 6h, 7i	8-14s,7-12p,7-11d,6-10f,5-7g, 6h, 7i
	Scheme (2.23)	
FS[a]	$6p, 6d, 5f$	$7p, 7d, 6f$
IH: P_m	$6p, 6d, 5f$	$9s, 7p, 7d, 6f$
IH: P	8-13s,6-10p,6-10d,5-8f,5-6g, 6h, 7i	9-13s,7-11p,7-11d,6-9f,5-6g, 6h

[a]Some determinants had to be moved from P to Q to achieve convergence of the FS calculations. These include p^2, pd and pf in scheme (2.22) for E122 and p^2, f^2 and pf in scheme (2.23) for both elements.

The second scheme starts with the [rare gas]$(ns)^2$ state of M^{2+}, adding two electrons to obtain states of the neutral and singly ionized species,

$$M^{2+}[(0)\ \text{sector}] \rightarrow M^{+}[(1)\ \text{sector}] \rightarrow M[(2)\ \text{sector}]. \tag{2.23}$$

It should be noted that the experimental ground state of Th^{2+} is $6d^2$ rather than $7s^2$ [61]; this is also the ground state determined by scheme (2.22). The closed-shell $7s^2$ state is, however, used as reference in the sequence (2.23). The calculated ground state of $E122^{2+}$ is $8s^2$.

The structure of the model space P in the Fock-space method and of P_m and P in the intermediate Hamiltonian approach is shown in Table 2.4. All determinants constructed from the orbitals listed in the table constitute the relevant space. P_m is a subspace of P in the IH-FSCC approach. Convergence difficulties of the FSCC formalism in sector (2) made it necessary to use an incomplete model space [62, 63], moving certain determinants from P to Q. The IH calculations employ much larger P spaces, which are always complete (i.e., include *all* combinations of relevant orbitals). Orbital selection was determined primarily on the basis of orbital energies.

The basis for both atoms was taken from the universal basis set of Malli *et al.* [64]. It consists of Gaussian-type orbitals, with exponents given by the geometric series

$$\zeta_n = \alpha \times \beta^{(n-1)}, \quad \alpha = 106\,111\,395.371\,615, \quad \beta = 0.486\,752\,256\,286. \tag{2.24}$$

A $35s30p25d20f11g9h9i7k7l$ basis was used, with n values of 1-35 for s, 5-34 for p, 9-33 for d, 13-32 for f, 21-31 for g, 22-30 for h and i, and 23-29 for k and l orbitals. The 62 external electrons were correlated in each atom. Virtual orbitals with energies above 100 a.u. were discarded. The atomic masses were 232.038 for Th and 306 for E122.

The ionization potentials and lower excitation energies of Th and its ions are reported in Tables 2.5 and 2.6. Very good agreement with experiment [61] is obtained: the average error of the 51 Fock-space energies at all ionization levels is 0.062 eV. The intermediate Hamiltonian approach reduces the average error to 0.051 eV. This level of accuracy is obtained in spite of the complicated interactions between different electronic configurations, which lead to a rather dense spectrum.

Table 2.5 *Fock-space and intermediate Hamiltonian transition energies of thorium compared with experiment [61] (eV)*

Transition		expt.	FS	IH
	Th. Ground state $6d^27s^2\ ^3F_2$			
→ Th^+ $6d7s^2$	$^2D_{3/2}$ [a]	6.537	6.497	6.521
→ $6d^27s^2$	$^3P_0$	0.317	0.486	0.450
	$^3F_3$	0.355	0.345	0.369
	$^3P_2$	0.457	0.592	0.538
	$^3P_1$	0.479	0.680	0.669
	$^3F_4$	0.615	0.623	0.624
	$^1D_2$	0.902	1.062	1.008
→ $5f6d7s^2$	$^3H_4$	0.966	1.048	1.062
→ $6d^27s^2$	$^1G_4$	1.005	1.155	1.070
→ $5f6d7s^2$	$^3F_2$	1.021	1.079	1.087
	$^1G_4$	1.290	1.361	1.371
	$^3G_3$	1.304	1.464	1.462
	Th^+. Reference state $6d^27s\ ^2D_{3/2}$ [a]			
→ $5f_{5/2}7s^2$	$^2F_{5/2}$	0.326	0.217	0.234
→ $6d_{5/2}7s^2$	$^2D_{5/2}$		0.512	0.510
→ $5f_{7/2}7s^2$	$^2F_{7/2}$	0.808	0.713	0.734
	Th^{3+}. Ground state $5f\ ^2F_{5/2}$			
→ Th^{4+}	$^1S_0$	28.752	28.759	28.734
→ $5f$	$^2F_{7/2}$	0.536	0.536	0.537
→ $6d$	$^2D_{3/2}$	1.140	1.167	1.151
	$^2D_{5/2}$	1.796	1.827	1.811
→ $7s$	$^2S_{1/2}$	2.868	2.831	2.815
average error			0.082	0.067

[a] The ground state of Th^+ is $6d^27s\ ^4F_{3/2}$, which cannot be reached in the Fock-space scheme used here. The IP of Th and excitations of Th^+ are therefore shown for the $6d7s^2\ ^2D_{3/2}$ state, which lies 0.23 eV above the ground state.

Few calculations of Th levels have been reported. Küchle *et al.* [65] give several *LS* term-averaged transition energies for the atom and the monocation using a pseudopotential and basis designed for molecular applications. Their results are less accurate than ours due to the limitations of the basis.

The ionization potentials and low excitation energies calculated for E122 are shown in Table 2.7. More values may be found in [60]. Intermediate Hamiltonian values for E122 and its monocation were calculated by the Dirac-Coulomb and Dirac-Coulomb-Breit schemes, to obtain the effect of the Breit interaction (2.2). The Breit term contribution is small (0.01–0.04 eV) for transitions not involving *f* electrons but increases to 0.07–0.1 eV when *f* orbital occupancies are affected, as observed above (Section 2.3.1). The ground state is predicted to be $8s^28p7d$, in agreement with early Dirac-Fock(-Slater) calculations [55–57], and not the $8s^28p^2$ configuration obtained by density functional theory [58]. The separation of the

Table 2.6 *Fock-space and intermediate Hamiltonian transition energies of Th^{2+} compared with experiment [61] (eV). The ground state is 6d5f $^3H_4$*

Transition		expt.	FS	IH	Transition		expt.	FS	IH
		odd levels					even levels		
→ Th^{3+} $5f$	$^2F_{5/2}$	18.325	18.473	18.336	→ $6d^2$	$^3F_2$	0.008	0.015	−0.019
→ $6d5f$	$^3F_2$	0.063	0.075	0.053		$^3F_3$	0.503	0.515	0.478
→ $5f7s$	$^3F_3$	0.313	0.345	0.291		$^1D_2$	0.580	0.585	0.536
	$^3F_2$	0.394	0.427	0.373		$^3P_0$	0.631	0.630	0.621
→ $5f6d$	$^1G_4$	0.395	0.398	0.390	→ $6d7s$	$^3D_1$	0.685	0.687	0.624
	$^3H_5$	0.557	0.564	0.568	→ $6d^2$	$^3F_4$	0.811	0.824	0.786
	$^3F_3$	0.598	0.640	0.604	→ $6d7s$	$^3D_2$	0.890	0.901	0.844
	$^3G_3$	0.627	0.759	0.707	→ $6d^2$	$^3P_1$	0.976	0.980	0.976
	$^1D_2$	0.780	0.786	0.769	→ $6d7s$	$^3D_3$	1.234	1.250	1.171
→ $5f7s$	$^3F_4$	0.783	0.796	0.754	→ $6d^2$	$^3P_2$	1.294	1.302	1.281
	$^3F_3$	0.930	0.988	0.920		$^1G_4$	1.307	1.319	1.282
→ $5f6d$	$^3D_1$	0.982	1.113	1.063	→ $7s^2$	$^1S_0$	1.483	1.444	1.394
	$^3G_4$	1.009	1.097	1.060					
	$^3H_6$	1.046	1.053	1.056					
	$^3F_4$	1.114	1.206	1.158					
	$^3D_2$	1.262	1.378	1.337					
	$^3D_3$	1.332	1.471	1.415					
	$^3P_1$	1.379	1.501	1.457					
	$^3P_0$	1.393	1.508	1.468					
	$^3G_5$	1.398	1.547	1.491					

average error, all levels: FS – 0.050 eV, IH – 0.041 eV

levels is not large: the lowest $8p^2$ and $7d^2$ states appear just 0.16 and 0.35 eV, respectively, above the ground state. It should be noted that the ground state of thorium is $7s^26d^2$, and the increased relativistic stabilization of the p orbital of E122 changes the relative energy of the configurations. The first excited state of E122 is $8p^2$, whereas the corresponding Th level is quite high. Similar phenomena occur in the ions. The ground state of Th^+ is $6d^27s$, while the lowest level of $E122^+$ is predicted to be $7d8s^2$. Th^{2+} has a $6d5f$ ground state, with $6d^2$ less than 0.01 eV away; the accuracy of the current method is not sufficient to decide between these two states. The lowest level of $E122^{2+}$ is $8s^2$, with all low excited states up to 4 eV having an $8snl$ configuration.

Quantum electrodynamic (QED) effects are not included in the present work. A recent calculation of these effects for s electrons [66] gave estimates of about 0.04 eV for the ionization potential of the Tl^{2+} $6s$ electron (self energy 0.05 eV, vacuum polarization −0.01 eV) and 0.06 eV for the E111 $7s$ electron (self energy 0.09 eV, vacuum polarization −0.03 eV). Since the p electrons responsible for the transitions discussed in this work exhibit much weaker penetration into the nucleus, QED effects here are expected to be considerably smaller, at most 0.01–0.02 eV, within the error limits of the method ($\sim$ 0.05 eV) estimated by comparing calculated and experimental results for thorium (Tables 2.5 and 2.6).

Table 2.7 Fock-space and intermediate Hamiltonian transition energies of eka-thorium (element 122) (eV)

Config.	J	FS	IH	
		DCB	DC	DCB
E122. Ground state $8s^2 7d_{3/2} 8p_{1/2}$ $J = 2$				
$\rightarrow$ E122$^+$ $7d_{3/2}$	3/2	5.651	5.613	5.595
$\rightarrow$ $8p_{1/2}^2$	0	0.185	0.157	0.162
$\rightarrow$ $7d_{3/2}^2$	2	0.348	0.385	0.353
$\rightarrow$ $7d_{3/2}8p_{1/2}$	1	0.662	0.651	0.636
$\rightarrow$ $7d_{3/2}7d_{5/2}$	3	0.860	0.891	0.856
$\rightarrow$ $7d_{5/2}8p_{1/2}$	2	0.875	0.872	0.862
$\rightarrow$ $7d_{5/2}8p_{1/2}$	3	0.955	0.954	0.940
$\rightarrow$ $7d_{3/2}7d_{5/2}$	4	1.012	1.028	0.988
	2	1.036	1.030	0.996
	1	1.154	1.144	1.113
E122$^+$. Ground state $8s^2 7d_{3/2}$				
$\rightarrow$ E122^{2+} $8s^2$	0	11.332	11.288	11.301
$\rightarrow$ $8s^2 6f_{5/2}$	5/2	0.262	0.342	0.261
$\rightarrow$ $8s^2 7d_{5/2}$	5/2	0.663	0.658	0.653
$\rightarrow$ $8s^2 8p_{1/2}$	1/2	0.696	0.644	0.681
$\rightarrow$ $8s^2 6f_{7/2}$	7/2	0.967	1.059	0.970
E122^{2+}. Ground state $8s^2$				
$\rightarrow$ E122^{3+} $8s$	1/2	20.483		20.379
$\rightarrow$ $8s6f_{5/2}$ $[^3F_2]$	2	1.952		1.969
$[^3F_3]$	3	2.021		2.041
$\rightarrow$ $8s7d_{3/2}$ $[^3D_1]$	1	2.339		2.358
$[^3D_2]$	2	2.448		2.464
E122^{3+}. Ground state $8s$				
$\rightarrow$ E122^{4+}	0	27.139		27.135
$\rightarrow$ $6f_{5/2}$	5/2	0.877		0.879
$\rightarrow$ $6f_{7/2}$	7/2	1.688		1.691
$\rightarrow$ $7d_{3/2}$	3/2	2.325		2.326
$\rightarrow$ $7d_{5/2}$	5/2	3.287		3.287

2.3.3 Rn-like actinide ions

An interesting case for study involves the Rn- or Fr-like actinide ions. The importance of the spectra for at least two systems, U^{5+} and Th^{3+}, was discussed recently [67]. Among other properties, Hanni *et al.* [67] analyzed the difference between U^{5+} and Th^{3+} polarizabilities, ascribing it to differences in electronic spectra, particularly of highly excited metastable states. To check this assumption, precise measurements or calculations of excited states of the Rn- and Fr-like ions energy levels are necessary. Knowing the electronic spectra of these species would provide a better understanding of important physical and chemical properties of actinides. The few available theoretical calculations of Rn-like actinides [68–72] are mostly limited to the ground state energy properties, and treat electron correlation and/or relativity at low order. The experimental results are collected in the NIST Atomic Spectra Database [73].

The relativistic IHFSCC approach was used in 2001 to calculate the spectra of neutral Xe and Rn atoms, obtaining unprecedented and still unsurpassed accuracy, with an average error of 0.6% for the lowest excitation energies (about 20 per atom) [74]. This accuracy allowed predictions, e.g., for the unobserved 8*s* states of Rn. It was demonstrated that only the combination of using large active spaces with all-order treatment of dynamic correlation in the framework of a high-quality relativistic Hamiltonian can yield such high level of accuracy.

Recently, using improved methodology, we reported the calculated spectra of the four lightest Rn-like actinides [75]. To assess the accuracy of the results we compare the calculated and experimental spectra of La^{3+}, the lightest Xe-like lanthanide, and the only homologue of the studied actinides with many experimentally observed excited energy levels. Starting from the closed shell rare-gas ground state M, the singly excited states in the $(1, 1)$ sector are obtained by the Fock-space scheme

$$\mathrm{M}[(0,0)\ \mathrm{sector}] \rightarrow \begin{matrix} \mathrm{M}^{+}[(1,0)\ \mathrm{sector}] \\ \mathrm{M}^{-}[(0,1)\ \mathrm{sector}] \end{matrix} \rightarrow \mathrm{M}^{*}[(1,1)\ \mathrm{sector}]. \tag{2.25}$$

The DCB Hamiltonian is used, and correlation is included at the CCSD level. The universal basis set of equation 2.24 is augmented until the transition energies converge satisfactorily. Two basis sets were used: basis A included $35s30p25d20f11g9h9i7k7l$ Gaussians, and basis B consisted of $37s33p25d23f14g12h11i6k$ Gaussians, with more diffuse functions but omitting the *l* Gaussians because of computational limitations. Both were considerably larger than the basis used in our FSCC calculations of neutral xenon and radon [74], and go to higher angular momentum. The basis orbitals are left uncontracted. Virtual orbitals with energies higher than 100 hartree in basis A and 120 hartree in basis B are omitted. Core electrons, left uncorrelated at the CCSD stage, include the 10-electron Ne-like core for La ions and 28 electrons (KLM shells, 1*s*–3*d* orbitals) for the actinides so that 44 electrons in La^{3+} and 58 electrons in the Rn-like actinides are correlated. The total number of active orbitals was 160 in basis A and 190 in basis B. A point to note is that the lowest-energy virtual orbitals in the Rn-like actinides (except Ac^{3+}) are of *d* and *f* symmetry, in contrast to the neutral Xe and Rn systems, making the inclusion of high-*l* virtual orbitals more important.

Our preliminary second order (PT2) calculations, as well as results of Safronova and Safronova [76], show that at least the first 52 excited states in Rn-like Ac, Th, Pa, and U ions are predominantly mixtures of the $[\mathrm{Xe}]4f^{14}5d^{10}6s^{2}6p^{5}nl$ and $[\mathrm{Xe}]4f^{14}5d^{10}6s6p^{6}nl$ states. The model spaces were constructed with this in mind. The spaces for both basis sets included 6*s* and 6*p* as valence holes, from which electrons are excited. The valence particles, to which electrons are transferred, were in basis A 5*f*, 6*d*, 7*s* and 7*p*, with the main P_m space including excitations from 6*p* to 5*f* and 6*d*; other excitations are in the intermediate space P_i. This gives a total of 55 states. The *n* values are lower by one in the La ion calculations. In the larger basis B, used here for La only, the valence particles were 6–8*s*, 6–8*p*, 4–6*d*, 4–6*f*, and 5*g*, yielding 912 determinants in *P*, which give rise to 172 *J* levels. Obviously,

Table 2.8 *La^{3+} ionization potential and excitation energies in first and second order and by XIH-FSCC method with basis sets A and B (cm^{-1}). DCB Hamiltonian. The mean absolute errors (MAE) are for the 14 known excitation energies. Experiment from the NIST Atomic Spectra Database [73]*

Level (*J*)	PT1(A)	PT2(A)	XIH(A)	XIH(B)	expt.
IP	399 746	436 985	404 161	404 107	402 900
$5p_{3/2}4f_{5/2}(1)$	155 493	152 368	142 953	142 689	143 354.7
$5p_{3/2}4f_{5/2}(2)$	158 439	155 243	145 630	145 332	145 949.0
$5p_{3/2}4f_{7/2}(5)$	163 004	156 998	148 723	148 425	
$5p_{3/2}4f_{5/2}(4)$	164 350	157 972	149 368	149 142	
$5p_{3/2}4f_{7/2}(3)$	163 151	159 991	149 744	148 947	149 927.1
$5p_{3/2}4f_{5/2}(3)$	169 304	163 586	153 365	153 140	153 339.1
$5p_{3/2}4f_{7/2}(4)$	173 456	168 163	157 093	157 087	
$5p_{3/2}5d_{3/2}(0)$	158 941	180 675	158 122	156 816	156 100.3
$5p_{3/2}5d_{3/2}(1)$	161 401	183 332	160 487	159 193	158 412.6
$5p_{3/2}4f_{7/2}(2)$	178 577	170 428	160 784	160 494	160 486.4
$5p_{3/2}5d_{5/2}(2)$	166 165	188 652	165 031	163 770	162 867.6
$5p_{3/2}5d_{5/2}(4)$	166 072	186 901	165 564	164 503	163 693.3
$5p_{3/2}5d_{3/2}(3)$	168 058	189 301	167 002	166 008	165 070.7
$5p_{3/2}5d_{3/2}(2)$	172 125	194 197	169 752	168 943	167 921.7
$5p_{1/2}4f_{5/2}(3)$	186 510	182 122	169 979	169 414	
$5p_{3/2}4f_{7/2}(3)$	189 204	186 134	172 756	172 362	
$5p_{1/2}4f_{7/2}(4)$	189 776	185 775	173 663	173 260	
$5p_{3/2}5d_{5/2}(3)$	177 664	201 793	175 687	174 805	173 335.5
$5p_{3/2}4f_{5/2}(2)$	195 829	188 809	177 415	177 026	175 013.8
$5p_{1/2}5d_{3/2}(2)$	189 754	213 314	186 896	185 712	184 885.7
MAE	8 615	19 159	1 420	853	

the higher levels obtained are not meaningful; we report the 10 lowest even-parity and 10 lowest odd-parity states.

The La^{3+} levels are known experimentally and help to assess the quality of the basis sets and model spaces. The ionization potential and 20 excitation energies are shown and compared with experiment [73] in Table 2.8. In addition to the XIH values in the two basis sets described above, we list the first-order (PT1) energies, obtained by diagonalizing the model space P with Q excluded, and the second-order (PT2) values, obtained from the 2-body terms in the first CC iteration. The PT1 and PT2 results are in the spirit of the CASSCF [77] and CASPT2 [78] methods, respectively; however, orbitals were not optimized for individual states, as done in CASSCF and CASPT2 calculations. No comparison with these methods is therefore claimed; the first- and second-order values are presented for comparison with the all-order coupled cluster method using the same orbitals. The PT1 excitation energies have a mean absolute error larger than 1 eV. The second order does even worse, with a mean absolute error (MAE) of 2.4 eV. The XIH values are much more accurate, with MAEs of 0.18 eV for basis A and 0.11 eV for basis B. Note that the B calculations involved a more extensive P space than A, which is probably responsible in large part for the improvement relative to A. Because of technical difficulties, we could not use basis B

Table 2.9 *IP and excitation energies of Ac^{3+} (cm^{-1})*

Level	PT1		PT2		XIH-FSCCSD	
	DC	DCB	DC	DCB	DC	DCB
IP	357 853	357 637	393 455	393 267	363 103	362 899
$6p_{3/2}6d_{3/2}(0)$	142 666	142 399	163 618	163 388	142 659	142 527
$6p_{3/2}6d_{3/2}(1)$	145 832	145 549	166 279	165 946	145 687	146 452
$6p_{3/2}6d_{5/2}(2)$	151 220	150 924	170 245	171 772	150 908	150 721
$6p_{3/2}6d_{3/2}(3)$	151 603	151 333	169 189	170 981	151 073	150 901
$6p_{3/2}5f_{5/2}(1)$	158 925	157 980	165 450	164 448	151 300	150 537
$6p_{3/2}6d_{5/2}(4)$	151 630	151 291	172 891	172 571	152 396	152 178
$6p_{3/2}6d_{3/2}(2)$	156 023	155 734	178 161	177 911	154 560	154 346
$6p_{3/2}5f_{5/2}(2)$	163 589	162 600	168 244	171 426	155 259	154 446
$6p_{3/2}5f_{5/2}(4)$	167 849	166 904	169 562	168 863	157 664	156 880
$6p_{3/2}5f_{7/2}(5)$	167 589	166 527	169 053	168 144	158 369	157 493
$6p_{3/2}6d_{3/2}(3)$	161 040	160 720	184 703	184 383	160 133	159 865
$6p_{3/2}5f_{7/2}(3)$	170 008	168 976	174 711	173 661	160 716	159 858
$6p_{3/2}5f_{5/2}(3)$	173 427	172 470	177 011	179 416	162 167	161 318
$6p_{3/2}7s_{1/2}(2)$	163 953	163 791	187 658	187 477	162 604	162 521
$6p_{3/2}6d_{5/2}(1)$	165 296	165 150	187 949	187 429	163 514	163 424
$6p_{3/2}5f_{5/2}(4)$	177 752	176 729	182 474	181 581	167 302	166 388
$6p_{3/2}5f_{7/2}(2)$	183 697	182 717	185 642	187 438	171 655	170 760
$6p_{1/2}7s_{1/2}(1)$	178 166	177 836	195 255	195 345	173 620	173 369
$6p_{3/2}7p_{1/2}(1)$	197 544	197 480	223 504	223 726	198 077	198 124
$6p_{3/2}5f_{5/2}(2)$	199 056	198 995	225 138	225 145	198 873	198 905

for the actinides; basis A, which gave highly satisfactory results, was used in all actinide calculations reported here.

We chose Ac^{3+} to assess the contribution of the Breit term. Table 2.9 compares energies of the DC and DCB Hamiltonians. The Breit effect is significant and varies considerably for different states, between 100–900 cm^{-1}. Its inclusion in calculations for these elements is therefore essential. The inadequacy of the PT2 values, discussed above, appears again here. These values are quite different from the all-order energies, which are closer to the first-order results. Calculated IPs and excitation energies for the Rn-like ions Th^{4+}, Pa^{5+}, and U^{6+} appear in Table 2.10. To our knowledge, these were the first reported excitation energies for these species with high-level inclusion of electron correlation. Based on the La^{3+} results, we believe the actinide excitation energies are correct to ~0.2 eV.

2.3.4 Electronic Spectrum of Superheavy Elements Nobelium (Z=102) and Lawrencium (Z=103)

As is well known, the effect of relativity increases when we go to superheavy elements. This term is usually applied to elements with atomic numbers above 100 (trans-fermium elements). The spectroscopic study of superheavy atoms presents a severe challenge to the experimentalist. An important relativistic effect involves changes in the level ordering,

Table 2.10 *XIH-FSCC IPs and excitation energies of* Th^{4+}, Pa^{5+}, *and* U^{6+} *(*cm^{-1}*). DCB Hamiltonian*

Level	Th^{4+}	Pa^{5+}	U^{6+}
IP	474 498	594 827	723 529
$6p_{3/2}5f_{5/2}(1)$	134 611	114 261	89 173
$6p_{3/2}5f_{5/2}(2)$	139 281	119 591	95 101
$6p_{3/2}5f_{5/2}(4)$	141 711	121 712	96 872
$6p_{3/2}5f_{7/2}(5)$	143 341	124 727	101 429
$6p_{3/2}5f_{5/2}(3)$	145 490	126 086	101 663
$6p_{3/2}5f_{7/2}(3)$	147 198	128 671	105 537
$6p_{3/2}5f_{7/2}(4)$	153 249	135 153	112 217
$6p_{3/2}5f_{7/2}(2)$	157 311	138 471	114 599
$6p_{3/2}6d_{3/2}(0)$	162 504	181 428	200 521
$6p_{3/2}6d_{3/2}(1)$	165 864	185 856	205 528
$6p_{3/2}6d_{3/2}(3)$	171 749	192 127	212 124
$6p_{3/2}6d_{3/2}(2)$	172 331	193 087	213 275
$6p_{3/2}6d_{5/2}(4)$	174 222	196 043	217 686
$6p_{3/2}6d_{5/2}(2)$	176 615	198 608	220 401
$6p_{3/2}6d_{5/2}(3)$	183 289	206 231	228 781
$6p_{3/2}6d_{5/2}(1)$	196 048	222 286	246 790
$6p_{3/2}7s_{1/2}(2)$	199 042	237 396	277 194
$6p_{1/2}7s_{1/2}(1)$	203 910	241 463	281 319
$6p_{1/2}5f_{5/2}(3)$	204 656	195 456	181 752
$6p_{3/2}5f_{7/2}(3)$	209 954	202 154	189 679

leading sometimes to a ground state configuration that differs from that of lighter atoms in the same group and, consequently, to different chemistry.

Although certain chemical properties of these elements may be elucidated in single-atom experiments [79, 80], spectra can be measured only in sizable samples. The first such study of a superheavy atom [81] used 2.7×10^{10} atoms of ^{255}Fm with a half life of 20.1 h, and was accompanied and guided by MCDF predictions of spectral energies. Other superheavy actinides have shorter lifetimes, on the order of seconds, and spectroscopic measurements for some of them were undertaken by a collaboration based at GSI [82]. The low production rates of the atoms and their short lifetimes necessitate reliable prediction of the position of transition lines, to avoid the need for broad wavelength scans. High level theoretical studies are very crucial for line identification of the superheavy elements and are still scarce.

In preparation to spectroscopic measurements for No and Lr, which have lifetimes of a few seconds, by a collaboration based at GSI [82], we undertook calculation of the expected spectra of these atoms [83, 84]. The accuracy of the predicted spectra for these elements was estimated by applying the same method to ytterbium [83] and lutetium [84], their lighter homologues, where experimental transition energies are available.

Large, converged basis sets ($37s31p26d21f16g11h6i$) and P spaces (up to $8s6p6d4f2g1h$) were used in the framework of the IH-FSCC method. Many electrons (42 for No, 43 for Lr) were correlated so that any core polarizations effects were included. The mean absolute error for the 20 lowest excitation energies was 0.04 eV for Yb, 0.05 eV for Lu. The calculated IP of No was 6.632 eV, in agreement with the semiempirically extrapolated value of 6.65(7) eV [85].

The transition energies and amplitudes of No are shown in Table 2.11. The energies were obtained by IH-FSCC, and their expected accuracy is $\sim$ 800 cm^{-1}. The amplitudes were calculated by relativistic CI and have lower accuracy, which is, however, sufficient for the purpose of identifying the strongest lines. The simulated spectrum is shown in Figure 2.2. The salient feature of the spectrum is a strong line at 30,100$\pm$800 cm^{-1}, with an amplitude of $A = 5 \times 10^8 s^{-1}$. There are other lines, with amplitudes at least one order of magnitude lower.

Table 2.12 shows the transition energies of lawrencium. Its ground state ($7s^27p\ ^2P_{1/2}$) is different from that of lutetium ($6s^25d\ ^2D_{3/2}$), as relativity pushes the $7p$ orbital below the $6d$. The QED corrections to the transition energies are small, below 30 cm^{-1}. This small contribution reflects the fact that the $7s$ population does not change for the transitions reported. Some excitations involving holes in the $7s$ shell were calculated by the RCI method; they exhibit larger QED effects, between 200–400 cm^{-1}. The prime region for observing transitions in the planned experiment was between 20,000 and 30,000 cm^{-1}. Our calculations predict several excitations with large transition amplitudes in this region. The strongest lines in the range of the experiment will correspond to $7p \rightarrow 8s$ at 20,100 cm^{-1} and $7p \rightarrow 7d$ at 28,100 cm^{-1}. The $7p \rightarrow 9s$ transition at 30,100 cm^{-1} is also dipole allowed, but the very different spatial distribution of the two orbitals is expected to make it weaker than the other two. The transition amplitudes are shown in Table 2.12. Note that some excited states, in particular those with a single $7s$ electron, have large contributions from several configurations. Thus, the first two states in Table 2.12 have RCI coefficients between 0.4–0.5 for each of the $7s7p_{1/2}7p_{3/2}$, $6d_{3/2}6d_{5/2}7s$, and $7s6d^2_{5/2}$ configurations, and their assignment

Table 2.11 *RCI electric dipole transition amplitudes of the strongest transitions of nobelium. τ is the lifetime of the upper level*

λ(Å)	upper state		τ(s)	lower state		A(s^{-1})
2 365	$7s8p$	$^1P_1$	2.9×10^{-8}	$7s^2$	$^1S_0$	3.2×10^7
2 457	$7s8p$	$^3P_1$	2.9×10^{-8}	$7s^2$	$^1S_0$	1.0×10^7
3 327	$7s7p$	$^1P_1$	2.0×10^{-9}	$7s^2$	$^1S_0$	5.0×10^8
4 103	$7s9s$	$^3S_1$	1.2×10^{-8}	$7s7p$	$^3P_1$	1.8×10^7
4 484	$7s7d$	$^3D_2$	4.5×10^{-8}	$7s7p$	$^3P_1$	1.4×10^7
5 140	$7s9s$	$^3S_1$	1.2×10^{-8}	$7s7p$	$^3P_2$	4.2×10^7
5 663	$7s7d$	$^3D_3$	6.1×10^{-8}	$7s7p$	$^3P_2$	1.7×10^7
6 168	$7s8s$	$^3S_1$	1.4×10^{-8}	$7s7p$	$^3P_0$	1.1×10^7
6 832	$7s8s$	$^3S_1$	1.4×10^{-8}	$7s7p$	$^3P_1$	3.3×10^7
7 679	$7s7d$	$^1D_2$	8.0×10^{-8}	$7s7p$	$^3P_1$	1.2×10^7
8 171	$7s8p$	$^3P_0$	3.7×10^{-8}	$7s6d$	$^3D_1$	1.6×10^7
10 290	$7s8s$	$^3S_1$	1.4×10^{-8}	$7s7p$	$^3P_2$	2.8×10^7
15 427	$7s8s$	$^1S_0$	8.9×10^{-8}	$7s7p$	$^1P_1$	1.1×10^7
18 235	$7s8p$	$^3P_0$	3.7×10^{-8}	$7s8s$	$^3S_1$	1.1×10^7

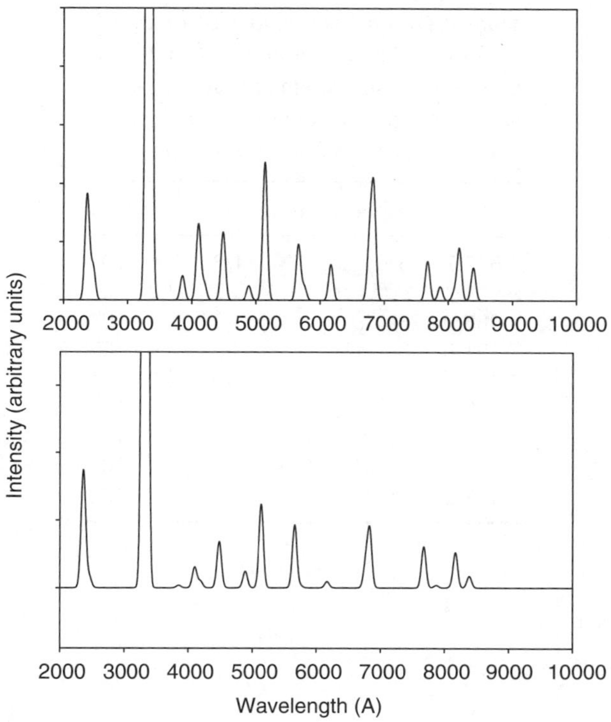

Figure 2.2 *Simulated E1 spectrum of No, assuming equal population of all excited levels. The lower panel is corrected for the total lifetime of the levels*

is somewhat arbitrary. The simulated spectrum, obtained by convolution with a Gaussian function with 20 Å full width at half maximum, is shown in Figure 2.3. The two states with the largest RCI transition amplitudes are outside the range of the planned experiment. They are dominated by the $6d^2 7s$ and $7s7p^2$ configurations, which cannot at present be included in the P space. Consequently, these states do not appear in the FSCC calculations, and their energies may have larger errors than states obtained by FSCC. The transitions at 20,100 and 28,100 cm^{-1} carry the next highest amplitudes, and are the most likely to be observed.

2.3.5 The Levels of U^{4+} and U^{5+}: Dynamic Correlation and Breit Interaction

Our first experience with f^2 levels involved the $4f^2$ states of Pr^{3+} and the $5f^2$ levels of U^{4+} [86]. Excellent agreement with experiment was obtained, with an MAE of 222 cm^{-1} for the Pr^{3+} levels and 114 cm^{-1} for U^{4+}. The error for Pr^{3+} was four times smaller than that of the MCDF calculation [87], demonstrating the more comprehensive inclusion of dynamic correlation by CC. More recently, a larger basis set was employed, extending the analysis to include the excited states of U^{5+} and the $5f^2$, $5f7s$, and $5f6d$ states of U^{4+}, which are important in interpreting the electronic spectrum of the neutral UO_2 molecule [25]. Results are presented and compared with experiment [88] and other calculations in Table 2.13. It should be noted that we use one set of canonical SCF spinors for all states in

Table 2.12 *RCI amplitudes of E1 transitions to the $7s^27p_{1/2}$ ground state of Lr. The upper levels are designated by the dominant electron configurations; other configurations may contribute substantially*

λ (Å)	upper level	J	$A(s^{-1})$
2 637.7	$6d_{3/2}6d_{5/2}7s$	1/2	3.6×10^8
2 911.3	$6d^2_{5/2}7s$	3/2	2.2×10^8
2 988.9	$7s^28d_{3/2}$	3/2	9.4×10^6
3 151.8	$6d^2_{3/2}7s$	3/2	8.6×10^6
3 319.5	$7s^29s$	1/2	6.0×10^5
3 559.2	$7s^27d_{3/2}$	3/2	3.5×10^7
3 616.2	$7s7p_{1/2}7p_{3/2}$	1/2	2.7×10^6
4 306.4	$7s7p^2_{1/2}$	1/2	1.4×10^7
4 967.5	$7s^28s$	1/2	2.7×10^7

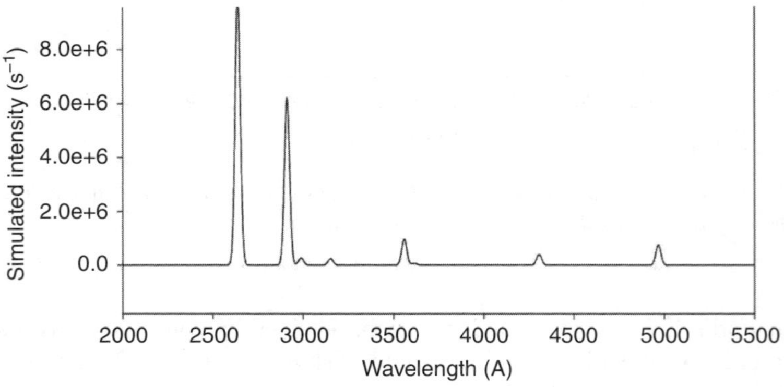

Figure 2.3 *Simulated E1 spectrum of Lr*

both IHFSCC and CASPT2 calculations, whereas the CASPT2 of Gagliardi *et al.* [89] uses common spinors only for states for the same spin and symmetry so that their orbitals are better adapted to individual states under consideration.

Comparing the different methods with the experimental data of Kaufman and Radziemski [88], the excitation energies of the U^{5+} ion appear to be best described by the XIHFSCCSD scheme (eXtrapolated Intermediate Hamiltonian Fock-Space Coupled Cluster Singles Doubles excitations), which within the large UBS basis set gives an MAE relative to the experimental data of 1,650 cm^{-1} without and 651 cm^{-1} with the Breit interaction.

Looking at the U^{5+} ion, we see that errors become larger for the more diffuse orbitals. While errors for 5*f* excitations are 100–200 cm^{-1}, the 6*d*, 7*s* and 7*p* excitations are 500–1000 cm^{-1}off. This occurs for the U^{4+} too. Comparing with the experimental data of Wyart *et al.* [90], the mean absolute error of the $5f^2$ states is 357 cm^{-1}, but it goes up to MAE of

Table 2.13 *U^{5+} and U^{4+} ionization potentials and excitation energies (cm^{-1}) [25]*

state		Exp. [90]	DCB-CASPT2 (5f6d7s)	DCB-XIH FSCC	DC-CASPT2 (5f6d7s)	DC-XIH FSCC
			U^{5+} states			
$6p^6 5f_{5/2}$(IP)	5f	–	–	508 183	–	507 326
$^2F_{7/2}$	5f	7 609	8 226	7 598	8 384	7 833
$^2D_{3/2}$	6d	91 000	95 309	90 562	92 989	89 564
$^2D_{5/2}$	6d	100 511	105 871	100 107	103 619	99 245
$^2S_{1/2}$	7s	141 448	144 946	140 211	142 206	139 062
$^2P_{1/2}$	7p	193 390	–	192 351	–	190 993
$^2P_{3/2}$	7p	215 886	–	215 112	–	213 698
MAE			–	651	–	1 650
			U^{4+} states			
$^3H_4$(IP)	$5f^2$	–	402 654	381 074	401 337	380 220
$^3F_2$	$5f^2$	4 161	3 773	4 202	3 742	4 190
$^3H_5$	$5f^2$	6 137	6 631	6 070	6 746	6 275
$^3F_3$	$5f^2$	8 983	8 897	8 974	8 989	9 147
$^3F_4$	$5f^2$	9 434	9 779	9 404	9 892	9 586
$^3H_6$	$5f^2$	11 514	12 486	11 420	12 676	11 780
$^1D_2$	$5f^2$	16 465	15 106	16 554	15 196	16 785
$^1G_4$	$5f^2$	16 656	17 391	16 630	17 599	16 937
$^3P_0$	$5f^2$	17 128	15 556	17 837	15 546	17 840
$^3P_1$	$5f^2$	19 819	18 426	20 441	18 500	20 570
$^1I_6$	$5f^2$	22 276	21 089	22 534	21 306	22 812
$^3P_2$	$5f^2$	24 652	23 539	24 991	23 753	25 315
$^1S_0$	$5f^2$	43 614	43 361	45 611	43 483	45 765
$^3H_4$	5f6d	59 183	65 821	57 161	63 221	56 289
$^3F_2$	5f6d	59 640	65 172	57 324	62 542	56 475
$^3G_3$	5f6d	63 053	68 182	61 331	65 353	60 510
$^1G_4$	5f6d	65 538	72 154	63 336	69 659	62 641
$^3F_3$	5f6d	67 033	71 826	64 485	69 537	64 141
$^3H_5$	5f6d	67 606	75 044	65 755	72 542	65 052
$^3F_2$	5f7s	94 070	97 573	91 410	94 548	90 411
$^3F_3$	5f7s	94 614	98 083	91 941	95 059	90 965
$^3F_4$	5f7s	101 612	105 500	98 921	102 614	98 168
$^1F_3$	5f7s	102 407	105 987	99 713	103 108	98 967
$^3G_3$	5f7p	139 141	–	138 614	–	137 582
$^3F_2$	5f7p	140 462	–	139 502	–	138 380
$^3G_4$	5f7p	146 926	–	145 150	–	143 970
$^3D_3$	5f7p	147 170	–	146 413	–	145 613
$^3F_3$	5f7p	156 493	–	156 024	–	155 028
MAE	$5f^2$		825	357	814	514
MAE	5f6d		6 024	2 110	3 467	2 824
MAE	5f7s		3 610	2 680	657	3 548
MAE	5f7p		–	898	–	1 924
MAE	all		–	1 191	–	1 738

2,110 cm^{-1} for transitions to $5f^16d^1$ states, and to 2,680 cm^{-1} for $5f^17s^1$ transitions. The overall MAE is 1,191 cm^{-1}. The CASPT2 method gives similar errors for the U^{5+} ion, but much larger differences from experiment for U^{4+}.

Table 2.13 shows that the effect of the Breit interaction is much larger in CASPT2 than in the all-order FSCC calculation. Similar large effects are observed in the results of the first FSCC iteration, equivalent to a second order perturbation calculation. The MAEs of atomic excitation energies calculated by CASPT2 increase strongly upon inclusion of the Breit interaction, indicating that the relatively good performance of the CASPT2 method based on the DC Hamiltonian may be due in part to cancellation of errors. We also investigated [25] the convergence of the CASPT2 energies with the systematic enlargement of the CAS, as they may be relevant in discussing the molecular results. It was found that inclusion of the *6d* orbital in the CAS is important for the quantitatively correct description of the intrashell excitations, while the effect of the *7p* orbital is less pronounced. While the inclusion of *6d* in the CAS is desirable in molecular calculations, it may often be prohibitively expensive.

2.3.6 Relativistic Coupled Cluster Approach to Actinide Molecules

The theoretical study of molecules containing such heavy atoms as the actinides presents serious methodological and computational difficulties. In addition to their sheer size, as measured by the number of electrons, these molecules exhibit a dense manifold of levels, resulting from the atomic *5s*, *6d*, *7s*, and *7p* orbitals, which are close in energy. Additional complications arise from the strong relativistic effects, both scalar and spin-orbit, and from the large electron correlation; as discussed above, relativity and correlation are non-additive and must be treated simultaneously. As an example, the electron affinities of UO_2^{2+}, which give the state energies of UO_2^+, are drawn in Figure 2.4. The first column shows spin-free, correlation-free levels; spin is included in the second column, correlation in the third, and the last column describes simultaneous inclusion of spin and correlation [25].

The treatment of actinide-containing molecules by DC-IHFSCC is described in more detail in the contribution of Saue and Visscher to this volume, and only a few comments are made here. The application of this approach was made possible by the incorporation of the FSCC scheme in the DIRAC relativistic program package [21], extended more recently to include IHFSCC. The first published application addressed the NpO_2^+, NpO_2^{2+}, and PuO_2^{2+} species [23]. Calculated excitation energies of PuO_2^{2+} were mostly within a few hundred wave numbers of experiment, significantly better than values given by SD+Q+SO [91] and CASPT2+SO [92] calculations. Extensive, detailed calculations on UO_2 and its monocation [25] helped to shed light on the complicated spectra of these species. Further discussion of the UO_2 spectrum may be found in the work of Real *et al.* [26].

While CC methods may be the best quantum chemistry can offer, they are also the most expensive, limiting the size of systems treatable by them. Exploration of larger systems became possible in recent years with the advent of schemes where only the active center of a large compound is treated by IHFSCC, while most of the compound is handled by simpler methods; the active center is then embedded in the potential created by its environment. This approach was applied to the spectrum of CUO with 4 rare gas atoms, simulating the effect of rare gas matrices on the spectra of the molecule [29] and to the spectrum of uranyl in the $Cs_2UO_2Cl_4$ crystal [30].

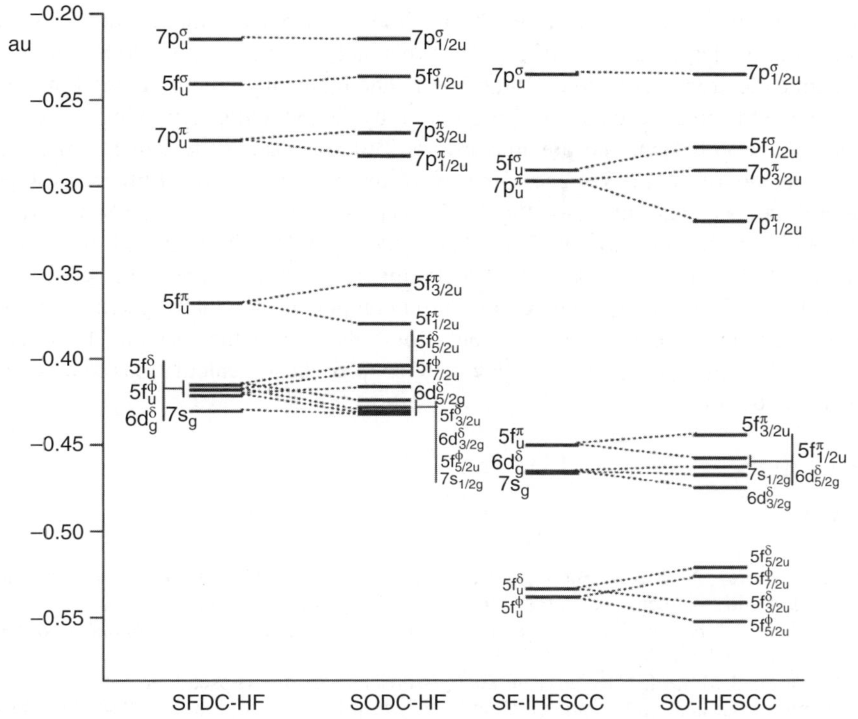

Figure 2.4 *UO_2^{2+} electron affinities, giving levels of UO_2^+ [25]*

2.4 Summary and Conclusion

The no-virtual-pair Dirac-Coulomb-Breit Hamiltonian, correct to second order in the fine-structure constant α, provides the framework for four-component methods, the most accurate approximations in electronic structure calculations for heavy atomic and molecular systems, including actinides. Electron correlation is taken into account by the powerful coupled cluster approach. The density of states in actinide systems necessitates simultaneous treatment of large manifolds, best achieved by Fock-space coupled cluster; to avoid intruder states, which destroy the convergence of the CC iterations, while still treating a large number of states simultaneously, intermediate Hamiltonian schemes are employed.

This chapter describes the IH-FSCC method and its implementation in the framework of the Dirac-Coulomb-Breit Hamiltonian. Applications to representative atomic and molecular actinide systems, as well as some lanthanides and eka-actinides, are reviewed. Most transition energies calculated agree with experimental data within a few hundredths of an eV, and reliable predictions are given for other, yet unknown, energies. Calculations of atomic excitation energies show that transitions involving f electrons display significant effects of the Breit term ($\sim$0.1 eV), with second order contributions much larger than the all-order results. Inclusion of high-l functions, up to i orbitals (l=6), is also important for these transitions.

The main advantages of the IHFSCC method include the simultaneous economical determination of large numbers of energy levels and the treatment of both dynamic and nondynamic electron correlation to high order. The main shortcoming is the limitation to states obtained from a closed shell configuration by adding and/or removing two electrons at most. We hope to continue previous efforts [36] and extend the scope of the method to higher sectors of the Fock space. Other work in progress includes further development of the Hilbert-space and mixed-sector IHCC [37], as well as the double FSCC formalism mentioned in the introduction, which will include higher QED terms and make possible the treatment of highly ionized species. A promising recent development is the combined wavefunction-DFT treatment, where the central component of a large system is treated by IHFSCC, with an embedding potential describing the rest of the system. This opens the possibility of bringing the power of relativistic coupled cluster calculations to much larger systems than before.

References

[1] L. R. Morss, N. M. Edelstein, and J. Fuger, Eds., "The Chemistry of the Actinide and Transactinide Elements," Springer, Dordrecht, The Netherlands, 4th ed. (2010).

[2] D. Wang, W. F. van Gunsteren, and Z. Chaia, "Recent advances in computational actinoid chemistry," Chem. Soc. Rev. **41**, 5836 (2012).

[3] X. Cao and M. Dolg, "Relativistic energy-consistent ab initio pseudopotentials as tools for quantum chemical investigations of actinide systems," Coord. Chem. Rev. **250**, 900 (2006).

[4] B. M. Roberts, V. A. Dzuba, and V. V. Flambaum, Parity nonconservation in Fr-like actinide and Cs-like rare-earth-metal ions," Phys. Rev. A **88**, 012510 (2013).

[5] Z.A. Adam, "Actinides and life's origins," Astrobiology **7**, 852 (2007).

[6] E. Eliav and U. Kaldor, "Four-component electronic structure methods," in "Relativistic Methods for Chemists," eds. M. Barysz and Y. Ishikawa (Springer, 2010), p. 279.

[7] E. Eliav and U. Kaldor, "Relativistic four-component multireference coupled cluster methods: Towards a covariant approach," in "Recent Progress in Coupled Cluster Methods: Theory and Applications," eds. J. Pittner, P. Charsky, and J. Paldus (Springer, 2010), p. 113.

[8] I. Lindgren, "Relativistic Many-Body Theory: A new field theoretical approach," Springer (2011).

[9] I. Lindgren, "A relativistic coupled cluster approach with radiative corrections," in: "Many-Body Methods in Quantum Chemistry," ed. U. Kaldor, Lecture Notes in Chemistry Vol. 52 (Springer-Verlag, Heidelberg, 1989) p. 293; "Towards a relativistic many-body procedure," Nucl. Instrum. Methods **B31**, 102 (1988).

[10] I.P. Grant, "Relativistic self-consistent fields," Proc. R. Soc. London A **262**, 555 (1961).

[11] S. Salomonson, I. Lindgren, and A.-M. Mårtensson, "Numerical many-body perturbation calculations on Be-like systems using a multi-configurational model space," Phys. Scr. **21**, 351 (1980).

[12] S.A. Blundell, W.R. Johnson, Z.W. Liu, and J. Sapirstein, "Relativistic all-order equations for helium," Phys. Rev. A **39**, 3768 (1989); "Relativistic all-order calculations of energies and matrix elements for Li and Be^+," Phys. Rev. A **40**, 2233 (1989); S.A. Blundell, W.R. Johnson, and J. Sapirstein, "High-accuracy calculation of the $6s_{1/2} \rightarrow 7s_{1/2}$ parity-nonconserving transition in atomic cesium and implications for the standard model," Phys. Rev. Lett. **65**, 1411 (1990); "Relativistic all-order calculations of energies and matrix elements in cesium" Phys. Rev. A **43**, 3407 (1991).

[13] Z.W. Liu and H.P. Kelly, "Atomic many-body perturbation method based on multiconfiguration Dirac-Fock wave functions," Phys. Rev. A **43**, 3305 (1991).

[14] S. Salomonson and P. Öster, "Relativistic all-order pair functions from a discretized single-particle Dirac Hamiltonian," Phys. Rev. A **40**, 5548 (1989).
[15] E. Eliav, U. Kaldor, and Y. Ishikawa, "Open-shell relativistic coupled-cluster method with Dirac-Fock-Breit wave functions: Energies of the gold atom and its cation," Phys. Rev. A **49**, 1724 (1994).
[16] E. Eliav, S. Shmulyian, U. Kaldor, and Y. Ishikawa, "Transition energies of lanthanum, actinium, and eka-actinium (element 121)," J. Chem. Phys. **109**, 3954 (1998).
[17] U. Kaldor and E. Eliav, "High-accuracy calculations for heavy and super-heavy elements," Adv. Quantum Chem. **31**, 313 (1998).
[18] B. P. Das, K. V. P. Latha, B. K. Sahoo, C. Sur, R. K. Chaudhuri, and D. Mukherjee, "Relativistic and correlation effects in atoms," J. Theor. Comp. Chem. **4**, 1 (2005).
[19] E. Eliav and U. Kaldor, "The Relativistic four-component coupled-cluster method for molecules: Spectroscopic constants of SnH_4," Chem. Phys. Lett. **248**, 405 (1996).
[20] L. Visscher, T. J. Lee, and K. G. Dyall, "Formulation and implementation of a relativistic unrestricted coupled-cluster method including noniterative connected triples," J. Chem. Phys. **105**, 8769 (1996).
[21] L. Visscher, E. Eliav, and U. Kaldor, "Formulation and implementation of the relativistic fock-space coupled cluster method for molecules," J. Chem. Phys. **115**, 9720 (2001).
[22] E. Eliav, U. Kaldor, and B. A. Hess, "The relativistic Fock-space coupled-cluster method for molecules: CdH and its ions," J. Chem. Phys. **108**, 3409 (1998).
[23] I. Infante, A. S. P. Gomes, and L. Visscher, "On the performance of the intermediate Hamiltonian Fock-space coupled-cluster method on linear triatomic molecules: The electronic spectra of NpO_2, NpO_2, and PuO_2," J. Chem. Phys. **125**, 074301 (2006).
[24] A. S. P. Gomes, C. R. Jacob, and L. Visscher, "Calculation of local excitations in large systems by embedding wave-function theory in density-functional theory," Phys. Chem. Chem. Phys. **10**, 5353 (2008).
[25] I. Infante, E. Eliav, L. Visscher, and U. Kaldor, "Fock-space coupled cluster study on the electronic structure of the UO_2, UO_2^+, U^{4+} and U^{5+} species," J. Chem. Phys. **127**, 124308 (2007).
[26] F. Real, A. Gomes, L. Visscher, V. Vallet, and E. Eliav, "Benchmarking electronic structure calculations on the bare UO_2^{2+} ion: How different are single and multireference electron correlation methods?," J. Phys. Chem. A **113**, 12504 (2009).
[27] F. Ruiperez, C. Danilo, F. Real, J.-P. Flament, V. Vallet, and U. Wahlgren, "An ab initio theoretical study of the electronic structure of UO_2^+ and $[UO_2(CO_3)_3]^{5-}$," J. Phys. Chem. A **113**, 1420 (2009).
[28] P. Tecmer, A. S. P. Gomes, U. Ekstroma, and L. Visscher, "Electronic spectroscopy of UO_2^{2+}, NUO^+ and NUN: an evaluation of time-dependent density functional theory for actinides," Phys. Chem. Chem. Phys. **13**, 6249 (2011).
[29] P. Tecmer, H. van Lingen, A. S. P. Gomes, and L. Visscher, "The electronic spectrum of $CUONg_4$ (Ng= Ne, Ar, Kr, Xe): New insights in the interaction of the CUO molecule with noble gas matrices," J. Chem. Phys. **137**, 084308 (2012).
[30] A. S. P. Gomes, C. R. Jacob, F. Real, L. Visscher, and V. Vallet, "Towards systematically improvable models for actinides in condensed phase: the electronic spectrum of uranyl in $Cs_2UO_2Cl_4$ as a test case," Phys. Chem. Chem. Phys. **15**, 15153 (2013).
[31] T. Fleig, L.K. Sorensen, and J. Olsen, " A relativistic 4-component general-order multi-reference coupled cluster method: initial implementation and application to HBr," Theo. Chem. Acc. **118**, 347 (2007).
[32] S. Pal and D. Mukherjee, "Use of cluster expansion methods in the open-shell correlation problem," Adv. Quantum Chem. **20**, 292 (1989).
[33] U. Kaldor, "The Fock-space coupled cluster method: Theory and application," Theor. Chim. Acta **80**, 427 (1991).
[34] Y. Ishikawa, R. C. Binning, and H. Sekino, "Relativistic many-body perturbation theory calculations on Be, Ne^{6+}, Ar^{14+} and Ne," Chem. Phys. Lett. **160**, 206 (1989); Y. Ishikawa, "Relativistic Dirac-Fock and many-body perturbation calculations on He, He-like ions, Ne,

and Ar," Phys. Rev. A **42**, 1142 (1990); "Dirac-Fock Gaussian basis calculations: Inclusion of the Breit interaction in the self-consistent field procedure," Chem. Phys. Lett. **166**, 321 (1990); Y. Ishikawa and H. M. Quiney, "Relativistic many-body perturbation-theory calculations based on Dirac-Fock-Breit wave functions," Phys. Rev. A **47**, 1732 (1993); Y. Ishikawa and K. Koc, "Relativistic many-body perturbation theory based on the no-pair Dirac-Coulomb-Breit Hamiltonian: Relativistic correlation energies for the noble-gas sequence through Rn (Z=86), the group-IIB atoms through Hg, and the ions of Ne isoelectronic sequence," Phys. Rev. A **50**, 4733 (1994).

[35] I. Lindgren and J. Morrison, "Atomic Many-Body Theory," second addition. Springer-Verlag, Berlin (1986).

[36] S. R. Hughes and U. Kaldor, "Relativistic coupled cluster calculations for closed shell atoms," Chem. Phys. Lett. **194**, 95 (1992); "The coupled cluster method with full inclusion of single, double and triple excitations applied to high sectors of the fock space," ibid. **204**, 339 (1993); "Fock-space coupled-cluster method: The (1,2) sector," Phys. Rev. A **47**, 4705 (1993); "The Fock-space coupled cluster method: Electron affinities of the five halogen elements with consideration of triple excitations," J. Chem. Phys. **99**, 6773 (1993); "The coupled cluster method in high sectors of the Fock space," Intern. J. Quantum Chem. **55**, 127 (1995).

[37] A. Landau, E. Eliav, Y. Ishikawa, and U. Kaldor, "Mixed-sector intermediate Hamiltonian Fock-space coupled cluster approach," J. Chem. Phys. **121**, 6634 (2004).

[38] A. Landau, E. Eliav, and U. Kaldor, "Intermediate Hamiltonian Fock-space coupled-cluster Method," Chem. Phys. Lett. **313**, 399 (1999); A. Landau, E. Eliav, and U. Kaldor, "Intermediate Hamiltonian Fock-space coupled-cluster method," Adv. Quantum Chem. **39**, 172 (2001).

[39] J.-P. Malrieu, Ph. Durand, and J.-P. Daudey, "Intermediate Hamiltonians as a new class of effective Hamiltonians," J. Phys. A **18**, 809 (1985).

[40] E. Eliav, M. J. Vilkas, Y. Ishikawa, and U. Kaldor, "Extrapolated intermediate Hamiltonian coupled cluster approach: Theory and pilot application to electron affinities of alkali atoms," J. Chem. Phys. **122**, 224113 (2005).

[41] E. Eliav, M. J. Vilkas, Y. Ishikawa, and U. Kaldor, "Ionization potentials of alkali atoms: Towards meV accuracy," Chem. Phys. **311**, 163 (2005).

[42] D. Mukhopadhyay, B. Datta, and D. Mukherjee, "The construction of a size-extensive intermediate Hamiltonian in a coupled-cluster framework," Chem. Phys. Lett. **197**, 236 (1992).

[43] E. Eliav, A. Borschevsky, R. K. Shamasundar, S. Pal, and U. Kaldor, "Intermediate Hamiltonian Hilbert space coupled cluster method: Theory and pilot application," Int. J. Quantum Chem. **109**, 2909 (2009).

[44] E. Eliav, U. Kaldor, and Y. Ishikawa, "Ionization potentials and excitation energies of the alkali-metal atoms by the relativistic coupled cluster method," Phys. Rev. A **50**, 1121 (1994).

[45] I. Goidenko, L. Labzowsky, E. Eliav, U. Kaldor, and P. Pyykkö, "QED corrections to the binding energy of the eka-radon (Z=118) negative ion," Phys. Rev. A **67**, 020102(R) (2003).

[46] DIRAC, a relativistic ab initio electronic structure program, Release DIRAC11.0 (2011), written by H. J. Aa. Jensen, T. Saue, and L. Visscher with contributions from V. Bakken, E. Eliav, T. Enevoldsen, T. Fleig, O. Fossgaard, T. Helgaker, J. Laerdahl, C. V. Larsen, P. Norman, J. Olsen, M. Pernpointner, J. K. Pedersen, K. Ruud, P. Salek, J. N. P. van Stralen, J. Thyssen, O. Visser, and T. Winther. (http://dirac.chem.vu.nl).

[47] P. Pyykkö, "Relativistic effects in structural chemistry," Chem. Rev. **88**, 563 (1988).

[48] E. Eliav, U. Kaldor, and Y. Ishikawa, "Transition energies of barium and radium by the relativistic coupled cluster method," Phys. Rev. A **53**, 3050 (1996).

[49] C. E. Moore, "Atomic Energy Levels," Natl. Bur. Stand. (U.S.) Circ. No. 467 (U.S. GPO, Washington, DC), Vol. II (1952) and Vol. III (1958).

[50] D. R. Lide (Ed.), "Handbook of Chemistry and Physics," 74th Ed., (CRC Press, Boca Raton FL, 1993).

[51] S. H. Vosko, J. B. Lagowski, I. L. Mayer, and J. A. Chevary, "Theoretical study of even- and odd-parity states in La^- and Ac^-: Evidence for the uniqueness of La^-," Phys. Rev. A **43**, 6389 (1991).

[52] W. C. Martin, R. Zalubas, and L. Hagan, "Atomic Energy Levels – The Rare-Earth Elements," Natl. Bur. Stand. Ref. Data Series, NBS Circ. No. 60 (U.S. GPO, Washington, DC, 1978).

[53] E. Eliav, U. Kaldor, P. Schwerdtfeger, B. A. Hess, and Y. Ishikawa, "The ground state electron configuration of element 111," Phys. Rev. Lett. **73**, 3203 (1994).
[54] J. B. Mann and J. T. Waber, "SCF relativistic Hartree-Fock Calculations on the superheavy element 118-131," J. Chem. Phys. **53**, 2397 (1970).
[55] B. Fricke, W. Greiner, and J. T. Waber, "The continuation of the periodic table up to $Z = 172$. The chemistry of superheavy elements," Theor. Chim. Acta **21**, 235 (1971).
[56] B. Fricke and G. Soff, "Dirac-Slater calculations for the elements Z=100, fermium, to Z=173," At. Data Nucl. Data Tables **19**, 83 (1977).
[57] C. C. Lu, T. A. Carlson, F. B. Malik, T. C. Tucker, and C. W. Nestor, "Relativistic Hartree-Fock-Slater eigenvalues, radial expectation values, and potentials for atoms, $2 \leq Z \leq 126$," At. Data **3**, 1 (1971).
[58] K. Umemoto and S. Saito, "Electronic configurations of superheavy elements," J. Phys. Soc. Japan **65**, 3175 (1996).
[59] A. Landau, E. Eliav, Y. Ishikawa, and U. Kaldor, "Benchmark calculations of electron affinities of the alkali atoms sodium to eka-francium (element 119)," J. Chem. Phys. **115**, 2389 (2001).
[60] E. Eliav, A. Landau, Y. Ishikawa, and U. Kaldor, "Electronic structure of eka-thorium (element 122) compared with thorium," J. Phys. B **35**, 1693 (2002).
[61] M. S. Fredi, "The Chemistry of the Actinide Elements," ed J. J. Katz, G. T. Seaborg, and L. R. Morss, 2nd ed., vol. 1 (Chapman and Hall, New york) p. 1196; VizieR astronomical database, http://vizier.u-strasbg.fr; S. Köhler, F. Albus, R. Deißenberger, N. Erdmann, H. Funk, H.-U. Hasse, G. Herrmann, G. Huber, H.-J. Kluge, M. Nunnemann, G. Passler, P. M. Rao, J. Riegel, N. Trautmann, and F.-J. Urban in *Resonance Ionization Spectroscopy 1994*, ed. H.-J. Kluge H-J, J. E. Parks, and K. Wendt (AIP Press, New York) p. 377.
[62] G. Hose and U. Kaldor, "Diagrammatic many-body perturbation theory for general model spaces," J. Phys. B **12**, 3827 (1979); "A general model space diagrammatic peturbation theory," Phys. Scr. **21**, 357 (1980).
[63] L. Meissner, S. A. Kucharski, and R. J. Bartlett, "A multireference coupled-cluster method for special classes of incomplete model spaces," J. Chem. Phys. **91**, 6187 (1989); L. Meissner and R. J. Bartlett, "A general model-space coupled-cluster method using a Hilbert-space approach," J. Chem. Phys. **92**, 571 (1990).
[64] G. L. Malli, A. B. F. Da Silva, and Y. Ishikawa, "Universal Gaussian basis set for accurate ab initio relativistic Dirac-Fock calculations," Phys. Rev. A **47**, 143 (1993).
[65] W. Küchle, M. Dolg, H. Stoll, and H. Preuss, "Energy-adjusted pseudopotentials for the actinides. Parameter sets and test calculations for thorium and thorium monoxide," J. Chem. Phys. **100**, 7535 (1994).
[66] L. Labzowsky, I. Goidenko, M. Tokman, and P. Pyykkö, "Calculated self-energy contributions for an ns valence electron using the multiple-commutator method," Phys. Rev. A **59**, 2707 (1999).
[67] M. E. Hanni, S. R. Lundeen, C. W. Fehrenbach, "Spectroscopy of high-L Rydberg levels of Fr-like U and Th Ions," http://meetings.aps.org/link/BAPS.2010.DAMOP.C5.7.
[68] M. S. Saxena, "Self-consistent field with exchange for U^{6+}," J. Phys. B **4**, 433 (1971).
[69] K. Rashid, M. Z. Saadi, and M. Yasin, "Dirac-Fock total energies, ionization energies, and orbital energies for uranium ions (U I to U XCII)," At. Data Nucl. Data Tables **40**, 365 (1988).
[70] W. J. Liu, W. Küchle, and M. Dolg, "Ab initio pseudopotential and density-functional all-electron study of ionization and excitation energies of actinide atoms," Phys. Rev. A **58**, 1103 (1998).
[71] X. Y. Cao and M. Dolg, "Theoretical prediction of the second to fourth actinide ionization potentials," Mol. Phys. **101**, 961 (2003).
[72] G. C. Rodrigues, P. Indelicato, J. P. Santos, P. Patte, and F. Parente, "Systematic calculation of total atomic energies of ground state configurations," At. Data Nucl. Data Tables **86**, 117 (2004).
[73] Yu. Ralchenko, A. E. Kramida, J. Reader, and NIST ASD Team (2010), NIST Atomic Spectra Database, http://physics.nist.gov/asd.

[74] A. Landau, E. Eliav, Y. Ishikawa, and U. Kaldor, "Intermediate Hamiltonian Fock-space coupled-cluster method in the one-hole one-particle sector: Excitation energies of xenon and radon," J. Chem. Phys. **115**, 6862 (2001).
[75] E. Eliav and U. Kaldor, "Transition energies of Rn- and Fr-like actinide ions by relativistic intermediate Hamiltonian Fock-space coupled-cluster methods," Chem. Phys. **392**, 78 (2012).
[76] U. I. Safronova and M. S. Safronova, "Correlation and relativistic effects in actinide ions," Phys. Rev. A **84**, 052515 (2011).
[77] B.O. Roos, "The CASSCF method and its applications in electronic structure calculations," in: "Advances in Chemical Physics: Ab Initio Methods in Quantum Chemistry" vol. II, K. P. Lawley, Ed., John Wiley & Sons, Chichester, U.K., (1987), p. 399.
[78] B. O. Roos, K. Andersson, M. P. Fülscher, P.-Å. Malmqvist, L. Serrano-Andrés, K. Pierloot, and M. Merchán, "Multiconfigurational perturbation theory: applications in electronic spectroscopy," Adv. Chem. Phys. **43**, 219 (1996).
[79] M. Schädel (ed.), "The Chemistry of Superheavy Elements," Kluwer Academic Publishers (Dordrecht, 2003).
[80] V. Pershina and D. C. Hoffman, "The chemistry of the heaviest elements," in "Theoretical Chemistry and Physics of Heavy and Superheavy Elements," Ed. U. Kaldor and S. Wilson, Kluwer Academic Publishers (Dordrecht, 2003) p. 55.
[81] M. Sewtz, H. Backe, A. Dretzke, G. Kube, W. Lauth, P. Schwamb, K. Eberhardt, C. Grüning, P. Thörle, N. Trautmann, P. Kunz, J. Lassen, G. Passler, C. Z. Dong, S. Fritzsche, and R. G. Haire, "First observation of atomic levels for the element fermium (Z=100)," Phys. Rev. Lett. **90**, 163002 (2003).
[82] H. Backe *et al.*, "Towards optical spectroscopy of the element nobelium (Z = 102) in a buffer gas cell," Eur. Phys. J D **45**, 99 (2007).
[83] A. Borschevsky, E. Eliav, M. J. Vilkas, Y. Ishikawa, and U. Kaldor, "Predicted spectrum of atomic nobelium," Phys. Rev. A **75**, 042514 (2007).
[84] A. Borschevsky, E. Eliav, M. J. Vilkas, Y. Ishikawa, and U. Kaldor, "Transition energies of atomic lawrencium," Eur. Phys. J D **45**, 115 (2007).
[85] J. Sugar, "Revised ionization energies of the neutral actinides," J. Chem. Phys. **60**, 4103 (1974).
[86] E. Eliav, U. Kaldor, and Y. Ishikawa, "The relativistic coupled cluster method: Intrashell excitations in the f^2 shells of Pr^{3+} and U^{4+}," Phys. Rev. A, **51**, 225 (1995).
[87] Z. Cai, V. Meiser Umar, and C. Froese Fischer, "Large-scale relativistic correlation calculations: Levels of Pr^{3+}," Phys. Rev. Lett. **68**, 297 (1992).
[88] V. Kaufman and L. F. Radziemski, Jr., "The sixth spectrum of uranium (U VI)," J. Opt. Soc. Am. **66**, 599 (1976).
[89] L. Gagliardi, M. C. Heaven, J. W. Krogh, and B. O. Roos, "The electronic spectrum of the UO_2 molecule," J. Am. Chem. Soc. **127**, 86 (2005).
[90] J. F. Wyart, V. Kaufman, and J. Sugar, "Analysis of the spectrum of four-times-ionized uranium (U5)," Phys. Scr. **22**, 389 (1980).
[91] L. Maron, T. Leininger, B. Schimmelpfennig, V. Vallet, J. L. Heully, C. Teichteil, O. Gropen, and U. Wahlgren, "Investigation of the low-lying excited states of PuO_2^{2+}," Chem. Phys. **244**, 195 (1999).
[92] C. Clavaguera-Sarro, V. Vallet, D. Maynau, and C. J. Marsden, "Can density functional methods be used for open-shell actinide molecules? Comparison with multiconfigurational spin-orbit studies," J. Chem. Phys. **121**, 5312 (2004).

3

Relativistic All-Electron Approaches to the Study of f Element Chemistry

Trond Saue[1] *and Lucas Visscher*[2]

[1]*Laboratoire de Chimie et Physique Quantiques, Université Paul Sabatier (Toulouse III)*
[2]*Theoretical Chemistry, Amsterdam Center for Multiscale Modeling, VU University Amsterdam*

3.1 Introduction

In this chapter we will provide a critical review of the use of 2- and 4-component relativistic Hamiltonians combined with all-electron methods and appropriate basis sets for the study of lanthanide and actinide chemistry. These approaches provide in principle the more rigorous treatment of the electronic structure but typically demand large computational resources due to the large basis sets that are required for accurate energetics. A complication is furthermore the open-shell nature of many systems of practical interest that make "black box" application of conventional methods impossible. Especially for calculations in which electron correlation is explicitly considered one needs to find a balance between the appropriate treatment of the multi-reference nature of the wave function and the practical limitations encountered in the choice of an active space. For density functional theory (DFT) calculations one needs to select the appropriate density functional approximation (DFA) on basis of assessments for lighter elements because little or no high-precision experimental information on isolated molecules is available for the f elements. This increases the demand for reliable theoretical ("benchmark") data in which all possible errors due to the inevitable approximations are carefully checked. In order to do so we need to understand how f elements differ from the more commonly encountered main group elements and also from the d elements with which they of course share some characteristics.

Computational Methods in Lanthanide and Actinide Chemistry, First Edition. Edited by Michael Dolg.

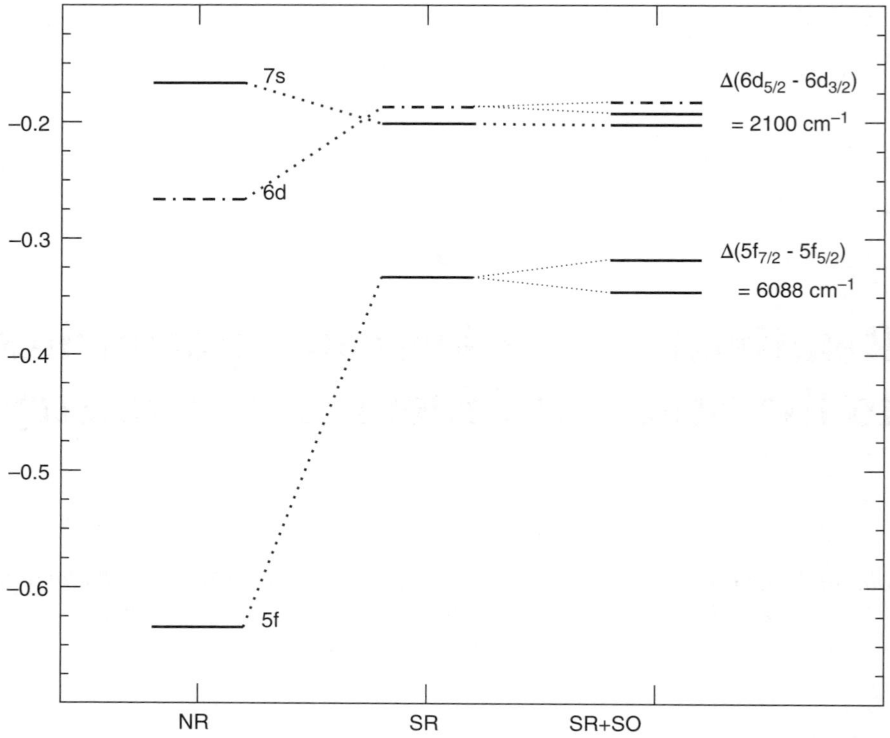

Figure 3.1 *Evolution of valence orbital energies (in E_h) of the uranium atom in the electronic ground state $[Rn]5f^3 6d^1 7s^2$ calculated by the Hartree-Fock method at the non-relativistic (NR) level, then variationally adding scalar relativistic (SR) effects and finally spin-orbit (SO) effects*

It is instructive to start out by considering what we can learn about f element chemistry from simple atomic calculations. In Figure 3.1 we demonstrate how relativistic effects influence the valence orbital energies of the neutral uranium atom in the electronic ground state $[Rn]5f^3 6d^1 7s^2$. We make a distinction between *scalar relativistic effects*, associated with the relativistic mass increase of the electron due to its high speed in the vicinity of heavy nuclei, and *spin-orbit effects*, associated with the interaction of the electron spin with the magnetic field induced by charges (e.g., nuclei and other electrons) in *relative motion*. The effect of scalar relativity is seen to be quite dramatic: The binding energy of the $5f$ electron is reduced from 0.634 E_h (17.3 eV) to 0.333 E_h (9.07 eV), making it much more available for chemical interaction, whereas the splitting between the $7s$ and $6d$ level is reduced from 0.100 E_h (21893 cm^{-1}) to -0.014 E_h (-3100 cm^{-1}), even changing sign. When we add spin-orbit interaction, the $5f$ level is split by 0.028 E_h (6088 cm^{-1}) to the $5f_{5/2}$ and $5f_{7/2}$ components, whereas the $6d$ level is split by 0.010 E_h (2100 cm^{-1}) to the $6d_{3/2}$ and $6d_{5/2}$ components. Clearly the inclusion of relativistic effects is crucial for a proper description of the chemistry of f elements.

In Figure 3.2 and Figure 3.3 we trace energies and expectation values $\langle r \rangle$, respectively, of the outer orbitals of 4f and 5f elements. These have been obtained from 4-component

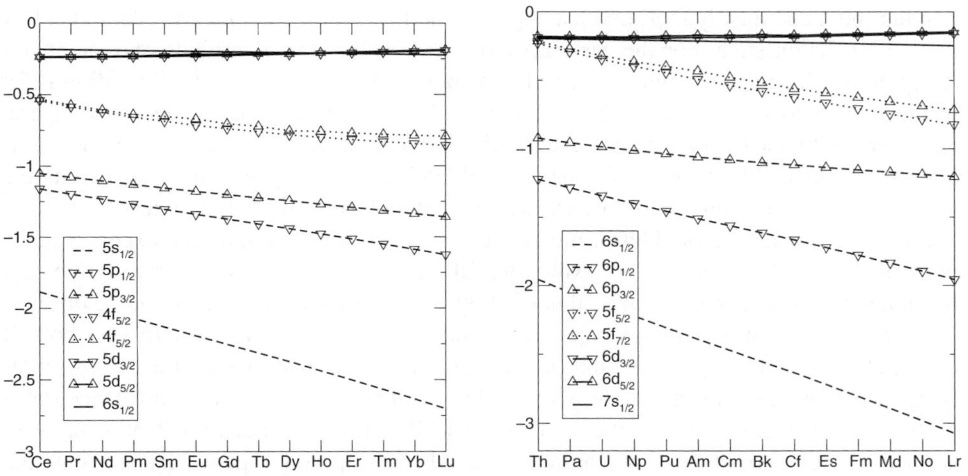

Figure 3.2 *Orbital energies (in E_h) of (left) the 4f elements cerium through lutetium and (right) the 5f elements thorium through lawrencium from 4-component relativistic Hartree-Fock calculations averaging over the $(n-2)f^i(n-1)d^1ns^2$ $(i = 1, 14)$ valence configuration of the neutral atom*

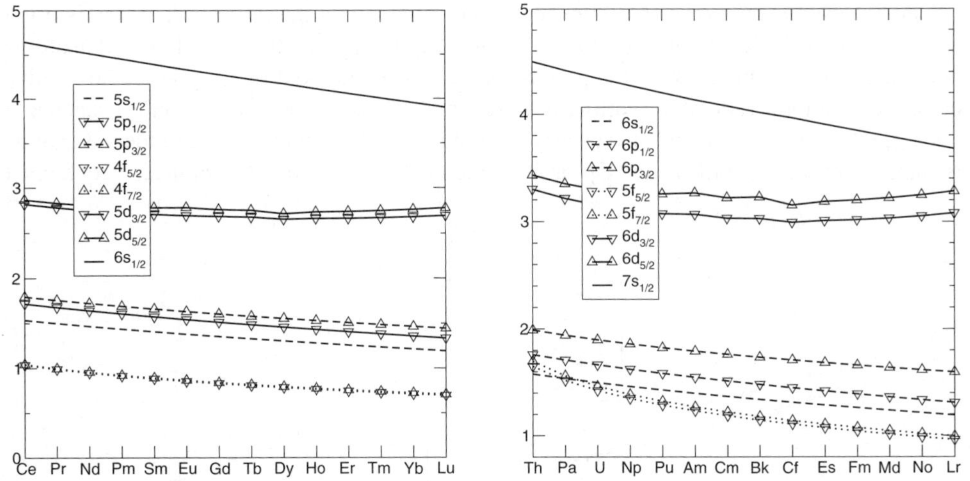

Figure 3.3 *Orbital radial expectation values $< r >$ (in a_0) of (left) the 4f elements cerium through lutetium and (right) the 5f elements thorium through lawrencium from 4-component relativistic Hartree-Fock calculations averaging over the $(n-2)f^i(n-1)d^1ns^2$ $(i = 1, 14)$ valence configuration of the neutral atom*

relativistic Hartree-Fock calculations averaging over the $(n-2)f^i(n-1)d^1ns^2$ $(i = 1, 14)$ valence configuration of the neutral atom. This configuration does not correspond to the ground electronic one for all f elements, but gives access to information about the valence (n–2)f, (n–1)d and ns orbitals. Beginning with the orbital energies of the 4f elements we see that the 4f orbitals are energetically well separated from the valence 5d and 6s orbitals, making them chemically inert and leading to a chemistry of the lanthanides dominated by the +3 oxidation state. For the early actinides we see that the 5f orbitals may be chemically active, leading to a wide range of possible oxidation states; plutonium allows, for instance, all oxidation states between +1 and +8. For the late actinides the energetic separation between 5f and 6d/7s increases and the chemistry reverts to that of the lanthanides. This leads to the difficulty of chemical separation of lanthanides and late actinides encountered, for instance, in the treatment of nuclear waste. A more technical point to note is that crossing of $(n-1)d$ and ns orbital levels along both the 4f and 5f series may lead to convergence problems in SCF calculations not exploiting full atomic symmetry, as is typically the case for molecular codes, since numerical noise may induce artificial mixing of s and d orbitals.

Looking now at orbital radial expectation values (see Figure 3.3) we see that for both 4f and 5f elements the ns, $(n-1)d$ and $(n-2)f$ are radially well separated, the latter orbitals lying inside the $(n-1)p$ and $(n-1)s$ levels. This somewhat ambivalent core/valence status of the f orbitals leads to f element complexes typically having an open f-shell but being very atomic in nature. An interesting feature to note when comparing Figure 3.2 and Figure 3.3 is that the 5f orbitals of the actinides shows a much larger spin-orbit splitting than the 6d orbitals, ranging from 3700 to 23600 cm^{-1}, whereas the difference in radial expectation values between the spin-orbit components is significantly larger for the 6d orbitals. The explanation is that since the 5f orbitals are more core-like than the 6d orbitals, it takes much more energy to deform them.

The steady decrease of 4f radial expectation values along the lanthanide series is often associated with the lanthanide contraction. However, it is a perfectly normal trend that atoms become smaller along a row in the periodic table. As pointed out by Lloyd [1], the relative contraction of ionic radii of +3 cations is *larger* for the 3d elements ($Sc^{3+}-Ga^{3+}$) than the 4f ones ($La^{3+}-Lu^{3+}$). Also, what is clearly seen in Figure 3.4 is that the size of the +3 cations is dictated by the size of 5s and 5p orbitals rather than 4f. When the pioneer geochemist

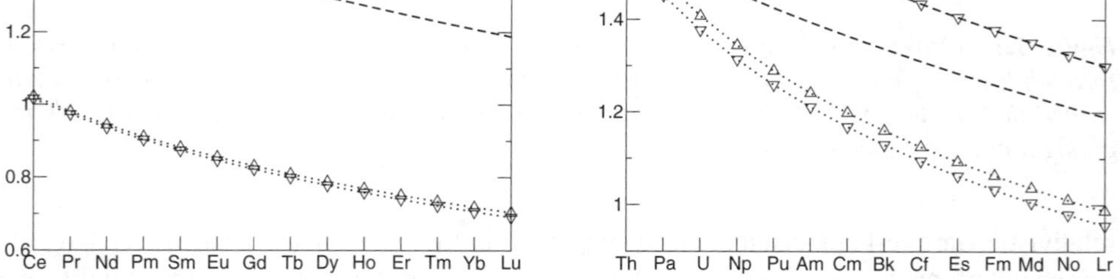

Figure 3.4 *Orbital radial expectation values $<r>$ (in a_0) of (left) the 4f elements cerium through lutetium and (right) the 5f elements thorium through lawrencium from 4-component relativistic Hartree-Fock calculations averaging over the $(n-2)f^i$ $(i=1,14)$ valence configuration of the +3 cations*

Victor Moritz Goldschmidt coined the term 'lanthanide contraction' in 1925 [2], it was with respect to a *vertical* rather than a horizontal trend in the periodic table: The covalent radius of silver is larger than that of copper, as expected when descending a column of the periodic table, whereas the covalent radius of gold is *smaller* than that of silver. This broken trend can be explained by noting that when going from copper to silver s, p and d shells are filled, whereas when going from silver to gold an f shell, with different screening properties than the preceeding shells, is also filled. An elegant theoretical study by Bagus *et al.* [3] shows that the expected increase of atomic size down the column of noble metal is restored when the combined effects of relativity and the lanthanide contraction are suppressed.

The one-electron picture is broken to some extent when including electron correlation. Especially for the early actinides, but also the first true f-element, cerium, configurations in which the $(n-1)d$ orbitals are occupied are energetically close or lower than the dominant $(n-2)f^ins^2$ valence configuration found in most of the lanthanides $(i=3-7, 9-14)$ and the later actinides $(i=6-7, 9-14)$ [4]. For atoms these different configurations are non-interacting due to the difference in parity of d and f-orbitals, however, breaking of the inversion symmetry by complex formation induces mixing and the need to treat all valence orbitals in a balanced fashion. As we will discuss later, this has implications both for the choice of basis sets and for the choice of a suitable electron correlation method. We will first start with a discussion of the available relativistic Hamiltonians that form a prerequisite for the treatment of electronic structure of heavy elements. We employ SI-based atomic units throughout this chapter.

3.2 Relativistic Hamiltonians

3.2.1 General Aspects

In this section we will provide a brief but self-contained overview of relativistic Hamiltonians for all-electron calculations. More detailed discussions can be found in [5–7] and references therein. In the non-relativistic domain the electronic Hamiltonian has the form

$$\hat{H} = V_{NN} + \sum_i \hat{h}(i) + \frac{1}{2}\sum_{i\neq j} \hat{g}(i,j); \quad V_{NN} = \frac{1}{2}\sum_{A\neq B} \frac{Z_A Z_B}{R_{AB}} \tag{3.1}$$

where V_{NN} is the classic repulsion of fixed nuclei. The one- and two-electron operators are taken to be

$$\hat{h} = \hat{h}_0 + V_{eN}; \quad \hat{g}(1,2) = \frac{1}{r_{12}} \tag{3.2}$$

where appears the free-particle operator $\hat{h}_0$ (kinetic energy), the interaction with the scalar potential of fixed nuclei $V_{eN} = -e\phi_{nuc}$ and the instantaneous two-electron Coulomb interaction. The free-particle operator has the same form as the classical free-particle energy, and quantization is obtained by the heuristic substitution

$$\left.\begin{array}{lcl} E & \rightarrow & i\dfrac{\partial}{\partial t} \\ \mathbf{p} & \rightarrow & \hat{\mathbf{p}} = -i\boldsymbol{\nabla} \end{array}\right\} \quad \Rightarrow \quad E = \frac{p^2}{2m} \quad \rightarrow \quad i\frac{\partial}{\partial t}\psi = \frac{\hat{p}^2}{2m}\psi \tag{3.3}$$

An important observation is that the relativistic electronic Hamiltonian has exactly the same generic form (3.1) as the non-relativistic one. This feature becomes perhaps even more manifest when expressing the electronic Hamiltonian on second-quantized form

$$\hat{H} = \sum_{p,q} h_{pq} p^\dagger q + \frac{1}{4} \sum_{p,q,r,s} \langle pq \,||\, sr \rangle \, p^\dagger q^\dagger rs \tag{3.4}$$

where $p^\dagger$ and q are creation- and annihilation operators, respectively, with respect to a specific set of one-electron functions (orbitals). These are combined with integrals h_{pq} and $\langle pq \,||\, sr\rangle$ over the one- and two-electron operators, respectively, the latter being on anti-symmetrized form

$$h_{pq} = \langle p|\hat{h}|q\rangle; \quad \langle pq||sr\rangle = \langle pq|sr\rangle - \langle pq|rs\rangle \tag{3.5}$$

In the non-relativistic case these integrals are real, and this also holds true when scalar relativistic effects are added. With the inclusion of spin-orbit interaction, complex algebra becomes mandatory (except for high-symmetry cases [8]) and fewer integrals will be zero due to symmetry. Yet the overall structure of the Hamiltonian is unchanged from the non-relativistic to the relativistic domain, and this generic form, combined with ansatzes for the wavefunction, suffices to generate the working formulas for the various methods of quantum chemistry. An important consequence is that whereas the choice of method (y) and basis set (N) combine to give a computation cost of N^y, the choice of Hamiltonian only contributes a prefactor, independent of system size. There are, however, some technical issues and challenges associated with the occurence of negative-energy solutions of the relativistic one-electron Hamiltonian that will be addressed in the following.

However, before going into a detailed discussion of various relativistic Hamiltonians we will introduce an alternative form of the electronic Hamiltonian (3.4), which is useful for wavefunction-based correlation methods. It is obtained by switching to a particle-hole formalism and then introducing normal ordering. In the second-quantization formalism creation and annihilation operators refer to some specific set of (orthonormal) orbitals, and Slater determinants in Hilbert space translate into occupation-number vectors in Fock space. The annihilation operators in equation 3.4 by definition give zero when acting on the vacuum state

$$p\,|0\rangle = 0; \quad \forall p \tag{3.6}$$

which by default is the empty occupation-number (ket) vector. The creation operators are simply the hermitian conjugates of the annihilation operators and accordingly satisfy

$$\langle 0|\, p^\dagger = 0; \quad \forall p. \tag{3.7}$$

From the above mentioned relations it is easy to see that the vacuum expectation value of the electronic Hamiltonian (3.4) is zero. The particle-hole formalism implies a redefinition of the vacuum state. Since correlation energy is defined with respect to the Hartree-Fock energy, we redefine the vacuum state as being the occupation-number vector corresponding to the converged HF determinant, the Fermi vacuum. This leads to a redefinition of creation and annihilation operators as well. For the moment we restrict attention to closed-shell systems for which we may distinguish occupied and virtual orbitals, indexed by i, j, k, l and a, b, c, d, respectively. Creation operators $i^\dagger$ of occupied orbitals as well as annihilation operators a of virtual orbitals give zero with respect to the HF vacuum and are now both annihilation operators, whereas their conjugates i and $a^\dagger$ are creation operators of holes and particles, respectively. The electronic Hamiltonian can now be recast as

$$\hat{H} = E^{HF} + \hat{H}_N \tag{3.8}$$

where the first term is the HF energy. The second term is the normal-ordered Hamiltonian

$$\hat{H}_N = \sum_{p,q} F_{pq}\{p^\dagger q\} + \frac{1}{4}\sum_{p,q,r,s} \langle pq \,||\, sr\rangle \{p^\dagger q^\dagger rs\} \tag{3.9}$$

where appears matrix elements F_{pq} over the Fock operator. The curly brackets indicate normal ordering, that is, all creation operators with respect to the HF vacuum are moved to the left of annihilation operators, as if they anti-commuted. The expectation value of the normal-ordered Hamiltonian with respect to the HF vacuum is therefore manifestly zero and so in correlated theories it gives direct access to the correlation energy.

3.2.2 Four-Component Hamiltonians

From the discussion above it becomes clear that the major difference between non-relativistic and relativistic quantum chemistry comes from the choice of one- and two-electron operators of the electronic Hamiltonian. To get some idea of the underlying physics, a brief review of the basics of relativity will be necessary. The special theory of relativity is special in that it restricts attention to transformation between inertial frames, that is, reference frames connected by constant velocity $\mathbf{v}$. The principle of relativity states that the laws of physics have the same form in all such frames. Whereas this principle has been known since the times of Galilei, the truly revolutionary aspect of the 1905 paper of Einstein [9] was the second postulate, stating the invariance of the speed of light c in all inertial frames. This implied that the Lorentz transformation, at that time already known from electrodynamics, applies to all domains of physics. Let us consider two inertial frames – K and K' – which we for simplicity, but without loss of generality, let coincide at $t = t' = 0$, when a flash of light is emitted from the origin. The invariance of the speed of light then implies that the *interval*

$$s = \sqrt{c^2t^2 - x^2 - y^2 - z^2} = \sqrt{c^2t'^2 - x'^2 - y'^2 - z'^2} \tag{3.10}$$

connecting the space and time coordinates (x, y, z, t) $((x', y', z', t'))$ of the lightfront in frame K (K') at a later time is zero. The resulting plasticity of space and time manifests itself through phenomena such as length contraction and time dilation, which are admittedly remote from everyday experience.

It is convenient to gather space and time coordinates in a vector, *4-position*

$$r_\mu = (\mathbf{r}, ict)\,; \quad r_\mu r_\mu = r^2 - c^2t^2 \tag{3.11}$$

whose norm is simply related to the interval and therefore conserved in Lorentz transformations. Other 4-vectors, with the same transformation properties, can be generated from 4-position by appropriate manipulations. A useful 4-vector is the 4-gradient

$$\partial_\mu = \left(\boldsymbol{\nabla}, -\frac{i}{c}\frac{\partial}{\partial t}\right) \tag{3.12}$$

4-velocity is obtained as a time derivative, but since absolute time is abolished in the theory of relativity, a Lorentz invariant *proper time* τ is generated from the infinitesimal interval

$$ds = \sqrt{c^2dt^2 - dx^2 - dy^2 - dz^2} = c\gamma^{-1}dt = cd\tau. \tag{3.13}$$

This relation contains the Lorentz factor

$$\gamma = \frac{1}{\sqrt{1 - v^2/c^2}} \tag{3.14}$$

which can be considered as a diagnostic of relativistic effects since it goes towards infinity as the particle speed v approaches the speed of light c. When the the relative speed v/c is small, the Lorentz factor is essentially unity, indicating a non-relativistic regime. The non-relativistic limit is obtained by letting the speed of light c tend towards infinity. The Lorentz factor also suggests that c is an upper speed limit for any interaction since $v > c$ makes the factor purely imaginary.

With the introduction of proper time and use of the chain rule 4-velocity is obtained as

$$v_\mu = \frac{dr_\mu}{d\tau} = \gamma\frac{dr_\mu}{dt} = \gamma(\mathbf{v}, ic); \quad v_\mu v_\mu = -c^2 \tag{3.15}$$

The length of this 4-vector is simply the speed of light, thus by construction Lorentz invariant. Proceeding we multiply 4-vector by Lorentz invariant mass m to obtain 4-momentum

$$p_\mu = (\gamma m\mathbf{v}, i\gamma mc) \tag{3.16}$$

The space part of 4-moment is relativistic momentum $\mathbf{p} = \gamma m\mathbf{v}$, which is often re-written $\mathbf{p} = M\mathbf{v}$, where appears relativistic mass $M = \gamma m$, which increases with the speed of the particle. The content of the time part of 4-momentum is less obvious, but quantification of 4-momentum, $\hat{p}_\mu = -i\partial_\mu$, suggests that it is proportional to energy so that we may write

$$p_\mu = (\mathbf{p}, \frac{i}{c}E); \quad p_\mu p_\mu = p^2 - \frac{E^2}{c^2} = -mc^2 \tag{3.17}$$

The relativistic energy of a free particle can thereby be expressed as

$$E = \pm\sqrt{m^2c^4 + c^2p^2} \in \left\langle -\infty, -mc^2\right] \cup \left[+mc^2, +\infty\right\rangle \tag{3.18}$$

Upon taking the square root, both signs must be considered, leading to a positive- and a negative-energy branch separated by a gap of $2mc^2$. In classical mechanics free particles of negative energy are not known, and since the energy can not change discontinuosly we can discard the negative-energy branch. The positive energy of relativistic free particles can then be related to that of non-relativistic ones through the expansion

$$E = +mc^2\sqrt{1+\frac{p^2}{m^2c^2}} = mc^2 + \frac{p^2}{2m} - \frac{p^4}{8m^3c^2} + \dots \tag{3.19}$$

The second term is seen to be the non-relativistic free-particle energy. Further terms to the right correspond to relativistic corrections and vanish when the speed of light goes to infinity. The first term, corresponding to the rest mass, then goes to infinity as well, but can be removed by simply resetting the origin of the energy scale.

In quantum mechanics things are not so simple since the energy can, as the name implies, change in leaps. The negative-energy branch can therefore not be ignored. Also, quantification of the energy expression (3.18) is far from straightforward since the direct heuristic substitution leads to the square-root of a Laplacian. Dirac [10] achieved a factorization of the squared energy expression (3.17) without square roots at the price of introducing a 4×4 matrix operator

$$\left(\hat{h}_0 - i\frac{\partial}{\partial t}\right)\psi = 0; \quad \hat{h}_0 = \beta mc^2 + c\,(\boldsymbol{\alpha}\cdot\mathbf{p}) = \begin{bmatrix} +mc^2 & c\,(\boldsymbol{\sigma}\cdot\mathbf{p}) \\ c\,(\boldsymbol{\sigma}\cdot\mathbf{p}) & -mc^2 \end{bmatrix} \tag{3.20}$$

with corresponding 4-component vector solutions

$$\psi = \begin{bmatrix} \psi^L \\ \psi^S \end{bmatrix} = \begin{bmatrix} \psi^{L\alpha} \\ \psi^{L\beta} \\ \psi^{S\alpha} \\ \psi^{S\beta} \end{bmatrix} \tag{3.21}$$

The Dirac matrices $\boldsymbol{\alpha}$ and β are defined in terms of the 2×2 Pauli spin matrices as well as identity and null matrices

$$\boldsymbol{\alpha} = \begin{bmatrix} \mathbf{0}_{2\times 2} & \boldsymbol{\sigma} \\ \boldsymbol{\sigma} & \mathbf{0}_{2\times 2} \end{bmatrix}; \quad \beta = \begin{bmatrix} I_{2\times 2} & 0_{2\times 2} \\ 0_{2\times 2} & -I_{2\times 2} \end{bmatrix} \tag{3.22}$$

The free-particle Dirac Hamiltonian $\hat{h}_0$ has the spectrum (3.18). To prevent positive-energy electrons from descending down into the negative-energy continuum, Dirac postulated that all negative-energy solutions were occupied and thus not available, due to the Pauli exclusion principle. This state of affairs does not hinder excitations from the negative-energy branch if sufficient energy ($\sim 2mc^2$) is provided. Since the negative-energy electrons are assumed to be not directly observable, such excitations will lead to the observation of an electron, but also the positively charged hole it leaves behind, identified as the positron, the anti-particle of the electron. The free-particle Dirac equation does not specify charge (as it is not needed) and therefore applies equally well to electrons and positrons.

Electric and magnetic fields are introduced in terms of their potentials through the principle of minimal electromagnetic coupling [11]

$$p_\mu \rightarrow p_\mu - qA_\mu \tag{3.23}$$

where appears the 4-potential A_μ

$$A_\mu = \left(\mathbf{A}, \frac{i}{c}\phi\right) \tag{3.24}$$

as well as particle charge q. For electrons one sets $q = -e$ such that the substitution (3.23) takes the form

$$\mathbf{p} \rightarrow \pi = \mathbf{p} + e\mathbf{A}; \quad E \rightarrow E + e\phi \tag{3.25}$$

For relativistic electons in the presence of the electrostatic potential ϕ_{nuc} of clamped nuclei the Dirac Hamiltonian accordingly reads

$$\hat{h}_D = \hat{h}_0 + V_{eN}; \quad V_{eN} = -e\phi_{nuc} \tag{3.26}$$

To align the zero of the relativistic energy scale with that of the non-relativistic one, the rest mass of the electron is subtracted through the substitution

$$\beta \rightarrow \beta^{'} = \beta - I_{4\times 4} \tag{3.27}$$

which completes the construction of the one-electron part of the relativistic electronic Hamiltonian.

It should be emphasized that (3.26) is the form of the one-electron Hamiltonian in the frame of clamped nuclei. In the frame of the electron an additional term would appear, describing the magnetic field induced by the moving nuclei, which couples to electron spin, thus representing spin-orbit coupling induced by the motion of nuclei relative to the electron.

Atomic solutions of h_D take the form

$$\psi(r, \theta, \phi) = \begin{bmatrix} R^L(r)\chi_{\kappa,m_j}(\theta, \phi) \\ iR^S(r)\chi_{-\kappa,m_j}(\theta, \phi) \end{bmatrix} \tag{3.28}$$

where R^X $(X = L, S)$ are real scalar radial functions and χ_{κ,m_j} complex 2-component angular functions. The angular functions are eigenfunctions of $\hat{j}^2$, where $\hat{\mathbf{j}}$ is total angular momentum

$$\hat{\mathbf{j}} = \hat{\mathbf{l}} + \hat{\mathbf{s}}, \tag{3.29}$$

reflecting spin-orbit coupling. However, the specification of quantum numbers j and m_j does not fully specify an angular function since orbital angular momentum l and spin $s = \frac{1}{2}$ can generally couple to both $j = l - \frac{1}{2}$ and $j = l + \frac{1}{2}$. The two possibilities are distinguished by the quantum number κ, which are eigenvalues of the operator $\hat{\kappa} = -\left[(\boldsymbol{\sigma}\cdot\mathbf{p}) + I_{2\times 2}\right]$ and take the values

$$\kappa = \begin{cases} l; & \kappa > 0 \\ -(l+1); & \kappa < 0 \end{cases} \tag{3.30}$$

as illustrated in Table 3.1.

Table 3.1 Quantum numbers for relativistic atomic orbitals

	$s_{1/2}$	$p_{1/2}$	$p_{3/2}$	$d_{3/2}$	$d_{5/2}$	$f_{5/2}$	$f_{7/2}$
j	1/2	1/2	3/2	3/2	5/2	5/2	7/2
κ	-1	+1	−2	+2	−3	+3	−4

No simple analytic form can be given to the fully relativistic two-electron interaction. Since two electrons are charges in relative motion, the electrostatic (charge-charge) interaction is supplemented by magnetic (current-current) interactions. Moreover, since the interactions are not instantaneous the complete history of the two moving particles is required for the specification of their interaction. The two-electron interaction is therefore given as a perturbation expansion in orders of c^{-2}. The zeroth-order term is the Coulomb term

$$g^{Coul}(i,j) = \frac{I_{4\times 4} \cdot I_{4\times 4}}{r_{ij}} \tag{3.31}$$

whereas the first-order term is the Breit term which splits into the Gaunt term and a gauge-dependent term

$$g^{Breit}(i,j) = g^{Gaunt}(i,j) + g^{gauge}(i,j); \quad g^{Gaunt}(i,j) = -\frac{c\alpha_i \cdot c\alpha_j}{c^2 r_{ij}} \tag{3.32}$$

The two-electron operator is given in the nuclear frame and not in the reference of either electron. The spin-orbit coupling due to the relative motion of elecrons therefore splits into two parts: The total interaction is the coupling of the spin of a selected reference electron with the magnetic field induced by a second electron. The spin-same orbit (SSO) and spin-other orbit (SOO) contributions arise from the motion of the reference electron and the other electron, respectively, relative to the nuclear frame and are carried by the Coulomb and Gaunt terms, respectively. For most molecular application it suffices to include the Coulomb term only, thus defining the Dirac-Coulomb Hamiltonian, but for the accurate calculation of molecular spectra the Gaunt term should be included as well.

3.2.3 Two-Component Hamiltonians

The negative-energy solutions are a troublesome aspect of the Dirac Hamiltonian and clearly of limited interest in chemical applications. Over the years, much effort has therefore been spent on eliminating the positronic degrees of freedom of the Dirac Hamiltonian. Most such efforts start from the Dirac equation in the molecular field

$$\begin{bmatrix} V_{eN} & c(\boldsymbol{\sigma} \cdot \mathbf{p}) \\ c(\boldsymbol{\sigma} \cdot \mathbf{p}) & V_{eN} - 2mc^2 \end{bmatrix} \begin{bmatrix} \psi^L \\ \psi^S \end{bmatrix} = \begin{bmatrix} \psi^L \\ \psi^S \end{bmatrix} E \tag{3.33}$$

and seek to generate a 2-component Hamiltonian h_{++}, which reproduces the positive-energy spectrum of the parent Hamiltonian. This can be accomplished by a unitary block diagonalization [12]

$$U^\dagger \begin{bmatrix} h_{LL} & h_{LS} \\ h_{SL} & h_{SS} \end{bmatrix} U = \begin{bmatrix} h_{++} & 0 \\ 0 & h_{--} \end{bmatrix} \tag{3.34}$$

or, equivalently, by elimination of the small components followed by renormalization of the transformed large components. The transformation can accordingly be expressed as seen in Heully *et al.* [13]

$$U = W_1 W_2; \quad W_1 = \begin{bmatrix} 1 & -R^\dagger \\ R & 1 \end{bmatrix}; \quad W_2 = \begin{bmatrix} \Omega_+ & 0 \\ 0 & \Omega_- \end{bmatrix}; \quad \begin{matrix} \Omega_+ = \left(1 + R^\dagger R\right)^{-1/2} \\ \Omega_- = \left(1 + R R^\dagger\right)^{-1/2} \end{matrix} \tag{3.35}$$

where the transformations W_1 and W_2 assure decoupling and renormalization, respectively. There is an infinite number of unitary transformations that will bring the parent Hamiltonian on block-diagonal form. This becomes evident by taking into consideration that after a complete diagonalization of the 4-component Hamiltonian, arbitrary unitary transformations can be carried out within the $\{++\}$ and $\{--\}$ blocks separately. The above form of the unitary transformation is dictated by the desire to maintain the resulting positive-energy 2-component functions as close to the original large component solutions as possible.

The identification of the operator R becomes clear when considering the effect of the unitary transformation on the orbitals

$$U^\dagger \begin{bmatrix} \psi^L \\ \psi^S \end{bmatrix} = \begin{bmatrix} \Omega_+ \left(\psi^L + R^\dagger \psi^S\right) \\ \Omega_- \left(\psi^S - R\psi^L\right) \end{bmatrix} \tag{3.36}$$

For positive-energy solutions we want the lower components to be zero, which implies

$$\psi_+^S = R\psi_+^L \tag{3.37}$$

and leads to 2-component positive-energy solutions of the form

$$\psi_+ = \frac{1}{\sqrt{1 + R^\dagger R}} \left(\psi^L + R^\dagger \psi^S\right) = \frac{1}{\sqrt{1 + R^\dagger R}} \left(\psi^L + R^\dagger R \psi^L\right) = \sqrt{1 + R^\dagger R}\,\psi^L, \tag{3.38}$$

suggesting that they can be expanded in the original large component basis. The operator R is thereby identified as the exact coupling between the large and small component of the positive-energy solutions of the Dirac equation (3.33), which is given by

$$R = \left(2mc^2 - V_{eN} + E\right)^{-1} c\,(\boldsymbol{\sigma} \cdot \mathbf{p}) \tag{3.39}$$

These results show not only that the coupling, and thus the separation of positive- and negative-energy solutions, depends explicitly on the external potential, but also on the energy. The latter feature implies that the proper 2-component Hamiltonian can only be generated after solving the parent 4-component problem and led to the development of several approximate 2-component relativistic Hamiltonian. They may be classified according to whether they are generated in one or several steps:

1. One-step decoupling
   (a) **Pauli Hamiltonian:** Using the approximate decoupling

$$R = \frac{1}{2mc}\left[1 + \frac{E - V}{2mc^2}\right]^{-1} (\boldsymbol{\sigma} \cdot \mathbf{p}) \sim \frac{1}{2mc} (\boldsymbol{\sigma} \cdot \mathbf{p}) \tag{3.40}$$

and retaining terms only to $O(c^{-2})$ gives the Pauli Hamiltonian

$$\hat{h}^{\text{Pauli}} = V + T \underbrace{-\frac{p^4}{8m^3c^2}}_{\text{mass-velocity}} \underbrace{+\frac{1}{8m^2c^2}(\nabla^2 V)}_{\text{Darwin}} \underbrace{+\frac{1}{4m^2c^2}\boldsymbol{\sigma} \cdot [(\boldsymbol{\nabla} V) \times \mathbf{p}]}_{\text{spin-orbit}} \tag{3.41}$$

It features three terms providing relativistic corrections to the non-relativistic Hamiltonian. The mass-velocity term can be recognized from equation 3.18 as the first-order relativistic correction to the kinetic energy. The Darwin term is a relativistic correction to the potential energy arising from *Zitterbewegung*, the highly oscillatory motion of the relativistic electron superimposed on its mean position. One interpretation of the *Zitterbewegung* is that in the vicinity of an electron its field is sufficiently strong to allow the creation of an electron-positron pair. The positron annihilates the original electron and the new electron takes over, leading to a jumping motion of the electron(s). In the presence of a single point nucleus the spin-orbit term takes the more familiar form

$$\hat{h}^{SO} = \frac{Z}{2m^2c^2r^3}\mathbf{s} \cdot \mathbf{l} \tag{3.42}$$

from which the name of the interaction arises. However, it should be carefully noted that spin-orbit interaction is due to magnetic induction and that the orbital angular momentum operator $\mathbf{l}$ appearing in the operator expression represents the *relative* motion of electron and nucleus.

The Pauli Hamiltonian has no lower bound and is therefore not recommended for variational calculations. It is nowadays mostly used in low-order perturbation calculations on light atoms ($Z < 40$).

   (b) **Regular Hamiltonians:** Using the approximate decoupling

$$R = \frac{c}{2mc^2 - V}\left[1 + \frac{E}{2mc^2 - V}\right]^{-1} (\boldsymbol{\sigma} \cdot \mathbf{p}) \sim \frac{c}{2mc^2 - V} (\boldsymbol{\sigma} \cdot \mathbf{p}) \tag{3.43}$$

and ignoring renormalization gives the Zeroth-Order Regular Approximation (ZORA) Hamiltonian [14–16]

$$\hat{H}^{\text{ZORA}} = V + \frac{1}{2m} (\boldsymbol{\sigma} \cdot \mathbf{p}) \frac{2mc^2}{2mc^2 - V} (\boldsymbol{\sigma} \cdot \mathbf{p}) \tag{3.44}$$

The second term can be thought of as an effective kinetic energy operator that goes to the non-relativistic one when $V \to 0$. Proper renormalization gives the Infinite Order Regular Approximation (IORA) [17], often approximated by scaled ZORA [16], which improves on ZORA.

2. Multi-step decoupling
   Use of the exact *free-particle* decoupling

$$R = \left(2mc^2 + E\right)^{-1} c\left(\boldsymbol{\sigma} \cdot \mathbf{p}\right) \tag{3.45}$$

   maintains the kinetic energy operator on the relativistic square-root form (3.18) and thereby provides regularized and variationally stable H, but no exact decoupling.
   (a) Subsequent decoupling transformations in orders of the potenial defines the **Douglas-Kroll-Hess (DKH) Hamiltonian** to a given order. This sequence of transformations was first proposed by Douglas and Kroll [18], but the name of Bernd Hess has later been associated with this Hamiltonian due to his introduction of an effective momentum basis [19, 20], thus paving the way for the use of the DKH Hamiltonian in molecular calculations.
   (b) Iterating the coupling equation of the free-particle transformed Hamiltonian to obtain the coupling correct through some odd order $2k - 1$ in c^{-1} and then perform a *single* unitary transformation defines the **Barysz, Sadlej and Snijders (BSS) Hamiltonian** [21] to order $2k$.

The past decade has witnessed the development of **eXact 2-Component (X2C) Hamiltonians** [22–25]. This development was a result of two important realizations: i) that the cost of solving the 4-component *one-electron* problem, explicitly or implicitly, is minor compared to solving the molecular *many-electron* problem, which is usually at hand, and ii) that the X2C Hamiltonian can be generated using matrix algebra, thus eliminating the need to program up integrals over complicated operator expressions. The basic algorithm for the construction of the X2C Hamiltonian is therefore to solve the parent 4-component equation on matrix form, extract the coupling R from the solutions, then construct the transformation U and finally h^{X2C}. The X2C Hamiltonian reproduces exactly the positive-energy spectrum of the parent 4-component Hamiltonian. The algorithm also has the advantage of providing an explicit representation of the transformation U. Any property operator can therefore be transformed on the fly, thus avoiding picture change errors (see below).

All of the above procedures provide an approximate block diagonalization of the starting 4-component one-electron Hamiltonian from which a 2-component Hamiltonian may be extracted

$$H^{2c} = \left[U^{\dagger} H^{4c} U\right]_{++} \tag{3.46}$$

but do not automatically provide two-electron or property operators to be used in conjuction with the one-electron Hamiltonian. Property operators Ω^{4c} must be subjected to the same transformation as the Hamiltonian, that is

$$\Omega^{2c} = \left[U^{\dagger} \Omega^{4c} U\right]_{++}. \tag{3.47}$$

Use of the approximate expression

$$\Omega^{2c} \approx \left[\Omega^{4c}\right]_{LL} \tag{3.48}$$

leads to *picture change errors* [26–29], which for properties sampling the wavefunction in the vicinity of nuclei may be larger than the relativistic effects.

The appropriate 2-component two-electron operator should likewise be constructed as

$$g^{2c}(1,2) = \left[(U^{\dagger}(1) \otimes U^{\dagger}(2))g^{4c}(1,2)(U(1) \otimes U(2))\right]_{++}. \tag{3.49}$$

However, the straightforward transformation on matrix form of the two-electron operator in this manner will lead to calculations that are more expensive than a full 4-component calculation [30], since one needs to generate and transform the full set of two-electron integrals of the 4-component calculation. One approximate approach consists of combining the untransformed two-electron Coulomb operator with Atomic Mean-Field Integrals (AMFI) [31]. The latter provide two-electron spin-orbit corrections and are generated as mean-field representations of the 2-component two-electron spin-orbit operators, analogous to the two-electron part of the Fock matrix, for the individual atoms of the molecule. For correlated calculations in the framework of X2C, another scheme, the Molecular Mean-Field Approach [32], consists of solving the 4-component relativistic Hartree-Fock problem for the molecule and then construct the decoupling transformation U, which block diagonalizes the converged Fock matrix. The resulting 2-component Fock operator on matrix form is then used together with the untransformed two-electron Coulomb operator at the correlated level, providing molecular mean-field two-electron spin-orbit corrections, as can be seen from equation 3.9. Since typically only valence electrons are correlated the approximations to two-electron integrals affect only valence electrons, keeping an unapproximated 4-component description for core electrons.

3.2.4 Numerical Example

To illustrate the performance of the various Hamiltonians discussed in this section we present in Table 3.2 orbital energies of the uranium atom obtained with different relativistic Hamiltonians. We have carried out Hartree-Fock calculations optimizing the average energy of the complete set of determinants generated from the $[Rn]5f^{3}6d^{1}7s^{2}$ ground state configuration of the uranium atom. For all calculations we employed an uncontracted 33s29p20d13f4g2h Gaussian (large component) basis [33], corresponding to the cc-pVTZ level, and the DIRAC molecular code [34].

Reference values for the various 2-component relativistic Hamiltonians are provided by the 4-component Dirac-Coulomb Hamiltonian, but we have also included orbital energies obtained with the Dirac-Coulomb-Gaunt (DCG) Hamiltonian. As already mentioned, the Gaunt term brings in spin-other-orbit (SOO) interaction. Since spin-orbit interaction induced by other electrons will oppose the one induced by nuclei we see from Table 3.3 that the spin-orbit splitting of orbital levels is overall reduced. However, one should note that the Gaunt term also modifies $s_{1/2}$ levels.

Turning now to the 2-component relativistic Hamiltonians we observe that the eXact 2-Component (X2C) Hamiltonians reproduces the DC orbital energies quite well, although there is a difference of about 10 E_h for the $1s_{1/2}$ level. This discrepancy stems entirely from picture change errors in the two-electron operator since the one-electron X2C Hamiltonian by construction reproduces *exactly* the positive-energy spectrum of the parent 4-component Hamiltonian. In the present calculations the one-electron X2C Hamiltonian

Table 3.2 *Orbital energies of the uranium atom obtained with different relativistic Hamiltonians and Hartree-Fock calculations averaging over the* $[Rn]5f^3 6d^1 7s^2$ *ground state configuration*

	DCG	DC	X2C	DKH2	DKH1	ZORA	scZORA
$1s_{1/2}$	−4262.599	−4281.813	−4272.178	−4253.946	−4568.402	−4890.081	−4267.639
$2s_{1/2}$	−804.292	−806.637	−804.996	−802.931	−840.315	−829.339	−804.400
$2p_{1/2}$	−773.067	−777.035	−775.649	−774.270	−791.143	−799.722	−775.573
$2p_{3/2}$	−633.274	−635.783	−635.010	−635.027	−634.978	−651.542	−634.900
$3s_{1/2}$	−206.265	−206.730	−206.350	−205.894	−214.216	−208.368	−206.214
$3p_{1/2}$	−192.463	−193.251	−192.949	−192.624	−196.579	−194.945	−192.940
$3p_{3/2}$	−159.897	−160.378	−160.206	−160.220	−160.067	−161.622	−160.178
$3d_{3/2}$	−138.721	−139.070	−138.997	−139.024	−138.568	−140.214	−138.982
$3d_{5/2}$	−132.183	−132.426	−132.367	−132.393	−131.938	−133.477	−132.350
$4s_{1/2}$	−54.250	−54.355	−54.259	−54.140	−56.332	−54.425	−54.223
$4p_{1/2}$	−48.048	−48.232	−48.161	−48.077	−49.085	−48.334	−48.159
$4p_{3/2}$	−39.454	−39.554	−39.515	−39.522	−39.437	−39.633	−39.508
$4d_{3/2}$	−29.688	−29.744	−29.734	−29.743	−29.590	−29.817	−29.730
$4d_{5/2}$	−28.100	−28.130	−28.123	−28.132	−27.980	−28.197	−28.119
$4f_{5/2}$	−15.207	−15.202	−15.211	−15.220	−15.089	−15.247	−15.210
$4f_{7/2}$	−14.802	−14.786	−14.795	−14.803	−14.676	−14.828	−14.792
$5s_{1/2}$	−12.582	−12.603	−12.582	−12.553	−13.081	−12.587	−12.573
$5p_{1/2}$	−10.098	−10.136	−10.122	−10.103	−10.320	−10.133	−10.122
$5p_{3/2}$	−8.077	−8.095	−8.088	−8.091	−8.049	−8.094	−8.087
$5d_{3/2}$	−4.347	−4.352	−4.353	−4.356	−4.305	−4.356	−4.353
$5d_{5/2}$	−4.040	−4.041	−4.042	−4.045	−3.995	−4.044	−4.041
$5f_{5/2}$	−0.350	−0.346	−0.349	−0.350	−0.321	−0.349	−0.349
$5f_{7/2}$	−0.323	−0.318	−0.321	−0.322	−0.294	−0.321	−0.321
$6s_{1/2}$	−2.135	−2.139	−2.135	−2.130	−2.234	−2.134	−2.133
$6p_{1/2}$	−1.338	−1.344	−1.342	−1.339	−1.371	−1.343	−1.342
$6p_{3/2}$	−0.983	−0.985	−0.984	−0.985	−0.968	−0.984	−0.984
$6d_{3/2}$	−0.193	−0.193	−0.193	−0.194	−0.181	−0.193	−0.193
$6d_{5/2}$	−0.183	−0.183	−0.184	−0.184	−0.173	−0.184	−0.184
$7s_{1/2}$	−0.202	−0.202	−0.202	−0.202	−0.211	−0.202	−0.202

Table 3.3 *Spin-orbit splitting of orbital levels of the uranium atom obtained with different relativistic Hamiltonians and Hartree-Fock calculations averaging over the* $[Rn]5f^3 6d^1 7s^2$ *ground state configuration*

	DCG	DC	X2C	DKH2	DKH1	ZORA	scZORA
2p	139.793	141.252	140.638	139.244	156.165	148.179	140.672
3p	32.565	32.874	32.743	32.404	36.512	33.324	32.762
3d	6.538	6.644	6.630	6.631	6.631	6.737	6.632
4p	8.594	8.678	8.645	8.555	9.648	8.701	8.651
4d	1.588	1.614	1.611	1.611	1.611	1.620	1.612
4f	2.021	2.041	2.034	2.012	2.271	2.038	2.035
5p	0.307	0.312	0.311	0.311	0.310	0.312	0.312
5d	0.307	0.312	0.311	0.311	0.310	0.312	0.312
5f	0.027	0.028	0.028	0.028	0.027	0.028	0.028
6p	0.797	0.795	0.793	0.790	0.862	0.791	0.791
6d	0.009	0.010	0.010	0.010	0.008	0.010	0.010

was supplemented by the 2-electron Coulomb term as well as an atomic spin-orbit correction provided by the AMFI package [35]. The latter employs orbitals generated from a scalar relativistic second-order Douglas-Kroll-Hess (DKH2) atomic HF calculation for the generation of a mean-field spin-same-orbit operator at the first-order DKH1 level. The same computational setup has been employed for the DKH1 and DKH2 Hamiltonians, whereas 2-electron spin-orbit corrections enter the ZORA Hamiltonian by the introduction of the mean-field potential in the approximate coupling R (3.43). The DKH1 is obtained by a free-particle decoupling transformation and is clearly an insufficient approximation, giving errors in the range of 0.25–0.70 eV for the valence uranium orbitals. DKH2 brings about clear improvement but introduces somewhat larger errors than X2C for core orbitals. The ZORA Hamiltonian gives errors that are even larger than DKH1 in the core region, but in contrast to DKH1 provides an excellent description of relativistic valence orbitals, which therefore has lead to its widespread use in relativistic molecular calculations. Finally, we note that with the introduction of approximate renormalization with the scaled ZORA Hamiltonian, the deficiencies in the core region are largely eliminated.

3.3 Choice of Basis Sets

A prerequisite for the application of all-electron approaches is the availability of a reliable basis set for the particular f element and the chosen Hamiltonian. Ideally one would like to be able to assess the size of basis set errors by being able to benchmark the results with a basis set large enough to exclude significant errors due to truncation effects. In atomic calculations one may do so by choosing a very large universal basis set. In this approach a set of exponents is generated via the recurrence relation

$$\alpha_i = \beta\alpha_{i-1}(i = 2, n) \tag{3.50}$$

in which only the first exponent α_1, the scale factor β, and the number of exponents n is to be specified. For each l-value different subsets of the generated exponents are taken, allowing for straightforward extension with additional functions needed in correlation calculations. While the Universal Gaussian Basis (UGB) basis set of Malli and coworkers [36] was initially designed for atomic Hartree-Fock calculations with the Dirac-Coulomb Hamiltonian, it has been widely applied also in correlated calculations. This universal approach leads to basis set sizes too large to be feasible in molecular calculations, and more economical representations of the orbitals have therefore been developed.

Molecular applications are typically done with element-specific basis sets in which the energy is minimised for each of the elements individually. As discussed above, one should here first decide upon the Hamiltonian that is to be used because the picture-change (Section 3.2.3) associated with the transformation from the Dirac to a 2-component Hamiltonian changes the shape of the wavefunction and the optimal set of exponents needed to describe it. In addition, one should define the nuclear model and the radius of the nucleus [37]. The latter provides an upper limit on the exponent values, in contrast to the conventional point nucleus approximation, which yields a singularity at the nuclear position that is impossible to describe with Gaussian (GTO) or Slater type (STO) basis functions. After having defined the Hamiltonian and nuclear model, one may then find optimal exponents by minimising the total electronic energy as a function of basis set exponents. For DFT calculations the procedure is relatively straightforward because the electron-electron coalescence needs not be described explicitly, reducing the need for functions with high values of angular momentum. Nevertheless, one needs to add polarisation functions to describe deformation of the atomic orbitals upon molecule formation, something that can only be tested in molecular calculations. For the STO choice of basis, energy-optimised sets [38] are available for the ZORA Hamiltonian in the ADF program package [39]. For the more common choice of GTOs, a number of energy-optimized sets are available. For the Dirac-Coulomb Hamiltonian sets covering the whole periodic table were provided by Faegri [40] and are described as "better than double zeta quality." These are dual family basis sets, that is, they are constructed with the constraint of having equal exponents for the s- and d-sets and for the p- and f-functions allowing for economical use in 4-component calculations. The Faegri sets do not contain polarisation functions, and different choices have been made by users of these sets, making their accuracy in molecular calculations somewhat difficult to assess. More extended sets in the spirit of the non-relativistic correlation-consistent basis sets have been developed by Dyall [33, 41] who has optimised the exponents under the constraint of a double (DZ), triple (TZ), or quadruple zeta (QZ) description of the valence orbitals. Although developed for the Dirac-Coulomb Hamiltonian, these basis sets can also be used in X2C calculations in which the small component basis set is only used to generate the transformed Hamiltonian matrix. For the Douglas-Kroll-Hess family of methods, only the picture-transformed upper component of the wavefunction needs to be described, which can either be done using the large component sets of the above basis sets or by a dedicated energy optimisation. Roos and coworkers [42, 43] used the Faegri sets as a starting point for basis sets for the scalar DKH2 Hamiltonian and added polarisation and correlation functions (up to l=6). These atomic natural orbitals (ANO) sets allow for high flexibility in choosing an appropriate subset of contracted functions, but predefined choices of double, triple, or quadruple zeta are available as well. A somewhat smaller and thereby more economical alternative for this type of Hamiltonian is provided by the Sapporo group [44, 45] who employed the third-order scalar Douglas-Kroll Hamiltonian to optimise segmented contracted sets including valence polarisation and correlation functions. Pantazis and Neese have reported compact basis sets for the lanthanides [46] and actinides [47] optimized for use in scalar relativistic DFT calculations based on the DKH2 or ZORA Hamiltonians.

In all cases one should be aware of the fact that exponents and contraction coefficients are obtained for a particular choice for electronic configuration of the atom. Due to the importance of the 6d in actinides, and the 5d in early lanthanides, the configuration of choice is either (Faegri, Noro [40, 45]) the $s^2d^1f^{n-1}$ configuration or an average of different configurations (Dyall, Roos [33, 43]). Taking just the $s^2d^0f^n$ ground state configuration of the later actinides does not allow for optimisation of the important $5d$ or $6d$ orbitals that play a role in chemical bonding and excited states. Another issue is the fact that exponents are typically constrained to be equal for the two components of a shell that is split by spin–orbit coupling. This is most important for the 6p and 7p orbitals in which the difference between the radial wavefunctions is appreciable, but less so for the 4f and 5f orbitals in which this difference is small (see Figure 3.3). Spin-orbit coupling is also important in defining contraction coefficients, because it is difficult to find a balanced description if the radial functions of both components of the spin-orbit shell need to be equal. For Hamiltonians in which spin–orbit coupling is only considered in a second stage, contractions are based on the scalar version of the Hamiltonian and the spin–orbit operator is calibrated to give good energies for the splitting. Extending the basis by adding (or decontracting) tight functions may in such cases have an adverse effect [48] and worsen the results rather than improving them. In 4-component Dirac calculations, in which elimination of spin-orbit coupling is more difficult, the basis set is typically left uncontracted. The same is true for the Slater functions used in the ADF ZORA basis sets, in which uncontracted sets are used for both scalar and spin-orbit calculations.

With the availability of different alternatives for DZ, TZ, and QZ series of basis sets and the possibility to reach high accuracy with uncontracted universal sets, the availability of basis sets for all-electron calculations of heavy elements starts to approach the situation for lighter elements in which systematic studies on basis set convergence are carried out and basis set limit extrapolation formulas have been proposed. This is, however, only true for valence correlation because correlating the (chemically uninteresting) deep core orbitals will give unpredictable results with any of these sets. The lack of functions specifically optimised to correlate electrons in orbitals below the 4d (lanthanides) or 5d (actinides) subshell will give rise to serious basis set superposition errors when such calculations are attempted. This makes it hard to determine basis set limit results even for small molecules as the influence of the core-valence electron correlation can not be fully addressed with the available sets. Also for high-accuracy valence correlation calculations, it would be desirable to be able to use systematically optimised 5Z or 6Z sets. These issues are, however, minor compared to the choice of an appropriate electronic structure method to carry out such calculations as we will discuss in the next section.

3.4 Electronic Structure Methods

In this chapter we focus on methods suitable for an all-electron treatment of f elements. In such calculations the aim is either to calculate molecular properties or to achieve benchmark accuracy. Depending on the application, one may be able to neglect or approximate spin-orbit coupling effects, which is nowadays possible with all Hamiltonians described above. Regarding the calculation of molecular properties [49], we note that the application of all-electron methods is typically required when the property operator probes the wave function in the vicinity of the nuclei. This is the case for calculation of Nuclear Magnetic Resonance (NMR) parameters, Electric Field Gradients (EFG), or Mössbauer shifts. For calculation of "valence" properties such as molecular dipole moments or polarizabilities, the use of pseudo-orbitals is less of a problem and the computationally cheaper effective core potential methods (discussed by Cao and Weigand elsewhere in this volume) are often a reasonable alternative [50]. For the second objective, reaching benchmark accuracy, all-electron methods have an advantage because they offer the possibility of systematic improvement of the results by extension of basis sets and the correlation of more electrons. The latter is more difficult with pseudopotential approaches because the core-valence correlation effects are part of the parametrisation. One typically can choose between a "large core" and a "small core" definition with accompanying basis sets that are developed to provide accurate results. A fundamental problem associated with the pseudo-orbital approximation is the change of the virtual orbital levels that are used to describe electron correlation in wavefunction-based approaches. As explained by Dyall [51], this necessarily leads to a decrease of the HOMO-LUMO gap and an overestimation of correlation effects. This can be remedied by parametrising on energy differences [52–54] rather than valence orbital shape. For DFT approaches, we note that the effective core potentials used in computational chemistry are typically derived by parametrising on basis of correlated atomic calculations, which may not be fully transferable to the Kohn-Sham ansatz used in DFT. An interesting aspect, which is rarely discussed, is the influence of smoothening out the oscillations of the electron density in the core region on the calculated generalised gradient terms in the exchange-correlation density functionals. This will lead to differences in the calculated exchange-correlation energies relative to all-electron approaches that should cancel out in the calculation of molecular bond energies or other valence energy differences.

Some insight into the methodological challenges in theoretical studies of f elements can already be gathered from looking at the uranyl double cation UO_2^{2+} which is a closed-shell molecule important in the treatment of nuclear waste. In this molecule the uranium atom has formal oxidation state $+VI$, but from projection analysis [55] at the HF level we find a charge of $+2.94$ and configuration $5f^{2.25}\ 6d^{1.16}\ 7s^{0.98}$ of the uranium atom in the molecule. Both features indicate a considerable covalent contribution to bonding, and indeed both $5f$ and $6d$ orbitals are found to participate in bonding with the oxygen ligands. What is perhaps even more remarkable is that the population of the subvalence $6p$ orbitals is only 5.64 (and not 6.00). This is a manifestation of the so-called 6p-hole [56], due to overlap with the oxygen ligands, and mostly located to the $6p_{3/2,1/2}$ orbitals. The lower virtual molecular orbitals are atomic $5f$ orbitals. As one moves to the right along the actinide series, these levels get filled in the homologous actinyl species, leading to manifolds of open-shell states. Due to the atomic nature of the active orbitals, the lower electronic states of a given actinyl species will have quite similar spectroscopic constants, in conformity with the concept of superconfigurations introduced by Field in spectroscopic studies of simple lanthanide molecules [57].

In conclusion we see that for actinides there may be a very large number of valence orbitals (5f, 6d, 7s, and perhaps even 7p) leading very important dynamical correlation effects preferably to be handled by a size extensive correlation method. The 6s and 6p orbitals are highly polarisable and may have to be included in the correlation treatment as well. Partially occupied 5f and 6d shells lead to a high density of multiconfigurational states. These are heavy elements and require a non-perturbative relativistic treatment. In addition much of the chemistry takes place in solution, so solvent effects should be accounted for. The large number of valence electrons to correlate combined with the strong multiconfigurational character of the wavefunction impose restrictions hardly met by any existing theoretical method. For the lighter lanthanides the methodological requirements are somewhat less severe. Yet, in a 1988 review in the *Handbook on the Physics and Chemistry of Rare Earths*, C. K. Jørgensen asked the rhetorical question 'Is Quantum Chemistry feasible?' and with special regard to these elements, he answered 'Sorry, not today; perhaps next century' [58]. Fortunately the situation has improved considerably since then.

Looking at technical limitations of the available electronic structure methods, we note that most techniques were originally developed for applications on light elements. The large number of f orbitals and the possibility of participation of d orbitals make f elements challenging. One is often faced with an electronic configuration in which more than 10 orbitals need to be placed in an active space to allow for variable occupation. Although such spaces can be treated with complete active space (CAS) models used in the CASSCF and CASPT2 approaches that are described below, the calculations do become expensive.

For multi-reference coupled cluster approaches, such a large number of open shells is usually prohibitive because they typically assume a simple reference wavefunction from which the states of interest can be generated by single or at most double excitations. Another consequence of open-shell configurations in the high density of states within the same energy range. A recurring theme in calculations is the determination of the precise electronic ground state of a compound, and how the energy differences between close-lying states are modified by ligation and solvation. For such applications it is essential to be able to calculate the energy of a multitude of states without introducing a bias for a particular state or configuration. An added requirement is the ability to analyse the results in a meaningful way, in particular for actinides the spin-orbit coupling can be so strong that decomposition of the wavefunction in terms of spin multiplets is difficult, in particular in methods that include the spin-orbit coupling terms already in the orbital generation step.

3.4.1 Coupled Cluster Approaches

3.4.1.1 Single-Reference Coupled Cluster Methods

A very useful concept in wavefunction-based correlation theories is the wave operator $\hat{\Omega}$ which, when acting on some suitable model function Ψ_0, gives the correlated, exact wave function Ψ

$$\Psi = \hat{\Omega}\Psi_0. \tag{3.51}$$

The equation for the exact wavefunction

$$\hat{H}\Psi = \hat{H}\hat{\Omega}\Psi_0 = E\hat{\Omega}\Psi_0 \tag{3.52}$$

can be simply transformed into an equation from which the exact energy E is obtained from the model function through the action of an *effective Hamiltonian*

$$\hat{H}_{\mathit{eff}}\Psi_0 = E\Psi_0; \quad \hat{H}_{\mathit{eff}} = \hat{\Omega}^{-1}\hat{H}\hat{\Omega}, \tag{3.53}$$

assuming that the inverse of the wavefunction exists. Perturbation theory can be formulated by partitioning the Hamiltonian as $\hat{H} = \hat{H}_0 + \hat{H}_1$, where Ψ_0 is an eigenfunction of H_0 and then expanding the wave operator in orders of the perturbation H_1. As an example, Møller-Plesset perturbation theory is obtained by setting H_0 equal to the Fock operator.

Coupled cluster theory emerged from a deep analysis of the structure of the wave operator, taking into account in particular the constraint of size extensivity, that is, the correct scaling of the correlation energy with system size. It was found that an exponential form was required to assure separability of non-interacting fragments. In its simplest form, the single-reference coupled cluster *ansatz* is accordingly

$$\Psi^{CC} = e^{\hat{T}}\Phi_0 \tag{3.54}$$

where Φ_0 is a single determinant, typically optimised with HF (Φ^{HF}), and $\hat{T} = \sum_\mu t_\mu \hat{\tau}_\mu$ is a linear combination of excitation operators $\hat{\tau}_\mu$. An important feature of the CC *ansatz* is that it is size extensive also when the excitation manifold $\hat{T}$ is truncated, in contrast to

Configuration Interaction (CI). The excitation manifold $\hat{T}$ is usually restricted to single and double excitations

$$\hat{T} = \hat{T}_1 + \hat{T}_2 = \sum_{i}^{occ.} \sum_{a}^{virt.} t_i^a a^\dagger i + \sum_{i,j}^{occ.} \sum_{a,b}^{virt.} t_{ij}^{ab} a^\dagger b^\dagger ij, \tag{3.55}$$

using perturbation theory to include the effect of triple excitations.

The wave equation for the CC wavefunction

$$\hat{H}\Phi^{CC} = \hat{H}e^{\hat{T}}\Phi_0 = E^{CC}e^{\hat{T}}\Phi_0 \tag{3.56}$$

can be transformed into a wave equation for the model function

$$\overline{H}\Phi_0 = E^{CC}\Phi_0; \quad \overline{H} = e^{-\hat{T}}\hat{H}e^{\hat{T}} \tag{3.57}$$

where the similarity-transformed Hamiltonian $\overline{H}$ is seen to constitute an effective Hamiltonian. The CC energy energy is obtained from projection with the model function

$$E^{CC} = \left\langle \Phi_0 \left| \overline{H} \right| \Phi_0 \right\rangle \tag{3.58}$$

whereas the CC amplitude equations are obtained by projection with excited determinants

$$\left\langle \Phi_\mu \left| \overline{H} \right| \Phi_0 \right\rangle = 0; \quad \Phi_\mu = \hat{\tau}_\mu \Phi_0 \tag{3.59}$$

The CC cluster amplitudes are coupled through these equations, but an important simplifying feature is that a Baker-Campbell-Haussdorf (BCH) expansion of $\overline{H}$ truncates exactly to quartic order

$$\overline{H} = e^{-\hat{T}}\hat{H}e^{\hat{T}} = \hat{H} + \left[\hat{H},\hat{T}\right] + \frac{1}{2!}\left[\left[\hat{H},\hat{T}\right],\hat{T}\right] + \frac{1}{3!}\left[\left[\left[\hat{H},\hat{T}\right],\hat{T}\right],\hat{T}\right] + \frac{1}{4!}\left[\left[\left[\left[\hat{H},\hat{T}\right],\hat{T}\right],\hat{T}\right],\hat{T}\right]$$

This follows from commutation of the excitation operators amongst each other and the fact that every new nested commutator eliminates one of the maximally four general indices of the electronic Hamiltonian.

As is clear from equation (3.55), the single-reference CC parametrisation assumes a distinction between occupied and virtual orbitals in the model wavefunction and is thereby most suited for molecules that are qualitatively well described by a single determinant, i.e., closed shell or high-spin open shell systems. This restriction prohibits application of the standard coupled cluster approach to most lanthanide complexes, because the 4f shell usually retains its atomic open shell character and requires a multi-determinant model wavefunction. Studies with coupled cluster methods therefore usually concern closed shell complexes with elements from the beginning or the end of the lanthanide series. The ^8S high-spin ground state of a $4f^7$ configuration can also be studied, but only if the environment consists of closed shell ligands. For actinides the situation is somewhat different, because the $5f$ is more extended and can get (formally) oxidised in molecular complexes. This is the case in the most studied actinide unit, the uranyl ion $[OUO]^{2+}$, in which the $5f$ shell is formally unoccupied at oxidation state VI of uranium. Explicit calculation shows a large HOMO-LUMO gap that is little affected by ligation or spin-orbit coupling, allowing coupled cluster studies with standard methods. The bare uranyl ion has been extensively studied with both all-electron and pseudo potential methods, and benchmark data for its bond distance and the vibrational frequencies have been determined [59–63].

Somewhat larger systems are still within reach of all-electron coupled cluster calculations, but the steep scaling of the computational cost prohibits studies of more than 10 atoms if both spin-obit coupling and core-valence correlation is to be explicitly included. The overview of recent calculations with the coupled cluster and other approaches in the chapter by Patzschke *et al.* in this volume gives an idea what is currently feasible. Systems that can be easily treated with single reference coupled cluster theory also include triatomic actinides in which only a single electron resides in the $5f$ shell such as $[ONpO]^{2+}$ or $[OUO]^{+}$. These ions correspond, however, to less common oxidation states of, respectively, uranium and neptunium. Study of the more common $5f^2$ species $[ONpO]^{+}$ is already more difficult for the single reference coupled cluster approach [64], and the same holds true for the neutral OUO molecule that has been subject of some debate relating to its precise ground state in rare gas matrices [65–68]. The reason is the close spacing of states arising from the $5f^2$, $5f^1 7s^1$, and $5f^1 d6^1$ configurations, which can be mixed in molecular complexes. This is specific for early actinides as can be seen from Figure 3.5 in which the HF lower virtual orbital energies, relative to the HOMO, for a $5f^0$ configuration are plotted.

For later actinides the increased nuclear charge makes the $5f$ levels drop sufficiently far below the $6d$ and $7s$ to decrease the configurational mixing. This makes the electron correlation problem similar to that of lanthanides for which it is not so much the interconfigurational mixing, but rather the difficulties in describing multiplet states with a simple reference wavefunction that prohibits single reference coupled cluster treatments. In both cases the number of systems that can be treated reliably can be extended using multireference coupled cluster techniques. In the next section we will in particular discuss the

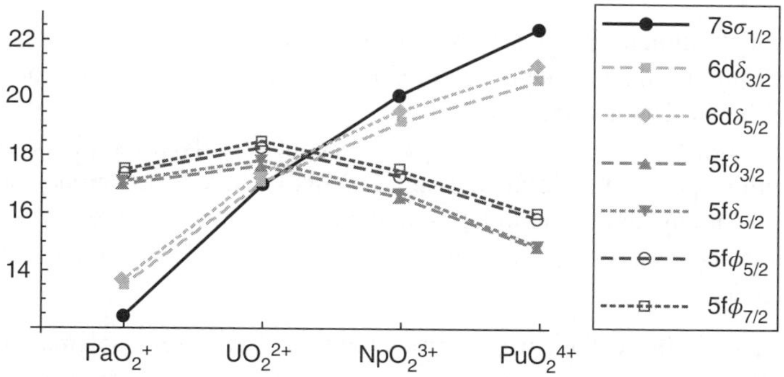

Figure 3.5 *Orbital energies (in eV) for virtual 5f, 6d, and 7s orbitals, relative to HOMO, in iso-electronic actinyl ions. Calculations were done with the ZORA Hamiltonian, using the Hartree-Fock model and the TZ2P basis in ADF*

Fock-space coupled cluster (FSCC) approach that has been mostly applied to actinides. A more extensive discussion can be found in Chapter 2 by Eliav and Kaldor.

3.4.1.2 Multiference Cluster Methods

A more typical situation in f element chemistry is that no single reference determinant provides a good starting point for correlated calculations so that the appropriate model function is a linear combination of several determinants

$$\Psi_0 = \sum_{\mu} \Phi_\mu C_{\mu 0} \tag{3.60}$$

spanning a model space defined by a projection operator P. We may now distinguish inactive (core) orbitals i, j, k, l, occupied in all determinants, and active (valence) orbitals u, v, w, x present in some, but not all determinants. A complete active space (CAS) contains the full set of determinants generated from distributing the active electrons in all possible ways amongst the active orbitals, consistent with whatever symmetries are imposed on the wavefunction, and lends itself better to size extensive approaches than incomplete spaces. We once again invoke a wave operator on exponential form and define an effective Hamiltonian. In an internally contracted approach, the expansion coeffients $C_{\mu 0}$ are kept fixed, whereas in a relaxed approach the CC energy and the expansion coefficients $C_{\mu K}$ are found by projecting the wave equation (3.53) with reference determinants within the model space

$$\sum_{\nu} \left\langle \Phi_\mu \left| \hat{H}_{\text{eff}} \right| \Phi_\nu \right\rangle C_{\nu K} = E C_{\mu K}; \quad \mu, \nu \in P \tag{3.61}$$

leading to an effective CI problem. The excitation amplitudes are found by projecting with determinants from the complementary space $Q = 1 - P$

$$\sum_{\nu} \left\langle \Phi_\mu \left| \hat{H}_{\text{eff}} \right| \Phi_\nu \right\rangle C_{\nu K} = 0; \quad \mu \in Q, \nu \in P \tag{3.62}$$

However, two problems, which breaks the nice structure observed in single-reference CC, immediately appear:

1. An excited determinant Φ_μ outside the model space can in principle be generated from any of the reference determinants, yielding more amplitudes than equations and a severe problem of linear dependence.
2. In single-reference CC the excitation operators contain only creation operators (particle or hole) with respect to the Fermi vacuum. In MRCC there is no unique choice of Fermi vacuum, but for any choice annihilation operators will appear in the excitation manifold so that the BCH-expansion of the similarity-transformed Hamiltonian will not truncate to quartic order.

Various flavours of MRCC have been formulated to tackle these problems. State-universal MRCC (SUMRCC) employs separate wave operators for each reference determinant [69]

$$\Psi_K^{SUMRCC} = \sum_{\mu} \hat{\Omega}_\mu \Phi_\mu C_{\mu K} \tag{3.63}$$

and then generates the complete set of states corresponding to the dimension of the model space. This eliminates linear dependencies, but only partially solves the problem of non-commuting cluster operators.

An alternative approach is obtained by noting that the Hamiltonian on second quantized form does not refer to the number of electrons. Thus, in order to calculate some specific open-shell state or a manifold of such states, one may find some state, possibly with a different number N of electrons, which can be described by a simple HF determinant that will define the Fermi vacuum and the (0,0) hole-particle *sector* in the ensuing calculations. In order to generate the leading determinants of the target state(s), one then successively adds holes and particles. Adding an electron (particle) to the active virtual orbitals leads to the reference determinants with $N+1$ electrons of the (0,1) sector, whereas removing an electron (adding a hole) from the active occupied orbitals leads to the reference determinants with $N-1$ electrons of the (1,0) sector. From both sectors one may then proceed to the (1,1) sector providing reference determinants appropriate for the description of excited states dominated by single excitations. The wave operator is accordingly an exponential of an excitation manifold on the form [70]

$$\Psi_K^{VUMRCC} = \sum_\mu \exp\left(\hat{T}\right) \Phi_\mu C_{\mu K}; \quad \hat{T} = \sum_{h,p} T^{(h,p)}. \tag{3.64}$$

and the model space contains all reference determinants in all sectors up to and including the target sector. Since the active occupied and virtual orbitals are commonly denoted valence orbitals, this approach is known as valence-universal MRCC (VUMRCC). If the wave operator is written as a simple exponential of $\hat{T}$ all amplitudes from the different sectors couple and have to be solved for simultaneously. If instead the wave operator is written as a *normal-ordered* exponential, $\hat{\Omega} = \left\{\exp\left(\hat{T}\right)\right\}$ the amplitudes can be solved in a hierarchical manner [71, 72], but at the price of reduced orbital relaxation. For single excitation energies one would then first carry out a single-reference CC calculation in the (0,0) sector, then solve the (1,0) and (0,1) sectors separately starting from the converged $T^{(0,0)}$ amplitudes and finally set up and diagonalize the effective Hamiltonian in the (1,1) sector to obtain the target states.

A severe problem affecting both state universal and valence universal approaches is the possible presence of *intruder states* [73], which are states dominated by determinants that are not in the model space. If these states become near-degenerate with one or more target states, the effective Hamiltonian becomes ill-conditioned so that the amplitude equations can not be solved. A possible way out is to drop the condition that the effective Hamiltonian should provide as many exact eigenvalues as there are states in the model space. Rather, one may separate the model space into the 'true' model space P_m and then a buffer space P_i that is described in an uncorrelated fashion. This approximation is known as the intermediate Hamiltonian approach [74–76] and makes it possible to obtain converged results for large model spaces. Interpretation of the results should be done with some care, however, as states that are primarily generated from the P_i model space will not be accurate. For actinides, the intermediate Hamiltonian coupled cluster method [75] has made it possible to treat molecules with two electrons outside a closed shell, using an active space that comprises both the $5f$ and the $6d$ orbitals. In this way one may reach very high accuracy in the calculation of atomic spectra, as is discussed in more detail by Eliav and Kaldor in Chapter 2.

For molecular applications, the choice of Fock-space sectors is more critical because a change of charge will affect the nature of the orbitals and ease with which the equations can be solved. An illustrative example is the calculation of the low-lying electronic states of the CUO molecule [77]. These are at first sight best defined as excitations from a closed shell reference determinant, thus utilising the (1,1) sector of Fock space. Obtaining these states does, however, also require finding the solutions of the (0,1) or electron attachment sector of Fock space. Although these solutions can be obtained in small or medium-sized basis sets, this becomes difficult for extended basis sets that contain sufficiently diffuse functions. The reason is the appearance of Rydberg orbitals with energies below the localised 5*f* orbitals that are needed to build the sought after localised model states. In this case it turned out to be advantageous to define the states of the neutral molecules as double electron-attached (0,2) states for the $[CUO]^{2+}$ dication. With this definition the equations to be solved comprise the (0,0) sector, the (0,1) sector of states of CUO^{+} and finally the desired (0,2) sector of CUO. Apart from yielding smoother convergence, this choice also has the advantage or removing a bias for the reference model wavefunction and for allowing mixing between this model state and other states of neutral CUO, something that is not possible when utilising the (1,1) sector. A disadvantage of this scheme is the use of orbitals that are optimised for a different charge state than one is interested in. Because the coupled cluster approach only yields a limited amount of orbital relaxation, this may also introduce errors in the final result. This problem of VUMRCC methods will become more severe when technological advances allow application of the method to larger molecules, for the current applications to atoms and highly symmetric molecules a change in number of electrons will change the extent of the orbitals but not their composition (as this is largely determined by symmetry). This may make state-specific (see Ivanov *et al.* [78] for a recent review) approaches competitive, even though the advantages of simultaneously obtaining many of the states of a dense manifold is then lost.

3.4.2 Multi-Reference Perturbation Theory

A computationally less expensive alternative to coupled-cluster theory is provided by multi-reference perturbation theory (MRPT). The currently most widely used form of MRPT in f element chemistry is second-order Complete Active Space Perturbation Theory (CASPT2) [79, 80], and we shall focus on this method in this section.

It will be useful to review certain features of Rayleigh-Schrödinger perturbation theory (RSPT): We start from a partitioning of the Hamiltonian into a zeroth-order part H_0 and a first-order part H_1

$$H = H_0 + H_1 \tag{3.65}$$

and focus on a particular solution Ψ_0, generally the ground-state solution. Expanding the wave equation in orders of the perturbation we obtain to first and second order

$$\left(H_0 - E_0^{(0)}\right)\left|\Psi_0^{(1)}\right\rangle + \left(H_1 - E_0^{(1)}\right)\left|\Psi_0^{(0)}\right\rangle = 0 \tag{3.66}$$

$$\left(H_0 - E_0^{(0)}\right)\left|\Psi_0^{(2)}\right\rangle + \left(H_1 - E_0^{(1)}\right)\left|\Psi_0^{(1)}\right\rangle = E_0^{(2)}\left|\Psi_0^{(0)}\right\rangle \tag{3.67}$$

Projecting with the zeroth-order solution $\Psi_0^{(0)}$ we obtain

$$E_0^{(1)} = \left\langle \Psi_0^{(0)} \left| H_1 \right| \Psi_0^{(0)} \right\rangle \tag{3.68}$$

$$E_0^{(2)} = \left\langle \Psi_0^{(0)} \left| H_1 \right| \Psi_0^{(1)} \right\rangle \tag{3.69}$$

provided we use intermediate normalization

$$\left\langle \Psi_0 \mid \Psi_0^{(0)} \right\rangle = 1 \tag{3.70}$$

and assume that $\Psi_0^{(0)}$ is a solution of H_0 with energy $E^{(0)}$. In conventional RSPT it is assumed that we dispose of the complete set of solutions of the zeroth-order Hamiltonian from which the target function Ψ_0 is constructed. Starting for instance from the canonical Hartree-Fock orbitals conventional Møller-Plesset theory is derived by choosing the Fock operator $\hat{F}$ as zeroth-order Hamiltonian. However, the situation changes if we wish to work with a non-canonical set of HF orbitals, for instance localized orbitals. The set of Slater determinants constructed from these orbitals are not solutions of the Fock operator. We can make the HF determinant Φ_0 a solution of H_0 by expressing it as

$$H_0 = P\hat{F}P + Q\hat{F}Q; \quad P = \left|\Phi_0\right\rangle\left\langle\Phi_0\right|, \quad Q = 1 - P \tag{3.71}$$

In order to generate the perturbation corrections to $\Psi_0^{(0)}$ we do not need the other solutions to H_0, just the orthogonal complement $\left\{\Psi_{m\neq 0}^{(0)}\right\}$, which is not required to form an orthogonal set in itself. Expanding the first-order wave equation in this set we then obtain a linear set of equations

$$\sum_n \left(H_{0;mn} - E_0^{(0)} S_{mn}\right) C_n^{(1)} = -H_{1:m0}; \quad \begin{array}{ll} H_{i;mn} & = \left\langle \Psi_m^{(0)} \left| H_i \right| \Psi_n^{(0)} \right\rangle \\ S_{mn} & = \left\langle \Psi_m^{(0)} \mid \Psi_n^{(0)} \right\rangle \end{array} \tag{3.72}$$

for the first-order expansion coefficients $C_m^{(1)}$.

In the CASPT2 method the zeroth-order wavefunction is taken to be a CASSCF wavefunction. It is not required that the solution is fully optimized, allowing thus to use solutions of a multi-root CASSCF, but it is assumed to be a solution of a CAS-CI in the same orbital set, so that it is orthogonal to the remaining CAS-CI space. Projectors P and Q are introduced as above (3.71) to make it a solution of a generalized Fock operator. To first-order only single and double excitations out of the reference CASSCF solution connect to the reference, but since these excitations connect to the remaining space, further decompositon of the Q projector is introduced to keep matrices H_0 and S_0 above of managable size. The final choice for the zeroth-order Hamiltonian is therefore

$$H_0^{CASPT2} = PFP + Q_K F Q_K + Q_{SD} F Q_{SD} + Q_{TQ} F Q_{TQ} \tag{3.73}$$

where Q_K is a projector of the complementary CAS-CI space, Q_{SD} a projector of single and double excitations out of the reference CASSCF solution, and Q_{TQ} a projector of the remaining space. A pilot implementation of CASPT2 based on the 4-component DC Hamiltonian has been reported [81], but the bulk of relativistic CASPT2 calculations are carried out using a scalar relativistic Hamiltonian such as DKH2. Spin-orbit corrections can then be added in a second perturbative step [82] in which off-diagonal spin-orbit matrix elements are determined between CASSCF wavefunctions, but the more accurate CASPT2 energies are taken to specify the diagonal values of a configuration interaction (CI) matrix. Diagonalization then provides spin-orbit coupled wavefunctions and energies. Provided that the basis of interacting spin-free states is complete and properly assigned when building the CI matrix [83], this efficient procedure is applicable to both weakly and strongly coupled cases.

Just as in the closed-shell case eigenvalues of the generalized Fock operator can be associated with ionization potentials and electron affinities for inactive and virtual orbitals, respectively. Since electrons can both enter and leave active orbitals, eigenvalues of the generalized Fock operator for such orbitals will be an average over ionization potentials and electron affinities, a feature of the CASPT2 zeroth-order Hamiltonian that lead to systematic errors. In a revised choice of H_0 shifts have been introduced to correct for this [84]. CASPT2 is also prone to intruder states. Both shortcomings can be curbed by introducing some two-electron interaction amongst the active orbitals [85]. This has lead to the development of n-electron valence perturbation theory (NEVPT2) [86, 87], which has nice formal properties, but whose actual performance in theoretical f element chemistry needs assessment.

3.4.3 (Time-Dependent) Density Functional Theory

A rather different approach compared to the ones discussed above is to the use of density functional theory (DFT) in conjunction with an all-electron method. Because of the more favourable computational scaling DFT is often used to describe large complexes in which only one or a few atoms belong to the class of f elements. Because the performance of the various classes of density functional approximations is well documented for light elements, we will focus our attention on the electronic structure of the f element.

For favourable cases, like those discussed above in conjunction with the single reference coupled cluster approach, description in terms of a single Kohn-Sham determinant is possible, and molecular structures and related properties can be obtained in the same way as in non-relativistic approaches. By comparison with reference calculations (MP2 or CC) and experiment it appears that the usual generalised gradient approximation (GGA) and hybrid functionals also perform sufficiently well for structures and energetics, respectively (see Schrechkenbach *et al.* [88] for some illustrative examples). Given the fact that most applications concern molecules in the condensed phase, the challenge is to account for environmental effects in an economical and reliable manner. The simplest approach to deal with solvation is to use continuum models for which the parametrisation of the f element is rather straightforward (requiring just a reasonable atomic size), but also explicit solvent models are in use.

In many cases, just applying ground state DFT is not sufficient because one would also like to investigate the relative energies of the excited states. This is nowadays usually done with time-dependent DFT (TDDFT), but the older "delta-SCF" method in which an excited state is calculated by selecting a different orbital occupation is also still used. In fact, the small HOMO-LUMO gaps that are characteristic for many actinide complexes often lead to non-aufbau occupations in the KS wavefunction, even when optimising on the lowest energy state. This indicates that none of the available functionals are flawless, as such a situation should not occur in exact DFT. Such problems are also manifest when applying TDDFT, with excited states sometimes appearing lower in energy than the reference state. This is especially true when turning to hybrid functionals as this introduces more severe triplet instabilities than present in a pure DFT approach. The analysis (and prevention) of such instabilities is difficult due to the spin-orbit coupling that blurs the distinction between singlet and triplet states, especially in actinides. Both problems indicate that more research is necessary, both on the parametrisation of existing DFAs as well as on possibly new functional forms that could be better suited for f elements. From a recent benchmark study (Tecmer *et al.* [89]) it appears that range-separated hybrids such as CAMB3LYP yield the best agreement with reference coupled cluster data already for relatively small triatomic molecules, but more research is necessary to establish whether this also holds for a wider class of compounds.

3.5 Conclusions and Outlook

In this chapter we have attempted to discuss the most important ingredients for an all-electron treatment of f elements. These can be subdivided into Hamiltonians, basis sets, and electronic structure methods, following the extended Pople-style diagram of quantum chemistry in which relativity is added next to the basis set and method axes to define a theoretical model chemistry. Of these three ingredients we can say that the Hamiltonian axis is well-described nowadays, with a clear hierarchy of Hamiltonians that approximate the reference Dirac-Coulomb-Breit Hamiltonian. For practical calculations the X2C approach appears to offer the best compromise between accuracy and computational efficiency. On the basis set axis we now start to see competing hierarchies of basis sets that each cover the full periodic table, including all f elements. These basis sets are complemented by very large even-tempered sets that can be used to reach very high precision in atomic calculations. The axis that, as usual, demands the most attention is that of electronic structure methods. Here the challenge remains to develop methods that can handle large active spaces and produce many energies in an accurate and efficient manner. The CASPT2 approach is the most generally applicable method but can be complemented by MRCC methods for molecules with only a few open shell electrons. DFT is well suited to optimise structure and energetics of molecules with a sufficiently large HOMO-LUMO gap, but it needs further benchmarking and development for use in determination of excited state energies and structures.

Acknowledgments

TS would like to acknowledge stimulating discussions with Debashis Mukherjee (Calcutta).

References

[1] Lloyd DR. On the Lanthanide and 'Scandinide' Contractions. J Chem Ed. 1986;63:502.

[2] Goldschmidt VM, Barth T, Lunde G. Geochemische Verteilungsgesetze der Elemente. V. Isomorphie und Polymorphie der Sesquioxyde. Die Lanthaniden-Kontraktion und Ihre Konsequenzen. Norske Vidensk Akad Skrifter I Mat Naturv Kl. 1925;7:1.

[3] Bagus PS, Lee YS, Pitzer KS. Effects of relativity and of the lanthanide contraction on the atoms from hafnium to bismuth. Chem Phys Lett. 1975;33:408.

[4] Kramida A, Yu Ralchenko, Reader J, and NIST ASD Team; 2013. NIST Atomic Spectra Database (ver. 5.1), [Online]. Available: http://physics.nist.gov/asd [2014, January 22]. National Institute of Standards and Technology, Gaithersburg, MD.

[5] Saue T. Relativistic Hamiltonians for Chemistry: A Primer. ChemPhysChem. 2011;12:3077.

[6] Dyall KG, Fægri K. Introduction to Relativistic Quantum Chemistry. Oxford University Press; 2007.

[7] Reiher M, Wolf A. Relativistic Quantum Chemistry: The Fundamental Theory of Molecular Science. Wiley-VCH; 2009.

[8] Visscher L. On the construction of double group molecular symmetry functions. Chem Phys Lett. 1996 Apr;253(1-2):20–26.

[9] Einstein A. Zur Elektrodynamik bewegter Körper. AnnPhys. 1905;17:891.

[10] Dirac PAM. The quantum theory of the electron. Proc Roy Soc London A. 1928;117:610–624.

[11] Gell-Mann M. The Interpretation of the New Particles as Displaced Charge Multiplets. Nuovo Cimento Suppl. 1956;4:848.

[12] Foldy LL, Wouthuysen SA. On the Dirac theory of spin 1/2 particles and its non-relativistic limit. PhysRev. 1950;78:29.

[13] Heully JL, Lindgren I, Lindroth E, Lundqvist S, Mårtensson-Pendrill AM. Diagonalisation of the Dirac Hamiltonian as a basis for a relativistic many-body procedure. J Phys B: At Mol Opt Phys. 1986;19:2799.

[14] Chang C, Pelissier M, Durand P. Regular two-component Pauli-like effective Hamiltonians in Dirac theory. PhysScr. 1986;34:394.

[15] van Lenthe E, Baerends EJ, Snijders JG. Relativistic regular two-component Hamiltonians. J Chem Phys. 1993;99:4597.

[16] van Lenthe E, Baerends EJ, Snijders JG. Relativistic total energies using regular approximations. J Chem Phys. 1994;101:9783.

[17] Dyall KG, van Lenthe E. Relativistic regular approximations revisited: An infinite-order relativistic approximation. J Chem Phys. 1999;111:1366–1372.

[18] Douglas M, Kroll NM. Quantum electrodynamical corrections to the fine structure of helium. AnnPhys. 1974;82:89–155.

[19] Hess BA. Applicability of the no-pair equation with free particle projection operators to atomic and molecular structure calculations. PhysRevA. 1985;32:756.

[20] Hess BA. Relativistic electronic-structure calculations employing a two-component no-pair formalism with external-field projection operators. PhysRevA. 1986;33:3742.

[21] Barysz M, Sadlej A, Snijders JG. Nonsingular Two/One-Component Relativistic Hamiltonians Accurate Through Arbitrary High Order in α. Int J Quant Chem. 1997;65:225.

[22] Dyall K. Interfacing relativistic and nonrelativistic methods. I. Normalized eliminiation of the small component in the modified Dirac equation. J Chem Phys. 1997;106:9618.

[23] Iliaš M, Jensen HJA, Kellö V, Roos BO, Urban M. Theoretical study of PbO and the PbO anion. Chem Phys Lett. 2005;408:210.

[24] Kutzelnigg W, Liu W. Quasirelativistic theory equivalent to fully relativistic theory. J Chem Phys. 2005;123:241102.

[25] Iliaš M, Saue T. An infinite-order two-component relativistic Hamiltonian by a simple one-step transformation. J Chem Phys. 2007;126:064102.

[26] Baerends EJ, Schwarz WHE, Schwerdtfeger P, Snijders JG. Relativistic atomic orbital contractions and expansions: magnitudes and explanations. J Phys B: At Mol Opt Phys. 1990;23:3225.

[27] Kellö V, Sadlej AJ. Picture change and calculations of expectation values in approximative relativistic theories. Int J Quant Chem. 1998;68:159–174.

[28] Dyall KG. Relativistic Electric and Magnetic Property Operators for Two-Component Transformed Hamiltonians. Int J Quant Chem. 2000;78:412.

[29] Pernpointner M, Schwerdtfeger P. Accurate nuclear quadrupole moments of the gallium isotopes ^{69}Ga and ^{71}Ga within the PCNQM model. Chem Phys Lett. 1998;295:347.

[30] Seino J, Nakai H. Local unitary transformation method for large-scale two-component relativistic calculations: Case for a one-electron Dirac Hamiltonian. J Chem Phys. 2012;136(24):244102.

[31] Heß BA, Marian CM, Wahlgren U, Gropen O. A mean-field spin-orbit method applicable to correlated wave functions. Chem Phys Lett. 1996;251:365.

[32] Sikkema J, Visscher L, Saue T, Iliaš M. The molecular mean-field approach for correlated relativistic calculations. J Chem Phys. 2009;131:124116.

[33] Dyall KG. Relativistic double-zeta, triple-zeta, and quadruple-zeta basis sets for the actinides Ac-Lr. Theor Chem Acc. 2007;117:491–500. The basis sets are available from the DIRAC web site http://dirac.chem.sdu.dk/basisarchives/dyall/index.html.

[34] DIRAC, a relativistic ab initio electronic structure program, Release DIRAC13 (2013), written by L. Visscher, H. J. Aa. Jensen, R. Bast, and T. Saue, with contributions from V. Bakken, K. G. Dyall, S. Dubillard, U. Ekström, E. Eliav, T. Enevoldsen, E. Faßhauer, T. Fleig, O. Fossgaard, A. S. P. Gomes, T. Helgaker, J. K. Lærdahl, Y. S. Lee, J. Henriksson, M. Iliaš, Ch. R. Jacob, S. Knecht, S. Komorovský, O. Kullie, C. V. Larsen, H. S. Nataraj, P. Norman, G. Olejniczak, J. Olsen, Y. C. Park, J. K. Pedersen, M. Pernpointner, K. Ruud, P. Sałek, B. Schimmelpfennig, J. Sikkema, A. J. Thorvaldsen, J. Thyssen, J. van Stralen, S. Villaume, O. Visser, T. Winther, and S. Yamamoto (see http://www.diracprogram.org).

[35] Schimmelpfennig B; 1996. Program AMFI, Stockholm, Sweden.

[36] DaSilva A, Malli G, Ishikawa Y. Relativistic Universal Gaussian-basis set for Dirac-Fock-Coulomb and Dirac-Fock-Breit SCF Calculations on heavy-atoms. Chem Phys Lett. 1993 FEB 19;203(2-3):201–204.

[37] Visscher L, Dyall KG. Dirac-Fock atomic electronic structure calculations using different nuclear charge distributions. At Data Nucl Data Tables. 1997;67:207.

[38] van Lenthe E, Baerends E. Optimized Slater-type basis sets for the elements 1-118. J Comp Chem. 2003;24:1142–1156.

[39] te Velde G, Bickelhaupt FM, Baerends E, Fonseca Guerra C, van Gisbergen SJA, Snijders JG, *et al.* Chemistry with ADF. J Comp Chem. 2001;22(9):931–967.

[40] Faegri Jr. K. Relativistic Gaussian basis sets for the elements K - Uuo. Theor Chem Acc. 2001;105:252. See also http://folk.uio.no/knutf/.

[41] Gomes AP, Dyall KG, Visscher L. Relativistic double-zeta, triple-zeta, and quadruple-zeta basis sets for the lanthanides La-Lu. Theor Chem Acc. 2010;127:369–381. The basis sets are available from the DIRAC web site http://dirac.chem.sdu.dk/basisarchives/dyall/index.html.

[42] Roos B, Lindh R, Malmquist PA, Veryazov V, Widmark PO, Borin AC. New Relativistic Atomic Natural Orbital Basis Set for Lanthanides Atoms with Applications to Ce Diatom and LuF_3. J Phys A. 2008;112:11431.

[43] Roos BO, Lindh R, Malmqvist PÅ, Veryazov V, Widmark PO. New relativistic ANO basis sets for actinide atoms. Chem Phys Lett. 2005;409:295–299.

[44] Sekiya M, Noro T, Miyoshi E, Osanai Y, Koga T. Relativistic correlating basis sets for lanthanide atoms from Ce to Lu. J Comp Chem. 2006 MAR;27(4):463–470.

[45] Noro T, Sekiya M, Osanai Y, Koga T, Matsuyama H. Relativistic correlating basis sets for actinide atoms from Th-90 to Lr-103. J Comp Chem. 2007 DEC;28(16):2511–2516.

[46] Pantazis DA, Neese F. All-Electron Scalar Relativistic Basis Sets for the Lanthanides. J Chem Theory Comput. 2009;5(9):2229–2238.

[47] Pantazis DA, Neese F. All-Electron Scalar Relativistic Basis Sets for the Actinides. J Chem Theory Comput. 2011;7(3):677–684.

[48] Visscher L, Saue T. Approximate relativistic electronic structure methods based on the quaternion modified Dirac equation. J of Chem Phys. 2000;113:3996–4002.

[49] Saue T. Post Dirac-Hartree-Fock Methods - Properties. In: Schwerdtfeger P, editor. Relativistic Electronic Structure Theory. Part 1. Fundamentals. Amsterdam: Elsevier; 2002. p. 332.

[50] Dolg M, Cao X. Relativistic Pseudopotentials: Their Development and Scope of Applications. Chem Rev. 2012;112:403–480.

[51] Dyall K. Formal analysis of effective core potential methods. J Chem Inf Comput Sci. 2001;41:30–37.

[52] Hellmann H. A New Approximation Method in the Problem of Many Electrons. J Chem Phys. 1935;3:61–61.

[53] Topp W, Hopfield J. Chemically Motivated Pseudopotential for Sodium. Physical Review B. 1973 Feb;7(4):1295–1303.

[54] Dolg M, Cao X. Accurate Relativistic Small-Core Pseudopotentials for Actinides. Energy Adjustment for Uranium and First Applications to Uranium Hydride. The Journal of Physical Chemistry A. 2009;113(45):12573–12581.

[55] Dubillard S, Rota JB, Saue T, Fægri K. Bonding analysis using localized relativistic orbitals: Water, the ultrarelativistic case and the heavy homologues H2X (X=Te, Po, eka-Po). J Chem Phys. 2007;124:154307.

[56] Pyykko P, Laaksonen L. Relativistically parameterized extended Hueckel calculations. 8. Double-.zeta. parameters for the actinoids thorium, protactinium, uranium, neptunium, plutonium, and americium and an application on uranyl. J Phys Chem. 1984;88(21):4892–4895.

[57] Field RW. Diatomic Molecule Electronic Structure beyond Simple Molecular Constants. Ber Bunsenges Phys Chem. 1982;86:771.

[58] Jørgensen CK. Influence of rare earths on chemical understanding and classification. In: jr KAG, Eyring L, editors. Handbook on the Physics and Chemistry of Rare Earths. vol. 11. Elsevier Science; 1988.

[59] de Jong W, Visscher L, Nieuwpoort W. On the bonding and the electric field gradient of the uranyl ion. J Mol Struct (Theochem). 1999 Jan;458(1–2):41–52.

[60] Straka M, Dyall K, Pyykko P. Ab initio study of bonding trends for f(0) actinide oxyfluoride species. Theor Chem Acc. 2001;106:393–403.

[61] Pierloot K, van Besien E. Electronic structure and spectrum of UO_2^{2+} and $UO_2Cl_4^{2-}$. J Chem Phys. 2005;123:204309.

[62] Réal F, Gomes ASP, Visscher L, Vallet V, Eliav E. Benchmarking electronic structure calculations on the bare UO_2^{2+} ion: how different are single and multireference electron correlation methods? J Phys Chem A. 2009;113:12504–12511.

[63] Wei F, Wu G, Schwarz WHE, Li J. Geometries, electronic structures, and excited states of UN_2, NUO^+, and UO_2^{2+} : a combined CCSD(T), RAS/CASPT2 and TDDFT study. Theor Chem Acc. 2011;129:467–481.

[64] Infante I, Gomes ASP, Visscher L. On the performance of the intermediate Hamiltonian Fock-space coupled-cluster method on linear triatomic molecules: The electronic spectra of NpO_2^+, NpO_2^{2+}, and PuO_2^{2+}. J Chem Phys. 2006;125:074301.

[65] Zhou M, Lester Andrews, Ismail, Nina, Colin Marsden. Infrared Spectra of UO_2, UO_2^+, and UO_2^- in Solid Neon. J Phys Chem A. 2000;104:5495–5502.

[66] Li J, Bursten B, Andrews L, Marsden C. On the electronic structure of molecular UO_2 in the presence of Ar atoms: Evidence for direct U-Ar bonding. J Am Chem Soc. 2004;126:3424–3425.

[67] Gagliardi L, Roos BO, Malmqvist P, Dyke J. On the electronic structure of the UO_2 molecule. J Phys Chem A. 2001;105:10602–10606.

[68] Infante I, Andrews L, Wang X, Gagliardi L. Noble Gas Matrices May Change the Electronic Structure of Trapped Molecules: The $UO_2(Ng)_4$ [Ng = Ne, Ar] Case. Chem Eur J. 2010;16:12804–12807.

[69] Jeziorksi B, Monkhorst HJ. Coupled-cluster method for multideterminantal reference states. Phys Rev A. 1981;24:1668.

[70] Mukherjee D, Moitra RK, Mukhopadhyay A. Applications of a non-perturbative many-body formalism to general open-shell atomic and molecular problems: calculation of the ground and the lowest π-π^* singlet and triplet energies and the first ionization potential of trans-butadiene. Mol Phys. 1977;33:955–969.

[71] Mukherjee D. On the hierarchical equations of the wave-operator for open-shell systems. Pramana. 1979;12:203–225.

[72] Lindgren I, Mukherjee D. On the connectivity criteria in the open-shell coupled-cluster theory. PhysRep. 1987;151:93.
[73] Schucan TH, Weidenmüller HA. The Effective Interaction in Nuclei and Its Perturbation Expansion: An Algebraic Approach. Ann Phys. 1972;73:108–135.
[74] Malrieu JP, Durand P, Daudey JP. Intermediate Hamiltonians as a new class of effective Hamiltonians. J Phys A. 1985;18:809.
[75] Landau A, Eliav E, Kaldor U. Intermediate Hamiltonian Fock-space coupled-cluster method. Chem Phys Lett. 1999;313:399–403.
[76] Landau A, Eliav E, Kaldor U. Intermediate Hamiltonian Fock-space coupled-cluster method. AdvQuantChem. 2001;39:171–188.
[77] Tecmer P, van Lingen H, Gomes ASP, Visscher L. The electronic spectrum of $CUONg_4$ (Ng = Ne, Ar, Kr, Xe): New insights in the interaction of the CUO molecule with noble gas matrices. J Chem Phys. 2012;137:084308.
[78] Ivanov VV, Lyakh DI, Adamowicz L. State-specific multireference coupled-cluster theory of molecular electronic excited states. Annu Rep Prog Chem, Sect C: Phys Chem. 2011;107: 169–198.
[79] Andersson K, Malmqvist P, Roos BO, Sadlej AJ, Wolinski K. Second-order perturbation theory with a CASSCF reference function. J Phys Chem. 1990;94:5483.
[80] Andersson K, Malmqvist P, Roos BO. Second-order perturbation theory with a complete active space self-consistent field reference function. J Chem Phys. 1992;96:1218.
[81] Abe M, Nakajima T, Hirao K. The relativistic complete active-space second-order perturbation theory with the four-component Dirac Hamiltonian. J Chem Phys. 2006;125:234110.
[82] Malmqvist PÅ, Roos BO, Schimmelpfennig B. The restricted active space (RAS) state interaction approach with spin-orbit coupling. Chem Phys Lett. 2002;357:230.
[83] Sánchez-Sanz G, Barandiarán Z, Seijo L. Energy level shifts in two-step spin–orbit coupling ab initio calculations. Chem Phys Lett. 2010;498:226–228.
[84] Ghigo G, Roos BO, Malmqvist P. A modified definition of the zeroth-order Hamiltonian in multiconfigurational perturbation theory (CASPT2). Chem Phys Lett. 2004;396:142–149.
[85] Dyall KG. The choice of a zeroth-order Hamiltonian for second-order perturbation theory with a complete active space self-consistent-field reference function. J Chem Phys. 1995;102:4909.
[86] Angeli C, Cimiraglia R, Evangelisti S, Leininger T, Malrieu JP. Introduction of n-electron valence states for multireference perturbation theory. J Chem Phys. 2001;114:10252.
[87] Angeli C, Cimiraglia R, Malrieu JP. n-electron valence state perturbation theory: A spinless formulation and an efficient implementation of the strongly contracted and of the partially contracted variants. J Chem Phys. 2002;117:9138.
[88] Schreckenbach G, Shamov GA. Theoretical Actinide Molecular Science. Acc Chem Res. 2010;43:19–29.
[89] Tecmer P, Gomes ASP, Ekström U, Visscher L. Electronic spectroscopy of UO_2^{2+}, NUO^+ and NUN: an evaluation of time-dependent density functional theory for actinides. PCCP. 2011;13:6249–6259.

4

Low-Lying Excited States of Lanthanide Diatomics Studied by Four-Component Relativistic Configuration Interaction Methods

Hiroshi Tatewaki,[1] *Shigeyoshi Yamamoto*[2] *and Hiroko Moriyama*[1]

[1]*Graduate School of Natural Sciences, Nagoya City University*
[2]*School of International Liberal Studies, Chukyo University*

4.1 Introduction

Lanthanide compounds generally have significant photoluminescent and magnetic properties, which arise from the characteristics of the localized 4*f* open-shell electrons. [1] The fluorescence of Ln^{3+} ions such as Tb^{3+} in La_2O_2S and Eu^{3+}in Y_2O_2S is used in flat color panels, and Nd^{3+} is used in the YAG laser, $Y_3Al_5O_{12}$. The Nd present in $Nd_2Fe_{14}B$ provides the strongest permanent magnetic material known, because the six distinct iron and two distinct neodymium sites are aligned in the same direction. The intensity of luminescence of some lanthanides in the visible or near-infrared region is sensitive to the nature of the metal-ion environment; for this reason these transitions are applied in bioprobes. [2–4] Other biomedical applications are based on the high magnetic moments of lanthanides, and paramagnetic lanthanide complexes are used as NMR shift reagents [5] or as magnetic resonance imaging (MRI) agents. [2–5]

The behavior of the *f* electrons of lanthanides is not fully understood, even for the diatomics. The oxides and fluorides are considered to have the simplest electronic structure of the diatomics, because of the strong electronegativity of the oxygen and fluorine atoms. We have analyzed the electronic structure of the ground state and excited states of cerium

Computational Methods in Lanthanide and Actinide Chemistry, First Edition. Edited by Michael Dolg.

monoxide (CeO), cerium monofluoride (CeF), europium monofluoride (EuF), and gadolinium monofluoride (GdF). The electronic structure of these molecules has previously been studied by methods such as ligand field theory, [6–8] molecular orbital theory with/without spin-orbit configuration interaction (SOCI), [9–13] and four-component relativistic theory. [14–29] Although theory and computer programs have developed rapidly in recent decades, the behavior of *f* electrons in lanthanide compounds remains a challenging subject.

In the present work we discuss the electronic structure of diatomics containing lanthanide atoms. Section 4.2 discusses the method used in the calculations. Section 4.3 treats the ground state of CeO and LnF (Ln = Ce, Gd), and Section 4.4 discusses the excited states of these molecules.

4.2 Method of Calculation

4.2.1 Quaternion Symmetry

We used the DIRAC program suite. Time-reversal symmetry [30] and Abelian point groups, including C_{2v}, are fully exploited in the DIRAC program with the help of quaternion algebra. [31] We briefly summarize quaternion algebra for the case of C_{2v}.

The molecular spinors are expanded in terms of the four quaternion units $(1, \check{i}, \check{j}, \check{k})$. Two-dimensional contour maps of the large components are created in this work for the molecular spinors so to illustrate the nodal structure, and we review the relation between the quaternion representation and the normal four-component complex representation. [32] Each quaternion unit belongs to one of the boson irreducible representations (boson irreps) of C_{2v} provided that the small components are neglected.

$$\begin{aligned}
&\text{quaternion } 1\text{: } L^{\alpha} \text{ Real-part } (\mathrm{A}_1 \text{ irrep})\\
&\text{quaternion } \check{i}\text{: } L^{\alpha} \text{ Imaginary-part } (\mathrm{A}_2 \text{ irrep})\\
&\text{quaternion } \check{j}\text{: } L^{\beta} \text{ Real-part } (\mathrm{B}_1 \text{ irrep})\\
&\text{quaternion } \check{k}\text{: } L^{\beta} \text{ Imaginary-part } (\mathrm{B}_2 \text{ irrep})
\end{aligned}$$

Here, '*L*' denotes the large component. The large component can then be written as

$$C_{2v}: \begin{pmatrix} \mathrm{A}_1 + i\mathrm{A}_2 \\ \mathrm{B}_1 + i\mathrm{B}_2 \end{pmatrix} \text{ [32]} \tag{4.1}$$

In the DIRAC program, molecular spinors are expressed as a sum of regular spherical harmonics (R^l_m). [33] As an example the *p* functions are expressed as follows:

$$\begin{aligned}
R^1_0 &= \sqrt{\frac{3}{4\pi}}z\\
R^1_{+1} &= -\frac{1}{\sqrt{2}}\left(\sqrt{\frac{3}{4\pi}}x + i\sqrt{\frac{3}{4\pi}}y\right)\\
R^1_{-1} &= \frac{1}{\sqrt{2}}\left(\sqrt{\frac{3}{4\pi}}x - i\sqrt{\frac{3}{4\pi}}y\right)
\end{aligned} \tag{4.2}$$

We shall write these functions associated with $\exp(-\zeta r^2)$ as p_0, p_{+1}, p_{-1}, respectively. A similar notation is adopted for the *d* and *f* functions.

Consider a $\Omega = 1/2$ molecular spinor having $C_{\infty v}$ symmetry as an example. Its large component is expressed as a linear combination of the following atomic spinors:

$$s_{1/2,1/2} = \begin{pmatrix} s \\ 0 \end{pmatrix}, \quad p_{1/2,1/2} = \begin{pmatrix} -\sqrt{\frac{1}{3}}p_0 \\ \sqrt{\frac{2}{3}}p_{+1} \end{pmatrix}, \quad p_{3/2,1/2} = \begin{pmatrix} \sqrt{\frac{2}{3}}p_0 \\ \sqrt{\frac{1}{3}}p_{+1} \end{pmatrix}, \quad \cdots \tag{4.3}$$

According to equation (4.1), p_0 and p_{+1} above have nonzero components (A_1) and (B_1+iB_2) respectively; 1, $\check{i}$, $\check{j}$ and $\check{k}$ are represented with the spherical harmonics. For the case $\Omega = -3/2$, the atomic spinor basis is:

$$p_{3/2,-3/2} = \begin{pmatrix} 0 \\ p_{-1} \end{pmatrix}, \quad d_{3/2,-3/2} = \begin{pmatrix} \sqrt{\frac{4}{5}}d_{-2} \\ -\sqrt{\frac{1}{5}}d_{-1} \end{pmatrix}, \quad d_{5/2,-3/2} = \begin{pmatrix} \sqrt{\frac{1}{5}}d_{-2} \\ \sqrt{\frac{4}{5}}d_{-1} \end{pmatrix}, \quad \cdots \tag{4.4}$$

Expressions (4.3) and (4.4) conform to the symmetry of equation (4.1). In the following sections we give the Mulliken gross atomic populations (GAOP) [34] of the respective molecular spinors in terms of the spherical harmonics.

4.2.2 Basis Set and HFR/DC Method

We first performed Hartree-Fock-Roothaan [35, 36] calculations with a Dirac-Coulomb Hamiltonian (HFR/DC) for CeO^+ and LnF^+ (Ln = Ce, Gd) as preparation for correlated calculations, since the correlation effects are well described by the spinors generated for the molecular cations. [18–20, 25–29] This is because the virtual spinors of the cation see a V^{N-1} potential, whereas virtual spinors of a neutral system see a V^N potential, which is not appropriate for describing the electron correlation effects of the neutral system. The basis set of the Ln atoms are primitive Gaussian-type functions (pGTF) as proposed by Koga *et al.*; [37] these are slightly modified, as will be shown for particular molecules. The pGTFs for $p_{1/2}, d_{3/2}, f_{5/2}$ are also used for $p_{3/2}, d_{5/2}, f_{7/2}$. The basis functions for O and F should also be those of Koga *et al.* [38] The small components are generated by the strict kinetic balance condition. [39, 40] In all molecular calculations we used the Gaussian nuclear model. [41] We also use the no-virtual-pair approximation. [42–47] All calculations were performed at the experimental internuclear distances of the respective molecules. The $C_{\infty v}$ double group is used in the framework of the four-component relativistic theory.

4.2.3 GOSCI and RASCI Methods

Using the LnL (L = O, F) molecular spinors, we performed four-component relativistic general open-shell CI (GOSCI) calculations, [48] and restricted active space CI (RASCI) calculations. [49, 50] These methods were implemented in the DIRAC program. [51] The RASCI module, modified by one of the chapter authors (Yamamoto), was used. [52] In molecular calculations of LnO or LnF, the molecular spinors very often have atomic character. Below we write an atomic spinor in square brackets such as [6*s*] and a molecular spinor in parentheses such as (6*s*).

In the following CI calculations the Ln (4*f*), (5*d*), and (6*s*) spinors are treated as the valence spinors. In this book, the GOSCI spaces for CeF and CeO are generated with the (4*f*), (5*d*), and (6*s*) spinors, and for GdF are generated with a direct product of $(4f^7)\otimes(5d,6s)^2$ so as to reduce the CI space. The RASCI spaces are generated with the three molecular spinor sets, i.e., RAS1 (active core), RAS2 (valence), and RAS3 (virtual). Details will be explained in the molecular calculations.

4.3 Ground State

In a study of the ionization potentials of LnF, Gotkis [53] predicted the existence of two types of LnF, $[Ln^{2+}(4f^n)F^-]$ $(6s)^1$ (Ln = Pr, Nd, Pm, Sm, Eu, Yb), and $[Ln^{3+}(4f^{n-1})F^-]$ $(6s)^2$ (Ln = La, Ce, Gd, Ho, Er, Yb). Both types of LnFs have spherical F^- moiety and one/two (6*s*) electrons polarized away from F^-; in the (6*s*) molecular spinor, the Ln [6*s*] atomic spinor mixes with the Ln [5*d**] atomic spinors. Schofield [54] analyzed the enthalpy and demonstrated the stable existence of MO^+ ions (Metal atom), including CeO^+. Several studies have been published [9, 11, 13, 16] of LnO and LnF using the molecular orbital method. Properties of the ground state such as bond distance (R_e) and the dissociation energy (D_e) were calculated [9, 11, 13] by the configuration interaction with substitution of singles and doubles method (CISD), [9] the coupled-cluster singles and doubles including perturbative treatment of triple excitations method (CCSD(T), [9, 11] and relativistic DFT. [13] Nevertheless we are unaware of any work that considers the characteristics of the ground state in the detail.

4.3.1 CeO Ground State

Prior to the correlated calculations we performed HFR/DC calculations for CeO^+. For Ce we constructed a (25*s*20*p*15*d*10*f*2*g*) basis functions from (25*s*18*p*15*d*10*f*) primitive Gaussian-type functions (pGTF), proposed by Koga *et al.* [37], by augmenting two *p*-type (2×*p*) GTFs of Huzinaga *et al.* [55] as the correlation functions. We first determined the exponent of the correlating *g*-type function to be 2.45, so that the resulting mean value of $r(\langle g|r|g\rangle)$ should match $\langle 4f|r|4f\rangle$, and secondly we have split this exponent 2.45 by dividing/multiplying by $\sqrt{2}$ and have obtained two *g*-type GTFs. The basis function for O is (12*s*8*p*2*d*). The *s* and *p* pGTFs are those of Koga *et al.*, [38] and the 2×*d* are those of Huzinaga *et al.* [55] All calculations were performed at the experimental internuclear distance (R_e), [56] which is 3.44 au.

We performed HFR/DC calculations for CeO^+, taking $[Ce^{4+}(1s^2.......5s^25p^6)$ $O^{2-}(1s^22s^22p^6)]^{2+}(valence^1)$ as its electronic configuration. We also performed HFR/DC calculations for CeO^{2+}, with $[Ce^{4+}(1s^2.......5s^25p^6)O^{2-}(1s^22s^22p^6)]^{2+}$. In Table 4.1 we set out the spinor energies and GAOPs of the CeO^+, the 25th – 32nd spinors, composed mainly of Ce [5*s*], [5*p*], [5*d*], O [2s], [2*p*], and spinor energies and GAOPs of the 33rd – 48th spinors composed of the Ce [4*f*], [5*d*], [6*s*], and [6*p*] spinors, where one electron is distributed among the latter 16 spinors and their time-reversal partners (TRP), namely among 16 Kramers pairs (KP).

Table 4.1 *GAOPs for CeO^{+}* [a)]

Spinor number	Spinor energy /au	Occ.	Ω	GAOPs of Ce				GAOPs of O		Character
				[*s*]	[*p*]	[*d*]	[*f*]	[*s*]	[*p*]	
25	−2.16582	1	1/2	0.97	0.00	0.00	0.00	0.02	0.01	Ce 5s
26	−1.55648	1	1/2	0.01	0.50	0.01	0.00	0.44	0.04	Ce *p* + O *sp*
27	−1.39425	1	1/2	0.00	0.91	0.00	0.00	0.07	0.01	Ce *p* + O *sp*
28	−1.31930	1	−3/2	0.00	0.99	0.00	0.00	0.00	0.01	Ce 5*p*
29	−1.18442	1	1/2	0.01	0.50	0.03	0.00	0.43	0.02	Ce *pds* + O *sp*
30	−0.66022	1	1/2	0.00	0.01	0.15	0.04	0.00	0.79	O *p* + Ce *dfp*
31	−0.65824	1	−3/2	0.00	0.01	0.15	0.04	0.00	0.79	O *p* + Ce *dfp*
32	−0.64907	1	1/2	0.01	0.05	0.21	0.05	0.03	0.65	O *ps* + Ce *dfps*
	sum(25-32)	8		1.00	2.97	0.56	0.14	0.98	2.32	
	2×sum	16		2.00	5.94	1.12	0.28	1.96	4.64	
33	−0.54700	1/32	1/2	0.00	0.00	0.12	0.88	0.00	0.00	Ce *fd*
34	−0.54582	1/32	5/2	0.00	0.00	0.00	1.00	0.00	0.00	Ce4*f*
…	…	…	…	…	…	…	…	…	…	…
39	−0.53598	1/32	5/2	0.00	0.00	0.05	0.95	0.00	0.00	Ce *fd*
…	…	…	…	…	…	…	…	…	…	…
42	−0.43523	1/32	1/2	0.85	0.06	0.07	0.01	0.00	0.02	Ce *sdpf*
43	−0.39618	1/32	1/2	0.00	0.17	0.62	0.09	0.00	0.12	Ce *dpf* + O *p*
…	…	…	…	…	…	…	…	…	…	…
48	−0.19489	1/32	1/2	0.84	0.09	0.06	0.02	0.01	−0.01	Ce *spdf*
	sum(33-48)	1/2		1.71	2.49	4.41	6.96	0.01	0.41	

a) HFR/DC total energy = −8935.81773 au

We next performed RASCI calculations for CeO^{2+}, CeO^{+}, and CeO, with (i) RAS1 composed of 8 active core molecular spinors and their TRPs (25th – 32nd) having characteristics of the Ce[5*s*], [5*p*] and O[2*s*], [2*p*] atomic spinors, and the maximum number of holes allowed is 2; (ii) RAS2 composed of 15 active molecular spinors and their TRPs (33rd – 47th) having characteristics of the Ce[4*f*], [5*d*], [6*s*], and [6*p*] atomic spinors; (iii) RAS3 composed of 23 molecular spinors and their TRPs (48th – 70th). For the RASCI calculations, the CeO^{2+} HFR/DC spinors are used for CeO^{2+}, whereas the CeO^{+} HFR/DC spinors are used for CeO^{+} and CeO. We denote this RASCI scheme by RASCI(8|15|23).

Table 4.2 sets out the electron configurations of the molecules and the atoms, the total energies of the molecules and atoms, and the dissociation energies (D_e) for the respective species. The calculated values of D_e (4.89 eV for CeO^{2+}, 6.23 eV for CeO^{+}, 5.97 eV for CeO) and in fact the experimental values (8.83 eV for CeO^{+}, 8.18 eV for CeO) [54] confirm that these species are long-lived. The calculated D_e values are smaller than experiment, but the order of values of D_e ($CeO^{+} > CeO$) is reproduced successfully by RASCI. Table 4.3 shows holes in RAS1, GAOPs in RAS2, particles in RAS3, and the most important configuration state functions (CSF). CeO^{2+} has 64 electrons in 32 closed-shell KPs. Tables 4.2 and 4.3 indicate that the ground states of CeO^{2+}, CeO^{+}, and CeO are well described by a single configuration of $|1,\underline{1}\ldots.32,\underline{32}|$, $|1,\underline{1}\ldots.32,\underline{32},34|$, and $|1,\underline{1}\ldots.32,\underline{32},34,\underline{42}|$, where n is the number of a spinor and $\underline{n}$ is the TRP of n. The 24 KPs (1st – 24th) are treated as frozen core in the RASCI calculations.

Table 4.2 *Total energies and dissociation energies of* $[Ce^{3+}O^{-}]^{2+}$, $[Ce^{3+}O^{-}]^{2+}(4f^1)$, *and* $[Ce^{3+}O^{-}]^{2+}(4f^1 6s^1)$ *obtained by RASCI (8|15|23)*

Molecule			Atom/Ion			
Configuration	TE[a)] /au	D_e / eV	Configuration	TE[b)] /au	Configuration	TE[c)]/au
$[Ce^{3+}O^{-}]^{2+}_{0}$	−8935.636174	4.89	$[Ce]^{4+}(4f^2)_4$	−8860.584040	$[O]^{4+}(2p^4)_2$	−74.872523
$[Ce^{3+}O^{-}]^{2+}(4f^1)_{5/2}$	−8936.127910	6.23	$[Ce]^{4+}(4f^1 5d^2)_{7/2}$	−8861.026467	$[O]^{4+}(2p^4)_2$	−74.872523
$[Ce^{3+}O^{-}]^{2+}(4f^1 6s^1)_2$	−8936.314017	5.97	$[Ce]^{4+}(4f^1 5d^1 6s^2)_4$	−8861.222001	$[O]^{4+}(2p^4)_2$	−74.872523

a) MOs for (8|15|23) of CeO^{2+} are closed-shell HFR/DC MOs for CeO^{2+}. MOs for (8|15|23) of CeO^+ and CeO are HFR/DC MOs for CeO^+ obtained by distributing one electron in the 4*f*, 5*d*, 6*s*, and 6*p*-like spinors (abbreviated as 1/32).
b) AOs for (4|15|20) of Ce^{2+} and Ce^+ are HFR/DC AOs for Ce^{2+} obtained by distributing two electrons in the 4*f*, 5*d*, 6*s*, and 6*p* spinors (2/32) and AOs for (4|15|20) of Ce are HFR/DC AOs of Ce obtained by distributing four electrons in the 4*f*, 5*d*, 6*s*, and 6*p* spinors (4/32).
c) AOs for (1|3|3) of O are HFR/DC AOs obtained by distributing four electrons in the 2*p* spinors (4/6).

Table 4.3 *Holes, GAOPs, particles and dominant CSFs in RASCI (8|15|23) for* CeO^{2+}, CeO^+, *and CeO*

Exptl.		RASCI(8|15|23)											
Species	Ω	Conf.	Ω	Hole in RAS1		GAOPs in RAS2					Particle in RAS3		Dominant CSF
				Ce	O	[6*s*]	[6*p*]	[5*d*]	[4*f*]	O	Ce	O	
CeO^{2+}		$5s^2 6p^6$	0	−0.09	−0.11	0.00	0.05	0.05	0.06	0.01	0.01	0.01	88%|....|[a)]
CeO^+		*f*	5/2-1	−0.08	−0.08	0.00	0.02	0.09	1.02	0.01	0.01	0.01	90%|....34|
CeO	2	*fs*	2-1	−0.09	−0.09	0.76	0.10	0.15	1.03	0.02	0.11	0.02	78%|....34,42|

Although we have not shown the results, the 1st – 32nd spinors of CeO^{2+} resemble those of CeO^+. CeO^{2+} has the electronic configuration $[Ce^{2.4+}(5s^{2.0}5p^{5.9}5d*^{1.2}4f*^{0.4})O^{0.4-}(2s^{2.0}2p^{4.4})]^{2+}$, which can be abbreviated as $[Ce^{3+}(5s^2 5p^6 5d*^1)O^{-}(2s^2 2p^5)]^{2+}$ or simply as $[Ce^{3+}(5d*^1)O^{-}]^{2+}$. This differs significantly from $[Ce^{4+}(5s^2 5p^6)O^{2-}(2s^2 2p^6)]^{2+}$ which was assumed at the beginning of the study. Approximately 1.6 electrons of the six O (2*p*) electrons are back-donated into Ce through the polarization functions. The asterisks in this notation indicate that electrons have been brought into the Ce moiety through the [4*f*] and [5*d*] polarization functions, described as (4*f**) and (5*d**).

Experimentally, Ce^{2+} has the electronic configuration $[5p^6 4f^2]$ [57] and O is $[2p^4]$. [58] CeO^{2+} is formed as

$$Ce^{2+}[5s^2 5p^6 4f^2] + O[2s^2 2p^4] \rightarrow [Ce^{3+}(5s^2 5p^6 5d*^1)O^{-}(2s^2 2p^5)]^{2+}. \quad (4.5)$$

The two Ce [4*f*] electrons move into O so as to construct the closed-shell. A stable molecular ion, CeO^{2+}, is formed as a result; the 1.6 electrons of the six electrons in the O (2*p*) spinors are back-donated to Ce^{4+} through the (4*f**) and (5*d**) molecular spinors as stated above.

Experimentally, Ce^{+}/Ce has the electronic configuration $[5p^6 4f^1 5d^2]/[5p^6 4f^1 5d^1 6s^2]$ [57]. Together with [4*f*] and [6*s*] GAOPs in Table 4.3, the sums of GAOPs in the 10th and 11th rows in Table 4.1 indicate that CeO^{+}/CeO is formed as

$$Ce^{+}[5s^2 5p^6 4f^1 5d^2] + O[2s^2 2p^4] \rightarrow [Ce^{3+}(5s^2 5p^6 5d^{*1})O^{-}(2s^2 2p^5)]^{2+}(4f^1) \quad (4.6)$$
$$Ce[5s^2 5p^6 4f^1 5d^1 6s^2] + O[2s^2 2p^4] \rightarrow [Ce^{3+}(5s^2 5p^6 5d^{*1})O^{-}(2s^2 2p^5)]^{2+}(4f^1 6s^1). \quad (4.7)$$

In CeO^{+}/CeO, the two valence electrons $[5d^2]/[5d^1 6s^1]$ move into O. The CeO^{2+} ionic core in CeO^{+}/CeO has electronic configuration $[Ce^{2.6+}(5s^{2.0}5p^{5.9}5d^{*1.1}4f^{*0.3})O^{0.6-}(2s^{2.0}2p^{4.6})]^{2+}$. This configuration is similar to that of CeO^{2+}. The dissociation energies in Table 4.2 indicate that the characteristics of the chemical bonds of these molecules are governed by the bondings in the ionic core, CeO^{2+}. We also observe that the 33th – 42nd valence spinors have almost no components of the oxygen atomic spinors. A detailed discussion now follows of the nature of the chemical bond in the CeO^{2+} core, using the CeO^{+} GAOPs. Contour maps of the 26th – 32nd spinors are set out in Figure 4.1. See color plate section. They show that the 26th spinor is bonding and the 29th spinor is antibonding; the 26th spinor has no nodal planes between the Ce and O nuclei, whereas the 29th spinor has one nodal plane between them; both have a nodal plane of Ce $[5p_0]$ that passes through the Ce nucleus. Since the 28th spinor is 99% composed of Ce $[5p_{-3/2}]$, the 28th spinor energy (-1.3193 au) is expected to be close to the atomic spinor energy of Ce $[5p_{-3/2}]$ placed in the field of the O^{-} ion. In fact, Moriyama *et al.* [29] have shown that the Ce $[5p_{-3/2}]$ spinor energy in the field produced by the negative point charge of -2.29 is -1.3120 au, which is much higher than the free gaseous Ce^{3+} value (-1.8653 au) and is close to the 28th spinor energy (-1.3193 au). The facts that the 26th spinor energy is lower than the 28th by 0.237 au, and that the 29th spinor energy is higher by 0.135 au than the 28th, suggest that bonding is predominant over antibonding. This contradicts the usual view that antibonding orbitals are more strongly antibonding than bonding orbitals are bonding. This unusual feature of CeO is due to the large shifts in spinor energies as a result of the effective charge of the other atom. Overall, the $O^{-}(2s)$ and $Ce^{3+}(5p)$ electrons are responsible for the covalent bond.

We next discuss the characteristics of the 30th – 32nd spinors, formed mainly by the O [2*p*] spinors. As illustrated above, the two valence Ce^{2+} electrons move into the O [2*p*] spinors, and more than half of them are back-donated through the [5*d**] and [4*f**] spinors. This situation is written as $[Ce^{2.6+}(5s^{2.0}5p^{5.9}5d^{*1.1}4f^{*0.3})O^{0.6-}(2s^{2.0}2p^{4.6})]^{2+}$, or in short as $[Ce^{3+}(5s^2 5p^6 5d^{*1})O^{-}(2s^2 2p^5)]^{2+}$. The GAOPs in Table 4.1 show considerable mixing of the O [2*p*] and Ce [5*d**] and [4*f**]. Nevertheless the 30th – 32nd spinor energies have almost constant values of -0.65 au, close to the -0.65 au value of the $O^{-}[2p]$ atomic spinor energies [29] in the field generated by the positive point charge (+1.79) at R_e. The Ce [5*d**] and [4*f**] spinors are used for the O 2*p* electrons to penetrate into the Ce core and for the electrons to move behind the Ce nucleus. The maps of the O (2*p*)-like spinors (30th – 32nd) in Figure 4.1 support this assertion. As an example, the 32nd spinor shows that the O $[2p_{1/2}]$ electron penetrates into the Ce region and lies on the other side of the Ce nucleus using the $[5d^{*}_{1/2}]$ and $[4f^{*}_{1/2}]$ polarization functions. The GAOPs and figures show that the bulk of the charge clouds of the spinors (30th – 32nd) is at the O moiety, and the O (2*p*) spinors form the ionic bond, further stabilizing the molecule.

In CeO^{+}/CeO, one/two valence electrons move in the field generated by $[Ce^{3+}(5s^{2.0}5p^{5.9}5d^{*1.1}4f^{*0.3})O^{-}(2s^{2.0}2p^{4.6})]^{2+}$. The valence electrons in the 34th and 42nd spinors

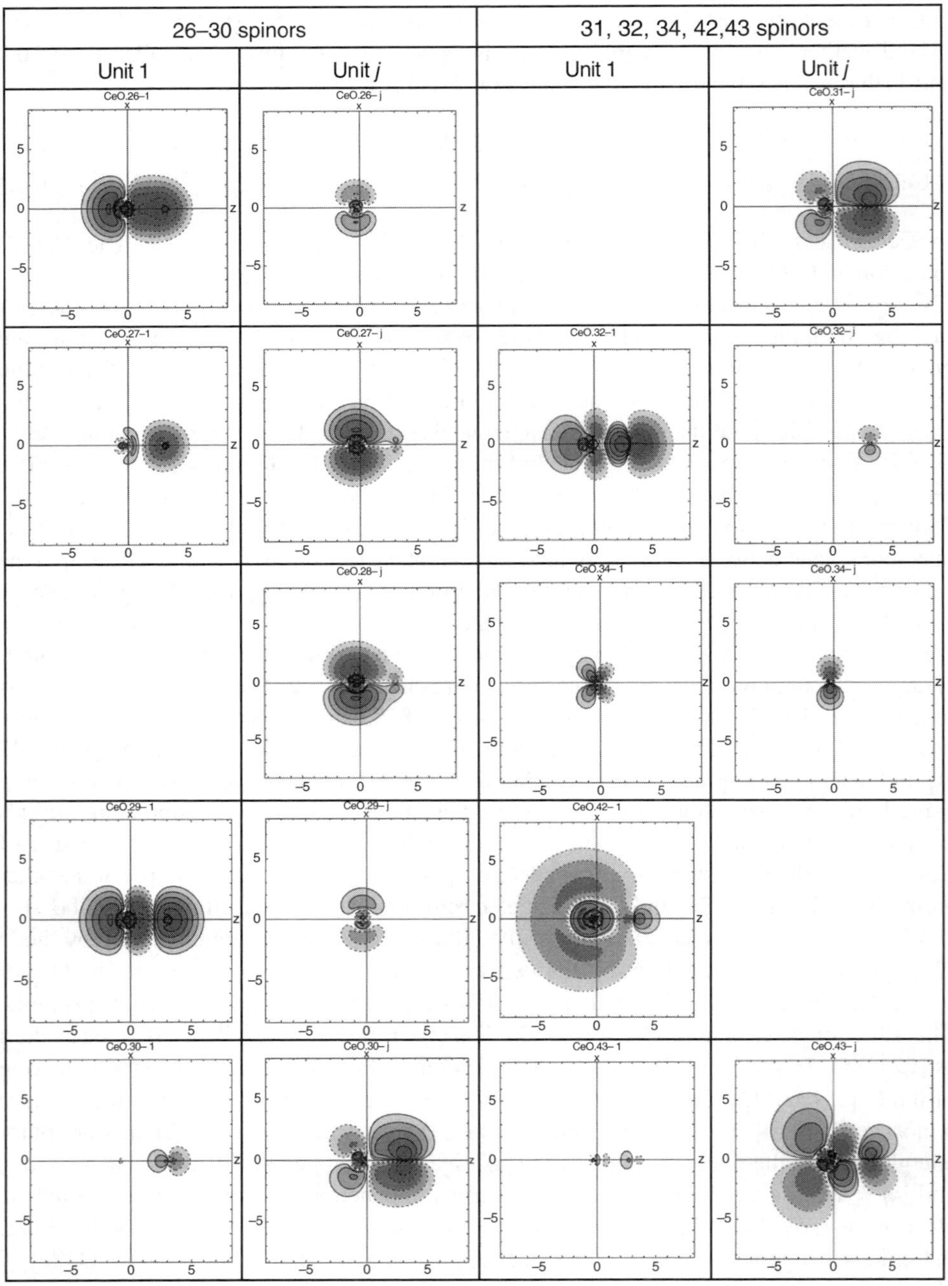

Figure 4.1 *Contour maps of the large components of the CeO^+ spinors. The quaternion units 1 and $\check{j}$ correspond, respectively, to the L^α real part of A_1 and to the L^β real part B_1. Coordinates run from -8.0 to 8.0 au. The CeO^+ nuclear distance is taken from the experimental R_e value (3.44 au) of neutral CeO. Solid and broken lines are contours of positive and negative values, respectively. Their values are ± 0.0125, ± 0.025, ± 0.05, ± 0.1, ± 0.2. See color plate section*

strengthen the binding by 1.1–1.3 eV. Figure 4.1 indicates that a considerable amount of electron density of the 34th ($4f$) and 42nd spinors ($6s$) lies behind the Ce^{3+} core (the opposite side of O^-), as Gotkis [53] suggested.

It follows that CeO is a charge-transfer molecule with strong ionic and covalent bonds in the CeO^{2+} core, to which two outer electrons are attached. Dolg *et al.* [9] pointed out that Ce[$5d$] and O[$2p$] form σ and π bonds, but they did not mention chemical bonds composed of Ce [$5p$] and O[$2s$].

4.3.2 CeF Ground State

The basis set used for Ce in CeF is the same as in CeO. The set for F is ($13s8p2d$). The s and p pGTFs are those of Koga *et al.* [38] One diffuse s-type GTF and $2\times d$ of Huzinaga *et al.* are augmented. [55] The HFR/DC calculations for CeF^+ and CeF^{2+} were performed first at the experimental CeF internuclear distance (R_e) of 3.87 au, [59] assuming a $[Ce^{4+}(5s^25p^6)F^-(2s^22p^6)]^{3+}$ ($valence\text{-}spinor^m$: $m = 1-2$) configuration. Table 4.4 shows the spinor energies and the GAOPs of the CeF^+ 25th – 32nd spinors composed mainly of Ce [$5s$], [$5p$], [$5d$], F [$2s$], [$2p$], and those of the 33rd – 48th spinors composed of Ce [$4f$], [$5d$], [$6s$], and [$6p$] atomic spinors, where two electrons are distributed among the latter 16 spinors and their TRPs.

We then performed RASCI calculations for CeF^{2+}, CeF^+, and CeF. We adopted a slightly different RASCI scheme from those of CeO^{2+}, CeO^+, and CeO in order to reduce the CI

Table 4.4 *GAOPs for $CeF^{+a)}$*

Spinor number	Spinor energy /au	Occ.	Ω	GAOPs of Ce				GAOPs of F		Character
				[s]	[p]	[d]	[f]	[s]	[p]	
25	−2.14003	1	1/2	0.99	0.00	0.00	0.00	0.01	0.00	Ce 5*s*
26	−1.70480	1	1/2	0.01	0.07	0.01	0.00	0.90	0.01	F *s* + Ce *dpf*
27	−1.40455	1	1/2	0.00	0.97	0.00	0.00	0.02	0.01	Ce 5*p*
28	−1.30284	1	−3/2	0.00	1.00	0.00	0.00	0.00	0.00	Ce 5*p*
29	−1.28809	1	1/2	0.00	0.92	0.00	0.00	0.05	0.03	Ce *p* + F *sp*
30	−0.77412	1	1/2	0.00	0.02	0.05	0.01	0.00	0.90	F *p* + Ce *dpf*
31	−0.77323	1	1/2	0.00	0.03	0.06	0.01	0.01	0.88	F *p* + Ce *dpf*
32	−0.77169	1	−3/2	0.00	0.01	0.04	0.01	0.00	0.94	F *p* + Ce *dpf*
	sum(25−32)	8		1.01	3.01	0.16	0.05	0.99	2.78	
	2×sum	16		2.02	6.02	0.32	0.10	1.98	5.56	
33	−0.52002	2/32	5/2	0.00	0.00	0.01	0.99	0.00	0.00	Ce 4*f*
34	−0.51967	2/32	1/2	0.00	0.00	0.11	0.89	0.00	0.00	Ce *fd*
…	…	…	…	…	…	…	…	…	…	…
39	−0.51098	2/32	−3/2	0.00	0.00	0.11	0.89	0.00	0.00	Ce *fd*
…	…	…	…	…	…	…	…	…	…	…
42	−0.45229	2/32	1/2	0.26	0.02	0.63	0.08	0.00	0.02	Ce *dsfp* + F *p*
43	−0.44851	2/32	1/2	0.49	0.02	0.43	0.06	−0.01	0.01	Ce *sdfp* + F *sp*
…	…	…	…	…	…	…	…	…	…	…
48	−0.22605	2/32	1/2	0.19	0.85	0.15	0.05	−0.25	0.01	Ce *psdf* + F *sp*
	sum(33–48)	1		1.19	2.99	4.97	7.03	−0.29	0.10	

a) HFR/DC total energy = −8960.398957 au

dimension: (i) RAS1 composed of 8 active core molecular spinors and their TRPs (25th – 32nd), which are the same as the CeO series; (ii) RAS2 composed of 13 active molecular spinors and their TRPs (33rd – 45th), reduced by the two spinors compared to the CeO case where the Ce [4*f*], [5*d*], and [6*p*] atomic spinors mix; (iii) RAS3 composed of 25 active molecular spinors and their TRPs (46th – 70th), where the 46th and 47th spinors and their TRPs are those excluded from RAS2. For the RASCI calculations, the CeF^{2+} HFR/DC spinors are used for CeF^{2+}, and the CeF^{+} HFR/DC spinors are used for CeF^{+} and CeF. We denote this RASCI scheme by RASCI(8|13|25).

Table 4.5 sets out the electron configurations, total energies and dissociation energies (D_e). The calculated values of D_e for CeF^{2+}, CeF^{+}, and CeF are 6.22 eV, 5.60 eV, and 5.44 eV, respectively. The experimental D_e value for CeF is 6.03 eV. [60] Table 4.6 shows holes in RAS1, GAOPs in RAS2, particles in RAS3, and the most important CSFs. Table 4.6 indicates that the ground states are well described by a single configuration for CeF^{2+} and CeF^{+}, but not for CeF. The 1st – 32nd spinors of CeF^{2+} are similar to

Table 4.5 *Total energies and dissociation energies of* $[Ce^{4+}F^{-}]^{3+}(4f)^1$, $[Ce^{4+}F^{-}]^{3+}(5d^1 4f^1)$, *and* $[Ce^{4+}F^{-}]^{3+}(4f^1 5d^1 6s^1)$ *obtained by RASCI(8|13|25)*

Molecule			Atom/Ion			
Configuration	TE$^{a)}$ /au	D_e / eV	Configuration	TE$^{b)}$ /au	Configuration	TE$^{c)}$ /au
$[Ce^{4+}F^{-}]^{3+}(4f^1)_{5/2}$	−8960.324376	6.22	$[Ce]^{4+}(4f^2)_4$	−8860.584040	$[F]^{5+}(2p^5)_2$	−99.511871
$[Ce^{4+}F^{-}]^{3+}(5d^1 4f^1)_{4/2}$	−8960.744062	5.60	$[Ce]^{4+}(4f^1 5d^2)_{7/2}$	−8861.026467	$[F]^{5+}(2p^5)_2$	−99.511871
$[Ce^{4+}F^{-}]^{3+}(4f^1 5d^1 6s^1)_{7/2}$	−8960.933787	5.44	$[Ce]^{4+}(4f^1 5d^1 6s^2)_4$	−8861.222001	$[F]^{5+}(2p^5)_2$	−99.511871

a) MOs for (8|13|25) of CeF^{2+} are HFR/DC MOs for CeF^{2+} obtained by distributing one electron in the 4*f*, 5*d*, 6*s*, and 6*p*-like spinors (abbreviated as 1/32). MOs for (8|13|25) of CeF^{+} and CeF are CeF^{+} HFR/DC MOs obtained by distributing two electrons in the 4*f*, 5*d*, 6*s*, and 6*p*-like spinors (2/32).
b) See footnote b) in Table 4.2
c) AOs for (1|3|3) of F are HFR/DC AOs obtained by distributing five electrons in the 2*p* spinor (5/6).

Table 4.6 *Holes, GAOPs, particles and dominant CSFs in RASCI (8|13|25) for* CeF^{2+}, CeF^{+}, *and CeF*

Exptl.		RASCI(8|13|25)											
Species	Ω	Conf.	Ω	Hole in RAS1		GAOPs in RAS2					Particle in RAS3		Dominant CSF
				Ce	F	[6*s*]	[6*p*]	[5*d*]	[4*f*]	F	Ce	F	
CeF^{2+}		*f*	5/2-1	−0.06	−0.04	0.01	0.00	0.08	1.02	0.00	0.01	0.00	92%|....33|$^{a)}$
CeF^{+}		*df*	4/2-1	−0.05	−0.01	0.01	0.00	0.96	1.10	0.00	0.03	0.01	78%|....33,40|
CeF	7/2	*sdf*	7/2-1	−0.06	0.01	0.94	0.03	0.93	1.13	0.00	0.07	0.01	26%|....36,43,$\underline{43}$|$^{b)}$

a)'....'≡ 1,$\underline{1}$,.........32,$\underline{32}$
b) The second and third dominant terms are 15% |....36,42,$\underline{43}$| and 15% |....36,$\underline{42}$,43|.

those of CeF^+ (although we have not shown the results). In fact the electronic configuration of CeF^{2+} is $[Ce^{2.5+}(5s^{2.0}5p^{6.0}5d^{*0.4}4f^{*0.1}4f^{1.0})F^{0.4-}(2s^{2.0}2p^{5.4})]^{2+}$ in terms of the CeF^{2+} HFR/DC spinors, and is $[Ce^{2.6+}(5s^{2.0}5p^{6.0}5d^{*0.3}4f^{*0.1}4f^{1.0})F^{0.6-}(2s^{2.0}2p^{5.6})]^{2+}$ in terms of the CeF^+ HFR/DC spinors. These are very close and can be abbreviated as $[Ce^{4+}(5s^25p^6)F^-(2s^22p^6)]^{3+}(4f^1)$; the electron in the (4*f*) spinor moves in the field generated by Ce^{4+} and F^-. This is quite different from [5*d**] in CeO^{2+}. The dication CeO^{2+} has electronic configuration $[Ce^{2.6+}(5s^{2.0}5p^{5.9}5d^{*1.1}4f^{*0.3})O^{0.6-}(2s^{2.0}2p^{4.6})]^{2+}$, abbreviated as $[Ce^{3+}(5s^25p^65d^{*1})O^-(2s^22p^5)]^{2+}$ or simply $[Ce^{3+}(5d^{*1})O^-]^{2+}$, where the [5*d**] spinors are used to describe the molecular O (2*p*) spinors. We therefore cannot exclude the electron in the [5*d**] orbitals from $Ce^{3+}O^-$ and we cannot write the CeO^{2+} as $[Ce^{4+}O^-]^{3+}(5d^{*1})$. CeF^{2+} is formed as

$$Ce^{2+}[5s^25p^64f^2] + F[2s^22p^5] \rightarrow [Ce^{4+}(5s^25p^6)F^-(2s^22p^6)]^{3+}(4f^1). \quad (4.8)$$

One of the two Ce [4*f*] electrons moves into the F [2*p*] spinors and makes up the closed-shell, giving a stable molecular ion, CeF^{2+}; the 0.4 electrons of the six 2*p* electrons in F are back-donated to Ce^{3+}. The 33rd spinor of CeF^{2+} consists almost entirely of [4*f*] ($\Omega = 5/2$). Tables 4.4, 4.5, and 4.6 indicate that CeF^+/CeF is formed as

$$Ce^+[5s^25p^64f^15d^2] + F[2s^22p^5] \rightarrow [Ce^{4+}(5s^25p^6)F^-(2s^22p^6)]^{3+}(4f^15d^1) \quad (4.9)$$

$$Ce[5s^25p^64f^15d^16s^2] + F[2s^22p^5] \rightarrow [Ce^{4+}(5s^25p^6)F^-(2s^22p^6)]^{3+}(4f^15d^16s^1). \quad (4.10)$$

Consider now the difference in the chemical bonds in CeO and CeF in terms of their cation spinors. In the CeO^+ 26th spinor, Ce [$5p_0$] mixes considerably with O [2*s*], while in CeF^+ F [2*s*] mixes slightly with Ce [$5p_0$]. See Tables 4.1 and 4.4. The contour maps of the CeF^+ spinors are given in Figure 4.2. See color plate section. The contours of the CeO^+ 26th spinor in the bonding region are denser than those of the CeF^+ 26th spinor, indicating a strong covalent bond in the CeO^{2+} core compared to that of the CeF^{2+} core.

We discuss further the characteristics of the 26th – 29th molecular spinors of CeO^+ and CeF^+ in terms of their spinor energies. Recall that the electron population of the CeO^{2+} core in CeO^+ is expressed as $[Ce^{3+}O^-]^{2+}$. The atomic Ce^{3+} spinor energies of [5*s*], [$5p_{1/2}$], and the degenerate [$5p_{3/2}$] spinors are respectively −2.7140 au, −1.9784 au, −1.8653 au, and −1.8653 au. The CeO^+ 28th spinor ($\Omega = -3/2$) is accurately located at Ce, as shown in Figure 4.1. The strength of the field generated by O^- on Ce^{3+} is estimated by comparing the [$5p_{3/2}$] spinor energy with the CeO^+ 28th spinor energy (−1.3193 au), as follows. O^- pushes up the value of Ce^{3+} [$5p_{3/2}$] by 0.5460 au (= −1.3193+1.8653). Addition of this value to the spinor energies of Ce^{3+}[5*s*], [$5p_{1/2}$], and the degenerate [$5p_{3/2}$] energies, gives values of −2.1680 au, −1.4324 au, −1.3193 au, and −1.3193 au. These can be compared to the CeO^+ 25th–28th spinor energies (−2.1658 au, −1.5565 au, −1.3943 au, −1.3193 au). The CeO^{2+} 26th spinor is stabilized more than expected (−0.12 au=−1.5565 au + 1.4324 au). This is due to the comparatively large mixing of Ce^{3+}[$5p_0$] with O^-[2*s*].

In CeF^+ the main component of the 26th spinor is F [2*s*] rather than Ce [$5p_0$] in the CeO 26th spinor. The atomic F^- spinor energies of [2*s*], [$2p_{1/2}$], and [$2p_{3/2}$] spinors are −1.0646 au, −0.1710 au, −0.1683 au, and −0.1683 au. The strength of the field generated by Ce^{3+} on F^- is estimated by comparing the F^-[$2p_{3/2}$] spinor energies with that of the 32nd spinor

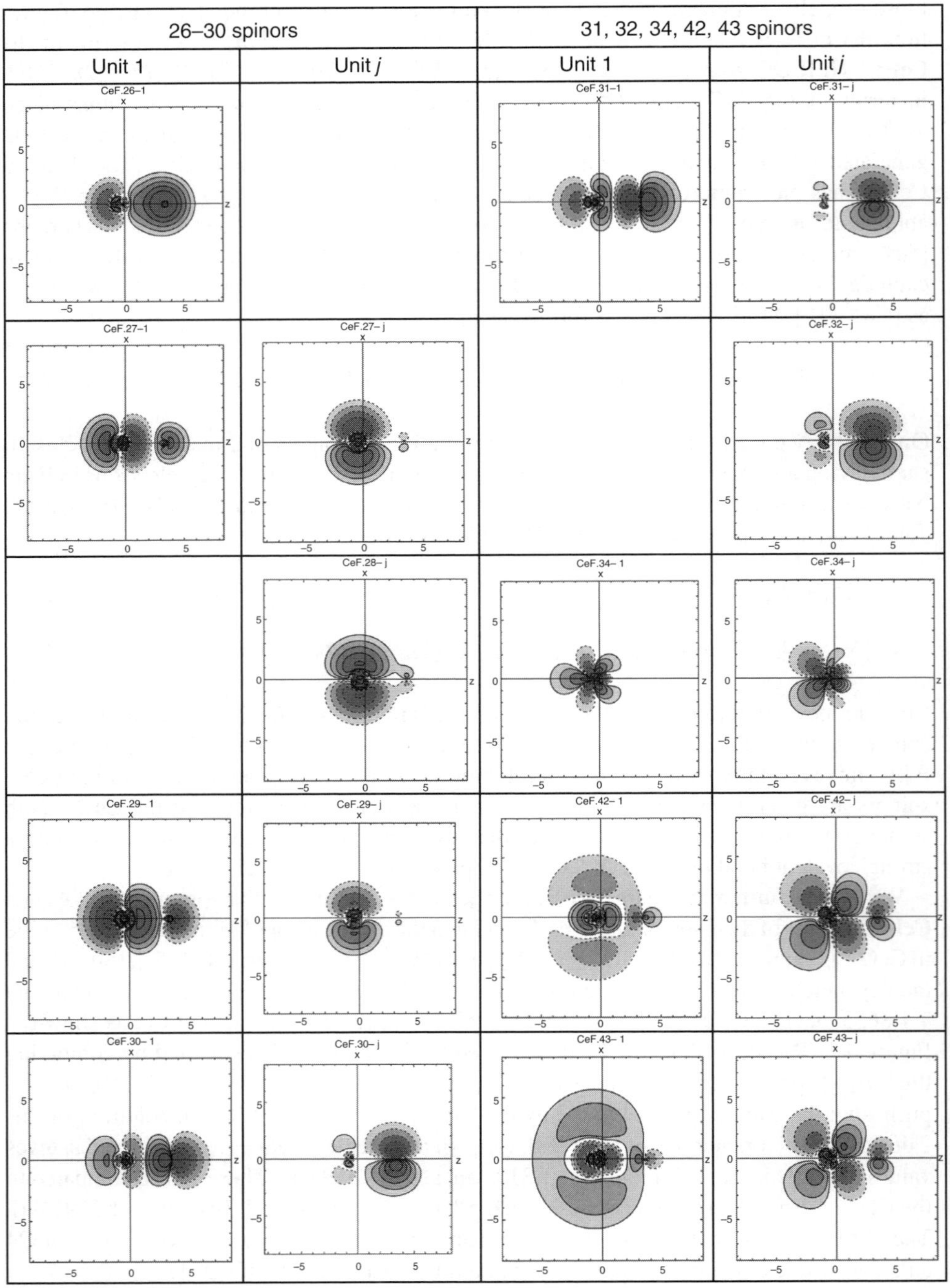

Figure 4.2 *Contour maps of the large components of CeF$^+$ spinors. The CeF$^+$ nuclear distance is taken from the experimental R_e value (3.87 au) of neutral CeF. The maps are drawn in the same manner as in Figure 4.1. See color plate section*

of CeF^+ (−0.7717 au); Ce^{3+} pushes down $F^-[2p_{3/2}]$ by −0.6034 au. Addition of this value to the spinor energies of F^- [2*s*], $[2p_{1/2}]$, and the degenerate $[2p_{3/2}]$ energies gives −1.6680 au, −0.7745 au, −0.7717 au, and −0.7717 au, which can be compared to the values for the 26th, 30th, 31st, and 32nd spinors (−1.7048 au, −0.7741 au, −0.7732 au, −0.7717 au). These shifted F^- spinor energies are close to the real molecular spinor energies except for the 26th, for which a slight further energy lowering (−0.04 au) is observed. This implies weak bonding in the CeF^+ 26th spinor relative to the CeO^+ 26th spinor. The CeF^+ 27th, 28th, and 29th spinors (spinor energies: −1.4046 au, −1.3028 au, −1.2881 au) are almost pure Ce [5*p*]. The differences in the molecular spinor energy between the 27th and 28th or between 27th and 29th spinors originate from the $Ce^{3+}[5p_{1/2}]$ and $[5p_{3/2}]$ spinors (spinor energy: −1.9784 au, −1.8653 au, −1.8653 au). We recall that $[p_{1/2,1/2}]$ is expressed as $-(1/3)^{1/2}p_0\alpha + (2/3)^{1/2}p_{+1}\beta$ and $[p_{3/2,1/2}]$ as $(2/3)^{1/2}p_0\alpha + (1/3)^{1/2}p_{+1}\beta$. See equation (4.3). The CeF^+ 27th and 29th spinors shown in Figure 4.2 hold atom-like [5*p*] spinors but the CeO^+ 27th and 29th do not, as shown in Figure 4.1. Overall, the contribution to the bonding of the 26th – 29th spinors composed of the Ce [5*p*] spinors and F [2*s*] spinors in CeF is smaller than that of the 26th – 29th spinors in CeO.

We compare briefly the characteristics of the 30th – 32nd spinors of CeO^+ and CeF^+, formed mainly of the $O^-[2p]$ spinors and $F^-[2p]$ spinors. The sum of the increments of the GAOPs (4.46/5.44) of ligand [2*p*] electrons in the 30th to 32nd spinors for CeO^+/CeF^+ compared to those (4/5) of atomic [2*p*] spinors is 0.46/0.44. This increased population explains the ionic bond produced by these spinors. We recall that CeO^+ back-donates 1.6 electrons from O to Ce, and CeF^+ back-donates 0.4 electrons from F to Ce through the [4*f**] and [5*d**] polarization functions. Moriyama *et al.* [29] have shown that the electric field generated by a positive point charge gives nearly the same spinor energies for the 30th - 32nd spinors as those of the molecular orbital calculations for CeO^+. We believe that the same is true for CeF^+ (notice almost the same spinor energies of the 30th to 32nd in Table 4.4). The O and F (2*p*) electrons penetrate into the Ce core, and a considerable part of the penetrated electrons is located behind the Ce core, through the polarization functions. There is, however, a difference between the CeO^+ 30th – 32nd spinors and the CeF^+ 30th – 32nd spinors. The CeO^+ spinors have somewhat bonding or antibonding characteristics, whereas the CeF^+ spinors retain atom-like features of the F [2*p*] spinors and the Ce [5*d*] spinors. We feel that the contribution to D_e by the ionic bond of CeO^+, which is covalent in some degree, is greater than that of the ionic bond of CeF^+.

4.3.3 Discussion of Bonding in CeO and CeF

We have discussed the stability of CeO^+ and CeF^+ using the CeO^{2+} and CeF^{2+} core. We have shown that Ce [5*s*], [5*p*], and O [2*s*] make the covalent bond stronger than do Ce [5*s*], [5*p*], and F [2*s*]. The same is true for the ionic bond composed of the O/F [2*p*] spinors, and the Ce [5*d**], [4*f**] spinors. In fact the experimental D_e value of CeO is 8.18 eV and of CeF is 6.03 eV, while the values calculated by RASCI are 5.97 eV and 5.44 eV, respectively. The calculated D_e values for CeO^+ (6.23 eV) and CeF^+ (5.60 eV) also support the reasoning above. The calculated values of D_e for CeO^{2+} (4.89 eV) and CeF^{2+} (6.22 eV) support strong bondings in CeO^{2+} and CeF^{2+}, but they fail to give the expected order of bond strengths ($CeO^{2+} > CeF^{2+}$). Further investigation is necessary.

4.3.4 GdF Ground State

We closely investigated the electronic structure of the ground states of EuF and GdF, when we prepared this chapter. We found that the electronic structure of the two molecules is very close. In order to save the space we treat only the GdF molecule because many spectra were observed for this molecule.

We constructed a ($25s21p16d10f2g$) basis function for Gd from the ($25s18p15d10f$) pGTFs of Koga *et al.* [37] by adding $3{\times}p$ ($2{\times}p$ correlating functions included) [55], $1{\times}d$ pGTFs to describe the $5d$ spinor flexibly, and $2{\times}g$ pGTFs for the electron correlation. The basis function for F is ($12s8p2d$), which is composed of ($12s8p$) [38] and $2{\times}d$ correlating functions of Huzinaga *et al.* [55] The HFR/DC calculations for GdF^{2+} and GdF^{+} were first performed at the calculated internuclear distance (R_e) of 3.7599 au, [26] whereas the experimental distance is 3.71 au. [8] We initially assumed the configuration $[Gd^{10+}(5s^2 5p^6)F^-(2s^2 2p^6)]^{9+}(4f^7)$ and $[Gd^{10+}(5s^2 5p^6)F^-(2s^2 2p^6)]^{9+}(4f^7 6s^1)$ for HFR/DC. Table 4.7 sets out the spinor energies and the GAOPs of the GdF^+ 25th – 32nd spinors, mainly composed of Gd[5*s*], [5*p*], [5*d*], F [2s], [2*p*], and the spinor energies and GAOPs of the 33rd – 48th spinors, composed of Gd [4*f*], [5*d*], [6*s*], and [6*p*] atomic spinors, where seven and one electrons are distributed among the seven ($4f$) spinors and a ($6s$) spinor and their TRPs, respectively.

We then performed GOSCI calculations for GdF^{2+}, GdF^{+}, and GdF. The GOSCI space is constructed as a direct product of the $(4f)^7$ and $(6s, 5d)^n$ full-CI spaces. In the RASCI

Table 4.7 *GAOPs for $GdF^{+a)}$*

Spinor number	Spinor energy /au	Occ.	Ω	GAOPs of Gd				GAOPs of F		Character
				[*s*]	[*p*]	[*d*]	[*f*]	[*s*]	[*p*]	
25	−2.535468	1	1/2	1.00	0.00	0.00	0.00	0.00	0.00	Gd 5s
26	−1.745830	1	1/2	0.00	0.25	0.00	0.00	0.73	0.01	F *sp* + Gd *p*
27	−1.629917	1	1/2	0.00	0.83	0.00	0.00	0.16	0.00	Gd *p* + F *s*
28	−1.480499	1	−3/2	0.00	1.00	0.00	0.00	0.00	0.00	Gd 5*p*
29	−1.457299	1	1/2	0.00	0.89	0.00	0.00	0.10	0.01	Gd *p* + F *sp*
30	−0.801817	1	1/2	0.01	0.02	0.07	0.01	0.01	0.88	F *ps* + Gd *dpfs*
31	−0.790217	1	1/2	0.00	0.01	0.04	0.01	0.00	0.94	F *p* + Gd *dfp*
32	−0.788329	1	−3/2	0.00	0.01	0.04	0.01	0.00	0.94	F *p* + Gd *dfp*
	sum(25–32)	8		1.01	3.01	0.15	0.03	1.00	2.78	
	2 × sum	16		2.02	6.02	0.30	0.06	2.00	5.56	
33	−0.475684	7/14	5/2	0.00	0.00	0.00	0.99	0.00	0.00	Gd 4f
…	…	…	…	…	…	…	…	…	…	…
36	−0.451103	7/14	−7/2	0.00	0.00	0.00	1.00	0.00	0.00	Gd 4f
…	…	…	…	…	…	…	…	…	…	…
39	−0.442355	7/14	1/2	0.00	0.00	0.00	1.00	0.00	0.00	Gd 4f
40	−0.315691	1/2	1/2	0.83	0.05	0.12	0.00	0.00	0.00	Gd *sdp*
…	…	…	…	…	…	…	…	…	…	…
48	−0.051468	0	1/2	0.13	0.71	0.17	0.00	−0.01	−0.01	Gd *pds* + F *sp*
	sum(33–48)	4		0.99	3.36	4.67	6.99	−0.01	−0.04	

a) HFR/DC total energy = −11373.720655 au

notation, this GOSCI scheme is denoted by $(0|(4f^m)\otimes(5d,6s)^n|0)$ ($m = 7, n = 0, 1, 2$). We symbolically noted this GOSCI scheme as (0|M|0) where M is the numbers of the Kramers pairs, which construct $(4f^m)\otimes(5d,6s)^n$ CI space. The GdF^{2+} HFR/DC spinors are used for the GdF^{2+}GOSCI calculations, and the GdF^+ HFR/DC spinors are used for GdF^+/GdF GOSCI. The GOSCI active space is composed of seven $(4f)$ spinors and six $(6s,5d)$ spinors and their TRPs.

In Table 4.8 the dominant electron configurations, total energies and dissociation energies (D_e) are set out. The calculated D_e values for GdF^{2+}, GdF^+, and GdF are respectively 3.68 eV, 4.50 eV, and 5.27 eV. The experimental D_e value for GdF (6.95 eV) [56] is favorably compared with the present GOSCI results. Table 4.9 shows the GAOPs in RAS2 and

Table 4.8 *Total energies and dissociation energies of* $[Gd^{10+}F^-]^{9+}(4f^7)(6s^0)$, $[Gd^{10+}F^-]^{9+}(4f^7)(6s^1)$ *and* $[Gd^{10+}F^-]^{9+}(4f^7)(6s^2)$ *obtained by RASCI(0|M|0) with M = 7, 7⊗6, 7⊗6 for the respective molecules*

Molecule			Atom/Ion			
Conf.	TE[a)] /au	D_e /eV	Conf.	TE[b)] /au	Conf.	TE[c)] /au
$[Gd^{10+}F^-]^{9+}(4f^7)(6s^0)$	−11373.770996	3.678	$[Gd]^{10+}(4f^7\,5d^1 6s^0)_2$	−11274.134075	$[F]^{5+}(2p^5)_2$	−99.501774
$[Gd^{10+}F^-]^{9+}(4f^7)(6s^1)$	−11374.215154	4.497	$[Gd]^{10+}(4f^7\,5d^1 6s^1)_{5/2}$	−11274.548125	$[F]^{5+}(2p^5)_2$	−99.501774
$[Gd^{10+}F^-]^{9+}(4f^7)(6s^2)$	−11374.421964	5.265	$[Gd]^{10+}(4f^7\,5d^1 6s^2)_2$	−11274.726719	$[F]^{5+}(2p^5)_2$	−99.501774

a) MOs for (0|7|0) of GdF^{2+} are HFR/DC MOs for GdF^{2+} obtained by distributing seven electrons in the 4*f*-like spinors (abbreviated as 7/14). MOs for (0|7⊗6|0) of GdF^+ and GdF are HFR/DC MOs of GdF^+ obtained by distributing seven electrons in the 4*f*-like spinors (7/14) and one electron in the 6s-like spinor (1/2).
b) AOs for (0|7⊗6|0) of Gd^{2+} are HFR/DC AOs for Gd^{2+} obtained by distributing seven electrons in the 4*f* spinors (7/7) and one electron in the 6s spinor (1/2). AOs for (0|7⊗6|0) of Gd^+ and Gd are HFR/DC AOs for Gd obtained by distributing seven electrons in the 4*f* spinors (7/14) and two electrons in the 6s spinor (2/2).
c) AOs for (0|3|0) of F are HFR/DC AOs obtained by distributing five electrons in the 2*p* spinors (5/6).

Table 4.9 *GAOPs and dominant CSFs in RASCI(0|M|0) for* GdF^{2+}, GdF^+, *and GdF*

Exptl.		RASCI(0|13|0)							
Species	Ω	Conf.	Ω	GAOPs in RAS2					Dominant CSF
				[6s]	[6*p*]	[5*d*]	[4*f*]	F	
GdF^{2+}		$(4f)^7(6s)^0$	7/2	0.00	0.00	0.01	6.98	0.01	$17\%|\ldots(f^7)_{7/2}|$[a)]
GdF^+		$(4f)^7(6s)^1$	4	0.83	0.05	0.13	6.98	0.01	$17\%|\ldots(f^7)_{7/2}40|$[b)]
GdF	7/2	$(4f)^7(6s)^2$	7/2	1.47	0.18	0.35	6.98	0.01	$15\%|\ldots(f^7)_{7/2}40,\underline{40}|$[c)]

a) $|\ldots(f^7)_{7/2}| \equiv |\ldots\underline{33},34,\underline{35},\underline{36},37,\underline{38},39|$
b) $|\ldots(f^7)_{7/2}40| \equiv |\ldots\underline{33},34,\underline{35},\underline{36},37,\underline{38},39,40|$
c) $|\ldots(f^7)_{7/2}40,\underline{40}| \equiv |\ldots\underline{33},34,\underline{35},\underline{36},37,\underline{38},39,40,\underline{40}|$

the most important CSFs. The weights of the important CSFs for all of the molecules are less than 20%, indicating their multiconfigurational character. The electronic configuration of GdF is written as $[Gd^{0.6+}(5s^{2.0}5p^{6.0}5d{*}^{0.3}4f{*}^{0.1}4f^{7.0}6s^{2.0})F^{0.6-}(2s^{2.0}2p^{5.6})]$. This can be abbreviated as $[Gd^{+}(5s^{2}5p^{6}4f^{7}6s^{2})F^{-}(2s^{2}2p^{6})]$ or $[Gd^{10+}(5s^{2}5p^{6})F^{-}(2s^{2}2p^{6})]^{9+}(4f^{7}6s^{2})$.

The contour maps of the molecular spinors for GdF^+ are set out in Figure 4.3. The figure shows the 26th – 29th spinors form the covalent bonds, and the 30th – 32nd spinor make the ionic bond as in CeO^+/CeO and CeF^+/CeF. Figures 4.1–4.3 show that in the bonding region the CeO^+ 26th spinor has the densest population and the GdF 26th spinor has denser population than that of CeF. Figures 4.1–4.3 also show that the CeO^+ spinors have somewhat stronger bonding or antibonding characteristics than those of the GdF^+ and CeF^+ in which the CeF^+ spinors retain atom-like features of the F [2*p*] spinors or the Ce [5*d*] spinors. These findings indicate that the order of D_e is CeO > GdF > CeF. Actually experimental D_es are 8.18 eV, 6.95 eV, and 6.03 eV for CeO, GdF, and CeF, respectively, while the calculated values are 5.97 eV, 5.23 eV, and 5.44 eV from CeO RASCI(8|15|23), GdF RASCI(013|0), and CeF RASCI(8|13|25). If the more correlated calculation is performed for GdF, a proper D_e for GdF would result. We summarize below the formation of GdF^{2+}, GdF^+, and GdF. GdF^{2+} is formed as

$$Gd^{2+}[5s^{2}5p^{6}4f^{7}5d^{1}] + F[2s^{2}2p^{5}] \rightarrow [Gd^{10+}(5s^{2}5p^{6})F^{-}(2s^{2}2p^{6})]^{9+}(4f^{7}) \tag{4.11}$$

and GdF^+/GdF as

$$Gd^{+}[5s^{2}5p^{6}4f^{7}5d^{1}6s^{1}] + F[2s^{2}2p^{5}] \rightarrow [Gd^{10+}(5s^{2}5p^{6})F^{-}(2s^{2}2p^{6})]^{9+}(4f^{7}6s^{1})/ \tag{4.12}$$

$$Gd[5s^{2}5p^{6}4f^{7}5d^{1}6s^{2}] + F[2s^{2}2p^{5}] \rightarrow [Gd^{10+}(5s^{2}5p^{6})F^{-}(2s^{2}2p^{6})]^{9+}(4f^{7}6s^{2}) \tag{4.13}$$

4.3.5 Summary of the Chemical Bonds, of CeO, CeF, GdF

We have investigated the electronic structure of the ground states for CeO, CeF, and GdF. The chemical bond is different between CeO and LnF as discussed above, but the bond has a common feature of having the stable dication core (LnX^{2+}). The ground state of CeO is written as $[Ce^{3+}([Xe]5d{*}^{1})O^{-}(2p^{5})]^{2+}(4f^{1})(6s^{1})$. The two fluorides can be written as $[Ce^{4+}([Xe])F^{-}(2p^{6})]^{3+}(4f^{1})(5d^{1}6s^{1})$, and $[Gd^{10+}([Xe])F^{-}(2p^{6})]^{9+}(4f^{7})(6s^{2})$. We found that (4*f*), (5*d*), (6*s*) in LnF almost consist of Ln atomic spinors. If the (4*f*) electrons in LnF are enclosed in the ionic core, the outer two valence electrons in (5*d*) and (6*s*) move in the field generated by the dication, LnX^{2+}. In CeO two valence electrons in (4*f*) and (6*s*) move in the field generated by CeO^{2+}. We, thus, expect the stable LnX^{2+} (X = O and F). We have no experimental D_es for LnX^{2+}. The calculated D_es for $[Ce^{3+}([Xe]5d{*}^{1})O^{-}(2p^{5})]^{2+}$, $[Ce^{3+}([Xe]4f^{1})F^{-}(2p^{6})]^{2+}$, and $[Gd^{3+}([Xe]4f^{7})F^{-}(2p^{6})]^{2+}$ are 4.89 eV, 6.22 eV, and 3.68 eV (see Tables 4.2, 4.5, and 4.8), respectively, indicating the existence of the stable LnX^{2+}.

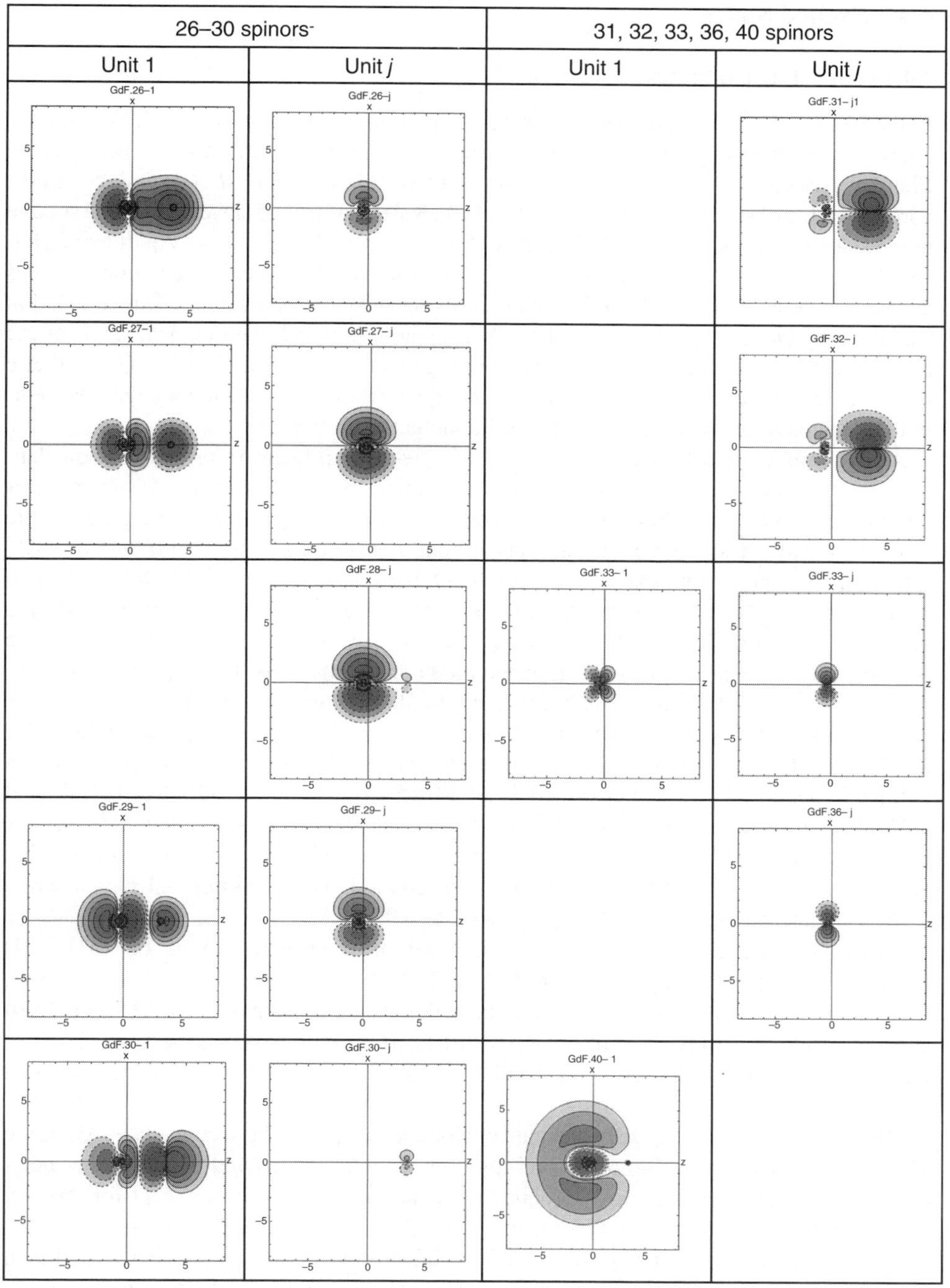

Figure 4.3 *Contour maps of the large components of GdF^+ spinors. The GdF^+ nuclear distance is the calculated R_e value (3.7599 au) of neutral GdF. The maps are drawn in the same manner as in Figure 4.1. See color plate section*

4.4 Excited States

4.4.1 CeO Excited States

In the late 1970s, molecular spectra of pure S-type stars in the range 1.513–2.752 eV were analyzed, and with the aid of laboratory studies, the lines were attributed to CeO. [61, 62] Many observations have since been performed. [7, 63–65] Linton *et al.* [64] gave an almost complete table of spectra in the range 0–3.546 eV. Kaledin *et al.* [65] investigated the spectra in the same range, using ligand field theory (LFT) as refined by Field. [6] Apart from LFT, which includes semi-empirical parameters, only a single theoretical calculation [10] has considered the CeO electronic excited states, looking at the lowest 16 states. This calculation was done by Dolg *et al.* [10] using the SOCI method, in which the diagonal terms of the Hamiltonian were replaced by the corresponding eigenvalues of the CISD + Q (quadratic excitation correction). [66] Dolg *et al.* used the quasirelativistic pseudopotential, and refer to their calculations as quasirelativistic pseudopotential CI (QRCI).

Analysis of the electronic structure of CeO yields significant information about lanthanide chemistry, because CeO is the first lanthanide oxide that contains 4*f* electrons. A number of excited states having the common electronic configuration as the ground state lies energetically near the ground state. The ground state of the Ce atom is $[4f^1 5d^1 6s^2]\ ^1G_4^o$; we use the symbolic notation $^1G_4^o$ for simplicity, but this state actually has a complicated configuration. [57] The ground state of the O atom is $[2s^2 2p^4]\ ^3P_2$. [58] The ground state of CeO is $[\mathrm{Ce}^{3+}(5s^2 5p^6 5d{*}^1)\mathrm{O}^-(2s^2 2p^5)]_0^{2+}(4f^1 6s^1)_2$, where the subscripts (0, 2) denote the electronic angular momentum projected on the molecular axis. In the present work we discuss the excited states below 0.56 eV using the RASCI module. [49, 50]

Using the CeO^+ HFR/DC molecular spinors, we performed RASCI calculations for the lowest 16 states having excitation energies below 0.560 eV. The RASCI space is defined in Section 4.3.1. We performed RASCI(0|15|23) in addition to RASCI(0|15|0) and RASCI(8|15|23) calculations. RASCI(0|15|0) is equivalent to the complete active space CI (CASCI) or GOSCI with two active electrons distributed among the 15 Kramers pairs, and we use the term RASCI(0|15|0) instead of GOSCI. The maximum CI dimension is 9,367,744 determinants in RASCI(8|15|23) for $\Omega = 0$.

Experimentally, all of the low-lying 16 states are considered to be $(4f^1 6s^1)$. [64] RASCI(0|15|0) gives the configuration $([4f^{0.5}][5d^{0.6}][6s^1])$ [29] for the 16th state, however, rather than $([4f^1][6s^1])$. [64] We therefore have doubts about the results of RASCI(0|15|0) and henceforth disregard the results of RASCI(0|15|0). Table 4.10 shows the lowest 16 excitation energies given by experiment, RASCI(0|15|23), and RASCI(8|15|23), together with the GAOPs.

RASCI(0|15|23) gives a fairly accurate excitation energy for the second state. All of the 16 low-lying states have $(4f^1 6s^1)$ configuration. Beyond the second excited state, however, RASCI(0|15|23) does not reproduce the experimental Ω values except for the 6th, 8th, and 9th states. We believe that the use of small numbers (23) of virtual spinors is the explanation of this discrepancy. We therefore added 80 virtual spinors to RAS3 and then performed the RASCI(0|15|103) calculation. This calculation then gives the same Ω values, and similar excitation energies for the respective states, as RASCI(0|15|23). For example, the excitation energies of the 2nd, 3rd, 15th, and 16th states obtained by RASCI(0|15|103) are 0.019 eV, 0.038 eV, 0.334 eV, 0.343 eV, which are very close to the RASCI(0|15|23)

Table 4.10 *CeO excitation energies and GAOPs of the lowest sixteen excited states lying 0 – 0.560eV above the ground state*

	Exptl				RASCI(0|15|23)						RASCI(8|15|23)					
No	Name	Ω	EE /eV	Conf	Ω	EE /eV	GAOPs in RAS2				Ω	EE /eV	GAOPs in RAS2			
							[6*s*]	[6*p*]	[5*d*]	[4*f*]			[6*s*]	[6*p*]	[5*d*]	[4*f*]
1	X_1	2	0.000	*fs*	2	0.000	0.67	0.09	0.12	0.98	2	0.000	0.76	0.10	0.15	1.03
2	X_2	3	0.010	*fs*	3	0.019	0.67	0.09	0.11	0.99	3	0.014	0.75	0.10	0.15	1.03
3	W_1	1	0.101	*fs*	0^-	0.042	0.71	0.07	0.19	0.88	1	0.061	0.76	0.10	0.21	0.96
4	W_2	2	0.113	*fs*	1	0.051	0.71	0.07	0.19	0.89	2	0.076	0.76	0.10	0.21	0.96
5	V_1	0^-	0.208	*fs*	0^+	0.080	0.70	0.07	0.20	0.88	0^-	0.092	0.77	0.10	0.25	0.91
6	V_2	1	0.232	*fs*	1	0.084	0.69	0.08	0.16	0.92	1	0.119	0.77	0.10	0.25	0.91
7	U_1	0^+	0.240	*fs*	2	0.088	0.69	0.08	0.15	0.93	0^+	0.136	0.76	0.10	0.25	0.91
8	X_3	4	0.253	*fs*	4	0.239	0.67	0.09	0.11	0.99	4	0.251	0.76	0.10	0.14	1.03
9	X_4	3	0.265	*fs*	3	0.253	0.67	0.09	0.11	0.99	3	0.266	0.75	0.10	0.15	1.03
10	W_3	3	0.324	*fs*	0^-	0.279	0.71	0.07	0.19	0.88	3	0.295	0.76	0.10	0.19	0.98
11	W_4	2	0.344	*fs*	1	0.288	0.71	0.07	0.19	0.88	2	0.313	0.76	0.10	0.20	0.97
12	V_3	2	0.429	*fs*	3	0.301	0.69	0.08	0.14	0.95	1	0.337	0.77	0.10	0.24	0.92
13	V_4	1	0.452	*fs*	2	0.306	0.70	0.08	0.16	0.92	0^-	0.339	0.77	0.10	0.25	0.90
14	T_1	0^-	0.474	*fs*	0^+	0.316	0.70	0.07	0.20	0.87	2	0.347	0.76	0.10	0.23	0.94
15	U_2	1	0.512	*fs*	2	0.340	0.68	0.08	0.15	0.94	1	0.418	0.76	0.10	0.24	0.92
16	U_3	0^+	0.553	*fs*	1	0.354	0.69	0.07	0.18	0.90	0^+	0.426	0.76	0.10	0.26	0.90

values of 0.019 eV, 0.042 eV, 0.340 eV, and 0.353 eV. The closeness of the results given by RASCI(0|15|23) and RASCI(0|15|103) indicates that the 23 RAS3 spinors are sufficient to describe the correlation effects of the two valence electrons; the small virtual spinor number of 23 in RAS3 is not the reason why RASCI(0|15|23) gives a poor description of the Ω values.

Based on the preliminary calculations above, we decided to implement RASCI(8|15|23) to take account of the electron correlation effects between the active core and the valence electrons. RASCI(8|15|23) accurately reproduces the experimental excitation energy for the first excited state. The calculation also gives the same Ω values as experiment except for the 12th – 14th states, although the calculated excitation energies are somewhat smaller than experiment. Overall, the features given by RASCI(8|15|23) are close to experiment. Dolg *et al.* [10] also calculated the lowest 16 states by QRCI; their results are better than the present ones. They used the Ce (5*s*) – (7*s*), (5*p*) – (7*p*), (5*d*) – (6*d*), (4*f*), and (5*f*) molecular orbitals (MOs) and the O (2*s*) – (4*s*), (2*p*) – (3*p*), and (3*d*) MOs in their QRCI, whereas in the present RASCI the Ce (5*s*) – (7*s*), (5*p*) –(8*p*), (5*d*) –(6*d*), (4*f*), and (5*f*) spinors and the (2*s*), (2*p*), and (3*p*) spinors of O are used. The CI spaces spanned by the Ce orbitals (spinors) are similar, but those spanned by the O orbitals (spinors) are not. In the reference 29 it is shown that the correlation effects arising from O (2*p*) are as important as those arising from Ce (5*s*) and (5*p*). The use of fewer O spinors in RASCI than in QRCI probably renders the results inferior to those of Dolg *et al.* [10]

Higher excited states than 0.56 eV is given in this Moriyama *et al.* [29]

4.4.2 CeF Excited States

Three excited states of the CeF molecule were observed by Bloch and coworkers. [59] The highest state is a red-degraded transition at 2.181 eV with $\Omega = 4.5$. The two lower states observed were at 0.087 eV ($\Omega = 4.5$) and 0.186 eV (tentatively assigned $\Omega = 3.5$) above the ground state. [59]

Two theoretical calculations can be found in the literature. One of these publications is a study by Wasada *et al.* [21] They performed CISD from a single reference CSF using a four-component DC Hamiltonian. In their CISD calculations, electron correlations among 17 electrons of Ce $(5p^6)(4f^1 5d^m 6s^n;\ n + m = 2)$ and $F(2s^2 2p^6)$ are considered. The calculated excitations are 0.319 eV, 0.518 eV, and 2.197 eV for the $\Omega = 4.5, 3.5, 4.5$ states, respectively. The two calculated lower excitation energies are considerably larger than the corresponding experimental values. The third calculated excitation energy is close to experiment, but the configuration obtained ($[4f^{1.0}][5d^{1.7}][6p^{0.3}]$) is different from that of experiment ($[4f^1][5d^1][6p^1]$).

The second theoretical calculation is by Tatewaki *et al.* [22] They used four-component relativistic multiconfigurational quasidegenerate perturbation theory to take account of correlation effects between the active core $[Ce^{3+}(4s^2 \ldots 5p^6)F^-(2s^2 2p^6)]^{2+}$ and three valence electrons in $(4f)$, $(5d)$, $(6s)$, $(6p)$, and intra-valence correlation effects. For the energy levels of the two excited states they obtained 0.144 eV and 0.351 eV. They did not reach conclusions about the higher excited states because of the limitations of the theory they used. A further study of the CeF excited states using larger RASCI and the f-shell Omega decomposition method (which is described below) is under way.

4.4.3 GdF Excited States

Kaledin *et al.* [8] made a detailed investigation of the GdF molecule, and characterized most of its excited states up to about 3.0 eV above the ground state. They also performed LFT calculations, which gave results consistent with experiment, although some uncertainties remained.

Several computational studies of the ground state of GdF [9, 12, 13, 17, 23–27] have been published. Studies of the excited states are less common, however. The chapter authors have systematically studied GdF. [24–27] We developed for this purpose a method for analyzing electronic configurations, including f electrons, called the "f shell Omega decomposition method."

4.4.3.1 f-Shell Omega Decomposition Method

We used the DIRAC program suite, in which the *jj*-coupling scheme is employed. For molecules containing f electrons, the electronic wavefunctions consist of many CSFs because of the near-degeneracy among the (f) spinors. In such molecular systems, numerous electronic configurations generated from the active spinors make contributions to the physical states, and it is scarcely possible to extract useful information merely by analyzing the individual CSFs. As an example, the GOSCI $(0|(4f^7)\otimes(5d,6s)^2|0)$ wavefunction for the GdF ground state ($\Omega = 7/2$) involves 14989 CSFs, and the CI weight is 0.1473 even at the maximum. This small CI weight indicates the multiconfigurational character of the GdF GOSCI wavefunction in the *jj*-coupling scheme. An index is needed that condenses the contributions from the individual Slater determinants. Accordingly, we introduced an index called the 'f-shell Omega component weight.' It is a series of numbers defined as

$$\text{weight}(\Omega_f) = \sum_{i \in \Omega_f} C_i^2 \tag{4.14}$$

where C_i is the CI coefficient of the i-th Slater determinant, and Ω_f denotes the electronic angular momentum projected onto the molecular axis generated from the n electrons of the $(4f)$ molecular spinors; in the GdF molecule, $n = 7$. The multiconfigurational appearance of the f-electron system is clearly reduced by the use of the f-shell Omega component weight. Addition of the electronic angular momentum generated from the two electrons of $(5d, 6s)$ to Ω_f yields the total electronic angular momentum around the molecular axis. Some examples of the f-shell Omega components, which are selected from those of all the excitation states below 3.0 eV [27] are listed in Table 4.11, together with GOSCI GAOPs of the respective states.

We here propose a terminology for molecular electronic states. As an example we shall consider the ground state of GdF, which is written as $^8\Sigma^-_{7/2}$ in the LS-coupling scheme. As shown in Table 4.11, the first state representing the $\Omega = 7/2$ ground state has a Ω_f weight of 1.00 at the $\Omega_f = 7/2$ column. The three states above this state are almost degenerate, and their Ω_f weights are close to 1.00, as shown in Table 4.11. Thus, the CSFs of these three states can be derived from the CSFs of the $\Omega = 7/2$ state. We call such a group of states a 'family.' The first state ($\Omega = 7/2$) is the origin of this family. We call such a state the 'root' of the family.

We now set out the four steps of the naming procedure. We assume that the electronic state is described as $^{\alpha}[\beta]_{\gamma}$ and explain how α, β, and γ are assigned.

1) Write the Ω value of the root of a family, Ω_{root}, as a superscript α.
2) Write the symbolized gross atomic orbital populations (GAOP) as β, such as $[(4f^7)(6s^2)]$; the GAOPs for the ground state are shown in Table 4.11 as $[4f^{7.0}]$ $[6s^{1.5}][5d^{0.4}][6p^{0.2}]$.
3) 'γ' is a subscript of the total angular momentum Ω of the corresponding family.

   The four lowest-lying states form a family, and this is designated as $^{7/2}[(4f^7)(6s^2)]_{\Omega}$ based on the nomenclature above. We also add the Ω_f value of the maximum Ω_f weight as a subscript to the $(4f^7)$ part, and we attach the angular momentum value (Ω_{sd}) of the (sd) part as its subscript. Consider now the four lowest states of GdF. These are almost degenerate and have almost the same GAOPs. In these states the Ω_f weights are almost 1.00, indicating that the wavefunctions of the three higher states are obtained by applying the lowering operators. These four states constitute the ground state family, and this family is expressed as $^{7/2}[(4f^7)_{\Omega_f}(6s^2)_{\Omega_{sd}}]_{\Omega}$, where the total angular momentum of each family member decreases by one from 7/2 to 1/2. This is our rule for naming a family. Based on the procedure described above, the first state ($\Omega = 7/2$) is designated as $^{7/2}[(4f^7)_{7/2}(6s^2)_{0/2}]_{7/2}$; we use the subscript $(2 \cdot \Omega_{sd})/2$ for the sd valence shell instead of Ω_{sd} because Ω_f takes a half integer value in this case. The second state ($\Omega = 5/2$) is designated as $^{7/2}[(4f^7)_{5/2}(6s^2)_{0/2}]_{5/2}$ as the maximum Ω_f weight appears at $\Omega_f = 5/2$; accordingly the angular momentum of the (sd) part is 0/2, and so on. Family members are thereby designated. We summarize the specification of the family as $^{\Omega\text{root}}[\text{symbolized GAOPs}]_{\Omega}$, where Ω_{root} is the highest Ω-value of the family, and the

Table 4.11 *Designation of the GdF excited states below 3.0 eV*

	Family name		EE	*f*-shell Omega component weight[d)]								GAOP[e)]		
SN[a)]	State designation[b)]	N[c)]	(eV)	−7/2	−5/2	−3/2	−1/2	1/2	3/2	5/2	7/2	[6*s*]	[5*d*]	[6*p*]
	$X^{7/2}[(f^7)(s^2)]_\Omega$													
1	$X^{7/2}[(f^7)_{7/2}(s^2)_{0/2}]_{7/2}$	1	0.00								**1.00**	1.47	0.35	0.18
2	$X^{7/2}[(f^7)_{5/2}(s^2)_{0/2}]_{5/2}$	1	0.00							1.00		1.47	0.35	0.18
3	$X^{7/2}[(f^7)_{3/2}(s^2)_{0/2}]_{3/2}$	1	0.00						0.99			1.47	0.35	0.18
4	$X^{7/2}[(f^7)_{1/2}(s^2)_{0/2}]_{1/2}$	1	0.00					0.99				1.47	0.35	0.18
	$a^{5/2}[(f^7)(s^1d^1)]_\Omega$													
5	$a^{5/2}[(f^7)_{7/2}(s^1d^1)_{-2/2}]_{5/2}$	2	0.46								**1.00**	0.80	1.13	0.07
6	$a^{5/2}[(f^7)_{5/2}(s^1d^1)_{-2/2} + ..]_{3/2}$	2	0.48							0.80	0.20	0.80	1.13	0.07
7	$a^{5/2}[(f^7)_{3/2}(s^1d^1)_{-2/2} + ..]_{1/2}$	2	0.49						0.62	0.36	0.02	0.80	1.13	0.07
	$b^{13/2}[(f^7)(s^1d^1)]_\Omega$													
8	$b^{13/2}[(f^7)_{-3/2}(s^1d^1)_{4/2} + ..]_{1/2}$	3	0.50		0.07	0.47	0.46					0.80	1.13	0.07
9	$b^{13/2}[(f^7)_{-1/2}(s^1d^1)_{4/2} + ..]_{3/2}$	3	0.51			0.14	0.54	0.32				0.80	1.14	0.07
10	$b^{13/2}[(f^7)_{1/2}(s^1d^1)_{4/2} + ..]_{5/2}$	3	0.53				0.24	0.56	0.20			0.80	1.14	0.07
11	$b^{13/2}[(f^7)_{3/2}(s^1d^1)_{4/2} + ..]_{7/2}$	2	0.54					0.37	0.53	0.10		0.80	1.14	0.07
12	$b^{13/2}[(f^7)_{3/2}(s^1d^1)_{6/2} + ..]_{9/2}$	1	0.55						0.53	0.43	0.04	0.80	1.14	0.06
13	$b^{13/2}[(f^7)_{5/2}(s^1d^1)_{6/2} + ..]_{11/2}$	1	0.57							0.74	0.26	0.80	1.14	0.06
14	$b^{13/2}[(f^7)_{7/2}(s^1d^1)_{6/2}]_{13/2}$	1	0.59								**1.00**	0.80	1.14	0.06
...														
	$E^{7/2}[(f^7)(s^15d^1 + d^2)]_\Omega$													
108	$E^{7/2}[(f^7)_{7/2}(s^1d^1 + d^2)_{0/2} + ..]_{7/2}$	17	2.50							0.03	**0.97**	0.57	0.94	0.47
109	$E^{7/2}[(f^7)_{5/2}(s^1d^1 + d^2)_{0/2} + ..]_{5/2}$	22	2.50						0.04	0.95		0.57	0.96	0.47
111	$E^{7/2}[(f^7)_{3/2}(s^1d^1 + d^2)_{0/2} + ..]_{3/2}$	26	2.50					0.04	0.95			0.56	0.97	0.46
112	$E^{7/2}[(f^7)_{1/2}(s^1d^1 + d^2)_{0/2} + ..]_{1/2}$	27	2.51			0.02	0.04	0.74	0.13	0.06		0.42	1.09	0.48
...														

a) The serial number in the entire CI space.
b) The 4*f*, 5*d*, and 6*s* are abbreviated as *f*, *d*, and *s* respectively.
c) The serial number in the respective OMEGA; please replace omega with capital greek.
d) *f*-shell Omega components with weight greater than 0.01 are listed. A bold-faced entry denotes a state of the highest weight in a family.
e) The GAOP of Gd [4*f*] is 6.98 for all states listed in this table.

total angular momentum of each family member decreases by one from Ω_{root} down to either 1/2 or 0.

4) We propose a rule for labeling a family in the *jj*-coupling scheme. A label *L* is placed before $^{\Omega\text{root}}$[configuration symbol]$_{\Omega}$, namely, $L^{\Omega\text{root}}$[configuration symbol]$_{\Omega}$. This rule is similar to that for a state in the *LS*-coupling scheme. [67] The label *X* denotes the family of the ground state. The label *A* denotes the family of the first excited state having the same Ω as the ground state. Subsequent families are labeled with *B*, *C*, *D*, and so on. The excited states for which Ω differs from the ground state are labeled with lowercase letters in alphabetical order (*a*, *b*, *c*, ...) as the excitation energy increases. (In the *LS*-coupling scheme, the state label is determined according to the spin multiplicity instead of Ω.)

The *f*-shell Omega decomposition method provides a systematic way to assign electronic spectra, especially for molecules containing heavy elements such as lanthanide monofluorides, in which the *jj*-coupling is dominant compared to the *LS*-coupling. In Table 4.11 we applied the *f*-shell Omega decomposition method to some example states of the GdF molecule, and give a unified terminology for the electronic states represented in the *jj*-coupling scheme. We give all the families and their excitation energies in Figure 4.4. See color plate section. For families having the highest Ω value, the same as the *X* family, solid lines are used; other families are connected by broken lines.

4.4.3.2 GdF Excited States Below 0.85 eV

Electronic transitions involving the low-lying excited states located below 0.85 eV will now be discussed. Kaledin *et al.* [8] found 10 forbidden bands around 0.548 eV (4416 cm^{-1}). We searched states in the range 0.01 eV to 0.85 eV and found 10 states. These are listed in Table 4.12 with the designations specified above and those of Kaledin *et al.* [8], with serial number (SN) in the entire CI space and serial number (N) in the respective Ω space, GAOPs, excitation energy (EE), and transition dipole moment. These 10 states are members of the two families, $a\ ^{5/2}[(f^7)(s^1d^1)]_{\Omega}$ and $b\ ^{13/2}[(f^7)(s^1d^1)]_{\Omega}$, concentrated in the energy region 0.46 eV to 0.59 eV. They all have small transition probabilities with the ground state, consistent with the fact that direct transitions from the ground state are not observed experimentally. Upon comparing the ground state population $[4f^7][6s^{1.5}][5d^{0.4}][6p^{0.2}]$ with the excited state populations $[4f^7][6s^{0.8}][5d^{1.1}][6p^{0.1}]$, it becomes clear that the difference of these populations corresponds to excitations from [6*s*] to [5*d*] atomic spinors. This is why the transition dipole moments are small. The increase in the calculated transition energies as the Ω value varies from 5/2, 3/2, 1/2, 1/2, 3/2, 5/2, 7/2, 9/2, 11/2, to 13/2 is consistent with the LFT calculations of Kaledin *et al.* [8]

4.4.3.3 GdF [15.8] and [18.3] States

Kaledin *et al.* [8] observed a strong band at 1.956 eV (15776.9 cm^{-1}), and designated it as [15.8]. They found another strong band at 2.264 eV (18261.9 cm^{-1}) and designated it as [18.3]. Table 4.13 sets out details of states that have a transition moment larger than 3.8 Debye. They are state designation, difference in $\Omega(\Delta\Omega)$ between the excited state and the ground state, serial number (SN) in the entire CI space, and serial number (N) in the respective Ω space, the excitation energy (EE), and the transition moment.

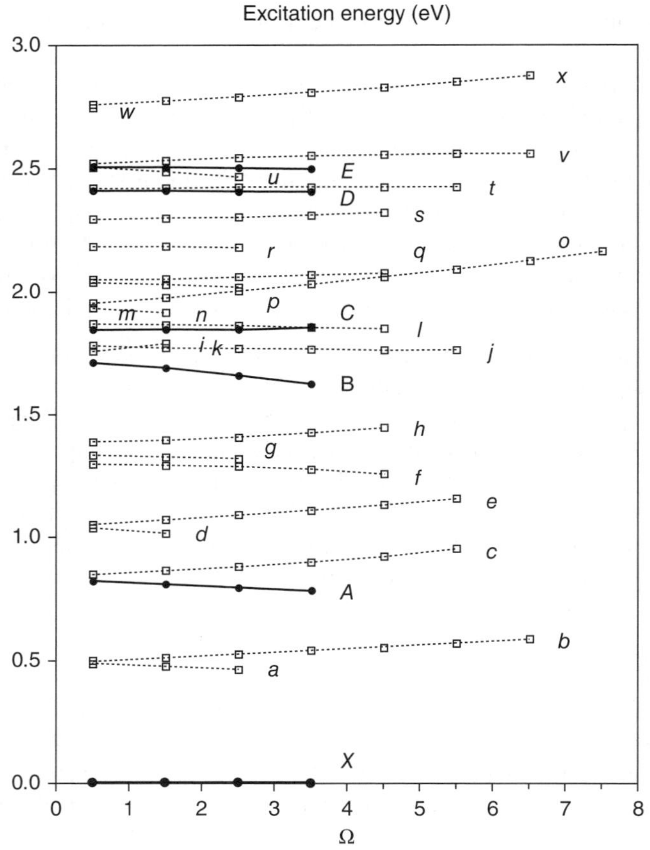

Figure 4.4 *Energy diagram of GdF. Vertical excitation energy of the GdF electronic state and family label. The ordinate denotes the vertical excitation energy in eV. The abscissa denotes the Ω value. Only positive values of Ω are treated. Solid circles and solid lines denote families having the same Ω as the X family. Open squares and broken lines denote families with Ω differing from the X family. See color plate section*

We shall start from [18.3], which has clear spectra. [8] The spectra of [18.3] of GdF arise from a single family, $E^{7/2}[(f^7)(d^1s^1+d^2)]_\Omega$ (SN=108, 109, 111, 112), and have degenerate excitation energies (T_v) as shown in Table 4.13. The spacing (0.005 eV) of the calculated ΔT_v values of [18.3] between $\Omega = 7/2$ (SN = 108) and $\Omega = 1/2$ (SN = 112) is comparable to that (0.0094 eV) of the experimental term values. [8] All transitions occur from the *E* family to the *X* family having the same Ω ($\Omega = 0$ transitions). These features suggest that the [18.3] states are $^8\Sigma^-$-like and that spin-orbit coupling is not dominant. This is consistent with the assignment of Kaledin *et al.* [8]

Compared to the [18.3] transitions, the [15.8] transitions are complicated. The five states (SN=79, 80, 82, 83, 84) form a family $q^{9/2}[(f^7)(d^1s^1+d^2)]_\Omega$. The three states (SN = 75, 76, 78) form another family, $p^{5/2}[(f^7)(d^1s^1+d^2)]_\Omega$. Linking the *q* and *p* families and the

Table 4.12 *GAOPs of the ground state and the ten low-lying excited states of the GdF molecule*

Designation					GAOP of Gd				EE/eV		Tr-M[e]/D
Theory	Experimental[a]	SN[b]	Ω	N[c]	[*f*]	[*s*]	[*d*]	[*p*]	$T_v^{d)}$	$T_v^{e)}$	
$X^{7/2}(f^7s^2)_{7/2}$	$^8\Sigma^-(f^7s^2)_{7/2}$	1	7/2	1	6.98	1.47	0.35	0.18	0.0	0.0	
$X^{7/2}(f^7s^2)_{52}$		2	5/2	1	6.98	1.47	0.35	0.18	0.000		
$X^{7/2}(f^7s^2)_{3/2}$		3	3/2	1	6.98	1.47	0.35	0.18	0.001		
$X^{7/2}(f^7s^2)_{1/2}$		4	1/2	1	6.98	1.47	0.35	0.18	0.001		
$a^{5/2}(f^7d^1s^1)_{5/2}$	$a^{10}\Delta(f^7d^1s^1)_{-5/2}$	5	5/2	2	6.98	0.80	1.13	0.07	0.464	0.245	(0.505)
$a^{5/2}(f^7d^1s^1)_{3/2}$	$a^{10}\Delta(f^7d^1s^1)_{-3/2}$	6	3/2	2	6.98	0.80	1.13	0.07	0.476	0.263	
$a^{5/2}(f^7d^1s^1)_{1/2}$	$a^{10}\Delta(f^7d^1s^1)_{-1/2}$	7	1/2	2	6.98	0.80	1.13	0.07	0.487	0.281	
$b^{13/2}(f^7d^1s^1)_{1/2}$	$a^{10}\Delta(f^7d^1s^1)_{1/2}$	8	1/2	3	6.98	0.80	1.13	0.07	0.500	0.301	
$b^{13/2}(f^7d^1s^1)_{3/2}$	$a^{10}\Delta(f^7d^1s^1)_{3/2}$	9	3/2	3	6.98	0.80	1.14	0.07	0.513	0.321	
$b^{13/2}(f^7d^1s^1)_{5/2}$	$a^{10}\Delta(f^7d^1s^1))_{5/2}$	10	5/2	3	6.98	0.80	1.14	0.07	0.526	0.343	(0.000)
$b^{13/2}(f^7d^1s^1)_{7/2}$	$a^{10}\Delta(f^7d^1s^1)_{7/2}$	11	7/2	2	6.98	0.80	1.14	0.07	0.540	0.366	(0.001)
$b^{13/2}(f^7d^1s^1)_{9/2}$	$a^{10}\Delta(f^7d^1s^1)_{9/2}$	12	9/2	1	6.98	0.80	1.14	0.06	0.554	0.391	(0.003)
$b^{13/2}(f^7d^1s^1)_{11/2}$	$a^{10}\Delta(f^7d^1s^1)_{11/2}$	13	11/2	1	6.98	0.80	1.14	0.06	0.570	0.417	
$b^{13/2}(f^7d^1s^1)_{13/2}$	$a^{10}\Delta(f^7d^1s^1)_{13/2}$	14	13/2	1	6.98	0.80	1.14	0.06	0.586	0.447	

a) LFT calculation (Ref. 8).
b) The serial number in the entire CI space.
c) The serial number in the respective OMEGA; please replace OMEGA with the capital Greek.
d) Calculated vertical excitation energy in eV from the ground state.
e) Transition dipole moment, with the ground state in Debye. Parentheses indicate that the transition is not strong enough to be observed.

Table 4.13 *GdF transitions of the [15.8] and [18.3] states to the ground states*

Designation		State							EE /eV			Tr-M^d) /D
Theoretical	Experimental	$\Delta\Omega$	SN′	Ω'	N′	SN″	Ω''	N″	$T_v'^{a)}$	$\Delta T_v^{b)}$	$T_0^{c)}$	
$q^{9/2}[f^7d^1s^1+..]_{\Omega'}$	[15.8]	+1	79	1/2	21	4	−1/2	1	2.049	2.048	1.956	4.226
$\rightarrow$	Designation	+1	80	3/2	20	4	1/2	1	2.054	2.053		4.641
$X^{7/2}[f^7s^2]_{\Omega''}$	is not given	+1	82	5/2	16	3	3/2	1	2.061	2.060		4.746
		+1	83	7/2	13	2	5/2	1	2.068	2.068		4.812
		+1	84	9/2	9	1	7/2	1	2.075	2.075		4.850
$p^{5/2}[f^7d^1s^1+..]_{\Omega'}$		−1	75	5/2	15	1	7/2	1	2.019	2.019		4.534
$\rightarrow$		−1	76	3/2	19	2	5/2	1	2.029	2.029		4.495
$X^{7/2}[f^7s^2]''_{\Omega}$		−1	78	1/2	20	3	3/2	1	2.036	2.035		4.148
		−1	79	−1/2	21	4	1/2	1	2.049	2.048		4.226
$E^{7/2}[f^7d^1s^1+..]_{\Omega'}$	[18.3]	0	108	7/2	17	1	7/2	1	2.499	2.499	2.264	4.537
$\rightarrow$	$^8\Sigma^-(?) \rightarrow$	0	109	5/2	22	2	5/2	1	2.502	2.501		4.503
$X^{7/2}[f^7s^2]_{\Omega''}$	$^8\Sigma^-(4f^76s^2)$	0	111	3/2	26	3	3/2	1	2.504	2.503		4.476
		0	112	1/2	27	4	1/2	1	2.505	2.504		3.849

a) Calculated vertical excitation energy in eV from the ground state.
b) Calculated vertical transition energy in eV.
c) Experimental excitation energy from the ground state (Ref. 8).
d) Transition dipole moment in Debye.

SN=79 state to the X family gives two spectra sequences ([15.8:$\Delta\Omega = +1$] and [15.8:$\Delta\Omega = -1$]). Table 4.13 shows that the [15.8:$\Delta\Omega = +1$] and [15.8:$\Delta\Omega = -1$] sequences lie in the narrow energy range 2.019 eV – 2.075 eV. This is consistent with the conclusion of the experimentalists [8] that "it is difficult to give a definitive configurational assignment for the [15.8] states as many Ω-components of the $^8\Pi(4f^7[^8S]5d6s)$ and $^{10}\Sigma^-(4f^7[^8S]5d^2)$ configurational states are predicted to fall in this energy region."

4.4.3.4 *GdF Transitions Between Excited States*

The selection rule gives three types of dipole transitions: $\Delta\Omega = \pm 1$ and 0. Kaledin *et al.* [8] observed $\Delta\Omega = +1$ transitions in the near-infrared region and $\Delta\Omega = 0$ transitions in the visible region, but they reported nothing about the $\Delta\Omega = -1$ case. Below we discuss only the $\Delta\Omega = +1$ case. The other cases are provided by Yamamoto and Tatewaki [26, 27].

Kaledin *et al.* [8] identified six bands in the near-infrared region (1.673 eV – 1.732 eV) as the transitions of $\Delta\Omega = +1$ listed in the 12th column of Table 4.14. In this energy region we searched for $\Delta\Omega = +1$ transitions having transition dipole moments larger than 2.5 D. We found 13 transitions, where the lower states are $a\ ^{5/2}[(f^7)(s^1d^1)]_\Omega$ or $b\ ^{13/2}[(f^7)(s^1d^1)]_\Omega$. We classified these 13 transitions into five sequences as shown in Table 4.14. Only the sequence from $(o \rightarrow b)$ is observed. This sequence begins from $\Delta T_v = 1.578$ eV, and the calculated transition energies are 0.2 eV smaller than experiment.

The transitions $(D \rightarrow a)$ have large transition moments and should be observed 0.3 eV above $(o \rightarrow b)$, but no spectra have been reported. It is possible that Kaledin *et al.* [8] overlooked these transitions in the visible region.

Table 4.14 *GdF transitions of* $\Delta\Omega = +1$ *to the* $b^{13/2}(4f^7 5d^1 6s^1)_\Omega$ *and* $a^{5/2}(4f^7 5d^1 6s^1)_\Omega$ *states*

Designation		State							Energetics /eV			Tr-M[d)]/D
Theoretical	Experimental	$\Delta\Omega$	SN′	Ω′	N′	SN″	Ω″	N″	$T_v{}'^{a)}$	$\Delta T_v^{b)}$	$\Delta T_0^{c)}$	
$o^{15/2}[f^7d^2+..]_{\Omega'} \rightarrow \underline{a}\,^{5/2}[f^7d^1s^1]_{\Omega''}$	Not observed	+1	72	1/2	19	$\underline{7}$	−1/2	2	1.955	1.467		4.773
$o^{15/2}[f^7d^2+..]_{\Omega'} \rightarrow b^{13/2}[f^7d^1s^1]_{\Omega''}$	$^{10}\Phi(f^7d^2)_{\Omega'}$	+1	73	3/2	18	8	1/2	3	1.978	1.478		4.808
	$\rightarrow$	+1	74	5/2	14	9	3/2	3	2.003	1.490	1.673	4.822
	$^{10}\Delta(f^7d^1s^1)_{\Omega''}$	+1	77	7/2	12	10	5/2	3	2.030	1.504	1.683	4.826
		+1	81	9/2	8	11	7/2	2	2.059	1.520	1.694	4.811
		+1	85	11/2	5	12	9/2	1	2.091	1.537	1.706	4.850
		+1	86	13/2	2	13	11/2	1	2.126	1.557	1.719	4.861
		+1	87	15/2	1	14	13/2	1	2.164	1.578	1.732	4.882
$D^{7/2}[f^7d^2+..]_{\Omega'} \rightarrow \underline{b}\,^{13/2}[f^7d^1s^1]_{\Omega''}$	Not observed	+1	98	1/2	24	$\underline{8}$	−1/2	3	2.407	1.907		3.373
$D^{7/2}[f^7d^2+..]_{\Omega'} \rightarrow a^{5/2}[f^7d^1s^1]_{\Omega''}$	Not observed	+1	97	3/2	23	7	1/2	2	2.407	1.919		3.754
		+1	96	5/2	19	6	3/2	2	2.406	1.931		4.096
		+1	99	7/2	15	5	5/2	2	2.407	1.943		4.708
$t^{11/2}[f^7d^2+..]_{\Omega'} \rightarrow \underline{b}\,^{13/2}[f^7d^1s^1]_{\Omega''}$	Not observed	+1	100	1/2	25	$\underline{8}$	−1/2	3	2.421	1.921		2.513

a), b), c), d) See a), b), c), d) in Table 4.13

4.4.3.5 Summary of GdF Excited States

We have performed relativistic GOSCI calculations for the excited states of GdF. To better compare the present computation of the electronic transitions with the experimental result of Kaledin *et al.*, [8] we have eliminated unimportant transitions with the aid of the calculated transition moments. We have successfully clarified the characters of the low-lying excited states of GdF. In particular, light has been shed on the characters of the [15.8] and [18.3] states. Transitions between the excited states have been classified according to the $\Delta\Omega$ values, however, only the results for $\Delta\Omega = 1$ are shown in this article. The observed spectra are ascribed to the transitions from higher families to the lower families of *a* and *b*, which lie ~0.5 eV above the ground state. Finally we add that the *f*-shell Omega decomposition method was also applied to EuF. [28] This clarifies the spectra that were not identified experimentally. [68]

4.5 Conclusion

The electronic structure of the CeO, CeF, and GdF molecules has been investigated using the four-component relativistic restricted active space configuration interaction (RASCI) and general open-shell configuration interaction (GOSCI) methods. CeO and LnF (Ln = Ce, Gd) are charge-transfer molecules that have in common a 64-electron closed-shell core bonded firmly by the covalent and ionic bonds composed of the Ln 5*s*, 5*p*, 5*d** and 4*f** and ligand 2*s*, and 2*p* spinors (here, 5*d** and 4*f**are polarization functions). The outer 4*f*-, 5*d*-, and 6*s*-like electrons moving around this core have weak-bonding or non-bonding characteristics. The excited states of CeO below 0.56 eV have been investigated by RASCI calculations. The GOSCI calculations were performed for GdF, for which transition dipole moments were also calculated. With the help of GAOPs and the *f*-shell Omega decomposition method, the GdF spectrum below 3.0 eV, determined experimentally by Kaledin *et al.* [8], has been almost completely explained.

References

[1] N. Kaltsoyannis and P. Scot, *The f elements* (Oxford Science Publications, 1999).
[2] S. Cotton, *Lanthanide and Actinide Chemistry* (John Wiley & Sons, Chichester, 2006).
[3] J.-C. G. Bünzli, Lanthanide luminescence for biomedical analyses and imaging, Chem. Rev. **110**, 2729–2755 (2010).
[4] K. Binnemans, Lanthanide-based luminescent hybrid materials, Chem. Rev. **109**, 4283–4374 (2009).
[5] S. Viswanathan, Z. Kovacs, K. N. Green, S. J. Ratnakar, and A.D. Sherry, Alternatives to gadolinium-based metal chelates for magnetic resonance imaging, Chem. Rev. **110**, 2960–3018 (2010).
[6] R. W. Field, Diatomic molecule electronic structure beyond simple molecular constants, Ber. Bunsenges. Phys. **86**, 771–779 (1982).
[7] L. A. Kaledin, J. E. McCord, and M. C. Heaven, Rotation-electronic deperturbation analysis of the 4*f* 6*s* configurational states of CeO, J. Mol. Spectrosc. **170**, 166–171 (1992).

[8] A. Kaledin, J. C. Bloch, M. C. McCarthy, E. A. Shenyavskaya, and R. W. Field, Laser spectroscopy of gadolinium monofluoride: ligand field assignments of states in the 0–3 eV range, J. Mol. Spectrosc. **176**, 148–161 (1996).

[9] M. Dolg and H. Stoll, Pseudopotential study of the rare earth monohydrides, monoxides and monofluorides, Theor. Chim. Acta **75**, 369–387 (1989).

[10] M. Dolg, H. Stoll, and H. Preuss, The low-lying electronic states of cerium monoxide CeO: ab initio calculations of using energy-adjusted psuedopotentials and spin-orbit Operators, J. Mol. Struct. **231**, 243–255 (1991).

[11] X. Cao, W. Liu, and M. Dolg, Molecular structure of diatomic lanthanide compounds, Sci. China Ser. B**45**, 91–96 (2002).

[12] S. G. Wang and W. H. E. Schwarz, Lanthanide diatomics and lanthanide contractions, J. Phys. Chem. **99**, 11687–11695 (1995).

[13] H. Heiberg, H. O. Gropen, J. K. Laerdahl, O. Swang, and U. Wahlgren, The performance of density functional theory for LnF (Ln=Nd, Eu, Gd, Yb) and YbH, Theor. Chem. Acc. **110**, 118–125 (2003).

[14] H. M. Quiney, H. Skaane, and I. P. Grant, Hyperfine and pt-odd effects in YbF doublet sigma, *J*.Phys. B **31**, L85–95 (1998).

[15] J. K. Laerdahl, K. Fægri, Jr., L. Visscher, and T. Saue, A fully relativistic Dirac–Hartree Fock and second-order Møller–Plesset study of the lanthanide and actinide contraction, J. Chem. Phys. **109**, 10806 (10 pages) (1998).

[16] Y. Wasada-Tsutsui, Y. Watanabe, and H. Tatewaki, Electronic structures of lanthanide monofluorides in the ground state: Frozen-core Dirac–Fock–Roothaan calculations, Int. J. Quantum Chem. **109**, 1874 -1885 (2009).

[17] M. Dolg, W. Liu, and S. Kalvoda, Performance of relativistic density functional and ab initio pseudopotential approaches for systems with high-spin multiplicities: Gadolinium diatomics GdX (X=H, N, O, F, P, S, Cl, Gd), Int. J. Quantum. Chem. **76**, 359–370 (2000).

[18] H. Moriyama, Y. Watanabe, H. Nakano, and H. Tatewaki, Electronic structure of LaF^+ and LaF from frozen-core four-component relativistic multiconfigurational quasidegenerate perturbation Theory, J. Phys. Chem. A**112**, 2683–2692 (2008).

[19] H. Moriyama, H. Tatewaki, Y. Watanabe, and H. Nakano, Molecular spinors suitable for four-component relativistic correlation calculations: studies of LaF^+ and LaF using multiconfigurational quasi-degenerate perturbation theory, Int. J. Quantum Chem. **109**, 1898 – 1904 (2009).

[20] H. Moriyama, Y. Watanabe, Y. H. Nakano, S. Yamamoto, and H. Tatewaki, Electronic structure of LaO based on frozen-core four-component relativistic multiconfigurational quasidegenerate perturbation theory, J. Chem. Phys. 132, 124310 (9 pages) (2010).

[21] Y. Wasada-Tsutsui, Y. Watanabe, and H. Tatewaki, Electronic structures and bonding of CeF: a frozen-core four-component relativistic configuration interaction study, J. Phys. Chem. A**111**, 8877–8883 (2007).

[22] H. Tatewaki, S. Yamamoto, Y. Watanabe, and H. Nakano, Electronic structure of CeF from frozen-core four-component relativistic multiconfigurational quasidegenerate perturbation theory, J. Chem. Phys. **128**, 214901 (8 pages) (2008).

[23] H. Tatewaki and O. Matsuoka, All-electron Dirac–Fock–Roothaan calculations on the electronic structure of the GdF molecule, J. Chem. Phys. **106**, 4558–4565 (1997).

[24] H. Tatewaki, Y. Watanabe, S. Yamamoto, and E. Miyoshi, Electronic structure of the GdF molecule by frozen-core four-component relativistic configuration interaction calculations, J. Chem. Phys. **125**, 044309 (9 pages) (2006).

[25] S. Yamamoto, H. Tatewaki, and T. Saue, Dipole allowed transitions in GdF: A four-component relativistic general open-shell configuration interaction study, J. Chem. Phys. **129**, 44505 (8 pages) (2008).

[26] S. Yamamoto and H. Tatewaki, Electronic spectra of GdF reanalyzed by decomposing state functions according to f-shell angular momentum, J. Chem. Phys. **134**, 164310 (11 pages) (2011).

[27] S. Yamamoto and H. Tatewaki, Assignment of electronic spectra of GdF by identifying families using the f-shell Omega decomposition method, Comput. Theoret. Chem. **980**, 37–43 (2012).

[28] S. Yamamoto and H. Tatewaki, and H. Moriyama, Electronic spectra of EuF studied by a four-component relativistic configuration interaction method, Theor. Chem. Acc. **131**, 1230 (9 pages) (2012).
[29] H. Moriyama, H. Tatewaki, and S. Yamamoto, Electronic structure of CeO studied by a four-component relativistic configuration interaction method, J. Chem. Phys. **138**, 224310 (8 pages) (2013).
[30] N. Rösch, Time-reversal symmetry, Kramers' degeneracy and the algebraic eigenvalue problem, Chem. Phys. **80**, 1–5 (1983).
[31] T. Saue and H. J. Aa Jensen, Quaternion symmetry in relativistic molecular calculations: The Dirac–Hartree–Fock method, J. Chem. Phys. **111**, 6211–6222 (1999).
[32] K. G. Dyall and K. Faegri, Jr., *Introduction to Relativistic Quantum Chemistry* (Oxford University Press, New York, 2007).
[33] T. Helgaker, P. Jørgensen, and J. Olsen, *Molecular Electronic-Structure Theory* (Wiley, Chichester, 2000).
[34] R. S. Mulliken, Electronic population analysis on LCAO–MO molecular wave functions. I, J. Chem. Phys. **23**, 1833–1840 (1955).
[35] C. C. J. Roothaan, New developments in molecular orbital theory, Rev. Mod. Phys. 23, 69–89 (1951).
[36] C. C. J. Roothaan and P. S. Bagus, in *Methods in Computational Physics* (Academic, New York, 1963), vol. 2, p. 215.
[37] T. Koga, H. Tatewaki, and O. Matsuoka, Relativistic Gaussian basis sets for molecular calculations: Cs–Hg, J. Chem. Phys. **117**, 7813–7814 (2002).
[38] T. Koga, H. Tatewaki, and O. Matsuoka, Relativistic Gaussian basis sets for molecular calculations: H–Xe, J. Chem. Phys. **115**, 3561–3565 (2001).
[39] Y. S. Lee and A. D. McLean, Relativistic effects on Re and De in AgH and AuH from all electron Dirac–Hartree–Fock calculations, J. Chem. Phys. **76**, 735–736 (1982).
[40] R. E. Stanton and S. Havriliak, Kinetic balance: A partial solution to the problem of variational safety in Dirac calculations, J. Chem. Phys. **81**, 1910–1918 (1984).
[41] L. Visscher and K. G. Dyall, Dirac–Fock atomic electronic structure calculations using different nuclear charge distributions, At. Data Nucl. Data Tables **67**, 207–224 (1997).
[42] G. E. Brown and D. G. Ravenhall, On the interaction of two electrons, Proc. Roy. Soc. London, A**208**, 552–559 (1951).
[43] J. Sucher, Foundation of the relativistic theory of many-electron atoms, Phys. Rev. A **22**, 348–362 (1980).
[44] B. A. Hess, Relativistic electronic-structure calculations employing a two-component no-pair formalism with external-field projection operators, Phys. Rev. A **33**, 3742–3748 (1986).
[45] Y. Watanabe, H. Nakano, and H. Tatewaki, Effect of removing the no-virtual pair approximation on the correlation energy of the He isoelectronic sequence. II. Point nuclear charge model, J. Chem. Phys. **132**, 124105 (7 pages) (2010).
[46] M. Bylicki, G. Pestka, and J. Karwowski, Relativistic Hylleraas configuration interaction method projected into positive-energy space, Phys. Rev. A **77**, 044501 (4 pages) (2008).
[47] H. Tatewaki and Y. Watanabe, Necessity of including the negative energy space in four-component relativistic calculations for accurate solutions, Chem. Phys. **389**, 58–63 (2011).
[48] O. Visser, L. Visscher, P. J. C. Aerts, and W. C. Nieuwpoort, Molecular open shell configuration interaction calculations using the Dirac–Coulomb Hamiltonian: The f^6-manifold of an embedded EuO_6^{9-} cluster, J. Chem. Phys. **96**, 2910–2919 (1992).
[49] J. Olsen, B. O. Roos, P. Jørgensen, and H. J. Aa. Jensen, Determinant based configuration interaction algorithms for complete and restricted configuration interaction spaces, J. Chem. Phys. **89**, 2185–2192 (1988).
[50] L. Visscher and W. C. Nieuwpoort, Relativistic and electron correlation effects on the $d - d$ spectrum of transition metal fluorides, Theor. Chim. Acta **88**, 447–472 (1994).
[51] DIRAC, A relativistic ab initio electronic structure program, release DIRAC08, written by L. Visscher, H. J. Aa. Jensen, and T. Saue, with new contributions from R. Bast, S. Dubillard, K. G. Dyall, U. Ekström, E. Eliav, T. Fleig, A. S. P. Gomes, T. U. Helgaker, J. Henriksson, M. Iliaš, Ch.

R. Jacob, S. Knecht, P. Norman, J. Olsen, M. Pernpointner, K. Ruud, P. Sałek, and J. Sikkema. (2008). (See http://dirac.chem.vu.nl).

[52] S. Yamamoto and H. Tatewaki, Excited states of PbF: A four-component relativistic study, J. Chem. Phys. **132**, 054303 (10 pages) (2010).

[53] I. Gotkis, Field-stimulated electron promotion from core 4f orbitals to out-of-core σ_{6s} Orbital Phenomenon in Simple Lanthanide Compounds, J. Phys. Chem. **95**, 6086–6095 (1991).

[54] K. Schofield, An Overlooked series of gas phase diatomic metal oxide ions that are long-lived, J. Phys. Chem. A **110**, 6938–6947 (2006).

[55] J. Andzelm, M. Kobukowski, E. Radzio-Andzelm, Y. Sakai, and H. Tatewaki, in *Gaussian Basis Sets for Molecular Calculations*, edited by S. Huzinaga (Elsevier, Amsterdam, 1984).

[56] K. P. Huber and G. Herzberg, *Molecular Spectra and Molecular Structure IV: Constants of Diatomic Molecules* (Van Nostrand Reinhold Company, 1979).

[57] W. C. Martin, R. Zalubas, and L. Hagan, *Atomic Energy Lev*els – *The Rare Earth Atoms*, Natl. Stand. Ref. Data Ser. Vol. 60 (US Natl. Bur. Stand., 1978), p. 47.

[58] C. E. Moore, *Atomic Energy Levels as Derived from the Analysis of Optical spectra*, Natl. Bur. Stand. (U. S.) Cir. 467 (U. S. Government Printing Office, 1949), p. 45.

[59] J. C. Bloch, M. C. McCarthy, R. W. Field, and L. A. Kaledin, Laser spectroscopy of cerium monofluoride: ligand field assignments of some 4f5d6p $\leftarrow$ 4f5d6s transitions, J. Mol. Spectrosc. **177**, 251–262 (1996).

[60] As quoted in *CRC Handbook of Chemistry and Physics, 70th* ed.; CRC Press, Inc.: Boca Raton, 1989; F-198 edited by R. C. Weast, D. R. Lide, M. J. Astle, and W. H. Beyer.

[61] S. Wyckoff and P. A. Wehinger, CeO: A new s-process molecules in S stars?, Astrophys. J. **212**, L139–141 (1977).

[62] R. F. Barrow, R. M. Clements, S. M. Harris, and P. P. Jensen, The Electronic spectra of Gaseous CeO, Astrophys. J. **229**, 439–447 (1979).

[63] C. Linton, M. Dulick, R. W. Field, P. Carette, and R. F. Barrow, Low lying electronic states of CeO, J. Chem. Phys. **74**, 189–191 (1981).

[64] C. Linton, M. Dulick, R. W. Field, P. Carette, P. C. Leyland, and R. F. Barrow, Electronic states of the CeO molecule: absorption, emission, and laser spectroscopy, J. Mol. Spectrosc. **102**, 441–497 (1983).

[65] L. A. Kaledin, J. E. McCord, and M. C. Heaven, Laser spectroscopy of CeO: characterization and assignment of states in the 0-3 eV range, J. Mol. Spectrosc. **158**, 40–61 (1993).

[66] S. R. Langhoff and E. R. Davidson, Configuration interaction calculations on the nitrogen molecule, Int. J. Quantum Chem. **8**, 61–72 (1974).

[67] W. T. Hill and C. H. Lee, *Light-Matter Interaction*, Wiley-VCH (2007), p. 128.

[68] L. V. Gurvich, Yu N. Dmitriev, L. A. Kaledin, A. I. Kobylyanskii, A. N. Kulikov, and E. A. Shenyavskaya, High-resolution electron spectra of diatomic molecules containing f elements: UO, GdO, GdF, EuF, Bull. Acad. Sci. USSR (Phys. Ser.) **53**, 75–79 (1989).

5

The Complete-Active-Space Self-Consistent-Field Approach and Its Application to Molecular Complexes of the f-Elements

Andrew Kerridge

Department of Chemistry, Lancaster University

5.1 Introduction

The electronic structure associated with elements present in the f-block of the periodic table, namely the lanthanides (Ln) and actinides (An), is both highly interesting and highly complex. This complexity, however, provides a significant challenge to quantum chemical methodologies and is, in essence, due to three phenomena:

- Strong electron correlation within the An/Ln valence orbital manifold
- Weak ligands fields experienced by the An/Ln ion
- The effects of relativity

The combined effects of these phenomena result in rich chemistry, with f-element compounds exhibiting novel bonding motifs absent from all other areas of the periodic table. Both the lanthanides and actinides behave as hard Lewis acids [1], resulting in complexes where the bonding is largely ionic in character. This ionic character is, however, more pronounced amongst the lanthanides. The 4f orbitals, with no radial node, experience strong nuclear attraction, resulting in a highly contracted, core-like character. The energetic stability associated with this has a profound effect on Ln chemistry: typically, the fourth ionization potential of a given lanthanide is greater than the sum of the first three [1], resulting in

Computational Methods in Lanthanide and Actinide Chemistry, First Edition. Edited by Michael Dolg.

chemistry dominated by the trivalent oxidation state. Nevertheless, electrons in the 4f manifold are strongly correlated and this, coupled with the fact that crystal field effects are weak, means that appropriate quantum chemical methodologies must be used in order to accurately model Ln complexes.

The situation is somewhat different in the actinides. The 5f shell has a single radial node and experiences more shielding from core electrons, partly due to the relativistic contraction of core electron orbitals in these very heavy elements. The 5f shell therefore tends to be less energetically stable and more spatially extended, allowing for a degree of covalent character in An-ligand bonding. The relative accessibility of the 5f electrons also results in a much greater range of oxidation states (particularly for the early actinides); the heptavalent oxidation state has been reported for Np and Pu [2, 3], while the hexavalent oxidation state is predominant in the aqueous chemistry of U [1]. The 5f shell is also similar in energy to the 7s and 6d orbitals, allowing for a large variation in bonding character. Finally the effects of relativity are more pronounced; in addition to the relativistic contraction of core orbitals, the effects of spin-orbit coupling are strong and must be considered in any quantitative description of An complexes.

In this chapter, the quantum chemical simulation of actinide and lanthanide complexes will be considered. The need for a multiconfigurational description of the wavefunction is discussed, and the complete-active-space self-consistent-field (CASSCF) approach, along with some related methods, is introduced and discussed. This approach, originally developed by Björn Roos, allows for the strong static correlation present in these complexes due to a combination of electron-electron interactions and weak crystal field splittings to be taken into consideration in a systematic manner. Extensions to this approach, which also account for dynamical correlation will also be considered. In the finally section, the application of the CASSCF approach will be illustrated with examples from the literature.

A useful tool for considering the quality of a quantum chemical calculation is the Pople diagram. Figure 5.1 shows a 3-dimensional version, suitable for use in the simulation of f-element complexes. The three axes of the Pople diagram represent the quantum chemical methodology, the chosen basis set, and the treatment of relativity, respectively. The main aim of this diagram is to ensure that a 'balanced' calculation is performed, i.e., relevant

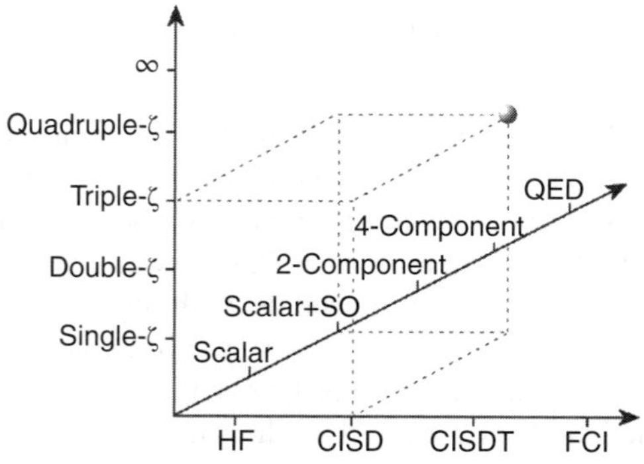

Figure 5.1 *Three dimensional Pople diagram. The three axes correspond to quantum chemical methodology, basis set, and treatment of relativity*

aspects of a calculation are given approximately equal consideration. Before we embark on the main aim of this chapter, we shall briefly consider appropriate choices of basis set and relativistic Hamiltonian.

5.1.1 Treatment of Relativistic Effects

Whilst relativistic effects are extremely important in the accurate modelling of f-element complexes [4], they are not the focus of this chapter. Here, we briefly consider some approximate approaches to incorporating relativistic effects. Bearing in mind that such effects must be implemented in the context of quantum chemical simulations, rendering approaches based on quantum electrodynamics (QED) effectively intractable, the most sophisticated approaches available are based on the replacement of the Schrödinger equation with the Dirac equation. Such approaches require a 4-component formalism in order to describe the particles under investigation, this being required since the formalism describes coupled particle-antiparticle pairs, with the effects of spin-orbit coupling treated explicitly. Four-component approaches are computationally intensive, and methods have been developed that effectively decouple the particle anti-particle pairs, retaining only the electronic components. Typically, such 2-component approaches still explicitly include the effects of spin-obit coupling and provide popular approaches to including relativistic effects. The most common implementations of these two-component methods employ either the Douglas-Kroll-Hess (DKH) Hamiltonian [5, 6] or the zeroth order regular approximation [7–9] (ZORA). In both of these approaches, further approximations can be made in order to eliminate spin-orbit coupling from the Hamiltonian. Such single-component approaches are said to treat scalar relativistic effects only, with the effects due to spin-orbit coupling sometimes incorporated in a *post-hoc* manner. Scalar relativistic effects can also be treated by the use of relativistic effective core potentials (RECPs), which replace core electrons with an appropriate potential experienced by the valence electrons. Such RECPs are designed so that, for example, the core contraction due to the relativistic mass increase is accurately incorporated.

5.1.2 Basis Sets

There are many different types of basis sets, of which the most popular are those designed by the Pople, Dunning, Ahlrichs, and Roos research groups. Assuming a well-defined basis set, multiconfigurational methods such as CASSCF require a high degree of flexibility since electron correlation is incorporated explicitly through the electronic wavefunction. As we will see, CASSCF-style approaches depend on linear combinations of different electronic configurations and so the chosen basis set must allow for a large number of configurations to be constructed. For this reason, quantitatively correct simulations typically require basis sets of at least 'triple-ζ' quality, meaning that each atomic valence orbital is constructed as a linear combination of three contracted Gaussian basis functions. Using such a basis set ensures that the configurational space is sufficiently large and flexible for correlation effects to be accurately modelled. When considering anionic systems, where valence electrons may be only loosely bound, the incorporation of diffuse functions is also recommended.

The use of basis sets as recommended above can result in very large number of basis functions, and it is common practice (depending on the properties of interest) to replace core electrons with an effective potential, a popular choice being the family of Stuttgart-Cologne energy-consistent pseudopotentials [10]. This approach significantly reduces the size of the basis set whilst having little effect on the flexibility of the valence space; as discussed in the previous section, for the case of the f-elements (as well as the heavier transition metal and main block elements), RECPs are designed so that the relativistic contraction of the core shells is well represented.

5.2 Identifying and Incorporating Electron Correlation

5.2.1 The Hartree Product Wavefunction

The simplest description of a many electron wavefunction is known as the Hartree product [11]. This wavefunction is constructed as a simple product of one electron wavefunction, or orbital. In the case of a two-electron system:

$$\Psi^{\mathrm{HP}}(\mathbf{x}_1, \mathbf{x}_2) = \chi_i(\mathbf{x}_1)\,\chi_j(\mathbf{x}_2) \tag{5.1}$$

Where $\chi_i(\mathbf{x}_n)$ is a spin-orbital occupied by electron n. The Hartree product corresponds to an independent particle approximation, which can be seen more clearly by considering the probability of simultaneously finding electron 1 at the point $\mathbf{r}_1$ and electron 2 at $\mathbf{r}_2$.

$$\begin{aligned} \mathrm{P}(\mathbf{r}_1, \mathbf{r}_2) &= \int \left|\Psi^{\mathrm{HP}}(\mathbf{x}_1, \mathbf{x}_2)\right|^2 \mathrm{d}s_1 \mathrm{d}s_2 \\ &= |\phi_i(\mathbf{r}_1)|^2\, |\phi_j(\mathbf{r}_2)|^2 \end{aligned} \tag{5.2}$$

Where we have integrated over the spin-degrees of freedom and $\phi_i(\mathbf{r}_n)$is a spatial orbital. Here, the probability of finding electron 1 at point $\mathbf{r}_1$ is independent of the position of electron 2, and the electronic motion is uncorrelated. Such an approach to building a many-body wavefunction is clearly only applicable in the case of non-interacting particles. Furthermore, the Hartree product is not antisymmetric under the exchange of space and spin coordinates of the two particles, and is therefore inappropriate for describing fermionic wavefunctions. This latter deficiency can be remedied by using a determinantal wavefunction.

5.2.2 Slater Determinants and Fermi Correlation

The Hartree product wavefunction described in equation 5.1 can easily be antisymmetrised:

$$\begin{aligned} \Psi^{\mathrm{SD}}(\mathbf{x}_1, \mathbf{x}_2) &= \frac{1}{\sqrt{2}}\left(\chi_i(\mathbf{x}_1)\,\chi_j(\mathbf{x}_2) - \chi_j(\mathbf{x}_1)\,\chi_i(\mathbf{x}_2)\right) \\ &= \frac{1}{\sqrt{2}}\begin{vmatrix} \chi_i(\mathbf{x}_1) & \chi_j(\mathbf{x}_1) \\ \chi_i(\mathbf{x}_2) & \chi_j(\mathbf{x}_2) \end{vmatrix} \\ &= |\chi_i \chi_j| \end{aligned} \tag{5.3}$$

The form of the wavefunction given in equation 5.3 is known as a Slater determinant [12] and can be generalised for an N-electron wavefunction, giving

$$\Psi^{\mathrm{SD}}(\mathbf{x}_1, \mathbf{x}_2, \ldots, \mathbf{x}_n) = \frac{1}{\sqrt{n!}} \begin{vmatrix} \chi_i(\mathbf{x}_1) & \chi_j(\mathbf{x}_1) & \cdots & \chi_k(\mathbf{x}_1) \\ \chi_i(\mathbf{x}_2) & \chi_j(\mathbf{x}_2) & \cdots & \chi_k(\mathbf{x}_2) \\ \vdots & \vdots & \ddots & \vdots \\ \chi_i(\mathbf{x}_n) & \chi_j(\mathbf{x}_n) & \cdots & \chi_k(\mathbf{x}_n) \end{vmatrix} = |\chi_i \chi_j \cdots \chi_k| \tag{5.4}$$

This form of the many-electron wavefunction, satisfying the antisymmetry principle, exhibits correlation between electrons of like spin. To see this, we take the two-electron wavefunction given in equation 5.3. For two electrons of unlike spin, i.e.,

$$\chi_i(\mathbf{x}_1) = \phi_i(\mathbf{r}_1)\,\alpha(s_1) \tag{5.5}$$

$$\chi_j(\mathbf{x}_2) = \phi_j(\mathbf{r}_2)\,\beta(s_2) \tag{5.6}$$

the probability of simultaneously finding electron 1 at the point $\mathbf{r}_1$ and electron 2 at $\mathbf{r}_2$ is given by

$$\begin{aligned} \mathrm{P}(\mathbf{r}_1, \mathbf{r}_2) &= \int \left|\Psi^{\mathrm{SD}}(\mathbf{x}_1, \mathbf{x}_2)\right|^2 \mathrm{d}s_1 \mathrm{d}s_2 \\ &= \frac{1}{2}\left(|\phi_i(\mathbf{r}_1)|^2 |\phi_j(\mathbf{r}_2)|^2 + |\phi_j(\mathbf{r}_1)|^2 |\phi_i(\mathbf{r}_2)|^2\right) \end{aligned} \tag{5.7}$$

and, since electrons are indistinguishable, we again see that the electronic motion is uncorrelated (this is made more obvious by setting $\varphi_i = \varphi_j$). In particular, it should be noted that $\mathrm{P}(\mathbf{r}_1, \mathbf{r}_1) \neq 0$, i.e., there is a non-zero probability of both electrons occupying the same point in space.

For two electrons of like spin, the integration over spin degrees of freedom produces a different probability distribution. Setting

$$\chi_i(\mathbf{x}_1) = \phi_i(\mathbf{r}_1)\,\alpha(s_1) \tag{5.8}$$

$$\chi_j(\mathbf{x}_2) = \phi_j(\mathbf{r}_2)\,\alpha(s_2) \tag{5.9}$$

gives

$$\begin{aligned} \mathrm{P}(\mathbf{r}_1, \mathbf{r}_2) &= \int \left|\Psi^{\mathrm{SD}}(\mathbf{x}_1, \mathbf{x}_2)\right|^2 \mathrm{d}s_1 \mathrm{d}s_2 \\ &= \frac{1}{2}(|\phi_i(\mathbf{r}_1)|^2 |\phi_j(\mathbf{r}_2)|^2 + |\phi_j(\mathbf{r}_1)|^2 |\phi_i(\mathbf{r}_2)|^2 \\ &\quad - \phi_i(\mathbf{r}_1)\,\phi_j^*(\mathbf{r}_1)\,\phi_j(\mathbf{r}_2)\,\phi_i^*(\mathbf{r}_2) - \phi_j(\mathbf{r}_1)\,\phi_i^*(\mathbf{r}_1)\,\phi_i(\mathbf{r}_2)\,\phi_j^*(\mathbf{r}_2)) \end{aligned} \tag{5.10}$$

Using equation 5.10, $\mathrm{P}(\mathbf{r}_1, \mathbf{r}_1) = 0$, demonstrating that the motion of the like-spin electrons *is* correlated and, for a given electron, the probability of finding another like-spin electron in its immediate vicinity is lower the that of an unlike-spin electron. This reduction in probability is known as a Fermi hole, and the correlation is known as Fermi or exchange correlation.

5.2.3 Coulomb Correlation

The determinantal form of the wavefunction discussed in Section 5.2.2 is used in the Hartree-Fock (HF) approach to solving the many-electron Schrödinger equation. The HF approach is considered to be an *uncorrelated* method, since it doesn't include Coulomb correlation, i.e., the correlation in electronic motion arising from the repulsive electrostatic electron-electron interactions. The correlation energy is therefore defined as the difference between the exact energy and the energy obtained by employing the Hartree-Fock approximation

$$E_C = E_{Exact} - E_{HF} \tag{5.11}$$

Coulomb correlation can be approximately partitioned into two components: the first, which has short range and is dependent on r_{ij}^{-1}, is known as dynamical correlation, whilst the second, which has long range and manifests itself due to near-degeneracy in orbital energies, is known as static correlation. Figure 5.2(a) illustrates angular correlation, where there is an energetic preference for electrons to be located on opposite sides of a nucleus. Figure 5.2(b) illustrates radial or *in-out* correlation, in which there is an energetic preference for electrons to be located at different distances from the nucleus. These are both forms of dynamical correlation, whereas Figure 5.2(c) illustrates an example of *left-right* correlation, whereby at large internuclear separation r_{AB}, there is an energetic preference for different electrons to be located near different nuclei. This is an example of static correlation, and it is extremely important in the simulation of f-element complexes to describe this form of correlation accurately.

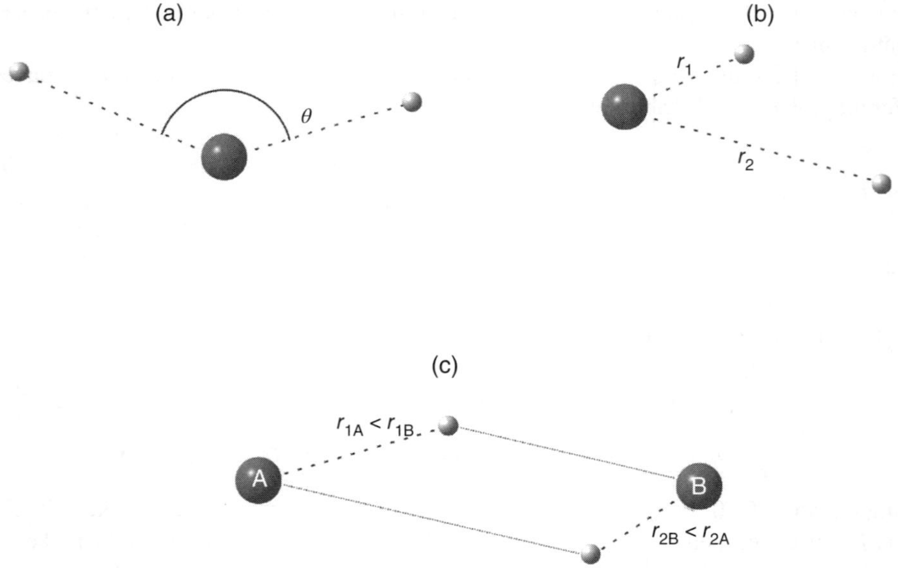

Figure 5.2 Examples of (a) angular correlation, (b) radial (in-out) correlation, and (c) left-right correlation. Nuclei and electrons are represented by dark and light spheres, respectively

5.3 Configuration Interaction and the Multiconfigurational Wavefunction

The weak crystal field experienced by f-block elements results in energetic near-degeneracy with respect to different 4f/5f configurations. This is a manifestation of static correlation and can be successfully modelled by replacing the monodeterminantal form of the wavefunction given by equation 5.4 with a multiconfigurational wavefunction, i.e., one in which several different electronic configurations can be incorporated using a linear combination of Slater determinants [12, 13]. In this section we will consider this 'configuration interaction' (CI) approach and its application to f-element complexes.

5.3.1 The Configuration Interaction Approach

Within the HF approximation, the many-electron wavefunction is described in terms of a single Slater determinant, but doesn't incorporate the effects of Coulomb correlation. However, during the optimisation of the HF wavefunction, a number of molecular spin-orbitals are produced. For a closed-shell, N-electron system, the $N/2$ lowest energy orbitals are occupied, whilst the remainder are unoccupied, or virtual, orbitals. These virtual orbitals can be used to construct excited configurations (which should not be confused with excited states of the system). In the complete basis set limit, we have a complete set of functions of $4N$ variables (the ground and excited configurations, each described by a different Slater determinant), and it can be shown [14] that *any* function of $4N$ variables can be expanded in terms of this complete set. We can therefore, in principle, construct the exact correlated N-electron wavefunction in terms of our HF reference configuration and a set of suitably weighted excited configurations. The CI approach gives us a mechanism for evaluating these weights, equal to the square of the corresponding CI coefficients. Using Dirac notation, we can write the CI wavefunction as

$$\left|\Psi^{\text{CI}}\right\rangle = c_0\left|\Psi_0\right\rangle + \sum c_i^a\left|\Psi_i^a\right\rangle + \sum c_{ij}^{ab}\left|\Psi_{ib}^{ab}\right\rangle + \sum c_{ijk}^{abc}\left|\Psi_{ijk}^{abc}\right\rangle + \cdots \tag{5.12}$$

Where $\left|\Psi_0\right\rangle = \left|\Psi^{\text{HF}}\right\rangle$, $\left|\Psi_i^a\right\rangle$ is a singly excited determinant formed by deoccupying orbital φ_i and occupying φ_a, and the summations run over occupied $(i, j, k, \ldots)$ or virtual $(a, b, c, \ldots)$ orbitals. Equation 5.12 makes explicit the fact that the CI wavefunction can be considered a series of corrections to the HF, or reference, wavefunction. The second term on the RHS of equation 5.12 corresponds to the correction made by singly excited determinants, the third term to that made by doubly excited determinants, and so on. Figure 5.3 gives examples of excited configurations that contribute to the CI wavefunction.

The CI coefficients, along with the total CI energy, are obtained by constructing and diagonalising the CI Hamiltonian matrix, $\mathbf{H}^{CI}$

$$\mathbf{H}^{\text{CI}} = \begin{bmatrix} \left\langle\Psi_0\middle|\hat{H}\middle|\Psi_0\right\rangle & \left\langle\Psi_0\middle|\hat{H}\middle|\Psi_{\text{S}}\right\rangle & \left\langle\Psi_0\middle|\hat{H}\middle|\Psi_{\text{D}}\right\rangle & \left\langle\Psi_0\middle|\hat{H}\middle|\Psi_{\text{T}}\right\rangle & \cdots \\ \left\langle\Psi_{\text{S}}\middle|\hat{H}\middle|\Psi_0\right\rangle & \left\langle\Psi_{\text{S}}\middle|\hat{H}\middle|\Psi_{\text{S}}\right\rangle & \left\langle\Psi_{\text{S}}\middle|\hat{H}\middle|\Psi_{\text{D}}\right\rangle & \left\langle\Psi_{\text{S}}\middle|\hat{H}\middle|\Psi_{\text{T}}\right\rangle & \cdots \\ \left\langle\Psi_{\text{D}}\middle|\hat{H}\middle|\Psi_0\right\rangle & \left\langle\Psi_{\text{D}}\middle|\hat{H}\middle|\Psi_{\text{S}}\right\rangle & \left\langle\Psi_{\text{D}}\middle|\hat{H}\middle|\Psi_{\text{D}}\right\rangle & \left\langle\Psi_{D}\middle|\hat{H}\middle|\Psi_{\text{T}}\right\rangle & \cdots \\ \left\langle\Psi_{\text{T}}\middle|\hat{H}\middle|\Psi_0\right\rangle & \left\langle\Psi_{\text{T}}\middle|\hat{H}\middle|\Psi_{\text{S}}\right\rangle & \left\langle\Psi_{\text{T}}\middle|\hat{H}\middle|\Psi_{\text{D}}\right\rangle & \left\langle\Psi_{\text{T}}\middle|\hat{H}\middle|\Psi_{\text{T}}\right\rangle & \cdots \\ \vdots & \vdots & \vdots & \vdots & \ddots \end{bmatrix} \tag{5.13}$$

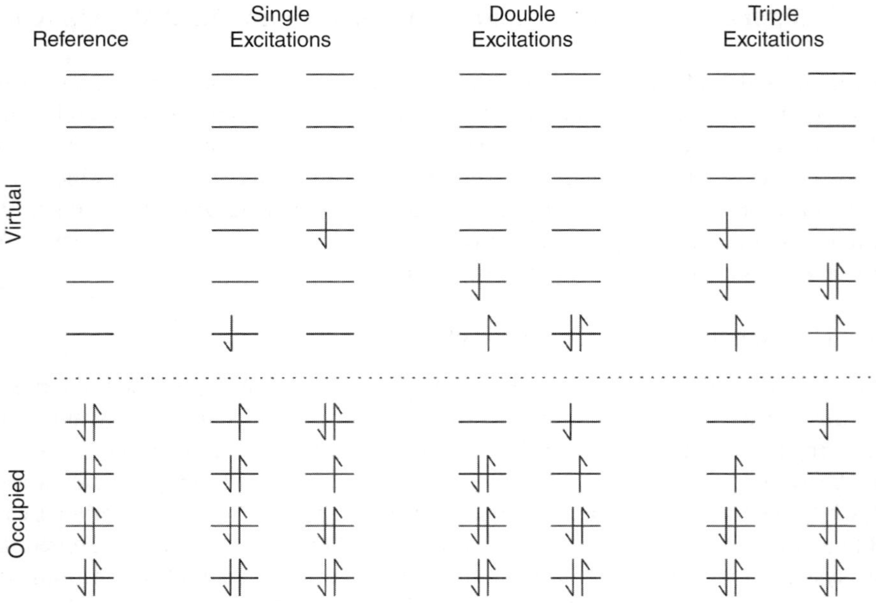

Figure 5.3 *Examples of singly, doubly, and triply excited configurations, relative to the HF reference for a singlet system*

where, for example, $\left\langle \Psi_S \middle| \hat{H} \middle| \Psi_D \right\rangle$ is the set of matrix elements between all singly and doubly excited determinants. These matrix elements can be evaluated in a relatively straightforward manner, taking advantage of Brillouin's theorem [15], $\left\langle \Psi_i^a \middle| \hat{H} \middle| \Psi_0 \right\rangle = 0$, and the Slater-Condon rules [12, 13].

In the limit of a complete basis set, diagonalisation of $\mathbf{H}^{CI}$ would give the exact energy of both the ground and excited states of an N-electron system. However, in reality, a truncated basis set is used. Expansion of the exact wavefunction in this truncated basis set is known as full CI (FCI), and whilst full CI gives an excellent description of both dynamical and static correlation, it is extremely expensive from a computational perspective, proving intractable for all but the smallest of systems. For this reason, truncated CI is more commonly applied. In truncated CI, only determinants differing from the reference by up to a maximum number of excitations are included. For example, including only singly and doubly excited configurations results in the CISD approximation, whereas also including triply excited determinants gives the CISDT approximation.

5.3.2 CI and the Dissociation of H_2

As an example of the application of the CI approach, we consider the dissociation of molecular hydrogen. Near the H_2 equilibrium geometry, the exact two electron wavefunction is well approximated by that obtained via application of the Hartree-Fock approach. Assuming

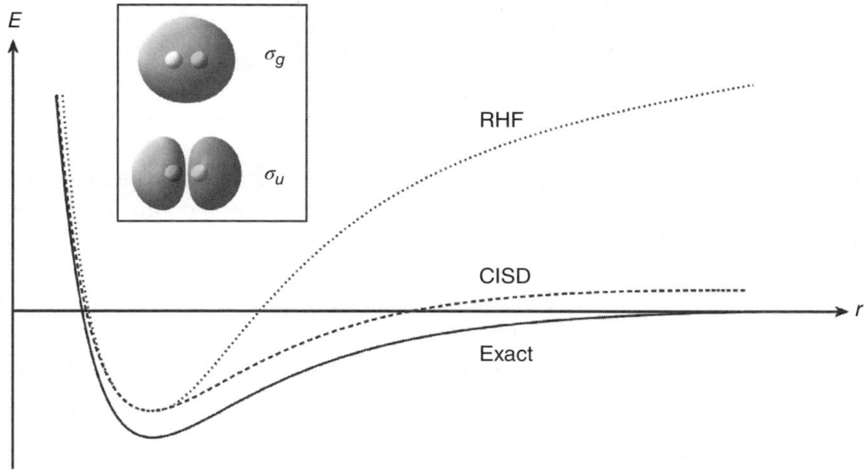

Figure 5.4 *Comparison of HF- and CISD-calculated potential energy surfaces to the exact surface for the dissociation of H_2. Inset: bonding and antibonding molecular orbitals of H_2. See color plate section*

a minimal basis set of 1s orbitals on each H centre, application of the HF approach yields two molecular orbitals (see Figure 5.4; See color plate section):

$$\sigma_g = N_g\left(1s_A + 1s_B\right) \tag{5.14}$$

$$\sigma_u = N_u\left(1s_A - 1s_B\right) \tag{5.15}$$

With the total wavefunction given by

$$|\Psi_0\rangle = |\sigma_g\sigma_g| \tag{5.16}$$

But, as $r_{AB} \to \infty$ and $\langle 1s_A \mid 1s_B\rangle \to 0$,

$$|\Psi_0\rangle \to \frac{1}{2}\left(|1s_A\alpha 1s_B\beta| + |1s_B\alpha 1s_A\beta| + |1s_A\alpha 1s_A\beta| + |1s_B\alpha 1s_B\beta|\right). \tag{5.17}$$

Whilst the first two terms on the right-hand side correspond to correct dissociation, i.e., $H_2 \to H+H$, the latter two correspond to incorrect *ionic* dissociation, i.e., $H_2 \to H^+ + H^-$. These spurious ionic terms correspond to both electrons being found on the same nucleus and therefore lead to a significant overestimation of the dissociation energy at the HF level.

The minimal basis description of H_2 allows for the construction of three excited configurations: two singly excited determinants $|\sigma_g\sigma_u|$ and $|\sigma_u\sigma_g|$, along with one doubly excited determinant of the form $|\Psi_D\rangle = |\sigma_u\sigma_u|$. From Brillouin's theorem, only the doubly excited configuration interacts directly with the reference and, in the limit $r_{AB} \to \infty$,

$$|\Psi_D\rangle \to \frac{1}{2}\left(-|1s_A\alpha 1s_B\beta| - |1s_B\alpha 1s_A\beta| + |1s_A\alpha 1s_A\beta| + |1s_B\alpha 1s_B\beta|\right) \tag{5.18}$$

Whilst $|\Psi_D\rangle$ suffers from the same incorrect long-range behaviour as $|\Psi_0\rangle$, the spurious ionic terms cancel when taking the linear combination, i.e.,

$$|\Psi_0\rangle + |\Psi_D\rangle \rightarrow |1s_A\alpha 1s_B\beta| + |1s_B\alpha 1s_A\beta| . \tag{5.19}$$

The equal contributions of $|\Psi_0\rangle$ and $|\Psi_D\rangle$ to the CI wavefunction might be expected since, as $r_{AB} \rightarrow \infty$, σ_g and σ_u become energetically degenerate. In general this is not the case, and the description of the H_2 electronic wavefunction can be written

$$|\Psi\rangle = c_0 |\Psi_0\rangle + c_D |\Psi_D\rangle \tag{5.20}$$

where c_0 and c_D are determined variationally. $|\Psi\rangle$ therefore accounts for the *static* correlation in H_2 and, as can be seen in Figure 5.4, gives a qualitatively correct description of the dissociation process.

5.3.3 Static Correlation and Crystal Field Splitting

In the absence of a crystal field, the seven 4f/5f orbitals of an f-element are degenerate. Whilst crystal field effects result in a large splitting of the d-shell in transition metal complexes, this is much less pronounced in f-element analogues; in particular, the lanthanides experience extremely weak crystal fields. In an octrahedral environment, the f manifold splits into two triply degenerate (t_{1u} and t_{2u}) and one non-degenerate (a_{2u}) level [16] and partial occupation of the 4f/5f manifold therefore results in several near degenerate electronic configurations, as can be seen schematically in Figure 5.5. In analogy with the H_2

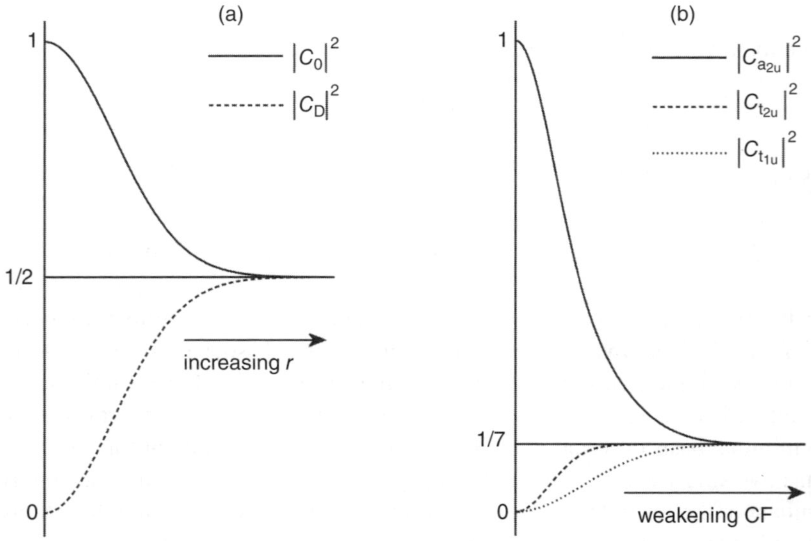

Figure 5.5 *CI coefficients for a) the reference and doubly excited configurations of H_2, and b) configurations incorporating different f-orbital occupation in a model f^1 system experiencing an octahedral crystal field*

example in the previous section, weak crystal field effects result in strong static correlation and therefore multiconfigurational character in the many-electron wavefunction. For this reason, the application of a multiconfigurational method is critical to the accurate modelling of open-shell f-element complexes.

5.3.4 Size Inconsistency and Coupled Cluster Theory

Consider two systems, A and B. Full CI is size consistent, meaning that

$$E_{\text{FCI}}(AB) = E_{\text{FCI}}(A) + E_{\text{FCI}}(B) \tag{5.21}$$

However, size-consistency is not maintained in truncated CI methods. CI expands the exact N-electron wavefunction in terms of Slater determinants using a linear ansatz:

$$|\Psi\rangle = \hat{C}\,|\Psi_0\rangle\,, \tag{5.22}$$

where

$$\hat{C} = 1 + \sum_i \hat{C}_i. \tag{5.23}$$

The exact wavefunction can, however, also be expanded using an exponential ansatz:

$$|\Psi\rangle = e^{\hat{T}}\,|\Psi_0\rangle\,, \tag{5.24}$$

where the cluster operator, $\hat{T}$, is defined as

$$\hat{T} = \hat{T}_1 + \hat{T}_2 + \hat{T}_3 + \cdots + \hat{T}_n \tag{5.25}$$

and $\hat{T}_i$ generates all determinants having i excitations from the HF reference. Expanding the wavefunction in this manner gives the coupled cluster (CC) approach [17] and, from equations 5.22 and 5.24, we see that full CI and full CC are equivalent. When we compare truncated CC to the equivalent CI calculation, however, we find that additional terms are incorporated in the former. If we consider only the double excitation operator $\hat{T}_2$ then the resulting CCD approximation gives:

$$\begin{aligned} |\Psi_{\text{CCD}}\rangle &= e^{\hat{T}_2}\,|\Psi_0\rangle \\ &= \left(1 + \hat{T}_2 + \frac{\hat{T}_2^2}{2!} + \frac{\hat{T}_2^3}{3!} + \cdots\right)|\Psi_0\rangle \end{aligned} \tag{5.26}$$

which, due to the presence of terms of the form $\hat{T}_2^n$, is size consistent. The corresponding CID approximation is given by

$$|\Psi_{\text{CID}}\rangle = \left(1 + \hat{T}_2\right)|\Psi_0\rangle \tag{5.27}$$

and, since terms of the form $\hat{T}_2^n$ are now absent, does not exhibit size consistency. Whilst CC approaches therefore outperform CI equivalents, they are computationally demanding. Furthermore, at least triple excitations must be included in order to include the effects of static correlation, resulting in the extremely expensive CCSDT approximation. Therefore, despite the limitations of truncated CI, advantage can be taken of the relative simplicity of constructing more advanced approaches based on the linear ansatz, and it is therefore used as a basis for multiconfigurational self-consistent-field (MCSCF) and other multireference techniques.

5.3.5 Computational Expense of CI and the Need for Truncation

Figure 5.6 compares the scaling of truncated CI approaches to that of the HF approximation and full CI. The latter scales factorially in the number of orbitals, rapidly becoming computational intractable. Whilst still computationally expensive, truncated CI approaches are tractable, however they provide only an approximate treatment of static and dynamical effects. If these two forms of correlation are instead treated separately, then the former can be treated by selecting key subsets of occupied and virtual orbitals within which full CI is performed. This dramatically reduces the number of configurations included in the CI expansion while ensuring that all static correlation effects are accurately described. This approach can be further improved by incorporating orbital optimisation, as opposed to utilising the set of canonical Hartree-Fock orbitals as a basis for the CI expansion. In essence, this application of full CI within an orbital subspace combined with orbital optimisation gives the complete-active-space self-consistent-field (CASSCF) approach and its variants.

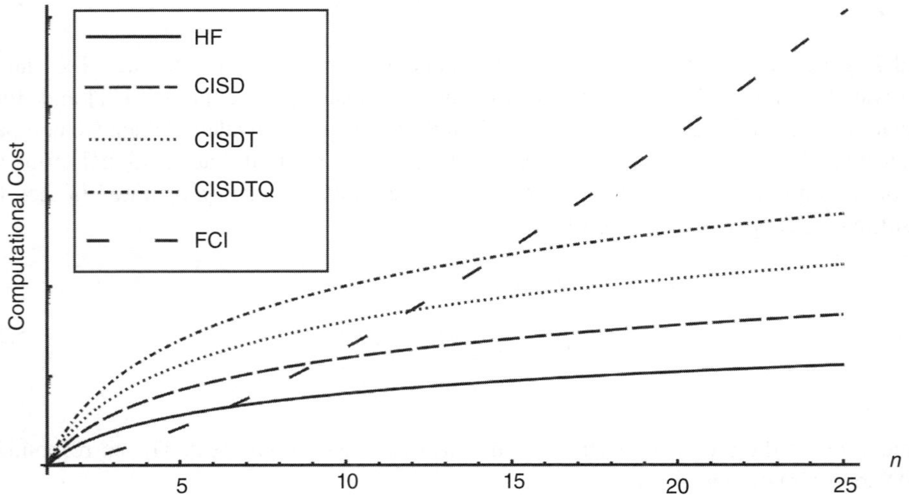

Figure 5.6 Computational scaling of post-HF methods as a function of n, the number of orbitals. Computational cost is plotted logarithmically

5.4 CASSCF and Related Approaches

The CASSCF approach is based on the concept that the orbital space of a system can be partitioned into a series of subspaces. As discussed in Section 5.3.5, the computational cost of FCI is so high that it is only practical for systems with a very small number of orbitals and electrons. In the CASSCF approach, an *active space* is defined, and it is within this space that full CI is performed. The active space typically encompasses some proportion (ideally, all) of the valence region of the system of interest, with deeper lying orbitals (for which the effects of static correlation have little impact with regard to chemical properties) deemed 'inactive.' The active space also includes a number of virtual orbitals, allowing for a CI expansion to be performed, with higher lying orbitals, occupied only in highly excited configurations that would not be expected to contribute significantly to the description of the static correlation within the system, deemed 'external.' Figure 5.7 summarises this partitioning into orbital subspaces.

5.4.1 The Natural Orbitals

A Hartree-Fock calculation produces a set of molecular orbitals, known as the canonical Hartree-Fock orbitals. These orbitals are the solution to the Hartree-Fock-Roothaan equations and, while they may be used as the basis for a CI expansion of the exact wavefunction, they are not optimal in the sense that a different choice of orbital basis may result in a CI expansion that converges more rapidly to the FCI limit. In 1955, Löwdin demonstrated that the optimal one-electron basis for the CI expansion of the exact wavefunction is the natural orbital basis [18]. In order to obtain the natural orbitals, we must first construct the first order reduced density matrix (RDM), defined as:

$$\gamma(\mathbf{x}_1, \mathbf{x}'_1) = \int \Psi(\mathbf{x}_1, \mathbf{x}_2, \cdots, \mathbf{x}_n)\Psi(\mathbf{x}'_1, \mathbf{x}_2, \cdots, \mathbf{x}_n)d\mathbf{x}_2 \cdots d\mathbf{x}_n \tag{5.28}$$

Diagonalising γ gives as eigenvectors the natural spin orbitals and as eigenvalues their corresponding occupancies. For an electronic state constructed from a monodeterminantal wavefunction, the natural occupancies take the values of 0, 1, or 2. However, for a multiconfigurational electronic state, the natural occupancies will differ from integer values, and these deviations can be taken as a measure of multiconfigurational character. As we shall see, the natural orbitals can be used in the construction of the active space required in a CASSCF calculation.

5.4.2 Optimisation of the CASSCF Wavefunction

The discussion in this section follows closely that in Roos [19]. In analogy to equation 5.12, we define the CASSCF state as a linear combination of configurations:

$$|0\rangle = \sum_m c_m |m\rangle \tag{5.29}$$

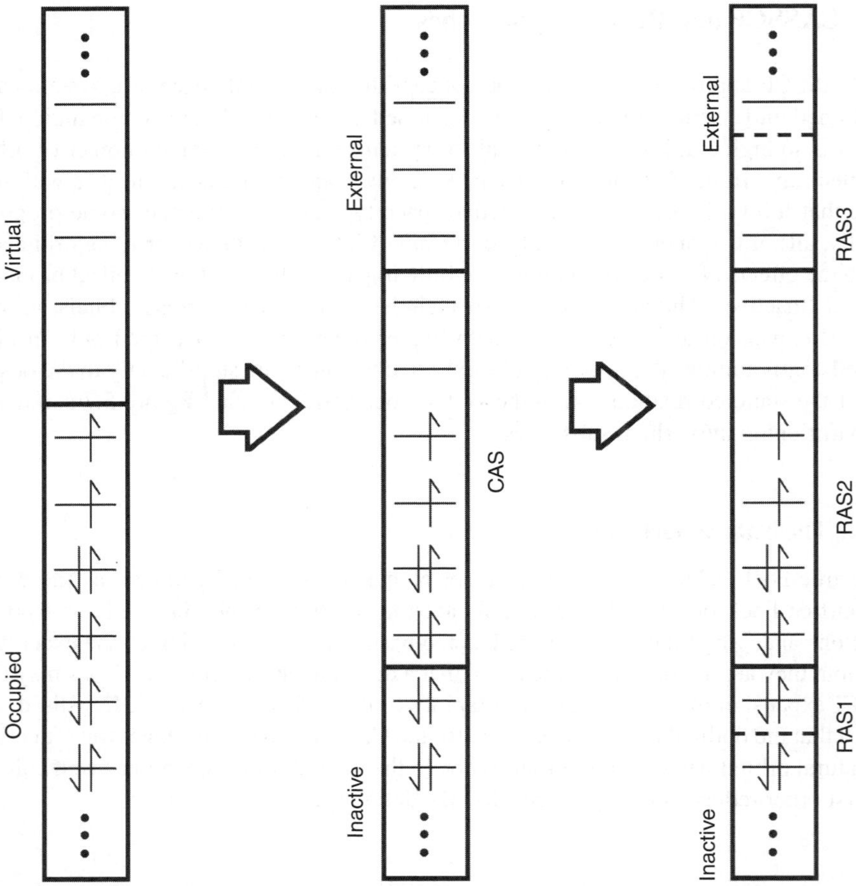

Figure 5.7 *Partitioning of the molecular orbital space into inactive, active, and external subspaces. The number of active subspaces differs between CASSCF, RASSCF, and GASSCF calculations*

where the $|m\rangle$ only differ in occupation of the active space orbitals. In a CASSCF wavefunction, both the molecular orbital (MO) and configuration interaction (CI) coefficients are optimised, and this optimisation can be described in terms of a unitary transformation of the non-optimal CASSCF state. We first consider the optimisation of the CI coefficients. We start by defining $|\mathrm{K}\rangle$, the *orthogonal complement* to $|0\rangle$:

$$|\mathrm{K}\rangle = \sum_m c_m^{\mathrm{K}} |m\rangle \tag{5.30}$$

The $|\mathrm{K}\rangle$ are expanded in the same basis as $|0\rangle$ with the CI coefficients c_m^{K} defined such that an orthonormal set is obtained, $\langle \mathrm{K} \mid \mathrm{L}\rangle = \delta_{\mathrm{KL}}$. A variational parameter $S_{\mathrm{K}0}$ can be assigned to each of the $|\mathrm{K}\rangle$ and the variation of our CASSCF state can be considered as a unitary transformation (or rotation) between $|0\rangle$ and $|\mathrm{K}\rangle$, with S_{K0} defining the magnitude of this rotation. In this manner we may define the anti-hermitian *replacement operator* $\hat{S}$:

$$\hat{S} = \sum_{\mathrm{K}} S_{\mathrm{K0}} \left(|\mathrm{K}\rangle \langle 0| - |0\rangle \langle \mathrm{K}| \right) \tag{5.31}$$

from which we can define the unitary operator $e^{\hat{S}}$.

The optimisation of the MO coefficients is implemented differently. Using second quantisation, a *substitution operator*, $\hat{E}$, can be defined:

$$\hat{E}_{ij} = \hat{a}^{\dagger}_{i\alpha}\hat{a}_{j\alpha} + \hat{a}^{\dagger}_{i\beta}\hat{a}_{j\beta} \tag{5.32}$$

Here, $\hat{a}^{\dagger}$ and $\hat{a}$ are creation and annihilation operators, respectively. $\hat{E}_{ij}$ therefore has the effect of substituting an electron in state i for one in state j. As was done for the CI coefficients, we now define an anti-hermitian operator, $\hat{T}$ in terms of the variational parameters T_{ij}:

$$\begin{aligned} \hat{T} &= \sum_{i,j} T_{ij}\hat{E}_{ij} \\ &= \sum_{i>j} T_{ij}\left(\hat{E}_{ij} - \hat{E}_{ji}\right) \end{aligned} \tag{5.33}$$

$\hat{E}_{ij} - \hat{E}_{ji}$ can also be considered as a replacement operator, and $T_{ij}\left(\hat{E}_{ij} - \hat{E}_{ji}\right)$ describes an orthogonal rotation between MOs i and j. The unitary operator $e^{\hat{T}}$ can therefore be used to describe rotations in the MO space.

We can now use the two exponential operators defined above to express a variation in the CI and MO coefficients of our CASSCF state $|0\rangle$:

$$|\Psi'\rangle = e^{\hat{T}}e^{\hat{S}}\,|0\rangle \tag{5.34}$$

$e^{\hat{T}}$ and $e^{\hat{S}}$ do not commute and the order defined above, while not affecting the final result, is chosen to simplify manipulation of resulting terms. We can define the energy of our transformed state $|\Psi'\rangle$ as:

$$\begin{aligned} E\left(\hat{T},\hat{S}\right) &= \left\langle \Psi' \middle| \hat{H} \middle| \Psi' \right\rangle \\ &= \left\langle 0 \middle| e^{-\hat{S}}e^{-\hat{T}}\hat{H}\, e^{\hat{T}}e^{\hat{S}} \middle| 0 \right\rangle \\ &= \left\langle 0 \middle| \hat{H} \middle| 0 \right\rangle \\ &\quad + \left\langle 0 \middle| \left[\hat{H},\hat{T}\right] + \left[\hat{H},\hat{S}\right] \middle| 0 \right\rangle \\ &\quad + \left\langle 0 \middle| \frac{1}{2}\left[\left[\hat{H},\hat{T}\right],\hat{T}\right] + \frac{1}{2}\left[\left[\hat{H},\hat{S}\right],\hat{S}\right] + \left[\left[\hat{H},\hat{T}\right],\hat{S}\right] \middle| 0 \right\rangle \\ &\quad + \ldots \end{aligned} \tag{5.35}$$

The first term on the right-hand side of equation 5.35 corresponds to the zeroth order energy of our untransformed CASSCF state. The second term gives the first derivatives of this energy with respect to the variational parameters T_{ij} and $S_{\mathrm{K}0}$, whilst the third term gives us the second derivatives, from which we can construct the Hessian. Considering the individual terms in the first derivative we have (for real wavefunctions):

$$\begin{aligned}\left\langle 0\left|\left[\hat{H},\hat{T}\right]\right|0\right\rangle &= \sum_{i>j}\left\langle 0\left|\left[\hat{H},\hat{E}_{ij}-\hat{E}_{ji}\right]\right|0\right\rangle T_{ij} \\ &= \sum_{i>j} g_{ij}^{\mathrm{MO}} T_{ij}\end{aligned} \tag{5.36}$$

$$\begin{aligned}\left\langle 0\left|\left[\hat{H},\hat{S}\right]\right|0\right\rangle &= \sum_{\mathrm{K}}\left(\left\langle 0\left|\hat{H}\right|\mathrm{K}\right\rangle + \left\langle \mathrm{K}\left|\hat{H}\right|0\right\rangle\right) S_{\mathrm{K}0} \\ &= \sum_{\mathrm{K}} 2\left\langle 0\left|\hat{H}\right|\mathrm{K}\right\rangle S_{\mathrm{K}0} \\ &= \sum_{i>j} g_{\mathrm{K}0}^{\mathrm{CI}} S_{\mathrm{K}0}\end{aligned} \tag{5.37}$$

where the g^{MO} and g^{CI} give the gradients in the energy with respect to variations in the MO and CI coefficients, respectively. Since these gradients vanish for an optimised wavefunction, equation 5.36 shows that all matrix elements of $\left[\hat{H},\hat{E}_{ij}-\hat{E}_{ji}\right]$ are zero for a set of optimised molecular orbitals, a generalisation of Brillouin's theorem. Equation 5.37, on the other hand, shows that an optimised CASSCF wavefunction does not interact through the Hamiltonian with the orthogonal complement $|\mathrm{K}\rangle$.

Given both the gradient and the Hessian, the optimised CASSCF wavefunction can, in principle, be obtained directly via the Newton-Raphson approach. For large configuration spaces however, such a direct approach becomes computationally intractable, and so a variety of iterative procedures have been developed in order to circumvent this problem. Another approach, known as the super-CI method [20], instead makes use of the generalised Brillouin theorem discussed above. We begin by defining a set of configurations corresponding to single excitations of our trial state $|0\rangle$:

$$\left|\Psi^{lm}\right\rangle = \left(\hat{E}_{lm}-\hat{E}_{ml}\right)|0\rangle \tag{5.38}$$

The super CI wavefunction is then defined as a linear combination of the reference state and these single excitations:

$$\left|\Psi^{\mathrm{SCI}}\right\rangle = |0\rangle + \sum_{l,m} c_{lm}\left|\Psi^{lm}\right\rangle. \tag{5.39}$$

For an optimised CASSCF wavefunction, the c_{lm} vanish, satisfying the generalised Brillouin theorem, and so the problem becomes one of incorporating the $\left|\Psi^{lm}\right\rangle$ into the trial state $\left|\Psi^{\mathrm{SCI}}\right\rangle$ so as to optimise the state. This can be done in an iterative manner. For example, the first order RDM, γ^{SCI}, can be constructed and diagonalised to generate a new set of trial orbitals, which would include contributions from the $\left|\Psi^{lm}\right\rangle$ to second order in c_{lm}. These new orbitals can then be used a basis for a CI expansion, and the new super CI wavefunction constructed in this manner (along with the corresponding CI expansion) would therefore be expected to have smaller contributions from the singly excited configurations. This process could be repeated until the wavefunction is optimised, at which point the c_{lm} vanish. The super-CI approach therefore optimises the wavefunction by systematically reducing its interaction with the singly excited configuration space.

5.4.3 Variants and Generalisations of CASSCF

CASSCF is a powerful approach to treating the effects of static correlation that commonly manifest themselves in complexes of the f-elements. However, it still suffers from the factorial scaling problem associated with full CI, resulting in severe limitations on the size of active space that can be considered. A typical upper limit corresponds to the explicit correlation of 16 electrons in an active space of 16 orbitals. We refer to such a calculation as a CASSCF (16,16) calculation and, in general, a calculation explicitly correlating n electrons in m orbitals is referred to as a CASSCF (n,m) calculation. In order to circumvent this computational bottleneck, approaches have been developed in order to expand the size of systems that can be considered with an MCSCF methodology.

Perhaps the best known variant of the CASSCF method is the restricted-active-space SCF (RASSCF) approach [21]. Here, the active space is divided into three subspaces labelled RAS1, RAS2, and RAS3 (see Figure 5.7). The RAS2 subspace is itself a complete active space, i.e., full CI is performed within RAS2. However, the other subspaces comprise orbitals, which are either doubly occupied (RAS1) or empty (RAS3) in the reference configuration. In addition to the full CI performed in the RAS2 subspace, truncated CI is performed amongst all three subspaces. This is achieved by specifying a maximum number of allowed holes in RAS1 and electrons in RAS3. This is understood most clearly by assuming RAS2 to be empty. Then, for example, specifying the maximum number of holes (electrons) in RAS1 (RAS3) to be three would correspond to performing a CISDT expansion within the active space. As with CASSCF, the RASSCF approach also allows for orbital optimisation.

The generalised-active-space SCF (GASSCF) method [22] is, as the name suggests, a generalisation of the CASSCF and RASSCF approaches. In the GASSCF approach, an arbitrary number of active subspaces can be defined. The total active space is then derived by specifying the *cumulative* minimum and maximum electron occupations over the subspaces, and it is straightforward to demonstrate that, using this definition, both CAS and RAS states are special cases of GAS states [22]. A closely related method, the occupation-restricted-multiple-active-space (ORMAS) SCF approach [23], also allows for an arbitrary number of active subspaces but, in contrast to GASSCF, considers minimum and maximum electron occupations for *each individual* subspace.

All methods discussed above achieve an increase in active space size by actually *reducing* the size of the configuration space, either by truncating the CI expansion or by defining disconnected subspaces between which excitations are forbidden. An alternative can be defined by taking advantage of the density matrix renormalisation group (DMRG) [24]. The DMRG-CASSCF method [25, 26] employs the DMRG algorithm as a substitute for the explicit diagonalisation of the extremely large Hamiltonians associated with large scale CASSCF calculations. The compact nature of the DMRG wavefunction allows for a significant increase in active space size, making calculations in which up to 30 electrons are explicitly correlated in a space of approximately 30 molecular orbitals, CASSCF (30,30), computationally tractable. This approach, however, is at its most efficient only when applied to (pseudo-)linear molecules, and so is currently not particularly well-suited to the study of f-element complexes.

5.5 Selection of Active Spaces

One of the most challenging aspects of a CASSCF calculation is the identification of a suitable active space. Much has been written on this subject, and the reader is directed to a recent contribution by Veryazov *et al.* [27] on the subject.

5.5.1 Chemical Intuition and Björn's Rules

There are many factors governing the choice of active space. An important consideration is the nature of the chemical problem that is to be addressed. For example, if a chemical reaction is to be studied, then all orbitals involved in bond breaking/formation must be considered. If, instead, excited state properties are of interest, for example in the simulation of absorption processes, then orbitals whose occupations differ significantly between the ground and excited state must be included. This can be particularly important in coordination complexes, where excitation processes may involve a change in the oxidation state of the metal ion. However, some general points with regard to the choice of active space orbitals can be made:

- Have most active orbitals paired. Typically, this pair of natural orbitals will consist of one strongly occupied orbital (i.e., one with occupation close to an integer value) and one weakly occupied 'correlating' orbital.
- Include bonding and antibonding orbitals of a bond that may be (partially) broken (see Section 5.3.2).
- Include conjugated and aromatic bonds in the active space.
- Leave orbitals associated with C-H bonds inactive.

Björn Roos attempted to develop rules to aid in the construction of active spaces, and some of these rules are briefly summarised here:

- Include 2s and 2p orbitals for light atoms such as Li, B, and C. For N, O, and F, only include 2p orbitals.
- Include two π-orbitals for each π-bond in planar molecules. If this is not possible, select active space based on energy.
- Include the valence d-shell of transition metals. For first row transition metals, a second d-shell may be needed in order to model changes in oxidation state.
- Where hybridisation is pronounced, include all orbitals with d-character.
- Include the 4f shell of the lanthanides, even when there is no evidence of participation in chemical bonding. 5d and 6s orbitals may also be important, although less so for strongly ionic complexes.

- The 5f, 6d, and 7s orbitals of the actinides should be included (although this is often impractical). Only 5f orbitals need be included for highly charged complexes.

5.5.2 Natural Orbital Occupations

The rules given in Section 5.5.1 serve as a useful starting point for the construction of active spaces. However, in many f-element complexes, the size of the active space generated using these rules is too large to be used in a CASSCF calculation. Another approach, which can be used in conjunction with chemical intuition, is to select an active space based on the natural orbital occupations. Here, a set of natural orbitals is generated using a post-HF method (which may be a CI or perturbational approach) and a criterion is imposed for inclusion in the active space. A typical criterion is that those natural orbital occupations that lie in the range 0.02–1.98 are included in the active space. The rationale behind this approach is that only natural orbitals with occupation numbers differing significantly from integer values contribute significantly to the static correlation of the system, since all other orbitals are either occupied or empty in all contributing configurations. This approach does not guarantee an appropriate active space, whose size may be very sensitive to variation in the selection criterion, but can be used to identify the most important orbitals in the active space.

5.5.3 RAS Probing

RAS probing is closely related to the approach described in Section 5.5.2. It is of most use when it is difficult to use chemical intuition in identifying the active space. In such cases, a RASSCF calculation is performed using a large number of orbitals (up to approximately 40) but a low order of excitation, considering only singly and doubly excited configurations. Again, natural orbital occupations are used in order to identify important orbitals and retain them in the active space. This approach can be used in an iterative manner in order to reduce the active space to a size to which CASSCF may be applied, although caution must be applied to ensure that results are robust to variation in the active space.

5.6 Dynamical Correlation

Whilst CASSCF and related methods give a qualitatively accurate description of static correlation, the effects of dynamic correlation are largely neglected. The inclusion of dynamical correlation is critical for the quantitatively correct simulation of f-element complexes. This can be recovered through the application of full CI but, as already discussed, this method is intractable for all but very small systems. In fact, CI expansions converge on the full CI limit very slowly. The Coulomb cusp condition specifies a relationship between the two-electron wavefunction and its first derivative when the interelectronic separation is equal to zero:

$$\left.\frac{\partial \Psi}{\partial r}\right|_{r_{12}=0} = \left.\frac{\Psi}{2}\right|_{r_{12}=0} \tag{5.40}$$

As discussed in Section 5.2.3, dynamical correlation is important when r_{12} is small and this is manifested in a significant difference between the exact two-electron wavefunction, which

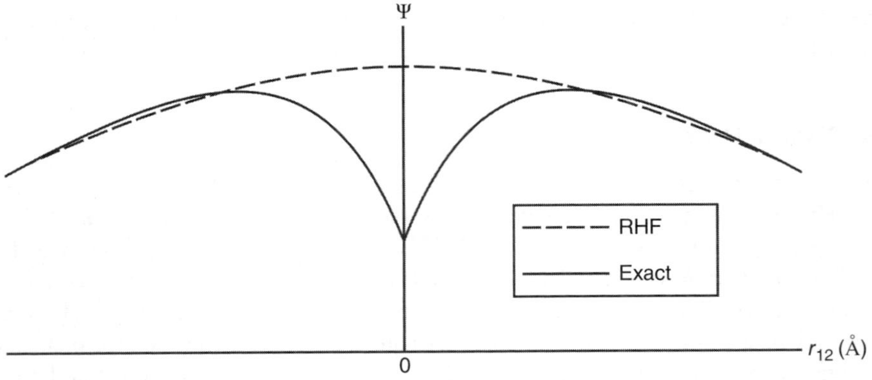

Figure 5.8 Comparison of exact and uncorrelated two-electron wavefunctions in the vicinity of $r_{12} = 0$. Only the former satisfies the Coulomb cusp condition

satisfies equation 5.40, and the uncorrelated wavefunction, which does not. This difference is shown in Figure 5.8.

Configuration interaction can only incorporate r_{12} dependence into the wavefunction via terms of the form r_{12}^2. This results in slow convergence to the exact wavefunction and consequently, the energy associated with dynamical correlation. For this reason, other approaches to recovering dynamical correlation have been developed. These approaches are applied after a CASSCF or related calculation has been performed in order to recover static correlation.

5.6.1 Multireference Configuration Interaction

Multireference configuration interaction (MRCI) [28] attempts to recover dynamical correlation by considering only the most important configurations. Taking the results of an MCSCF calculation, the leading configurations are taken as reference configurations for a series of truncated CI expansions in the full orbital space. Due to the truncated expansion, MRCI is significantly cheaper than full CI, but still becomes extremely expensive for all but moderately small systems. Typically, no configurations with more than two or three excitations from the references are considered.

5.6.2 Multireference Second Order Perturbation Theory

There are two popular approaches to recovering dynamical correlation through second order perturbation theory. Whilst both take a CASSCF wavefunction as their starting point, their definition of the zeroth order Hamiltonian, $\hat{H}_0$, required for the perturbational treatment, differ.

The first approach is known as complete active space second order perturbation theory [29] (CASPT2). The CASPT2 approach can be considered a multireference analogue to Møller-Plesset perturbation theory, a monodeterminantal method used to recover the correlation energy. In CASPT2 $\hat{H}_0$ is chosen so that it has the CASSCF state as an eigenfunction, and the effective Fock operator, $\hat{F}$, is defined so that it is diagonal in the active, inactive, and external subspaces. Furthermore, its diagonal elements correspond to orbital energies for inactive and external orbitals. This approach can be developed so that the second order correction to the energy can be obtained via construction of the first order correction to the wavefunction, and reduces to the Møller-Plesset form for a monodeterminantal wavefunction.

It can be shown that within the CASPT2 approach, the reference state directly interacts only with those states that differ from it through either single or double excitation. It is not uncommon to find that the zeroth order energy of one or more of these excited states can be similar to, or even below, the energy of the reference. Such 'intruder states' cause the perturbational approach to fail, sometimes in dramatic fashion, and must be eliminated through either a redefinition of the active space or via the application of level-shifting techniques [30].

The intruder state problem can be avoided by employing second order n-electron valence state perturbation theory [31] (NEVPT2). NEVPT2 again takes a CASSCF state as a reference function, but uses a more advanced definition of $\hat{H}_0$, a modified Dyall Hamiltonian. This Hamiltonian behaves like the true Hamiltonian inside the CAS subspace, explicitly including all two-electron interactions. NEVPT2 can also be considered as a generalisation of the Møller-Plesset approach, and the improved definition of $\hat{H}_0$ eliminates the intruder state problem.

5.7 Applications

We finish this chapter with some example applications from the literature. These serve to illustrate the fundamental questions in f-element chemistry that can be probed with the CASSCF method.

5.7.1 Bonding in Actinide Dimers

A series of An dimers (An=Ac-U) have been studied using the CASSCF methodology [32, 33]. The bonding in An_2 is rather complicated, in contrast to the transition metal analogues Mo_2 and W_2, which both exhibit sextuple bonds, and arises from the aforementioned near degeneracy of the An 5f, 6d, and 7s orbitals. In a seminal demonstration of the CASSCF methodology, Gagliardi and Roos speculated that the electronic structure U_2 could be interpreted as being consistent with a quintuple U-U bond [32]. This bond was composed of three traditional two-electrons bonds, formed from uranium $7s_\sigma$ and $6d_\pi$ orbitals, and four 'one electron bonds,' formed from $6d_\sigma$, $6d_\delta$, $5f_\pi$, and $5f_\delta$ orbitals, giving a total of ten electrons in bonding orbitals.

Table 5.1 summarises some CASSCF calculated properties of An dimers. These data reveal an increase in bond order as one traverses the actinide series, with a concomitant reduction in bond length and increase in binding energy up to Pa_2. U_2, exhibiting the same bond order as the protactinium analogue, nevertheless shows a dramatic decrease in binding energy. This was rationalised in terms of stronger occupation of bonding orbitals in Pa_2 as well as greater contribution of the more strongly interacting 6d orbitals to the Pa-Pa bond [33].

Table 5.1 Selected properties of the An_2 bond. Data from Roos and van Besien [33]

System	Bond order	r_{An-An} (Å)	Bond energy (eV)
Ac_2	2	3.63	1.2
Th_2	4	2.76	3.3
Pa_2	5	2.37	4.0
U_2	5	2.43	1.2

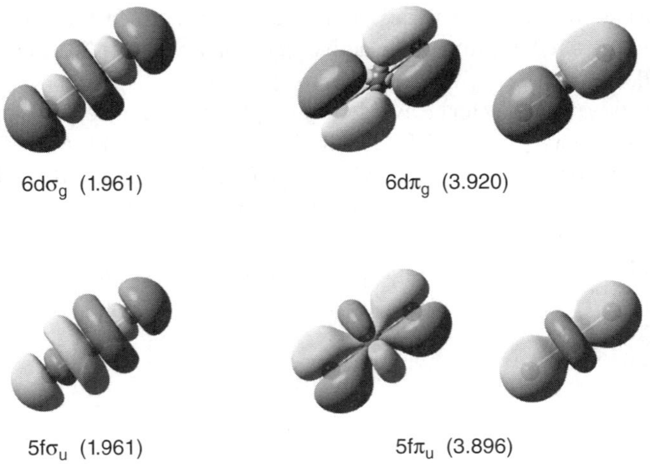

Figure 5.9 Strongly occupied natural orbitals and corresponding occupation numbers of $[UO_2]^{2+}$ obtained from a CASSCF (12,12) calculation. Data from Pierloot and van Besien [34]. See color plate section

5.7.2 Covalent Interactions in the U-O Bond of Uranyl

The uranyl dication $[UO_2]^{2+}$ is ubiquitous in the aqueous chemistry of uranium and, again in contrast to the transition metal analogues $[MoO_2]^{2+}$ and $[VO]^{2+}$, which exhibit O=M=O bond angles of approximately 110°, is linear, with $D_{\infty h}$ symmetry. This linear motif is also found in other synthesized actinyls, $[AnO_2]^{+/2+}$ (An=U-Am). These complexes are characterised by short, chemically inert, covalent An-O bonds. Pierloot and van Besien have performed CASSCF calculations on uranyl [34], confirming the description of bonding established by previous work [35]. Figure 5.9 shows the strongly occupied natural orbitals and corresponding occupation numbers obtained from a CASSCF (12,12) calculation. See color plate section.

In contrast to the An_2 dimers considered previously, no δ- or φ-type bonding interactions are possible via O 2p donation, simplifying the bonding picture considerably. In this complex the covalent bonds formed by donation from the –yl oxygen 2p orbitals into the empty uranium 5f and 6d orbitals can be clearly seen, and the bonding characterised in terms of a series of three-centre electron-pair bonds. In total, there are twelve electrons occupying bonding orbitals, which can interpreted as representing two U-O triple bonds.

Whilst the uranyl bonding orbitals have primarily oxygen 2p and uranium 5f and 6d character, other contributions can be significant: the 5fσ_u orbital has an antibonding admixture of the uranium pseudocore 6p orbital, resulting in energetic destabilisation due to the *pushing up from below* mechanism [36]. The contributions from U 5f, 6p, and 6d orbitals to the uranyl U-O bond illustrates the complexity of bonding in f-element compounds and exemplifies the need for a carefully selected active space.

5.7.3 Covalency and Oxidation State in f-Element Metallocenes

Since the 1968 synthesis of uranocene, $UCOT_2$, (COT = η^8-C_8H_8) by Streitwieser and Müller-Westerhoff [37], the f-element metallocenes have been the subject of intense interest from the f-element community. Since each COT ligand requires two electrons in order to attain aromatic stability, it is reasonable to assume that they may form sandwich complexes with elements in the +4 oxidation state. The stability of actinocenes, $AnCOT_2$, has its origins in the availability of the An 5f and 6d orbitals to participate in δ-bonding with the π_2 ligand orbitals. The transition metals, with only d-orbitals able to participate in bonding, and the lanthanides, with the chemical inertness of the 4f shell combined with a prevalent +3 oxidation state, are significantly less likely to form analogous complexes. One exception is cerium, which is known to form tetravalent complexes, and $CeCOT_2$ has been successfully synthesized [38, 39]. The participation of the Ce 4f$_\delta$ orbitals to bonding in cerocene can be seen in Figure 5.10. See color plate section. It was suggested that $CeCOT_2$ may actually be a molecular analogue of a Kondo system, with an open shell trivalent cerium centre antiferromagnetically coupled to the ligand subsystem [40]. The high (D_{8h}) symmetry implies that such an electronic structure would be implicitly multiconfigurational, and both MCSCF and CASSCF studies have demonstrated that, while there is a degree of ambiguity in the interpretation of the ground state of this complex, there is significant multiconfigurational character, with nonnegligible contributions from both Ce(III) and Ce(IV)

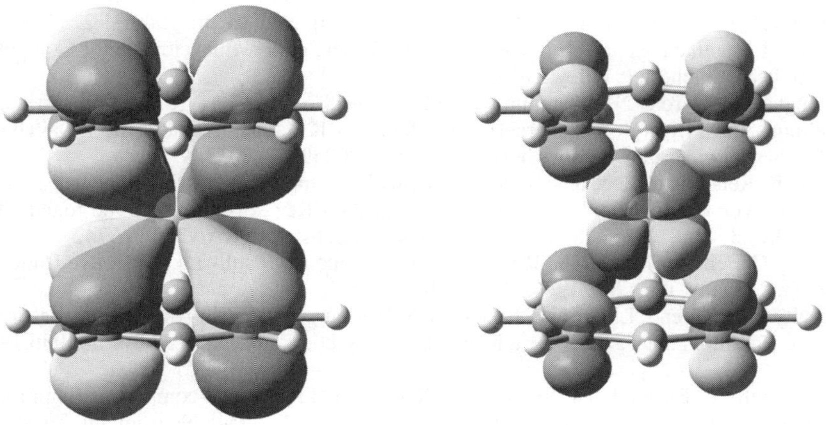

Figure 5.10 *Bonding and antibonding MOs of cerocene, exhibiting 4f$_\delta$ contributions. See color plate section*

configurations [41–45]. This strong multiconfigurational character has also been found in other Ce complexes [46, 47].

CASSCF calculations, when combined with subsequent bonding analysis, can also shed light on the origins of covalency in f-element metallocenes. Although there is a strong similarity in the electronic structure of cerocene and the plutonium analogue $PuCOT_2$, topological analysis of the electron density reveals that the latter exhibits enhanced covalent character [44, 48]. A related study of the actinocenes (An=Th-Cm) has demonstrated that covalent character reaches a maximum between Pa and Pu, and that the variation in covalency is almost entirely due to differing contributions from actinide $5f_\delta$ orbitals [49].

5.8 Concluding Remarks

The CASSCF methodology is a powerful quantum chemical approach, with characteristics that make it ideally suited to the study of f-element complexes. As opposed to simpler monodeterminantal methodologies, careful planning and consideration is required in order to derive meaningful results. However, upon the identification of a suitable active space, the quantitative evaluation of a range of chemical properties becomes possible. Current computational and methodological restrictions limit the applicability of CASSCF to relatively small systems containing tens of atoms, but promising related approaches are in development. As these approaches mature, so the scope for their application will increase.

References

[1] Kaltsoyannis N, Scott P. The f elements. Oxford: Oxford University Press; 1999.
[2] Krot NN, Gelman AD. Production of neptunium and plutonium in heptavelent state. Dokl Chem. 1967;177:124–9.
[3] Spitsyn VI, Gelman AD, Krot NN, Mefodiyeva MP, Zakharova FA, Komkov YA, *et al.* Heptavalent State of Neptunium and Plutonium. J Inorg Nucl Chem. 1969;31:2733–45.
[4] Dolg M. Lanthanides and Actinides. In: Schaefer HF, editor. Encyclopedia of Computational Chemistry. Chichester: Wiley; 2002.
[5] Douglas M, Kroll N. Quantum electrodynamical corrections to the fine structure of helium. Ann Phys (N Y) [Internet]. 1974 [cited 2012 Jun 19];155:89–155. Available from: http://www.sciencedirect.com/science/article/pii/0003491674903339.
[6] Hess B. Relativistic electronic-structure calculations employing a two-component no-pair formalism with external-field projection operators. Phys Rev A [Internet]. 1986 [cited 2012 Jun 19];33(6):3742–8. Available from: http://pra.aps.org/abstract/PRA/v33/i6/p3742_1.
[7] Chang C, Pelissier M, Durand P. Regular Two-Component Pauli- Like Effective Hamiltonians in Dirac Theory. Phys Scr. 1986;34:394–404.
[8] Heully J, Lindgren I, Lindroth E, Lundqvist S, Mårtensson-Pendrill AM. Diagonalisation of the Dirac Hamiltonian as a basis for a relativistic many-body procedure. J Phys B. 1986;19:2799–815.
[9] Van Lenthe E, Baerends EJ, Snijders JG. Relativistic regular two-component Hamiltonians. J Chem Phys [Internet]. 1993 [cited 2014 Feb 24];99(6):4597. Available from: http://link.aip.org/link/JCPSA6/v99/i6/p4597/s1&Agg=doi.
[10] Stuttgart-Cologne energy-consistent pseudopotentials [Internet]. Available from: http://www.tc.uni-koeln.de/PP/index.en.html.

[11] Hartree DR. The wave mechanics of an atom with a non-Coulomb central field. Part I: theory and methods. Proc Camb Philol Soc. 1928;24:89–110.
[12] Slater JC. The theory of complex spectra. Phys Rev. 1929;34:1293.
[13] Condon E. The Theory of Complex Spectra. Phys Rev [Internet]. 1930 Oct;36(7):1121–33. Available from: http://link.aps.org/doi/10.1103/PhysRev.36.1121.
[14] Szabo A, Ostlund NS. Modern Quantum Chemistry: Introduction to Electronic Structure Theory. New York: Macmillan; 1982.
[15] Brillouin L. Perturbation problems and self consistent fields. J Phys le Radium. 1932;3:373–89.
[16] Friedman H, Choppin G, Feuerbacher D. The shapes of the f orbitals. J Chem Educ [Internet]. 1964 [cited 2012 Sep 10];(9):354–8. Available from: http://pubs.acs.org/doi/abs/10.1021/ed041p354.
[17] Cizek J. On the Correlation Problem in Atomic and Molecular Systems. Calculation of Wavefunction Components in Ursell-Type Expansion Using Quantum-Field Theoretical Methods. J Chem Phys [Internet]. 1966 [cited 2014 Feb 25];45(11):4256. Available from: http://link.aip.org/link/?JCP/45/4256/1&Agg=doi.
[18] Löwdin P. Quantum theory of may-particle systems. I. Physical interpretations by means of density matrices, natural spin-orbitals, and convergence problems in the method of configurational interaction. Phys Rev [Internet]. 1955 [cited 2012 Jun 19];376(17):1474–89. Available from: http://prola.aps.org/abstract/PR/v97/i6/p1474_1.
[19] Roos BO. Multiconfigurational (MC) Self-Consistent-Field (SCF) Theory. In: Widmark P-O, editor. European Summer School in Quantum Chemistry, Book II. Lund: Lund University; 2009.
[20] Grein F, Chang TC. Multiconfiguration wavefunctions obtained by application of the generalized Brillouin theorem. Chem Phys Lett. 1971;12:44.
[21] Malmqvist P-Å, Rendell A, Roos BO. The Restrlcted Active Space Self-Consistent-Field Method, Implemented with a Split Graph Unitary Group Approach. J Phys Chem. 1990;(4).
[22] Ma D, Li Manni G, Gagliardi L. The generalized active space concept in multiconfigurational self-consistent field methods. J Chem Phys [Internet]. 2011 Jul 28 [cited 2012 Mar 1];135(4):044128. Available from: http://www.ncbi.nlm.nih.gov/pubmed/21806111.
[23] Ivanic J. Direct configuration interaction and multiconfigurational self-consistent-field method for multiple active spaces with variable occupations. I. Method. J Chem Phys [Internet]. 2003 [cited 2014 Feb 25];119(18):9364. Available from: http://link.aip.org/link/JCPSA6/v119/i18/p9364/s1&Agg=doi.
[24] White SR. Density Matrix Formulation for Quantum Renormalization Groups. Phys Rev Lett. 1992;69(19):2863.
[25] Zgid D, Nooijen M. The density matrix renormalization group self-consistent field method: orbital optimization with the density matrix renormalization group method in the active space. J Chem Phys [Internet]. 2008 Apr 14 [cited 2014 Jan 22];128(14):144116. Available from: http://www.ncbi.nlm.nih.gov/pubmed/18412432.
[26] Ghosh D, Hachmann J, Yanai T, Chan GK-L. Orbital optimization in the density matrix renormalization group, with applications to polyenes and beta-carotene. J Chem Phys [Internet]. 2008 Apr 14 [cited 2014 Feb 6];128(14):144117. Available from: http://www.ncbi.nlm.nih.gov/pubmed/18412433.
[27] Veryazov V, Malmqvist PÅ, Roos BO. How to Select Active Space for Multiconfigurational Quantum Chemistry. Int J Quantum Chem. 2011;111:3329–38.
[28] Lischka H, Shepard R, Brown FB, Shavitt I. New implementation of the graphical unitary group approach for multireference direct configuration interaction calculations. Int J Quantum Chem [Internet]. 2009 Jun 19;20(S15):91–100. Available from: http://doi.wiley.com/10.1002/qua.560200810.
[29] Anderson K, Malmqvist P-Å, Roos BO, Sadlej AJ, Wolinskit K. Second-Order Perturbation Theory with a CASSCF Reference Function. J Phys Chem. 1990;(7):5483–8.
[30] Forsberg N, Malmqvist P-Å. Multiconfiguration perturbation theory with imaginary level shift. Chem Phys Lett [Internet]. 1997 Aug;274(1-3):196–204. Available from: http://linkinghub.elsevier.com/retrieve/pii/S0009261497006696.
[31] Angeli C, Cimiraglia R, Evangelisti S, Leininger T, Malrieu J-P. Introduction of n-electron valence states for multireference perturbation theory. J Chem Phys [Internet]. 2001 [cited

2014 Jan 20];114(23):10252. Available from: http://link.aip.org/link/JCPSA6/v114/i23/p10252/s1&Agg=doi.
[32] Gagliardi L, Roos. Quantum chemical calculations show that the uranium molecule U2 has a quintuple bond. Nature. 2005;433(February):848–51.
[33] Roos BO, Malmqvist P-A, Gagliardi L. Exploring the actinide-actinide bond: theoretical studies of the chemical bond in Ac2, Th2, Pa2, and U2. J Am Chem Soc [Internet]. 2006 Dec 27;128(51):17000–6. Available from: http://www.ncbi.nlm.nih.gov/pubmed/17177451.
[34] Pierloot K, van Besien E. Electronic structure and spectrum of UO2 2+ and UO2Cl4 2-. J Chem Phys [Internet]. 2005 Nov 22 [cited 2011 Oct 10];123(20):204309. Available from: http://www.ncbi.nlm.nih.gov/pubmed/16351258.
[35] Zhang Z, Pitzer RM. Application of Relativistic Quantum Chemistry to the Electronic Energy Levels of the Uranyl Ion. J Phys Chem A [Internet]. 1999 Aug;103(34):6880–6. Available from: http://pubs.acs.org/doi/abs/10.1021/jp991867q.
[36] Kaltsoyannis N. Computational Study of Analogues of the Uranyl Ion Containing the -N=U=N-Unit. Inorg Chem. 2000;39:6009–17.
[37] Streitwieser A, Müller-Westerhoff U. Bis(cyclooctatetraenyl)uranium (uranocene). A new class of sandwich complexes that utilize atomic f orbitals. J Am Chem Soc. 1968;90(26):7364.
[38] Greco A, Cesca S, Bertolini G. New pi-cyclooctatretraenyl and pi-cyclopentdienyl complexes of cerium. J Organomet Chem. 1976;113:321–30.
[39] Streitwieser A, Kinsley SA, Rigsbee JT, Fragala IL, Ciliberto E, Rosch N. Photoelectron spectra and bonding in cerocene, bis(pi-[8]annulene)cerium(IV). J Am Chem Soc. 1985;107:7786–8.
[40] Neumann C, Fulde P. Is there a molecular analogue of a Kondo singlet state? 1989;278:277–8.
[41] Dolg M, Fulde P, Ku¨chle W, Neumann C-S, Stoll H. Ground state calculations of di-π-cyclooctatetraene cerium. J Chem Phys [Internet]. 1991 [cited 2013 Jul 4];94(4):3011. Available from: http://link.aip.org/link/JCPSA6/v94/i4/p3011/s1&Agg=doi.
[42] Dolg M, Fulde P, Stoll H, Preuss H, Chang A, Pitzer RM. Formally tetravalent cerium and thorium compounds?: a configuration interaction study of using energy-adjusted quasirelativistic ab initio pseudopotentials. Chem Phys. 1995;195:71–82.
[43] Kerridge A, Coates R, Kaltsoyannis N. Is cerocene really a Ce(III) compound? All-electron spin-orbit coupled CASPT2 calculations on M(eta(8)-C8H8)2 (M = Th, Pa, Ce). J Phys Chem A [Internet]. 2009 Mar 26 [cited 2011 Oct 7];113(12):2896–905. Available from: http://www.ncbi.nlm.nih.gov/pubmed/19256474.
[44] Kerridge A. Oxidation state and covalency in f-element metallocenes (M = Ce, Th, Pu): a combined CASSCF and topological study. Dalton Trans [Internet]. 2013 Sep 26 [cited 2013 Nov 6];42:16428–36. Available from: http://www.ncbi.nlm.nih.gov/pubmed/24072035.
[45] Mooßen O, Dolg M. Two interpretations of the cerocene electronic ground state. Chem Phys Lett [Internet]. Elsevier B.V.; 2014 Feb [cited 2014 Feb 26];594:47–50. Available from: http://linkinghub.elsevier.com/retrieve/pii/S0009261414000311.
[46] Coates R, Coreno M, DeSimone M, Green JC, Kaltsoyannis N, Kerridge A, *et al.* A mystery solved? Photoelectron spectroscopic and quantum chemical studies of the ion states of CeCp3+. Dalt Trans [Internet]. 2009 Aug 14 [cited 2011 Jul 14];5943–53. Available from: http://www.ncbi.nlm.nih.gov/pubmed/19623382.
[47] Kerridge A, Kaltsoyannis N. All-electron CASPT2 study of Ce(η8–C8H6)2. Comptes Rendus Chim [Internet]. 2010 Jun [cited 2011 Aug 29];13(6-7):853–9. Available from: http://linkinghub.elsevier.com/retrieve/pii/S1631074810000251.
[48] Kerridge A, Kaltsoyannis N. Are the ground states of the later actinocenes multiconfigurational? All-electron spin-orbit coupled CASPT2 calculations on An(eta(8)-C8H8)2 (An = Th, U, Pu, Cm). J Phys Chem A [Internet]. 2009 Jul 30;113(30):8737–45. Available from: http://www.ncbi.nlm.nih.gov/pubmed/19719318.
[49] Kerridge A. f-orbital covalency in the actinocenes (An=Th-Cm): multiconfigurational studies and topological analysis. RSC Adv [Internet]. 2014 [cited 2014 Feb 26];4(24):12078–86. Available from: http://pubs.rsc.org/en/Content/ArticleLanding/2014/RA/c3ra47088a.

6

Relativistic Pseudopotentials and Their Applications

Xiaoyan Cao and Anna Weigand

Theoretical Chemistry, University of Cologne

6.1 Introduction

Computational studies of f element chemistry are still a challenge because of significant correlation and relativistic effects due to the large number of electrons. Furthermore, the partially occupied f, d, s, and/or p shells cause a high density of low-lying states, which also complicate calculations. One possibility to deal with these difficulties is the effective core potential (ECP) approach, which is one of the most successful and applied approximations in relativistic quantum chemistry for heavy elements. The idea of ECPs is based on the chemically intuitive restriction of the explicit calculations to the chemical important valence electrons, while the essentially inert atomic cores, i.e., the nucleus and the electrons in the inner shells, are replaced by a suitable parametrized (relatively simple) one-electron operator acting on the remaining valence electrons, i.e., the effective core 'potential' [1]. In contrast to the alternative approximate relativistic all-electron (AE) methods as the Douglas–Kroll–Hess (DKH) [2–5] approximation, ECPs save significant computation time and therefore often allow for calculations, which are not feasible at the AE level, e.g., investigations of large complexes even including more than one heavy element. Additionally, they concentrate the computational effort on the chemically relevant valence part allowing for accurate correlation methods and large active spaces. However, one should not forget that the restriction to valence electrons is a compromise between the required computational effort and the accuracy of the results. Thus, it is important to choose the ECP core carefully (see Section 6.2.3). If a small core is chosen and care

Computational Methods in Lanthanide and Actinide Chemistry, First Edition. Edited by Michael Dolg.

has been taken in the adjustment, the error produced by the ECP approximation is almost always smaller than the error due to the ab initio electron correlation or density functional procedure [6].

Besides the computational savings, ECPs have the advantage that they allow for the implicit inclusion of relativistic effects, even of the Breit interaction or quantum electrodynamic (QED) corrections, by simple parametrizations to relativistic AE data. Furthermore, ECPs permit the usage of smaller basis sets and thus the basis set superposition error is less significant compared to AE calculations. Even the difficulties due to open shells may be avoided by applying ECPs, if these open shells are included in the core system as it is the case for the 4f-in-core [7–9] and 5f-in-core [10–12] pseudopotentials (PP) for lanthanides and actinides, respectively. However, these PPs can only be applied, if the f orbitals do not participate significantly in chemical bonding (see Section 6.3.1).

Scalar-relativistic or spin-averaged ECPs can be applied in unchanged nonrelativistic quantum chemistry and essentially all modern quantum chemistry electronic structure packages allow their usage, e.g., GAMESS [13], GAUSSIAN [14], MOLCAS [15], MOLPRO [16], or TURBOMOLE [17]. Moreover, they are available in solid state codes using Gaussian basis functions as CRYSTAL [18]. Spin–orbit (SO) effects can also be accounted for using various strategies, which range from a simple perturbative treatment subsequent to scalar-relativistic calculations to the rigorous variational inclusion already from the beginning of the calculations [19]. Even the unforeseen application of ECPs developed within a wavefunction-based framework to density functional theory (DFT) lead to reasonable results [20].

There are two main lines of ECPs, i.e., the model potential (MP) technique, which utilizes valence orbitals with a nodal structure corresponding exactly to those of the AE valence orbitals, and the PP scheme, which uses valence orbitals exhibiting a simplified nodal structure with respect to the AE valence orbitals, i.e., the so-called pseudovalence orbitals. This chapter will only focus on the PP approach, while the chapter from Barandiáran and Seijo will deal with MPs.

One further distinguishes ECPs by the kind of their adjustment, i.e., energy-consistent PPs (see Section 6.3.1) and shape-consistent PPs/MPs (see Section 6.3.2). Furthermore, ECPs are categorized by the size of their core, e.g., one differs between f-in-valence small-core [21, 22] and f-in-core large-core PPs (LPP) [7–12] for the f elements. Finally, the accuracy of the underlying AE reference data determines the ECP type, e.g., for early actinides scalar-relativistic Wood–Boring (WB) [22] or relativistic multiconfiguration Dirac–Hartree–Fock (MCDHF) [23, 24] small-core PPs (SPP) are available.

In the following we will emphasize only those modern variants of the PP approach, which experience a widespread use for f elements. At first, we will summarize the most important concepts and equations of the PP method. We will then describe in detail the two different kinds of PP adjustment, i.e., the energy- and shape-consistent technique. Next we will briefly comment on the valence basis sets for PP calculations. Then we will show some characteristic applications, and finally, we will give a short conclusion and outlook demonstrating the possibilities and limitations of the PP approach.

For further reading the minireview on the PP approximation by Schwerdtfeger [6], the rigorous recent review on relativistic PPs by Dolg and Cao [19], and the relatively complete list of older reviews given in the latter are recommended.

6.2 Valence-Only Model Hamiltonian

In ECP theory an effective model Hamiltonian only acting on the explicitly treated valence electrons is searched. There are several choices for the formulation of such a valence-only model Hamiltonian, i.e., four-, two-, or one-component approaches and explicit or implicit relativistic treatment [20]. Nonrelativistic, scalar-relativistic, and quasirelativistic ECPs use a formally nonrelativistic valence-only model Hamiltonian implicitly including relativistic effects [19]

$$\hat{H}_{\mathrm{v}} = -\frac{1}{2}\sum_{i}^{n_{\mathrm{v}}} \Delta_{\mathrm{i}} + \sum_{i<j}^{n_{\mathrm{v}}} \frac{1}{r_{\mathrm{ij}}} + \sum_{I}^{N} \hat{V}_{\mathrm{cv}}^{\mathrm{I}}(i) + \sum_{I<J}^{N} \frac{Q_{\mathrm{I}}Q_{\mathrm{J}}}{R_{\mathrm{IJ}}} + \hat{V}_{\mathrm{CPP}}. \tag{6.1}$$

The Hamiltonian as well as all following formulas are given in atomic units. The subscripts c and v denote core and valence, respectively, and the Hamiltonian is given for a molecule with n_{v} valence electrons and N cores with effective core charges Q. r_{ij} and R_{IJ} denote interelectronic and internuclear distances, respectively. The individual terms of equation 6.1 are the kinetic energy of the valence electrons, the Coulomb interaction between the valence electrons, the superposition of N atomic ECPs $\hat{V}_{\mathrm{cv}}^{\mathrm{I}}$, the point charge Coulomb repulsion between cores/nuclei, and a core-polarization potential (CPP) $\hat{V}_{\mathrm{CPP}}$, which will be further described in Section 6.2.2. If required, additive corrections accounting for deviations from the point charge approximation for the repulsion between cores/nuclei may be included, e.g., corrections for mutually penetrating cores [19]. For a neutral system the number of valence electrons corresponds to

$$n_{\mathrm{v}} = n - \sum_{I}^{N} (Z_{\mathrm{I}} - Q_{\mathrm{I}}), \tag{6.2}$$

where n is the total number of electrons and Z_{I} as well as Q_{I} denote the nuclear charge and the core charge of core I, respectively.

Neglecting $\hat{V}_{\mathrm{CPP}}$ in equation 6.1 for a moment, it is usually assumed that all relativistic effects are described by a suitable parametrization of the ECPs $\hat{V}_{\mathrm{cv}}^{\mathrm{I}}$, i.e., it is sufficient to apply the nonrelativistic kinetic energy operator as well as the nonrelativistic Coulomb interaction between the valence electrons [19]. Besides the relativistic contributions, the ECP $\hat{V}_{\mathrm{cv}}^{\mathrm{I}}$ accounts for all interactions of the valence electrons with the nucleus and the (removed) core electrons, and it is given by [19]

$$\hat{V}_{\mathrm{cv}}^{\mathrm{I}}(i) = -\sum_{i}^{n_{\mathrm{v}}} \frac{Q_{\mathrm{I}}}{r_{\mathrm{iI}}} + \sum_{i}^{n_{\mathrm{v}}} \hat{V}_{\mathrm{PP}}^{\mathrm{I}}(i). \tag{6.3}$$

The leading term corresponds to the Coulomb attraction between the valence electrons and the cores, and the expression $\hat{V}_{\mathrm{PP}}^{\mathrm{I}}$ of the second term is the one-electron pseudopotential operator. This operator keeps the valence electrons out of the core and in the valence space, i.e., it is repulsive in the short-range and attractive in the long-range [6]. The main task for the PP development is now to find an analytical form, i.e., a suitable parametrization, for $\hat{V}^{\mathrm{I}}_{\mathrm{PP}}$, which is able to compensate for all errors introduced by the simplification of the original Hamiltonian and at the same time easily applicable [19].

6.2.1 Pseudopotentials

The pseudopotential approximation was originally introduced by Hellmann already in 1935 for a semiempirical treatment of the valence electron of potassium [25]. However, it took until 1959 for Phillips and Kleinman from the solid state community to provide a rigorous theoretical foundation of PPs for single valence electron systems [26]. Another decade later in 1968 Weeks and Rice extended this method to many valence electron systems [27, 28]. Although the modern PPs do not have much in common with the PPs developed in 1959 and 1968, respectively, these theories prove that one can get the same answer as from an AE calculation by using a suitable effective valence-only model Hamiltonian and pseudovalence orbitals with a simplified nodal structure [19].

The analytical form applied for PPs nowadays is the semilocal ansatz (local in the coordinate r, but nonlocal in spherical angle coordinates θ and ϕ [6]), which goes back to Abarenkov and Heine working in the field of solid states [29, 30] and was introduced a few years later to quantum chemistry by Schwarz [31] as well as Kahn and Goddard [32]. In this ansatz besides the r-dependency of the PP also a l-dependency, i.e., a dependency on the angular momentum quantum number l, is taken into account. In the case of nonrelativistic and scalar-relativistic, i.e., one-component, PPs in equation 6.3 the following semilocal ansatz for the PP $\hat{V}^{\mathrm{I}}_{\mathrm{PP}}$ is used [32]

$$\hat{V}^{\mathrm{I}}_{\mathrm{PP}}(i) = V^{\mathrm{I}}_{\mathrm{L}}(r_{\mathrm{iI}}) + \sum_{l=0}^{l_{\mathrm{max}}} V^{\mathrm{I}}_{\mathrm{l}}(r_{\mathrm{iI}})\hat{P}^{\mathrm{I}}_{\mathrm{l}}(i). \tag{6.4}$$

This semilocal PP consists of a sum of local potentials $V^{\mathrm{I}}_{\mathrm{l}}(r_{\mathrm{iI}})$ acting separately on each angular momentum symmetry $0 \leq l \leq l_{\mathrm{max}}$ present in the core and a common local potential $V^{\mathrm{I}}_{\mathrm{L}}(r_{\mathrm{iI}})$, which acts on all angular momentum symmetries $l > l_{\mathrm{max}}$ not included in the core [19]. If l_{max} is taken large enough, the leading local term $V^{\mathrm{I}}_{\mathrm{L}}$ can be avoided [6].

The second term in equation 6.4 contains the angular momentum projection operator $\hat{P}^{\mathrm{I}}_{\mathrm{l}}$ based on spherical harmonics $|lm, I\rangle$

$$\hat{P}^{\mathrm{I}}_{\mathrm{l}}(i) = \sum_{m=-l}^{l} |lm, I\rangle\langle lm, I|. \tag{6.5}$$

Since there are no core functions in equation 6.4, the pseudovalence orbitals belonging to the lowest Hartree–Fock (HF) or Kohn–Sham solutions for each angular momentum l are thus nodeless [6].

In the case of relativistic two-component PPs, the inclusion of SO effects requires some modification of the analytical PP form. At the Dirac–Hartree–Fock (DHF) level, the degeneracy of the orbitals is reduced and depends in addition to n and l also on the total angular momentum quantum number j implying a semilocal PP with a lj-dependency [33]

$$\hat{V}^{\mathrm{I}}_{\mathrm{PP}}(i) = V^{\mathrm{I}}_{\mathrm{LJ}}(r_{\mathrm{iI}}) + \sum_{l=0}^{l_{\max}} \sum_{j=|l-1/2|}^{l+1/2} V^{\mathrm{I}}_{\mathrm{lj}}(r_{\mathrm{iI}})\hat{P}^{\mathrm{I}}_{\mathrm{lj}}(i). \tag{6.6}$$

Here, the projection operator $\hat{P}^{\mathrm{I}}_{\mathrm{lj}}$ is set up with spinor spherical harmonics $|ljm, I\rangle$

$$\hat{P}^{\mathrm{I}}_{\mathrm{lj}}(i) = \sum_{m=-j}^{j} |ljm, I\rangle\langle ljm, I|. \tag{6.7}$$

Ermler *et al.* showed that equation 6.6 may be rewritten as the sum of a spin-averaged (SA) scalar-relativistic and a SO term [34]

$$\hat{V}^{\mathrm{I}}_{\mathrm{PP}}(i) = \hat{V}^{\mathrm{I}}_{\mathrm{PP,SA}}(i) + \hat{V}^{\mathrm{I}}_{\mathrm{PP,SO}}(i), \tag{6.8}$$

where

$$\hat{V}^{\mathrm{I}}_{\mathrm{PP,SA}}(i) = \sum_{l=0}^{l_{\max}} \frac{lV^{\mathrm{I}}_{\mathrm{l},|\mathrm{l}-1/2|}(r_{\mathrm{iI}}) + (l+1)V^{\mathrm{I}}_{\mathrm{l,l}+1/2}(r_{\mathrm{iI}})}{2l+1}\hat{P}^{\mathrm{I}}_{\mathrm{l}}(i) \tag{6.9}$$

and

$$\hat{V}^{\mathrm{I}}_{\mathrm{PP,SO}}(i) = \sum_{l=1}^{l_{\max}} \frac{V^{\mathrm{I}}_{\mathrm{l,l}+1/2}(r_{\mathrm{iI}}) - V^{\mathrm{I}}_{\mathrm{l},|\mathrm{l}-1/2|}(r_{\mathrm{iI}})}{2l+1}\left[l\hat{P}^{\mathrm{I}}_{\mathrm{l,l}+1/2}(i) - (l+1)\hat{P}^{\mathrm{I}}_{\mathrm{l,l}-1/2}(i)\right]. \tag{6.10}$$

The spin-averaged PP, which can be calculated from a two-component PP by equation 6.9, may be applied in scalar-relativistic (one-component) calculations using standard quantum chemistry program packages as GAUSSIAN [14] or MOLPRO [16].

In order to become computationally efficient also for molecular calculations and to handle PPs by standard quantum chemistry codes, the potentials $V^{\mathrm{I}}_{\mathrm{L}}$ and $V^{\mathrm{I}}_{\mathrm{l}}$ in the one-component case (see equation 6.4) as well as $V^{\mathrm{I}}_{\mathrm{LJ}}$ and $V^{\mathrm{I}}_{\mathrm{lj}}$ in the two-component case (see equation 6.6) are usually represented by linear combinations of k radial Gaussian functions multiplied by powers of the electron–core distance r_{iI} [35]

$$V^{\mathrm{I}}_{\mathrm{m}}(r_{\mathrm{iI}}) = \sum_{k} A^{\mathrm{I}}_{\mathrm{km}} r_{\mathrm{iI}}^{n_{\mathrm{km}}} \mathrm{e}^{-a^{\mathrm{I}}_{\mathrm{km}} r^2_{\mathrm{iI}}} \quad \text{with} \quad \mathrm{m} = \mathrm{L}, \mathrm{l}, \mathrm{LJ}, \mathrm{lj}. \tag{6.11}$$

The parameters $A^{\mathrm{I}}_{\mathrm{km}}$, n_{km}, and $a^{\mathrm{I}}_{\mathrm{km}}$ are optimized within the PP adjustment procedure (see Section 6.3). The powers n_{km} of the electron–core distance are restricted to the values -2, -1, and 0, because these values proved to accurately cover the behavior of $\hat{V}^{\mathrm{I}}_{\mathrm{PP}}$ at the origin [35].

6.2.2 Approximations

At least five approximations have to be made in order to work with a computationally practical valence-only model Hamiltonian as in equation 6.1. To compensate for the errors introduced by these approximations, the valence-only Hamiltonian usually contains free parameters, e.g., in the pseudopotentials $\hat{V}^{\mathrm{I}}_{\mathrm{PP}}$, which are adjusted to accurately reproduce AE or experimental data [19].

The first approximation is connected with the choice of the core and valence subsystems or the so-called core–valence separation. From a quantum mechanical point of view the partitioning of a many-electron system into subsystems is not possible, because electrons are indistinguishable. However, this partitioning can approximately be achieved within an independent-particle framework, as the HF or DHF level [35–37], using an orbital-based definition of the core and valence subsystems. Due to the core–valence separation, the core–valence correlation and, if uncorrelated reference data are chosen (as is most often the case), also the core–core correlation is neglected. Of course the assumption of valence-only methods is that core correlation effects can be omitted [19], but due to this neglect bond lengths and energies as well as ionization potentials (IPs) are at least slightly affected [38]. Obtained bond lengths and energies are too long and too low, respectively, because the electron–electron repulsion is overestimated, and calculated IPs are too small, since core correlation stabilizes the atom. Therefore, especially for LPPs, it may become important to include the dynamic core-polarization, i.e., core–valence correlation, by the addition of a suitable CPP $\hat{V}_{\mathrm{CPP}}$ [39, 40] to the valence-only Hamiltonian (see equation 6.1).

Second, the atomic cores are assumed to be transferable for the atom and molecule regardless of the electronic state, i.e., they are inert and remain unchanged, which corresponds to a freezing of the core orbitals, the so-called frozen-core (FC) approximation [36]. Therefore, care has to be taken that all low-energy configurations of the neutral atom and its low-charged ions, which might become important in chemical processes, are considered in the PP adjustment [41]. However, only energy-consistent PPs are adjusted to more than one reference configuration, which constitutes an advantage over shape-consistent PPs (see Section 6.3). Clearly, the appropriate choice of the core is also for this approximation an important factor and will be discussed in detail in Section 6.2.3. Besides the usage of several reference configurations within the PP parametrization, CPPs [39, 40] can be used to correct for the FC approximation, because they also account for static core-polarization, which arises from the deformation of the atomic cores under the field of the other cores and all valence electrons in the system at the HF level [19].

Meyer and coworkers [39, 40] proposed in the framework of AE calculations for alkali and alkaline earth elements the use of an effective CPP of the form

$$\hat{V}_{\mathrm{CPP}} = -\frac{1}{2}\sum_{I}\alpha_{\mathrm{I}}\hat{\vec{f}}_{\mathrm{I}}^{\,2} \tag{6.12}$$

with

$$\hat{\vec{f}}_{\mathrm{I}} = -\sum_{i}\frac{\vec{r}_{\mathrm{iI}}}{r^{3}_{\mathrm{iI}}}\mathrm{F}\left(r_{\mathrm{iI}}, \delta^{\mathrm{I}}_{\mathrm{e}}\right) + \sum_{J\neq I} Q_{\mathrm{J}}\frac{\vec{R}_{\mathrm{JI}}}{R^{3}_{\mathrm{JI}}}\mathrm{F}\left(R_{\mathrm{JI}}, \delta^{\mathrm{I}}_{\mathrm{c}}\right). \tag{6.13}$$

Here, α^{I} denotes the dipole polarizability of the core I and $\hat{\vec{f}}_{\mathrm{I}}$ is the electric field at this core generated by the valence electrons (at relative positions $\vec{r}_{\mathrm{iI}}$) and all other cores/nuclei (with charges Q_{J} at relative positions $\vec{R}_{\mathrm{JI}}$). Since the validity of the underlying multipole expansion breaks down for small distances from the core I, the electric field $\hat{\vec{f}}_{\mathrm{I}}$ has to be multiplied by a cutoff factor $\mathrm{F}\left(r_{\mathrm{iI}}, \delta_{\mathrm{e}}^{\mathrm{I}}\right)$ for electron–nuclear terms

$$\mathrm{F}\left(r_{\mathrm{iI}}, \delta_{\mathrm{e}}^{\mathrm{I}}\right) = \left(1 - \mathrm{e}^{-\delta_{\mathrm{e}}^{\mathrm{I}} r_{\mathrm{iI}}^{2}}\right)^{n_{\mathrm{e}}} \tag{6.14}$$

and $\mathrm{F}\left(R_{\mathrm{JI}}, \delta_{\mathrm{c}}^{\mathrm{I}}\right)$ for nuclear–nuclear terms

$$\mathrm{F}\left(R_{\mathrm{JI}}, \delta_{\mathrm{c}}^{\mathrm{I}}\right) = \left(1 - \mathrm{e}^{-\delta_{\mathrm{c}}^{\mathrm{I}} R_{\mathrm{IJ}}^{2}}\right)^{n_{\mathrm{c}}}. \tag{6.15}$$

The exponents n_{e} and n_{c} for the electronic and nuclear contributions are usually one or two, and the cutoff parameters $\delta_{\mathrm{e}}^{\mathrm{I}}$ and $\delta_{\mathrm{c}}^{\mathrm{I}}$ are adjusted to suitable reference data, e.g., IPs. This ansatz for CPPs was adapted to energy-consistent LPPs by Fuentealba *et al.* [42] and proved to be quite successful, e.g., in calculations using 4f-in-core [43] and 5f-in-core LPPs [11, 44], respectively.

The third approximation is the replacement of the core electrons in an AE FC Hamiltonian by an ECP $\hat{V}_{\mathrm{cv}}^{\mathrm{I}}$ modeling the real nonlocal HF potential [19]. The atomic effective one-electron operators representing the ECPs are usually cast into a computationally convenient form with adjustable parameters allowing for an implicit treatment of relativistic effects [19].

Fourth, the valence-only model Hamiltonian given in equation 6.1 assumes a classical point charge Coulomb repulsion between cores/nuclei. However, for very large cores this approximation might be too crude and mutually penetrating or overlapping cores need to be accounted for by additional corrections [19].

Finally, in contrast to MPs the PP scheme introduces the so-called pseudovalence orbital transformation, i.e., atomic core and virtual orbitals are mixed into the valence orbitals in order to make these radially smooth and nodeless for the energetically lowest solution in each angular symmetry [41]. Although the pseudovalence orbitals possess in the chemically inert core region a simplified nodal structure, their shapes in the chemically important valence region as well as their orbital energies should be very similar to the AE case. The advantage of the simplified radial nodal structure is that compact basis functions, required to describe the nodal structure in the core region, are avoided, and thus considerable savings with respect to the one-particle basis sets are achieved using PPs [20]. However, pseudovalence orbitals tend to give too large valence correlation energies and too large multiplet splittings [45, 46], because the exchange integrals are overestimated. In practice, the accuracy of correlation energies from PP calculations [47, 48] are not worse than that of MP calculations [49, 50] and especially correlation contributions to energy differences as binding energies are well described due to the modern PP parametrization.

6.2.3 Choice of the Core

The appropriate choice of the ECP core is, besides the choice of the reference data, the most important decision underlying the construction of an ECP. The size of the core is an important factor for both the accuracy (small cores preferable) as well as the efficiency (large cores preferable) of the ECP [19]. Thus, a compromise is needed, where the errors due to the two approximations depending on the core size, i.e., the core–valence separation and the FC approximation, are still acceptable.

Table 6.1 *Range of the FCEs (in eV) for configurations with fixed 4f occupations for 4, 12, and 30 valence electron systems of Ce*

	4 Valence electrons	12 Valence electrons	30 Valence electrons
valence space	$4f^1 5d^1 6s^2$	$4f^1 5s^2 5p^6 5d^1 6s^2$	$4s^2 4p^6 4d^{10} 4f^1 5s^2 5p^6 5d^1 6s^2$
FC	1s–4d, 5s, 5p	1s–4d	1s–3d
FCEs for $4f^1$	0.007–0.416	0.000–0.001	0.000
FCEs for $4f^0$	0.845–5.037	0.364–0.466	0.002
FCEs for $4f^2$	0.824–1.238	0.180–0.223	0.001–0.002

The FCEs correspond to relative DHF energies based on the DC Hamiltonian of the 2J+1-weighted average of all J levels belonging to a nonrelativistic configuration [51] with respect to the value for the Ce [Xe]$4f^1 5d^1 6s^2$ ground state configuration and were taken from Dolg and Cao [41]. Considered are 9, 12, and 6 configurations of the neutral atom and its cations for the fixed $4f^1$, $4f^0$, and $4f^2$ occupations, respectively.

In the case of heavy elements, more than one choice of the core is possible and the related accuracy has to be further investigated. For this purpose AE FC HF and DHF calculations provide a good impression of the FC errors (FCE), which result for a specific core size [19]. However, one should not forget that additional FCEs arise at the correlated level.

The most complex systems to model by ECPs are lanthanides and actinides, since their valence shells have three different main quantum numbers n, i.e., $(n-2)$f, $(n-1)$d, and ns, and thus three radial density maxima at different distances from the nucleus [19]. In order to demonstrate the possibilities and difficulties in choosing the ECP core for f elements, three different core definitions for Ce will be discussed. Table 6.1 lists FCEs for configurations with fixed 4f occupations for 4, 12, and 30 valence electron systems of Ce taken from Dolg and Cao [41]. The FCEs correspond to relative energies from AE DHF calculations based on the Dirac–Coulomb (DC) Hamiltonian using the finite difference program GRASP [51] between several configurations with respect to the value for the Ce [Xe]$4f^1 5d^1 6s^2$ ground state configuration. Considered are 9, 12, and 6 configurations of the neutral atom and its cations for the fixed $4f^1$, $4f^0$, and $4f^2$ occupations, respectively, including excitations and ionization from the 5d and 6s shell.

From a chemical point of view, the valence electron system of Ce consists of four electrons ($4f^1 5d^1 6s^2$), which corresponds to a core–valence separation according to orbital energies. However, FCEs of up to 5.037 eV arise using this core definition. Especially, if the 4f occupation number (with respect to the ground state) changes, the FCEs become significantly large. This is due to the fact that the 5s and 5p semicore orbitals are more diffuse than the compact 4f shell, and thus they experience a significant change of the effective nuclear charge, if the 4f occupation changes [41]. The associated relaxation of the 5s and 5p orbitals is not possible, if they are included in the ECP core, and this leads to large FCEs. Furthermore, a weak dependence on the 5d occupation is observed, because this shell still has a noticeable radial overlap with the 5s and 5p semicore orbitals [41]. Therefore, a much better choice is to include the 5s and 5p shells in the valence space leading to ECPs with 12 valence electrons. In this case the dependence of the FCEs on the 5d occupation is negligible (at most 0.001 eV), whereas a noticeable, however, systematic dependence on the 4f occupation is still present. These findings can be explained by the radial overlap between the 4f valence and the 4s, 4p, and 4d core shells [41]. The best choice with respect to the FCEs (at most

0.002 eV) is a small Ce^{30+} core, which treats all shells with $n \geq 4$ explicitly. It is therefore strongly advocated to separate the ECP core using spatial rather than energetic criteria.

Table 6.1 also shows that the range of the FCEs for a fixed 4f occupation, i.e., the FCEs for configurations with the same 4f occupation, are relatively small ($4f^0$: 0.102 eV, $4f^2$: 0.043 eV), if a 12 valence electron system is considered. This was the motivation to derive PPs for f elements including the open f shell with a fixed occupation number into the core. Thus, for each valency, corresponding to a fixed f occupation, a f-in-core LPP was adjusted (see Section 6.3.1).

6.3 Pseudopotential Adjustment

Among the PPs one distinguishes energy- and shape-consistent PPs. Although the former are simultaneously adjusted to total AE valence energies of all chemical important configurations of the neutral atom and its low-charged ions, the latter rely on quantities defined within an effective one-electron picture, i.e., orbital shapes in the spatial valence region and orbital energies, of one specific reference state [19]. Thus, the energy-consistent PPs better account for the transferability of the core orbitals, i.e., the FC approximation (see Section 6.2.2). Both methods will be described in the following two sections.

6.3.1 Energy-Consistent Pseudopotentials

The method of energy-consistent PPs was developed during the last three decades from the semiempirical PP approach, where the free parameters were optimized to fit the experimental atomic spectrum analogous to the original work of Hellmann [25]. However, such a semiempirical adjustment is restricted to single valence electron systems, where valence correlation is absent and the Schrödinger equation can be solved almost exactly using finite difference methods or very large basis sets [19]. The problem for many-electron systems is that the occurring valence correlation cannot be described sufficiently accurate within the PP parametrization [19]. Since especially for f elements it is important to choose a medium or small core due to the FC approximation (see Section 6.2.3), the semiempirical adjustment is not appropriate to produce accurate and transferable PPs for these elements [19]. Furthermore, for f elements, especially actinides, the required accurate experimental atomic reference data as excitation energies and IPs are often missing or incomplete. However, the idea to fit PPs to such quantum mechanical observables, instead of relying on quantities defined in an approximate one-particle picture as orbital shapes and energies, was adopted in the ab initio framework of fitting energy-consistent PPs [52]. In contrast to semiempirical energy-adjusted PPs in the case of ab initio PPs, the term energy-consistent was introduced [19].

The energy adjustment consists of three steps. First, the reference configurations, *LS* states, or *J* levels *I* of the neutral atom and its low-charged ions are chosen and their total energies E_I^{AE} are determined using an AE method. Next the AE total valence energies $E_I^{AE,V}$ are calculated by subtracting the AE core energy from the total energies E_I^{AE}. Finally, the free parameters of the PP, i.e., the coefficients and exponents of the Gaussians in equation 6.11, are adjusted by a least-squares fit to the total valence energies of the reference [52, 53]

$$\sum_I \omega_\mathrm{I} \left(E_\mathrm{I}^\mathrm{PP} - E_\mathrm{I}^\mathrm{AE,V} \right)^2 := \min. \tag{6.16}$$

Here, E_I^PP denotes the total valence energy for the Ith configuration/LS state/J level from the PP calculation using the same coupling scheme and correlation treatment as for the calculation of the AE total valence energy $E_\mathrm{I}^\mathrm{AE,V}$. The weight factors ω_I are typically chosen to be equal for all nonrelativistic configurations independent from the number of the underlying LS states or J levels [19].

The available energy-consistent PPs and corresponding valence basis sets (see Section 6.4) for lanthanides and actinides are listed in Table 6.2 and are also compiled on the Cologne PP website [54] and partially on the Environmental Molecular Sciences Laboratory (EMSL)

Table 6.2 *Energy-consistent relativistic PPs and valence basis sets for lanthanides and actinides*

Elements	Method	Core	n_c	n_v	Basis sets	Ref.
La–Lu	WB	1s–3d	28	29–43	(14s13p10d8f6g)/[6s6p5d4f3g], gen., ANO p-VQZ reducible to p-VTZ and p-VDZ	[21,58]
					(14s13p10d8f6g)/[10s8p5d4f3g], seg., ANO	[59]
	WB	1s–4f^n	46–60	11	(7s6p5d)/[5s4p3d], seg.	[7,8]
					(8s7p6d)/[6s5p5d,5s4p4d], (7s6p5d)/[6s5p4d,5s4p4d], seg.; sol.	[60]
					3f2g	[44]
La–Yb	WB	1s–4f^{n+1}	47–60	10	(7s6p5d)/[5s4p3d], seg.	[7,8]
Ce–Nd	WB	1s–4f^{n-1}	46–48	12	(6s5p4d,7s6p5d)/p-VXZ, X = D, T, Q, seg.; +2f1g; sol.	[9]
Ac–Lr	WB	1s–4f	60	29–43	(12s11p10d8f)/[8s7p6d4f], seg.	[22]
					(14s13p10d8f6g)/[6s6p5d4f3g], gen., ANO p-VQZ reducible to p-VTZ and p-VDZ	[61]
					(14s13p10d8f6g)/[10s9p5d4f3g], seg., ANO	[62]
Ac–Lr	WB	1s–5f^n	78–92	11	(6s5p4d,7s6p5d,8s7p6d)/p-VXZ, X = D, T, Q, seg.; +2f1g; sol.	[10]
Pu–No	WB	1s–5f^{n+1}	84–92	10	(6s5p4d,7s6p5d)/p-VXZ, X = D, T, Q, seg.; +2f1g; sol.	[11]
Th–Cf	WB	1s–5f^{n-1}	78–86	12	(6s5p4d,7s6p5d)/p-VXZ, X = D, T, Q, seg.; +2f1g; sol.	[11]
Pa–Am	WB	1s–5f^{n-2}	78–82	13	(6s5p4d,7s6p5d)/p-VXZ, X = D, T, Q, seg.; +2f1g; sol.	[12]
U–Am	WB	1s–5f^{n-3}	78–81	14	(6s5p4d,7s6p5d)/p-VXZ, X = D, T, Q, seg.; +2f1g; sol.	[12]
Ac–U	DHF/ DC+B	1s–4f	60	29–43	(14s13p10d8f6g)/[6s6p5d4f3g], gen., ANO p-VQZ reducible to p-VTZ and p-VDZ	[23,24]

Listed are the elements, for which the PPs are available, the method used to calculate the AE reference data, the chosen core, the number of core electrons n_c, the number of valence electrons n_v, the available basis sets, and the references for both PPs and basis sets. In the case of the basis set description, the following abbreviations are used: generalized contracted (gen.), atomic natural orbital (ANO), polarized valence double-, triple-, and quadruple-zeta quality (p-VXZ with X = D, T, Q), segmented contracted (seg.), and subsets of the basis sets are applicable to solid state calculations (sol.).

database [55]. Although the scalar-relativistic PPs relying on AE WB reference data have been derived using equation 6.16, for the recently optimized relativistic PPs using AE MCDHF reference data based on the DC Hamiltonian perturbatively including the Breit interaction (DC+B) an additional adjustable parameter $\Delta E_{\rm shift}$ was used

$$\sum_{I} \omega_{\rm I} \left(E_{\rm I}^{\rm PP} - E_{\rm I}^{\rm AE,V} + \Delta E_{\rm shift} \right)^2 := \min. \tag{6.17}$$

The global valence energy shift $\Delta E_{\rm shift}$ was originally introduced in the case of the PP adjustment for 3d transition metals [56]. Typically this shift amounts to less than 1% of the ground state total valence energy, but it can improve the accuracy of the parametrization by one or two orders of magnitude. Whereas the restriction to $\Delta E_{\rm shift} = 0$ in the original method (see equation 6.16) implied that, e.g., the ground state valence energy equals the sum of all IPs leading from the neutral atom to the core system, this is not the case for the new fitting procedure. Here only the sum of all IPs leading from the neutral atom to the most highly ionized system included in the reference data is reproduced correctly. The shift can also be viewed as a shift of the AE core energy. Since the bare core position relative to the valence states is not expected to be overly relevant for chemical processes, this shift changing the reference energies can be justified [57]. Moreover, it is obvious that the quantities of interest as the electron affinity, IPs, and excitation energies, i.e., all possible energy differences between configurations included as references, remain unchanged [1]. Furthermore, thanks to this energy shift the adjustment to higher ionized states even with holes in core and semicore orbitals becomes possible.

In the case of the scalar-relativistic WB PPs, one distinguishes between f-in-valence SPPs [21, 22] and f-in-core LPPs [7–12]. The LPPs include the open f shell in the PP core and thus avoid its difficult correlation treatment. Since the occupation of the f shell depends on the oxidation state of the lanthanide/actinide, f-in-core LPPs are adjusted for fixed f occupations, i.e., they are available for di- ($n_{\rm v} = 10$), tri- ($n_{\rm v} = 11$), and tetravalent ($n_{\rm v} = 12$) lanthanides as well as di- ($n_{\rm v} = 10$), tri- ($n_{\rm v} = 11$), tetra- ($n_{\rm v} = 12$), penta- ($n_{\rm v} = 13$), and hexavalent ($n_{\rm v} = 14$) actinides (see Table 6.2). Certainly, the large size of the core is a crude approximation connected with a loss of accuracy, but the lanthanide LPPs were quite successfully applied during the last 24 years [41, 63–65]. The actinide LPPs have been developed nearly 20 years later due to the well-known occasional stronger involvement of the 5f orbitals in chemical bonding. However, they also yield satisfactory results if they are applied to molecules, where the 5f orbitals do not significantly contribute to chemical bonding. If this assumption is valid, it should be explicitly tested, e.g., in single-point HF calculations with an explicit treatment of the 5f shell using the corresponding SPP. If the LPP and SPP 5f occupations obtained from Mulliken population analyses differ by less than 0.5 electrons, the application of the LPP should yield reasonable results. In the case of actinide fluorides, the 5f occupations differ by at most 0.1 electrons for AnF_2/AnF_3/AnF_4 (except for ThF_3/PaF_3, where the trivalent oxidation state is not preferred/unstable), 0.3 electrons for AnF_5, and 0.9 electrons for AnF_6 [12], which explains the good results using the di-, tri-, and tetravalent, the acceptable results using the pentavalent, and the bad results using the hexavalent LPPs. Thus, the 5f-in-core approximation reaches its limit for the hexavalent oxidation state and therefore these LPPs should only be used to preoptimize structures. The larger deviations in the 5f occupations of AnF_5 and AnF_6

compared to those of $AnF_2/AnF_3/AnF_4$ are due to the fact that the higher the assumed oxidation state, the less probable it is, since ionization energies increase with increasing positive charge [12].

The originally adjusted 4f-in-core LPPs for di- and trivalent lanthanides [7] did not allow for any 4f orbital occupation. This leads to some systematic errors, especially at the beginning of the lanthanide series, where the 4f orbitals are still diffuse and their participation in chemical bonding is not completely negligible. Therefore, in 1993 the f part of these LPPs was modified to allow for 4f orbital contribution in chemical bonding [8], i.e., the coefficients and exponents of the f potentials were readjusted to reference configurations, where the 4f occupation (in the valence space) amounts to one. These improved f parts better describe partial 4f occupations due to some mixing of the 4f shell into the molecular orbitals, i.e., the 4f occupation number q modeled with the LPPs might be $m \leq q < m+1$, where m corresponds to the fixed 4f occupation of the PP core. In the case of the 5f-in-core LPPs for actinides, the f part was modified once again [10]. Here, the f potentials V_f consist of two types of potentials V_1 and V_2, which are linear combined as follows

$$V_f = \left(1 - \frac{m}{14}\right) V_1 + \frac{m}{14} V_2. \tag{6.18}$$

Here, m is the number of electrons in the 5f orbitals and V_1 and V_2 model 5f shells, which can and respectively cannot accommodate an additional electron. In the case of the trivalent LPPs, V_1 is the exact potential for Ac ($5f^0$, $m = 0$) and was adjusted to the four reference configurations An^{10+} $5f^{m+1}$, $5f^m6f^1$, $5f^m7f^1$, and $5f^m8f^1$, whereas V_2 is exact for Lr ($5f^{14}$, $m = 14$) and was only adjusted to the last three of these reference configurations. Thus, the new f potentials allow for a decreasing participation of the 5f orbitals in chemical bonding along the actinide series, i.e., the 5f shell can completely, partially, and not at all contribute to chemical bonding for Ac, Th–No, and Lr, respectively.

In contrast to the LPPs for the more rigorous SPPs also SO operators [58, 61] are available. However, it has to be noted that these SO operators are effective valence SO operators, i.e., they have to be applied in SO–CI calculations for the valence electrons (Ln: 4f/5d/6p; An: 5f/6d/7p), where the semicore shells (Ln: 4p/4d/5p; An: 5p/5d/6p) are frozen in their scalar-relativistic form, or a corresponding perturbative treatment. If the SO contributions of the semicore shells cannot be neglected, the recently developed MCDHF/DC+B SPPs [23, 24] suitable for a variational two-component treatment or SO–CI calculations including excitations from semicore shells should be applied. However, these SPPs are still under construction and only available for Ac–U so far.

6.3.2 Shape-Consistent Pseudopotentials

Shape-consistent PPs are adjusted to AE valence orbitals φ_v and orbital energies ϵ_v [66, 67], i.e., the pseudovalence orbital φ_p is required to retain the correct radial distribution of charge given by the φ_v in the valence region outside a critical radius r_c and the corresponding PP pseudovalence orbital energy ϵ_p is set equal to the ϵ_v. For a given lj one requires that

$$\varphi_{p,lj}(r) = \begin{cases} \varphi_{v,lj}(r) & \text{for} \quad r \geq r_c \\ f_{lj}(r) & \text{for} \quad r < r_c \end{cases} \qquad \text{and} \qquad \epsilon_{p,lj} = \epsilon_{v,lj}. \tag{6.19}$$

The auxiliary function f_{lj} is required to be radially nodeless and smooth in the core region ($r < r_{\mathrm{c}}$). Except for the normalization and continuity conditions for the $\varphi_{\mathrm{p,lj}}(r)$, the choice of r_{c} as well as the choice of f_{lj} is within certain limits arbitrary and a matter of experience.

With a given $\varphi_{\mathrm{p,lj}}$ and $\epsilon_{\mathrm{v,lj}}$, the pseudopotential $V_{\mathrm{lj}}^{\mathrm{PP}}(r)$ can be determined by the radial Fock equation

$$\left[-\frac{1}{2}\frac{\mathrm{d}^2}{\mathrm{d}r^2} + \frac{l(l+1)}{2r^2} + V_{\mathrm{lj}}^{\mathrm{PP}}(r) + \hat{W}_{\mathrm{p,lj}}(\{\varphi_{\mathrm{p',l'j'}}\})\right] |\varphi_{\mathrm{p,lj}}(r)\rangle = \epsilon_{\mathrm{v,lj}} |\varphi_{\mathrm{p,lj}}(r)\rangle. \quad (6.20)$$

The first two terms in the parentheses are the radial kinetic energy operator and the term $\hat{W}_{\mathrm{p,lj}}$ stands for an effective valence Coulomb and exchange potential for $\varphi_{\mathrm{p,lj}}$. Relativistic effects are implicitly included in $V_{\mathrm{lj}}^{\mathrm{PP}}$, since the AE reference calculation explicitly describes these effects. Repeating this procedure for each *lj*-set, the resulting potentials $V_{\mathrm{lj}}^{\mathrm{PP}}$ are tabulated on a grid and are usually fitted by means of a least-squares criterion to a linear combination of Gaussian functions according to

$$\hat{V}^{\mathrm{PP}}(\vec{r}_{\mathrm{i}}) = -\frac{Q}{r} + \sum_{lj}\left(\sum_{k} A_{\mathrm{lj,k}} r^{n_{\mathrm{lj,k}}-2} \mathrm{e}^{-\alpha_{\mathrm{lj,k}} r^2}\right) \hat{P}_{\mathrm{lj}}. \quad (6.21)$$

The separation into scalar-relativistic and SO potentials, $\hat{V}_{\mathrm{PP,SA}}$ and $\hat{V}_{\mathrm{PP,SO}}$, is the same as for energy-consistent PPs (see equation 6.8).

Concerning the auxiliary function f_{l} in equation 6.19, one choice used for a very popular set of PPs for main group and transition elements published by Hay and Wadt [68, 69] is

$$f_{\mathrm{l}}(r) = r^{\mathrm{b}}\left(a_0 + a_1 r + a_2 r^2 + a_3 r^3 + a_4 r^4\right) \quad (6.22)$$

with $b = l + 3$ in the nonrelativistic and $b = \lambda + 2$ in the relativistic case, where

$$\lambda + 1 = \frac{1}{2}(1 - \delta_{0,1}) + \sqrt{l(l+1) + \frac{1}{4}(1 + \delta_{1,0})^2 + (\alpha Z)^2} \quad (6.23)$$

and α is the fine structure constant. For relativistic s orbitals the choices of $b = \lambda + 3$ and f_0 as sixth degree polynomial have been found to lead to smoother s pseudovalence orbitals. The five coefficients a_{i} are determined by requiring that

1. $\varphi_{\mathrm{p}}(r)$ remains normalized.
2. $f_{\mathrm{l}}(r)$ and its first three derivatives match φ_{v} and its first three derivatives at r_{c}.

For subsequent usage in molecular valence-only calculations compact valence basis sets were generated, i.e., the pseudovalence orbitals were fitted by using a nonlinear least-squares procedure, similar to the one for fitting the potentials, to a linear combination of Gaussian functions

$$\varphi_{\mathrm{p,l}} = \sum_{i} C_{\mathrm{i}} r^{\mathrm{l}} \mathrm{e}^{-\alpha_{\mathrm{i}} r^2}. \quad (6.24)$$

A closer look at equation 6.20 will tell us about the singularity problem in the PP for a radial node of the pseudovalence orbital φ_{p}. Most shape-consistent PPs are therefore derived for

Table 6.3 *Shape-consistent relativistic PPs and valence basis sets for lanthanides and actinides*

Elements	Method	Core	n_c	n_v	Basis set	Ref.
La	DHF/DC	1s–4d, $5s^2 5p^6$	54	3	(3s3p4d)	[70, 71]
	DHF/DC	1s–4d	46	11	(9sp5d)/[3sp3d]	[72]
	CG	1s–4d, $5s^2 5p^6$	54	3	(3s3p3d)	[68]
	CG	1s–4d	46	11	(5s5p3d)	[69]
Ce–Lu	DHF/DC	1s–4d, $5s^2 5p^6$	54	4–17	(6s6p6d6f)	[73]
	DHF/DC	1s–4d	46	12–25	(6sp3d7f)/[4sp2d2f]	[74]
Ac–Pu	DHF/DC	1s–5d	78	11–16	(5s5p4d4f)	[75]
Am–Lr	DHF/DC	1s–5d	78	17–25	(6s7p6d5f)	[76, 77]
U–Pu	CG	1s–5d	78	14–16		[78, 79]

Listed are the elements for which the PPs are available, the method used to calculate the AE reference data, the chosen core, the number of core electrons n_c, the number of valence electrons n_v, the basis set, and the references for both PPs and basis sets. The abbreviation CG stands for the Cowan–Griffin method.

positive ions, which are chosen in such a way that this problem does not occur. For lanthanides and actinides one or more pseudovalence orbitals may have a radial node when small cores are used. Therefore, only large-core shape-consistent PPs are available for lanthanides and actinides (see Table 6.3).

On the other hand a small core has to be used in case of lanthanides and actinides for accurate calculations. A possible solution to the singularity problem for the small-core shape-consistent PPs is attempted in the so-called generalized relativistic effective core potential (GRECP) approach of Titov, Mosyagin, and coworkers [80–83], where the PP is interpolated in a vicinity of the pseudospinor node [81]. If more than one pseudospinor per *lj* is available, more than one radial PP for this *lj* combination can be derived. The GRECP approach therefore employs the idea of separating the orbital space of a heavy atom into three regions: inner core, outer core, and valence, which are treated differently. The idea to partition into these three regions appeared already in 1985 in the work of Andzelm *et al.* [84] on the MPs of Huzinaga type, however the approach is currently not further pursued.

The GRECP operator to be used in atomic calculations is written in the form [81]

$$
\begin{aligned}
\Delta V_{\mathrm{GRECP}} = V_{\mathrm{n_v,LJ}}(r) + \sum_{l=0}^{L} \sum_{j=|l-1/2|}^{l+1/2} \Bigg\{ & \left[V_{\mathrm{n_v,lj}}(r) - V_{\mathrm{n_v,LJ}}(r) \right] \hat{P}_{\mathrm{lj}} \\
& + \sum_{n_{oc}} \left[V_{\mathrm{n_{oc},lj}}(r) - V_{\mathrm{n_v,lj}}(r) \right] \hat{\tilde{P}}_{\mathrm{n_{oc},lj}} \\
& + \hat{\tilde{P}}_{\mathrm{n_{oc},lj}} \sum_{\mathrm{n_{oc}}} \left[V_{\mathrm{n_{oc},lj}}(r) - V_{\mathrm{n_v,lj}}(r) \right] \\
& - \sum_{n_{oc},n'_{oc}} \hat{\tilde{P}}_{\mathrm{n_{oc},lj}} \left[\frac{V_{\mathrm{n_{oc},lj}} + V_{\mathrm{n'_{oc},lj}}}{2} - V_{\mathrm{n_v,lj}}(r) \right] \hat{\tilde{P}}_{\mathrm{n'_{oc},lj}} \Bigg\}
\end{aligned}
\tag{6.25}
$$

Here, n_v and n_{oc} are the principle quantum numbers for valence and outer core orbitals, respectively. The maximum L and J quantum numbers are related by $J = L + 1/2$. $\hat{P}_{\mathrm{lj}}$ is the projector on the spinor spherical harmonics $|ljm\rangle$

$$\hat{P}_{\mathrm{lj}} = \sum_{m=-j}^{m=j} |ljm\rangle\langle ljm|, \tag{6.26}$$

whereas $\hat{\hat{P}}_{\mathrm{n_{oc},lj}}$ is the projector on the outer core pseudospinors

$$\hat{\hat{P}}_{\mathrm{n_{oc},lj}} = \sum_{m=-j}^{m=j} |\widetilde{n_{\mathrm{oc}}, ljm}\rangle\langle \widetilde{n_{\mathrm{oc}}, ljm}|. \tag{6.27}$$

Obviously, besides the standard semilocal form of the PPs, additional nonlocal terms are used in GRECPs to take into account the difference between the effective potentials acting on the outer core and valence electrons with the same l and j quantum numbers.

The GRECP operator is often rewritten as a sum over spin-free (averaged) and spin-dependent parts [81] for application in molecular calculations. The parametrized form of the GRECPs contains a relatively large number of parameters, which easily can exceed 100 for heavier elements. Parametrizations are available on the website of the Titov group [85] for 18 atoms including one lanthanide and four actinide atoms, i.e., Yb, Th, U, Pu, and Am. Since the GRECP ansatz is at present not supported by most of the standard quantum chemistry codes, applications of GRECPs are still scarce and were performed mainly by the authors of the method [86–91].

6.4 Valence Basis Sets for Pseudopotentials

For the accuracy of PP calculations the applied valence basis sets are as important as the PPs themselves. Therefore, these basis sets should be optimized very carefully. Furthermore, only a valence basis set corresponding to the PP under consideration will provide a reliable description of the pseudovalence orbitals, which exhibit different radial shapes in the spatial core region for different PPs [19]. Thus, even valence basis sets corresponding to PPs for the same element and with the same core size are not transferable. If one needs a more extended basis set than provided, the original basis set can just be augmented by adding diffuse functions.

Tables 6.2 and 6.3 list all valence basis sets corresponding to the available energy- and shape-consistent PPs, respectively. For the energy-consistent PPs often more than one basis set is available, since these PPs have been supplemented by several high quality valence basis sets in addition to the basis sets published together with the PP. For example in the case of the actinide SPPs, in additional to the original (12s11p10d8f)/[8s7p6d4f] basis sets [22], in 2003 (14s13p10d8f6g)/[6s6p5d4f3g] atomic natural orbital (ANO) basis sets were adjusted [61], because the former basis sets were not completely satisfactory, e.g., they were only adjusted to one configuration, therefore they are not flexible enough to describe states with different 5f occupations. Moreover, in the case of the lanthanide LPPs for trivalent oxidation states, in 2005 several valence basis sets applicable to solid state calculations were adjusted [60], i.e., in these basis sets the most diffuse exponents for each angular symmetry were fixed to 0.15 in order to avoid linear dependencies in solid state calculations.

6.5 Selected Applications

Due to the singularity problem of shape-consistent SPPs (see Section 6.3.2) only shape-consistent LPPs are available for f elements, leading to the limited usage of shape-consistent PPs. Compared to this the energy-consistent PPs, where both large-core and small-core options are available for lanthanides and actinides, have been widely used in quantum chemistry calculations.

In case of the energy-consistent f-in-valence SPPs [21, 22] the most widely used sets are adjusted to AE WB reference data [92]. The recently developed highly accurate MCDHF/DC+B adjusted SPPs for Ac, Th, Pa, and U [23, 24] have only been applied in a few test calculations so far. For example, multireference averaged coupled-pair functional (MRACPF) calculations of the first to fourth IPs agree very well with corresponding AE calculations using the second-order Douglas–Kroll–Hess Hamiltonian and the Breit–Pauli SO operator in first-order perturbation theory (DKH2+BP) [2–5], i.e., the mean absolute deviations (m.a.d.) are all below 0.05 eV (see Tables 6.4 and 6.5). Compared to the available experimental data [95–102] the mean absolute errors (m.a.e.) of the PP MRACPF results amount to 0.07, 0.24, and 0.58 eV for IP_1–IP_3, respectively. For IP_4 of U the calculated results (PP MRACPF: 32.62 eV; AE DKH2+BP MRACPF: 32.61 eV) are much lower than the experimental value (36.70±0.99 eV). Therefore, for IP_4 the m.a.e. of the calculated results are up to 2.23 eV. However, the very large deviations of the PP and AE calculated values most likely cannot be completely attributed to the deficiencies of the calculations, but rather suggest that the experimental value [100] might be somewhat too high. More rigorous SPP Fock-space coupled-cluster (FSCC) calculations are in excellent agreement with AE FSCC results [93, 94], i.e., the m.a.d. amount at most to 0.03 eV, and compared to the experimental data the disagreements are less than 0.15 eV. Moreover, some applications by using the MCDHF/DC+B adjusted PPs to UH [23], U^{4+}/U^{5+} [103], and UO_2^{2+} halide complexes [104] have been published.

Table 6.4 *Calculated IP_1 and IP_2 of Ac, Th, Pa, and U (in eV)*

	IP_1					IP_2				
	PP		AE			PP		AE		
An	ACPF	FSCC	DKH2	FSCC	exp.	ACPF	FSCC	DKH2	FSCC	exp.
Ac	5.38	5.30	5.44	5.32	5.38	11.59	11.87	11.59	11.90	11.75 ±0.03
Th	6.29		6.27		6.31	12.41		12.46		
Pa	6.12		6.10		5.89 ±0.12	11.63		11.64		
U	6.17		6.17		6.19	11.89		11.94		11.59 ±0.37
m.a.e.	0.07		0.08			0.24		0.26		
m.a.d.	0.03					0.03				

Energy-consistent MCDHF/DC+B adjusted PP MRACPF and FSCC calculations [24] in comparison to AE results [24, 93] and experimental data [95–100]. Basis sets: PP MRACPF and AE DKH2+BP MRACPF extrapolated basis sets, PP FSCC (16s15p12d10f8g7h7i), AE FSCC (35s26p21d16f10g6h5i). m.a.e. and m.a.d. refer to experimental and AE DKH2+BP MRACPF values, respectively.

Table 6.5 *Calculated IP_3 and IP_4 of Ac, Th, Pa, and U (in eV)*

	IP_3					IP_4				
	PP		AE			PP		AE		
An	ACPF	FSCC	DKH2	FSCC	exp.	ACPF	FSCC	DKH2	FSCC	exp.
Ac	17.29	17.51	17.28	17.51		44.52	44.98	42.64	44.99	
Th	18.11		18.18		18.33 ±0.05	28.27	28.66	28.28	28.65	28.65 ±0.02
Pa	18.64		18.70			30.59		30.57		
U	18.86		18.88		19.80 ±0.25	32.62		32.61		36.70 ±0.99
m.a.e.	0.58		0.54			2.23		2.23		
m.a.d.	0.04					0.01				

Analogous to Table 6.4. AE FSCC basis sets (35s26p21d16f10g6h5i) for Ac IP_3 [93] and (35s30p25d20f11g9h9i7k7l) for IP_4 of Ac and Th [94]. Experimental data taken from [100–102]. The m.a.d. for IP_4 excludes Ac, where the application of the BP Hamiltonian leads to much too large SO contributions.

In the case of energy-consistent LPPs, calibration studies have been published for lanthanide tri- and tetrafluorides [9, 44] and actinide di-, tri-, tetra-, penta-, and hexafluorides [10–12, 44], respectively. For trifluorides comparing to SPP state-averaged multiconfiguration self-consistent field (MCSCF) bond lengths, bond angles, and ionic binding energies LPP+CPP HF results deviate only by 0.005/0.01 Å (0.2/0.5%), 0.2/2.1° (0.2/1.9%), and 0.14/0.24 eV (0.3/0.5%) for lanthanides/actinides, respectively. Furthermore, the comparison of LPP+CPP coupled cluster calculations with single and double excitation operators and a perturbative estimate of triple excitations [CCSD(T)] to AE or SPP data using various correlation methods show an excellent agreement [44].

6.5.1 DFT Calculated M–X (M = Ln, An; X = O, S, I) Bond Lengths

Although the ab initio PPs were adjusted to wavefunction-based AE HF or DHF reference data, they also perform quite well when combined with various DFT methods. By applying different density functionals together with PPs to actinoids Averkiev *et al.* [105] have found that MPW3LYP, B3LYP, M06, and M05 produced as accurate or even more accurate bond energies and IPs than AE complete active space second-order perturbation theory (CASPT2), CCSD, and CCSD(T) methods. Kovács *et al.* [106] evaluated trends in the bond distances and dissociation enthalpies of AnO and AnO_2 (An = Th–Lr) at the DFT level and demonstrated the very good performance of the B3LYP exchange–correlation functional in conjunction with SPPs. Pereira *et al.* [107] published SPP DFT studies on neutral and monocationic AnS (An = Ac–Cm), where it was found that both B3LYP and MPW1PW91 produced ionization energies in good agreement with the experimental values, i.e., the mean difference amounts to 22 kJ/mol with the larger deviations occurring for Np (65 kJ/mol) and Cm (35 kJ/mol). For ThS and CmS, DFT yields binding energies, which are within 15 kJ/mol of the SPP CCSD(T) values.

However, special care should be taken for the calculated molecular geometries at the DFT level. It is well known that the crystal and ionic radii of Eu^{III} (1.09, 0.95 Å) and Am^{III}

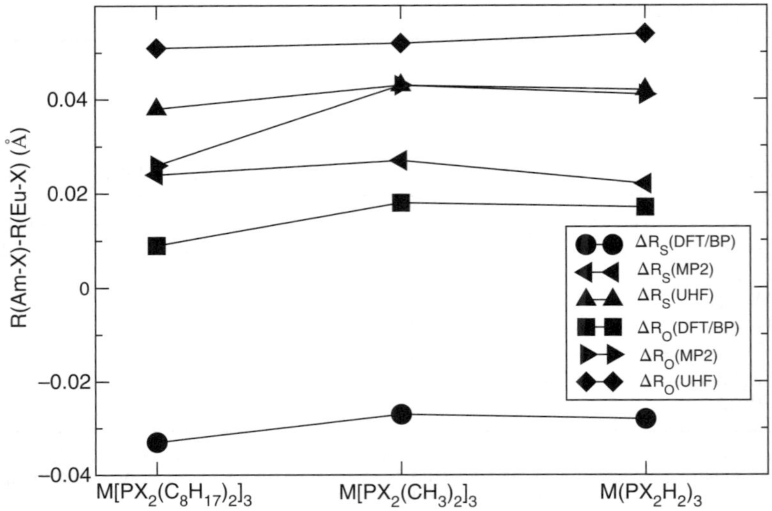

Figure 6.1 *Averaged M–X (M = Eu^{III}, Am^{III}; X = S, O) bond length differences obtained by using SPPs [109]*

(1.12, 0.98 Å) for sixfold coordination [108] are about 0.03 Å larger for Am^{III} than for Eu^{III}. A very similar finding with a 0.02 Å difference holds for eightfold coordination [108]. It might thus be expected that ionic compounds of Am^{III} exhibit slightly longer metal–ligand distances than corresponding Eu^{III} systems. A deviation from this pattern, i.e., shorter bonds in Am^{III} than in Eu^{III} complexes, might be interpreted to be related to a higher covalency in the former systems. Figure 6.1 shows the averaged M–S and M–O (M = Eu^{III}, Am^{III}) bond length differences for selected complexes, where M is sixfold coordinated, obtained by using SPPs combined with DFT/BP, unrestricted Hartree–Fock (UHF), and second-order Møller–Plesset perturbation theory (MP2) methods, respectively [109]. In contrast to the ab initio data as well as the data for M–O bond distances, the DFT/BP calculated Am^{III}–S bonds are shorter than the Eu^{III}–S bonds. The shorter Am^{III}–S bonds at the DFT level are also reported by Manna and Ghanty [110] for trivalent lanthanides with 1,10-phenanthroline-2,9-dicarboxylic acid based ligands. Dolg *et al.* found that the Eu f population is significantly higher at the DFT/BP level than at the UHF or MP2 level, which leads to the longer Eu^{III}–S bonds [109]. Moreover, they compared bond distances of EuI_3 and AmI_3 obtained with SPPs as well as LPPs at the DFT/BP level to those from UHF, MP2, CCSD(T), and multireference configuration interaction (MRCI) calculations (see Table 6.6). Obviously the Eu^{III}–I bond distance evaluated at the SPP DFT/BP level is by more than 0.1 Å longer than the recommended experimental value [111, 112] and the correlated ab initio results. As a consequence the order of the bond lengths is different at the DFT and ab initio level when SPPs are applied, i.e., at the SPP DFT level Eu^{III}–I bonds are longer than Am^{III}–I bonds. It should be noted that this finding is not due to a PP defect, i.e., very similar results are obtained in relativistic AE calculations using DKH2 Hamiltonian [113]. Therefore, Dolg *et al.* concluded that the shorter Am^{III}–S (or I) compared to the Eu^{III}–S (or I) bonds should not be interpreted as a higher covalency in the former systems, but rather as a consequence

Table 6.6 *Averaged M–I (M = Eu^{III}, Am^{III}) bond distances (in Å) [109]*

	DFT/BP	MRCI	CCSD(T)	MP2	UHF
EuI_3	2.968/2.880	2.861/2.891	2.841/2.854	2.822/2.885(2.847[a])	2.915/2.926
AmI_3	2.929/2.931	2.910/2.948	2.883/2.909	2.862/2.937(2.904[a])	2.961/2.983
Δ	−0.039/0.051	0.049/0.057	0.042/0.055	0.040/0.052(0.057[a])	0.046/0.057

The first and second values refer to SPPs and LPPs, respectively. The recommended experimental value for EuI_3 is 2.831 ± 0.022 Å [111, 112]. [a]Extended basis sets with up to g functions were used for Eu, Am, and I.

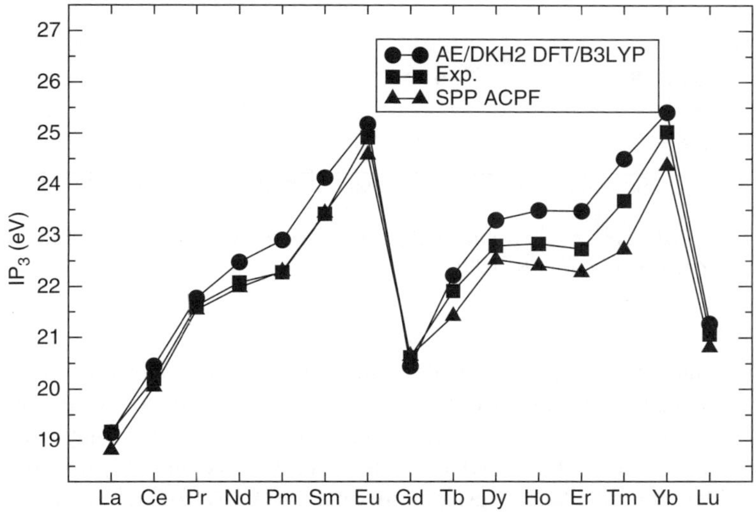

Figure 6.2 *IP_3 for lanthanides calculated at the AE/DKH2 DFT/B3LYP [119] and the SPP ACPF [115] level in comparison to experimental results [116]*

of the so-called delocalization error of many commonly used density functionals [114] as well as the underestimation of the repulsive interaction of electrons with equal spins within the f shell (no DFT functional adjusted to f elements) [109].

The too-strong binding of the electrons in the f shell of Eu compared to that of Am at the DFT level is also obvious from results for the atomic third IPs, i.e., the energy difference between M^{2+} f^7 $^8S_{1/2}$ and M^{3+} f^6 ^7F (M = Eu, Am). Using f-in-valence SPPs, the DFT/BP calculated IP_3 for Eu (26.00 eV) is 1.42 and 1.08 eV larger than the basis set extrapolated result of complete active space self-consistent field (CASSCF) with subsequent ACPF calculations (24.58 eV) [115] and the experimental value (24.92 eV) [116], respectively. For Am the DFT/BP calculated IP_3 (22.03 eV) is 0.31 eV lower than the ACPF result (22.34 eV) [115]. A substantial improvement of the DFT result for Eu was obtained when a self-interaction correction was included [117, 118]. Moreover, Pantazis *et al.* [119] published IP_3 values for lanthanides at the AE DFT/B3LYP level using the DKH2 Hamiltonian. The obtained results are significantly larger than the experimental data [116] except for La and Gd, where the f occupation is not changed when one electron is removed from La^{2+} and Gd^{2+} (see Figure 6.2). A very similar overestimation is observed when compared to

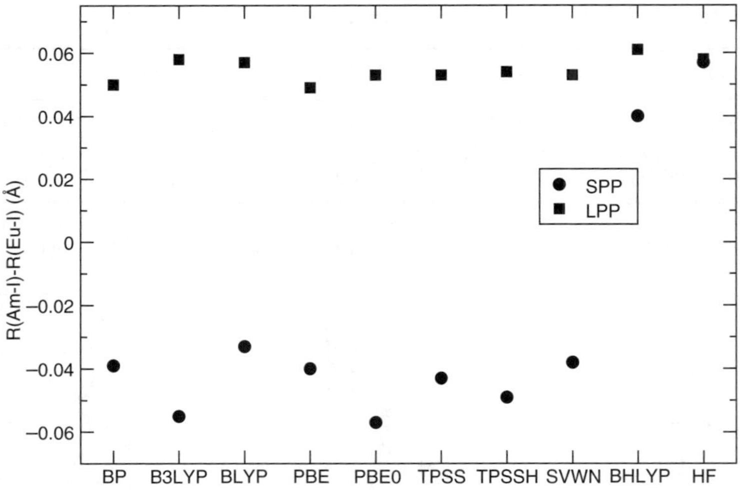

Figure 6.3 *M–I (M = Eu^{III}, Am^{III}) bond length differences for MI_3 from SPP and LPP calculations using different density functionals [109]*

SPP ACPF results [115]. It is worthwhile to mention that the longer Eu^{III}–I compared to the Am^{III}–I bond distance caused by DFT is avoided by using f-in-core LPPs (see Table 6.6 and Figure 6.3). It can be seen from Figure 6.3 that when the SPP is applied, except for BHLYP with its 50% admixture of HF exchange, all selected commonly used density functionals, i.e., SVWN for the local density approximation (LDA), BP, BLYP, and PBE for the generalized gradient approximation (GGA), BHLYP, B3LYP, and PBE0 for hybrid functionals as well as TPSS and TPSSH for meta-GGA and hybrid-meta-GGA functionals, failed to reproduce the correct order of the Eu^{III}–I and Am^{III}–I bond distances. In contrast to the SPP values all LPP values show the same order of the Eu^{III}–I and Am^{III}–I bond distances in agreement with the ab initio HF results. It is noteworthy that the HF results for the LPP and SPP treatment are very close to each other.

Figure 6.4 shows the obtained M–O (M = Ln, An) bond distances for lanthanide [120] and actinide [106] monoxides at the SPP DFT/B3LYP level. It was found that due to the different singly occupied orbitals for Eu and Am, i.e., seven Eu 4f orbitals [121] and six Am 5f orbitals as well as the Am 7s orbital [122], the Eu–O bond is significantly longer than the Am–O bond.

6.5.2 Lanthanide(III) and Actinide(III) Hydration

The lanthanide(III) hydration was studied at the DFT and MP2 level combined with LPPs by Ciupka and coworkers [123] as well as at the SPP DFT level by Kuta and Clark [124]. It was found that both SPP and LPP produced very similar Ln–ligand distances in good agreement with experimental evidence [125] (see Table 6.7). Moreover, the DFT calculated Ln^{III}–O bond lengths are systematically longer than the results obtained from the LPP MP2 calculations and extended X-ray absorption fine structure (EXAFS), i.e., comparing to EXAFS results the m.a.d. for SPP DFT/B3LYP, LPP DFT/B3LYP, and LPP DFT/BP86 is 0.040,

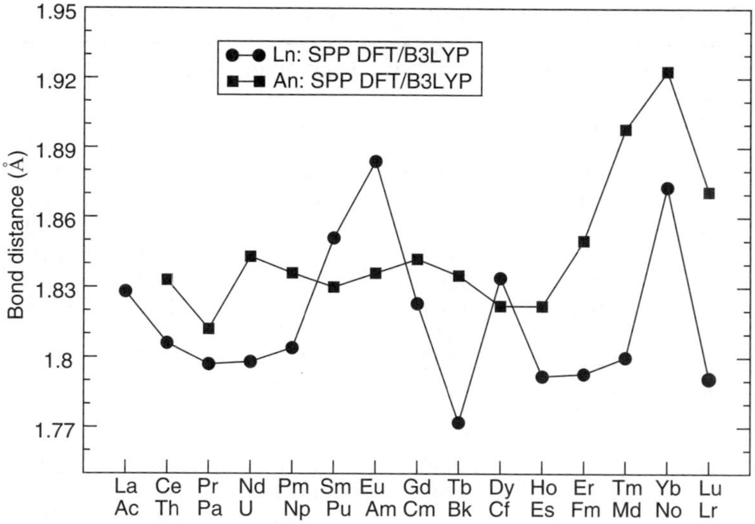

Figure 6.4 *Calculated bond distances for LnO (Ln = La–Lu) [120] and AnO (An = Ac–Lr) [106]*

Table 6.7 *Mean Ln^{III}–O distances (in Å) for $[Ln^{III}(H_2O)_{CN}]^{3+}$*

Ln^{III}	CN	[a]SPP B3LYP	[b]LPP B3LYP	[b]LPP BP86	[b]LPP MP2	[c]EXAFS
La	9	2.62	2.626	2.618	2.591	2.600
Ce	9	2.59	2.606	2.598	2.571	2.570
Pr	9	2.57	2.587	2.581	2.551	2.550
Nd	9	2.55	2.570	2.564	2.535	2.525
Pm	9	2.54	2.554	2.549	2.518	
Sm	9	2.52	2.539	2.535	2.503	2.490
Eu	8	2.47	2.484	2.477	2.455	
	9	2.51	2.523	2.521	2.488	2.470
Gd	8	2.45	2.471	2.463	2.441	
	9	2.50	2.510	2.508	2.475	2.455
Tb	8	2.45	2.456	2.449	2.427	
	9	2.50	2.496	2.495	2.462	2.440
Dy	8	2.43	2.443	2.436	2.414	
	9		2.483	2.484	2.449	2.425
Ho	8	2.41	2.430	2.423	2.401	
	9		2.470	2.472	2.436	2.405 (8.9)
Er	8	2.40	2.418	2.411	2.389	
	9	2.45	2.458	2.461	2.424	2.390 (8.9)
Tm	8	2.39	2.406	2.400	2.378	
	9	2.44	2.447	2.451	2.413	2.375 (8.8)
Yb	8	2.38	2.395	2.388	2.366	
	9	2.43	2.435	2.441	2.402	2.360 (8.7)
Lu	8	2.37	2.387	2.381	2.359	2.345 (8.2)
	9	2.42	2.427	2.435	2.395	
m.a.d.		0.040	0.053	0.051	0.020	

Experimentally determined coordination numbers are given in parentheses in the last column. [a]Data taken from [124]; [b]data taken from [123]; [c]data taken from [125].

Table 6.8 *Gibbs free energies of hydration $-\Delta G_H^0$ (in kJ/mol) for [$Ln^{III}(H_2O)_{CN}]^{3+}$*

Ln^{III}	CN	[a]SPP B3LYP	[b]LPP B3LYP	[b]LPP BP86	[b]LPP MP2	[c]Exp.	[d]Exp.
La	9	3205	3105	3088	3136	3061	3145
Ce	9	3253	3135	3117	3165	3112	3200
Pr	9	3321	3167	3147	3196	3156	3245
Nd	9	3340	3196	3176	3225	3183	3280
Pm	9	3378	3225	3203	3253	3210	3250
Sm	9	3376	3252	3230	3281	3228	3325
Eu	8	3365	3291	3273	3307		
	9	3385	3281	3258	3309	3279	3360
Gd	8	3446	3315	3296	3331		
	9	3454	3302	3280	3331	3292	3375
Tb	8	3423	3342	3322	3356		
	9	3441	3327	3303	3355	3331	3400
Dy	8	3456	3367	3346	3382		
	9		3350	3326	3379	3337	3425
Ho	8	3511	3392	3371	3407		
	9		3373	3349	3403	3382	3470
Er	8	3498	3417	3395	3432		
	9	3522	3397	3373	3426	3404	3495
Tm	8	3542	3441	3418	3457		
	9	3556	3418	3394	3448	3431	3515
Yb	8	3569	3466	3443	3482		
	9	3566	3442	3417	3471	3473	3570
Lu	8	3588	3489	3465	3502	3488	3515
	9	3592	3463	3438	3490		
m.a.d.		92	34	56	15		

[a]Data taken from [124]; [b]data taken from [123]; [c]data taken from [126]; [d]data taken from [127]. The m.a.d. are calculated according to the averaged values of both experimental results. For Dy and Ho the ΔG_H^0 for CN = 8 are used to determine the m.a.d. for SPP DFT/B3LYP. For all others the ΔG_H^0 with respect to the CN determined by EXAFS are used (see Table 6.7).

0.053, and 0.051 Å, respectively. The m.a.d. is reduced to 0.020 Å when the MP2 method is applied.

Table 6.8 lists the calculated Gibbs free energies of hydration ΔG_H^0 for lanthanide(III) ions. The LPP MP2 calculated results agree very well with the experimental data [126, 127], i.e., the m.a.d. is only 15 kJ/mol. Among all applied DFT methods the B3LYP functional combined with LPPs produced the best results compared to the experimental data, i.e., the m.a.d. are 92, 34, and 56 kJ/mol for SPP DFT/B3LYP, LPP DFT/B3LYP, and LPP DFT/BP86, respectively. A closer look at the results calculated at the LPP MP2 level shows that the coordination numbers (CN) of water to Ln^{III} affect the ΔG_H^0 slightly for Eu, Gd, Tb, Dy, Ho, and Er, i.e., the differences of obtained ΔG_H^0 for CN = 8, 9 are at most 6 kJ/mol. For Tm, Yb, and Lu the differences are increased up to 12 kJ/mol.

Compared to lanthanides(III) the actinide(III) hydration has not received enough attention of both experimentalists and theoreticians. The calculated An^{III}–O (An = Ac–Lr) bond lengths for $[An^{III}(H_2O)_{CN}]^{3+}$ (CN = 8, 9) [128, 129] as well as available experimental data [130–136] are shown in Table 6.9. In analogy with the Ln^{III}–O the calculated An^{III}–O bond distances are found to decrease almost linearly with respect to the nuclear charge,

Table 6.9 *Mean An^{III}–O distances (in Å) for $[An^{III}(H_2O)_{CN}]^{3+}$*

An^{III}	CN	[a]SPP B3LYP	[b]LPP BP86	LPP MP2	Exp.
Ac	9		2.699		
Th	9		2.682		
Pa	9		2.656		
U	9		2.636		[c]2.52 (9.1)
Np	9		2.617		[c]2.50 (9),[c]2.52 (10)
Pu	9		2.599		[c]2.49 (10),[d]2.51 (9)
Am	9		2.582	2.557	[d]2.48 (10)
Cm	9	2.537	2.566	2.542	[c]2.45 (10),[e]2.47 (9) [f]2.48 (8.5)
Bk	9		2.551		[g]2.43 (9)
Cf	8		2.495		
	9		2.537		[h]2.42 (8.5)
Es	8		2.482		
	9		2.524		
Fm	8		2.469		
	9		2.513		
Md	8		2.455		
	9		2.502		
No	8		2.443		
	9		2.490		
Lr	8		2.431		
	9		2.478		

Experimentally determined coordination numbers are given in parentheses in the last column.
[a]Data taken from [128]; [b]data taken from [129]; [c]data taken from [130]; [d]data taken from [131]; [e]data taken from [132]; [f]data taken from [133]; [g]data taken from [134]; [h]data taken from [135] and [136].

e.g., the slope/correlation coefficients obtained from the linear fitting of LPP DFT/BP86 data are −0.013/0.9941 and −0.016/0.9927 for lanthanides and actinides, respectively. By applying linear fitting to the ionic radii of Ln^{III} and An^{III} reported by Shannon [108], a very similar fitting quality for M^{III}–O (M = Ln, An) bonds has been obtained, i.e., the slope/correlation coefficients are −0.012/0.9956 and −0.017/0.956 for Ln^{III} and An^{III}, respectively. Therefore, a similar accuracy of the LPP DFT/BP86 calculated Ln^{III}–O and An^{III}–O bond lengths has been expected. However, compared to the available experimental data for U, Np, Pu, Am, Cm, Bk, and Cf, the theoretical An^{III}–O bond lengths are about 0.1 Å longer, i.e., the deviations are about 0.05 Å larger than the m.a.d. for the Ln^{III}–O bond lengths (see Tables 6.7 and 6.9).

Using DFT and MP2 combined with LPPs, the Gibbs free energies of hydration for actinide(III) ions have been calculated by Wiebke and coworkers [129] (see Table 6.10). It can be seen that the single-point corrected LPP MP2 results are about 152 kJ/mol lower than the LPP DFT/BP86 results. There are only two experimental values available, i.e., −3205 kJ/mol for U and −3235 kJ/mol for Pu [137]. Compared to LPP MP2 results, the experimental data are by 41 and 6 kJ/mol lower for U and Pu, respectively. The differences between the ΔG^0_H obtained from a model by David and Vokhmin [138] and the DFT calculated results are very significant, i.e., the deviations amount up to 338 (Lr) and 238 kJ/mol

Table 6.10 Gibbs free energies of hydration $-\Delta G^0_H$ (in kJ/mol) for $[An^{III}(H_2O)_{CN}]^{3+}$

An^{III}	CN	^{a}Model	^{b}LPP BP86	^{b}LPP MP2	^{c}LPP MP2	^{d}Exp.
Ac	9	2806	2897	3044		
Th	9	2885	2938	3088		
Pa	9	2921	2971	3124		
U	9	3045	3008	3164		3205
Np	9	3073	3043	3202		
Pu	9	3134	3072	3229		3235
Am	9	3159	3100	3252	3207	
Cm	9	3225	3133	3284	3235	
Bk	9	3340	3159	3310		
Cf	8		3218	3334		
	9	3444	3187	3338		
Es	8		3243	3360		
	9	3498	3212	3363		
Fm	8		3267	3384		
	9	3546	3234	3386		
Md	8		3294	3411		
	9	3585	3256	3408		
No	8		3318	3436		
	9	3623	3279	3431		
Lr	8		3343	3461		
	9	3644	3306	3459		

^{a}Data taken from [138]; ^{b}data taken from [129], corrections of −106 kJ/mol for DFT/BP86 and −179.9 kJ/mol for MP2 from cluster cycle [123] added; ^{c}geometries taken from the fifth column of Table 6.9, single-point energies corrected by using aug-cc-pVQZ basis sets for O and H; ^{d}data taken from [137].

(Ac) for LPP DFT/BP86 and LPP MP2, respectively. We notice that the accuracy of the model of David and Vokhmin depends on the experimental data, e.g., ionic radii and An^{III}–O distances.

6.5.3 Lanthanide(III) and Actinide(III) Separation

The chemical separation of lanthanides and actinides is a very difficult and long-standing problem, e.g., for the work-up of spent nuclear fuel and nuclear waste [139]. More than 15 years ago it was experimentally found by Zhu *et al.* [140] that purified Cyanex301 (sulfur-donor ligand), which contains as a main component bis(2,4,4-trimethylpentyl)dithiophosphinic acid (HBTMPDTP), exhibits a very high separation factor ($\approx$5900) between Am^{III} and Eu^{III} for the liquid–liquid extraction from a slightly acidic aqueous solution (pH $\approx$ 3–4) to kerosene as organic phase. Besides sulfur-donor ligands there are nitrogen- and oxygen-donor ligands used in lanthanide and actinide extraction processes. The extraction efficiency and Am^{III} selectivity are ordered as sulfur- > nitrogen- > oxygen-donor ligands. Due to the large size of the extraction complexes, the theoretical investigations have used exclusively DFT so far [141, 142] and have been in addition sometimes restricted to truncated model systems [143].

In 2010 one of the present authors published a LPP DFT/BP86 study on the separation of Am^{III}/Cm^{III} from Eu^{III} with Cyanex301 containing mainly HBTMPDTP (denoted as HL)

[141]. It was shown that the neutral complexes ML_3, where L acts as a bidentate ligand and the metal cation is coordinated by six sulfur atoms, are most likely the most stable extraction complexes. For explaining the high selectivity of Cyanex301 for Am^{III}/Cm^{III} over Eu^{III} the change of the Gibbs free energy ΔG_e of the following extraction equation was calculated

$$M^{3+}_{aq} + 3HL_{org} \xrightarrow{\Delta G_e} ML_{3,org} + 3H^+_{aq}. \tag{6.28}$$

In equation 6.28 the initial state supposes an aqueous ion with the HL residing in the organic phase, while in the final state the ion is complexed by three L^- anions in the organic phase after release of three protons. The obtained ΔG_e of the extraction reactions agree with the thermodynamical priority for Am^{3+} and Cm^{3+} [141]. It was found that the Gibbs free energies of hydration for M^{3+} play an important role for the high selectivity of Cyanex301 for Am^{III}/Cm^{III} over Eu^{III}, i.e., in the gas phase the calculated values of ΔG_g showed a thermodynamical priority for Eu^{3+}, whereas this priority is reverted for the aqueous solution. By using Gibbs free energies of hydration for Eu^{3+} and Am^{3+} calculated at the MP2 level (see Tables 6.8 and 6.10) ΔG_e of 68.1 and 46.5 kJ/mol were obtained for Eu^{3+} and Am^{3+}, respectively, in good agreement with the experimental findings [140] (Eu: 63.3 kJ/mol; Am: 44.1 kJ/mol). Using SPP DFT Keith and Batista [142] performed a detailed thermodynamic examination of the selective extraction of Am^{3+} from Eu^{3+} by two CF_3^- substituted diaryldithiophosphinic acids. They found that in order to obtain reasonable values of ΔG_e, it is crucial to accurately describe the Gibbs free energies of hydration for M^{3+}. The extraction factor is primarily due to the binding free energies of the ligands to the metals and is not dependent on side reactions or complicated solvent effects.

Using DFT combined with SPPs, Lan and coworkers [144] studied the complexation of Am^{III} and Eu^{III} with tetradentate nitrogen-donor ligands, i.e., 6,6'-bis(5,6-dialkyl-1,2,4-triazin-3-yl)-2,2'-bipyridines (BTBPs). The structures and stabilities of the inner-sphere BTBPs complexes were explored in the presence of various counterions such as NO_3^-, Cl^-, and ClO_4^-. They found that the changes of Gibbs free energies play an important role for Am^{III}/Eu^{III} separation. Moreover, the reactions $M(NO_3)_3(H_2O)_4 \rightarrow ML(NO_3)_3$ and $[M(NO_3)(H_2O)_7]^{2+} \rightarrow [ML_2(NO_3)]^{2+}$ are found to be probably the dominant ones in the Am^{III}/Eu^{III} separation process [145]. Roy *et al.* [146] studied the structures and stabilities of the aqueous phase complexes $[M^{III}(DTPA)\text{-}H_2O]^{2-}$ (M = Nd, Am) as well as the changes in the Gibbs free energy for complexation in the gas phase and aqueous solution through SPP DFT calculations, where DTPA denotes diethylenetriamine-N,N,N',N",N"-pentaacetic acid. The calculated changes of the Gibbs free energy favor the formation of $[Am^{III}(DTPA)\text{-}H_2O]^{2-}$ over $[Nd^{III}(DTPA)\text{-}H_2O]^{2-}$. Furthermore, all bonding analyses show the important role of the electrostatic and covalent interactions of the oxygen atoms with the nitrogen chelates providing an additional, yet small, covalent interaction.

Manna and Ghanty [50] investigated the complexation behavior of preorganized 1,10-phenanthroline-2,9-dicarboxylic acid (PDA)-based ligands, monothiodicarboxylic acids (TCA/TCA1), and dithiodicarboxylic acid (THIO) with lanthanide(III) and actinide(III) ions using DFT combined with SPPs. It was shown that the complexation energy in terms of selectivity for actinides over lanthanides is at most with TCA1, where the metal is complexed by the oxygen atoms. Within the framework of the Pearson's Hard-Soft-Acid-Base (HSAB) principle it was proven that the presence of softer nitrogen atoms in the phenanthroline moiety, which also act as donors to the metal ion, has a profound influence in changing the soft nature of the actinide ions, which in turn bind with the hard oxygen atoms in a stronger way as the valence isoelectronic lanthanide ions.

6.6 Conclusions and Outlook

Modern relativistic pseudopotentials are invaluable tools for accurate quantum chemical calculations of f element systems, i.e., by far more than 1,000 investigations were performed using pseudopotentials of each of the most popular sets [19]. However, the pseudopotential approximation should not be applied to investigate properties, where the polarized core-density, the nodal structure of the valence orbitals, core–valence correlation, or core-overlap effects become important [6]. If the pseudopotential core is chosen appropriately small to minimize the frozen-core and other errors, both shape-consistent and energy-consistent pseudopotentials are able to compete in accuracy with all-electron approaches for structural and electronic valence properties and save at the same time computational effort. Therefore, the pseudopotential method is essential except for atoms and diatomic molecules, where highly accurate all-electron correlation techniques can achieve spectroscopic accuracy. The main field of future applications will be large molecules containing several heavy atoms, e.g., clusters or infinite systems as surfaces or the solid state. However, even if pseudopotentials are applied, quantum chemical investigations of large systems such as the extraction complexes used for the lanthanide and actinide separation often have to be carried out by means of density functional theory or the calculations have to be restricted to truncated model systems. Thus, also the construction of large-core pseudopotentials with corresponding core-polarization potentials is an important goal.

Today the main emphasis seems to be the adjustment of highly accurate small-core pseudopotentials with rigorous all-electron calculations as reference data, e.g., Dirac–Hartree–Fock calculations based on the Dirac–Coulomb(–Breit) Hamiltonian or Hartree–Fock calculations using the Douglas–Kroll–Hess Hamiltonian. Sometimes even quantum electrodynamic effects and the finite nucleus are taken into account, which cannot routinely be included using all-electron methods. Since the inclusion of spin–orbit contributions at the self-consistent field level will become more and more standard, reliable two-component instead of scalar-relativistic one-component pseudopotentials are now primarily developed. For example, in the energy-consistent approach, new small-core two-component pseudopotentials are adjusted to multiconfiguration Dirac–Hartree–Fock finite nucleus reference data based on the Dirac–Coulomb–Breit Hamiltonian. Although such two-component pseudopotentials are already available for all main group, transition metal, and even superheavy elements, they have only been parametrized for Ac, Th, Pa, and U in the case of the f elements so far. Thus, one future project is to complete this kind of pseudopotentials for lanthanides and actinides.

References

[1] X. Cao and M. Dolg, Relativistic pseudopotentials, in M. Barysz and Y. Ishikawa (Eds.), *Relativistic methods for chemists. Challenges and advances in computational physics*, vol. 10, Springer, Berlin, pp. 215–278 (2010).

[2] M. Douglas and N. M. Kroll, Quantum electrodynamical corrections to the fine structure of helium, *Ann. Phys.*, **82**, 89–155 (1974).
[3] B. A. Hess, Relativistic electronic-structure calculations employing a two-component no-pair formalism with external-field projection operators, *Phys. Rev. A*, **33**, 3742–3748 (1986).
[4] G. Jansen and B. A. Hess, Revision of the Douglas-Kroll transformation, *Phys. Rev. A*, **39**, 6016–6017 (1989).
[5] A. Wolf, M. Reiher, and B. A. Hess, The generalized Douglas-Kroll transformation, *J. Chem. Phys.*, **117**, 9215–9226 (2002).
[6] P. Schwerdtfeger, The pseudopotential approximation in electronic structure theory, *Chem. Phys. Chem.*, **12**(17), 3143–3155 (2011).
[7] M. Dolg, H. Stoll, A. Savin, and H. Preuss, Energy-adjusted pseudopotentials for the rare earth elements, *Theor. Chim. Acta.*, **75**, 173–194 (1989).
[8] M. Dolg, H. Stoll, and H. Preuss, A combination of quasirelativistic pseudopotentials and ligand field calculations for lanthanoid compounds, *Theor. Chim. Acta*, **85**, 441–450 (1993).
[9] M. Hülsen, A. Weigand, and M. Dolg, Quasirelativistic energy-consistent 4f-in-core pseudopotentials for tetravalent lanthanide elements, *Theor. Chem. Acc.*, **122**, 23–29 (2009).
[10] A. Moritz, X. Cao, and M. Dolg, Quasirelativistic energy-consistent 5f-in-core pseudopotentials for trivalent actinide elements, *Theor. Chem. Acc.*, **117**, 473–481 (2007).
[11] A. Moritz, X. Cao, and M. Dolg, Quasirelativistic energy-consistent 5f-in-core pseudopotentials for divalent and tetravalent actinide elements, *Theor. Chem. Acc.*, **118**, 845–854 (2007).
[12] A. Moritz and M. Dolg, Quasirelativistic energy-consistent 5f-in-core pseudopotentials for pentavalent and hexavalent actinide elements, *Theor. Chem. Acc.*, **121**, 297–306 (2008).
[13] http://www.msg.chem.iastate.edu/gamess (accessed Oct. 19, 2013).
[14] http://www.gaussian.com (accessed Oct. 19, 2013).
[15] http://www.molcas.org (accessed Oct. 19, 2013).
[16] http://www.molpro.net (accessed Oct. 19, 2013).
[17] http://www.turbomole.com (accessed Oct. 19, 2013).
[18] http://www.crystal.unito.it (accessed Oct. 19, 2013).
[19] M. Dolg and X. Cao, Relativistic pseudopotentials: their development and scope of applications, *Chem. Rev.*, **112**, 403–480 (2012).
[20] M. Dolg, Relativistic effective core potentials, in P. Schwerdtfeger (Ed.), *Relativistic electronic structure theory, part 1: fundamentals*, vol. 11, Elsevier, Amsterdam, pp. 793–862 (2002).
[21] M. Dolg, H. Stoll, and H. Preuss, Energy-adjusted ab initio pseudopotentials for the rare earth elements, *J. Chem. Phys.*, **90**, 1730–1734 (1989).
[22] W. Küchle, M. Dolg, H. Stoll, and H. Preuss, Energy-adjusted pseudopotentials for the actinides. Parameter sets and test calculations for thorium and thorium monoxide, *J. Chem. Phys.*, **100**(10), 7535–7542 (1994).
[23] M. Dolg and X. Cao, Accurate relativistic small-core pseudopotentials for actinides. Energy-adjustment for uranium and first applications to uranium hydride. *J. Phys. Chem. A*, **113**, 12573–12581 (2009).
[24] A. Weigand, X. Cao, T. Hangele, and M. Dolg, Relativistic small-core pseudopotentials for actinium, thorium, and protactinium, *J. Phys. Chem. A*, **118**, 2519–2530 (2014).
[25] H. Hellmann, A new approximation method in the problem of many electrons, *J. Chem. Phys.*, **3**, 61 (1935).
[26] J. C. Phillips and L. Kleinman, New method for calculating wave functions in crystals and molecules, *Phys. Rev.*, **116**(2), 287–294 (1959).
[27] J. D. Weeks, A. Hazi, and S. A. Rice, On the use of pseudopotentials in the quantum theory of atoms and molecules, *Adv. Chem. Phys.*, **16**, 283–342 (1969).
[28] J. D. Weeks and S. A. Rice, Use of pseudopotentials in atomic-structure calculations, *J. Chem. Phys.*, **49**, 2741–2755 (1968).
[29] V. Heine and I. V. Abarenkov, *Philos. Mag.*, New method for electronic structure of metals, **9**(99), 451–460 (1964).
[30] I. V. Abarenkov and V. Heine, The model potential for positive ions, *Philos. Mag.*, **12**(117), 529–537 (1965).

[31] W. H. E. Schwarz, Hellmann's pseudopotential method. 1. Theoretical basis, *Theor. Chim. Acta*, **11**(4), 307–324 (1968).
[32] L. R. Kahn and W. A. Goddard, Ab initio effective potentials for use in molecular calculations, *J. Chem. Phys.*, **65**, 3826–3853 (1976).
[33] Y. S. Lee, W. C. Ermler, and K. S. Pitzer, Ab initio effective core potentials including relativistic effects. I. Formalism and applications to the Xe and Au atoms, *J. Chem. Phys.*, **67**, 5861–5876 (1977).
[34] W. C. Ermler, Y. S. Lee, P. A. Christiansen, and K. S. Pitzer, Ab initio effective core potentials including relativistic effects. A procedure for the inclusion of spin-orbit coupling in molecular wavefunctions,*Chem. Phys. Lett.*, **81**, 70–74 (1981).
[35] L. R. Kahn, P. Baybutti, and D. G. Truhlar, Ab initio effective core potentials: Reduction of all-electron molecular structure calculations to calculations involving only valence electrons, *J. Chem. Phys.*, **65**(10), 3826–3853 (1976).
[36] V. Fock, M. Vesselov, and M. Petrashen, *Zh. Eksp. Teor. Fiz.*, **10**, 723 (1940).
[37] S. Huzinaga, Effective Hamiltonian method for molecules, *J. Mol. Struct.* (Theochem), **80**, 51–73 (1991).
[38] H. Partridge, in P. v. Ragué Schleyer *et al.* (Eds.), *Encyclopedia of computational chemistry*, vol. 1, Wiley & Sons, Chichester, pp. 581–591 (1998).
[39] W. Müller, J. Flesch, and W. Meyer, Treatment of intershell correlation effects in ab initio calculations by use of core polarization potentials. Method and application to alkali and alkaline earth atoms, *J. Chem. Phys.*, **80**(7), 3297–3310 (1984).
[40] W. Müller and W. Meyer, Ground-state properties of alkali dimers and their cations (including the elements Li, Na, K) from ab initio calculations with effective core polarization potentials, *J. Chem. Phys.*, **80**(7), 3311–3320 (1984).
[41] M. Dolg and X. Cao, The relativistic energy-consistent ab initio pseudopotential approach and its application to lanthanide and actinide compounds, in K. Hirao and Y. Ishikawa (Eds.), *Recent advances in computational chemistry*, vol. 6, World Scientific, New Jersey, pp. 1–35 (2004).
[42] P. Fuentealba, H. Preuss, H. Stoll, and L. v. Szentpály, A proper account of core-polarization with pseudopotentials: single valence-electron alkali compounds, *Chem. Phys. Lett.*, **89**, 418–422 (1982).
[43] Y. Wang and M. Dolg, Pseudopotential study of the ground and excited states of Yb_2, *Theor. Chem. Acc.*, **100**, 124–133 (1998).
[44] A. Weigand, X. Cao, J. Yang, and M. Dolg, Quasirelativistic f-in-core pseudopotentials and core-polarization potentials for trivalent actinides and lanthanides: molecular test for trifluorides, *Theor. Chem. Acc.*, **126**, 117–127 (2010).
[45] B. Pittel and W. H. E. Schwarz, Correlation energies from pseudopotential calculations, *Chem. Phys. Lett.*, **46**, 121–124 (1977).
[46] C. Teichteil, J. P. Malrieu, and J. C. Barthelat, Non-empirical pseudopotentials for molecular calculations. 2. Basis set extensions and correlation effects on X_2 molecules (X = F, Cl, Br, I), *Mol. Phys.*, **33**, 181–197 (1977).
[47] M. Dolg, On the accuracy of valence correlation energies in pseudopotential calculations, *J. Chem. Phys.*, **104**, 4061–4067 (1996).
[48] M. Dolg, Valence correlation energies from pseudopotential calculations, *Chem. Phys. Lett.*, **250**, 75–79 (1996).
[49] M. Klobukowski, Atomic correlation energies from effective-core-potential and model-potential calculations, *Chem. Phys. Lett.*, **172**, 361–366 (1990).
[50] L. Seijo, Z. Barandiarán, and S. Huzinaga, Atomic valence correlation energies from ab initio model potential calculations, *Chem. Phys. Lett.*, **192**, 217–220 (1992).
[51] G. K. Dyall, I. P. Grant, C. T. Johnson, F. A. Parpia, and E. P. Plummer, GRASP - a general-purpose relativistic atomic structure program, *Comput. Phys. Commun.*, **55**, 425–456 (1989); modified for pseudopotential calculations/optimizations by M. Dolg and B. Metz.
[52] M. Dolg, U. Wedig, H. Stoll, and H. Preuss, Energy-adjusted ab initio pseudopotentials for the first row transition elements, *J. Chem. Phys.*, **86**, 866–872 (1987).

[53] M. Dolg, H. Stoll, H. Preuss, and R. M. Pitzer, Relativistic and correlation effects for element 105 (Hahnium Ha). A comparative study of M and MO (M = Nb, Ta, Ha) using energy-adjusted ab initio pseudopotentials, *J. Phys. Chem.*, **97**, 5852–5859 (1993).
[54] http://www.tc.uni-koeln.de/PP/index.en.html (accessed Oct. 26, 2013).
[55] http://bse.pnl.gov/bse/portal (accessed Oct. 26, 2013).
[56] M. Dolg, Improved relativistic energy-consistent pseudopotentials for 3d-transition metals, *Theor. Chem. Acc.*, **114**, 297–304 (2005).
[57] K. A. Peterson, D. Figgen, M. Dolg, and H. Stoll, Energy-consistent relativistic pseudopotentials and correlation consistent basis sets for the 4d elements Y–Pd, *J. Chem. Phys.*, **126**, 124101-1–124101-12 (2007).
[58] X. Cao and M. Dolg, Valence basis sets for relativistic energy-consistent small-core lanthanide pseudopotentials, *J. Chem. Phys.*, **115**, 7348–7355 (2001).
[59] X. Cao and M. Dolg, Segmented contraction scheme for small-core lanthanide pseudopotential basis sets, *J. Mol. Struct.* (Theochem), **581**, 139–147 (2002).
[60] J. Yang and M. Dolg, Valence basis sets for lanthanide 4f-in-core pseudopotentials adapted for crystal orbital ab initio calculations, *Theor. Chem. Acc.*, **113**, 212–224 (2005).
[61] X. Cao, M. Dolg, and H. Stoll, Valence basis sets for relativistic energy-consistent small-core actinide pseudopotentials, *J. Chem. Phys.*, **118**, 487–496 (2003).
[62] X. Cao and M. Dolg, Segmented contraction scheme for small-core actinide pseudopotential basis sets, *J. Mol. Struct.* (Theochem), **673**, 203–209 (2004).
[63] M. Dolg and H. Stoll, Electronic structure calculations for molecules containing lanthanide atoms, in K. A. Gschneidner Jr. and L. Eyring (Eds), *Handbook on the physics and chemistry of rare earths*, vol. 22, Elsevier, Amsterdam, pp. 607–729 (1996).
[64] X. Cao and M. Dolg, Relativistic energy-consistent ab initio pseudopotentials as tools for quantum chemical investigations of actinide systems, *Coord. Chem. Rev.*, **250**, 900–912, (2006).
[65] M. Dolg and X. Cao, Computational methods: lanthanides and actinides, in E. I. Solomon, R. A. Scott, and R. B. King (Eds), *Computational inorganic and bioinorganic chemistry*, Wiley, Chichester, pp. 503–515 (2009).
[66] P. A. Christiansen, Y. S. Lee, and K. S. Pitzer, Improved ab initio effective core potentials for molecular calculations, *J. Chem. Phys.*, **71**, 4445–4450 (1979).
[67] P. Durand and J. C. Barthelat, A theoretical method to determine atomic pseudopotentials for electronic structure calculations of molecules and solids, *Theor. Chim. Acta*, **38**, 283–302 (1975).
[68] P. J. Hay and W. R. Wadt, Ab initio effective core potentials for molecular calculations. Potentials for the transition metal atoms Sc to Hg, *J. Chem. Phys.*, **82**, 270–283 (1985).
[69] P. J. Hay and W. R. Wadt, Ab initio effective core potentials for molecular calculations. Potentials for K to Au including the outermost core orbitals, *J. Chem. Phys.*, **82**, 299–310 (1985).
[70] R. B. Ross, J. M. Powers, T. Atashroo, W. C. Ermler, L. A. LaJohn, and P. A. Christiansen, Ab initio relativistic effective potentials with spin-orbit operators. IV. Cs through Rn, *J. Chem. Phys.*, **93**, 6654–6670 (1990).
[71] R. B. Ross, J. M. Powers, T. Atashroo, W. C. Ermler, L. A. LaJohn, and P. A. Christiansen, Erratum, *J. Chem. Phys.*, **101**, 10198–10199 (1994).
[72] W. J. Stevens, M. Krauss, H. Basch, and P. J. Jasien, Relativistic compact effective potentials and efficient, shared-exponent basis sets for the third-, fourth-, and fifth-row atoms, *Can. J. Chem.*, **70**, 612–630 (1992).
[73] R. B. Ross, S. Gayen, and W. C. Ermler, Ab initio relativistic effective potentials with spin-orbit operators. V. Ce through Lu, *J. Chem. Phys.*, **100**, 8145–8155 (1994).
[74] T. R. Cundari and W. J. Stevens, Effective core potential methods for the lanthanides, *J. Chem. Phys.*, **98**, 5555–5565 (1993).
[75] W. C. Ermler, R. B. Ross, and P. A. Christiansen, Ab initio relativistic effective potentials with spin-orbit operators. VI. Fr through Pu, *Int. J. Quant. Chem.*, **40**, 829–846 (1991).

[76] C. S. Nash, B. E. Bursten, and W. C. Ermler, Ab initio relativistic potentials with spin-orbit operators. VII. Am through element 118, *J. Chem. Phys.*, **106**, 5133–5142 (1997).
[77] C. S. Nash, B. E. Bursten, and W. C. Ermler, Erratum, *J. Chem. Phys.*, **111**, 2347–2347 (1999).
[78] P. J. Hay, Ab initio studies of excited-states of polyatomic-molecules including spin-orbit and multiplet effects – the electronic states of UF_6, *J. Chem. Phys.*, **79**, 5469–5482 (1983).
[79] P. J. Hay and R. L. Martin, Theoretical studies of the structures and vibrational frequencies of actinide compounds using relativistic effective core potentials with Hartree-Fock and density functional methods: UF_6, NpF_6, and PuF_6, *J. Chem. Phys.*, **109**, 3875–3881 (1998).
[80] N. S. Mosyagin, A. V. Titov, and Z. Latajka, Generalized relativistic effective core potential: Gaussian expansion of potentials and pseudospinors for atoms Hg through Rn, *Int. J. Quant. Chem.*, **63**, 1107–1122 (1997).
[81] A. V. Titov and N. S. Mosyagin, Generalized relativistic effective core potential: Theoretical grounds, *Int. J. Quant. Chem.*, **71**, 359–401 (1999).
[82] A. V. Titov and N. S. Mosyagin, The generalized relativistic effective core potential method: Theory and calculations, *Russ. J. Phys. Chem.*, **74**, S376–S387 (2000).
[83] A. V. Titov, N. S. Mosyagin, T. A. Isaev, and A. N. Petrov, Accuracy and efficiency of modern methods for electronic structure calculation on heavy- and superheavy-element compounds, *Phys. At. Nucl.*, **66**, 1152–1162 (2003).
[84] J. Andzelm, E. Radzio, Z. Barandiarán, and L. Seijo, New developments in the model potential method, *J. Chem. Phys.*, **83**, 4565–4572 (1985).
[85] http://www.qchem.pnpi.spb.ru/recp (accessed Aug. 28, 2013).
[86] N. S. Mosyagin, A. N. Petrov, and A. V. Titov, The effect of the iterative triple and quadruple cluster amplitudes on the adiabatic potential curve in the coupled cluster calculations of the ground electronic state of the Yb dimer, *Int. J. Quant. Chem.*, **111**, 3793–3798 (2011).
[87] A. N. Petrov, L. V. Skripnikov, A. V. Titov, and R. J. Mawhorter, Centrifugal correction to hyperfine structure constants in the ground state of lead monofluoride, *Phys. Rev. A*, **88**, 010501-1–010501-5 (2013).
[88] R. A. Evarestov, M. V. Losev, A. I. Panin, N. S. Mosyagin, and A. N. Petrov, Electronic structure of crystalline uranium nitride: LCAO DFT calculations, *Phys. Stat. Sol. B*, **245**, 114–122 (2008).
[89] A. N. Petrov, N. S. Mosyagin, A. V. Titov, A. V. Zaitsevskii, and E. A. Rykova, Ab initio study of Hg–Hg and E112–E112 van der Waals interactions, *Phys. At. Nucl.*, **72**, 396–400 (2009).
[90] A. V. Zaitsevskii, C. v. Wüllen, and A. V. Titov, Relativistic pseudopotential model for superheavy elements: applications to chemistry of eka-Hg and eka-Pb, *Rus. Chem. Rev.*, **78**, 1173–1181 (2009).
[91] A. Zaitsevskii, C. v. Wüllen, E. A. Rykova, and A. V. Titov, Two-component relativistic density functional theory modeling of the adsorption of element 114 (eka-lead) on gold, *Phys. Chem. Chem. Phys.*, **12**, 4152–4156 (2010).
[92] J. H. Wood and A. M. Boring, Improved Pauli Hamiltonian for local-potential problems, *Phys. Rev. B*, **18**, 2701–2711 (1978).
[93] E. Eliav, S. Shmulyian, U. Kaldor, and Y. Ishikawa, Transition energies of lanthanum, actinium, and eka-actinium (element 121), *J. Chem. Phys.*, **109**, 3954–3958 (1998).
[94] E. Eliav and U. Kaldor, Transition energies of Rn- and Fr-like actinide ions by relativistic intermediate Hamiltonian Fock-space coupled-cluster methods, *Chem. Phys.*, **392**, 78–82 (2012).
[95] J. Rossnagel, S. Raeder, A. Hakimi, R. Ferrer, N. Trautmann, and K. Wendt, Determination of the first ionization potential of actinium, *Phys. Rev. A*, **85**, 012525-1–012525-6 (2012).
[96] W. C. Martin, L. Hagan, J. Reader, and J. Sugar, Ground levels and ionization potentials for lanthanide and actinide atoms and ions, *J. Phys. Chem. Ref. Data*, **3**, 771–780 (1974).
[97] N. Erdmann, M. Nunnemann, K. Eberhardt, G. Herrmann, G. Huber, S. Koehler, J. V. Kratz, G. Passler, J. R. Peterson, N. Trautmann, and A. Waldek, Determination of the first ionization potential of nine actinide elements by resonance ionization mass spectroscopy (RIMS), *J. Alloys Compd.*, **271**, 837–840 (1998).
[98] J. Sugar, Ionization energies of the neutral actinides, *J. Chem. Phys.*, **59**, 788–791 (1973).

[99] A. Waldek, N. Erdmann, C. Gruening, G. Huber, P. Kunz, J. V. Kratz, J. Lassen, G. Passler, and N. Trautmann, RIMS measurements for the determination of the first ionization potential of the actinides actinium up to einsteinium, *AIP Conf. Proc.*, **584**, 219–224 (2001).
[100] J. Blaise and J. F. Wyart, Energy levels and atomic spectra of actinides, in *International tables of selected constants*, vol. 20, CNRS, Paris (1992).
[101] J. F. Wyart and V. Kaufman, Extended analysis of doubly ionized thorium (Th III), *Phys. Scr.*, **24**, 941–952 (1981).
[102] P. F. A. Klinkenberg, Spectral structure of trebly ionized thorium, Th IV, *Physica B+C*, **151**, 552–567 (1988).
[103] A. Weigand, X. Cao, V. Vallet, J.-P. Flament, and M. Dolg, Multi-configuration Dirac–Hartree–Fock adjusted energy-consistent pseudopotential for uranium: spin–orbit configuration interaction and Fock-space coupled-cluster study of U^{4+} and U^{5+}, *J. Phys. Chem. A*, **113**, 11509–11516 (2009).
[104] G. S. Groenewold, M. J. van Stipdonk, J. Oomens, W. A. de Jong, G. L. Gresham, and M. E. McIlwain, Vibrational spectra of discrete UO_2^{2+} halide complexes in the gas phase, *Int. J. Mass Spec.*, **297**, 67–75 (2010).
[105] B. B. Averkiev, M. Mantina, R. Valero, I. Infante, A. Kovacs, D. G. Truhlar, and L. Gagliardi, How accurate are electronic structure methods for actinoid chemistry?, *Theor. Chem. Acc.*, **129**, 657–666 (2011).
[106] A. Kovacs, P. Pogany, and R. J. M. Konings, Theoretical study of bond distances and dissociation energies of actinide oxides AnO and AnO_2, *Inorg. Chem.*, **51**, 4841–4849 (2012).
[107] C. C. L. Pereira, C. J. Marsden, J. Marcalo, and J. K. Gibson, Actinide sulfides in the gas phase: experimental and theoretical studies of the thermochemistry of AnS (An = Ac, Th, Pa, U, Np, Pu, Am and Cm), *Phys. Chem. Chem. Phys.*, **13**, 12940–12958 (2011).
[108] R. D. Shannon, Revised effective ionic-radii and systematic studies of interatomic distances in halides and chalcogenides, *Acta Cryst.*, **A32**, 751–767 (1976).
[109] M. Dolg, X. Cao, and J. Ciupka, Misleading evidence for covalent bonding from $Eu^{III}X$ and $Am^{III}X$ density functional theory bond lengths, *J. Electron Spectrosc. Relat. Phenom.*, **194**, 8–13 (2014).
[110] D. Manna and T. K. Ghanty, Complexation behavior of trivalent actinides and lanthanides with 1,10-phenanthroline-2,9-dicarboxylic acid based ligands: insight from density functional theory, *Phys. Chem. Chem. Phys.*, **14**, 11060–11069 (2012).
[111] M. Hargittai, The molecular-geometry of gas-phase metal-halides, *Coord. Chem. Rev.*, **91**, 35–88 (1988).
[112] A. Kovacs and R. J. M. Konings, Structure and vibrations of lanthanide trihalides: An assessment of experimental and theoretical data, *J. Phys. Chem. Ref. Data*, **33**, 377–404 (2004).
[113] M. Dolg, Segmented contracted Douglas-Kroll-Hess adapted basis sets for lanthanides, *J. Chem. Theory Comput.*, **7**, 3131–3142 (2011).
[114] A. J. Cohen, P. Mori-Sánchez, and W. Yang, Challenges for density functional theory, *Chem. Rev.*, **112**, 289–320 (2012).
[115] X. Cao and M. Dolg, Theoretical prediction of the second to fourth actinide ionization potentials, *Mol. Phys.*, **101**, 961–969 (2003).
[116] W. C., Martin, R. Zalubas, and L. Hagan, (1978) *Atomic energy levels – the rare earth elements*, NSRDS-NBS 60, Washington, DC.
[117] W. Liu and M. Dolg, Benchmark calculations for lanthanide atoms: Calibration of ab initio and density-functional methods, *Phys. Rev. A*, **57**, 1721–1728 (1998).
[118] W. Liu, W. Küchle, and M. Dolg, Ab initio pseudopotential and density-functional all-electron study of ionization and excitation energies of actinide atoms, *Phys. Rev. A*, **58**, 1103–1110 (1998).
[119] D. A. Pantazis and F. Neese, All-electron scalar relativistic basis sets for the lanthanides, *J. Chem. Theory Comput.*, **5**, 2229–2238 (2009).
[120] Z. J. Wu, W. Guan, J. Meng, and Z. M. Su, Density functional studies of diatomic LaO to LuO, *J. Cluster Science*, **18**, 444–458 (2007).

[121] M. Dolg, H. Stoll, and H. Preuss, Ab initio pseudopotential study of europium monoxide EuO: $^8\Sigma^-$ ground state and $^8\Sigma^-$ first excited state, *Chem. Phys.*, **148**, 219–227 (1990).
[122] A. Kovacs, R. J. M. Konings, J. Raab, and L. Gagliardi, A theoretical study of AmO_2 and CmO_n ($n = 1, 2$), *Phys. Chem. Chem. Phys.*, **10**, 1114–1117 (2008).
[123] J. Ciupka, X. Cao, J. Wiebke, and M. Dolg, Computational study of lanthanide(III) hydration, *Phys. Chem. Chem. Phys.*, **12**, 13215–13223 (2010).
[124] J. Kuta and A. E. Clark, Trends in aqueous hydration across the 4f period assessed by reliable computational methods, *Inorg. Chem.*, **49**, 7808–7817 (2010).
[125] P. D'Angelo, A. Zitolo, V. Migliorati, G. Chillemi, M. Duvail, P. Vitorge, S. Abadie, and R. Spezia, Revised ionic radii of lanthanoid(III) ions in aqueous solution, *Inorg. Chem.*, **10**, 4572–4579 (2011).
[126] F. David, V. Vokhmin, and G. Ionova, Water characteristics depend on the ionic environment. Thermodynamics and modelisation of the aquo ions, *J. Mol. Liq.*, **90**, 40–62 (2001).
[127] Y. Marcus, Thermodynamics of solvation of ions. Part 5.– Gibbs free energy of hydration at 298.15 K, *J. Chem. Soc. Faraday Trans.*, **87**, 2995–2999 (1991).
[128] T. Yang and B. E. Bursten, Speciation of the curium(III) ion in aqueous solution: a combined study by quantum chemistry and molecular dynamics simulation, *Inorg. Chem.*, **45**, 5291–5301 (2006).
[129] J. Wiebke, A. Moritz, X. Cao, and M. Dolg, Approaching actinide (+III) hydration from first principles, *Phys. Chem. Chem. Phys.*, **9**, 459–465 (2007).
[130] B. Brendebach, N. Banik, C. M. Marquardt, J. Rothe, M. Denecke, and H. Geckeis, X-ray absorption spectroscopic study of trivalent and tetravalent actinides in solution at varying pH values, *Radiochim. Acta*, **97**, 701–708 (2009).
[131] P. Allen, J. Bucher, D. K. Shuh, N. M. Edelstein, and I. Craig, Coordination chemistry of trivalent lanthanide and actinide ions in dilute and concentrated chloride solutions, *Inorg. Chem.*, **39**, 595–601 (2000).
[132] S. Skanthakumar, M. Antonio, R. Wilson, and L. Soderholm, The curium aqua ion, *Inorg. Chem.*, **46**, 3485–3491 (2007).
[133] P. Lindqvist-Reis, C. Apostolidis, J. Rebizant, A. Morgenstern, R. Klenze, O. Walter, T. Faenghänel, and R. G. Haire, The structures and optical spectra of hydrated transplutonium ions in the solid state and in solution, *Angew. Chem. Int. Ed.*, **46**, 919–922 (2007).
[134] M. R. Antonio, L. Soderholm, C. W. Williams, J. P. Blaudeau, B. E. Bursten, Neptunium redox speciation, *Radiochim. Acta*, **89**, 17–25 (2001).
[135] E. Galbis, J. Hernández-Cobos, C. den Auwer, C. L. Naour, D. Guillaumont, E. Simoni, R. R. Pappalardo, and E. Sánchez-Marcos, Solving the hydration structure of the heaviest actinide aqua ion known: the californium(III) case, *Angew. Chem. Int. Ed.*, **49**, 3811–3815 (2010).
[136] R. Revel, C. den Auwer, C. Madic, F. David, B. Fourest, S. Hubert, J.-F. Le Du, and L. R. Morssposition, First investigation on the L edges of the ^{249}Cf aquo ion by X-ray absorption spectroscopy, *Inorg. Chem.*, **38**, 4139–4141 (1999).
[137] Y. Marcus, A simple empirical-model describing the thermodynamics of hydration of ions of widely varying charges, sizes, and shapes, *Biophys. Chem.*, **51**, 111–127 (1994).
[138] F. H. David and V. Vokhmin, Thermodynamic properties of some tri- and tetravalent actinide aquo ions, *New J. Chem.*, **27**, 1627–1632 (2003).
[139] K. L. Nash, A review of the basic chemistry and recent developments in trivalent f-elements separations, *Solv. Extr. Ion Exch.*, **11**, 729–768 (1993).
[140] Y. Zhu, J. Chen, and R. Jiao, Extraction of Am(III) and Eu(III) from nitrate solution with purified cyanex301, *Solvent. Extr. Ion. Exch.*, **14**, 61–68 (1996).
[141] X. Cao, D. Heidelberg, J. Ciupka, and M. Dolg, First-principles study of the separation of Am^{III}/Cm^{III} from Eu^{III} with cyanex301, *Inorg. Chem.*, **49**, 10307–10315 (2010).
[142] J. M. Keith and E. R. Batista, Theoretical examination of the thermodynamic factors in the selective extraction of Am^{3+} from Eu^{3+} by dithiophosphinic acids, *Inorg. Chem.*, **51**, 13–15 (2012).
[143] A. Bhattacharyya, T. K. Ghanty, P. K. Mohapatra, and V. K. Manchanda, Selective americium(III) complexation by dithiophosphinates: a density functional theoretical validation for

covalent interactions responsible for unusual separation behavior from trivalent lanthanides, *Inorg. Chem.*, **50**, 3913–3921 (2011).

[144] J. H. Lan, W. Q. Shi, L. Y. Yuan, Y. L. Zhao, J. Li, and Z. F. Chai, Trivalent actinide and lanthanide separations by tetradentate nitrogen ligands: a quantum chemistry study, *Inorg. Chem.*, **50**, 9230–9237 (2011).

[145] J. H. Lan, W. Q. Shi, L. Y. Yuan, Y. X. Feng, Y. L. Zhao, and Z. F. Chai, Thermodynamic study on the complexation of Am(III) and Eu(III) with tetradentate nitrogen ligands: a probe of complex species and reactions in aqueous solution, *J. Phys. Chem. A*, **116**, 504–511 (2012).

[146] L. E. Roy, N. J. Bridges, and L. R. Martin, Theoretical insights into covalency driven f element separations, *Dalton Trans.*, **42**, 2636–2642 (2012).

[147] D. Manna and T. K. Ghanty, Complexation behavior of trivalent actinides and lanthanides with 1,10-phenanthroline-2,9-dicarboxylic acid based ligands: insight from density functional theory, *Phys. Chem. Chem. Phys.*, **14**, 11060–11069 (2012).

7

Error-Balanced Segmented Contracted Gaussian Basis Sets: A Concept and Its Extension to the Lanthanides

Florian Weigend

Karlsruhe Institute of Technology

7.1 Introduction

Application of quantum chemical methods to molecules of chemical interest requires expansion of (one-particle) wavefunctions by basis functions. In this way, tasks are reduced mainly to integral evaluations and matrix operations and thus tailored to the abilities of contemporary computer architectures, which allows for calculations of systems consisting of several hundred atoms. But then, results obtained depend not only on the quality of the method, for instance (post-) Hartree-Fock (HF) or density functional theory (DFT) methods, and the terms included in the Hamiltonian, like scalar relativity or spin-orbit coupling, but also on the quality of the basis set expansion. This chapter is dedicated to the last issue. Clearly, higher accuracy in the basis set expansion leads to higher computational costs. Unfortunately this is not true the other way round; so far best results at given costs basis sets have to be designed and optimized carefully. This is particularly true for the heavier elements with large numbers of electronic shells.

Concerning economy, a proven way is the employment of 'contracted' Gaussian functions, combinations of primitive Gaussian functions with fixed expansion coefficients, as integral evaluation is much easier for Gaussian than for Slater functions, albeit the latter may be physically more reasonable. Usually one contracted function is used for each core

Computational Methods in Lanthanide and Actinide Chemistry, First Edition. Edited by Michael Dolg.

shell and X functions for each valence shell, resulting in 'X-tuple-zeta valence' bases. For many computer programs in quantum chemistry, it is economically advantageous to use so-called segmented contracted sets, for which atomic shells are not fitted as a whole, but by segments. This scheme, which allows but also requires certain flexibility in the choice of the number of contracted functions, is explained in detail in Section 7.2. The valence basis sets have to be extended by functions of higher angular quantum numbers for description of polarization and/or electron correlation, or, to be more specific, for the description of the so-called 'correlation cusp.' [1] For the description of correlation there are well known schemes for the consistent extension, [2] derived from the contributions of functions to the total correlation energy in post-HF methods. Often those functions are also used for orbital polarization. This effect results from the presence of the neighbor atoms in molecules and has to be covered by the basis also for HF and DFT treatments. For simple cases, e.g., the 2p elements, employing just the same functions for polarization and correlation works well, but not in general; sometimes more, less, or other functions are needed to obtain similar errors for compounds all across the periodic table ('error balancing'). The basic considerations for this error-balancing are discussed in Section 7.3. Complexity of development of this kind of bases increases for heavier elements, a particular case being lanthanides. In Section 7.4 we discuss the special aspects arising for lanthanides from the 'missing' inner electrons when using effective core potentials, from the presence of the partly occupied f shell, the flexibility of the 6s/5d/4f occupation, and the comparably close energetic distance of the 6p shell.

7.2 Core and Valence Shells: General and Segmented Contraction Scheme

Using Gaussian-type functions (GTF) for atomic basis sets requires contraction of primitive functions (PGTF) to contracted ones (CGTF). The smaller number of basis functions N resulting from the contraction leads to lower memory requirements (Fock and density matrices scale like N^2) and to shorter computation times, in particular for diagonalizing the Fock matrix, which scales like N^3. Further, convergence of the SCF procedure is significantly better for contracted sets. One distinguishes between two types of contraction schemes, general and segmented. The general scheme is conceptually simpler and briefly sketched first.

One starts with a set of PGTFs, e.g., a series of Gaussians with even-tempered or well-tempered exponents. [3] For obtaining the contraction coefficients, in the simplest case an atomic HF calculation is carried out. The CGTFs are identical to the resulting orbitals, that is, the contraction coefficients are the expansion coefficients obtained in this atomic calculation. In this way, one would obtain N CGTFs from N PGTFs. Using such a basis of course would not reduce the effort in molecular treatments; the number of CGTFs has to be reduced. The minimum needed are the CGTFs representing occupied orbitals, e.g., two s and one p CGTF for the 2p elements. This would yield the correct result for the atoms, but the flexibility is by far too low to account for changes resulting from the neighbor atoms in molecules. Thus, the CGTFs representing the valence shell have to be modified by releasing the one/two/…/X-1 of the most diffuse functions (those with the smallest exponents, ζ) from the contraction. In this way the valence shell is described by two/three/…*X* GTFs, and the resulting bases are termed "double/triple/ …/*X*-tuple zeta valence bases." The core shells are described by one CGTF (see above). The diffuse *X* functions are removed from them, because they are still present as PGTFs for the valence shell and further do only weakly contribute to the inner shells. So, for a triple zeta valence basis for phosphorous, for example, altogether five s-type CGTFS and four p-type CGTFs are needed, for which in case of the widely used bases by Dunning [4] 15 different s-type and 9 different p-type PGTFs are required, which is summarized in the contraction scheme (15s9p)/[5s4p]{13,13,13,1,1/7,7,1,1}. Note that the first number, 15s, is the number of *different* s-type PGTFs, of which 13 occur in three different CGTFs. Thus, if the identity of exponents is *not* exploited in integral calculations, as it is the case for many quantum chemical program systems, integrals over of $3 \cdot 13 + 2 \cdot 1 = 41$ s-type PGTFs and 16 p-type PGTFs have to be calculated.

It has to be noted that the choice of PGTFs to be used for an *X*-tuple zeta valence basis requires some care: atomic shells with cardinal quantum number n show n-1 radial nodes. Most relevant for chemical bonds is the part beyond the last radial node, and here the flexibility is needed. Thus, the PGTFs have to be chosen in a way so that the *X* most diffuse PGTFs describe exactly this part of the valence shell. This means that their expansion coefficients in the atomic calculation have the same sign, while the exponent of the $(X+1)^{st}$ PGTF (in the sequence of increasing exponents), which represents the region just before the last radial node, has to be chosen in a way that the respective expansion coefficient is of opposite sign. In this way, *X* determines (roughly) the ratio of exponents (smaller for larger *X*, larger for smaller *X*) and thus the total number of PGTFs.

Within the general contraction scheme, each PGTF except of the *X* most diffuse contributes to each CGTF of the same angular quantum number, which is shown in Figure 7.1 for the s-type CGTFs of phosphorous for the cc-pVTZ basis set. [4] In the segmented contraction scheme in contrast, one exploits the similarity of radial shape of atomic orbitals of different cardinal quantum numbers, which opens the possibility for significant savings. For instance, the radial shape of the 1s orbital is nearly the same as that of the regions between the core and the first radial node of the higher s orbitals, the radial shape of the outer part of the 2s orbital is very similar to that of the region between the first and the second radial node of the higher s orbitals, and so on. See Figure 7.1. The same is true for the other angular quantum numbers. In this way, orbitals are described by CGTFs representing 'segments' for the regions between the nodes, as shown in Figure 7.1. It is obvious, that for a given segment only, CGTFs of exponents within a certain range of values are needed, for instance, only those with large exponents for the most inner segment. This dramatically reduces the numbers of PGTFs contributing to CGTFs, the so-called 'contraction lengths.' The number of CGTFs is the same for segmented and generally contracted bases, at least for the example in Figure 7.1: five s-type CGTFs are needed (one for the 1s and the inner parts of the 2s and 3s shell, one for the outer part of the 2s and the middle part of the 3s, and three for the outer part of the 3s shell), and similarly for the p shells four p-type CGTFs (one for the 2p shell and the inner part of the 3p shell, three for the outer part of the 3p shell). A reasonable contraction scheme for a segmented contracted triple zeta valence basis [5] for P turned out to be (14s9p)/[5s4p]{7,3,2,1,1/6,1,1,1}. The much smaller total number of PGTFs of segmented bases compared to general contracted bases (e.g., for the s-space 14 versus 41) dramatically reduces the computational effort in the integral evaluation for most programs (e.g., ORCA [6] or TURBOMOLE [7]), as it formally scales with the fourth power of the

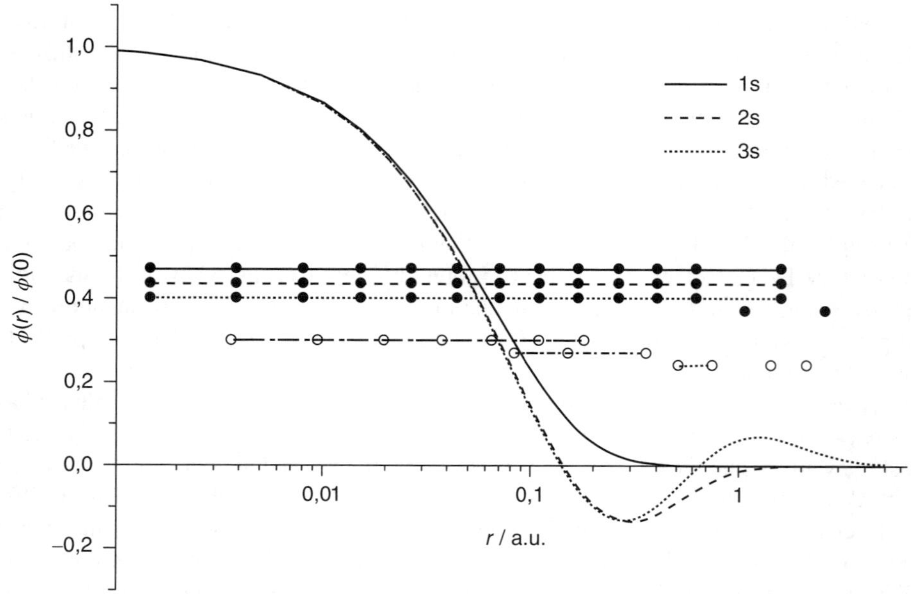

Figure 7.1 *Amplitudes of atomic s orbitals of P and respective PGTFs and CGTFs for triple zeta valence bases according to the generalized contraction scheme (cc-pVTZ, filled circles) and the segmented scheme (def2-TZVP, open circles). Each PGTF is represented by a circle, the r-position is given by half of the full with at half maximum (FWHM/2) of the respective Gaussian. CGTFs consisting of more than one PGTF are represented by solid/dashed/ dotted lines connecting the respective PGTFs. In the generalized scheme the same PGTFs (with different coefficients) are used to fit the 1s orbital (solid), the 2s orbital (dashed), and the 3s orbital (dotted), the latter together with two PGTFs. In the segmented scheme the first CGTF (7 PGTFs) describes the 1s orbital and the inner segment of both the 2s and the 3s orbital, the second CGTF (3 PGTFs) the outer segment of the 2s and the middle segment of the 3s including corrections in the region of the first radial node and the third CGTF together with two PGTFs the outer part of the 3s orbital, including corrections in the region of the second radial node*

number of PGTFs. Exceptions are program systems that exploit the identity of PGTFs in case of the generally contracted bases (e.g., Molpro [8]).

The ‘multi-use’ of segments requires a simultaneous optimization of exponents (or their logarithms) and contraction coefficients, e.g., by repeatedly forming gradients of the (HF-) energy with respect to all basis set parameters followed by a relaxation step, until an energy minimum is reached. It remains the task to find start values for the optimization, i.e., the number of PGTFs per CGTF and reasonable initial values for the exponents and coefficients. Here, it is best to start with an uncontracted, well-tempered basis and determine the segments by the change of the sign in the expansion coefficients of the valance shell in an atomic calculation. Next, for the first segment (the CGTF representing the 1s shell and the most inner parts of the other s shells), one takes the expansion coefficients of the 1s orbital as contraction coefficients. The coefficients for the second segment are taken from the expansion coefficients of the 2s orbital (those between the first and the second change of the sign). This is continued for the further s shells and in the same way for the shells with higher angular quantum number. As is evident from Figure 7.1, a problem occurs for the description of the regions around the radial nodes. Here, the outer part of the *n*s orbital and the respective part of the (*n*+1)s orbital differ greatly. The amplitude of the further approximates zero like exp(-*r*), or, if modeled by Gaussians like exp($-r^2$), the latter crosses the *r*-axis. For the accurate description in these regions, the CGTFs usually have to be extended by a steep PGTF (which has a contraction coefficient of opposite sign), at least if one is interested in accurate total energies. For heavier elements or higher accuracy, it is sometimes necessary to extend the strict scheme and to even add further CGTFs, as positions of radial nodes also are only similar but not identical for shells of different cardinal quantum numbers. The bases obtained this way finally are optimized by minimizing the energy with respect to all parameters.

This procedure is much less strict than that for generally contracted sets, and for the same accuracy concerning total energies, more PGTFs with different exponents (but much less PGTFs in total) are required. Nevertheless, if properly done, the increase is moderate. For instance, for P the (core and valence) part of the generally contracted basis cc-pVQZ [4] has the contraction scheme (16s11p)[6s5p]{13,13,13,1,1,1/8,8,1,1,1} (16 different s-type CGTFs, overall 42 s-type PGTFs, and 19 p-type PGTFs, see above) and shows an error in the atomic HF energy of 0.44 mE_H; the segmented contracted basis QZVP [9] is of scheme (20s14p)/[9s6p]{10,3,1,1,1,1,1,1,1/8,2,1,1,1,1}. The error is even somewhat smaller; 0.26 mE_H.

7.3 Polarization Functions and Error Balancing

So far, we have only discussed the construction of basis sets for the core and the valence shells. For the use in quantum chemistry functions with higher angular quantum number are also needed, which describe polarization effects due to the neighboring atoms and are also needed for post-HF correlation treatments in order to describe the correlation cusp. For the latter a strict scheme for the consistent extension of valence bases with polarization functions exists, which is based on the energy contribution of functions with higher angular quantum number to the correlation energy. [2] If one considers this quantity for the valence shell of, for example, a p element, it turns out that the largest additional contribution arises from a d function with an exponent similar to that of the maximum of the valence-p shell. Adding this function to the DZV set yields a reasonable balance between the flexibility of the valence shell and the description of the correlation. The next two contributions, which are of similar size, are obtained by either adding an f function (of similar exponent as the d function) or by splitting the d function in two d functions that are larger/smaller than the original one by a factor of ca. $\sqrt{3}$. This 2d1f set added to a triple zeta valence basis yields a reasonable balance in the above sense as does similarly adding a 3d2f1g set to a quadruple zeta valence basis.

Despite not being designed for this purpose, the same sets often are used for the description of polarization effects. For many cases this works well, as polarization also requires functions of higher angular quantum numbers with exponents similar to that of the shell to be polarized. Also the numbers of functions to be added to the *X*-tuple zeta valence bases added for consistent description of correlation turned out to be reasonable also for polarization, at least for the main group elements. Nevertheless, there are many examples, where polarization requires more or less or different functions than correlation. Such cases are most easily found when testing bases 'in situ' during their development for the purpose they are intended to be used for further. This is the idea behind so-called 'def2-XVP' bases [10], X = DZ (or S, for historical reasons), TZ, QZ, developed in Karlsruhe some years ago. A very large part of quantum-chemical applications focuses on structure optimizations and bond energy calculations at HF, DFT, and economic correlated treatments like HF+MP2 (Hartree-Fock plus second-order Møller-Plesset perturbation theory) for compounds all across the periodic table. Thus the partly already existing [5], [11] basis sets constructed following the concepts discussed above were tested at levels HF, DFT (with the functional BP86 [12]), and MP2 in calculations of bond energies, dipole moments, and structure parameters for a set of more than 300 molecules, which represents nearly each element in its common oxidation states. The main goal was to obtain similar errors (within a method and a basis set type X) all across the periodic table. This is a helpful property in praxis, and, as for segmented contracted sets a strict construction principle does not apply anyway, it appears most reasonable to rather focus on the property of consistent errors. The size of typical errors was oriented at that for first row main group compounds treated with basis sets like cc-pVXZ or XVP, that is for bond energies per atom at HF or DFT level ca. 20/5/1 kJ/mol for X = DZ(S)/TZ/QZ and ca. 50/20/5 kJ/mol at MP2 level, which serves also as orientation for higher correlated levels. It turned out (once more) that at MP2 level even for the comparable low accuracy requirements, more effort is needed concerning polarization functions than for DFT in several cases. Therefore providing two sets, a larger for HF+MP2 and a smaller for DFT, appeared to be reasonable. Errors in total energies were of lower interest for TZV and DZV sets, the typical amount to several ten mE_H for TZV bases and a few hundred mE_H for DZV bases per atom. QZV bases in contrast show errors of below 1 mE_H (up to Kr), which is less than for the generally contracted sets. [4] The prefix 'def2' was chosen to distinguish the error-balanced sets from their predecessors, the 'def-SVP' and 'def-TZVP' bases. Here the prefix 'def' means 'default,' as these bases were taken by the TURBOMOLE input generator, if nothing else is specified. The main corrections/extensions needed to reach error balance are briefly summarized now.

For the s elements a comparably large number of polarization functions is needed, even for HF and DFT at DZV or TZV level, if one wants to stay within the above error limits. The subsequent p orbitals are energetically in reach and thus often partly occupied, for the heavier s elements this holds also for the d orbitals and for the 6s and 7s elements additionally for the f orbitals. The respective functions are determined best in restricted open-shell HF calculations with the respective occupations, e.g., s^1p^1 for the optimization of p functions of Mg. Thus, for error-consistency in the case of s elements, comparably large basis sets are needed, for instance for the triple zeta basis for Mg two diffuse p and two diffuse d sets and additionally one steep d set for the polarization of the 2p shell. For Ba apart from a diffuse p and a {4,1}d set, a {4}f set had to be added even to the double zeta basis. These functions are partly occupied, for instance in case of BaO by ca. 0.1 electrons. Neglecting the f set changes the equilibrium distance for BaO (at DFT level) by ca. 9 pm and the bond energy by ca. 50 kJ/mol, which is not tolerable at all.

For the 3p elements, restriction to the polarization of the valence shell leads to non-tolerable errors; the polarization of the 2p shell with a steep d set is essential, otherwise bond energies, already at HF or DFT level, are off by ca. 3 kJ/mol, which is too high for quadruple zeta bases and may be critical already for triple zeta bases. Note that no extension of the description of the valence shell (by more valence and polarization functions) will be able to correct this error. Thus, if solely the valence shell description is stepwise improved, convergence to an energetic limit will be observed, of course, but one must be aware that this is not the basis set limit, for total as well as for bond energies. For the heavier p elements, this problem does not occur, because suited d functions are already present as they are needed for the description of the occupied d shell. The transition metals, on the other hand, require less effort concerning polarization of the d shell than main group elements for the p shell, in particular for DFT. At double zeta level accuracy matching that of s and p elements is achieved without any polarization of the d orbitals. At triple zeta level one f function is sufficient.

7.4 Considerations for Lanthanides

For heavy elements relativistic effects are important and have to be incorporated into basis set development. In case of all-electron treatments, scalar relativistic effects and also spin-orbit coupling are treated by decoupling the four-component Dirac equation in the way suggested by Douglas, Kroll, and Hess (DKH) [13], or more recent with the exact decoupling method, X2C [14] ('exact' refers only to the one-electron part), or alternatively with the 'zeroth order regular approximation', ZORA. [15] Achieving accurate total energies with segmented sets for the lanthanides with those elements is not easy; errors of several E_H in the total energy and thus also considerable errors for the inner shells seem to be tolerable for some developers. [16] Of course, for energy differences most of these errors cancel, but nevertheless, there is still some space for improvement, in particular, as a main reason for using all-electron methods (and not effective core potentials, see below) as the accessibility of inner orbitals, which then of course should be accurately described. This space (partly) was bridged by Dolg, [17] whose bases show errors in total energies of several hundred mE_H, which is in the range of def2-SVP bases (for the lighter elements, for which they are currently available). The inner orbitals are described with reasonable accuracy by these bases, as evident from very small errors in the higher ionization potentials. The bases are tested for the trihalides LnX_3 (Ln=La-Lu, X=F, Cl, Br, I) thus representing at least the most typical oxidation state. The errors in bond energies (listed only with respect to the experiment, but not to the basis set limit) are reasonably small, but it has to be noted that the diversity of the test set is much smaller than for the error-balanced basis sets [10] discussed above. Both, these bases as well as the predecessors, [16] are not obtained by the rather elaborate way described above, that is, not by simultaneously optimizing exponents and coefficients in this way fitting segments as well a possible, but by contracting a fixed set of exponents with the coefficients of the atomic orbitals. As evident from the above considerations, in this way only a few PGTFs can be contracted, as already around the first radial node the first segment of the higher shells of a given angular quantum number is slightly different from the first shell of this angular quantum number. Consequently, those bases are loosely contracted with only one contraction per angular quantum number covering the steepest four to six PGTFs. Here most probably will still be some space for improvements towards more compact segmented contracted bases following the above mentioned ideas, but currently for all-electron calculations with segmented contracted sets of Gaussians, the bases by Dolg [17] or, if one is not that much interested in inner orbitals, also those of similar size by Neese [16] are a reasonable choice.

A pragmatic, economic, and proven alternative to all-electron calculations is the employment of effective core potentials (ECPs) for heavier elements. The description of inner electrons by an effective core potential leads to significant computational savings, but even more important is the possibility of very economically model relativistic effects, scalar effects as well as spin-orbit coupling. The def2-series discussed above (segmented contracted, error-balanced, for levels DZ, TZ, QZ with two sets of polarization functions for DFT and MP2) also included basis sets for the use in connection with ECPs for the elements beyond Kr, namely scalar-relativistic Wood-Boring ECPs [18] for the s and d elements and Dirac-Fock ECPs [19] for the p elements. Meanwhile also for the s and d elements beyond Kr bases for the use with Dirac-Fock ECPs [20] following the above ideas are available, termed 'dhf-XVP', [21] for consistency also the def2-bases for the heavier p elements received the label 'dhf' (dhf = Dirac-Hartree-Fock). For all dhf-bases, patches for self-consistent calculations of the spin-orbit coupling are available, [21] which are needed for the splitting of the p shells into the spatially more extent $p_{3/2}$ and the more compact $p_{1/2}$ shell, and similarly for the d shells.

For the lanthanides currently available scalar relativistic ECPs are of Wood-Boring type, on one hand small-core ECPs covering the inner 28 electrons, thus leaving the shells 4s, 5s, 6s, 4p, 5p, 4d, and 4f for the explicit treatment, [22] on the other hand large-core ECPs that also cover the 4s4p4d shells and the (open) 4f shell. [23] The latter exists in different variants for most elements for the cases f^n, f^{n+1}, and f^{n-1} (n = 1 for Ce, etc.). These ECPs allow for comparably unproblematic convergence of SCF procedures, but the f occupation is fixed by the ECP. The small-core ECPs have the full flexibility in the occupation of the valence shell, but this is sometimes paid with convergence problems. Nevertheless, when extending our series of error-balanced segmented contracted bases, we assigned priorities to the higher flexibility and focused on basis sets for the small-core ECPs. For these ECPs, reasonable bases of around triple zeta quality were already available from Cao and Dolg [24] (*CD*), which served as a valuable starting point for our developments of error-balanced lanthanide bases of DZV, TZV, and QZV quality. [25]

For the development of basis sets, in particular segmented contracted basis sets, ECPs are a certain challenge, as the inner orbitals, which are used for the description of the inner part of outer orbitals, now are missing, but the inner part of the outer orbitals is not zero, moreover its structure is non-physical, as shown in Figure 7.2. Nevertheless, the segmented scheme still can be used at least partly. The shape of the respective segments is also similar for the different orbitals. As evident from Figure 7.2 for the triple zeta valence basis of Eu, the s orbitals (obtained from simultaneous optimization of exponents and coefficients) are rather loosely contracted; only the steepest five PGTFs form a CGTF, and the remaining nine are left as PGTFs. An attempt was made to describe the second segment with a CGTF (like for P, see Figure 7.1), but this came along with a significant increase of PGTFs with exponents overlapping in the two CGTFs, which turned out to lead to higher computational costs in the end. Similar is true for the other angular quantum numbers and for the double and quadruple zeta basis sets. Details are given below and are also provided by Gulde *et al.* [25].

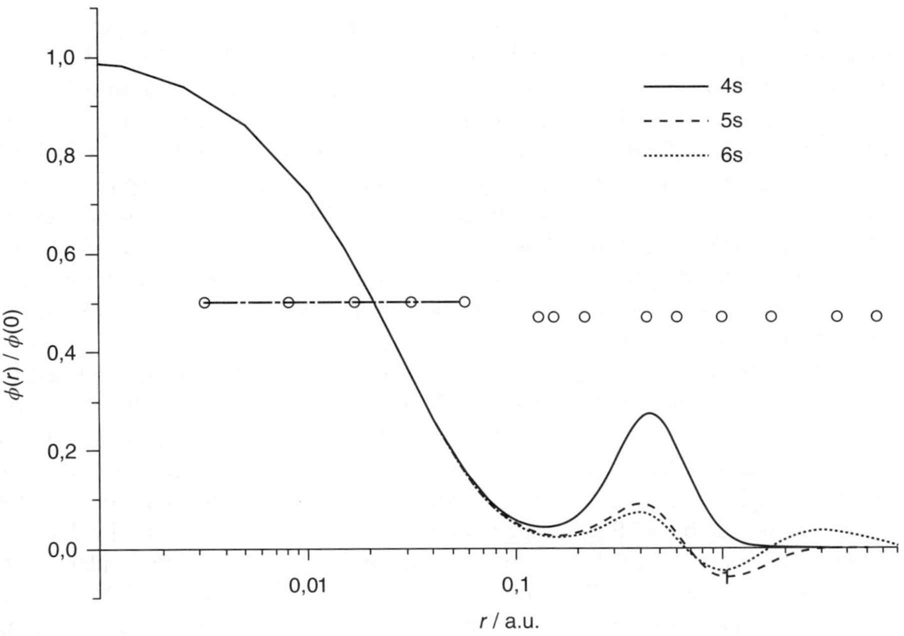

Figure 7.2 *Amplitudes of atomic 4s, 5s, and 6s orbitals of Eu obtained with a Wood-Boring-ECP covering the inner 28 electrons and respective PGTFs and CGTFs for a segmented contracted triple zeta valence basis. See also Figure 7.1*

Following the ideas sketched above, a representative test set and reference bases are needed for determining the number and kind of further functions for balanced errors. For the test set the following molecules were chosen: CeF, CeH_2, CeO, CeF_3, PrCl, PrH_2, PrF_3, NdCl, NdH_2, NdO, NdF_3, PmF, PmO, PmH_2, SmF, SmH_2, $SmCl_3$, Eu_2, EuCl, EuF_2, EuH_2, Gd_2, GdF, GdF_2, GdH_2, GdF_3, TbF, DyF, DyF_2, $DyCl_3$, HoO, HoF_3, $ErCl_2$, ErF_3, TmCl, TmF_2, Yb_2, YbCl, YbH_2, Lu_2, LuF, Lu_2O, Lu_2N, LuF_2, LuH_3, LuF_3, $LuCl_3$, $LuBr_3$, LuI_3. The reference bases are of even-tempered type, typically (31s25p18d16f6g2h1i). The factor between subsequent exponents was set to $10^{-1/4}$, which is very close to the complete basis but still numerically stable. For the non-lanthanide atoms in the molecules of the test set, QZVPP bases were taken as reference. With these bases and the respective ECPs, [22] HF, DFT(BP86), and MP2 calculations were carried out for the test set with fixed structure parameters, yielding reference bond energies and dipole moments. Differences of results obtained with the def2-XVP bases to the results obtained with those reference bases are termed 'errors' in the following, usually mean value and standard deviation obtained for the test set are given.

It turned out that for the *CD* bases, errors in bond energies are in the same range as those obtained with def2-TZVP bases (MP2: def2-TZVPP) for the rest of the periodic table. Further it became evident that the comparably large g set of the CD bases be reduced for most cases, as shown below. For dipole moments in contrast, errors of *CD* bases were very large, e.g., for HF 0.561±0.533, which is below double zeta quality. The reason for this turned out to be a missing diffuse p function.

Thus, the TZVP and TZVPP sets were derived from the *CD* bases by adding diffuse p sets, by slightly changing the contraction pattern of the p set from {61111111} to {6311111} for Ce-Dy and to {63111111} for Er-Lu, and by reducing the polarizing g set from {411} to {1} for TZVP and to {21} for TZVPP. Exponents and contraction coefficients were re-optimized at unrestricted HF level for the respective HF ground state (usually $s^2d^1f^{n-1}$, except for Nd, Sm, Eu, Yb). The resulting sets are of very similar quality as *CD* concerning bond energies and in the typical TZVP/PP range for dipole moments. The double zeta valence sets SVP and SV(P) were derived from the triple zeta valence bases by removing one CGTF from the p/d/f set of TZVP, keeping the contraction pattern of the s set unchanged and re-optimizing the entire basis. For the SVP basis, the same 1g set as for the TZVP was used; for the lower polarized SV(P) set, it was omitted.

For QZVP/PP sets, not only energy differences but also total (HF) energies are desired to be close to the limit. Errors for first and second row elements are much smaller than 1 mE_H. For heavier elements errors about 1-2 mE_H are achieved. It was decided to slightly ease criteria for lanthanides due to the higher number of electrons and the increased complexity due to the occupied f shell and allow for errors of ca. 3-4 mE_H. In order to fulfill these requirements the number of PGTFs had to be increased compared to TZVP by 2/4/1/2 for s/p/d/f sets. The number of CGTFs was regularly increased by one for each angular quantum number. Polarization sets of QZVP are identical to that of TZVPP. For QZVPP the original set from *CD* was taken plus one 1h set, which was roughly optimized by maximization of the MP2 correlation energy, as the higher polarized sets are intended mainly for the use in correlated treatments. The resulting contraction patterns are listed in Table 7.1.

The most important results, the basis set errors in bond energies per atom at levels DFT(BP86) and MP2, are shown in Figure 7.3. As intended, one observes a reduction of typical errors when switching from a smaller to a larger basis. The only exception is the

Table 7.1 *Contraction patterns for the basis sets def2-SV(P) to def2-QZVPP, the reference bases, and the bases by Cao and Dolg, CD [24]. N denotes the number of basis, functions in the basis of spherical harmonics (AO) or Cartesian functions (CAO)*

basis	spdf	ghi		N(AO/CAO)	
		P	PP	P	PP
SV(P)/P	(14s12p9d7f)/[10s6p4d3f] {5111111111/531111/ 6111/511}	–	(1g)/[1g] {1}	69/82	78/97
TZVP/PP	(14s14p10d8f)/[10s7p5d4f] {5111111111/6311111/ 61111/5111}	(1g)/[1g] {1}	(3g)/[2g] {21}	93/116	101/131
QZVP/PP	(16s18p11d10f)/[11s8p6d5f] {61111111111/93111111/ 611111/61111}	(3g)/[2g] {21}	(6g1h)/[3g1h] {411/1}	118/151	138/187
CD	(14s13p10d8f)/[10s8p5d4f] {5111111111/61111111/ 61111/5111}	(6g)/[3g] {411}		114/149	
reference	(31s25p18d16f)/[31s25p18d16f]	(6g2h1i)/[6g2h1i]		397/534	

change from SV(P) (without polarizing g) to SVP (with one polarizing g) at DFT level, which does not improve results. Further, the typical errors obtained with the def2-bases for the lanthanides match well to those obtained previously for other elements with def2-bases. Errors (absolute of the mean value plus the standard deviation) of the SV(P) basis amount to 10-20 kJ/mol at DFT (and HF) level and thus are smaller than typical errors of the method; for MP2 this is reached at triple zeta valence quality. With triple zeta valence bases (HF, DFT) and quadruple zeta valence bases (MP2) 2–5 kJ/mol are reached, which is sufficient for most practical applications. Further, at DFT level the larger polarization set does not provide significant advantage over the smaller set, but at MP2 level the larger set is recommendable. For DFT increasing the basis mainly reduces the standard deviation, as already at double zeta valence level mean errors are small; for MP2 in contrast, also mean errors systematically become smaller for larger bases.

An important advantage of small-core ECPs compared to large-core ECPs is the possibility to cover different occupations of the f shells, as mentioned above. This of course requires a certain flexibility of the basis set, otherwise the description becomes biased towards the cases with occupations same as for the atoms. A detailed look at bond energies of PrCl, $ErCl_2$ PmF, TmCl, and TmF_2, which have one additional f electron compared to the atoms, showed that for DFT (and HF) these compounds behave well. All errors are as small as requested above, but MP2 results reveal a different picture: errors for PmF and PrCl with SVP/TZVPP/QZVPP bases amount to ca. 3/10/40 kJ/(mol·atom) and thus are still tolerable, but for TmCl we get 9.4/23/84 kJ/(mol·atom), which in particular for the QZVPP basis is more than desired. These problems, which only occur at MP2 level for the heavier lanthanides, can be solved by partial decontraction of f/g shells to 511111/2211. The errors at MP2 level for these modified QZVPP sets amount to 3.6/2.5/3.1 kJ/(mol·atom) for $ErCl_2$/TmCl/TmF_2. We thus maintained the contraction scheme for consistency and costs; nevertheless, in case of changing f occupations in heavy lanthanides the partial decontraction is advisable.

Also errors of dipole moments and equilibrium distances are in the same range as for the other elements. Quadruple and also triple zeta valence bases yield very small typical errors of below 0.08 (QZVP) (0.15, TZVP) Debye for both HF and DFT. For HF also at double zeta valence level errors are very reasonable, ca. 0.3 Debye. This error almost completely arises from standard deviation, thus, nearly no systematic errors occur. At DFT level in contrast, a systematic underestimation by ca. 0.4 Debye is observed, which together with a standard deviation of 0.3 Debye (similar to that of HF), amounts to comparably large typical errors of ca. 0.7 Debye. The reason for this is rather found in the SVP basis set for the non-lanthanide atoms, which are too rigid to describe the change (compared to the atom) of orbitals at the bond partners, which usually are of higher electronegativity and thus show a negative partial charge leading to spatial extension of orbitals.

Structure parameters were investigated for those diatomic systems that turned out to be worst cases for bond energies; these are CeF, Gd_2, Lu_2, NdO, PmF, PmO, SmF, and TmCl. Among these, worst cases for errors in distances are the dimers, with errors of -11 to $+7$ pm for double zeta valence bases, -2 to $+2$ pm for triple zeta valence bases and less than 1 pm for quadruple zeta valence bases. Compared to ‘usual’ compounds, for metal dimers the minima of the potential curves are rather shallow, so small inaccuracies in the description of the potential curve cause large changes in distances. Therefore, omitting the two dimers from the statistics may lead to a characterization of higher practical relevance showing errors

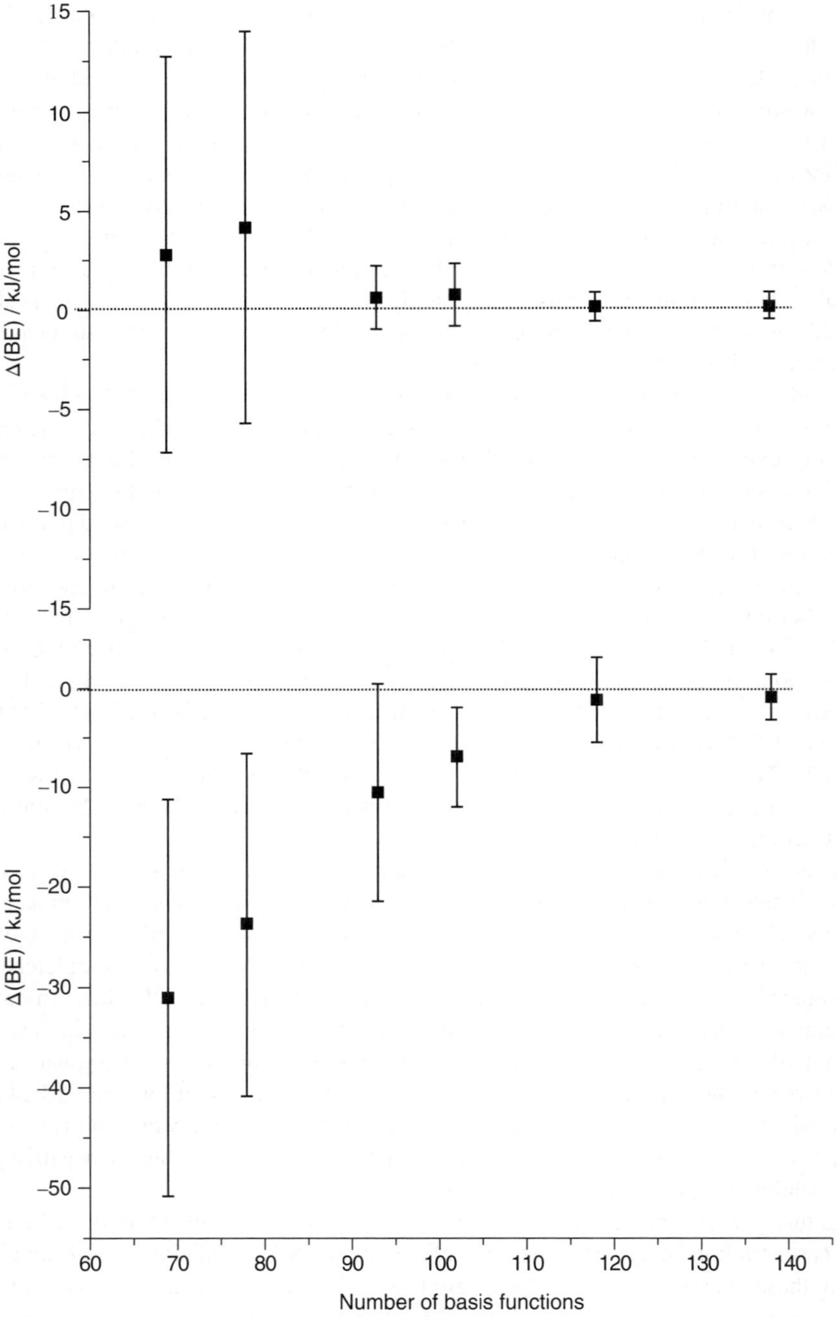

Figure 7.3 *Mean values and standard deviation of errors in bond energies for SV(P), SVP, TZVP, TZVPP, QZVP, and QZVPP basis sets (from left to right, characterized by their number of spherical harmonic basis functions) for the molecules of the test set at levels DFT(BP86), upper part and MP2, lower part*

typically amounting to 2-3 pm for double zeta, 0.5-0.8 pm for triple zeta and ca. 0.2-0.3 pm for quadruple zeta valence bases. This probably still is a conservative estimation, as here only the worst cases were considered. Notably, in contrast to bond energies, MP2 is not significantly more critical than HF or DFT in case of bond lengths.

For more detailed statistics and data for single compounds we refer to the publications about the def2-bases for the lanthanides, [25] and for the other elements, [10] as well as to that about the dhf-bases. [21] All def2-bases are available from the internet at http://www.cosmologic.de/basis-sets/basissets.php.

References

[1] Klopper W and Kutzelnigg W. 1987. Møller-Plesset calculations taking care of the correlation cusp. *Chem. Phys. Lett.* **134**, 17–22.

[2] Ahlrichs R, Scharf P, and Jankowski K. 1985. The impact of higher polarization basis functions on molecular ab-initio results. 2. The ground states of CO, N2, NO, O2 and F2. *Chem. Phys.* **98**, 381–386.

[3] Huzinaga S and Klobukowski M. 1985. Well-tempered basis sets for the atoms K through Xe. *Chem. Phys. Lett.* **120**, 509–512.

[4] Woon TE and Dunning TH. 1993. Gaussian basis sets for use in correlated molecular calculations. III. The atoms aluminum through argon. *J. Chem. Phys* **98**, 1358–1371.

[5] Schäfer A, Huber C, and Ahlrichs R. 1994. Fully optimized contracted Gaussian basis sets of triple zeta valence quality for Atoms Li to Kr. *J. Chem. Phys* **100**, 5829–5835.

[6] Neese F. 2012. The ORCA program system. *WIREs Comput Mol Sci* **2**, 73–78.

[7] a) Furche F, Ahlrichs R, Hättig C, Klopper W, Sierka M, and Weigend F. 2013. Turbomole. WIREs *Comput. Mol. Sci.* **4**, 91–100; b)TURBOMOLE Version 6.5, TURBOMOLE GmbH 2013. TURBOMOLE is a development of University of Karlsruhe and Forschungszentrum Karlsruhe 1989–2007, TURBOMOLE GmbH since 2007.

[8] Werner HJ, Knowles PJ, Knizia G, Manby FR, and Schütz M. 2012. Molpro: a general-purpose quantum chemistry program package. WIREs Comput Mol Sci 2, 242–253.

[9] Weigend F, Furche F, and Ahlrichs R. 2003. Gaussian basis sets of quadruple zeta valence quality for atoms H-Kr. *J. Chem. Phys.* **119**, 12753.

[10] Weigend F, and Ahlrichs R. 2005. Balanced basis sets of split valence, triple zeta valence and quadruple zeta valence quality for H to Rn: Design an assessment of accuracy. *Phys. Chem. Chem. Phys.* **7**, 3297–3305.

[11] Schäfer A, Horn H, and Ahlrichs R. 1992. Fully Optimized Contracted Gaussian Basis Sets for Atoms Li to Kr. *J. Chem. Phys.* **97**, 2571–2577.

[12] a) Becke, AD. 1988. Density functional exchange-energy approximation with correct asymptotic behavior. *Phys Rev. A* **38**, 3098–3100; b) Perdew JP. 1986. Density-functional approximation for the correlation-energy of the inhomogeneous electron-gas. *Phys. Rev. B* **33**, 8822–8824.

[13] Hess BA. 1986. Relativistic electronic-structure calculations employing a 2-component no-pair formalism with external-field projection operators. *Phys. Rev.* A **33**, 3742–3748.

[14] Kutzelnigg W, and Liu WJ. 2005. Quasirelativistic theory equivalent to fully relativistic theory. *Chem. Phys.***123**, 241102.

[15] van Lenthe E, Baerends J, and Snijders JG. 1993. Relativistic regular 2-component Hamiltonians. *Chem. Phys.* **99**, 4597–4610.

[16] Pantatzis AD, and Neese F. 2009. All-Electron Scalar Relativistic Basis Sets for the Lanthanides. *J. Chem. Theory Comput.* **5**, 2229–2238.

[17] Dolg M. 2011. Segmented Contracted Douglas-Kroll-Hess Adapted Basis Sets for Lanthanides. *J. Chem. Theory Comput.* **7**, 3131–3142.

[18] Andrae A, Häußermann U, Dolg M, Stoll H, and Preuss H. 1990. Energy-adjusted ab initio pseudopotentials for the 2nd and 3rd row transition-metals. *Theor. Chim. Acta* **77**, 123–141.

[19] a) Metz B, Stoll H, and Dolg M. 2000. Small-core multiconfiguration-Dirac-Hartree-Fock-adjusted pseudopotentials for post-d main group elements: Application to PbH and PbO. *J. Chem. Phys.* **113**, 2563–2569; b) Peterson KA, Figgen D, Goll E, Stoll H, Dolg M. 2003. Systematically convergent basis sets with relativistic pseudopotentials. II. Small-core pseudopotentials and correlation consistent basis sets for the post-d group 16–18 elements. *J. Chem. Phys.* **119**, 11113.

[20] Peterson KA, Figgen D, Dolg M, and Stoll H. 2007. Energy-consistent relativistic pseudopotentials and correlation consistent basis sets for the 4d elements Y-Pd. *J. Chem. Phys.* **126**, 124101; b) Figgen D, Peterson KA, Dolg M, Stoll H. 2009. Energy-consistent pseudopotentials and correlation consistent basis sets for the 5d elements Hf-Pt. *J. Chem. Phys.* **130**, 164108.

[21] Weigend F, and Baldes A. 2010. Segmented contracted basis sets for one- and two-component Dirac-Fock effective core potentials. *J. Chem. Phys.* **133**, 174102.

[22] Dolg M, Stoll H, and Preuss H. 1989. Energy-adjusted ab initio pseudopotentials for the rare-earth elements. *J. Chem. Phys.* **90**, 1730–1734.

[23] Dolg M, Stoll H, Savin A, and Preuss H. 1989. Energy-adjusted pseudopotentials for the rare-earth elements. *Theor. Chim. Acta* **75**, 173–194.

[24] Cao X, and Dolg M. 2002. Segmented contraction scheme for small-core lanthanide pseudopotential basis sets *J. Molec. Struct. Theochem.* **581**, 139–147.

[25] Gulde R, Pollak P, and Weigend F. 2012. Error-Balanced Segmented Contracted Basis Sets of Double-zeta to Quadruple-zeta Valence Quality for the Lanthanides. *J. Chem. Theor. Comput.* **8**, 4062–4068.

8

Gaussian Basis Sets for Lanthanide and Actinide Elements: Strategies for Their Development and Use

Kirk A. Peterson[1] *and Kenneth G. Dyall*[2]

[1]*Department of Chemistry, Washington State University*
[2]*Dirac Solutions*

8.1 Introduction

Molecules containing f-block elements, particularly the actinides, are currently of very strong interest due to their role in the nuclear fuel cycle and waste remediation efforts. In particular, those containing actinide elements demonstrate a rich chemistry due to the existence of numerous stable oxidation states of the metal together with partially filled 6d and 5f shells, especially amongst the early members of the block. For experimental spectroscopic studies this can lead to complicated, congested spectra due to the high density of states, while the theoretical work is very challenging due to the large number of electrons, strong relativistic effects, and often multireference character due to nearly degenerate 5f, 6d, 7s, and 7p orbitals. In contrast to the actinides, lanthanides are generally found in just their +3 oxidation states, but they can still exhibit some of the same challenges as the actinides due to partially filled 4f shells. Fundamental studies on small molecular systems involving f-block elements have been an active area for both experiment and theory, since the results and insights arising from these studies can often be used for understanding the chemistry of f-block elements in complex molecular environments.

In carrying out ab initio quantum chemistry calculations it is now well known that the choice of basis set can be nearly as important as the electronic structure method that is

Computational Methods in Lanthanide and Actinide Chemistry, First Edition. Edited by Michael Dolg.

used. For the main group elements, families of accurate Gaussian basis sets have been available for some time, which in many cases can be used to estimate the complete basis set (CBS) limit for density functional and wavefunction-based approaches. Basis sets for the transition metals have also received considerable attention, and families of accurate Gaussian basis sets suitable for extrapolation are available for these elements as well. The lanthanides and actinides have received less attention, partly because the challenges of developing basis sets are greater than for the transition metals. This chapter aims to outline the challenges of the f block for the development of basis sets and present considerations and strategies for developing well-balanced, accurate basis sets. The existing basis sets for the lanthanide and actinide elements are enumerated and evaluated, and some results presented for illustration.

8.2 Basis Set Design

8.2.1 General Considerations

The design of a basis set depends on many factors: which Hamiltonian is to be used, which orbitals need to be represented, how correlation is to be treated (wavefunction or DFT), which properties need to be represented, whether contraction is used and by what contraction method, what accuracy level is required, what nuclear model is used. The process of developing a basis set starts with generating a suitable set of primitive functions. This process usually involves generating a primitive set that represents the atomic occupied orbitals and perhaps the most important low-lying virtual orbitals; adding primitives for polarization of the atom in the molecular environment, which is necessary for both DFT and wave function correlation methods; adding primitives for correlation of various orbitals or shells; and adding primitives for properties.

Of crucial importance in designing a basis set is the issue of linear dependence, because this affects the numerical stability of the atomic or molecular calculations that use the basis sets. Linear dependence in the primitive set can be controlled by the use of even-tempered or well-tempered basis sets, which minimize the linear dependence by construction. However, such basis sets tend to be larger than energy-optimized basis sets, where linear dependence problems can become significant as the basis set size increases.

Contraction of the basis set can alleviate or remove entirely any problems with linear dependence. Of course this depends on how the contraction is done and how the basis set is to be used. If a general contraction is used for all basis functions, as is common for ANO sets, there are no linear dependence problems, and even the addition of a few primitives for extra flexibility does not usually cause problems. Segmented contractions in which the inner core is contracted and the rest of the basis is largely left uncontracted could suffer from linear dependence if the primitive basis has linear dependence, whereas segmented contractions in which all or most shells are represented by a contraction are less likely to suffer from linear dependence.

When primitives for correlation are added on top of an SCF contraction, as in the correlation-consistent style of basis set, the correlating set usually includes higher angular momentum functions that are not represented in the SCF set, and functions of lower angular momentum that may already be represented in the SCF set. Care must be taken with the latter, to ensure that the functions added for correlation or valence flexibility do not create a set that is linearly dependent on the SCF occupied functions, particularly if multiple shells are correlated.

A further issue is the relation between the correlating functions and the SCF functions, particularly when primitives are used for correlation. For efficiency in sets like the correlation-consistent sets, the primitive functions added to an SCF orbital set for correlation in the occupied symmetries are normally taken from the SCF set, because this requires less work in the integral code. However, it is usually the case that the optimal exponents for the correlating functions do not match very well with those of the SCF primitive functions: the ratio between the correlating exponents is often larger than that in the SCF basis set, except perhaps for the outer few SCF primitives, which tend to be more widely spaced. Some compromise must usually be made: for example, to choose the SCF set to match the optimal correlating functions, as has been done for the correlation consistent sets, or to choose the correlating functions from the SCF set that maximize the correlation contribution, as done in the Dyall basis sets. A third possibility, also used in the Dyall basis sets where the valence correlating functions overlap with the outer core orbitals is to replace the linearly dependent SCF functions with the valence correlating functions, and then reoptimize the SCF set with the correlating functions frozen. Such a strategy is useful if the basis is to be used uncontracted, because the tail of the outer core functions is not as important as the correlation of the valence shell. Other approaches include optimizing the SCF and the correlating exponents together, or replacing them with an even-tempered set.

8.2.2 Basis Sets for the f Block

When applying the principles of basis set design, the characteristics of the elements for which the basis sets are developed must be taken into account. The f series start out like the d-block transition metals, with the 6d orbital occupied for several of the early actinides, and the 5d occupied for La and Ce. These early elements even have some low-lying states in which the d orbital is multiply occupied. Further along the series, the f orbital is the dominant occupied open-shell orbital and the d is unoccupied, except in the middle of the block. The outer s orbital is doubly occupied in all of these elements. The chemistry of the lanthanides is largely (but by no means solely) that of the +3 oxidation state, whereas higher oxidation states are of importance in the early part of the actinide series, and the +3 oxidation state becomes dominant later in the series.

For the development of basis sets, the radial behavior of the orbitals is important, because it determines the range of exponents of the Gaussian functions that are used, and to what extent the exponent sets for different shells overlap, particularly those for electron correlation. Because the f shell is fairly compact, any basis set must cover a radial range that extends from that of the f shell to that of the outer valence s and d shells. The radial behavior is elaborated below.

Relativistic effects are also critical, particularly for the actinides, where both direct and indirect effects significantly change the radial behavior of the orbitals compared to that of the lanthanides. The relativistic effects are not negligible in the lanthanides, though, because they contribute a good fraction of the lanthanide contraction. The spin-orbit splitting is important for the 6p of the actinides and for the core shells of the same principal quantum number as the f shell.

The accessibility of the orbitals (or spinors) for bonding, and hence whether they are considered part of the valence, the outer core or the inner core, must be assessed on the basis of both the energetics and the radial extent of the orbitals. How the orbitals are partitioned is also relevant to the development of basis sets, which must describe not only the atomic behavior but the bonding behavior as well. For the energy, the eigenvalue is usually sufficient, because it gives an approximate measure of the ionization potential. For the radial extent, mean properties, although useful, are insufficient to portray the entire radial behavior, because the behavior at large distances depends on the angular momentum due the power of r in the radial functions (r^l). Thus, two orbitals of different angular momentum with the same mean radius or radial maximum will have different radial extents: the higher angular momentum function will extend further than the lower. Here we use three measures of the radial extent: the root-mean-square radius, r(rms), the position of the radial maximum, r(max), and the radius that contains 95% of the density, r(95%). The last of these is a measure of how far out the charge distribution extends. Together with the spinor eigenvalues, these quantities are plotted for the lanthanides and actinides in Figures 8.1–8.4, based on Dirac-Hartree-Fock (DHF) calculations on the (degeneracy-weighted) average energy of the $f^{n-1}d^1s^2$ configuration.

For the lanthanides, the rms radii and the 95% density radius are in shell order, i.e., 4d $<$ 4f $<<$ 5s $<$ 5p $<$ 5d $<<$6s. The radial maximum of the 4f is inside that of the 4d, but otherwise the shell order is observed. What is perhaps not obvious from these plots is that the radial maximum of the 6s is outside the 95% density radius of the 5p shell, whereas the radial maximum of the 5d is inside the 95% density radius of the 5p shell. If the radial maximum is taken as some measure of where the midpoint of a bond would be, this indicates that bonding with the 6s does not incur much repulsion of the ligand orbitals by the outer core (5s and 5p) of the lanthanide, whereas there would be somewhat more repulsion from bonding with the 5d. In any case, the 5d is substantially inside the 6s on all measures of radial extent.

The tail of the 4f (measured by the 95% density radius) extends almost as far as the tail of the 5s at the beginning of the block but pulls in further as Z increases. The tail of the 5p is always further out than the 4f. The 4f is consequently not very radially accessible for bonding, lying inside the outer core 5s and 5p shells, although energetically it is much higher than the 5s and 5p.

The actinides display similar behavior to the lanthanides for the rms radius, although the 5f is much closer to the 6s at the beginning. The tail of the 5f extends further out relative to the outer core than the tail of the 4f in the lanthanides – as far as the tail of the 6p at the beginning of the block, it pulls in as Z increases, though it remains in a similar range to the 6s tail. The somewhat larger extent of the 5f relative to the 4f probably contributes to its involvement in bonding in the early actinides, in the actinyl unit for example. It is also energetically more accessible than the 4f in the lanthanides.

The 6s and 6p spinors of the actinides have very similar radial behavior to the 5s and 5p of the lanthanides. The difference in the outer core shells is in the p_+ spinors: the $6p_+$ of the actinides is more extended than the $5p_+$ of the lanthanides, due to the much larger spin-orbit splitting of the p shell.

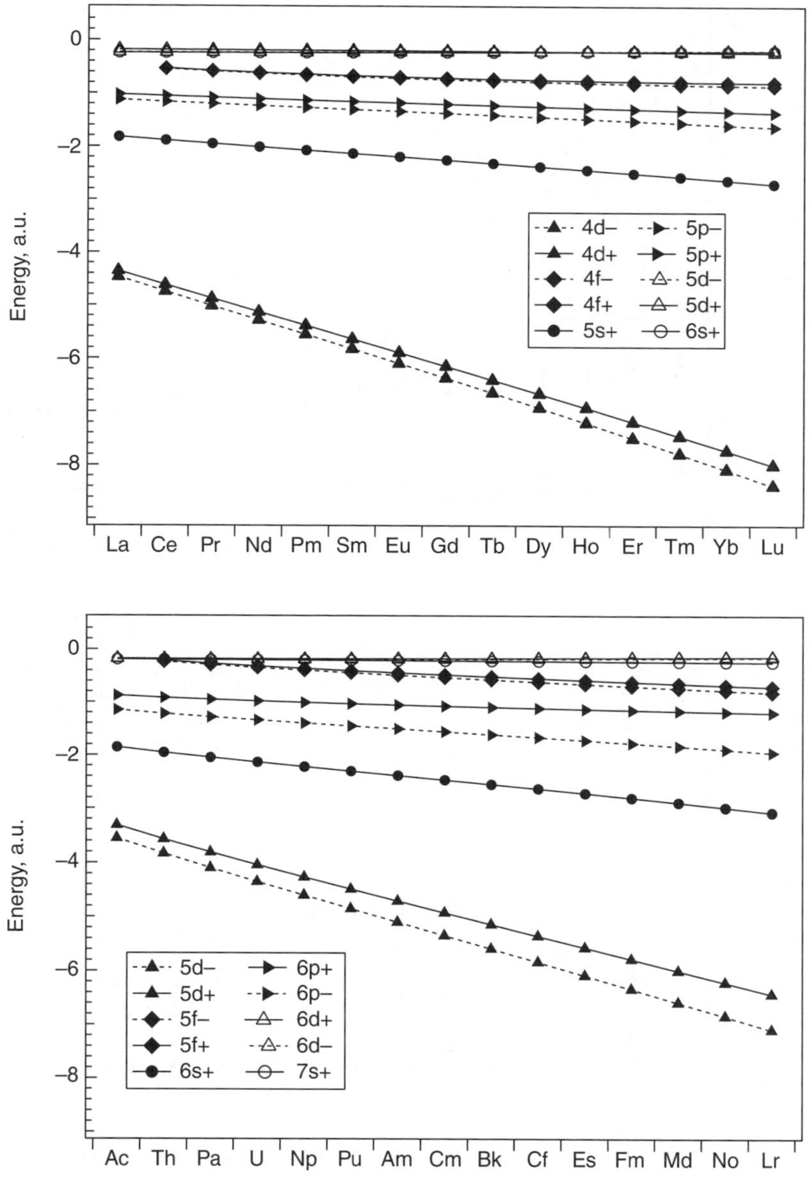

Figure 8.1 *Calculated DHF spinor eigenvalues for the lanthanide and actinide atoms*

The radial maximum of the 7s falls at about the same distance as the 95% density radius of the $6p_+$, and only a little outside that of the $6p_-$ and the 6s. As for the 5d in the lanthanides, the radial maximum of the 6d spinors in the actinides is inside the 95% density radius of the outer core spinors, but these spinors extend out almost as far as the 7s spinors at the 95% density radius. Bonding that involves any of the valence spinors is likely to involve a fair

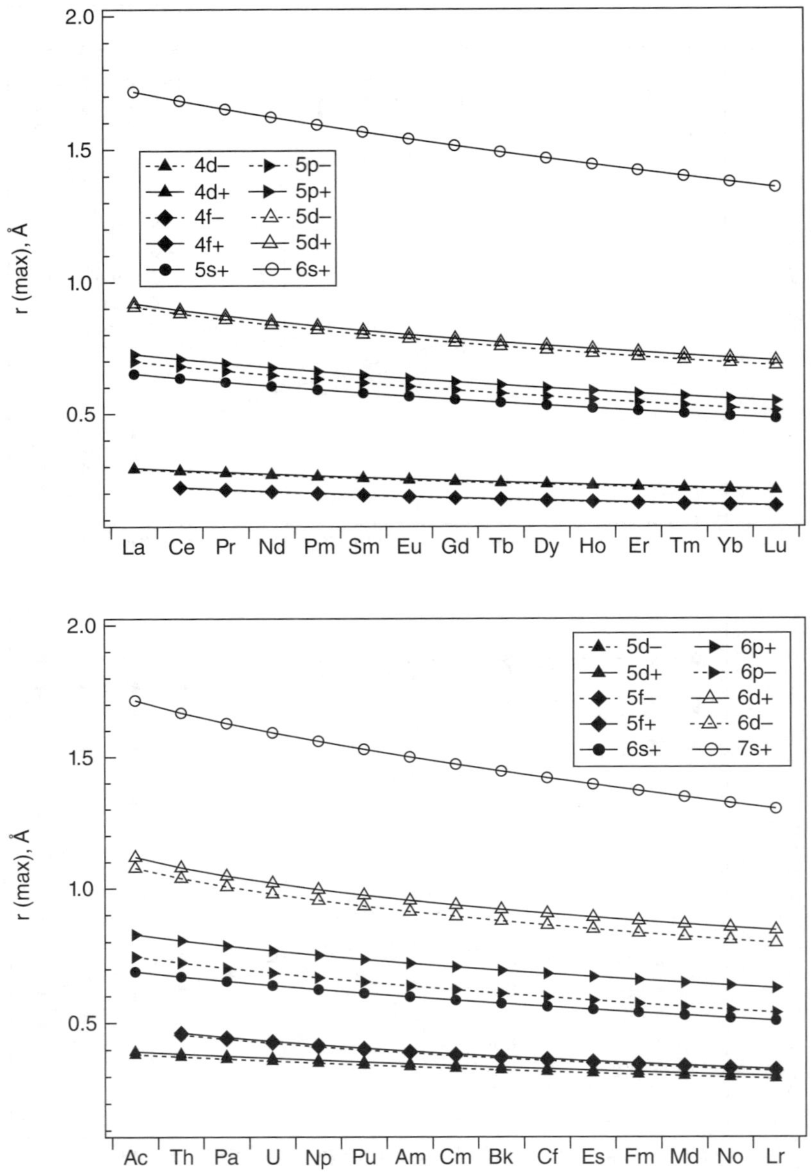

Figure 8.2 *Calculated radial maxima of the DHF spinors for the lanthanides and actinides*

amount of overlap with the 6p spinors as evidenced by the not insubstantial $6p_+$ hole in species like the actinyls.

Given the overlap of the f shell with the outer core shells and even the valence shell, as well as the rest of the shell of the same principal quantum number, basis sets for the lanthanides and actinides must cover the correlation of three shells: the shell that includes

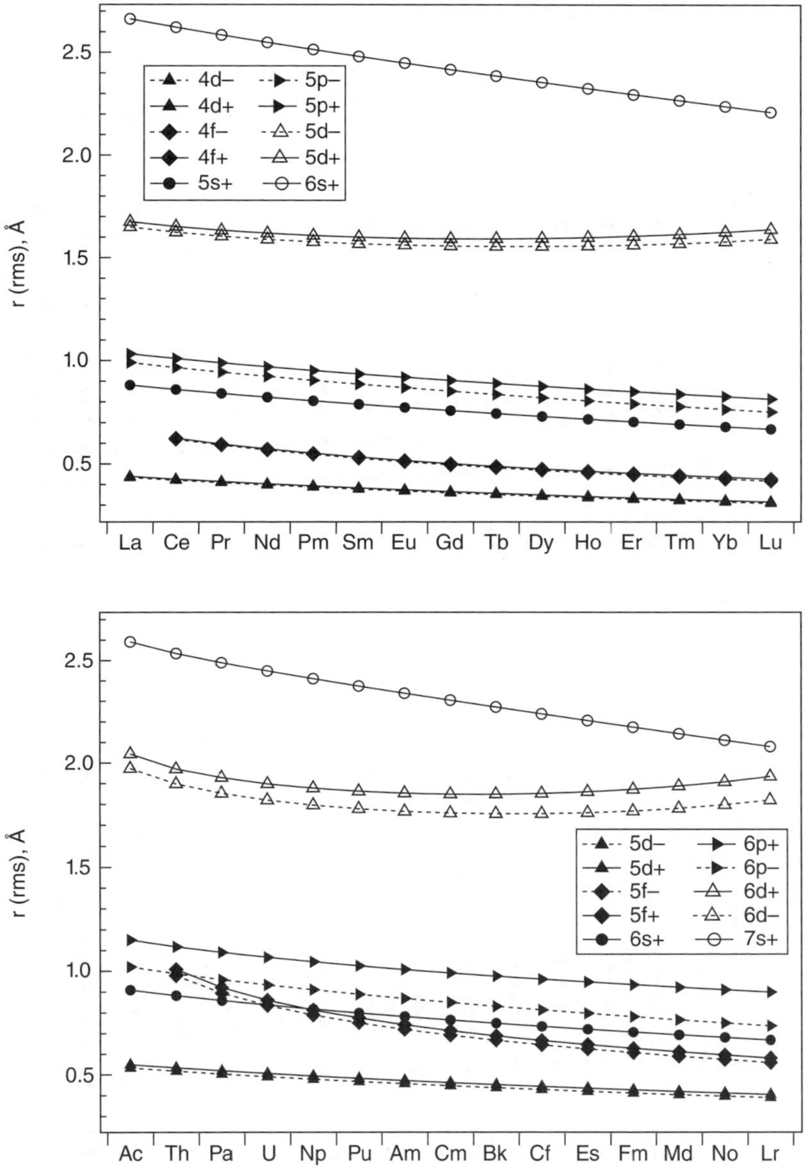

Figure 8.3 *Calculated RMS radii of the DHF spinors for the lanthanides and actinides*

the f orbital, the next shell up, which is the outer core shell, and the next shell up again, which is the valence shell. The changes in behavior along the block means that the basis set size could vary, as the overlap between correlating sets for the various shells changes.

The SCF basis set should cover the low-lying orbitals that are usually unoccupied in the ground state, which usually means the $(n+1)$d and the $(n+2)$p orbitals (where n=4 or 5, the

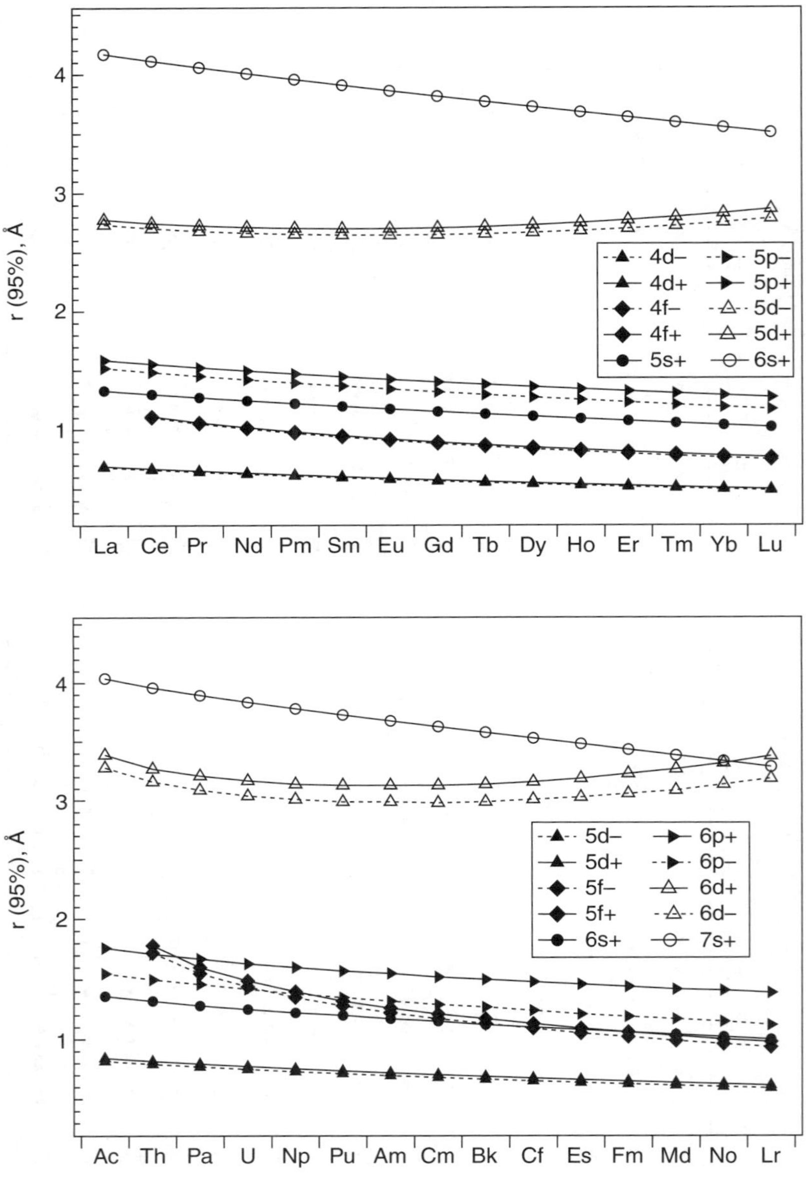

Figure 8.4 *Calculated radii that contain 95% of the DHF spinor densities*

principal quantum number of the f shell). The latter is unoccupied in all of these elements but Lr, whose ground state is $7s^2 7p^1$, rather than $7s^2 6d^1$; the former is occupied for four of the lanthanides (La, Ce, Gd, and Lu) and six of the actinides (Ac, Th, Pa, U, Np, and Cm). The main reason for including the (n+2)p is for polarization of the (n+2)s valence orbital, filling in the function space between the s and d sets in the outer valence region.

Correlating sets must provide for correlation of the *n*f orbital, which may be regarded as an inner valence orbital, the (*n*+1)s and p outer core orbitals, and the outer valence (*n*+1)d and (*n*+2)s orbitals. As there is considerable radial overlap of the *n*f with both the *n*d and the (*n*+1)s and p orbitals, there may be linear dependence between the optimal correlating functions in these sets. There is also some radial overlap between the outer valence orbitals and the outer core orbitals, and to a lesser extent with the f orbital, which could result in linear dependence issues.

As an example, the optimal g sets from the Dyall quadruple zeta basis sets for the actinides [1] are plotted in Figure 8.5, on a logarithmic scale for the exponents. The exponent sets for each group of spinors were optimized independently on the energy. The two g functions for correlation of the 5d (from a 3f2g1h set) have almost the same values as the inner two g functions for 5f correlation (3g2h1i set), so the former can be omitted. The g function for 6s6p correlation (from a 3d2f1g set) starts out reasonably different from the outer g function for 5f correlation at the beginning of the block but by the end of the block is very similar. A decision must be made about when to include this function and when to leave it out. The outer g function for 5f correlation overlaps with the inner g function for 7s6d correlation (from a 3f2g1h set). The rather erratic behavior of these g functions is due to a fairly flat potential energy surface in the exponent space so that a fairly large change in exponent results in a fairly small change in the energy. However, at the beginning of the block, a choice between the g functions must be made. As the g function for 5f correlation is more important energetically than the g function for 7s6d correlation, it is the former that is used in the early actinides.

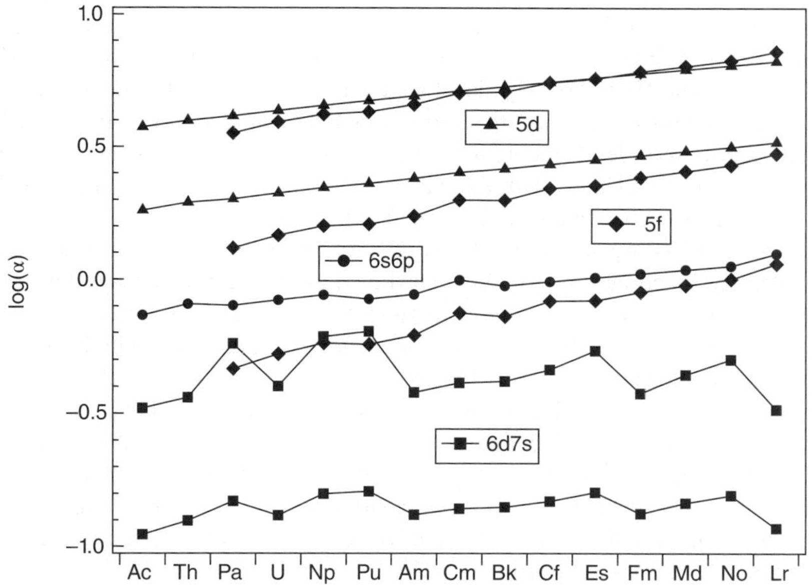

Figure 8.5 *Optimal g-type basis functions from the Dyall quadruple zeta basis sets for actinides*

Independent optimization and collation of different sets of exponents is not the only option, of course. Reoptimization of some of the exponents in one set with exponents taken from another and held fixed can be done, or collective optimization of all sets of exponents, or optimization of an even-tempered set that covers the range and number of exponents required are some other options. All of them represent some sort of compromise. As the basis sets get larger along a series of systematically designed sets, the overlap between the optimal sets of correlating functions also increases, but to compensate, the energy loss from a compromise decreases, as does the energy gain from locating the lowest energy for each set. The use of an even-tempered set might well be more effective and incur less linear dependence in very large basis sets.

8.3 Overview of Existing Basis Sets for Lanthanides and Actinide Elements

8.3.1 All-Electron Treatments

Currently, there is a variety of all-electron basis sets available for the lanthanides and actinides, of different classes and qualities.

In the first class are basis sets developed at the SCF level only. These sets cover a substantial part of the entire periodic table, and for our purposes cover both the lanthanides and the actinides (except where noted). Most of them are at least of triple zeta quality (again, except where noted). These sets include the universal basis sets of Malli *et al.* [2, 3] and Jorge *et al.*, [4], which are even-tempered basis sets; the well-tempered basis set of Huzinaga and Klobukowski [5], which was developed with the nonrelativistic Hamiltonian and covers the lanthanides; the two energy-optimized dual family sets of Fægri [6, 7], developed at the DHF level with a finite nucleus, and which are of better than double-zeta and better than triple-zeta quality; the nonrelativistic and relativistic (DHF) energy-optimized sets of Koga *et al.* [8, 9]; and the energy-optimized third-order Douglas-Kroll-Hess (DKH3) sets of Nakajima and Hirao [10, 11], developed with both point and finite nuclear models. These sets are primitive sets only; they do not include contraction coefficients, although they have been used to develop contracted basis sets. They also do not include functions for polarization or correlation, with the exception of the larger Fægri set [7], which has recommendations for extension to higher angular momentum.

In the second class are segmented basis sets. Some of these have been developed for use with density functional theory. As such, functions for correlation are not important, only functions for polarization of the density in a molecular environment. Among these are the sets of Pantazis and Neese for the lanthanides [12] and the actinides [13], which employ even-tempered primitive sets with g polarization functions, and include a segmented contraction in the core; and the set of Dolg for the lanthanides [14], which are based on energy-optimized primitives but have no functions with angular momenta higher than f-type. These contractions are made explicitly for the DKH2 and the ZORA Hamiltonians. Sekiya *et al.* have developed segmented contractions of correlating functions for the f shell and the valence s and d shells, using the DKH3 Hamiltonian. These are available both for the lanthanides [15] and the actinides [16] and can be added to SCF sets. They have also developed DKH3 segmented basis sets that include both SCF and correlating functions for the lanthanides [17] as part of the Sapporo basis set library. These basis sets include functions for correlating all shells from $n = 4$ up, and are systematically expanded in angular momentum from double-zeta through quadruple-zeta.

In the third class are generally-contracted basis sets. Roos *et al.* developed DKH2 ANO basis sets for the lanthanides [18] and the actinides [19] of double-zeta, triple-zeta, and quadruple-zeta quality, based on an extension of the Fægri primitive set to include up to h functions. These basis sets correlate the f shell, the outer core s and p shells, and the valence s and d shells. Although they do not have functions that explicitly correlate the 4s, 4p, and 4d orbitals of the lanthanides and the 5s, 5p, and 5d orbitals of the actinides, the functions for correlating the f shells are probably adequate for this purpose. Gomes *et al.* and Dyall developed Dirac-Hartree-Fock basis sets of double-zeta, triple-zeta, and quadruple-zeta quality for the lanthanides [20] and the actinides [1], with averaged SCF contractions for the occupied spinors and the valence d and p. Correlating functions are added as primitives, following the style of the correlation consistent sets to systematically expand the angular momentum range.

8.3.2 Effective Core Potential Treatments

There have been a number of basis sets for lanthanide and actinide elements previously reported in the literature that are based on relativistic effective core (ECP) potentials, or pseudopotentials (PP). These can be most easily categorized by the type of underlying ECP used: (a) shape consistent pseudopotentials, (b) energy consistent pseudopotentials, and (c) model potentials.

The shape consistent category includes the so-called CRENL and SBKJC PPs. The CRENL PPs of Ross *et al.* [21] for the lanthanide elements have 54-electron [Xe] cores, which leaves just the 6s, 5d, and 4f electrons in the valence. They are accompanied by a set of (6s6p6d6f) basis functions, but without any contractions. For the actinide atoms [22–24], the CRENL PPs are defined by a 78-electron core, leaving the 6s-7s, 6p, 6d, and 5f electrons in the valence. The sets for Ac-Pu consist of (5s5p4d4f) and are a little smaller than those for Am-Lr, (6s7p6d5f). For most of the angular momenta in the latter case, the exponents are shared and they were reportedly optimized for the atomic ground states, although two diffuse *p* functions were optimized for the lowest $5f^{n-1}7p^1$ states. Cartesian functions seem to be used throughout the CRENL sets. The second set of shape consistent PPs, in the SBKJC family, were reported for the lanthanide atoms by Cundari and Stevens [25]. These consist of a smaller (compared to CRENL) 46-electron core, which leaves the 5s-6s, 5p, 5d, and 4f electrons in the valence space. This is analogous to the 78-electron actinide CRENL PPs. The Cundari and Stevens PPs are accompanied by a [4s4p2d2f] basis set that was contracted from a (6s6p3d7f) primitive set. The s, p, and f functions were optimized for the $4f^{n-1}$ state of Ln^{+3} with the d's optimized for the $4f^{n-1}5d^1$ state of Ln^{+2}. The $4f^{n-1}6s^1$state of Ln^{+2} was also used to obtain an s function to describe the 6s orbital. The s and p exponents of these basis sets are shared. (Note that the occupation index n for the ions has been standardized to be consistent with that of the neutral atoms.)

Model potentials using the Cowan-Griffin Hamiltonian have been reported for the lanthanide elements, first by Sakai *et al.* [26] and then by Seijo *et al.* [27]. The latter group also included the actinides in their study. Both model potentials for the lanthanides consist of a 46-electron core, while the actinide potentials utilized a 78-electron core. Primitive basis sets supplied with these potentials were larger in the Seijo *et al.* cases compared to Sakai *et al.*, i.e., (14s10p9d8f) vs. (10s7p7d6f). The work of Sakai *et al.* [26] optimized their s, p, f functions for the $4f^n6s^2$ states while the d functions were optimized for $4f^{n-1}6s^25d^1$. They also added up to 2 g-type functions. The model potentials of Seijo *et al.* [27] included spin-orbit coupling at the Wood-Boring level and their basis set contraction coefficients, obtained as general contractions, were modified to reproduce the spin-orbit corrected valence orbitals.

Energy-consistent PPs for the lanthanide and actinide atoms have been developed by the Stuttgart/Köln groups. They have the smallest core definition of any PP mentioned to this point, 28 electrons for Ln and 60 electrons for An, where the valence includes all orbitals with $n \geq 4$ for Ln and $n \geq 5$ for An. The original PPs of this type [28, 29] were based on the quasi-relativistic Wood-Boring Hamiltonian. The An PPs were accompanied by contracted spdf basis sets [29], but both the Ln and An PPs are generally used with the more extensive basis sets subsequently reported by Cao and Dolg [30–33]. For both lanthanides and actinides these latter sets are based on (14s13p10d8f6g) primitives and are available in both general and segmented contracted forms. Most of the primitives were optimized for the f^ns^2 atomic states with diffuse d and p functions optimized for $f^{n-1}d^1s^2$ and $f^{n-1}s^2p^1$ states, respectively. The sets were contracted based on ANOs from primarily CASSCF calculations (a few elements utilized MRCI) and are designed for electron correlation with an (*n*-2)spd frozen core. Recently Gulde *et al.* [34] have reported new basis sets for the lanthanide atoms in the family of "def2" basis sets [35] that are based on the PPs of Dolg *et al.* [28], SV(P)/P, TZVP/PP, and QZVP/PP. Their development was based loosely on the previously mentioned sets of Cao and Dolg [32, 33], but with exponents and (segmented) contraction coefficients optimized at the UHF level for the atomic ground states. Several "f-in-core" PPs based on the Wood-Boring Hamiltonian have also been reported by Dolg and co-workers [36–39] for both the lanthanide and actinide elements. These PPs have also been accompanied by basis sets in segmented contraction schemes [40, 41]. These sets have been designed for solid state calculations, but with additional functions specified for use in molecules.

Recently, new energy-consistent, small-core (60 electron) PPs for the f-block elements adjusted to extensive multiconfigurational Dirac-Hartree-Fock reference data (with contributions from the Breit interaction) have been reported for Ac–U [42, 43]. In these cases the PPs were tested on both atomic and molecular calculations using large ANO basis sets. As in previous work the primitive sets were optimized primarily for the f^ns^2 states with diffuse d and p functions added. The ANO contractions were based on either CASSCF or MRCI averaged density matrices.

8.4 Systematically Convergent Basis Sets for the f Block

Given the complexities of elucidating the electronic structure of atoms and molecules involving f-block elements, it is not surprising that the development of extensive basis sets for this block has lagged behind that of the rest of the periodic table. The availability of basis sets that exhibit systematic convergence towards the CBS limit is, however, extremely important for accurate and reliable studies in ab initio thermochemistry and spectroscopy. Since the errors due to basis set and the electronic structure method can be strongly coupled, it is only in the CBS limit that the inherent accuracy of the latter is apparent. This ability to systematically eliminate sources of error in a given calculation is the cornerstone of composite methods, e.g. FPD [44], HEAT [45], Wn [46] that have been applied with so much success in accurate thermochemistry and spectroscopy. As discussed below, there have recently been efforts to develop sequences of basis sets for the lanthanide and actinide elements that could be used to accurately estimate CBS limits in molecular calculations.

8.4.1 All-Electron

There are a few candidates for systematically convergent all-electron basis sets for the f block. The ANO basis sets of Roos [18, 19] are available as double, triple, and quadruple zeta sets. For the purpose of extrapolation, these sets lack i functions at the QZ level. The Sapporo segmented basis sets for the lanthanides [17] have a systematic increase in the angular momentum with basis set cardinal number as is done in the correlation consistent basis sets, with 9s7p6d4f1g for DZ, 11s9p7d5f3g1h for TZ, and 13s11p9d7f5g3h1i for QZ. These sets may be suitable for extrapolation, as the number of primitives for each angular momentum increases with the set size, and so the coverage of the correlating space is increasing. Use of the uncontracted Fægri sets [6, 7] is another possibility, though the number of primitives needed in each angular space to gain a systematic expansion in the correlating space would be large. ANO contractions of this basis, as Roos *et al.* have done [18, 19] with the earlier Fægri basis sets, would reduce the size of the basis and allow a systematic expansion. Ultimately, though, any extrapolation with these basis sets would fall a little short because the density of coverage in exponent space is not increasing with the basis set size, due to the fact that the basis sets are extended from the existing sets by adding higher angular momentum but keeping the same exponent ratios. The final candidate is the Dyall basis sets [1, 20], which were designed along similar lines to the correlation consistent basis sets. These basis sets are described in more detail next.

The SCF basis sets were developed to include a representation of the $(n+2)$p and the $(n+1)$d orbitals for all elements. The d exponents were optimized for the $n\text{f}^{n-1}(n+1)\text{d}^1(n+2)\text{s}^2$ configuration, which is the ground configuration for several of the elements in each block. The rest of the exponents for the occupied orbitals were optimized on the ground configuration, which may be either $n\text{f}^{n-1}(n+1)\text{d}^1(n+2)\text{s}^2$ or $n\text{f}^{n}(n+2)\text{s}^2$. The exception here is Th, for which the ground configuration is $6\text{d}^2 7\text{s}^2$. In this case the ground configuration was used for all but the f functions, for which the $5\text{f}^1 6\text{d}^1 7\text{s}^2$ configuration was used.

The additional functions for the $(n+2)$p were optimized on the configuration in which the two $(n+2)$s electrons were replaced by two $(n+2)$p electrons. In addition to optimizing the exponents for this orbital, the exponents for the outermost antinode of the $(n+1)$p were reoptimized, which allows some inward relaxation of the exponents for this antinode. The result is that the outermost maximum of the $(n+2)$s and p and the $(n+1)$s and p are represented by N functions for an NZ basis set; that is, in a QZ basis set there are four functions whose coefficients are of comparable magnitude and of the same sign representing the outermost maximum of these shells.

Correlating functions for the nd, nf, $(n+1)$s and p, $(n+1)$d, and $(n+2)$s shells were optimized in MR-SDCI calculations. The ground state was used for the reference configuration for the nd, nf, $(n+1)$s and p shells; for the $(n+1)$d and $(n+2)$s shells, the $n\text{f}^{n-1}(n+1)\text{d}^1(n+2)\text{s}^2$ configuration was used for the reference. The correlating sets were chosen so that the double zeta basis sets had one function with one unit higher angular momentum than the shell correlated: a g function for the f shell, a d function for the (n+1)s and p shells, an f function for the (n+1)d shell. Functions from the occupied space were used for lower angular momentum where the exponents were close to the SCF exponents. Going to triple and quadruple zeta, one more unit of angular momentum was added for each basis set, with one extra function in each angular momentum compared to the next lowest basis set. Thus, for the f shell, the correlating set was 1g for the double-zeta basis, 2g1h for the triple-zeta basis, and 3g2h1i for the quadruple-zeta basis.

In order to avoid linear dependence, some decisions must be made about which exponents to include, because there is some overlap of the exponent sets for the triple and quadruple zeta sets. For example, one g function is needed for (n+1)sp correlation, three for nf correlation, and two for (n+1)d correlation in the quadruple-zeta basis set. The behavior is illustrated in Figure 8.1. In the actinides, the g function for 6s6p correlation lies between the outermost two g functions for 5f correlation for about half the block, but then comes close to the middle g function. The outermost g function for 5f correlation is very similar to the innermost g function for 6d7s correlation for the first half of the block, but becomes tighter as Z increases. In the lanthanides, the g function for 5s5p correlation is similar to the outermost g function for 4f correlation, whereas the g functions for 5d6s correlation do not overlap the 4f correlating g set at all. The composition of the correlating set must therefore change across the block. For both lanthanides and actinides, the g functions for nd correlation overlapped fairly well with the g functions for nf correlation, so it is not necessary to include these separately. The final contracted lanthanide basis set sizes for the elements where the 4f is occupied are 8s7p5d4f1g for DZ, 10s9p7d5f3g1h for TZ, and 12s11p8d7f5g3h1i for QZ without 4s4p4d correlation; for these shells an extra 1s1p1d1f for DZ, 2s2p2d for TZ and 3s3p3d2f for QZ are added to the basis sets. For the actinides, the final contracted basis sets sizes are 9s8p6d5f1g for DZ, 11s10p7d6f3g1h for TZ, and 13s12p9d7f5g3h1i for QZ for Pa, Np, Pu, Cf-Lr, with an extra g for U, Am, Cm, and Bk. With the addition of 5d correlation, additional primitives are included: 1s1p1d1f for DZ, 2s2p2d for TZ, 3s3p1d2f for QZ for all elements and an extra h function for Pa-Bk. The lack of extra f functions for 4d and 5d correlation in the TZ basis stems from the fact that the primitives added for 4f and 5f correlation are already sufficient.

Functions for the dipole polarization (polarizability) of the nf shell and of the (n+1) d shell were also optimized. For the double-zeta basis sets, a single function was optimized on the polarizability of the atom. The ratio of the polarizing exponent to the correlating exponent was then transferred to the higher basis sets to generate diffuse functions of high angular momentum. It was not felt necessary to add diffuse functions of lower angular momentum, because the outer s, p, and d shells are already fairly diffuse, and for the most part the chemistry of these elements involves the positive ions.

8.4.2 Pseudopotential-Based

With the possible exception of the new def2-style basis sets of Weigend and co-workers [34] for the lanthanide atoms as discussed above, there have not been any standard PP-based basis sets reported in the literature for the f-block elements that are amenable to reliable extrapolation to the complete basis set limit, e.g., correlation consistent basis sets. This is currently being mitigated by work in one of the authors' groups whereby the cc-pVnZ-PP basis set family (n=D, T, Q, etc.), which have been previously developed for both transition metals and post-*d*, main group elements (see [47] and [48] and references therein), is being extended to the *f*-block [49]. As in these previous works, the sets are optimized in conjunction with the newest available small-core, energy consistent PPs that were adjusted to atomic energy differences computed at the multiconfigurational Dirac-Hartree-Fock level of theory together with perturbational estimates of the Breit interaction. Such a PP has been previously reported for the uranium atom in 2009 by Dolg and Cao [42], and those for the remaining early actinide elements (Ac, Th, Pa) have recently been completed [43]. In particular, unlike the older Stuttgart PPs for the *f*-block [28, 29] the accompanying spin-orbit potentials in these new PPs have been constructed in a manner that makes them suitable for use in 2-component DHF, DFT, or CCSD(T) calculations. The actinide PPs subsume the 60 electrons that occupy atomic orbitals with principal quantum numbers ≤ 4. The new cc-pV*n*Z-PP basis sets for U range in size from DZ through QZ and include sets for valence electron correlation (here defined as the 6s6p5f6d7s electrons), as well as those appropriate for correlation of the 5s5p5d electrons (cc-pwCV*n*Z-PP). Optimization of analogous sets for Ac–Pa are in progress.

As with all correlation consistent basis sets, both all-electron (non-relativistic and relativistic) and PP-based, regular convergence to *both* the HF limit and the CBS limit of the correlation energy are the primary goals. In the case of U, the underlying HF primitive sets range in quality: (14s13p9d8f) for DZ, (17s16p11d10f) for TZ, and (20s17p12d11f) for QZ. In each case the Gaussian exponents were fully optimized with only the constraint that the ratio of two exponents in the same angular momentum must be greater than 1.6 to avoid coalescence and linear dependency problems. Either numerical derivatives in a conjugate gradient approach or downhill Simplex was used in the optimizations within the MOLPRO suite of ab initio programs [50]. In previous basis set work on the transition metal atoms [47, 51, 52] care was taken to avoid biasing the sets towards any one of the many low-lying electronic states by optimizing exponents for average HF energies of states with varying d-orbital occupations. The strategy was similar for the U atom, although not surprisingly a bit more complicated: the (*n*s) sets were optimized for the $5f^47s^2$ state, the (*n*p) sets for the average of the $5f^37s^27p^1$ and $5f^47s^17p^1$ states, the (*n*d) sets for the average of the $5f^36d^17s^2$ and $5f^36d^27s^1$ states, and finally the (*n*f) sets to the $5f^47s^2$ and $5f^36d^17s^17p^1$ states. The energetically lowest-lying electronic state for each configuration was used throughout and were fully symmetry equivalenced, which generally required several determinants in each symmetry type. In each case the resulting primitive sets were first contracted to [3s2p1d1f] using atomic orbital (AO) coefficients from state-averaged calculations involving the $5f^47s^2$ and $5f^36d^17s^2$ states, and then an additional *d* contraction from the 6*d* AO of the $5f^36d^17s^2$ state and a *p* contracted function from a 7*p* AO obtained by averaging the $5f^47s^17p^1$ and $5f^37s^27p^1$ states were added to yield a contracted [3s3p2d1f] basis set.

As discussed above, the optimization of correlating/polarization functions can be challenging for the *f*-block elements due to both the comparable radial extents of the valence and inner-valence orbitals and the large range of occupied orbital angular momenta (s, p, d, and f). The correlating functions for occupied angular momenta were represented as ANO contractions using averaged natural orbitals obtained from MRCI calculations on the $5f^47s^2$ and $5f^36d^17s^2$ states – [1s1p1d1f] for DZ, [2s2p2d2f] for TZ, and [3s3p3d3f] for QZ with the most diffuse exponent also uncontracted in each angular momentum. All 14 electrons down through the 6s orbital were correlated in all cases. The g-type and higher angular momentum functions were optimized separately for correlation of the 6s6p5f and 6d7s electrons using the $5f^47s^2$ and $5f^36d^17s^2$ states, respectively. In order to reduce the overlap of exponents, however, all of these correlating functions were optimized together at the MRCI level (iterating until consistent) with the same constraint on the exponents as mentioned above for the HF primitives. Somewhat typical shells of correlating functions were chosen: (a) 1g for 6s6p5f correlation for DZ, (b) 2g1h for 6s6p5f correlation and 1g for 6d7s correlation for TZ, and (c) 3g2h1i for 6s6p5f and 2g1h for 6d7s correlation for QZ. The final contracted set sizes corresponded to [5s5p4d3f1g], [6s6p5d4f3g1h], and [7s7p6d5f5g3h1i] for DZ, TZ, and QZ, respectively. Finally additional functions for correlating the inner 5s5p5d electrons – 1s1p1d1f for wCVDZ, 2s2p2d2f1g for wCVTZ, and 2s2p2d3f2g1h for wCVQZ – were optimized in the usual weighted core-valence scheme [53] for the $5f^47s^2$ state and added to the previously described valence sets to produce cc-pwCV*n*Z-PP basis sets. In the wCVQZ case there was significant overlap between the tightest g function in the valence set, so all the valence g's were reoptimized (for valence correlation) along with the 5s5p5d correlating g's. This approach is consistent with the construction of the valence sets, but of course different from that used in the Dyall sets [1, 20] where one of the functions would have been dropped due to linear dependency.

8.5 Basis Set Convergence in Molecular Calculations

In order to compare the basis set convergence characteristics of the U atom in comparison to the well-known behavior of lighter systems, a number of molecular HF and CCSD(T) calculations were carried out with the new cc-pV*n*Z-PP and cc-pwCV*n*Z-PP basis sets for U.

Figure 8.6 displays the basis set convergence characteristics, for both the HF energy and the frozen-core CCSD(T) correlation energy, of the UF_6, SF_6, and WF_6 molecules. In these calculations the aug-cc-pV*n*Z sequence of sets was used for F [54, 55], cc-pV$(n+d)$Z for S [56], and cc-pV*n*Z-PP for both W and U [47, 49]. The n=Q CCSD(T) equilibrium geometries were used throughout. In each case the basis set incompleteness error (BSIE), estimated using CBS limits obtained using the HF extrapolation formula of Karton and Martin [57] and a relation for the CCSD(T) correlation energy [58, 59], is plotted as a function of the basis set. The HF basis set convergence is observed to be very similar between all three species, with the BSIE at the DZ level slightly smaller for UF_6, but all three are nearly identical at the TZ and QZ levels. For the valence correlation energy (6s6p5f6d7s for U, 2s2p for F, 3s3p for S, 5d6s for W), even with the larger number of electrons correlated in the UF_6 case, the convergence rate is very similar to that of SF_6 and WF_6, with UF_6 being only slightly slower towards the CBS limit.

Figure 8.7 shows the convergence of the core-valence correlation contributions using cc-pwCV*n*Z-PP (heavy atoms) and aug-cc-pwCV*n*Z (light atoms) basis sets [53]. In this case results for UF_6 are compared to those of SeF_6 since the core in both cases corresponds to a set of 18 *spd* electrons. The convergence of the core-valence correlation *effect* [difference of core correlated and valence correlated CCSD(T) calculations] with basis set is shown for 5d/3d correlation and also for 5s5p5d/3s3p3d correlation (U/Se). The 1s electrons of F are not correlated in all cases. In each case the convergence is smooth, with the UF_6 results approaching the estimated CBS limit slightly faster. The key, however, is that the UF_6 results exhibit a very similar systematic convergence as the SeF_6 case, which is encouraging for the

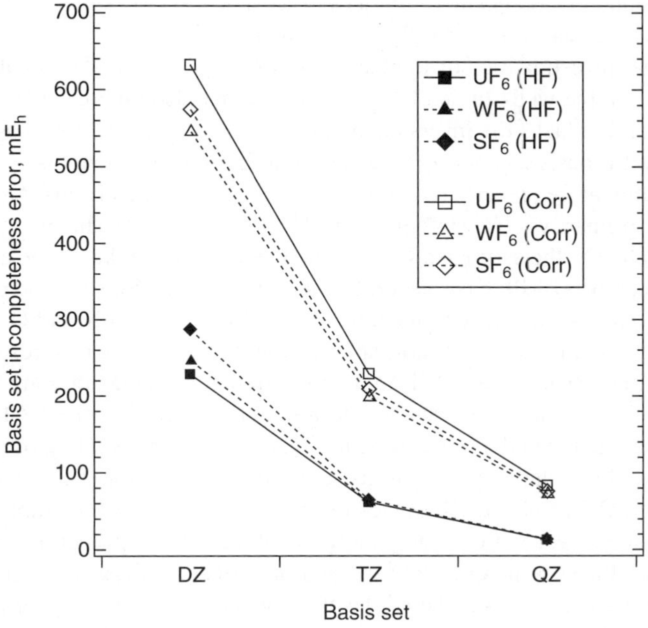

Figure 8.6 *Convergence of the HF energy and valence CCSD(T) correlation energy as a function of correlation consistent basis set. The values are given relative to estimated CBS limits*

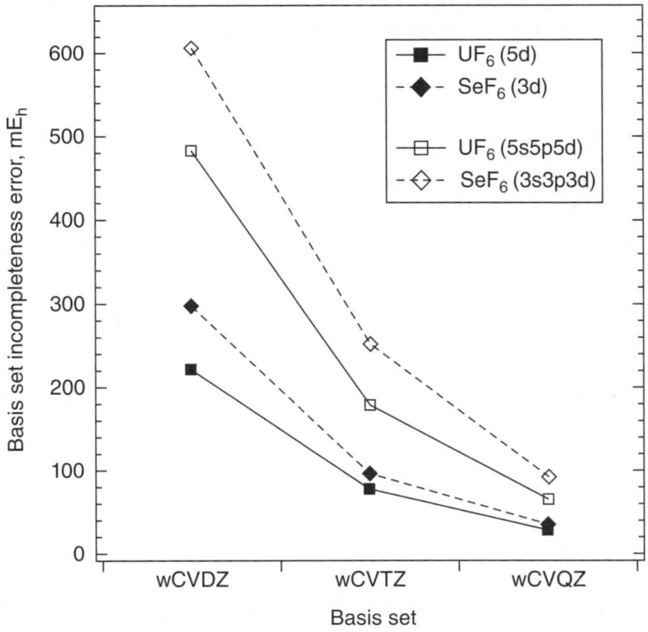

Figure 8.7 *Convergence of the CCSD(T) correlation energy for outer-core electrons using core-valence correlation consistent basis sets. Results are given relative to estimated CBS limits*

application of standard composite techniques to actinide thermochemistry and spectroscopy at least as far as the basis set approach is concerned.

For additional investigations of the basis set convergence, CCSD(T) calculations were also carried out on the well-studied UO_2^{+2} molecule, but also on NUO^+, which is isoelectronic and linear, but lacks the inversion symmetry of the uranyl cation. The two panels of Figure 8.8 show the basis set convergence of the equilibrium bond lengths as well as all three harmonic vibrational frequencies (ω_1 and ω_3 are the symmetric and antisymmetric stretches in UO_2^{+2}, ω_1 is primarily a UN stretch in NUO^+, ω_3 is a UO stretch, and ω_2 is the bend in both cases) relative to the estimated CBS limit. The latter was taken to be a CCSD(T) calculation with a cc-pVQZ-PP basis set on U with 3 i-type functions instead of 1 (exponents equal to the h-type functions) with the aug-cc-pV5Z basis set on N and O. Both the bond lengths and stretching frequencies show smooth, rapid convergence towards their respective CBS limits. At the DZ level the BSIE in the two stretching frequencies are relatively large at around 20–35 cm^{-1}, but this decreases to less than 10 cm^{-1} at the TZ level. The bond lengths are converged to under 0.005 Å with a TZ basis set. The bending frequency exhibits what is presumably a relatively large basis set superposition error (BSSE) at the DZ level, but both TZ and QZ exhibit BSIE values of just a few cm^{-1}. An attempt was also made to estimate the importance of each angular momentum to these properties by repeating the calculations with the U cc-pVQZ-PP basis set truncated at f-only, g-only, and h-only levels. Figure 8.9 shows these results at the HF level of theory using a full aug-cc-pVQZ basis set on the light atoms. Except for the bending frequency, very similar results were obtained using the aug-cc-pVTZ on N and O. At the HF level both UO_2^{+2} and NUO^+ show relatively minor contributions due to h and i functions. This is particularly true for NUO^+ where even g functions are relatively unimportant for the stretching vibrational frequencies at the HF level. For UO_2^{+2}, however, g functions are obviously more important than the higher angular

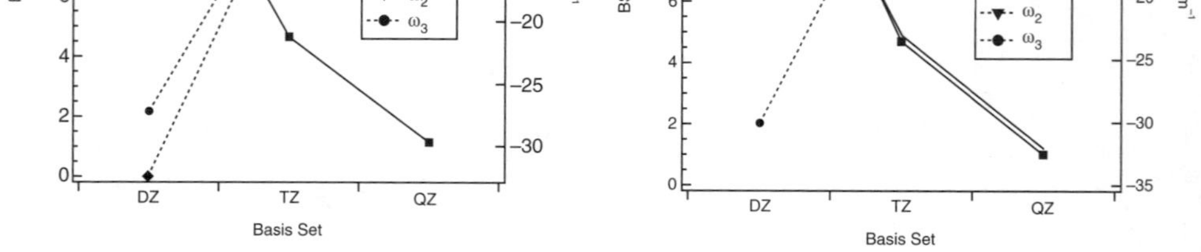

Figure 8.8 Convergence of frozen-core CCSD(T) bond lengths and harmonic frequencies with basis set for UO_2^{+2} relative to the cc-pVQZ-PP(3i)/aug-cc-pV5Z basis set result (U/O). Convergence of frozen-core CCSD(T) bond lengths and harmonic frequencies with basis set for NUO^+ relative to the cc-pVQZ-PP(3i)/aug-cc-pV5Z basis set result (U/O,N)

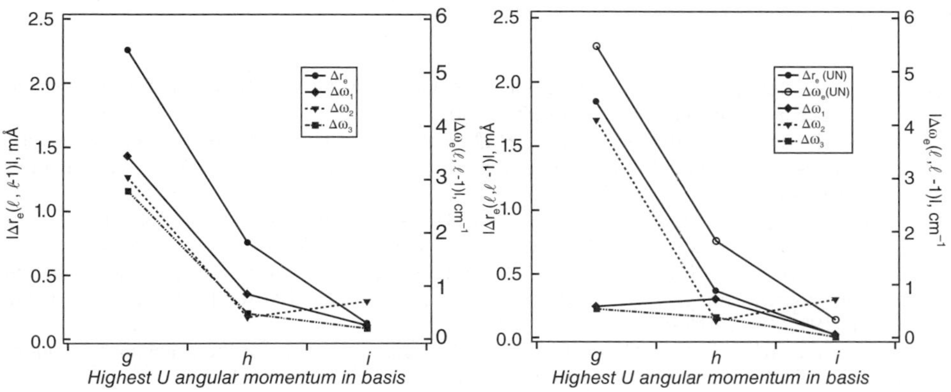

Figure 8.9 *Incremental effect on the HF spectroscopic properties of truncating the U cc-pVQZ-PP basis set to different level of angular momentum functions for UO_2^{+2}. The aug-cc-pVTZ basis set was used on O. The largest set replaced the single i-type function in cc-pVQZ-PP with 3. Incremental effect on the HF spectroscopic properties of truncating the U cc-pVQZ-PP basis set to different level of angular momentum functions for NUO^+. The aug-cc-pVTZ basis set was used on O and N. The largest set replaced the single i-type function in cc-pVQZ-PP with 3*

momentum functions for all three vibrational frequencies. Even so, overall these HF contributions are small and do not increase appreciably when the smaller aug-cc-pVTZ basis set is used on N and O. The situation is much more ambiguous, however, at the CCSD(T) level. Presumably due to BSSE, many of these same results were qualitatively different depending on which light atom basis set was used. Regardless of the light atom basis set, however, the bending frequency was strongly affected by the inclusion of *g* functions in the UO_2^{+2} case. Inclusion of *g* functions raised ω_2 by 25.0 and 11.2 cm^{-1} when aug-cc-pVTZ or aug-cc-pVQZ was used on O, respectively, at the CCSD(T) level of theory.

8.6 Conclusions

The f block presents considerable challenges for the development of basis sets, particularly if systematic convergence to the basis set limit is desired. These challenges arise from the necessity to correlate three shells, and from the changing behavior of the f shell in particular across the blocks. Accordingly, though there are quite a few basis sets available, both at the PP and the AE level, there are few that are suitable for systematic convergence studies. This deficiency, particularly for PPs, is being filled in by the development of new basis sets, for which sample calculations on U show the same convergence characteristics as calculations on main-group or transition-metal compounds with correlation consistent basis sets.

Acknowledgments

K.A.P gratefully acknowledges support from the Department of Energy, Basic Energy Sciences, Heavy Element program (DE-FG02-12ER16329).

References

[1] Dyall KG. Relativistic double-zeta, triple-zeta, and quadruple-zeta basis sets for the actinides Ac-Lr. Theor Chem Acc. 2007;117(4):491–500.
[2] Malli GL, da Silva ABF, Ishikawa Y. Highly accurate relativistic universal Gaussian basis set: Dirac–Fock–Coulomb calculations for atomic systems up to nobelium. J Chem Phys. 1994;101(8):6829–33.
[3] Malli GL, da Silva ABF, Ishikawa Y. Universal Gaussian basis set for accurate ab initio relativistic Dirac-Fock calculations. Phys Rev A. 1993;47(1):143–6.
[4] Jorge FE, de Castro EVR, Da Silva ABF. A universal Gaussian basis set for atoms cerium through lawrencium generated with the generator coordinate Hartree-Fock method. J Comp Chem. 1997;18(13):1565–9.
[5] Huzinaga S, Klobukowski M. Well-tempered Gaussian basis sets for the calculation of matrix Hartree—Fock wavefunctions. Chem Phys Lett. 1993;212(3–4):260–4.
[6] Faegri K. Relativistic Gaussian basis sets for the elements K-Uuo. Theor Chem Acc. 2001;105(3):252–8.
[7] Fægri K. Even tempered basis sets for four-component relativistic quantum chemistry. Chem Phys. 2005;311(1–2):25–34.
[8] Koga T, Tatewaki H, Matsuoka O. Relativistic Gaussian basis sets for molecular calculations: Tl-Lr. J Chem Phys. 2003;119(2):1279–80.
[9] Koga T, Tatewaki H, Shimazaki T. Chemically reliable uncontracted Gaussian-type basis sets for atoms H to Lr. Chem Phys Lett. 2000;328(4–6):473–82.
[10] Nakajima T, Hirao K. Accurate relativistic Gaussian basis sets determined by the third-order Douglas-Kroll approximation with a finite-nucleus model. J Chem Phys. 2002;116:8270.
[11] Tsuchiya T, Abe M, Nakajima T, Hirao K. Accurate relativistic Gaussian basis sets for H through Lr determined by atomic self-consistent field calculations with the third-order Douglas-Kroll approximation. J Chem Phys. 2001;115(10):4463–72.
[12] Pantazis DA, Neese F. All-Electron Scalar Relativistic Basis Sets for the Lanthanides. J Chem Theory Comp. 2009;5(9):2229–38.
[13] Pantazis DA, Neese F. All-Electron Scalar Relativistic Basis Sets for the Actinides. J Chem Theory Comp. 2011;7(3):677–84.
[14] Dolg M. Segmented Contracted Douglas–Kroll–Hess Adapted Basis Sets for Lanthanides. J Chem Theory Comp. 2011;7:3131–42.
[15] Sekiya M, Noro T, Miyoshi E, Osanai Y, Koga T. Relativistic correlating basis sets for lanthanide atoms from Ce to Lu. J Comp Chem. 2006;27(4):463–70.
[16] Noro T, Sekiya M, Osanai Y, Koga T, Matsuyama H. Relativistic correlating basis sets for actinide atoms from Th-90 to Lr-103. J Comp Chem. 2007;28(16):2511–6.
[17] Sekiya M, Noro T, Koga T, Shimazaki T. Relativistic segmented contraction basis sets with core-valence correlation effects for atoms ^{57}La through ^{71}Lu: Sapporo-DK-nZP sets (n = D, T, Q). Theor Chem Acc. 2012;131(7).
[18] Roos BO, Lindh R, Malmqvist P-A, Veryazov V, Widmark P-O, Borin AC. New Relativistic Atomic Natural Orbital Basis Sets for Lanthanide Atoms with Applications to the Ce Diatom and LuF_3. J Phys Chem A. 2008;112(45):11431–5.
[19] Roos BO, Lindh R, Malmqvist PA, Veryazov V, Widmark PO. New relativistic ANO basis sets for actinide atoms. Chem Phys Lett. 2005;409(4–6):295–9.
[20] Gomes AP, Dyall KG, Visscher L. Relativistic double-zeta, triple-zeta, and quadruple-zeta basis sets for the lanthanides La-Lu. Theor Chem Acc. 2010;127(4):369–81.
[21] Ross RB, Gayen S, Ermler WC. Ab initio relativistic effective potentials with spin–orbit operators. V. Ce through Lu. J Chem Phys. 1994;100(11):8145.
[22] Ermler WC, Ross RB, Christiansen PA. Ab initio relativistic effective potentials with spin-orbit operators. VI. Fr through Pu. Int J Quantum Chem. 1991;40(6):829–46.
[23] Nash CS, Bursten BE, Ermler WC. Erratum: "Ab initio relativistic effective potentials with spin-orbit operators. VII. Am through element 118" [J. Chem. Phys. 106, 5133 (1997)]. J Chem Phys. 1999;111(5):2347.

[24] Nash CS, Bursten BE, Ermler WC. Ab initio relativistic effective potentials with spin-orbit operators. VII. Am through element 118. J Chem Phys. 1997;106(12):5133.
[25] Cundari TR, Stevens WJ. Effective core potential methods for the lanthanides. J Chem Phys. 1993;98(7):5555.
[26] Sakai Y, Miyoshi E, Tatewaki H. Model core potentials for the lanthanides. Journal of Molecular Structure: THEOCHEM. 1998;451(1–2):143–50.
[27] Seijo L, Barandiaraìn Z, Harguindey E. The ab initio model potential method: Lanthanide and actinide elements. J Chem Phys. 2001;114(1):118–29.
[28] Dolg M, Stoll H, Preuss H. Energy-adjusted ab initio pseudopotentials for the rare earth elements. J Chem Phys. 1989;90(3):1730–4.
[29] Kuchle W, Dolg M, Stoll H, Preuss H. Energy-adjusted pseudopotentials for the actinides — parameter sets and test calculations for thorium and thorium monoxide. J Chem Phys. 1994;100(10):7535–42.
[30] Cao XY, Dolg M. Segmented contraction scheme for small-core actinide pseudopotential basis sets. Journal of Molecular Structure: THEOCHEM. 2004;673(1–3):203–9.
[31] Cao XY, Dolg M, Stoll H. Valence basis sets for relativistic energy-consistent small-core actinide pseudopotentials. J Chem Phys. 2003;118(2):487–96.
[32] Cao XY, Dolg M. Segmented contraction scheme for small-core lanthanide pseudopotential basis sets. Journal of Molecular Structure: THEOCHEM. 2002;581:139–47.
[33] Cao XY, Dolg M. Valence basis sets for relativistic energy-consistent small-core lanthanide pseudopotentials. J Chem Phys. 2001;115(16):7348–55.
[34] Gulde R, Pollak P, Weigend F. Error-Balanced Segmented Contracted Basis Sets of Double- to Quadruple- Valence Quality for the Lanthanides. J Chem Theory Comp. 2013;8(11):4062–8.
[35] Weigend F, Ahlrichs R. Balanced basis sets of split valence, triple zeta valence and quadruple zeta valence quality for H to Rn: Design and assessment of accuracy. Phys Chem Chem Phys. 2005;7:3297–305.
[36] Dolg M, Stoll H, Preuss H. A combination of quasirelativistic pseudopotential and ligand field calculations for lanthanoid compounds. Theor Chim Acta. 1993;85:441–50.
[37] Hülsen M, Weigand A, Dolg M. Quasirelativistic energy-consistent 4f-in-core pseudopotentials for tetravalent lanthanide elements. Theor Chem Acc. 2009;122(1–2):23–9.
[38] Moritz A, Cao X, Dolg M. Quasirelativistic energy-consistent 5f-in-core pseudopotentials for divalent and tetravalent actinide elements. Theor Chem Acc. 2007;118(5–6):845–54.
[39] Moritz A, Cao X, Dolg M. Quasirelativistic energy-consistent 5f-in-core pseudopotentials for trivalent actinide elements. Theor Chem Acc. 2007;117(4):473–81.
[40] Hülsen M, Dolg M, Link P, Ruschewitz U. Improved valence basis sets for divalent lanthanide 4f-in-core pseudopotentials. Theor Chem Acc. 2010;129(3–5):367–79.
[41] Yang J, Dolg M. Valence basis sets for lanthanide 4f-in-core pseudopotentials adapted for crystal orbital ab initio calculations. Theor Chem Acc. 2005;113(4):212–24.
[42] Dolg M, Cao X. Accurate Relativistic Small-Core Pseudopotentials for Actinides. Energy Adjustment for Uranium and First Applications to Uranium Hydride. J Phys Chem A. 2009;113(45):12573–81.
[43] Weigand A, Cao XY, Hangele T, Dolg M. Relativistic Small-core Pseudopotentials for Actinium, Thorium, and Protactinium. J. Phys. Chem. A 2014; 118(13):2519–30.
[44] Feller D, Peterson KA, Dixon DA. A survey of factors contributing to accurate theoretical predictions of atomization energies and molecular structures. J Chem Phys. 2008;129(20):204105.
[45] Harding ME, Vazquez J, Ruscic B, Wilson AK, Gauss J, Stanton JF. High-accuracy extrapolated ab initio thermochemistry. III. Additional improvements and overview. J Chem Phys. 2008;128(11):114111.
[46] Karton A, Rabinovich E, Martin JML, Ruscic B. W4 theory for computational thermochemistry: In pursuit of confident sub-kJ/mol predictions. J Chem Phys. 2006;125(14):144108.
[47] Figgen D, Peterson KA, Dolg M, Stoll H. Energy-consistent pseudopotentials and correlation consistent basis sets for the 5d elements Hf–Pt. J Chem Phys. 2009;130:164108.
[48] Peterson KA, Yousaf KE. Molecular core-valence correlation effects involving the post-d elements Ga–Rn: Benchmarks and new pseudopotential-based correlation consistent basis sets. J Chem Phys. 2010;133:174116.

[49] Peterson KA. Correlation consistent basis sets for actinides. I. The thorium and uranium atoms. In preparation.
[50] MOLPRO, version 2012.1, a package of ab initio programs, H.-J. Werner, P. J. Knowles, G. Knizia, F. R. Manby, M. Schütz, and others, see http://www.molpro.net.
[51] Peterson KA, Figgen D, Dolg M, Stoll H. Energy-consistent relativistic pseudopotentials and correlation consistent basis sets for the 4d elements Y-Pd. J Chem Phys. 2007;126:124101.
[52] Balabanov NB, Peterson KA. Systematically convergent basis sets for transition metals. I. All-electron correlation consistent basis sets for the 3d elements Sc-Zn. J Chem Phys. 2005;123:064107.
[53] Peterson KA, Dunning Jr. TH Accurate correlation consistent basis sets for molecular core-valence correlation effects. The second row atoms Al - Ar, and the first row atoms B - Ne revisited. J Chem Phys. 2002;117:10548–60.
[54] Kendall RA, Dunning Jr TH, Harrison RJ. Electron affinities of the first row atoms revisited. Systematic basis sets and wave functions. J Chem Phys. 1992;96:6796.
[55] Dunning Jr TH. Gaussian basis sets for use in correlated molecular calculations. I. The atoms boron through neon and hydrogen. J Chem Phys. 1989;90:1007.
[56] Dunning Jr TH, Peterson KA, Wilson AK. Gaussian basis sets for use in correlated molecular calculations. X. The atoms aluminum through argon revisited. J Chem Phys. 2001;114:9244.
[57] Karton A, Martin JML. Comment on: "Estimating the Hartree-Fock limit from finite basis set calculations" [Jensen F (2005) Theor Chem Acc 113, 267]. Theor Chem Acc. 2006;115:330–3.
[58] Martin JML. Ab initio total atomization energies of small molecules — towards the basis set limit. Chem Phys Lett. 1996;259:669.
[59] Feller D, Peterson KA, Hill JG. On the effectiveness of CCSD(T) complete basis set extrapolations for atomization energies. J Chem Phys. 2011;135:044102.

9

$4f$, $5d$, $6s$, and Impurity-Trapped Exciton States of Lanthanides in Solids

Zoila Barandiarán and Luis Seijo

Department of Chemistry, Autonomous University of Madrid

9.1 Introduction

The basic element of a large number of solid-state optical devices of technological interest is an imperfect crystal that incorporates a lanthanide ion as a substitutional impurity. Lanthanide-activated solid-state lasers, scintillators, and lamps reflect the ability of the lanthanide ions to absorb, transfer, exchange, and emit light under the influence of the host crystal, which deforms and multiplies the lanthanide's atomic electronic states in various ways and extents, until local electronic states arise in the imperfect crystal, which are rooted in the lanthanide and capable of administering a variety of energy-related processes.

Probably as a result of the considerable ionic nature that the lanthanide keeps while establishing bonds with nearest neighbors and interacting with the reminder of the host, the lanthanide free ion parentage becomes very visible as the local electronic states of the defect are studied. They can be associated with open-shell electronic structures where the atomic $4f$, $5d$, and $6s$ shells dominate the molecular orbital mixing as much as atomic-like $4f^N$, $4f^{N-1}5d$, and $4f^{N-1}6s$ open-shell configurations lead the multiconfigurational expansions. These characteristics are far from formal. Rather, they are basic characteristics that must be taken into account to favor efficient and increasing understanding of the materials. They are widely recognized by expert experimentalists and theoreticians alike because they are extremely useful to refer to groups of states associated to which significantly different

Computational Methods in Lanthanide and Actinide Chemistry, First Edition. Edited by Michael Dolg.

structural and optical behaviors can be ascribed. These discussions will become more clear throughout this chapter.

Doped lanthanides in solids are most stable as trivalent ions (Ln^{3+}). Some elements can be stabilized, in certain hosts, as divalent ions (Ln^{2+}) as well, but other oxidation states are rare [1–3]. The ground configuration is $[Xe]4f^N$ in the trivalent lanthanide series ($4f^N$ from now on) and the $4f$-shell occupancy uniformly grows from Ce^{3+} ($N = 1$) to Lu^{3+} ($N = 14$), leading to a number of electronic states, which is maximal in the middle of the series, varying from 14 Ce^{3+}(Yb^{3+}) to 91 Pr^{3+}(Tm^{3+}), 364 Nd^{3+}(Er^{3+}), 1001 Pm^{3+}(Ho^{3+}), 2002 Sm^{3+}(Dy^{3+}), 3003 Eu^{3+}(Tb^{3+}), and 3432 Gd^{3+}, as shown by Dieke and Carnall [1, 2]. This is also the ground configuration in the Ln^{2+} isoelectronic series, with some exceptions for the lighter lanthanides.

The impurity $4f$ shell can be described as a very internal open-shell whose electrons are hidden from the crystal environment behind the $5s$ and $5p$ shells, which are, to some extent, valence shells. This explains their secondary role in determining the bond lengths of ground and excited $4f^N$ states, as well as the little dependence of the $4f \rightarrow 4f$ spectra on the host, which results in a very close correspondence between the $4f \rightarrow 4f$ spectra in crystals and in gas phase, which is the basis of the success of the well-known parametric Crystal Field Theory [2, 4]. In contrast, the energy levels of the tight and internal $4f^N$ shell are very sensitive to electron-electron repulsion, electron correlation, and relativistic effects, which provoke the appearance of energy gaps between members of the $4f^N$ manifold, which are large enough to prevent non-radiative decay, making it possible for the existence of several metastable exited states, which can either luminesce, further absorb new photons prior to higher energy emissions, or originate a cascade of shorter energy emissions upon high energy absorption, leading, respectively, to proven f-f multiple luminescence, upconversion luminescence, or quantum cutting, as well as to other optical phenomena that contribute to their applied interest. Note that f is a common shorthand notation used to refer to $4f$ or to $4f^N$; disambiguation relies on the context. The same is true for d relative to $5d$ or $4f^{N-1}5d$. These terms will be used throughout this chapter.

In addition to the wealth of the $4f^N$ manifold and its intraconfigurational f-f transitions and processes, very different excited states and energy-related mechanisms appear as one electron is excited to the higher $5d$ shell leading to the $4f^{N-1}5d$ manifolds. Only at the edges of the series, where the number of $4f^N$ states and their energy range is small the $4f^{N-1}5d$ excited states lie clearly above in energy; otherwise, the two manifolds overlap to some extent. The $4f^N$ and $4f^{N-1}5d$ states have different parity and this has a strong impact on transition selection rules and, hence, transition intensities, when the impurity occupies centrosymmetric sites; it still affects the spectra when it is lost by low site symmetry. But, parity is not the only important difference between these two configurations. Other new characteristics accompany the $4f^{N-1}5d$ manifolds, which also become available to tune the material's optical behavior and response.

It is useful to see the $4f^{N-1}5d$ manifold as the result of the coupling of the $4f^{N-1}$ subshell with the $5d$ electron because each part conveys its own electronic structure characteristics. So, the $4f^{N-1}$ subshell carries its typical energy gaps due to f-f repulsions and spin-orbit coupling, commented above, whereas the $5d$ part imprints some transition metal character to the lanthanide through the typical, well-known sensitivity of the d shells to chemical environment, to the crystal environment here, which extends and splits further the complex $4f^{N-1}$ energy level structure upon $4f^{N-1} \otimes 5d$ coupling. Yet, another point of view is also needed: The two parts together, the $4f^{N-1}5d$ manifolds, resulting, basically, from single $4f \rightarrow 5d$ excitations from the $4f^N$ ground manifold, strongly modify the electron correlation balance in the latter. On the one hand, tight f-f electron pairs are broken and substituted by more distant f-d pairs, which has increasing importance across the series, particularly in its second half, where the so-called heavy lanthanides ($N > 7$) are reached. On the other hand, correlation with electrons of the valence shells of the impurity first neighbors, the ligands, becomes more important for the chemically more active $5d$ shell than for the shielded $4f$ shell, hence for the $4f^{N-1}5d$ manifolds. All of this affects the theoretical models applicable as much as models the materials local structure and optical properties, as will be discussed in this chapter.

The fast and intense emissions that may occur from some members of the $4f^{N-1}5d$ manifold, together with the possibility of tuning their wavelength and band shape by changing the chemical environment (changing the host, co-doping other impurities, modifying the site symmetry by doping a non-stoichiometric host, etc.) have widened the fields of applications of lanthanide ions in solids to areas of enormous societal interest among which, solid-state lighting and medical imaging can be singled out, and have provoked an increase in the number of studies focused on the $4f^{N-1}5d$ optical properties. Yet, their investigation and applicability face difficulties and/or limitations of various origins, which challenge the experimental research as much as the theoretical models, which can no longer be based on parametric crystal field theory nor disregard strong correlation and relativistic effects.

Mimicking the variation observed for Ln^{3+} in gas phase, the energy onset of the $4f^{N-1}5d$ spectrum in crystals increases sharply towards half/full occupation of the $4f^{N-1}$ subshell: Roughly, it grows from Ce^3+ ($4f^15d$) to Gd3+ ($4f^65d$), and, again, from Tb^3+ ($4f^75d$) to Lu^3+ ($4f^{13}5d$) (being the onset of Tb^{3+} higher than that of Ce^{3+}, in general). Hence, often the $4f \rightarrow 5d$ transitions are very high in energy, lying in the ultra-violet (UV) or vacuum UV, or simply higher than the strong host absorption, so that they cannot be observed except for a few Ln^{3+} ions doped in hosts that preferably create strong crystal fields and have large transparency windows or band gaps (insulators). Consistent with this, the number of works on Ln^{2+} has experienced an increase, since their $4f^{N-1}5d$ states lie much lower in energy than those of isoelectronic Ln^{3+}; however, Ln^{2+} ions are more unstable against oxidation than Ln^{3+}, which poses more challenges to the syntheses of the materials. In any case, the occurrence or the efficiency of the $4f^{N-1}5d$ luminescence can be also compromised by non-radiative decay due to different, not well-understood, quenching mechanisms. Their strong coupling with the host vibrations, particularly with the local ones, favors non-radiative decay. It is also favored when the energy gap between the potentially emitting $4f^{N-1}5d$ state and the $4f^N$ state lying next and lower to it in energy is not large enough to prevent multiphonon relaxation, a circumstance that may vary from one host to another, since the variation in chemical environment can shift the $4f^{N-1}5d$ states while the $4f^N$ levels remain basically unaffected. The $4f^{N-1}5d$ luminescence of phosphors and scintillators can also be quenched (and, eventually, favored) due to interaction with impurity-trapped excitons. This is the case of the so-called "anomalous" emissions characterized by extremely wide bands (resembling host bands) that peak at much lower energies than expected from the pumped $4f^{N-1}5d$ states. Impurity-trapped excitons are also believed to play a role in the host-to-activator energy transfer mechanisms that either favor or reduce the emission of scintillation photons from the $4f^{N-1}5d$ states of the active center upon ionizing radiation.

Excitons trapped at impurity centers in highly ionic crystals were first described by McClure and Pédrini [5] as excited states consisting of a bound electron-hole pair with the hole localized on the impurity and the electron on the nearby lattice sites, and a very short impurity–ligand bond length. The study of the electronic structure of some actinide and lanthanide impurities in crystals using methods of solid state quantum chemistry has revealed the existence of electronic states whose characteristics match this description. Analyses of their wavefunctions, their potential energy surfaces, and their local geometry reveal a significant charge leak outside the first coordination shell, which favors a significant impurity-ligand bond shrinkage, towards values associated with the oxidized impurity. Also apparent are the numerous crossings and avoided crossings that occur when impurity-trapped excitons coincide in energy with impurity states, particularly of the $4f^{N-1}5d$ and $4f^{N-1}6s$ manifolds. Their microscopic description is probably close to the limits of what can be studied using embedded-cluster approximations, hence, it requires considerable extensions of their framework. As a matter of fact, they are still a challenge for theoretical methods, which, so far, have not been able to produce an accurate description of the anomalous emissions observed experimentally, even though routes for progress from accumulated knowledge are foreseen.

Altogether, impurity states and impurity-trapped excitons define the realm of lanthanide activated solid-state materials. This is a realm where experiment and theory should meet but where the research work conducted is overwhelmingly experimental. Their structure and optical properties are complex and rich. They are a genuine challenge for quantum chemists. What is needed is not massive production of theoretical results, which follow experiments (which, in any case, would probably be very difficult to attain, given the pace of experimental work and sophistication of the theoretical methods applicable). What is needed is to answer basic questions that cannot be answered by experimental techniques alone so that their electronic structures are mastered beyond simple model and beyond empirical model descriptions, to the point where the intensive and constant search for new materials could count on the ability to predict, which is characteristic of *ab initio* quantum chemical methods when it is found how to stretch them to the limits of their capabilities.

9.2 Methods

The methods of choice must be adequate for manifolds of electronic states that are localized around a lanthanide ion in a solid host. The combination of a solid environment, a heavy element, and $4f$, $5d$, and other open-shells, demands the consideration of the effects of the solid host, the use of relativistic Hamiltonians up to spin-orbit coupling, the correct treatment of static and dynamic correlation, and handling large manifolds of quasi-degenerate excited states. We decided to use embedded-cluster wavefunction theory-based (EC-WFT) methods, with a two-component relativistic Hamiltonian to be used in two-steps, a multiconfigurational variational treatment of static correlation, and a multireference second-order perturbation theory treatment of dynamic correlation.

As an alternative, density functional theory methods with periodic boundary conditions (PBC-DFT) mean a better description of the solid host effects, and they may provide better ground state structures. Besides, they can be used in more extended solid defects that involve a larger number of atoms. They are not so adequate, however, to provide excited state structures and high quality absorption and emission transition energies. In consequence with this, we reserve the use of PBC-DFT methods to the study of the ground state structures of materials involving complex defects, like those resulting when a material doped with an optically active lanthanide ion is co-doped with other impurities in an attempt to control the luminescent properties of the active centers.

9.2.1 Embedded-Cluster Methods

Since the pioneering cluster calculation on the $KNiF_3$ solid of Shulman and Sugano [6] there has been a wide variety of proposals of procedures to handle relatively localized electronic states of a solid with a molecule-like Hamiltonian that includes the relevant solid host effects, depending on the type of solids and on methodological flavors (Green's functions, wavefunctions, density functional, etc.). A recent summary of practical methods can be found in Huang and Carter [7]. Here we describe our choice of embedded-cluster method, particularly useful in ionic materials.

9.2.1.1 Embedded-Cluster Hamiltonian

The embedded-cluster Hamiltonian we use is based on the group function theory for two or more groups of electrons without mutual correlation, as developed by McWeeny and Huzinaga [8–10]. If we assume this is the case for the interactions between a group of electrons localized on a particular region of the solid close to a number of atoms (which we call *cluster*) and other groups of electrons that are distributed over the rest of the solid (which we call *embedding host*), then, the ground state and excited state wavefunctions of the set of N_e^C cluster electrons under the effects of the embedding host can be approximately computed as the eigenfunctions of an embedded-cluster Hamiltonian $\hat{H}^{\mathrm{EC}}$ that is the sum of the Hamiltonian of the otherwise isolated cluster $\hat{H}^{\mathrm{IC}}$ and a host embedding Hamiltonian $\hat{H}^{\mathrm{emb}}$. The latter is a sum over the N_e^C cluster electrons of one-electron embedding operators $\hat{V}^{\mathrm{emb}}(i)$. $\hat{V}^{\mathrm{emb}}$ is known as the embedding potential.

$$\hat{H}^{\mathrm{EC}} = \hat{H}^{\mathrm{IC}} + \hat{H}^{\mathrm{emb}} = \hat{H}^{\mathrm{IC}} + \sum_{i}^{N_e^C} \hat{V}^{\mathrm{emb}}(i). \tag{9.1}$$

The isolated cluster Hamiltonian $\hat{H}^{\mathrm{IC}}$ can be either the non-relativistic many-electron Hamiltonian or a suitable relativistic choice, both in their all-electron versions or in any effective core potential version. We will discuss $\hat{H}^{\mathrm{IC}}$ later. The embedding potential acting on the cluster electrons reads:

$$\hat{V}^{\mathrm{emb}}(i) = -\sum_{\xi}^{N_n^H} \frac{Z_\xi}{r_{\xi i}} + \sum_{\mu}^{N_{oo}^H} \left[f_\mu \int \frac{\varphi_\mu^*(j)[2 - \hat{P}_{ij}]\varphi_\mu(j)}{r_{ij}} d\tau_j + B_\mu |\varphi_\mu\rangle\langle\varphi_\mu| \right]. \tag{9.2}$$

It includes the following interactions experienced by one cluster electron: First, the electrostatic interaction with the N_n^H nuclei ξ of the embedding host, and second, the Coulomb and exchange interactions with the embedding host electrons, whose charge density is represented by means of the N_{oo}^H occupied orbitals φ_μ (with respective fractional occupancies f_μ), and finally, the Pauli repulsion with the embedding host electrons. This term is positive for non-converged cluster orbitals (all constants B_μ are positive), and it prevents their collapse onto the host orbitals (and the over occupancy of the latter, so called orbital collapse); it is zero for fully converged cluster orbitals in a complete basis set.

Embedding AIMPs. The set of embedding host orbitals φ_μ in equation 9.2 can be any unitary transformation of the occupied host orbitals so that both delocalized and localized orbitals can be used. In an ionic crystal, host localized orbitals can often be found that only contain small mixing between orbitals of the individual ions, ϕ_k^ξ; in these cases, substituting the φ_μ set by the ϕ_k^ξ set can be a reasonable approximation. The AIMP embedded potential results from adopting such an approximation and substituting the Coulomb and exchange operators of the individual ions of the host by model potential representations of them [11, 12],

$$\hat{V}^{\text{emb}}(i) = \sum_{\xi}^{N_n^H} \hat{V}_\xi^{\text{AIMP}}(i).$$

Here, the embedding AIMP of the host ion ξ is:

$$\begin{aligned}\hat{V}_\xi^{\text{AIMP}}(i) = &-\frac{Q_\xi}{r_{\xi i}} + \frac{1}{r_{\xi i}} \sum_p C_p^\xi \exp(-\alpha_p^\xi r_{\xi i}^2) + \sum_p \sum_q |\chi_p^\xi\rangle A_{pq}^\xi \langle \chi_q^\xi| \\ &+ \sum_k B_k^\xi |\phi_k^\xi\rangle\langle\phi_k^\xi|. \end{aligned} \tag{9.3}$$

Here, Q_ξ is the ionic charge, $Q_\xi = Z_\xi - N_\xi^{\text{elec}}$. The second term is such that

$$\frac{1}{r_{\xi i}} \sum_p C_p^\xi \exp(-\alpha_p^\xi r_{\xi i}^2) \approx -\frac{N_\xi^{\text{elec}}}{r_{\xi i}} + 2\sum_k f_k \int \frac{\phi_k^{\xi *}(j)\phi_k^\xi(j)}{r_{ij}} d\tau_j, \tag{9.4}$$

with the C_p^ξ and α_p^ξ chosen to minimize the deviations; in this way, the second term corrects the point charge potential with the electrostatic contributions from the ionic electron density and $Q_\xi - \sum_p C_p^\xi \exp(-\alpha_p^\xi r_{\xi i}^2)$ is the effective ionic charge. Finally, the coefficients in the last terms are

$$A_{pq}^\xi = -\sum_r \sum_s S_{pr}^{-1} \sum_k f_k \int \frac{\chi_r^{\xi *}(i)\phi_k^{\xi *}(j)\phi_k^\xi(i)\chi_s^\xi(j)}{r_{ij}} d\tau_j d\tau_i S_{sq}^{-1}, \tag{9.5}$$

with the S matrix defined as

$$S_{pq} = \langle \chi_p^\xi | \chi_q^\xi \rangle, \tag{9.6}$$

so that the third term is the exchange interaction between the embedding electrons and the i-th cluster electron, in the form of a spectral representation (resolution of the identity). In equations 9.3, 9.5, and 9.6, the χ_p^ξ are a set of auxiliar functions; usually, these are the primitive Gaussians used to expand the ion orbitals ϕ_k^ξ.

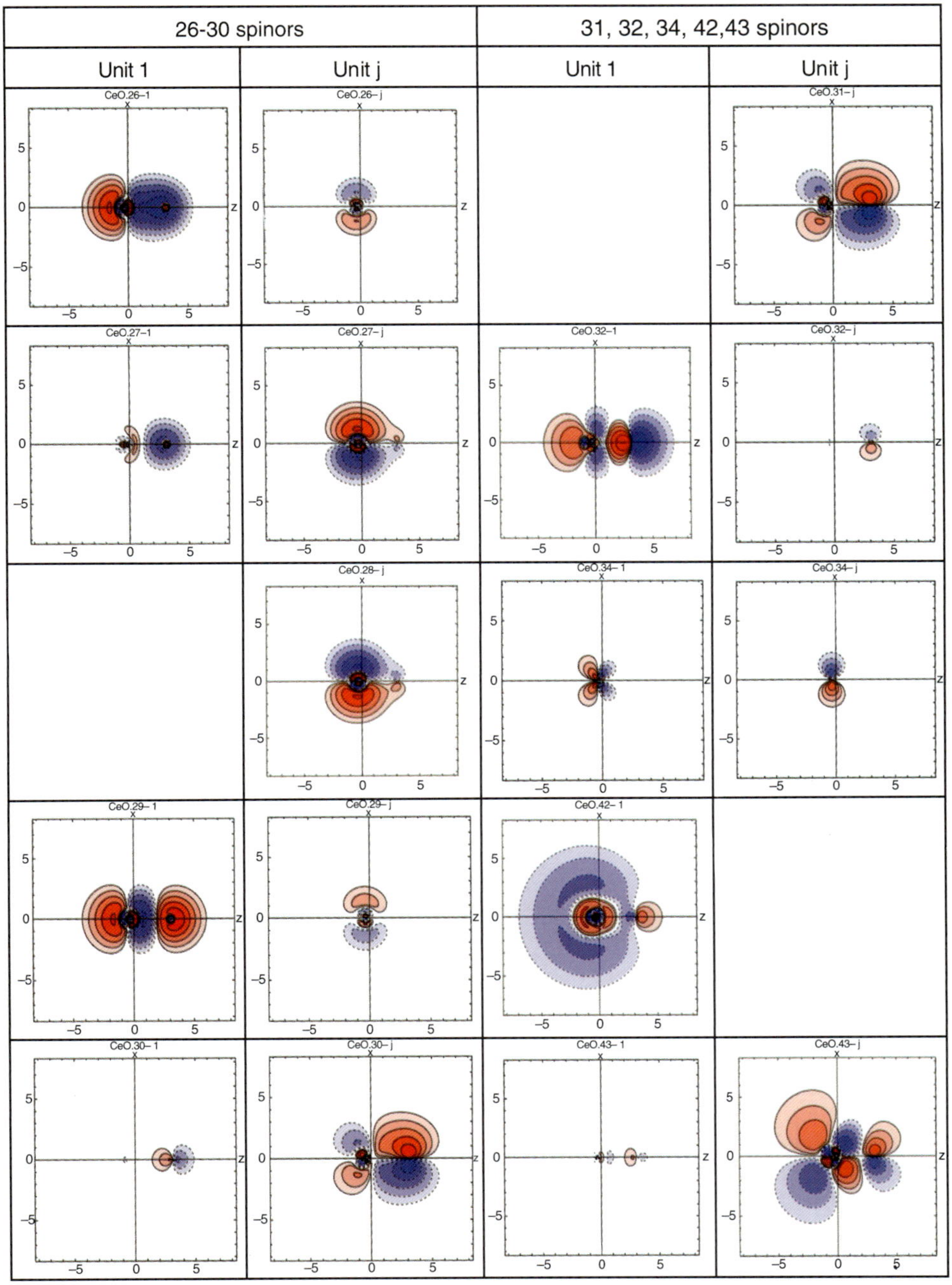

Figure 4.1 *Contour maps of the large components of the CeO^+ spinors. The quaternion units 1 and $\check{j}$ correspond, respectively, to the L^α real part of A_1 and to the L^β real part B_1. Coordinates run from -8.0 to 8.0 au. The CeO^+ nuclear distance is taken from the experimental R_e value (3.44 au) of neutral CeO. Solid and broken lines are contours of positive and negative values, respectively. Their values are ± 0.0125, ± 0.025, ± 0.05, ± 0.1, ± 0.2*

Computational Methods in Lanthanide and Actinide Chemistry, First Edition. Edited by Michael Dolg.

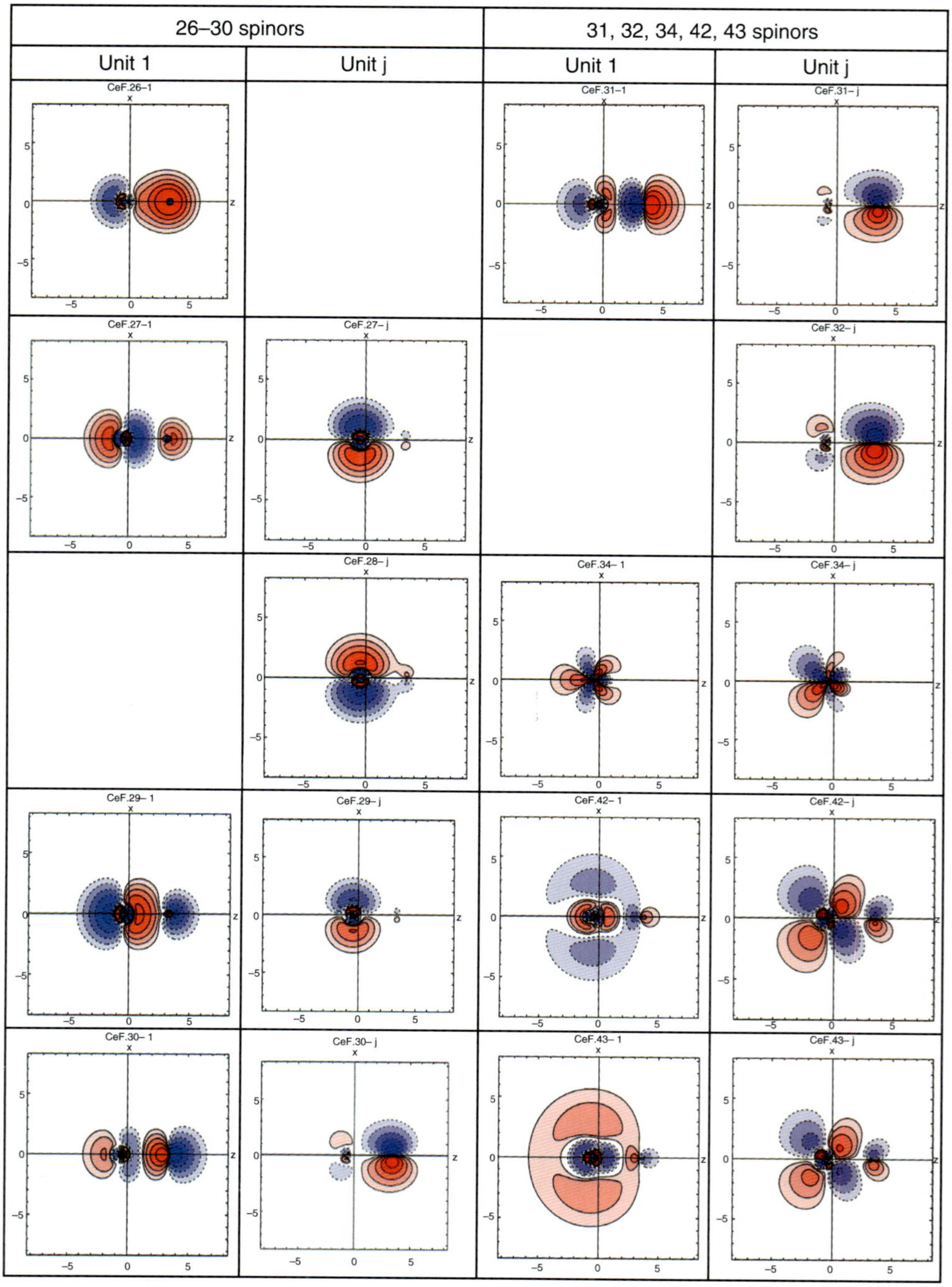

Figure 4.2 *Contour maps of the large components of CeF^+ spinors. The CeF^+ nuclear distance is taken from the experimental R_e value (3.87 au) of neutral CeF. The maps are drawn in the same manner as in Figure 4.1*

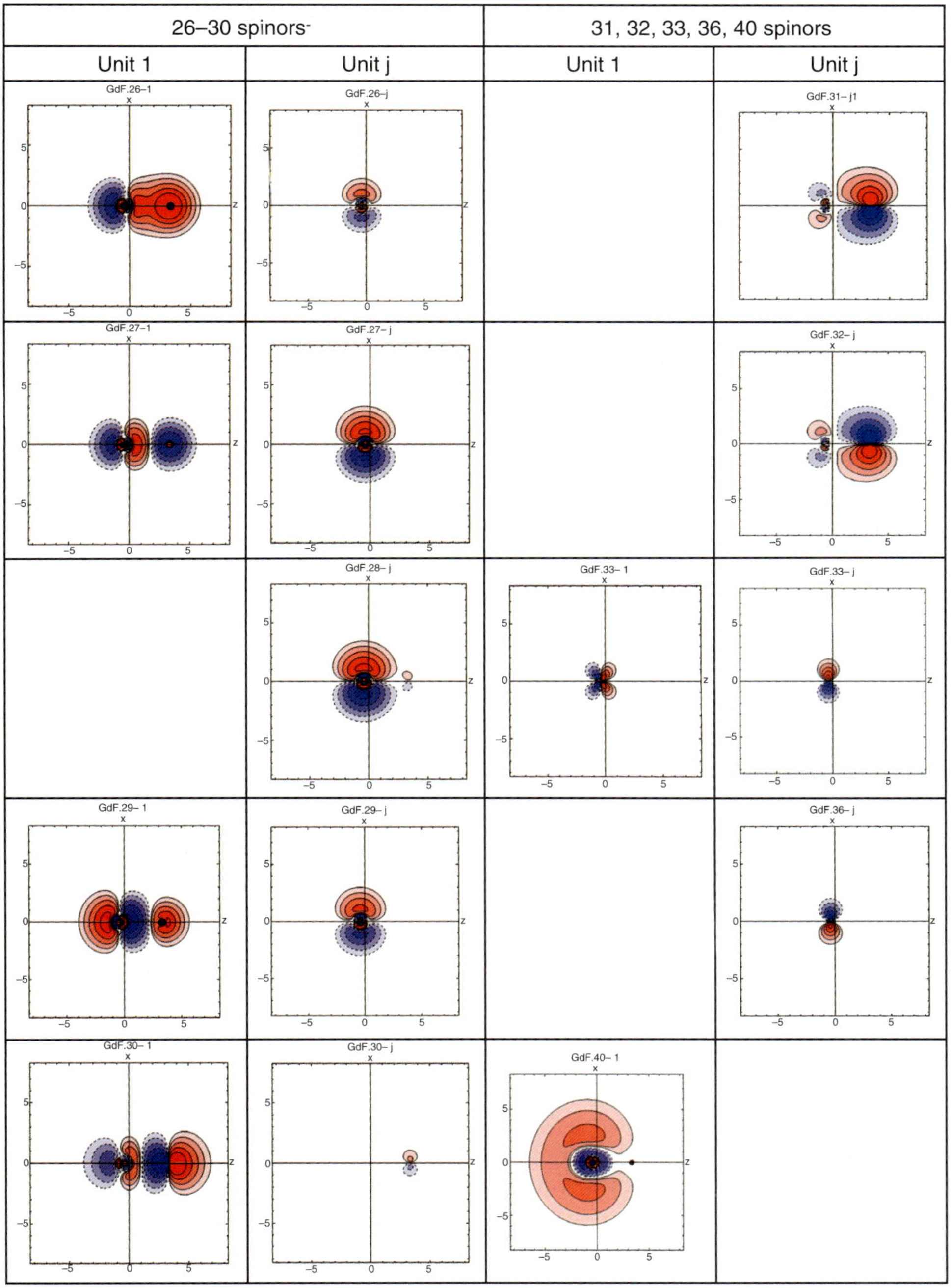

Figure 4.3 *Contour maps of the large components of GdF^+ spinors. The GdF^+ nuclear distance is the calculated R_e value (3.7599 au) of neutral GdF. The maps are drawn in the same manner as in Figure 4.1*

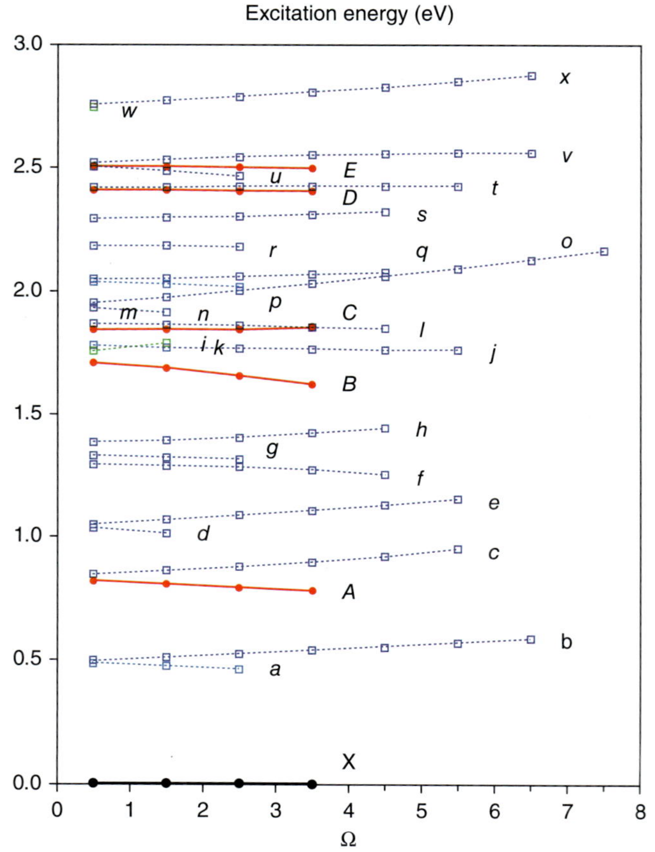

Figure 4.4 *Vertical excitation energy of the GdF electronic state and family label. The ordinate denotes the vertical excitation energy in eV. The abscissa shows Ω; only positive values of Ω are considered. Black line with solid circles denotes the ground state family and red lines with solid circles denote families having the same Ω as the X family. Blue broken lines with open squares denote families with Ω differing from the X family*

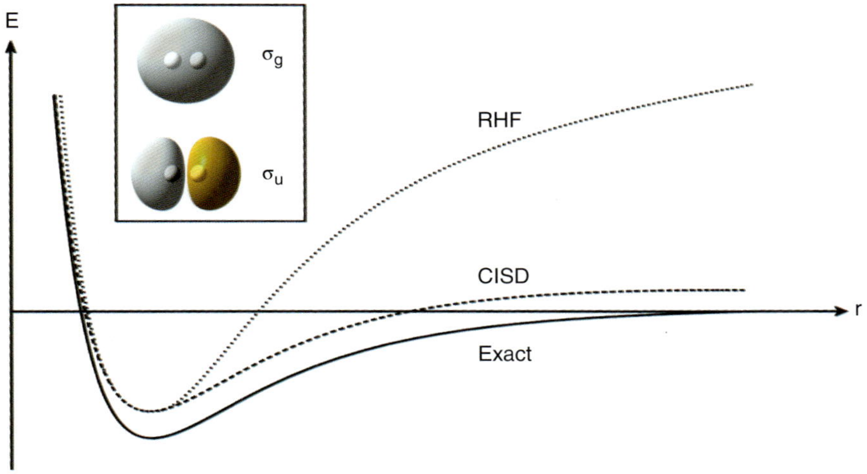

Figure 5.4 *Comparison of HF- and CISD-calculated potential energy surfaces to the exact surface for the dissociation of H_2. Inset: bonding and antibonding molecular orbitals of H_2*

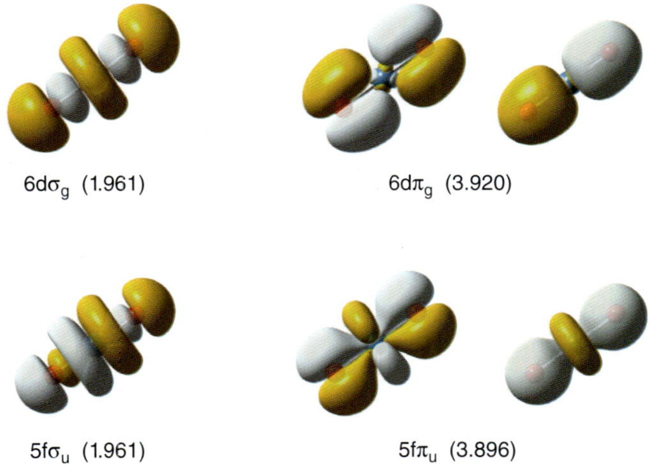

Figure 5.9 *Strongly occupied natural orbitals and corresponding occupation numbers of $[UO_2]^{2+}$ obtained from a CASSCF (12,12) calculation. Data from Pierloot and van Besien [34]*

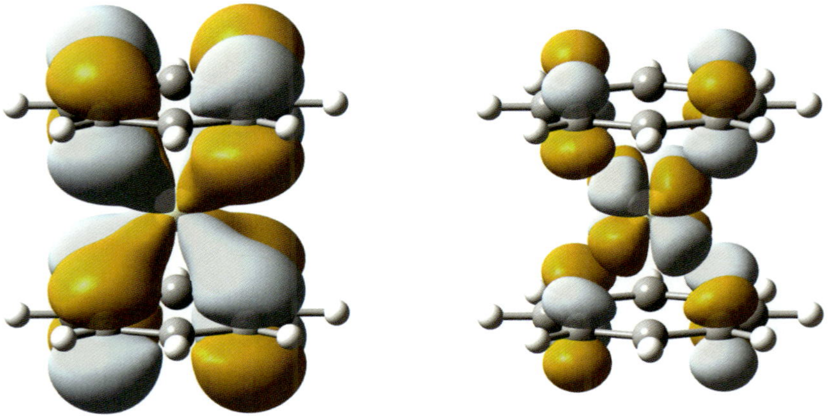

Figure 5.10 *Bonding and antibonding MOs of cerocene, exhibiting* $4f_\delta$ *contributions*

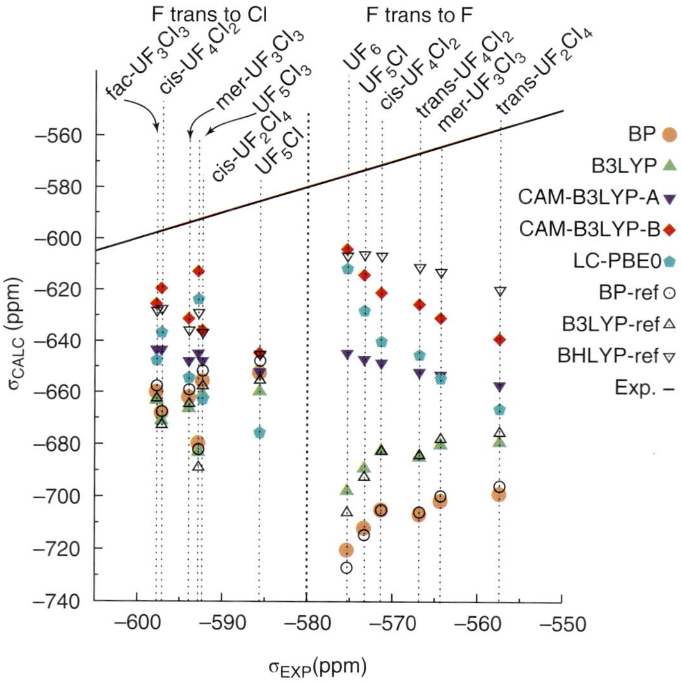

Figure 12.1 *Performance of different functionals in the calculation of* ^{19}F *shielding for the series of diamagnetic* U^{VI} *complexes* UF_nCl_{6-n} $(n = 1 - 6)$. *BP-ref, B3LYP-ref, and BHLYP-ref are DFT results employing a scalar relativistic effective core potential for U, from Straka and Kaupp [43]. The other data are from all-electron scalar ZORA calculations with NWChem from Aquino et al. [15]. CAM and LC indicates hybrid functionals with range-separated exchange. The straight black line indicates where calc = expt. Absolute shielding constants converted from experimental shifts taken from Straka and Kaupp [43]. Figure reprinted with permission from Aquino et al. [15].* © *2011 American Chemical Society*

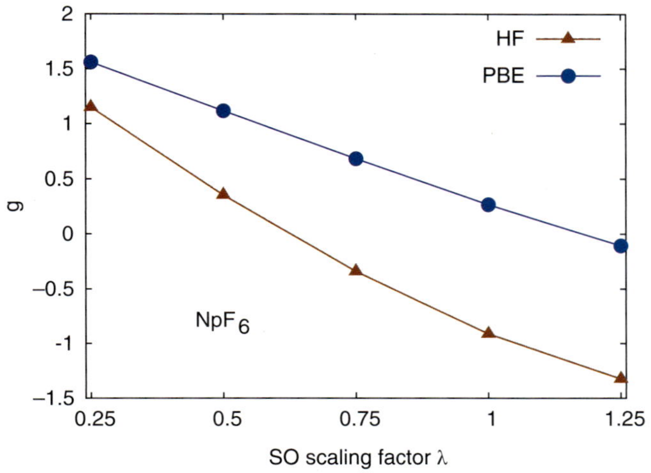

Figure 12.2 *Computed g-factors of NpF_6 using HF theory and a non-hybrid density functional (PBE). ZORA/MA, with SO integrals scaled by a factor λ. Expt. approx. $g = -0.6$. Data to prepare the figure were taken from Verma and Autschbach [16]*

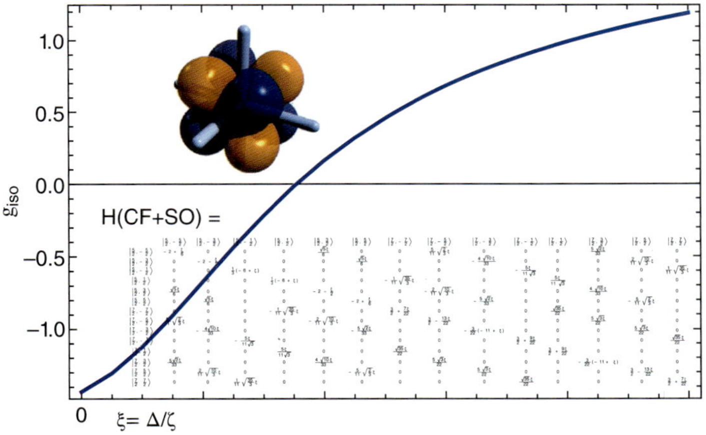

Figure 12.3 *g-factor of NpF_6 as a function of SO splitting ζ and CF splitting Δ for a $\mathcal{O}_h$ crystal field. Symbolic computation using software written by one of the authors (JA). For a description of the CF model see Notter and Bolvin [48]. The inset shows the unpaired Np orbital $5f_{xyz}$ from a scalar ZORA DFT calculation*

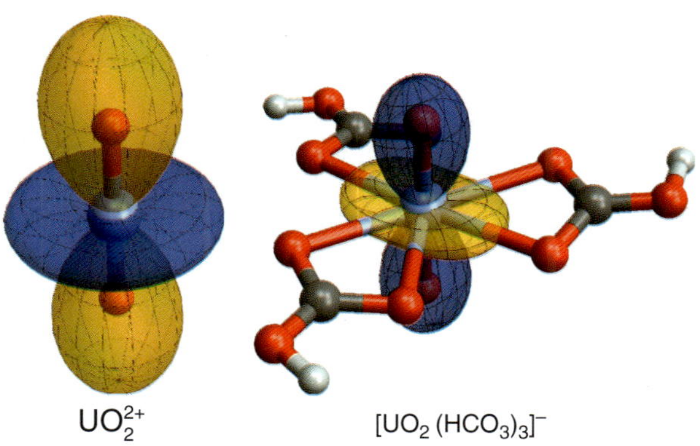

Figure 12.4 *EFG tensors for uranyl and a uranyl-carbonate complex. Polar plots of the field gradient in the direction of the field, scaled to 40 au / pm. Blue (dark shading) = positive, orange (light shading) = negative field gradient (SO X2C, V_{mp}, B3LYP). Reprinted with permission from Autschbach et al. [14]. © 2012 American Chemical Society*

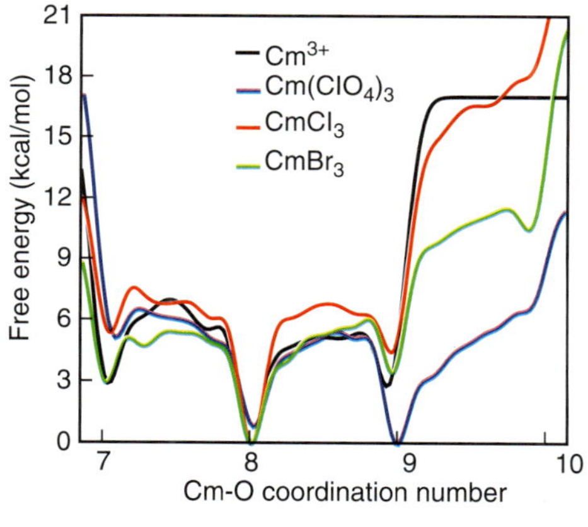

Figure 12.6 *Free energy profiles of Cm coordination with respect to O in Cm^{3+} (aq), Cm^{3+}+$3ClO_4^-$ (aq), Cm^{3+}+$3Cl^-$ (aq), and Cm^{3+}+$3Br^-$ (aq). Reprinted with permission from Atta-Fynn et al. [118]. © 2013 American Chemical Society*

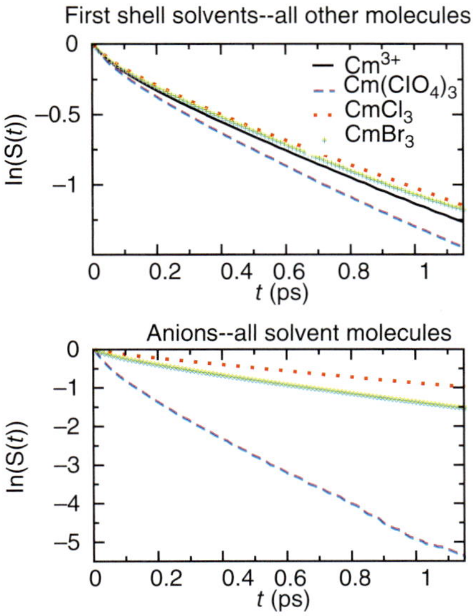

Figure 12.7 *HB survival autocorrelation function of Cm^{3+} (aq), Cm^{3+}+$3ClO_4^-$ (aq), Cm^{3+}+$3Cl^-$ (aq), and Cm^{3+}+$3Br^-$ (aq). The vertical axis is on a logarithmic scale. Reprinted with permission from Atta-Fynn et al. [118]. © 2013 American Chemical Society*

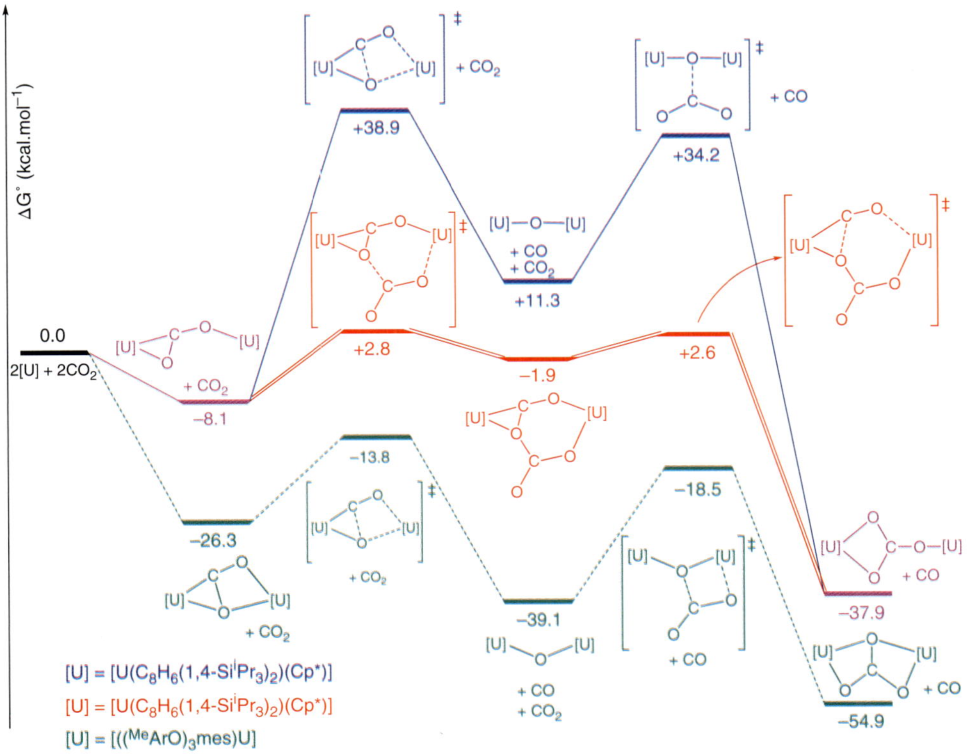

Figure 13.9 *Gibbs free energy profiles of the reactions of U(III) complexes and CO_2. The profile on the bottom concerns $[((^{Me}ArO)_3mes)U^{III}]$ while both other pathways concern $[U^{III}(C_8H_6(1,4\text{-}Si^iPr_3)_2)(Cp^*)]$. Both complexes are simplified as [U]*

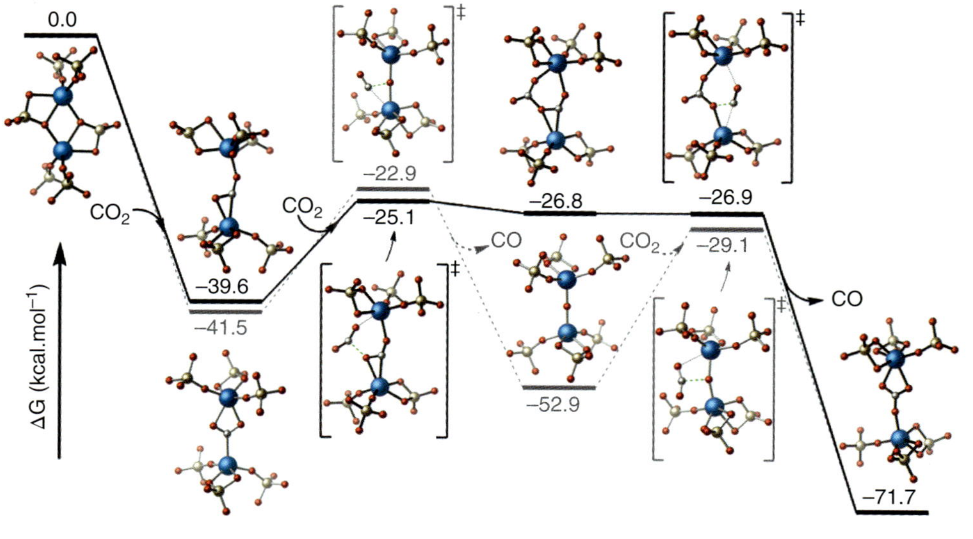

Figure 13.11 *Gibbs free energy profiles of the reactions between* $[U^{III}(OSi(OtBu)_3)_3]$ *and* CO_2

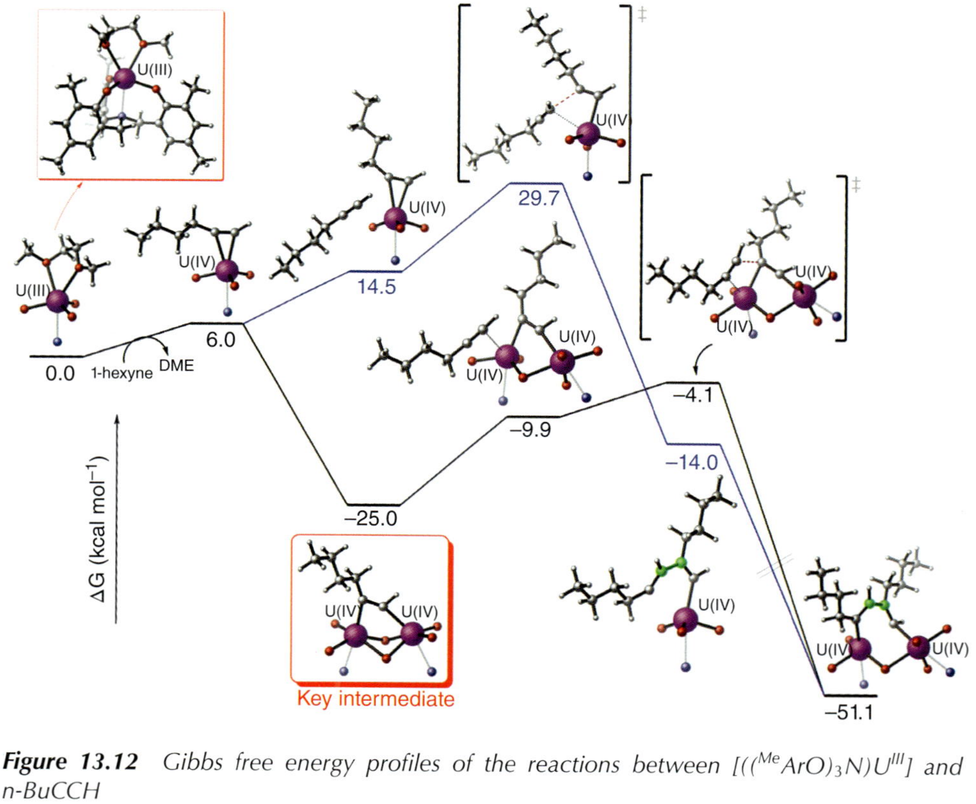

Figure 13.12 *Gibbs free energy profiles of the reactions between* $[((^{Me}ArO)_3N)U^{III}]$ *and n-BuCCH*

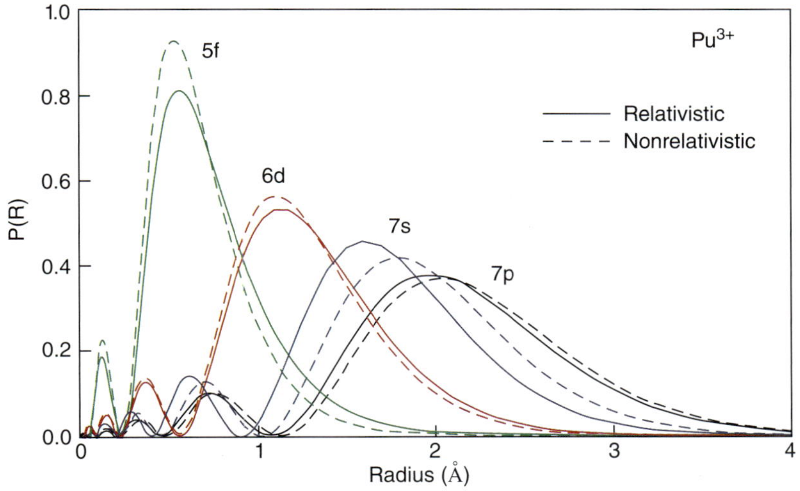

Figure 14.1 *Relativistic effect on the radial distribution function of Pu^{3+}. P. Jeffrey Hay Los Alamos Science No. 26, vol II, p. 371*

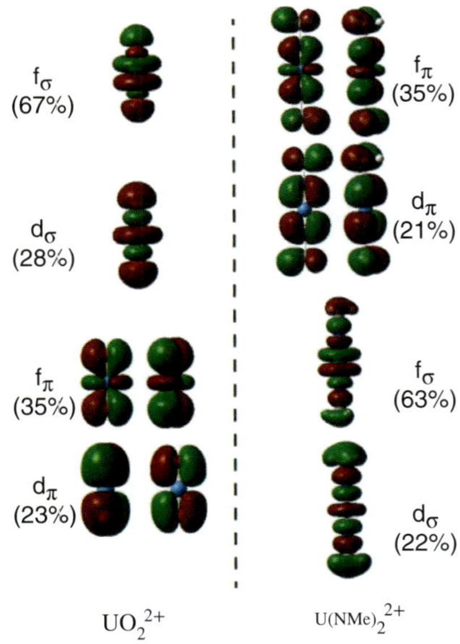

Figure 14.4 *Bonding orbitals of UO_2^{2+} and $U(NMe)_2^{2+}$. The column on the sides shows the type of uranium atomic orbital participating in each MO, along with its percentage contribution to the total MO*

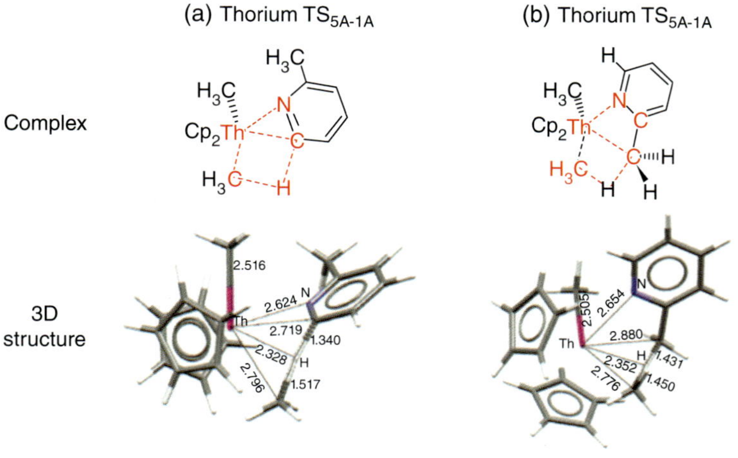

Figure 14.7 *Illustrated and 3D structures of transition states for the sp^2 and sp^3 C–H bond activation. Notice the close resemblance between the geometries of the thorium and uranium complexes. Reprinted with permission from Yang et al.[5] Copyright 2008, American Chemical Society*

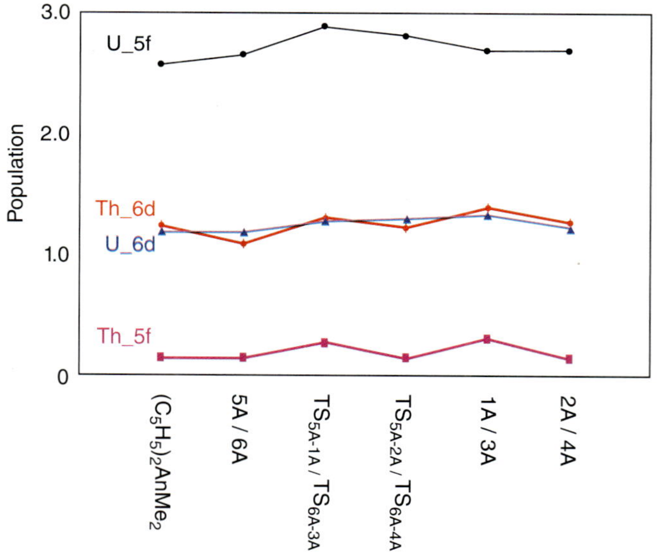

Figure 14.10 *Orbital population on the actinide metal center along reaction pathways. Reprinted with permission from Yang et al.[5] Copyright 2008, American Chemical Society*

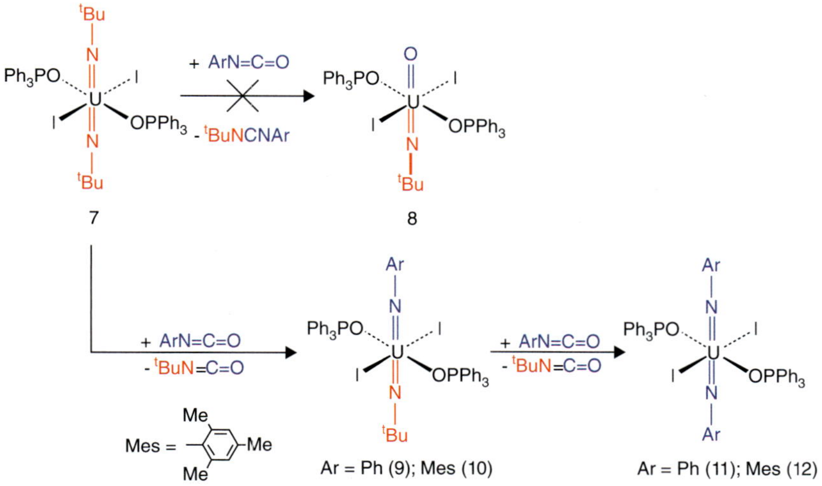

Scheme 14.2 *Potential reaction pathways bis(imido) uranium complex with aryl isocyanates. Reprinted with permission from Spencer et al.[47] Copyright 2008, American Chemical Society*

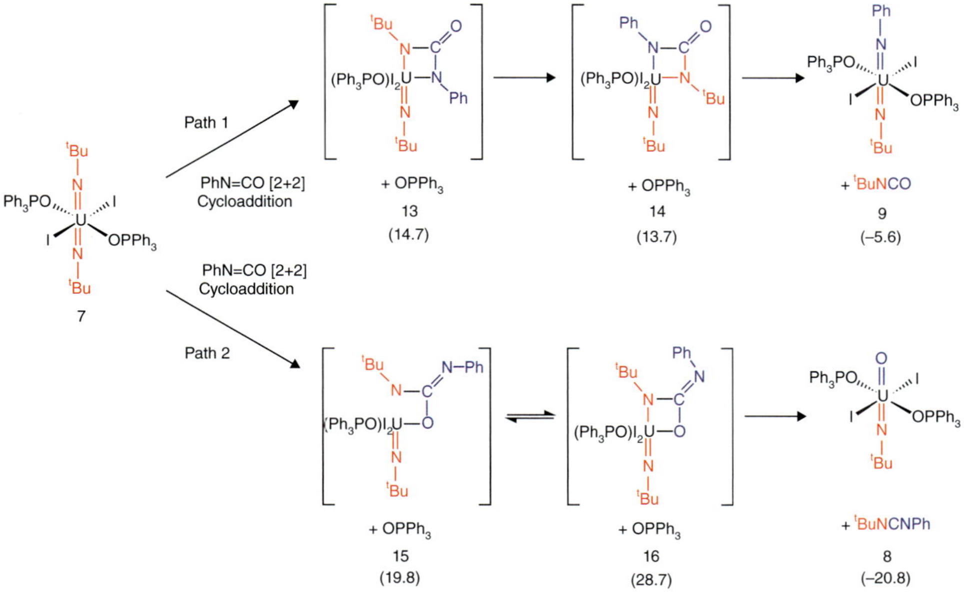

Scheme 14.3 *Relative free energies of the products and intermediates in the potential pathways for the formation of 8 and 9. The energies are provided in parentheses in kcal/mol. Reprinted with permission from Spencer et al.[47] Copyright 2008, American Chemical Society*

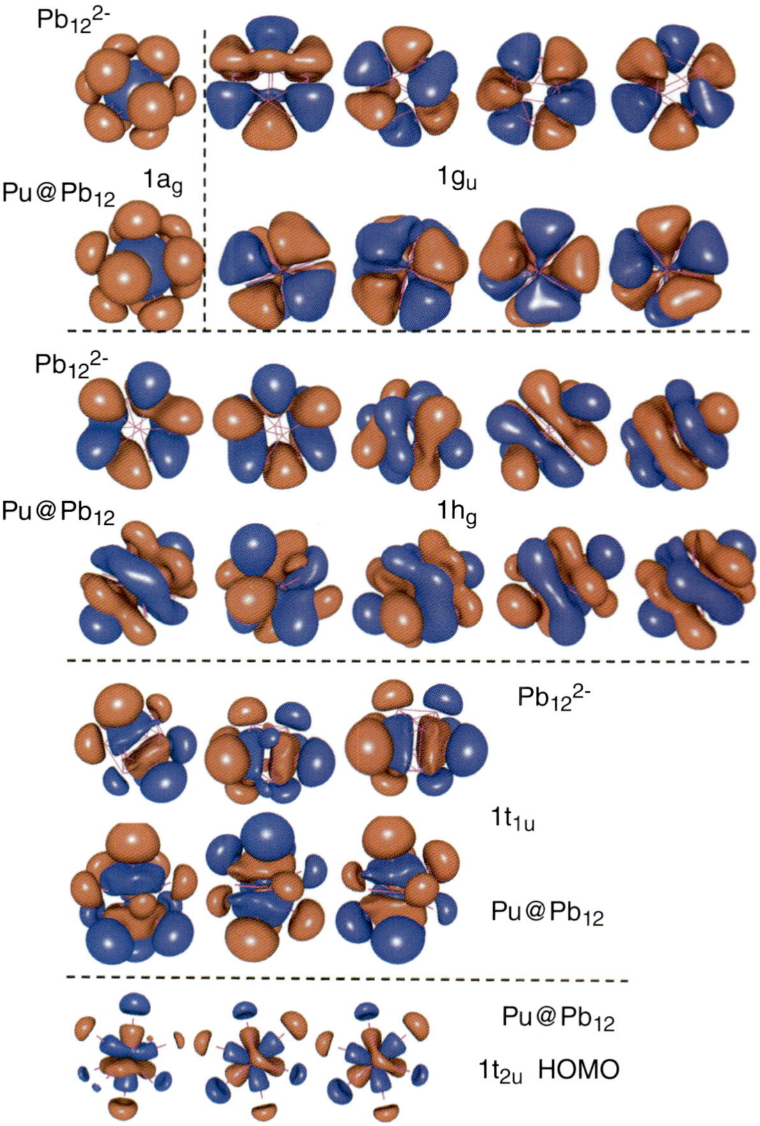

Figure 15.4 *Scalar relativistic valence molecular orbitals of* $Pu@Pb_{12}^{2-}$ (I_h) *and* Pb_{12}^{2-} (I_h). *Adapted from Dognon et al. [13]*

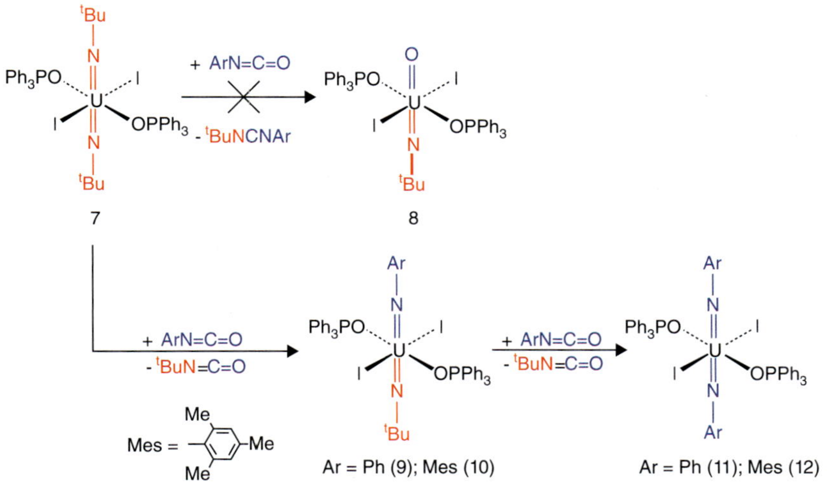

Scheme 14.2 *Potential reaction pathways bis(imido) uranium complex with aryl isocyanates. Reprinted with permission from Spencer et al.[47] Copyright 2008, American Chemical Society*

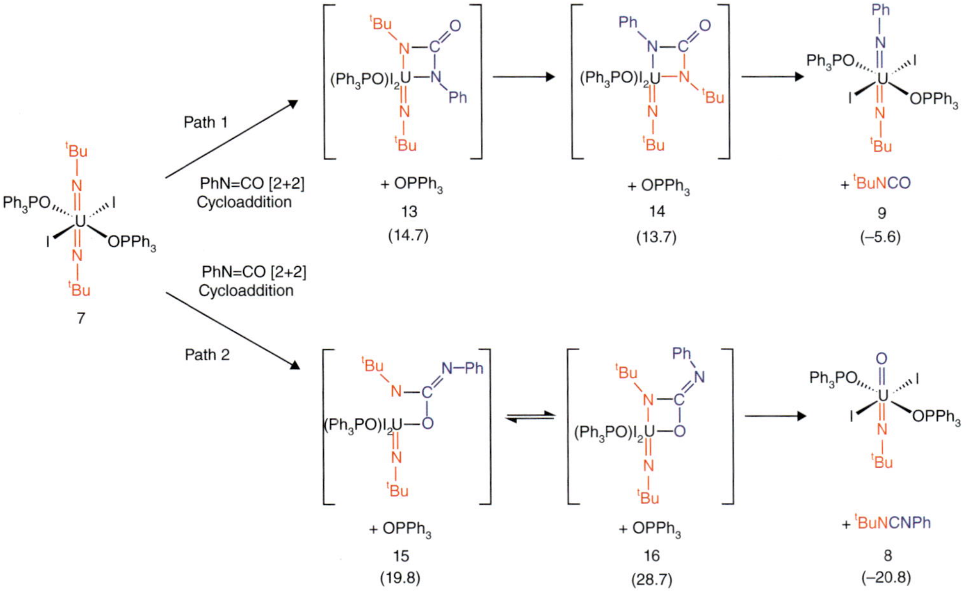

Scheme 14.3 *Relative free energies of the products and intermediates in the potential pathways for the formation of 8 and 9. The energies are provided in parentheses in kcal/mol. Reprinted with permission from Spencer et al.[47] Copyright 2008, American Chemical Society*

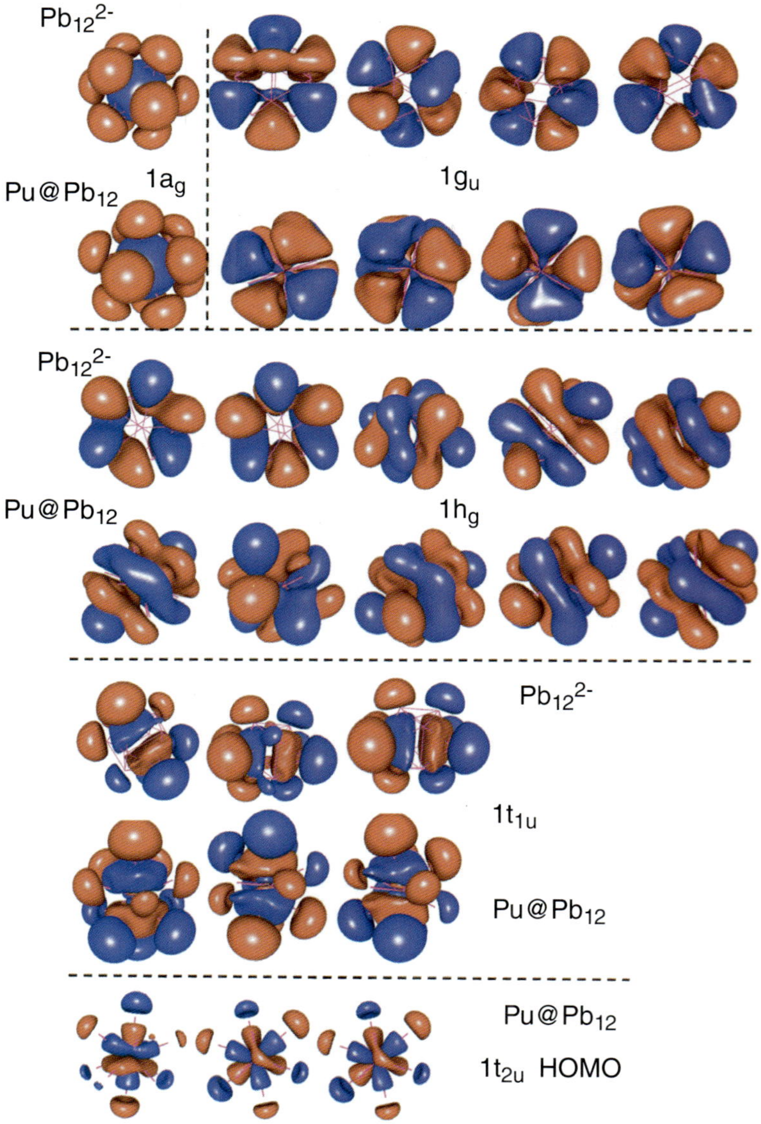

Figure 15.4 *Scalar relativistic valence molecular orbitals of* $Pu@Pb_{12}^{2-}$ (I_h) *and* Pb_{12}^{2-} (I_h). *Adapted from Dognon et al. [13]*

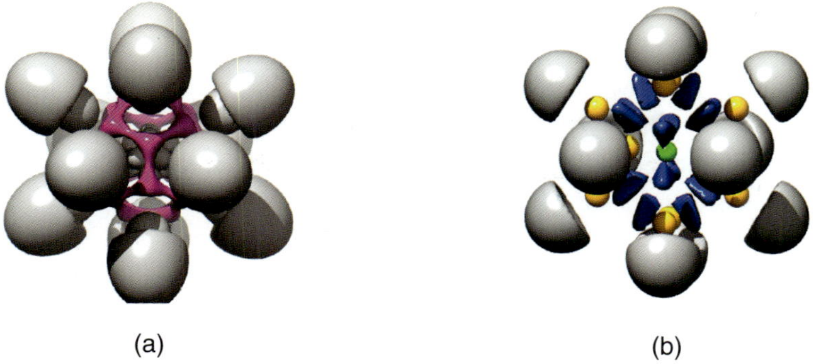

Figure 15.6 *ELI-D isosurfaces of the [Pu@Pb$_{12}$] (I_h) : magenta domains (a) visualize the interaction between the Pb cage and the Pu atom. Blue domains (b) are pELI-D isosurfaces (0.25) of summation over 5 MOs (1h_g) visualizing the 6p(Pb)-6d(Pu) orbital mixing. From scalar relativistic all-electron DFT/PBE*

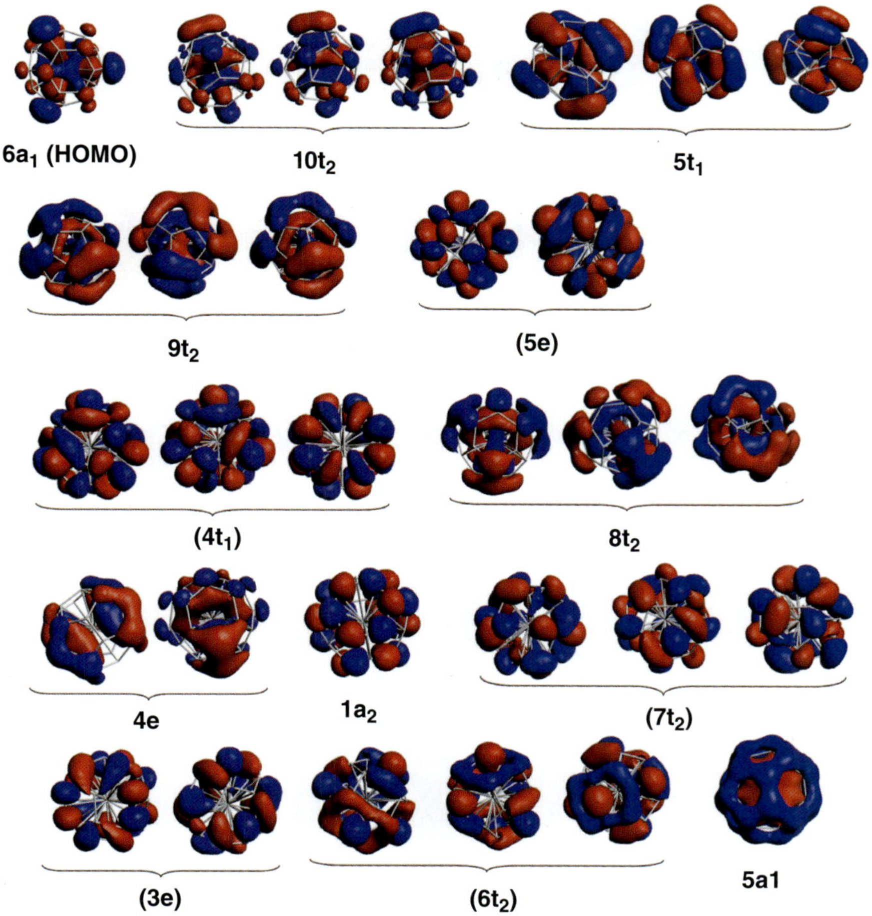

Figure 15.8 *Scalar-relativistic valence molecular orbitals of Pu^{4+}@C_{28} (T_d). Adapted from Dognon et al. [15]. The 'pure cage' orbitals (5e, 4t$_1$, 1a$_2$, 7t$_2$, 3e, 6t$_2$) are given in parentheses*

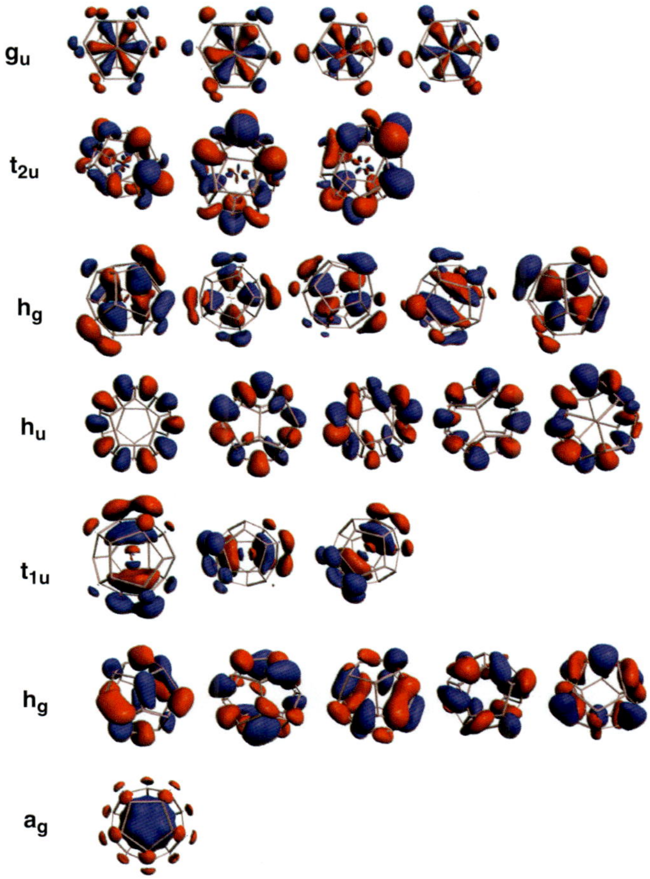

Figure 15.11 *Scalar relativistic valence molecular orbitals for* $[U@Si_{20}]^{6-}$ (I_h)*. Adapted from Dognon et al. [16]*

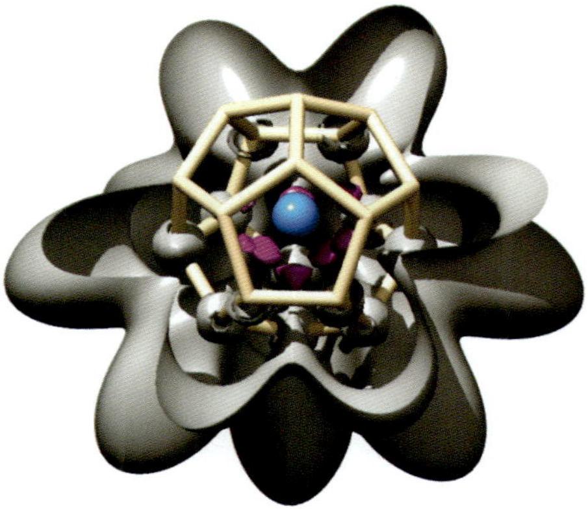

Figure 15.13 $[U@Si_{20}]^{6-}$ (I_h) *ELI-D isosurface (0.75). A cutting plane is used to show the U-Si basins (in magenta). U atom in blue*

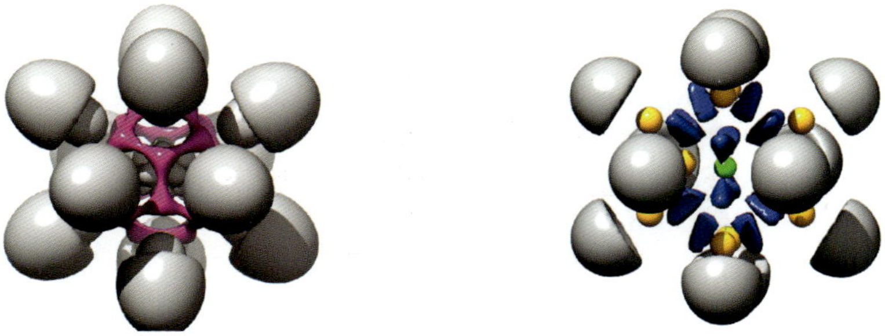

(a) (b)

Figure 15.6 *ELI-D isosurfaces of the* $[Pu@Pb_{12}]$ (I_h) *: magenta domains (a) visualize the interaction between the Pb cage and the Pu atom. Blue domains (b) are pELI-D isosurfaces (0.25) of summation over 5 MOs* $(1h_g)$ *visualizing the 6p(Pb)-6d(Pu) orbital mixing. From scalar relativistic all-electron DFT/PBE*

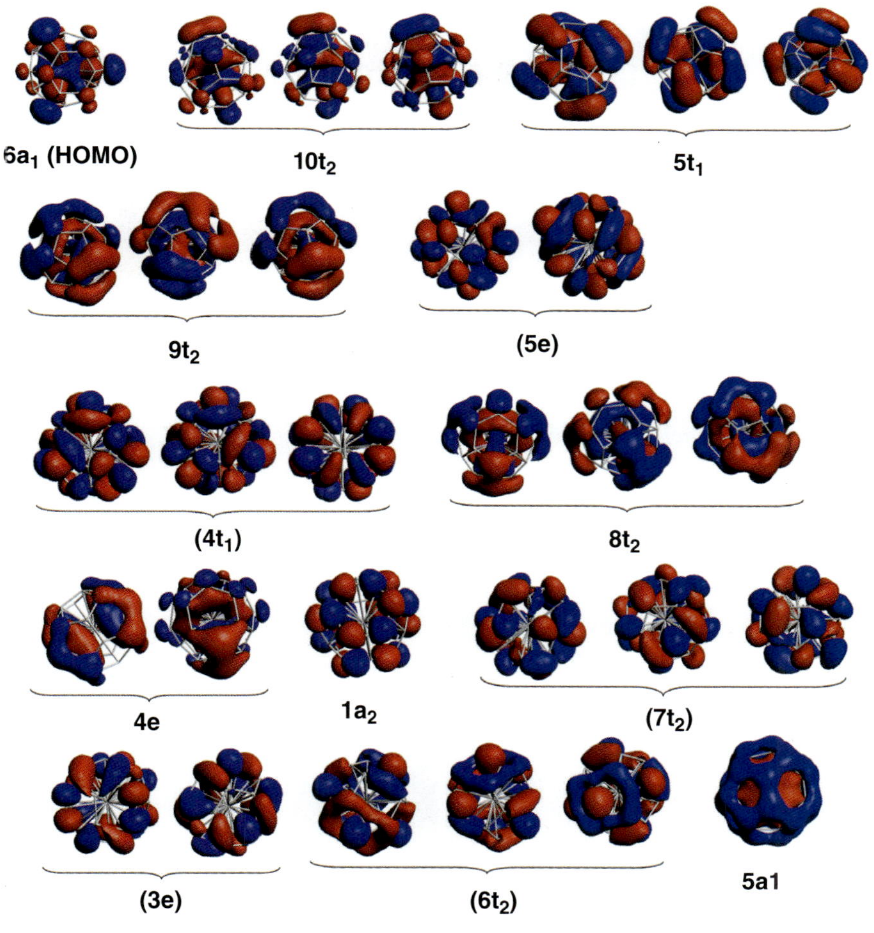

Figure 15.8 *Scalar-relativistic valence molecular orbitals of* $Pu^{4+}@C_{28}$ (T_d)*. Adapted from Dognon et al. [15]. The 'pure cage' orbitals (5e,* $4t_1$*,* $1a_2$*,* $7t_2$*, 3e,* $6t_2$*) are given in parentheses*

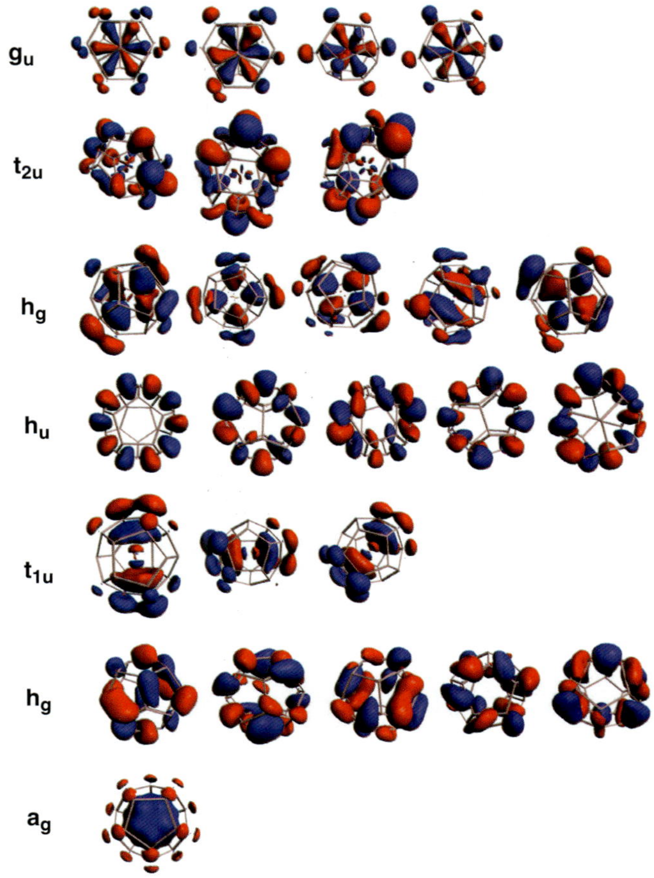

Figure 15.11 *Scalar relativistic valence molecular orbitals for [U@Si$_{20}$]$^{6-}$ (I_h). Adapted from Dognon et al. [16]*

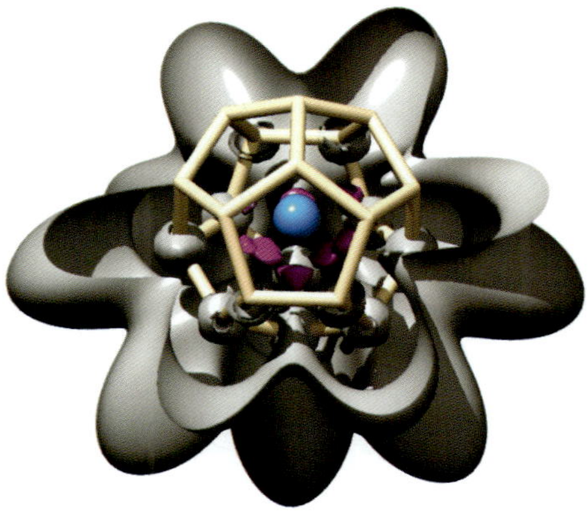

Figure 15.13 *[U@Si$_{20}$]$^{6-}$ (I_h) ELI-D isosurface (0.75). A cutting plane is used to show the U-Si basins (in magenta). U atom in blue*

Self-Consistent Embedded Ions. The calculation of the AIMP embedding potential (equation 9.3) demands the knowledge of the ionic orbitals ϕ_k^{ξ}. We compute them in an iterative self-consistent embedded ions calculation SCEI [13]. Let us take the $Lu_3Ga_5O_{12}$ garnet as an example. Here we identify four types of ions: Lu^{3+}, Ga_{oct}^{3+}, Ga_{tet}^{3+}, and O^{2-}. Starting with an initial representation of the $Lu_3Ga_5O_{12}$ embedding potential (e.g., that corresponding to a point charge representation of all the ions), we perform Hartree-Fock (HF) calculations on the individual embedded ions. With the orbitals and orbital energies of these ions, new embedding AIMPs are made for them after adopting $B_k^{\xi} = -2\varepsilon_k^{\xi}$, and they are used to update the AIMP embedding potential representation of $Lu_3Ga_5O_{12}$. Then, new HF calculations on the embedded ions are made. This SCEI cycle is stopped when the embedded-ion total and orbital energies of two consecutive iterations converge within given thresholds (e.g., 10^{-8} au). The embedding AIMPs of the ions are located at experimental sites of the respective garnets, within a cube made of $3 \times 3 \times 3$ unit cells (160 atoms per unit cell) surrounding the reference ion. The AIMP embedding potential of the garnet is completed with a set of $\sim 10^5$ additional point charges situated at lattice sites, generated by the zero-multipole method of Gellé and Lepetit [14], in order to closely reproduce the Ewald potential [15] within the cluster. The Evjen method [16] of fractional charges has been used for this purpose in our group in many occasions, but we found in a number of crystal lattices that the Evjen charges reproduce the Ewald potential within the cluster and its surroundings *with a constant added.* Although this is not a problem for a majority of structural and spectroscopic calculations, the effect of such a constant has to be corrected to calculate ionization. The method of Gellé and Lepetit [14] overcomes this problem in an elegant manner. An extension of the embedding AIMP method for multiatomic ions has been recently implemented [17].

Isolated Cluster Hamiltonian. Lanthanides are heavy elements and their ions have partially filled $4f$ shells in which spin-orbit coupling plays a key role. In consequence, the isolated cluster Hamiltonian $\hat{H}^{\mathrm{IC}}$ of equation 9.1 must be a relativistic Hamiltonian that includes spin-orbit coupling. We found the CGWB-AIMP Hamiltonian [18, 19] to be a reasonable choice. This Hamiltonian is the frozen-core *ab initio* model potential version [20] of the spin-orbit coupling extension of Wood and Boring [21] of the Cowan and Griffin Hamiltonian [22]. See Seijo and Barandiaran [23] for a detailed description. A good alternative is the all-electron second order (or higher order) Douglas-Kroll-Hess Hamiltonian [24, 25] with the atomic mean-field approximation AMFI for the spin-orbit coupling operator [26], or a frozen-core AIMP version of it [27].

9.2.1.2 *Embedded-Cluster Wavefunctions*

The embedded-cluster approximation is a natural way to focus the methodological effort on the local defect electronic structure, studying it using methods designed for the calculation of the ground and excited states of molecules containing heavy elements in gas phase. The size of the defect cluster to be singled out and the type of wavefunctions associated with it are basic decisions that must be guided by the chemical composition of the point defect and by the intrinsic open-shell, highly correlated nature of the electronic states of the substitutional impurity. The experience gathered on these lines is summarized next.

Cluster Size and Basis Set. When a quantum-mechanical embedding technique like that described above is used and the defect is local, the cluster can be quite small. In many applications it includes only the lanthanide impurity and its first neighbors: the ligands. Yet, even when the small impurity+ligands cluster is a reasonable choice, basis functions must be used at the neighboring atoms to respond to the strong orthogonality conditions [28] that must be fulfilled between the cluster wavefunctions and the frozen ion wavefunctions of the embedding, represented by the AIMP embedding potentials (last term in equation 9.3). For this reason, the outermost atomic orbitals of the (next) frozen embedding ions are customarily used in the cluster basis set, which avoids artificial cluster bond length shrinkage [12]. Alternatively, in cases where the ligands are tightly bound to surrounding ions, like in granates, or when impurity-trapped excitons are calculated, second neighbor cations should also be included in the cluster and contribute to the cluster molecular basis set. In addition to this, empty orbitals of the next neighbor cations and functions at interstices have been found to be necessary for a balanced representation of impurity states and impurity-trapped excitons [29–31].

In all cases, each contributing atomic basis set should be flexible enough so that sophisticated electron correlation methods like the ones described below can be used without contaminating basis set errors. Hence, at least quadruple-zeta plus polarization should be used.

Electron Correlation. Given the open-shell nature of the ground and excited states of lanthanide impurity ions in crystals, electron correlation is extremely important. Currently, electronic structure methods based on the use of multireference wavefunctions appear to be the fittest to respond to the requirements. Furthermore, their current evolution towards allowing more and more flexible definitions of the active space makes them even more adequate. Even though the methods used in the applications contained in this chapter are well known and their performance has been proven and documented in many highly correlated systems, we summarize here how to adapt them to the impurity lanthanide ion electronic structure demands.

The Multireference Space. The calculations of the embedded-cluster wavefunctions have a first step where multiconfigurational self-consistent field wavefunctions and energies are calculated using complete and/or restricted active spaces (CASSCF [32–34] and/or RASSCF [35, 36]). The lanthanide $4f$, $5d$, and $6s$ shells must be included in the active space for the calculation of the $4f^N$, $4f^{N-1}5d$, and $4f^{N-1}6s$ manifolds, N being the number of active electrons [37]. Impurity-trapped excitons may also occur and can be labeled as $4f^{N-1}\phi$, where ϕ stands for significantly more delocalized orbitals whose electron density spreads beyond first neighbors; these orbitals also have to be included in the active space. If the active space is complete, the previous choice may generate very large multiconfigurational expansions, which are still affordable in some common impurity ions. An active $5f$ shell is also necessary for the heavy lanthanides ($N > 7$) to account for large radial correlation effects in the $4f$ shell for interconfigurational transitions, as we comment below. In these cases, the use of restricted instead of complete active spaces is necessary given the size of the multireference space that might be generated [31]. The orbitals used in the multiconfigurational expansions are optimized for the average of all states with main $4f^N$, $4f^{13}5d^1$, and $4f^{13}6s^1$ configurational character of a given total spin and point symmetry irreducible representation. Although approximate, this practice is appropriate for the calculation of the local structure and spectroscopy of lanthanide ions doped in crystals. Very often the calculations are done using a subgroup of the actual site symmetry group (e.g., D_{2h} when the site symmetry is O_h). In these cases it is very important to restrict the orbital rotations to enforce actual site symmetry. Important *a posteriori* assignments of the wavefunctions to irreducible representations rest upon fulfillment of this restriction. Low symmetry sites or sites with no symmetry at all are simpler in these respects but are naturally far more demanding computationally. It is known that CASSCF wavefunctions converge far more readily than RASSCF ones. For the latter, the restriction to use only pure super-CI iterations ensures smooth convergence even though the number of iterations is usually very large.

Multi-State Second Order Perturbation Method. The state-average CAS/RASSCF states of lanthanide ions in gas phase and in solids are known to interact at second order perturbation level. Therefore, dynamic correlation must be computed using the multi-state multiconfigurational second order perturbation theory (MS-CASPT2 [38–41] or MS-RASPT2 [42], respectively). $N + 8$ electrons of the lanthanide (which includes the $5s$ and $5p$ closed shells) and the valence electrons of the ligands should be correlated at this level. The use of imaginary shifts prevents the appearance of intruder states [43]. The standard IPEA value (0.25 au) can also be used; this parameter has been introduced by Ghigo *et al.* [44] as a simple way to correct for systematic underestimations of CASPT2 transition energies from closed-shell ground states to open-shell excited states, although it is also recommended as a default option in other cases.

Radial Correlation in the 4*f* Shell. The effects of radial correlation on the ionization potential and $4f^N \rightarrow 4f^{N-1}5d$ transitions at the end of the lanthanide series were investigated recently by means of RASSCF + MS-RASPT2 studies of Yb^{2+} ($N = 14$) in gas phase and in the $SrCl_2$ solid [31]. The study shows the limitations of the second order perturbation treatment to account for the large contribution of differential radial correlation in the $4f$ shell when the number of f electron changes in the transition. The sample case chosen poses most stringent methodological requirements in the lanthanide series since the number of f electron pairs that are broken in the excitation is maximal.

In effect, it has been shown that the large energy gap of 25000 cm^{-1} (~29500 cm^{-1} without spin-orbit coupling) experimentally observed between the $4f^{14}$ ground state and the first excited state of the $SrCl_2$:Yb^{2+} material is mostly due to electron correlation [31]. A minimal multiconfigurational restricted active space (RASSCF) calculation that includes only the $4f^{14}$ ground and $4f^{13}5d$ open-shell excited configurations gives a very small gap (5400 cm^{-1}), which is increased by some 35600 cm^{-1} when electron correlation is included by means of second order perturbation theory. This correction is too large to be accurate at second order perturbation level. In effect, when a second f-shell is also included in the active space and single and double excitations to the $5d$, $6s$, and $5f$ shells are treated variationally, the (extended) RASSCF energy gap above the ground state and the electronic transitions increase by 22038±120 cm^{-1} and the RASPT2 correlation energy corrections become small (–721±571 cm^{-1}), as it is desirable for a second order perturbation. If both RASPT2 results are compared it is possible to conclude that the minimal RASPT2 calculation overestimates the interconfigurational transition energies by 14223±80 cm^{-1}, most of which (12700 cm^{-1}) are due to the lack of the second f-shell in the active space, which indicates an inaccurate calculation of the differential radial correlation between the $4f^{14}$ and

$4f^{13}5d$ configurations by second order perturbation theory. In order to establish practical and accurate procedures for the calculation of $4f \rightarrow 5d$ and other interconfigurational transitions at the end of the lanthanide series, the above mentioned and other RASSCF+RASPT2 calculations on the ionization potential of Yb^{2+} in gas phase and in $SrCl_2$ were benchmarked against coupled cluster (CCSD and CCSD-T) calculations and experiment. The results support that variational calculation of single and double $4f \rightarrow 5f$ excitations at the RASSCF stage prior to MS-RASPT2 can be a realistic, accurate, and feasible choice to model radial correlation effects in the heavy lanthanides ($N > 7$).

9.2.1.3 *Spin-orbit coupling*

Our choices of relativistic isolated cluster Hamiltonian allow us to write the embedded-cluster Hamiltonian of equation 9.1 as a sum of spin-free and spin-orbit coupling Hamiltonians,

$$\hat{H}^{\mathrm{EC}} = \hat{H}^{\mathrm{EC}}_{\mathrm{SF}} + \hat{H}^{\mathrm{EC}}_{\mathrm{SO}}. \tag{9.7}$$

Since the dynamic electron correlations are associated with the two-electron interactions in $\hat{H}^{\mathrm{EC}}_{\mathrm{SF}}$ and their treatment is much more cumbersome for a spin-dependent Hamiltonian like $\hat{H}^{\mathrm{EC}}_{\mathrm{SO}}$, it is very convenient to split the full calculation in two-steps, a first step with $\hat{H}^{\mathrm{EC}}_{\mathrm{SF}}$ in which all correlations are calculated (e.g., at RASSCF+RASPT2 level), and a second step with $\hat{H}^{\mathrm{EC}}$ where the spin-orbit couplings between eigenfunctions of the spin-free Hamiltonian are considered. Besides, since the spin-orbit coupling operator is a one-electron effective operator [26], the calculation of the latter couplings does not demand the use of dynamically correlated wavefunctions, but, instead, statically correlated wavefunctions like RASSCF are sufficient. A convenient way of using spin-orbit couplings calculated between statically correlated wavefunctions Φ_i^S, together with dynamically correlated energies of the spin-free Hamiltonian E_i^D, instead of the poorer statically correlated ones E_i^S, is the use of spin-free-state-shifting operators [45]:

$$\hat{H}^{\mathrm{EC}}_{\mathrm{sfss}} = \hat{H}^{\mathrm{EC}} + \sum_i |\Phi_i^S\rangle (E_i^D - E_i^S) \langle \Phi_i^S|. \tag{9.8}$$

The application of this two-step procedure for RASSCF statically correlated wavefunctions and RASPT2 dynamically correlated energies is shown by Malmqvist *et al.* [46]. Basically, it consists of computing the matrix of the spin-orbit coupling operator $\hat{H}^{\mathrm{EC}}_{\mathrm{SO}}$ in the basis of RASSCF wavefunctions, adding the RASPT2 energies to the diagonal elements, and diagonalizing it. In Paulovic *et al.* [47], the option of doing the same in the basis of the first-order wavefunctions of the MS-CASPT2 method—the modified RASSCF functions that diagonalize the MS-CASPT2 effective Hamiltonian—was followed. This is equivalent to supplementing the spin-orbit operator matrix with the full MS-CASPT2 effective Hamiltonian matrix (in the original RASSCF basis) prior to diagonalization, which prevents the anomalous results near avoided crossing points between states with equal spins and symmetries but different electronic natures (e.g., an impurity $4f^{N-1}5d$ state and an excitonic $4f^{N-1}\phi$ state) that can be found with the first approach [48].

9.2.2 Combined Use of Periodic Boundary Condition Methods and Embedded Cluster Methods

The strength of embedded-cluster methods as the ones described in Section 9.2.1 resides in their ability to produce accurate energies and properties of large manifolds of local excited states. Their weakness is in their limitations to handle structures of defects that involve many atoms. As an intermediate alternative for complex defects, periodic boundary conditions DFT methods can be used for ground state structural studies, and the resulting structures can be later used in embedded-cluster WFT methods for excited states. This has been proven to be extremely useful, e.g., to study effects of co-dopants on the local states of lanthanide ions [49, 50] and to help in determining the optically active defects among the list of potential defects present in a given lanthanide based phosphor [49–51]. The weaker point of this alternative appears when the structures of defects must be known in their excited states, e.g., in Stokes shifts calculations and in studies of excited state energy transfer mechanisms, non-radiative decays, etc.

9.2.3 Absorption and Emission Spectra

Using the computed bond lengths, breathing mode vibrational frequencies, energies of the adiabatic transitions, and absorption/emission electric dipole transition moments, it is possible to simulate absorption and emission spectra profiles. A good choice for it is the semiclassical time-dependent approach of Heller [52, 53], which follows the propagation of a vibrational wave packet of the original electronic state in the potential energy surface of the final electronic state. According to it, the intensity profile of an individual electronic absorption or emission (one electronic origin) reads

$$I(\omega) = C\,\omega^n \int_{-\infty}^{\infty} \exp(i\omega t)\,\langle \phi \mid \phi(t)\rangle\, dt, \tag{9.9}$$

where ω is the frequency of the absorbed or emitted radiation, $n = 1$ in absorption and $n = 3$ in emission, ϕ is the initial wave packet or vibrational wavefunction on the ground state, and $\phi(t)$ is its propagation in the final electronic state energy surface, which results from the vibrational time dependent Schrödinger equation

$$i\hbar\frac{\partial\phi(t)}{\partial t} = H\phi(t) \quad (\phi(0) \equiv \phi). \tag{9.10}$$

Since temperature, disorder, and other factors widen the experimental absorption and emission spectral lines, it is common and convenient to include an arbitrary widening factor Γ in the overlap between the initial and the propagated vibrational functions $\langle \phi \mid \phi(t)\rangle$. For instance, if only one breathing mode is taken into account, a common value of its vibrational frequency ω_k is used for the initial and final electronic states, and a harmonic approximation is assumed, the $\langle \phi \mid \phi(t)\rangle$ reduces to [54]

$$\langle \phi \mid \phi(t)\rangle = \exp\{-\frac{\Delta_k^2}{2}\left(1 - e^{-i\omega_k t}\right) - \frac{i\omega_k t}{2} - iE_0 t - \Gamma^2 t^2\}, \tag{9.11}$$

where Δ_k is the dimensionless displacement along the vibrational mode k, $\Delta_k = (\mu_k\omega_k/\hbar)^{1/2}\Delta Q_k$ (e.g., in a ML_6 octahedral moiety, the effective mass is $\mu_k = \mu_{a_{1g}} = m_L$

and the breathing mode distortion of the ML_6 moiety is given by $\Delta Q_k = \Delta Q_{a_{1g}} = \sqrt{6}\Delta d_{M-L}$), E_0 is the difference between the minima of the upper and lower energy surfaces, and Γ is the mentioned arbitrary damping factor whose value determines the width of the vibrational lines. Values of Γ ranging between 10 cm^{-1} and 200 cm^{-1} are common. The full absorption or emission intensity profile is calculated as the superposition of the profiles of the individual electronic origins, with weight factors having the same ratios as the oscillator strengths of the individual absorptions.

9.3 Applications

The results of some applications of the methods to the study of structural and optical properties of lanthanide ions in crystals are summarized here. They are organized to show their ability for giving insight and building a model of their electronic structure and interactions. We also focus on showing their capacity to predict optical properties, a very valuable characteristic on the line of search for new materials.

9.3.1 Bond Lengths

In addition to the expected lanthanide contraction of the $4f^N$ ground state bond length across the Ln^{2+} and Ln^{3+} series [55], an interesting result, which is most easily visible in highly symmetric crystals, is the configuration dependence of the bond length between the lanthanide impurity and the first neighbors of the crystal, the ligands. The dependence is so manifest that it contributes to the classification of large numbers of states into manifolds that share a common leading configuration and show parallel potential energy curves with common bond length values, with deviations as small as 10^{-3} Å. This can be observed in Figure 9.1, where the energy curves of the electronic states of Yb^{2+}-doped $CsCaBr_3$, calculated including electron correlation and excluding spin-orbit coupling, are plotted

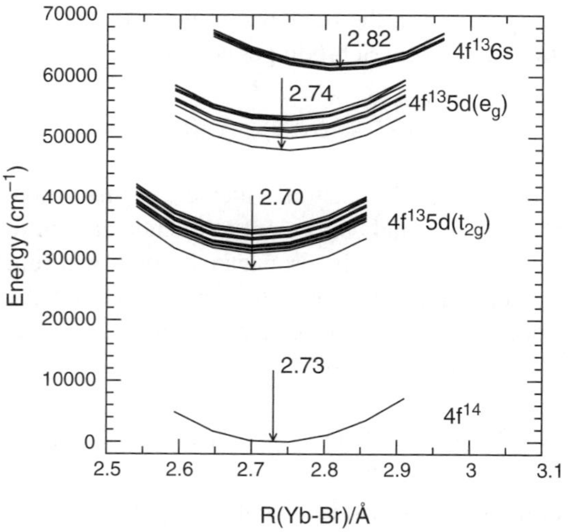

Figure 9.1 *Electronic states of Yb^{2+}-doped $CsCaBr_3$ without spin-orbit coupling*

against the Yb–Br distance of the octahedral $YbBr_6^{4-}$ embedded cluster. Furthermore, the bond length dependence of the leading electronic configuration is so systematic that the following relative values for octahedral sites:

$$R_e[4f^{N-1}d(t_{2g})] < R_e[4f^N] < R_e[4f^{N-1}d(e_g)] < R_e[4f^{N-1}6s],$$

and for cubic sites:

$$R_e[4f^{N-1}d(e_g)] < R_e[4f^N] < R_e[4f^{N-1}d(t_{2g})] < R_e[4f^{N-1}6s],$$

which are also illustrated in Figure 9.1, have been systematically obtained in O_h substitutional defects of Ln^{2+} [30, 56–58], Ln^{3+} [59–62], and actinide ions [23, 29, 37, 63, 64].

Predictions of the local geometry are important because local distortions affect the optical properties and direct measurements of them are very difficult. In effect, very little quantitative information is available on the local geometry of the defects around the lanthanide impurities and very often the undistorted local geometry of the perfect host is assumed when geometry dependent models are used. In principle, EXAFS experiments could give coordination numbers and bond distances between the impurity and its closest neighbors in the $4f^N$ ground state of the defect, however, the local geometry of excited states is even more difficult since it would require EXAFS measurements on a pumped excited state, a type of experiment which has not been done so far. Yet, it is possible to infer from experiments that the $4f^N$ and $4f^{N-1}5d^1$ manifolds have different nuclear equilibrium configurations and to estimate the absolute value of their shift out of the analysis of the absorption and emission band shapes. In this respect, it has been widely assumed that the bond distances between the lanthanide ion and the ligands are larger in the lowest $4f^{N-1}5d^1$ than in the $4f^N$ states. (See configuration coordinate diagrams in Figures 7, 2, 3, and 7 of these references [65–68], respectively, among many examples of this extended assumption.) This idea is probably based on the fact that Ln $5d$ orbitals have much larger radial extent than Ln $4f$ orbitals. Only exceptionally, it has been pointed out the possibility of an opposite shift like the one mentioned above [69].

Even though the calculated offsets between f^N and $f^{N-1}d(t_{2g})$ states were validated by the good agreement found between the theoretical and experimental $f \longleftrightarrow d(t_{2g})$ band profiles corresponding to the totally symmetric a_{1g} vibrational progression built on a single electronic origin [64, 70, 71] and on multiple electronic origins [37, 72] it was necessary to give an explanation to the unexpected trend. Hence, quantum chemical analyses of the bond lengths in f^N and $f^{N-1}d$ states of Ce^{3+}, Pr^{3+}, Pa^{4+}, and U^{4+} defects in chloride hosts were conducted by Barandiaran *et al.* [61] using the Constrained Space Orbital Variation method [73, 74]. The analyses showed the origin of the bond length shrinkage upon $4f \rightarrow 5d$ excitation: The inner $4f^N$ open-shell electrons are shielded from the ligands by the outer $5p^6$ closed-shell, whose interaction with the ligands determines the bond distance in states of the $4f^N$ configuration. When one electron is excited to lead to the $4f^{N-1}5d$ configuration, it crosses the $5p^6$ barrier and is exposed to covalent interactions with the ligands, which strengthen the bonds. At the same time, a $4f$ whole is left behind, which promotes charge transfer from the ligands, this contributing further to the bond length shrinkage. Finally, a large $5d(t_{2g}) - 5d(e_g)$ ligand field splitting, which is enhanced by covalency, decreases the $4f^{N-1}5d(t_{2g})$ bond lengths clearly below the $4f^N$ ones, and increases the $4f^{N-1}5d(e_g)$ ones clearly above, which leads to the final trends. Note that in cubic, rather than octahedral, O_h coordination the t_{2g} and e_g behaviors are opposite.

The study of the bond length shifts were also extended from the doped crystals to octahedral complexes of f element ions in liquid solutions and gas phase, which showed the generality of the bond length shortening upon $f^N \rightarrow f^{N-1}d(t_{2g})$ excitation [61]. Since no experimental proof of the actual sign of the bond length change upon excitation had been given, a quantum chemical simulation of spectroscopic experiments under high hydrostatic pressure followed that showed a continuous redshift of the lowest $f^N \rightarrow f^{N-1}d$ band of Cs_2NaYCl_6:Ce^{3+} with pressure in the range 1 bar to 26 kbar [75]. The pressure-induced redshift was shown to be a direct consequence of the bond length shrinkage upon the $f \rightarrow d(t_{2g})$ excitation, and the prediction called for the experimental detection of the pressure induced redshift as a way to proof (or reject) the bond length shortening. Shortly after, Valiente *et al.* found the predicted red shift in Ce^{3+}-doped $Cs_2NaLuCl_6$ [76].

The long bond lengths found for the $4f^{N-1}6s$ states are associated with their large radial extent, strong electron–electron repulsion within the cluster, and confinement by the host embedding, which shifts their energies to very high values, compared with the free ions, at the same time that their bond lengths are maximal, as discussed by Sanchez-Sanz *et al.* [58].

It should be noted that the trend, of bond lengths we have just described is obtained at the lowest methodological level where the energy curves are calculated at the CASSCF or RASSCF levels. Embedded-Cluster valence electron correlation adds a quite constant shrinkage of all bonds without altering the trend, and spin-orbit does not affect, in general, the local structure. Exceptions to this are the cases where manifolds with different dominant electron configuration interact due to either dynamic electron correlation or spin-orbit coupling. In these cases, configurational mixing leads to distortions of the energy curves, which may result in a range of bond length values comprising those of the interacting configurational manifolds and anharmonic curves associated with avoided crossings. An example of this can be seen in Figure 9.2 where the manifolds of spin-orbit free states presented in

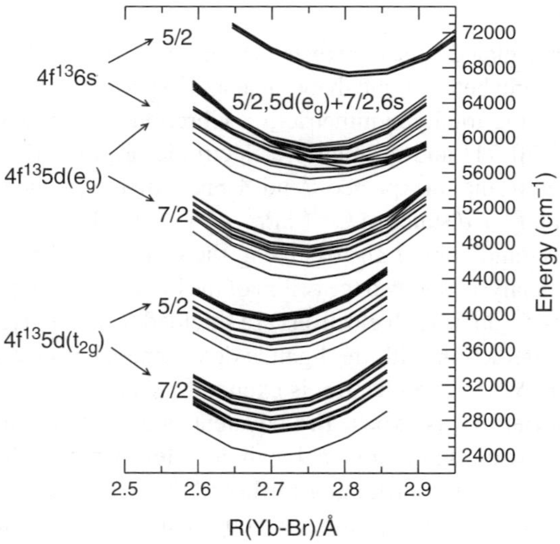

Figure 9.2 *Electronic states of Yb^{2+}-doped $CsCaBr_3$ with spin-orbit coupling*

Figure 9.1 appear to be split in two submanifolds by spin-orbit coupling. Each submanifold corresponds, basically, to the $J = 7/2$ and $J = 5/2$ components of the $4f^{13}$ subshell. Given that the spin-orbit free $4f^{13}5d(t_{2g})$ and $4f^{13}6s$ manifolds are close in energy (see Figure 9.1), their corresponding 5/2 and 7/2 subsets appear to be mixed and distorted, as can be observed in Figure 9.2 [56, 57].

9.3.2 Energy Gaps

The existence of energy gaps between excited states is characteristic of the lanthanide ions. This grants them very interesting optical properties such as multiple luminescence, upconversion, quantum cutting, which require the existence of several metastable excited states. The stability of a particular excited state of a lanthanide in a crystal depends primarily on the magnitude of the energy gap separating it from the next state of lower energy, relative to the value of the higher energy phonon of the material, since an important mechanism of non-radiative decay is multiphonon relaxation. Hence, it is interesting to study the origin of the energy gaps between excited states, and it is valuable to predict their occurrence.

Given that the $4f^N$ shell is very efficiently shielded from the crystalline environment, the energy gaps shown by the lanthanide ions in gas phase may be observed in the solids, where several excited states are metastable and capable to luminesce or absorb to higher excited states. However, the occurrence of metastable excited states of the $4f^{N-1}5d$ configuration is not so frequent. As a matter of fact, in many cases, the $4f^{N-1}5d$ levels coincide in energy with the numerous $4f^N$ states and decay through them. Yet, the $4f^{N-1}5d$ manifolds of heavy lanthanides may contain a number of metastable excited states, as deduced from the facts that multiple spontaneous emissions originating in different $4f^{N-1}5d$ states have been detected [77–79] and upconversion luminescence has been proven and interpreted in terms of energy transfer mechanisms where more than one $4f^{N-1}5d$ states are involved [80, 81].

Using the embedded-cluster methods described above multiple spontaneous emission has been predicted for Yb^{2+}-doped $CsCaBr_3$ crystals and the origin of the energy gaps existing below the potentially emitting electronic states has been analyzed by Sanchez-Sanz *et al.* [57]. Five large energy gaps are found that result from the following interactions: The interactions of the Yb^{2+} impurity with the ligands and with the embedding host are responsible for the splitting of the $4f^{13}5d$ levels into the $4f^{13}5d(e_g)$ and $4f^{13}5d(t_{2g})$ manifolds, separated by 19000 – 20000 cm^{-1} (Figure 9.1). Differential correlation between $4f$–$4f$ and $4f$–$5d$ pairs is responsible for shifting the whole $4f^{13}5d$ spectrum above the ground state by 25000 – 27000 cm^{-1} (Figure 9.1). Spin-orbit coupling within the $4f^{13}$ sub shell (which is found to amount about 10500 cm^{-1}), further splits the excited manifolds as observed in Figure 9.2. Correspondingly, the computed absorption spectrum profile, which is shown in Figure 9.3, shows four groups of separate bands, three of which could lie below the host absorption and are the bases for the prediction of multiple spontaneous $4f^{13}5d \rightarrow 4f^{14}$ emissions in this material. The predicted emissions at 23900, 26600, 34600, and 43900 cm^{-1} should be experimentally observable at low temperatures. The first, third, and fourth bands are slow, electric dipole forbidden emissions and can be described as spin-forbidden. The second band, with 400 ns emission lifetime, is an electric dipole allowed emission that cannot be described as spin-allowed, but, rather, as spin-enabled. The energy gaps below the states responsible for the slow bands—23900, 4600, 4000 cm^{-1}, respectively—are large compared to the maximum local phonon energies calculated: 185 cm^{-1}, which supports

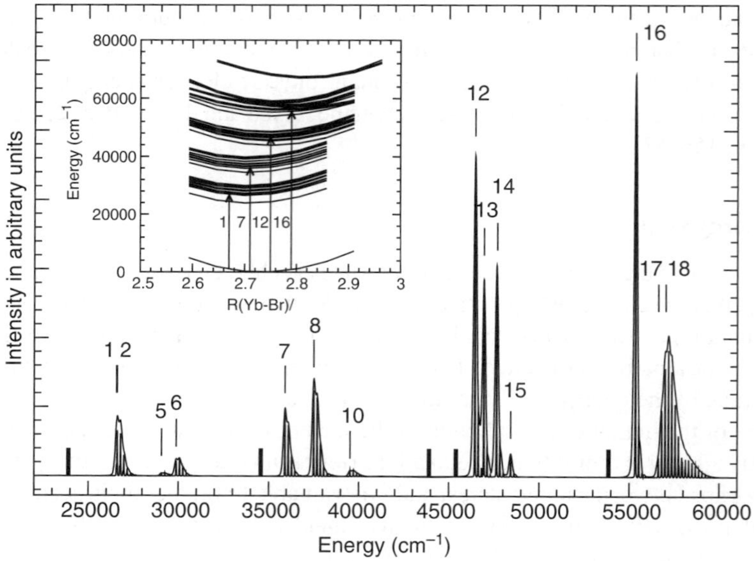

Figure 9.3 *Calculated absorption spectrum of $CsCaBr_3$:Yb^{2+} using different values for the line broadening parameter. The most intense electronic origins are indicated. Electric dipole forbidden transitions are indicated with vertical bars*

the stability of the three states against multiphonon relaxation. A smaller gap (2600 cm^{-1}) separates the states leading to the fast emission, the second emission band, which could result in a temperature-dependent competition between radiative and non-radiative decay, which could affect the relative intensities of the second and first emission bands, in line with experimental observations in similar systems. To the best of our knowledge the predicted multiple luminescence has not yet been investigated experimentally.

Similar energy gaps due to crystalline environment, electron correlation, and spin-orbit effects occur in other systems. The outcome in terms of stability of $4f^{N-1}5d$ excited states depends strongly on the particular impurity + host combination and demands comparable studies where all the methodological effects are included at once.

9.3.3 Impurity-Trapped Excitons

The combination of methods described above is flexible enough to allow for the appearance of unexpected excited states. This is what happened along a study of the electronic states of U^{4+} in the strong field hexafluoride environment of the Cs_2GeF_6 host. Whereas calculations had been set up to obtain the $5f^2$, $5f6d$, and $5f7s$ manifolds, the characteristics of the latter, namely, very short U–F bond lengths and unnormal, delocalized $7s$ shell, pointed at their impurity-trapped exciton nature as described in the Introduction by Ordejon *et al.* [29].

Figure 9.4(a) shows the energy curves obtained in the embedded cluster calculations including host effects, 68 valence electron correlation, and relativistic effects up to spin-orbit coupling [29]. Above the dense $5f^2$ and $5f6d(t_{2g})$ manifolds, and below the $5f6d(e_g)$ states, a set of levels appear from 62000 to 76000 cm^{-1} that can be identified with what

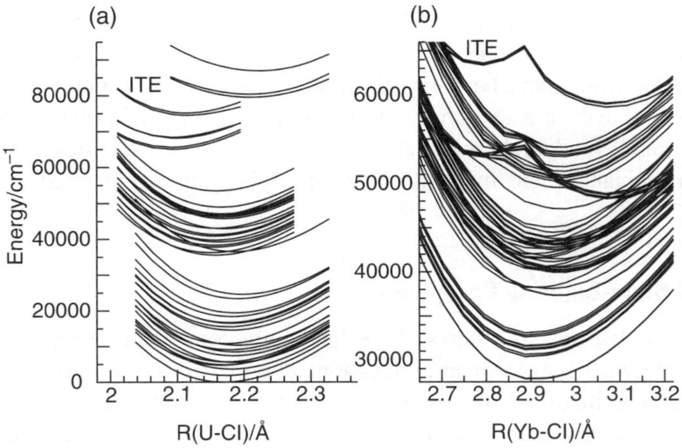

Figure 9.4 *Impurity-trapped excitons (ITE) of Cs_2GeF_6:U^{4+} (a), and $SrCl_2$:Yb^{2+} (b)*

McClure and Pédrini called impurity-trapped excitons [5]. Their wavefunctions are more delocalized than those of the impurity states, extending beyond the UF_6 moiety, and their bond lengths are very short. A significant U 7*s* – Cs 6*s* orbital mixing is observed, which strongly stabilizes the energy of these states. As a result of this charge delocalization, the bond lengths and a_{1g} vibrational frequencies become closer to the ones of U(V) defects than to U(IV) defects. The results suggested a model where the electron excited to the very diffuse U 7*s* orbital experiences such large electron repulsion in the small hexafluoride cage that significant charge spill occurs towards the second neighbor Cs sites, withdrawing charge from the cluster interior and allowing for the bond length shrinkage, all of which leads to the formation of impurity-trapped exciton states. This explains why in crystals like Cs_2ZrCl_6:U^{4+}, with larger hexachloride cage, localized $5f7s$ excited states appear at high energies with long bond lengths [63], following the bond lengths trend described above.

Impurity-trapped excitons with similar characteristics were also found in the fluorite-type $SrCl_2$ crystal doped with Yb^{2+}, in cubic chloride coordination [30]. They appear above the highest *d* manifold, $4f5d(t_{2g})$, at the short Yb–Cl distance side of double well energy curves. Splitting of all manifolds in two sets corresponding to the $4f^{13}$ 7/2 and 5/2 components (analogously to the $CsCaBr_3$:Yb^{2+} case above, shown in Figure 9.2), spread their interaction with impurity states as can be seen in Figure 9.4(b), which affected the high energy part of the absorption spectrum. The double-well energy curves result from avoided crossings between Yb-trapped exciton states (more stable at short Yb–Cl distances) and $4f^{13}6s$ impurity states (more stable at long Yb–Cl distances); the former are found to be delocalized states in which the impurity holds the excited electron in close lying interstices of the fluorite-type structure, located outside but next to the $YbCl_8$ moiety.

Impurity-trapped excitons are thought to be responsible for anomalous emissions in some Eu^{2+} and Yb^{2+} containing crystals [82]. They are also thought to participate in energy transfer and decay processes in scintillators. The widespread description of their electronic structure corresponds very well with the characteristics of the high-energy excitons we have found and described above, where the excited electron delocalizes symmetrically towards second neighbors or next interstices. However, work in progress on the electronic states of Yb^{2+}-doped CaF_2 and SrF_2 cubic defects show no trace of the low-lying, emitting impurity-trapped excitons, which have been experimentally observed and reported [5, 83–85], even though extensions to the cluster definition and basis sets are used. Further work in non-cubic defects is in progress and might shed light into the peculiar characteristics of their anomalous emissions. What makes them different from higher lying impurity-trapped excitons remains to be found.

9.3.4 Solid-State-Lighting Phosphors

The rich manifolds of excited states of the lanthanide ions in solids makes them suitable for a large variety of technological applications ranging from lasers and color displays to solar-cells and scintillators [86]. Among them, their use as phosphors in solid-state based energy efficient white illumination devices, or solid-state-lighting (SSL), has the potential of an especially large social impact due to the enormous energy savings that are expected from SSL [87]. The expected contributions from first-principles calculations have been recognized as key contributions to the Grand Challenges in SSL research: "moving from serendipitous discovery towards rational design" and "understand and control the microscopic pathways through which losses occur as electrons produce light" [87]. Here we will give an example of the potential of *ab initio* embedded-cluster wavefunction theory calculations to help such understandings and rational designs.

The most popular phosphor in commercial white light SSL devices nowadays is Ce^{3+}-doped YAG (Yttrium Aluminum Garnet, $Y_3Al_5O_{12}$). It converts the blue light emitted by a very efficient InGaN based light-emitting-diode (LED) [88] into yellow, which is mixed with the LED blue light to produce white light useful in illumination. The conversion is made by means of the local states associated with the $4f$ and $5d$ configurations of the Ce^{3+} impurity [89, 90]. The use of this phosphor produces a relatively bluish white light that prevented it from a wider use in general illumination, so that the search for phosphor materials with an emission red shifted (shifted to longer wavelengths) with respect to YAG:Ce^{3+}, which could be used in warm light illumination devices, was and still is an issue. As a matter of fact, controlling the color is considered to be one of the key issues governing the success of SSL technologies [91], as important as achieving high energy efficiencies. It is not strange, then, that a great deal of research in the field has been devoted to tailoring the phosphors color rendering indexes [89, 90, 92–95]. Co-doping YAG:Ce^{3+} was one of the lines of approach to this problem, and an initial success was achieved when it was found that co-doping Gd^{3+} and La^{3+} in YAG:Ce^{3+} red-shifted the yellow luminescence of Ce^{3+} whereas co-doping Ga^{3+} blue-shifted it [89, 92, 93]. However, an understanding of the reasons behind these red and blue shifts that could lead the search for new color-controlled phosphors was not achieved. In effect, both types of co-dopings increased the lattice constants, and the associated atomistic expansions could only explain the blue shift of Ga^{3+} co-doping: they lower the crystal-field splitting of the $5d$ shell of Ce^{3+} and rise the energy of the first $5d$ level with respect to the ground state of the $4f$ configuration. The same argument as applied to La^{3+} co-doping would also predict a blue shift rather than the detected red shift. Although many hypotheses were explored in order to explain these contradictory results, like assuming local atomistic contractions around Ce^{3+} upon La^{3+} co-doping in spite of the general lattice expansion [96], the issue remained an unsolved question for a long while, and *ab initio* embedded-cluster calculations provided convincing answers and its basic understanding. Here we summarize them.

SA-CASSCF/MS-CASPT2/*sfss*-SOCI *ab initio* embedded-cluster calculations with the spin-orbit Wood-Boring AIMP Hamiltonian on YAG:Ce^{3+} using D_2 local symmetry [97] identified the first three $4f \rightarrow 5d$ (blue) absorptions as transitions from the ground state to Γ_5 states of almost pure character of 2^2A, 3^2B_3, and 3^2A, respectively, and overestimated them 2300 and 3300 cm^{-1} with respect to experiment. The remaining two absorptions were predicted at 52500 and 54200 cm^{-1} and identified as Γ_5 states of dominant character 3^2B_2 and 3^2B_1; their energies make them to be hidden in the experiments by the absorptions of the pure host. The lowest $5d \rightarrow 4f$ emission takes place from the Γ_5 (2^2A) state and is made of two overlapping bands, each of them built on three electronic origins. The emission energies show overestimations of around 4000 cm^{-1}, which result in smaller Stokes shifts than the experimental ones. The calculations support the assignment of a wide emission band with two maxima observed just under 30000 cm^{-1} as $5d \rightarrow 4f$ transitions from higher 5*d* levels made by Blasse and Bril [89]: it is interpreted as an emission from the second 5*d* level, which is Γ_5 (3^2B_3). The shapes and relative intensities of the calculated absorption and emission band profiles agreed with the experiments and the assignments of the upper bands, which contradicted older interpretations, were later supported by further analysis of the experiments [98]. Such calculations of the absorption and emission spectra demanded previous structural optimizations in the ground 4*f* state (absorption) and in excited 5*d* states (emission), which were done at the MS-CASPT2 spin-orbit free level. Later, an optimization of the ground state structure, in PBC-DFT PBE calculations [99] lead to similar results (e.g., the two Ce-O distances of the D_2 CeO_8 moiety were 2.37 and 2.44 Å in MS-CASPT2 and 2.37 and 2.47 Å in DFT PBE) and EXAFS measurements of the ground state local structure gave 2.38 and 2.52 Å [100]. Neither the DFT calculations nor the EXAFS measurements could provide excited state structures, and EXAFS provided first neighbor coordination numbers and distances, but not detailed structural data with bond and torsion angles.

Combined PBC-DFT ground state structural studies and WFT SA-CASSCF/MS-CASPT2 excited state energy calculations were used to study the effects of Ga^{3+} co-doping [49] and La^{3+} co-doping [50] of YAG:Ce^{3+}. The structures and energetics of the large list of defects that result from such double dopings of the YAG 160 atom unit-cell gave structural expansions in all cases, in agreement with experiments. Besides, an important structural difference was found between both types of co-dopings: the Ga^{3+} co-dopants that substitute for Al^{3+} ions in four fold and six fold oxygen coordinated sites do not show a clearly preferable location with respect to the Ce^{3+} position, the La^{3+} co-dopants that substitute for Y^{3+} ions in eight fold oxygen coordinated sites do. Their most stable site is one with the shortest Ce^{3+}-La^{3+} distance. This structural difference will be responsible for the opposite shifts of Ga^{3+} and La^{3+} co-dopings. In effect, the WFT spectral calculations done with the most stable DFT structures lead to an increase of the energy difference between the lowest states of the 5*d* and 4*f* configurations after Ga^{3+} co-doping and to its decrease after La^{3+} co-doping, in agreement with the experimental observations, which explained the result but not the reason for it. The results were further analyzed in terms of energy differences between the 5*d* and the 4*f* energy centroids (averages of the energies of the individual states) and the ligand-field stabilizations; (of the lowest 5*d* and 4*f* states with respect to their respective energy centroids). The result was that the blue shift due to Ga^{3+} is determined by the 5*d* ligand field stabilization; it is smaller after co-doping because of the expansions it causes. This is in agreement with the original interpretations. However, the red shift of La^{3+} is determined by a $5d$ centroid energy stabilization, which is a bonding controlled factor not easily predictable and is stronger than the blue shift induced by the ligand-field effects associated with the atomistic expansions. These studies solved the long standing puzzle of the co-doping induced blue and red shifts of the yellow emission of YAG:Ce^{3+} solid-state-lighting phosphor. They also showed that *ab initio* calculations of this kind, which may be very demanding in terms of computational resources and time and on the theoretical background of the researchers, have a high potential for complementing the experiments in this complicated field, where simple models and empirical theories are bound to be very limited at some point.

Besides helping to understand experimental facts in the area of solid-state-lighting phosphors, *ab initio* calculations can be very helpful in the search for new color controlled phosphors, like the search for phosphors with an emission red-shifted with respect to YAG:Ce^{3+}. As an example, a series of SA-CASSCF/MS-CASPT2/RASSI-SO calculations with the DKH spin-orbit coupling Hamiltonian on the $(CeO_8)^{13-}$ cluster embedded in a cubic oxide lattice [101] and embedded in a set of 21 $A_3B'_2B''_3O_{12}$ garnets [102] (with some of the combinations of B''=Si,Al,Ga, B'=Al,Ga,Fe,Sc, and A=Mg,Ca,Fe,Mn,Y,Lu,Yb,Er,Gd,Ho,Dy,Tb,Sm,Nd) pinpointed the first-neighbor effects and the host effects on the energies of the 12 local states of main character Ce-$4f$ and Ce-$5d$. They showed that only two out of the six degrees of freedom of the effective D_2 field (resulting from bonding and embedding effects) can provide a red shift of the lowest $5d \rightarrow 4f$ emission. These are the cubic, breathing mode, a_{1g} symmetric bond stretching, and the tetragonal, $e_g(\theta)$ symmetric bond bending (see Figure 1 in Seijo and Barandiaran [102]). This result largely reduces the field of search for red phosphors based on Ce^{3+}-doped garnets. Furthermore, it was found that the unrelaxed host effect, i.e., the interactions between Ce^{3+} and the unrelaxed host (second and further distant neighbors), is largely responsible for the red shift of the first $4f \rightarrow 5d$ transition with respect to a purely cubic reference and plays a very important role in the differentiation of the values of the transition in different garnet families (Si, Al, and Ga based garnets). Interestingly, it is the large red shift created by the unrelaxed host effect that makes of Ce^{3+}-doped $Lu_2CaMg_2Si_3O_{12}$ an orange phosphor [94] with a larger wave length emission than YAG:Ce^{3+}.

References

[1] Dieke GH. Spectra and Energy Levels of Rare Earth Ions in Crystals. New York: Interscience Publishers; 1968.

[2] W T Carnall, G L Goodman, K Rajnak, and R S Rana. A Systematic Analysis of the Spectra of the Lanthanides Doped into Single Crystal LaF_3. Argonne Illinois: Argonne National Laboratory; 1988.

[3] Schilling G, Meyer G. Ternary Bromides and Iodides of Divalent Lanthanides and Their Alkaline-Earth Analoga of the Type AMX_3, and AM_2X_5. Z Anorg Allg Chem. 1996;622:759.

[4] Li CL, Reid MF. Correlation-crystal-field analysis of the $^2H(2)_{11/2}$ multiplet of Nd^{3+}. Phys Rev B. 1990;42:1903.

[5] McClure DS, Pédrini C. Excitons trapped at impurity centers in highly ionic crystals. Phys Rev B. 1985;32:8465.

[6] Sugano S, Shulman RG. Covalenci effects in KNiF3. III. Theoretical studies. Phys Rev. 1963;130:517.
[7] Huang P, Carter EA. Advances in Correlated Electronic Structure Methods for Solids, Surfaces, and Nanostructures. Ann Rev Phys Chem. 2008;59:261.
[8] McWeeny R. Some Recent Advances in Density Matrix Theory. Rev Mod Phys. 1960;32:335.
[9] Huzinaga S, Cantu AA. Theory of Separability of Many-Electron Systems. J Chem Phys. 1971;55:5543.
[10] Huzinaga S, McWilliams D, Cantu AA. Projection Operators in Hartree-Fock Theory. Adv Quantum Chem. 1973;7:187.
[11] Barandiarán Z, Seijo L. The ab initio model potential representation of the crystalline environment. Theoretical study of the local distortion on NaCl:Cu^+. J Chem Phys. 1988;89:5739.
[12] Seijo L, Barandiarán Z. The ab initio model potential method: A common strategy for effective core potential and embedded cluster calculations. In: Leszczyński J, editor. Computational Chemistry: Reviews of Current Trends. vol. 4. Singapore: World Scientific; 1999. p. 55–152.
[13] Seijo L, Barandiarán Z. *Ab initio* model potential study of local distortions around Cr^+ and Cr^{3+}. J Chem Phys. 1991;94:8158.
[14] Gellé A, Lepetit MB. Fast calculation of the electrostatic potential in ionic crystals by direct summation method. J Chem Phys. 2008;128:244716.
[15] Ewald PP. Die Berechnung optischer und elektrostatischer Gitterpotentiale. Ann Phys. 1921;369:253.
[16] Evjen HM. On the Stability of Certain Heteropolar Crystals. Phys Rev. 1932;39:675.
[17] Swerts B, Chibotaru LF, Lindh R, Seijo L, Barandiarán Z, Clima S, *et al.* Embedding fragment ab initio model potentials in CASSCF/CASPT2 calculations of doped solids: Implementation and applications. J Chem Theory Comput. 2008;4:586.
[18] Barandiarán Z, Seijo L. The ab initio model potential method. Cowan-Griffin relativistic core potentials and valence basis sets from Li (Z = 3) to La (Z = 57). Can J Chem. 1992;70:409.
[19] Seijo L. Relativistic ab initio model potential calculations including spin-orbit effects trhough the Wood-Boring Hamiltonian. J Chem Phys. 1995;102:8078.
[20] Huzinaga S, Seijo L, Barandiarán Z, Klobukowski M. The ab initio model potential method. Main group elements. J Chem Phys. 1987;86:2132.
[21] Wood JH, Boring AM. Improved Pauli hamiltonian for local-potential problems. Phys Rev B. 1978;18:2701.
[22] Cowan RD, Griffin DC. Approximate relativistic corrections to atomic radial wave functions. J Opt Soc Am. 1976;66:1010.
[23] Seijo L, Barandiarán Z. Relativistic Ab-Initio Model Potential Calculations for Molecules and Embedded Clusters. In: Schwerdtfeger P, editor. Relativistic Electronic Structure Theory: Part 2. Applications. Amsterdam: Elsevier; 2004. p. 417–475.
[24] Douglas M, Kroll NM. Quantum electrodynamical corrections to the fine structure of helium. Ann Phys (NY). 1974;82:89.
[25] Hess BA. Relativistic electronic-structure calculations employing a two-component no-pair formalism with external-field projection operators. Phys Rev A. 1986;33:3742.
[26] Hess BA, Marian CM, Wahlgren U, Gropen O. A mean-field spin-orbit method applicable to correlated wavefunctions. Chem Phys Lett. 1996;251:365.
[27] Tsuchiya T, Nakajima T, Hirao K, Seijo L. A third-order Douglas-Kroll ab initio model potential for the lanthanides. Chem Phys Lett. 2002;361:334.
[28] McWeeny R. Methods of molecular quantum mechanics. 2nd ed. London: Academic Press; 1989.
[29] Ordejón B, Seijo L, Barandiarán Z. Geometry and electronic structure of impurity-trapped excitons in Cs_2GeF_6:U^{4+} cyrstals. The $5f^17s^1$ manifold. J Chem Phys. 2007;126:194712.
[30] Sánchez-Sanz G, Seijo L, Barandiarán Z. Yb^{2+}-doped $SrCl_2$: Electronic structure of impurity states and impurity-trapped excitons. J Chem Phys. 2010;133:114509.
[31] Barandiarán Z, Seijo L. Radial correlation effects on interconfigurational excitations at the end of the lanthanide series: A restricted active space second order perturbation study of Yb^{2+} and $SrCl_2$:Yb^{2+}. J Chem Phys. 2013;138:074102.

[32] Roos BO, Taylor PR, Siegbahn PEM. A complete active space SCF method (CASSCF) using a density-matrix formulated super-CI approach. Chem Phys. 1980;48:157.
[33] Siegbahn PEM, Heiberg A, Almlöf J, Roos BO. The complete active space SCF (CASSCF) method in a Newton-Raphson formulation with application to the HNO molecule. J Chem Phys. 1981;74:2384.
[34] Siegbahn PEM, Heiberg A, Roos BO, Levy B. Comparison of the super-CI and the Newton-Raphson scheme in the complete active space SCF method. Phys Scr. 1980;21:323.
[35] J Olsen, BO Roos, P Jørgensen, Jensen JA. Determinant based configuration interaction algorithms for complete and restricted configuration interaction spaces. J Chem Phys. 1988;89:2185.
[36] Malmqvist PÅ, Rendell A, Roos BO. The restricted active space self-consistent-field method, implemented with a split graph unitary group approach. J Phys Chem. 1990;94:5477.
[37] Seijo L, Barandiarán Z. Ab initio theoretical studies on U^{3+} and on the structure and spectroscopy of U^{3+} substitutional defects in Cs_2NaYCl_6. $5f^26d^1$ manifold. J Chem Phys. 2003;118:5335.
[38] K Andersson, P-A Malmqvist, BO Roos, AJ Sadlej, Wolinski K. Second-order perturbation theory with a CASSCF reference function. J Phys Chem. 1990;94:5483.
[39] K Andersson, P-A Malmqvist, Roos BO. Second-order perturbation theory with a complete active space self-consistent field reference function. J Chem Phys. 1992;96:1218.
[40] Zaitsevskii A, JP Malrieu. Multi-partitioning quasidegenerate perturbation theory. A new approach to multireference Møller-Plesset perturbation theory. Chem Phys Lett. 1995;233:597.
[41] J Finley, P-A Malmqvist, BO Roos, Serrano-Andrés L. The multi-state CASPT2 method. Chem Phys Lett. 1998;288:299.
[42] P-A Malmqvist, K Pierloot, AR Moughal Shahi, CJ Cramer, L Gagliardi. The restricted active space followed by second-order perturbation theory method: Theory and application to the study of CuO_2 and Cu_2O_2 systems. J Chem Phys. 2008;128:204109.
[43] Forsberg N, Malmqvist P-A. Multiconfiguration perturbation theory with imaginary level shift. Chem Phys Lett. 1997;274:196.
[44] G Ghigo, BO Roos, P-A Malmqvist. A modified definition of the zeroth-order Hamiltonian in multiconfigurational perturbation theory (CASPT2). Chem Phys Lett. 2004;396:142.
[45] Llusar R, Casarrubios M, Barandiarán Z, Seijo L. Ab initio model potential calculations on the electronic spectrum of Ni^{2+}-doped MgO including correlation, spin-orbit and embedding effects. J Chem Phys. 1996;105:5321.
[46] Malmqvist P-A, Roos BO, Schimmelpfennig B. The RASSI approach with spin-orbit coupling. Chem Phys Lett. 2002;357:230.
[47] Paulovic J, Nakajima T, Hirao K, Lindh R, Malmqvist P-A. Relativistic and correlated calculations on the ground and excited states of ThO. J Chem Phys. 2003;119:798.
[48] Sánchez-Sanz G, Barandiarán Z, Seijo L. Energy level shifts in two-step spin-orbit coupling ab initio calculations. Chem Phys Lett. 2010;498:226.
[49] Muñoz-García AB, Seijo L. Structural, electronic, and spectroscopic effects of Ga codoping on Ce-doped yttrium aluminum garnet: First-principles study. Phys Rev B. 2010;82:184118.
[50] Muñoz-García AB, Pascual JL, Barandiarán Z, Seijo L. Structural effects and 4f-5d transition shifts induced by La codoping in Ce-doped yttrium aluminum garnet: First-principles study. Phys Rev B. 2010;82:064114.
[51] Wen J, Ning L, Duan CK, Chen Y, Zhang Y, Yin M. A Theoretical Study on the Structural and Energy Spectral Properties of Ce^{3+} Ions Doped in Various Fluoride Compounds. J Phys Chem C. 2012;116:20513.
[52] Heller EJ. Time-dependent approach to semiclassical dynamics. J Chem Phys. 1975;62:1544.
[53] Heller EJ. The semiclassical way to molecular spectroscopy. Acc Chem Res. 1981;14:368.
[54] Zink JI, Shin KS. In: Molecular distortions in excited electronic states determined from electronic and resonance raman spectroscopy. vol. 16 of Advances in Photochemistry. New York: Wiley; 1991. p. 119–214.
[55] Ordejón B, Seijo L, Barandiarán Z. Quantum chemical study of the lanthanide bond length contraction on Ln^{3+}-doped Cs_2NaYCl_6 crystals (Ln=Ce to Lu). J Chem Phys. 2003;119:6143.

[56] Sánchez-Sanz G, Seijo L, Barandiarán Z. Spin-forbidden and spin-enabled $4f^{14} \rightarrow 4f^{13}5d^1$ transitions of Yb^{2+}-doped $CsCaBr_3$. J Chem Phys. 2009;131:024505.
[57] Sánchez-Sanz G, Seijo L, Barandiarán Z. Energy gaps in the $4f^{13}5d^1$ manifold and multiple spontaneous emissions in Yb^{2+}-doped $CsCaBr_3$. J Phys Chem A. 2009;113:12591.
[58] Sánchez-Sanz G, Seijo L, Barandiarán Z. Energy shift of the $4f^{13}6s^1$ excited states of Yb^{2+} from gas phase to the $CsCaBr_3$ Solid. Spec Letters. 2010;43:393.
[59] Barandiarán Z, Seijo L. Quantum chemical analysis of the bond lengths in f^n and $f^{n-1}d^1$ states of Ce^{3+}, Pr^{3+}, Pa^{4+}, and U^{4+} defects in chloride hosts. J Chem Phys. 2003;119:3785.
[60] Tanner PA, Mak CSK, Edelstein NM, Murdoch KM, Liu G, Huang J, *et al.* Absorption and emission spectra of Ce^{3+} in elpasolite lattices. J Amer Chem Soc. 2003;125:13225.
[61] Barandiarán Z, Edelstein NM, Ordejón B, Ruipérez F, Seijo L. Bond lengths of 4f(1) and 5d(1) states of Ce^{3+} hexahalides. J Solid State Chem. 2005;178:464.
[62] Barandiarán Z, Seijo L. On the bond length change upon 4f(1)→5d(1) excitations in eightfold coordination: CaF_2:Ce^{3+} cubic defects. Theor Chem Acc. 2006;116:505.
[63] Barandiarán Z, Seijo L. Structure and spectroscopy of U^{4+} defects in Cs_2ZrCl_6: Ab initio theoretical studies on the $5f^2$ and $5f^16d^1$ manifolds. J Chem Phys. 2003;118:7439.
[64] Seijo L, Barandiarán Z. Structure and spectroscopy of Pa^{4+} defects in Cs_2ZrCl_6. An ab initio theoretical study. J Chem Phys. 2001;115:5554.
[65] Andres HP, Krämer K, Güdel HU. Optical absorption and luminescence spectroscopy of U^{3+} in K_2LaX_5 (X=Cl,Br,I). Phys Rev B. 1996;54:3830.
[66] Shen Y, Bray KL. Effect of pressure and temperature on 4f-4f luminescence properties of Sm^{2+} ions in MFCl crystals (M=Ba, Sr, and Ca). Phys Rev B. 1998;58:11944.
[67] Dereń PJ, Stręk W, Zych E, Drożdżyński J. Up-conversion in elpasolite crystals doped with U^{3+}. Chem Phys Lett. 2000;332:308.
[68] N M Khaidukov, M Kirm, S K Lam, D Lo, Mkhov VN, Zimmerer G. VUV spectroscopy of KYF_4 crystals doped with Nd^{3+}, Er^{3+} and Tm^{3+}. Opt Comm. 2000;184:183.
[69] Meijerink A, Blasse G. Luminescence properties of Eu^{2+}-activated alkaline-earth haloborates. J Lumin. 1989;43:283.
[70] Piehler D, Kot WK, Edelstein N. The 6d→5f fluorescence spectra of $PaCl_6^{2-}$ in a Cs_2ZrCl_6 crystal. J Chem Phys. 1991;94:942.
[71] Edelstein N, Kot WK, Krupa JC. The 5f→6d absorption spectrum of Pa^{4+}:Cs_2ZrCl_6. J Chem Phys. 1992;96:1.
[72] Karbowiak M, Simoni E, Drożdżyński JD, Hubert S. Acta Phys Pol A. 1996;90:367.
[73] Bagus PS, Hermann K, Bauschlicher CW. A new analysis of charge transfer and polarization for ligand-metal bonding: Model studies of Al_4CO and Al_4NH_3. J Chem Phys. 1984;80:4378.
[74] Bauschlicher CW, Bagus PS. The metal-carbonyl bond in $Ni(CO)_4$ and $Fe(CO)_5$: A clear-cut analysis. J Chem Phys. 1984;81:5889.
[75] Ruipérez F, Seijo L, Barandiarán Z. Prediction of pressure-induced redshift of $f^1 \rightarrow d(t_{2g})^1$ excitations in Cs_2NaYCl_6:Ce^{3+} and its connection with bond-length shortening. J Chem Phys. 2005;122:234507.
[76] R Valiente, F Rodríguez, J González, H U Güdel", R Martín-Rodríguez, L Nataf, M N Sanz-Ortiz, Krämer K. High pressure optical spectroscopy of Ce^{3+}-doped $Cs_2NaLuCl_6$. Chem Phys Lett. 2009;481:149.
[77] Witzke H, McClure DS, Mitchell B. The photoluminescence spectra of single crystals of $SrCl_2$:Yb^{2+}. In: Williams FE, editor. Luminescence of Crystals, Molecules, and Solutions. New York: Plenum Press; 1973. p. 598.
[78] Grimm J, Güdel HU. Five different types of spontaneous emission simultaneously observed in Tm^{2+} doped $CsCaBr_3$. Chem Phys Lett. 2005;404:40.
[79] Pan Z, Duan C, Tanner PA. Electronic spectra and crystal field analysis of Yb^{2+} in $SrCl_2$. Phys Rev B. 2008;77:085114.
[80] E Beurer, J Grimm, P Gerner, Güdel HU. New type of near-infrared to visible photon upconversion in Tm^{2+}-doped $CsCaI_3$. J Amer Chem Soc. 2006;128:3110.

[81] J Grimm, E Beurer, P Gerner, Güdel HU. Upconversion between 4f-5d excited states in Tm^{2+}-doped $CsCaCl_3$, $CsCaBr_3$, and $CsCaI_3$. Chem Eur J. 2007;13:1152.
[82] Dorenbos P. J Phys: Condens Matter. 2003;15:2645.
[83] B Moine, C Pédrini, D S McClure, Bill H. Fluorescence and photoionization processes of divalent Yb ions in SrF_2. J Lumin. 1988;40&41:299.
[84] B Moine, B Courtois, Pédrini C. Luminescence and photoionization processes of Yb^{2+} in CaF_2, SrF_2 and BaF_2. J Phys France. 1989;50:2105.
[85] MF Reid, PS Senanayake, J-P R Wells, G Berden, A Meijerink, AJ Salkeld, C-K Duan, RJ Reeves. Transient photoluminescence enhancement as a probe of the structure of impurity-trapped excitons in CaF_2:Yb^{2+}. Phys Rev B. 2011;84:113110.
[86] Weber MJ. Inorganic scintillators: today and tomorrow. J Lumin. 2002;100:35.
[87] Phillips JM, Burrows PE, Davis RF, Simmons JA, Malliaras GG, So F, *et al.* Basic Research Needs for Solid-State Lighting. Department of Energy, Office of Science; 2006. Http://science.energy.gov/bes/news-and-resources/reports/.
[88] Nakamura S, Fasol G. The blue laser diode: GaN based light emitters and lasers. Berlin: Springer; 1997.
[89] Blasse G, Bril A. Investigation of some Ce^{3+}-activated phosphors. J Chem Phys. 1967;47:5139.
[90] Jüstel T, Nikol H, Ronda C. New developments in the field of luminescent materials for lighting and displays. Angew Chem, Int Ed. 1998;37:3085.
[91] Brodrick J. Next-generation lighting initiative at the US Department of Energy: Catalyzing science into the marketplace. J Disp Technol. 2007;3:91.
[92] Tien TY, Gibbons EF, DeLosh RG, Zacmanidis PJ, Smith DE, Stadler HL. Ce^{3+} activated $Y_3Al_5O_{12}$ and some of its solid-solutions. J Electrochem Soc. 1973;120:278.
[93] Robertson JM, van Tol MW, Smits WH, Heynen JPH. Color shift of the Ce^{3+} emission in mono-crystalline epitaxially grown garnet layers. Philips J Res. 1981;36:15.
[94] Setlur AA, Heward WJ, Gao Y, Srivastava AM, Chandran RG, Shankar MV. Crystal chemistry and luminescence of Ce^{3+}-doped $(LuCaMg_2)Ca_2(Si,Ge)_3O_{12}$ and its use in LED based lighting. Chem Mater. 2006;18:3314.
[95] Shimomura Y, Honma T, Shigeiwa M, Akai T, Okamoto K, Kijima N. Photoluminescence and crystal structure of green-emitting $Ca_3Sc_2Si_3O_{12}$:Ce^{3+} phosphor for white light emitting diodes. J Electrochem Soc. 2007;154:J35.
[96] Pan YX, Wang W, Liu GK, Skanthakumar S, Rosenberg RA, Guo XZ, *et al.* Correlation between structure variation and luminescence red shift in YAG:Ce. J Alloys Compd. 2009;488:638.
[97] Gracia J, Seijo L, Barandiarán Z, Curulla D, Niemansverdriet H, van Gennip W. Ab initio calculations on the local structure and the 4f-5d absorption and emission spectra of Ce^{3+}-doped YAG. J Lumin. 2008;128:1248.
[98] Tanner PA, Fu L, Ning L, Cheng BM, Brik MG. Soft synthesis and vacuum ultraviolet spectra of YAG:Ce^{3+} nanocrystals: reassignment of Ce^{3+} energy levels. J Phys: Condens Matter. 2007;19:216213.
[99] Muñoz-García AB, Seijo L. Ce and La Single- and Double-Substitutional Defects in Yttrium Aluminum Garnet: First-Principles Study. J Phys Chem A. 2011;115:815.
[100] Ghigna P, Pin S, Ronda C, Speghini A, Piccinelli F, Bettinelli M. Local structure of the Ce^{3+} ion in the yellow emitting phosphor YAG:Ce. Opt Mater. 2011;34:19.
[101] Seijo L, Barandiarán Z. 4f and 5d levels of Ce^{3+} in D_2 8-fold oxygen coordination. Opt Mater. 2013;35:1932.
[102] Seijo L, Barandiarán Z. Host effects on the optically active 4f and 5d levels of Ce^{3+} in garnets. Phys Chem Chem Phys. 2013;15:19221.

10

Judd-Ofelt Theory — The *Golden* (and the Only One) Theoretical Tool of f-Electron Spectroscopy*

Lidia Smentek

Department of Chemistry, Vanderbilt University

10.1 Introduction

The year 2012 marked the golden anniversary of the Judd-Ofelt (J-O) theory as a tool used widely for the theoretical description of the spectroscopic properties of the lanthanide materials; in fact it is **the only** tool, as pointed out in the title of this chapter, that is used in *ab initio* type calculations. It is indeed a unique tool applicable for all of these investigations whose main aim is not to reproduce the measurements by a fitting procedure but rather to understand the physical mechanisms responsible for a plethora of properties observed for lanthanide systems. The golden anniversary of the J-O theory was celebrated by a special scientific session followed by a birthday ceremony organized at the Conference ICfE-8 in Udine, Italy. Its festive atmosphere was enhanced by the presence of the authors of the theory [1, 2].

Two separate papers authored by Brian Judd and George Ofelt, of which over the years have become the origin of the J-O Theory, were published independently on the same day, August 1, 1962, in two different journals, Physical Review and Journal of Chemical Physics [3, 4]. The research was not only performed independently on two distant shores of the country, but the scientists never met personally while working—at Berkeley, Brian Judd, and at Johns Hopkins, George Ofelt. To celebrate the fortieth anniversary of these world

*This presentation is dedicated to the memory of George S. Ofelt who passed away on June 5, 2014.

Computational Methods in Lanthanide and Actinide Chemistry, First Edition. Edited by Michael Dolg.

famous publications, their authors met for the first time at the conference in Lądek Zdrój, Poland in 2003 [5].

In a 2001 paper devoted to the general aspects of the J-O theory (of which I was one of the authors), it was written [6]:

"It is a common opinion that the task of reproduction of spectra of rare earth doped materials is reduced to the problem of adjusting at most three intensity parameters in accordance with the following expression

$$S_{f\leftarrow i} = \sum_{\lambda=2,4,6} \Omega_\lambda |\langle \Psi_f || U^{(\lambda)} || \Psi_i \rangle|^2 \tag{10.1}$$

For the vast majority of researchers that are involved in the spectroscopy of rare earth ions this technical point of view very often defines the well-known Judd-Ofelt theory [1, 2]; for some of them however, the Judd-Ofelt theory has a broader interpretation.

Without the definition of the physical model that characterizes the Judd-Ofelt theory, expression (10.1) is an algebraic scheme of the parametrization of the spectrum. From such a point of view, there is no reason to limit the terms in the summation to just those with even ranks. Actually, treating the expression in (10.1) in a formal way, it is possible to introduce in an ad hoc way completely new terms for λ odd without giving them any physical explanation. However it should be realized that such a technical improvement of the fitting procedure (more degrees of freedom, better adjustment) loses its identity as an extension of the original Judd-Ofelt theory as introduced by its authors in 1962.

When the parameters Ω_λ in (10.1) are determined in a semiempirical way, and the intensities are reproduced, one may conclude that the one particle parametrization scheme applied for the spectrum works well; and this is the success of the Judd-Ofelt theory.

The aim of this presentation is to answer the question about the physical mechanisms that contribute to the intensity parameters which are determined from (10.1) in a semiempirical way. The present discussion demonstrates the generality and universality of the Judd-Ofelt theory which, in fact, goes far beyond the initial limitations of its original derivation."

The parametrization of the f-spectra is the subject of the last part of this chapter. However, after more than ten years since this opinion quoted above was written, it is still valid. In fact it is even more important now when the properties of the lanthanide materials are really revolutionizing modern science, technology, and everyday life. In addition to technical problems of various applications of these materials there are so many effects monitored experimentally that are still not fully understood or theoretically described. Therefore, they are not further used for designing new equipment of desired properties, especially those crucial for modern applications. The best example is the correlation between the structure, including the geometry and morphology of a sample, and the observed spectroscopic patterns of lanthanide materials of ultra-small nano size. In fact not only from a theoretical point of view this is a novel and rather pioneering research. Although the ultra-small nano-samples of the lanthanide materials are synthesized in some laboratories, and their spectroscopic patterns are experimentally monitored, their physical origin is not verified precisely enough to establish a reliable theoretical model of their description. In comparison to the features of the lanthanide materials in bulk, the changes of spectra of nano samples caused by their size is a new phenomenon. To describe these observations a correlation between the spectroscopic activity of the lanthanide ion and its environment has to be established via (*ab initio*) calculations. The J-O Theory is a powerful tool that connects these two aspects of research, and it is used to monitor modifications of the sample when its size reaches the nano scale.

Another very important research method that requires deeper knowledge and understanding of the electronic structure of the lanthanides is addressed directly to their medical applications. There is a demand for theoretical investigations useful for predicting a proper architecture of organic chelates caging very toxic lanthanide ions. Such compounds are potentially the diagnostic and therapeutic tools to detect and cure the cancerous changes in various tissues [7–10].

The role of the J-O Theory cannot be overestimated not only in these two examples of research. Indeed, the $f \longleftrightarrow f$ electric dipole transitions, described by this theory, that are characteristic for the lanthanide materials, are the best probes used in many investigations, and consequently in various applications. In order to demonstrate the validity of the latter statement, prior to analyzing the details of the physical background and resulting theoretical model of Judd and Ofelt, it is illustrative to inspect the graphs below that present the rapid development of the lanthanide field of macroscopic and microscopic scales.

The first graph, Figure 10.1, presents the number of citations of the J-O Theory based on the statistics that started in 1990. (For almost 30 years between 1962 and 1990 the bars on the graph would be only slightly visible if preserving the same scale of the present graph.) The last bar on this illustration shows only citations from the first half of 2012 when the

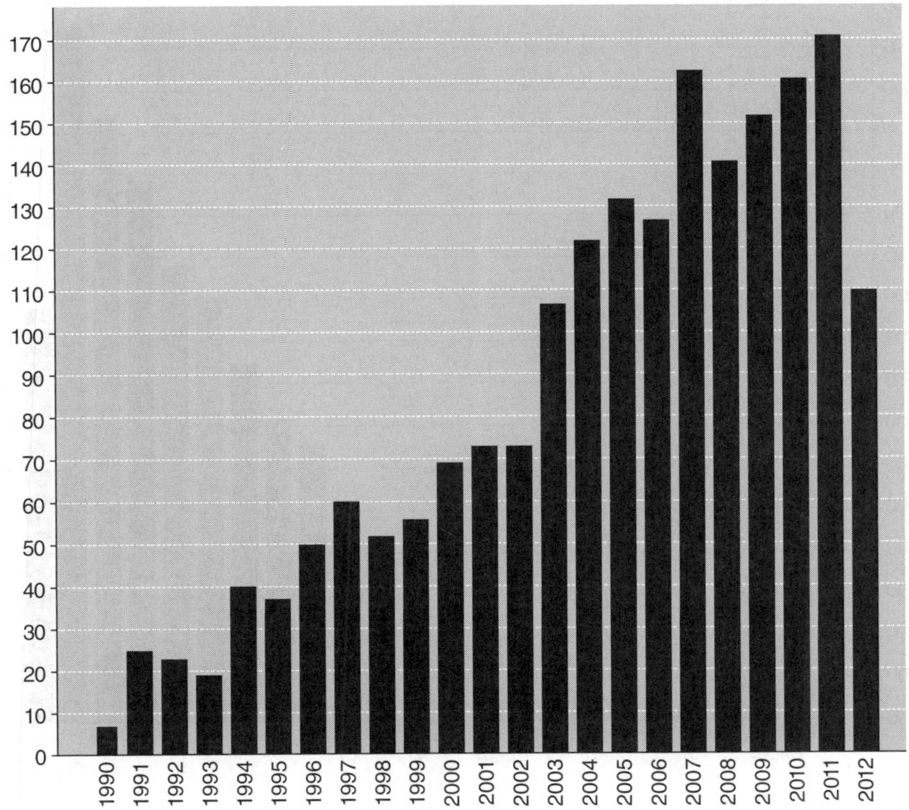

Figure 10.1 *Rapid increase of the number of citations of the Judd-Ofelt theory in scientific publications; the last bar shows the result for the first half of 2012*

results were presented at the golden anniversary celebration. Thus, this graph presents the importance of the J-O Theory expressed in numbers.

The importance of the field of the lanthanides is presented in Figure 10.2 where the numbers of citations of the publications that cited the J-O Theory are collected for the same period of time (again the last bar reflects the situation during the first half of 2012). It is seen that indeed the dynamics of this field of research is exponential and has the symptoms of a revolutionary development.

These are the numbers that evidently show the role of the J-O Theory in the field with undefined limits due to the broad applications of its achievements. In order to understand why this theory is so important, its physical background must be presented. It is possible to conclude briefly that the J-O Theory is a simple application of the outstanding beauty of tri-positive lanthanide ions, and in particular their unusual electronic structure. Its features are defined in the language of Racah algebra applied for the concept of effective tensor operators. The simplicity and clarity of this approach, including the well-known Judd-Ofelt parametrization scheme of the f-spectra based on (10.1), when successfully applied to very complex systems makes one wonder how is it possible that this tool works so well; in fact this query is its power.

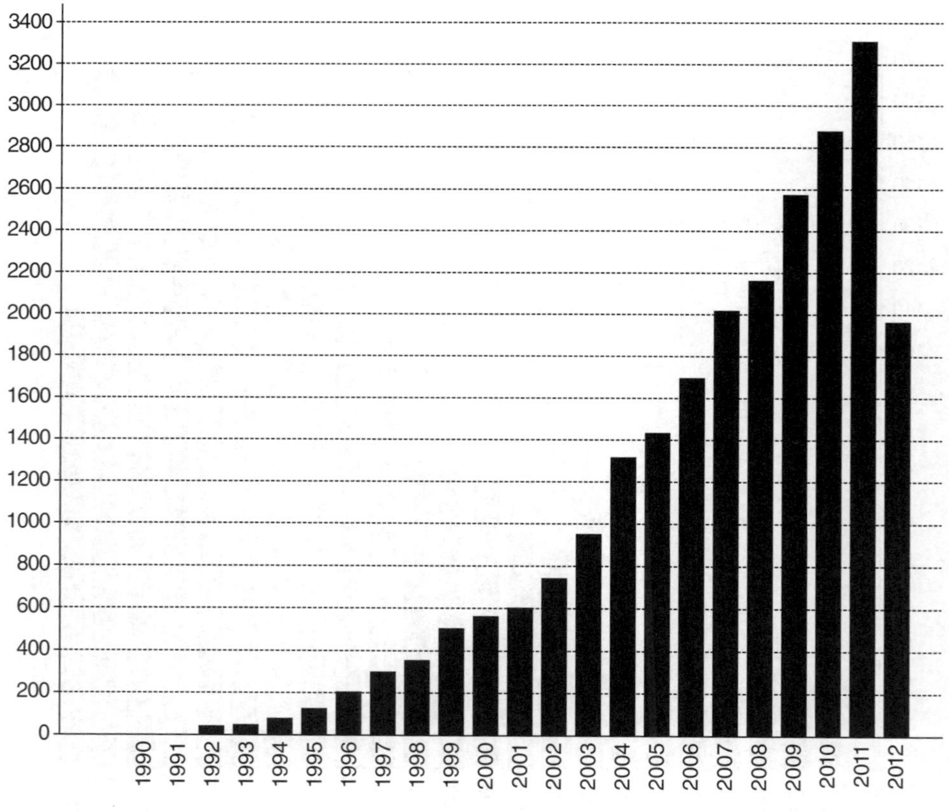

Figure 10.2 *Rapid increase of the importance of the field of the lanthanide materials as reflected by the numbers of citations of the publications that cite the J-O Theory*

In summary, the advantages and limitations of the Judd-Ofelt theory when applied in ab initio and also semi-empirical calculations are discussed. The model is presented in a standard version as introduced in 1962 by its authors, Brian Judd and George Ofelt, together with its extensions and modifications by the third-order contributions to the amplitude of the electric dipole f-f transitions. In addition to the standard non-relativistic approach, the relativistic version is also presented to enable a theoretical description of those transitions that are observed in various experiments but forbidden within the original Judd-Ofelt parameterization scheme of f-spectra. The presentation is concluded by practical remarks on the application of this theoretical tool, including the solution of the problem of the completeness of the radial basis sets of one-electron functions when the direct calculations are performed. In the case of semi-empirical calculations the meaning of the intensity parameters is clarified. Finally, an extensive list of important literature is provided to help the potential reader to understand the challenges of the theoretical description of the lanthanide ions, and most of all, to facilitate the application of existing theoretical models.

10.2 Non-relativistic Approach

It sounds like a slogan to state that the electronic structure of tri-positive lanthanide ions is special. Indeed, almost all publications on this subject start in this way. To follow this pattern it should be repeated again that the optically active electrons that occupy one-electron energy states of $4f$ symmetry are shielded from the perturbing environment by the closed shells of the $5s$ and $5p$ symmetry. At the same time, in the energy scale, they are the outermost, the so-called valence electrons, and therefore responsible for the spectroscopic properties. As a consequence of this property it is possible to assume that the tri-positive lanthanide ion embedded in any structure keeps its free-ionic identity, represented by the unperturbed Hamiltonian H_0, and its properties are only slightly modified by the environment that might be treated as a perturbation V. This is why the properties of systems with the lanthanide ions are theoretically described by perturbation theory (in practise the Rayleigh-Schrödinger approach) applied for the Hamiltonian [11],

$$\mathcal{H} = H_0 + \lambda V. \tag{10.2}$$

The unperturbed Hamiltonian, H_0, contains a part defined within the central field approximation (in practice the Hartree-Fock model), h_0, and also all such operators that are necessary for a proper description of the free ion, like the noncentral part of Coulomb interaction that is responsible for electron correlation effects, for example (since h_0 is defined within the single configuration approximation). V represents in general all physical mechanisms that are important for the description of an ion in a system but still are weak enough to be treated as a perturbation.

As mentioned, the most attractive properties of these systems used in their applications originate from the electric dipole $f \longleftrightarrow f$ transitions. They might be direct and observed as absorption or emission, or play a role as initial and final steps of sensitized luminescence, for example. In all of these cases the amplitude of the electric dipole transition is determined by the certain order of the time-dependent perturbation theory applied for the description of the interaction between matter and the radiation field. Each order of the perturbing expansion describes a process with a certain number of photons absorbed or emitted. In general, the transition rate is defined by matrix elements of the interaction Hamiltonian $\mathcal{H}_{\mathcal{I}}$ between both subsystems that connects one- electron states involved in the process [12],

$$\frac{1}{\tau} = \frac{2\pi}{\hbar^2}\sum_f | \langle f \mid \mathcal{H}_{\mathcal{I}} \mid i\rangle + \frac{1}{\hbar}\sum_{\ell} \frac{\langle f \mid \mathcal{H}_{\mathcal{I}} \mid \ell\rangle\langle \ell \mid \mathcal{H}_{\mathcal{I}} \mid i\rangle}{\omega_i - \omega_\ell} + \cdots +$$
$$\frac{1}{\hbar^{n-1}}\sum_{\ell_1}\sum_{\ell_2}\cdots\sum_{\ell_{n-1}} \frac{\langle f \mid \mathcal{H}_{\mathcal{I}} \mid \ell_1\rangle\langle \ell_1 \mid \mathcal{H}_{\mathcal{I}} \mid \ell_2\rangle\cdots\langle \ell_{n-1} \mid \mathcal{H}_{\mathcal{I}} \mid i\rangle}{(\omega_i - \omega_{\ell_1})(\omega_i - \omega_{\ell_2})\cdots(\omega_i - \omega_{\ell_{n-1}})} |^2$$
$$\times\, \delta(\omega_i - \omega_f) \tag{10.3}$$

In (10.3) $|i\rangle$ and $\langle f|$ denote the initial and final states of the radiative process while $\ell, \ell_1, \ldots$ are the so-called virtual intermediate states that are not physical states that take part in the process.This is indicated by the Dirac delta function that reflects the energy-conservation law, and it covers the energies of the initial and final states between which the physical transition occurs.

The J-O Theory describes one photon $f \longleftrightarrow f$ transitions, thus the first-order term from the above expansion is the starting point for further analysis.

In the electric dipole approximation, $\mathcal{H}_{\mathcal{I}}$ reduces to the electric dipole radiation operator, $D_{\varrho}^{(1)}$ in a tensorial form, a pair of photon creation and annihilation operators (their order depends on the type of the radiative process), and a pair of creation and annihilation operators that change one electron initial state of a matter into the final state due absorption/annihilation of a photon. The atomic part of the first-order term, when the free ionic system approximation and the electric dipole approximation are adopted, is defined by the matrix element,

$$\langle 4f^N \Psi_f^0 \mid D^{(1)} \mid 4f^N \Psi_i^0\rangle \tag{10.4}$$

where $\langle 4f^N \Psi_f^0 \mid$ and $\mid 4f^N \Psi_i^0\rangle$ are the solutions of the eigenvalue problem of H_0 from (10.2).

This is the transition amplitude between the final and initial states of the electron configuration $4f^N$ of a tri-positive lanthanide ion. Due to the parity requirements, this amplitude vanishes while a direct electric dipole transition between these energy states for many systems containing the lanthanide ions are observed. This means that in reality the parity of the states of $4f^N$ configuration is not well defined. Therefore, the theoretical model of the description of these traditions has to include components of opposite parities. This is realized by the higher-order corrections to the wavefunctions as defined within the time-independent perturbation approach. This is why in equation (10.2) the perturbing mechanism V is introduced to complement the total hamiltonian of the system. Thus, the Rayleigh-Schrödinger perturbation theory is applied and the crystal field potential is chosen as a perturbation to lower the spherical symmetry of a free ion represented by h_0. This perturbation allows the admixing to the functions of the zeroth order new components of an opposite parity. This is the main concept of the J-O Theory, which is based on the perturbation theory applied for the Hamiltonian,

$$H = H_0 + \lambda(PV_{cryst}Q + QV_{cryst}P) \tag{10.5}$$

with the tensorial definition of the crystal field potential V_{cryst} as follows,

$$V_{cryst} = \sum_{k,q}^{all} B_q^k \sum_i^N r_i^k C_q^{(k)}(\vartheta_i, \phi_i) \tag{10.6}$$

where B_q^k is the crystal field parameters (structural parameters), r^k is the radial part of the potential, and $C_q^{(k)}$ is the spherical tensor.

The partitioning of the Hamiltonian leads to the separation of the subspace P, spanned by the wavefunctions of the unperturbed Hamiltonian Ψ^0, from its orthogonal complement Q. Since H_0 describes the system within the single configuration approximation, then P projects onto the space spanned by the states of the $4f^N$ configuration. Q contains the projection operators built of the states of various excited configurations, among which those of the opposite parity to the parity of the states of the ground configuration are of main interest. Due to this construction, the perturbing operator in equation (10.6) represents the inter-shell interactions via crystal field potential. In this case V_{cryst} couples the states of the ground and excited configurations, and that is the origin of the standard J-O Theory. The remaining part of the perturbation that represents the intra-shell interactions, $PV_{cryst}P$, may be included within H_0, if the free ionic system approximation is not applied, and the so-called $j-j$ mixing is introduced. It should be mentioned that this formal partitioning of the space into two separate subspaces prevents possible count of various interactions twice.

Taking into account the perturbing influence of the crystal field potential upon the electronic structure of the lanthanide ion, the wavefunction defined up to the first order in perturbation V_{cryst} has the following form:

$$\Psi_i = \Psi_i^0 + \lambda \sum_{k \neq i} \frac{\langle \Psi_k^0 \mid QV_{cryst}P \mid \Psi_i^0 \rangle}{E_i^0 - E_k^0} \Psi_k^0. \tag{10.7}$$

For a non-vanishing matrix element in equation (10.7) it is evident that Ψ_k^0 has to belong to Q and Ψ_i^0 to P. Since $PQ = 0$, the summation in equation (10.7) is over the states of excited configurations, and therefore the energy denominators are always non-zero.

The first non-vanishing terms contributing to the transition amplitude are the terms of second order. These are the terms with one energy denominator in time-independent perturbation expansion, and they define the original version of the J-O Theory as introduced in 1962,

$$\Gamma^2 = \lambda \{ \langle \Psi_f^0 \mid D_\rho^{(1)} \mid \Psi_i^1 \rangle + \langle \Psi_f^1 \mid D_\rho^{(1)} \mid \Psi_i^0 \rangle \} + \theta(\lambda^m), \quad m \geq 2 \tag{10.8}$$

with the electric dipole transition represented by a tensor operator,

$$D_\rho^{(1)} = \sum_i^N r_i C_\rho^{(1)}(\vartheta_i, \phi_i). \tag{10.9}$$

The parity requirements for the non-vanishing matrix elements limit the set of excited configurations to those of the parity opposite to the parity of $4f^N$. This means that the set of excited configurations contains the following singly excited configurations (due to the one particle nature of dipole radiation and crystal field potential operators),

$$X \equiv 4f^{N-1}n'\ell', \quad for \quad \ell' = even\,(\equiv d, g, ...). \tag{10.10}$$

Finally, the transition amplitude is determined by two second-order terms that differ from each other by the order of appropriate operators and energy denominators,

$$\Gamma^2 = \sum_{Xx} \Big\{ \langle \Psi_f^0 \mid D_\rho^{(1)} \mid Xx\rangle\langle Xx \mid QV_{cryst}P \mid \Psi_i^0\rangle/(E_i^0 - E_{Xx}^0) + \langle \Psi_f^0 \mid PV_{cryst}Q \mid Xx\rangle\langle Xx \mid D_\rho^{(1)} \mid \Psi_i^0\rangle/(E_f^0 - E_{Xx}^0) \Big\}, \tag{10.11}$$

and x denotes the energy states of the excited configurations X that are of opposite parity to the parity of Ψ_f^0 and Ψ_i^0 (thus the matrix elements of $D_\varrho^{(1)}$ do not vanish). This is the physical background of the J-O Theory.

The form of the transition amplitude defined in equation (10.11) is not suitable for a practical application. The form of the products of matrix elements in each contributing term may suggest the so-called closure procedure over the complete basis sets of x, for each X, however the presence of the energy denominators prevents this simplification. This is why special steps have to be undertaken to finally perform the partial closure to derive the expression applicable for numerical calculations.

10.2.1 Standard Judd-Ofelt Theory and Its Original Formulation of 1962

The whole theoretical procedure applied to derive the famous expression of the J-O Theory is becoming clearer when the inter-shell unit tensor operators are introduced. They are defined by the reduced matrix elements in the following way:

$$\langle n\ell \,||\, u^{(k)}(n_1\ell_1, n_2\ell_2) \,||\, n'\ell'\rangle = \delta(n\ell, n_1\ell_1)\delta(n_2\ell_2, n'\ell'). \tag{10.12}$$

It is easily seen from this definition that such an operator is interpreted as coupled products of annihilation and creation operators of one-electron states. Therefore, due to the orthogonality of one electron states, the reduced matrix element is equal to 1 when to the right (and to this left) the same one electron state is created and annihilated.

In the terms of unit tensor operators, V_{cryst} has the following form:

$$QV_{cryst}P \equiv \sum_{t,p} B_p^t \sum_j \langle n'\ell' \mid r_j^t \mid 4f\rangle\langle \ell' \,||\, C^{(t)} \,||\, f\rangle u_{j,p}^{(t)}(n'\ell', 4f) \tag{10.13}$$

The electric dipole transition operator is expressed as follows:

$$D_\rho^{(1)} \rightarrow PD_\rho^{(1)}Q \equiv \sum_i \langle 4f \mid r_i \mid n'\ell'\rangle\langle f \,||\, C^{(1)} \,||\, \ell'\rangle u_{i,\rho}^{(1)}(4f, n'\ell'). \tag{10.14}$$

The tensorial form of each product of the unit tensor operators in the terms contributing to the transition amplitude (10.11) can be simplified by coupling the objects to a tensorial product following the relation,

$$u_{i,\rho}^{(1)}(4f,n'\ell')u_{j,p}^{(t)}(n'\ell',4f) = \sum_{k,q}(-1)^{1-t-q}[k]^{1/2}\begin{pmatrix}1 & t & k\\ \rho & p & -q\end{pmatrix}$$
$$[u_i^{(1)}(4f,n'\ell') \times u_j^{(t)}(n'\ell',4f)]_q^{(k)}, \qquad (10.15)$$

where $[k] = 2k+1$, and k is the rank of the tensorial product of two tensor operators.

When taking into account the orthogonality of one electron functions, it is evident that the matrix element of the operator at the right-hand side of equation (10.15) is not vanishing only if both tensor operators in the product act on the same coordinate. This is the condition that allows replacement the tensorial product by a single tensor operator that acts within the $4f^N$ shell and thus represents the effective interactions. In order to perform this derivation, the so-called commutator of the tensor operators is used (contraction of creation and annihilation operators), and consequently each tensorial product is reduced to a single effective operator; that is, it is effective in the sense that it represents the inter-shell interactions but formally acts only within the $4f^N$ shell,

$$\left[u_i^{(1)}(4f,n'\ell') \times u_i^{(t)}(n'\ell',4f)\right]_q^{(k)}$$
$$= (-1)^k[k]^{1/2}\begin{Bmatrix}t & k & 1\\ f & \ell' & f\end{Bmatrix} u_{i,q}^{(k)}(4f,4f). \qquad (10.16)$$

This is in fact the main point of the derivation of the expression for the transition amplitude of the standard Judd-Ofelt formulation. The procedure described here is applied to both terms in equation (10.11). As mentioned above, in general these two terms differ by the order of operators in the matrix elements and energy denominators. However they both are expressed by the same effective tensor operator, yet each is associated with a different energy denominator. This is the reason that the following approximations/assumptions are introduced in order to make the final expression even simpler:

1. The distance in an energy scale between the ground configuration $4f^N$ and all excited configurations $4f^{N-1}n'\ell'$ is large.
2. This energy difference is so large that it can be assumed that when "looking" from the ground configuration up the energy scale, the excited configurations are "seen" as degenerate and vice versa; the consequence is such that $E^0_{4f^N,i} - E^0_{Xx} \simeq E^0_{4f^N} - E^0_X$, and the same relation for the final state Ψ_f of $4f^N$, where $E^0_{4f^N}$ and E^0_X are the average energies of configurations.
3. In light of the above, it is possible to assume that

$$E^0_{4f^N,i} - E^0_{Xx} \simeq E^0_{4f^N,f} - E^0_X = E^0_{4f^N} - E^0_X.$$

These assumptions about the energy denominators are treated in some publications as a weak point of the J-O Theory. This is especially the case when they are erroneously interpreted as the assumption that the energy levels of the ground configurations are degenerate!

This interpretation is obviously wrong, and it must be wrong, since this theoretical model is established to describe the transitions between distinct energy levels of $4f^N$. It is important to point out that the assumptions about the energy denominators of the J-O Theory have to be understood as presented above. They are in fact introduced as approximations that are necessary to make *ab initio* calculations possible in practice. It should also be remembered that the role of such investigations is not to reproduce the measurements but to understand the mechanisms and establish their hierarchy of importance in order to improve a theoretical model via comparison of the magnitude of various contributions to the transition amplitude. At the same time, when the semiempirical realization of the J-O Theory is applied, all errors possibly introduced by the assumptions on the energy denominators are compensated for by the values of freely adjusted parameters.

Using the tools of Racah algebra, it is rather straightforward to derive the final expression for the amplitude of one photon electric dipole $f \longleftrightarrow f$ transitions. This is the original version that was defined by Judd and Ofelt in the papers from 1962:

$$\Gamma_{J-O} = 2\sum_{t,p}^{odd} B_p^t \sum_{\lambda,q}^{even} \sum_{\ell'}^{even} (-1)^q [\lambda]^{1/2} \begin{pmatrix} t & 1 & \lambda \\ p & \rho & -q \end{pmatrix} A_t^{\lambda}(\ell') R_{JO}^t(\ell')$$
$$\langle 4f^N \Psi_f^0 \mid U_q^{(\lambda)} \mid 4f^N \Psi_i^0 \rangle \quad (10.17)$$

where $U_q^{(\lambda)}$ is $N-$electron unit tensor operator, and the factor of 2 results from two terms in the perturbing expression of equation (10.15) that differ only by $(-1)^{\lambda}$ (if accepting the assumptions about the energy denominators). This condition requires the contributions to Γ_{J-O} to be limited to the effective operators with ranks $\lambda = even$.

The angular term is defined as

$$A_t^{\lambda}(\ell') = [k]^{1/2} \begin{Bmatrix} t & \lambda & 1 \\ f & \ell' & f \end{Bmatrix} \langle f \| C^{(1)} \| \ell' \rangle \langle \ell' \| C^{(t)} \| f \rangle. \quad (10.18)$$

The symmetry properties of the $6j-$ symbol limit the even values of rank λ to $2, 4, 6$. In addition, due to the triangular conditions for the non-vanishing $3j-$ symbols that determine the reduced matrix elements of spherical tensors in equation (10.18), the rank t of the crystal field potential operator has to be odd. Indeed, since ℓ' is even, and $f = 3$ for $4f$ electrons, the odd parts of the crystal field potential contribute to the transition amplitude, while the terms with even values of t contribute to the energy.

All the terms that depend on the principal quantum number n' of one electron excited states are collected in the radial term that consists of the product of two radial integrals and the appropriate energy denominator

$$R_{JO}^t(\ell') = \sum_{n'}^{exc.} \frac{\langle 4f \mid r \mid n'\ell' \rangle \langle n'\ell' \mid r^t \mid 4f \rangle}{\Delta(4f, n'\ell')} \quad (10.19)$$

Here r is the radial part of $D_{\varrho}^{(1)}$ (10.9) and r^t is the radial dependence of V_{cryst} (10.6). The summation in equation (10.19) is overall n' of the ℓ' states of a given excited configuration $4f^{N-1}n'\ell'$. The form of the energy denominator in this radial term, as the difference of the orbital energies, indicates that the virtual one electron excited states are generated for an average energy of a configuration and with the core orbitals frozen (only the excited orbital $n'\ell'$ is adjusted within the Hartree-Fock model for the frozen core orbitals of $4f^N$). The validity of such approximations is demonstrated in Chapter 21 of Wybourne and Smentek [13].

Thus, introducing the intensity as a square of the transition amplitude defined within the J-O theory we have, as presented in equation (10.1)

$$\mathcal{S}_{f\longleftarrow i} = \sum_{\lambda}^{2,4,6} \Omega_\lambda |\langle \Psi_f || U^{(\lambda)} || \Psi_i \rangle|^2$$

where Ω_λ, for $\lambda = 2, 4, 6$ are the so-called Judd-Ofelt parameters, the famous intensity parameters. These parameters contain (as seen from the above derivation) the angular terms, radial integrals, coefficients of coupling, and re-coupling of the angular momenta. The genealogy of the energy states, between which the transition is observed, is included within the reduced matrix element of the unit tensor operator. However, when Ω_λ's are adjusted by the fitting procedure they represent more physical mechanisms than the presented derivation may indicate, which is discussed below).

10.2.2 Challenges of *ab initio* Calculations

When analyzing the final expression for the transition amplitude defined in equation (10.17), the task of its evaluation seems to be straightforward. Indeed, for a particular case of the $f \longleftrightarrow f$ transitions there are only a few terms that have to be taken into account, namely those with the following values of various ranks: $\lambda = 2, 4, 6; t = 1, 3, 5, 7$ (at the most, since the ranks depend on the symmetry of the crystal field). In addition, one electron excited states have the following symmetry: $\ell' = 2(\equiv d), = 4(\equiv g)$. In equation (10.17) also there are summations over the components of certain tensor operators: q is the component of the unit tensor operator (the effective operator), and its values are determined by the triangle conditions with the components of the atomic levels (via the matrix element) between which the transition is occurring; p as a component of the crystal field potential depends on the symmetry of the system, and only such terms from all of $2t+1$ projections contribute, which denote the invariants of the symmetry operations of a given point group; and ϱ describes the polarization and has the values of $-1, 0$, and 1.

What are the problems then with the direct calculations of the transition amplitude? This question has its origin in the fact that the numerical procedures performed to interpret the experimental f-spectra always are based on the semi-empirical approach applied for equation (10.1). However, the choice of the semi-empirical procedure is made due to neither the preference nor simpler form of equation (10.1) when compared to the terms in equation (10.17). Indeed, there are objective, or rather technical reasons, why it is still impossible to perform *ab initio* calculations that would provide reliable results.

In equation (10.17), in addition to the angular term, radial integral and matrix element of the unit tensor operator there is the crystal field parameter B^t_p. This is the structural factor from the multipole expansion of the crystal field potential; the radial part from equation (10.6) is included in the radial term, as shown in equation (10.19). The limitation of the summation over t shows that these parameters are of odd rank. This fact is the source of a problem that is addressed separately below.

Another possible problem with a direct calculations based on equation (10.17) arises from the definition of the radial integral of the J-O model. The summation over the principal quantum number of one electron excited state to which the $4f$ electron is promoted covers the whole spectrum that consists of all function of ℓ' symmetry, discrete part and continuum. It is common practice to limit this summation and include only the excitations to the first one-electron state of d symmetry, $5d$, ignoring the rest of the spectrum. In the case of more advanced calculations, also the excitation to the $5g$ one-electron state is explicitly taken into account. It is possible to find such publications where the closure procedure is performed in (10.19) over the complete set of g one electron functions using the argument that g orbitals are not occupied in the ground configuration of the lanthanide ion. In such a procedure the energy denominators are assumed to be common for all members of the complete set of $g-$ symmetry, and therefore are treated as a common factor independent of n'. In all such cases however the evaluated values of the radial terms are quite far from being reliable. This aspect of numerical calculations is verified when the complete radial basis sets are used for their evaluation. In Table 21-7 of Wybourne and Smentek [13] the effect of the incompleteness of the radial basis set is easily seen in the case of the J-O standard radial term $R^3_{JO}(n'd)$, as an example. The main conclusions derived from the detailed analysis of the radial integrals and their sensitivity to the completeness of the radial basis sets demonstrate almost an equal importance of $d-$ and $g-$ excitations when included for all members of each spectrum. In order to avoid an unnecessary source of error, it is very important to address this result when *ab initio* calculations are performed.

10.2.2.1 Crystal Field Parameters

In general, the crystal field potential is represented by one-particle effective operators of even and odd ranks depending on the coordinates of the electrons of the central ion. The surrounding ligands are placed in fixed positions (within the so-called static approximation), and they define the symmetry of the environment described by the crystal field parameters. When the perturbing influence of the crystal field potential upon the energy of $4f^N$ configuration is evaluated within the single configuration approximation, only the terms with the even ranks give non-zero contributions. In addition, due to the triangular conditions for non-vanishing reduced matrix elements of spherical tensors, in the case of the energy the rank has only three values of 2, 4, and 6. Thus, in order to evaluate the energy of a system, the values of the crystal field parameters must be known.

It is common practice to determine the values of these crystal field parameters from the fitting procedure used for the reproduction of the measured values of energy levels. In most cases this procedure is stable, and it gives the values that are needed for the evaluation of the energy of such levels that are not obtainable from experiment.

The standard fitting procedure is based on the one-particle parametrization scheme. The discrepancy between the observed and calculated crystal field energy levels quite often requires extension of the set of standard parameters by, for example, spin-correlated and orbitally correlated parameters introduced by Newman and collaborators [14–17].

It should be pointed out however that the semi-empirical way of the determination of the crystal field parameters is the only reliable method. Indeed, a very extensive study on the character of the interactions between the central ion and the surrounding ligands, even when including the higher-order multipoles in the expansions used by Faucher and collaborators [18–22] did not lead to a reliable method of their direct evaluation. However, the authors of molecular dynamics simulation used for determining the actual positions of ligands (beyond the static approximation) claim that even the calculations in which only the point charges are taken into account give satisfactory results [23, 24]. In any case it should be realized that any comparison of the values of the structural parameters evaluated directly with those obtained from a fitting procedure is possible only for even ranks. This is the case because the parameters with t odd do not contribute to the energy, and therefore it is impossible to adjust their values. This is why it is impossible to evaluate directly the transition amplitude defined in (10.17). The lack of a reliable model for the evaluation of crystal field parameters is also the reason that very often, in order to derive the conclusions on the importance of various physical mechanisms affecting the transition amplitude, the relative intensities evaluated directly are discussed rather than the absolute values evaluated with poorly approximated crystal field parameters. This was the practice used in a series of publications devoted to the nature of the $f \longleftrightarrow f$ transitions analyzed in the language of the standard Judd-Ofelt model, its extensions by various third-order contributions, and also in the case of the relativistic versions of description of f-spectra (see Chapters 17 and 18 of Wybourne and Smentek [13], and references therein).

10.2.2.2 Radial Integrals

The angular parts of the effective operators of the transition amplitude defined in equation (10.17) have the same values for all of the lanthanide ions. The radial integrals, as carriers of specific features of the electronic structure of each ion, have to be evaluated individually. In order to find their values, the one electron functions have to be known. This task is easily solved by applying the H-F model and performing the calculations for the average energy of the configurations and with frozen core orbitals when generating the excited states. At the same time, the summation present in the definition of the radial integral (10.19) cannot be performed in a direct way as covering the discrete and continuum parts of the spectrum for a given symmetry of one electron state. This problem is solved by applying the so-called perturbed function approach originally introduced by I. Lindgren in the *Many Body Perturbation Theory* [25] and applied to the lanthanides by J. Morrison [26, 27]. The perturbed functions are used here to evaluate the radial integrals in an exact way, and for a complete radial basis sets of one electron excited functions. This means that in such a way the perturbing influence of all excited configurations of appropriate parity is taken into account via the inter-shell interactions that are represented by the perturbing operators of various physical mechanisms affecting the lanthanide ion [28] (see also Chapters 17 and 21 of Wybourne and Smentek [13]).

The original definition of the radial integral of the J-O Theory is of the form

$$R^t_{JO}(\ell') = \sum_{n'}^{exc.} \frac{\langle 4f \mid r \mid n'\ell' \rangle \langle n'\ell' \mid r^t \mid 4f \rangle}{\Delta(4f, n'\ell')}.$$

It is rewritten in a explicit form to group all those terms that depend on the principal quantum number n', namely

$$R^t_{JO}(\ell') = \int \left\{ \sum_{n'}^{exc.} \frac{\langle 4f \mid r^t \mid n'\ell' \rangle}{(\varepsilon_{4f} - \varepsilon_{n'\ell'})} P_{n'\ell'}(r') \right\} r' P_{4f}(r') dr'. \qquad (10.20)$$

It is seen that the term within the curly brackets is a linear combination of one-electron functions describing the excited orbitals, $P_{n'\ell'}$. This particular part of the radial integral, including the summation, defines a new function, the so-called perturbed function,

$$\varrho^t(4f \longrightarrow \ell'; r) = \sum_{n'}^{exc.} \frac{\langle 4f \mid r^t \mid n'\ell' \rangle}{(\varepsilon_{4f} - \varepsilon_{n'\ell'})} P_{n'\ell'}(r). \qquad (10.21)$$

Thus, each perturbed function contains all the first-order corrections that are due to singly excited configurations $4f^{N-1}n'\ell'$, for all n', and which are taken into account via the odd-rank part of the crystal field potential.

With the definition (10.21), the radial terms of the Judd-Ofelt approach are expressed now by single integrals that involve the perturbed functions, namely

$$R^t_{JO}(\ell') = \langle \varrho^t(4f \longrightarrow \ell') \mid r \mid 4f \rangle. \qquad (10.22)$$

The problem of performing the summation over the complete radial basis sets of functions of a given symmetry ℓ' is replaced by the task of finding the newly defined perturbed functions. They satisfy the following equation (for explicit derivation see Jankowski et al. [28] or Chapter 17 of Wybourne and Smentek [13]).

$$(\varepsilon_{4f} - h_0^{\ell'}) \varrho^t(4f \longrightarrow \ell'; r) = r^t P_{4f} - \sum_{n'}^{occ} P_{n'\ell'}(r) \langle n'\ell' \mid r^t \mid 4f \rangle \qquad (10.23)$$

with the operator defined as follows

$$h_0^{\ell'} = -\frac{1}{2}\frac{d^2}{dr^2} - \frac{Z}{r} + U(r) + \frac{\ell'(\ell'+1)}{2r^2}. \qquad (10.24)$$

Note that the summation in equation (10.23) is performed over the occupied one-electron states of a given symmetry. This means that in the case of excitations to d- orbitals only $3d$ and $4d$ one-electron functions have to be known. For g excitations the second term at the right hand side of equation (10.23) does not contribute, since there are no occupied orbitals of g symmetry in the case of the ground configuration of the trivalent lanthanide ions.

At the end of Chapter 21 of Wybourne and Smentek [13] the numerical results are presented and the radial integrals of standard J-O theory and also its extensions by the third-order contributions of various physical origins are presented for all of the lanthanide ions.

10.2.3 Problems with the Interpretation of the *f*-Spectra

There are two kinds of problems when an attempt is made to interpret the f-electron spectra using the J-O theory. One of them is conceptual. The standard realization of the Judd-Ofelt model is based on the assumption that only the central ion is perturbed by the environment, while in fact there is a mutual interaction between two subsystems, the central ion and the surrounding ligands. As a consequence the dynamic model has been introduced [29–32]. In this model the lanthanide ion plays a static role. The dipoles on the ligands, which result from the presence of the lanthanide ion, induce in turn the multi-poles on the central ion. This interaction between the multi-poles on the central ion and the dipoles on the ligands is the origin of additional contributions to the transition amplitude, and consequently to the intensity parameters Ω_λ,

$$\Omega_\lambda : T_{dyn} = A^L_{\lambda+1}\alpha_\lambda(L)\langle 4f \mid r^\lambda \mid 4f\rangle\langle f \mid\mid C^{(\lambda)} \mid\mid f\rangle U^\lambda_q, \tag{10.25}$$

where $A^L_{\lambda+1}$ is a structural parameter, and $\alpha_\lambda(L)$ is related to the frequency dependent polarizability of the ligand

$$\alpha_\lambda(L) = (k+1)(2k+3)^{1/2}\alpha'_L.$$

Here the isotropic polarizability is included while the corrections due to anisotropy were introduced by Mason [33, 34] and applied by Reid *et al.* [35]

It is interesting to note that the dynamic part of the transition amplitude is independent of any excited configuration and its evaluation is rather straightforward. Since the static and dynamic models contribute to the intensity parameters simultaneously it is possible to verify their relative magnitude, or at least their signs. The results of such an analysis [36] demonstrated that the radial terms of the second-order static (standard Judd-Ofelt) effective operators are negative for all the lanthanide ions, and the dynamic radial parts are positive; the angular parts are the same for all members of the lanthanide family.

The results of calculations presented in the literature indicate that the dynamic second-order contributions enhance the $f \longleftrightarrow f$ transition probabilities defined within the Judd-Ofelt approach [35]. There are a considerable number of examples for which the dynamic coupling model is even the dominant one [33, 34, 36–41]. It should be mentioned, however, that in spite of such an importance of the dynamic coupling model, the results of *ab initio* calculations performed at the second order still do not provide satisfactory agreement with experiment. In this sense the second-order model is not complete.

Here another source of a conceptual problem of the second order-approach appears. The standard formulation of the J-O theory, even if extended by the dynamic coupling model, is based on the single configuration approximation. This means that in such a description all the electron correlation effects are neglected and it is well known that the transition amplitude strongly depends on them. At this point also the spin-orbit interactions should be taken into consideration as possibly important in the description of the spectroscopic patterns of the lanthanides. In the case of all of these possibly important physical mechanisms there is a demand for an extension of the standard Judd-Ofelt formulation. The transition amplitude in equation (10.17) has to be modified by the third-order contributions that originate from various perturbing operators introduced in addition to the crystal field potential that plays a forcing role for the electric dipole one photon $f \longleftrightarrow f$ transitions. This extension possibly improves the description of the transitions when the direct calculations are performed. It should be mentioned that if the semi-empirical approach is applied, all of these additional contributions are included within the values of the adjusted intensity parameters Ω_λ, unless they are not represented by one particle effective operators.

Although the Judd-Ofelt theory is successful as a parametrization scheme used in the fitting procedure, it fails when used in the *ab initio*-type calculations. This means that its physical model is not rich enough to reproduce the subtleties of the electronic structure of f-electron systems.

The other nature of problems with the interpretation of f-spectra is rather technical, or numerical, and it originates from the formal limitations of the Judd-Ofelt model. This is the case of all such transitions that violate the selection rules of one particle parametrization scheme of f-spectra introduced by the J-O model.

When the intermediate coupling scheme is used for identification of the energy levels of the lanthanide ion, the matrix element of unit tensor operator $U^{(\lambda)}$ in the expression for the transition amplitude (10.17) (or the line strength) exists only if the triangular condition is satisfied, namely

$$\Delta(J\lambda J') \Rightarrow |J - J'| \leq \lambda \leq J + J' \ and \ \lambda \neq 0.$$

In particular, the one-photon electric dipole transitions $[^7F]_0 \longleftrightarrow [^5D]_0$ and $[^7F]_0 \longleftrightarrow [^5D]_1$ observed in Eu^{3+} in various hosts are not described by the standard Judd-Ofelt theory. In the first case the triangle condition would give $\lambda = 0$, and due to the orthogonality of the wavefunctions the contribution to the transition amplitude would vanish. In the second example, $\lambda = 1$, which is beyond the limitation of the ranks to even values. This is an unfortunate situation, not only because of a purely theoretical nature of problems with their description (see Smentek and Hess [42], and references therein), but rather because of their roles in many investigations based on experimentally observed spectroscopic patterns. For example, the electric dipole transition $[^5D]_0 \longrightarrow [^7F]_0$ is used as a benchmark providing a detailed insight into the structure, symmetry, and geometry of various compounds doped by Eu^{3+}. This interest arises from the fact that such a line is observed but yet not predicted by the J-O Theory. It is even more intriguing to observe several components attributed to this particular transition. Such a pattern is not theoretically expected, since initial and the final energy states involved in the process are not degenerate. At the same time it means that in such a case the crystal field potential is too weak to directly force the electric dipole transitions (in the sense of the Judd-Ofelt model), but yet different morphologies and local symmetries affect the optical centers in different ways resulting in more than one component of the emission line. This argument was used for identification of two diastereomeric forms of the $Eu : DOTA$, a tissue selective organic chelate for which two $0 \longrightarrow 0$ lines were observed; this conclusion was confirmed by the results of DFT calculations [7, 8]. This possibility to connect the macroscopic observations with the properties at a micro-world level is of special importance. The chelates like DOTA are the precursors for the majority of modern therapeutic agents used for early detection of cancerous cells and the agents applied in radio-immunotherapy as local centers of radiation when the lanthanide ion is replaced by its radioactive isotope.

The same transition, $[^7F]_0 \longleftrightarrow [^5D]_0$ of Eu^{3+}, is a useful tool also for finding the correlation between the structural properties and the spectroscopic patterns of ultra small nano-samples in order to design the compound and adjust its architecture to obtain a material with the desired properties. This is the subject of novel research devoted to the ultra-small nanosamples of Eu_2O_3 [43].

Such examples of applications of the tool of the lanthanide spectroscopy as mentioned here add impetus to the search for an extended J-O model, which would be applicable for interpretation of entire lanthanide spectra in various compounds of *macro* and *micro* scales.

10.3 Third-Order Contributions

The Judd-Ofelt model of the $f \longleftrightarrow f$ transitions extended by the third-order contributions to the transition amplitude is based on the double perturbation theory applied for the following Hamiltonian,

$$\begin{aligned} H = H_0 &+ \lambda(PV_{cryst}Q + QV_{cryst}P + QV_{cryst}Q) \\ &+ \mu(PVQ + QVP + QVQ) \end{aligned} \tag{10.26}$$

where V_{cryst} still plays the forcing role of these transitions, and V denotes an additional physical mechanism that possibly modifies the transition amplitude (for details see Chapter 17 of Wybourne and Smentek [13]). In order to break the limitations of a single configuration approximation of the standard J-O Theory, in the present approach V represents the non-central part of Coulomb interaction, which is responsible for electron correlation effects,

$$V_{corr} = \sum_{i<j} \frac{1}{r_{ij}} - u_{HF} \tag{10.27}$$

In this perturbing operator, u_{HF} is the centro-symmetric Hartree-Fock (H-F) potential. Similarly as in the case of the crystal field potential included as a perturbation, also here the inter-shell interactions via the V_{corr} are taken into account, thus the same partitioning of space in equation (10.26). However, for a certain order of three operators, $D_{\varrho}^{(1)}$, V_{cryst}, and V_{corr} in the final third-order contributions, it is also possible to take into account the interactions via the perturbing operator within the Q-space. This is why additional operators, $QV_{cryst}Q$ and QVQ, are included in the Hamiltonian of equation (10.26).

Following the standard procedure of double perturbation theory, the transition amplitude is now determined by the following contributions,

$$\begin{aligned} \Gamma = \lambda\{&\langle\Psi_f^0 \mid D_\rho^{(1)} \mid \Psi_i^{10}\rangle + \langle\Psi_f^{10} \mid D_\rho^{(1)} \mid \Psi_i^0\rangle\} \\ &+ \mu\{\langle\Psi_f^0 \mid D_\rho^{(1)} \mid \Psi_i^{01}\rangle + \langle\Psi_f^{01} \mid D_\rho^{(1)} \mid \Psi_i^0\rangle\} \\ &+ \lambda\mu\{\langle\Psi_f^{10} \mid D_\rho^{(1)} \mid \Psi_i^{01}\rangle + \langle\Psi_f^{01} \mid D_\rho^{(1)} \mid \Psi_i^{10}\rangle\} \\ &+ \lambda^2\{\ldots\} + \mu^2\{\ldots\}. \end{aligned} \tag{10.28}$$

The terms associated with λ define the second-order contributions, and they lead to the Judd-Ofelt theory; this part is known as presented in the previous section. The terms proportional to the perturbing parameter μ vanish due to the parity requirements. Indeed, Ψ_i^{01} and Ψ_f^{01} are the first-order corrections due to the even parity operator V_{corr}. The most interesting terms are those proportional to $\lambda\mu$ as completely new. They represent the interplay between both perturbing mechanisms, V_{cryst} and V_{corr}. The remaining third-order terms, proportional to λ^2, as a result of the crystal field perturbation used twice, do not provide any new aspect to the approach, and finally, the terms proportional to μ^2 vanish due to the parity requirements.

In general the third-order contributions to the transition amplitude that originate from the inter-shell interactions have the usual perturbing form, now with two energy denominators,

$$
\begin{aligned}
&{}^3\Gamma_{corr}(inter) \\
&= \sum_{Xx}\sum_{Bb}\left\{\frac{\langle\Psi_f^0 \mid PV_{corr}Q \mid Bb\rangle\langle Bb \mid D_\rho^{(1)} \mid Xx\rangle\langle Xx \mid QV_{cryst}P \mid \Psi_i^0\rangle}{(E_i^0 - E_{Xx}^0)(E_f^0 - E_{Bb}^0)}\right. \\
&\quad \left. + \frac{\langle\Psi_f^0 \mid PV_{cryst}Q \mid Xx\rangle\langle Xx \mid D_\rho^{(1)} \mid Bb\rangle\langle Bb \mid QV_{corr}P \mid \Psi_i^0\rangle}{(E_f^0 - E_{Xx}^0)(E_i^0 - E_{Bb}^0)}\right\}, \qquad (10.29)
\end{aligned}
$$

where $|Bb\rangle$ and $|Xx\rangle$ are the states of excited configurations of the same and opposite parities to the parity of $4f^N$ configuration, respectively. Due to the two particle character of Coulomb interaction in V_{corr}, the set of excited configurations taken into account in the Judd-Ofelt theory is extended at the third order by all doubly excited configurations of the same parity as the parity of $4f^N$ (included within $|Bb\rangle$).

The impact due to the interactions via V_{corr} and V_{cryst} within the Q space of excited configurations are represented by the third-order contributions of the form,

$$
\begin{aligned}
&{}^3\Gamma_{corr}(intra) \\
&= \sum_{Xx}\sum_{Yy}\left\{\frac{\langle 4f^N\Psi_f^0|D^{(1)}|Yy\rangle\langle Yy|QV_{corr}Q|Xx\rangle\langle Xx|V_{cryst}|4f^N\Psi_i^0\rangle}{(E_i^0 - E_{Yy}^0)(E_i^0 - E_{Xx}^0)}\right. \\
&\quad \left. + \frac{\langle 4f^N\Psi_f^0|V_{cryst}|Xx\rangle\langle Xx|QV_{corr}Q|Yy\rangle\langle Yy|D^{(1)}|4f^N\Psi_i^0\rangle}{(E_f^0 - E_{Yy}^0)(E_f^0 - E_{Xx}^0)}\right\} \\
&\quad + \sum_{Bb}\sum_{Yy}\left\{\frac{\langle 4f^N\Psi_f^0|D^{(1)}|Yy\rangle\langle Yy|QV_{cryst}Q|Bb\rangle\langle Bb|V_{corr}|4f^N\Psi_i^0\rangle}{(E_i^0 - E_{Yy}^0)(E_i^0 - E_{Bb}^0)}\right. \\
&\quad \left. + \frac{\langle 4f^N\Psi_f^0|V_{corr}|Bb\rangle\langle Bb|QV_{cryst}Q|Yy\rangle\langle Yy|D^{(1)}|4f^N\Psi_i^0\rangle}{(E_f^0 - E_{Yy}^0)(E_f^0 - E_{Bb}^0)}\right\} \qquad (10.30)
\end{aligned}
$$

where $|Xx\rangle$ and $|Yy\rangle$ are of the same parity, which is opposite to the parity of $4f^N$.

In summary, the perturbing influence of the following excited configurations is taken into account at the third-order analysis: $4f^{N-1}n'\ell'$ with $\ell' = even$, $4f^{N-1}n''\ell''$ with $\ell'' = odd$, and $4f^{N-2}n'\ell'n''\ell''$ with $\ell' + \ell'' = even$.

10.3.1 Third-Order Electron Correlation Effective Operators

The form of the effective operators depends on the intermediate configurations and on the order of operators in the triple products of matrix elements. To derive the tensorial form of the third-order terms, the J-O Theory assumptions about the energy denominators are adopted. These conditions mean that also now the radial basis sets of one electron functions are generated for the average energy of configuration and with frozen core orbitals in the case of the excited states. As a consequence, the energy denominators, determined by appropriate orbital energies, are included within the newly defined perturbed functions.

The two-particle nature of Coulomb interaction in equation (10.27) is the reason that among the third-order contributions to the transition amplitude, in addition to one particle effective operators (as in the standard J-O approach), two particle objects are also present. However, the numerical analysis based on *ab initio* calculations performed for all lanthanide ions, applying the radial integrals evaluated for complete radial basis sets (due to perturbed function approach), demonstrated that the contributions due to two-particle effective operators are relatively negligible [11, 44–58]. This is why here they are not presented in an explicit tensorial form (see for example Chapter 17 in [13]). At the same time it should be pointed out that two-particle effective operators, as the only non-vanishing terms, play an important role in determining the amplitude of transitions that are forbidden by the selection rules of second- and the third-order approaches. This is the only possibility, at least within the non-relativistic model, to describe the so-called special transitions like, $0 \longleftrightarrow 0$ in Eu^{3+}, for example, as discussed above.

The results of this extensive numerical analysis showed also that for almost all lanthanide ions the third-order electron correlation contributions to the transition amplitude are larger than the standard second- order terms. This relative magnitude of the third versus second-order terms is not in contradiction with the requirement that the perturbation expansion has to be convergent, since this condition is valid for the energy. The demanded convergency of the expansion of energy reflects that the applied Hamiltonian has been properly partitioned. Here, when the transition amplitude is evaluated with the functions of good quality in respect to the energy, the contributions of a certain order, rather than the corrections, are determined. This means that there is no *a priori* expectation that the higher-order terms contributing to the transition amplitude have to be smaller than those of a lower order.

Finally, the one particle effective operators that are defined up to the third order have the following form (based on the static model, as the original J-O Theory),

$$^{3}D(1) = \sum_{\lambda}^{even} {}^{1}T^{\lambda t}_{stat} U^{(\lambda)}(4f, 4f) \tag{10.31}$$

where

$$^{1}T^{\lambda t}_{stat} = \sum_{\ell'}^{even} [R^{t}_{JO}(\ell') - R^{t}_{HF}(\ell', f) + \frac{(N-1)}{2} R^{0}_{t}(\ell', f)] A^{\lambda}_{t}(\ell') \tag{10.32}$$

and the angular term $A^{\lambda}_{t}(\ell')$ is the same as in the J-O and is defined by equation (10.18). The radial terms in (10.32), defined in the terms of the perturbed functions, have the following interpretation:

$R^t_{JO}(\ell')$ represents the standard Judd-Ofelt theory (see 10.22), and the third-order radial terms:

$R^t_{HF}(\ell',f) =< \varrho^1(4f \to \ell')|r^t|\varrho_{HF}(4f \to f) > + < \varrho^t(4f \to \ell')|r^1|\varrho_{HF}(4f \to f) >$, arising from the H-F potential,

$R^0_t(\ell',f) =< \varrho^0(4f \to f)|r^1|\varrho^t(4f \to \ell') > + < \varrho^0(4f \to f)|r^t|\varrho^1(4f \to \ell') >$, caused by the Coulomb operator.

It is seen from equation (10.32) that in order to include the third-order electron correlation contributions it is enough to evaluate the radial integrals and modify the values of the J-O radial term. This is a very strong conclusion, since it indicates that when the intensity parameters of the one-particle parametrization scheme of f-spectra are evaluated via the fitting procedure, they also contain the major part of electron correlation effects. In this sense, the J-O theory, in its semi-empirical version, is more general than its original derivation might suggest. In fact, it should be mentioned that other physical mechanisms are also included when the intensity parameters Ω_λ are fitted, and this aspect is presented as a discussion of the parametrization scheme of the f-spectra at the end of this chapter.

Finally it should be concluded that one-particle parametrization scheme of the standard J-O Theory is preserved at the third order. The limitations caused by the original derivation based on the single configuration approximation are compensated in a perturbative way by the third-order electron correlation contributions analyzed here.

Since the inclusion of the higher-order terms caused by electron correlation effects does not change the original selection rules for the non-vanishing transition amplitude (although they modify the values in a tremendous way), it is worthwhile to generalize the description of the $f \longleftrightarrow f$ transitions by introducing a relativistic approach.

10.4 Relativistic Approach

In order to derive a relativistic version of the electric-dipole $f \longleftrightarrow f$ transitions the concept of Sandras and Beck [59] is applied to include new effects in an effective way. This means that every unit tensor operator $u^{(x)}$ analyzed in the non-relativistic approach has to be replaced prior to the partial closure by a double unit tensor operator $w^{(\kappa,k)x}$ that acts within the spin (κ) - orbital (k) space. The transformation is as follows,

$$r^x\langle\ell||C^{(x)}||\ell'\rangle u^{(x)}_\varrho(\ell,\ell') \Rightarrow \mathcal{A}R^x w^{(\kappa_1 k_1)x}_\varrho(s\ell,s\ell'), \tag{10.33}$$

where the double tensor operator is defied by its reduced matrix element,

$$\langle s\ell''||w^{\kappa_1 k_1}(s\ell,s\ell')||s\ell'''\rangle = \delta(\ell'',\ell)\delta(\ell',\ell''') \tag{10.34}$$

and $\mathcal{A}$ denotes a coefficient that is responsible for such replacement [60].

In this way the relativistic effects are taken into account in an effective way. At the same time the $S - L$ coupling, natural for the non-relativistic Schrödinger equation and the standard theory of tri-positive lanthanide ions, is preserved, instead of the $j - j$ basis of a relativistic Dirac equation. As a result of such replacements of all the operators, new angular terms appear and the radial integrals R^x in (10.33) are defined by the *small* and *large* components, the solutions of the Dirac equation, but they depend on both, ℓ and j quantum numbers [60–64].

From a physical point of view this relativistic model is also based on the perturbation approach, and at the second order, similarly as in the case of the standard J-O Theory, the crystal field potential plays the role of a mechanism that forces the electric dipole $f \longleftrightarrow f$ transitions. The only difference is that now the transition amplitude is in effectively relativistic form, as determined by the double tensor operator, but still of one particle nature. Furthermore, the same partitioning of space as in non-relativistic approach is valid here. The same requirements about the parity of the excited configurations are expected to be satisfied. As a final step of derivation of the effective operators, the coupling of double inter-shell tensor operators has to be performed. This procedure is based on the same rules of Racah algebra as presented in the case of the standard J-O theory. However, the coupling of the inter-shell double tensor operators consists of two steps, for spin and orbital parts separately. Thus, the rules presented in equations (10.15) and (10.16) have to be applied twice for orbital and spin momenta couplings, resulting in two $3j-$ and two $6j-$ coefficients.

Taking into account two second-order contributions that differ by the order of the operators [as in (10.11)] the final expression for the second-order transition amplitude, in fact the relativistic analog of the Judd-Ofelt theory, has the following form:

$$
\begin{aligned}
{}^2\Gamma^R = \sqrt{3}\sum_{tp}^{odd} B_p^t[t]^{1/2} \sum_{\kappa_1=0,1}\ \sum_{k_1\leq|\kappa_1-1|}^{\kappa_1+1}\ \sum_{\kappa_2=0,1}\ \sum_{k_2\leq|\kappa_2-t|}^{\kappa_2+t} \\
\sum_{\ell'}^{even} \mathcal{A}_{k_1k_2}^{\kappa_1\kappa_2}(t\ell') \sum_{\kappa_3=0,1}\ \sum_{k_3\leq|t-1|}^{t+1} a \sum_{\lambda\leq|\kappa_3-k_3|}^{\kappa_3+k_3} \\
[\lambda]^{1/2}\sum_q (-1)^{\kappa_3+k_3+t-q}\left[\kappa_3,k_3\right] \\
\begin{pmatrix} 1 & t & \lambda \\ \varrho & p & -q \end{pmatrix}
\begin{Bmatrix} \kappa_2 & \kappa_3 & \kappa_1 \\ \frac{1}{2} & \frac{1}{2} & \frac{1}{2} \end{Bmatrix}
\begin{Bmatrix} k_2 & k_3 & k_1 \\ \ell & \ell' & \ell \end{Bmatrix} \\
\begin{Bmatrix} \kappa_1 & k_1 & 1 \\ \kappa_2 & k_2 & t \\ \kappa_3 & k_3 & \lambda \end{Bmatrix} W_q^{(\kappa_3 k_3)\lambda}(4f,4f)
\end{aligned}
\tag{10.35}
$$

where $\mathcal{A}_{k_1k_2}^{\kappa_1\kappa_2}(t\ell')$ is defined by the angular and radial terms (for details see Chapter 18 in Wybourne and Smentek [13]). The factor $a \equiv a(\kappa_1\kappa_2k_1k_2;\kappa_3k_3)$ in (10.35) is equal 2 when the parity of appropriate ranks of operators is the same, $p(\kappa_1+k_1+\kappa_2+k_2) = p(\kappa_3+k_3)$; otherwise it vanishes.

The effective operators defined in (10.35) extend the standard Judd-Ofelt effective operators by the interactions via the crystal field potential within the spin part of the space. The reduction to the non-relativistic case is easily seen when all of the ranks of operators acting within the spin space are equal to zero. Indeed, setting $\kappa_1 = \kappa_2 = \kappa_3 = 0$ results in $k_1 = 1$ $k_2 = t$ and $k_3 = k$. In such a situation, the rank of effective unit tensor operator is even, the spin-dependent $6-j$ symbol in equation (10.35) is a number, $9-j$ is reduced to a number, and the remaining $6-j$ has the form

$$\begin{Bmatrix} t & k & 1 \\ \ell & \ell' & \ell \end{Bmatrix}$$

which, together with the $3-j$ symbol, gives exactly the angular part of standard the J-O effective operators (10.17). The contributions that are represented by the terms with $\kappa_1, \kappa_2, \kappa_3 = 1$ describe the effects that are new in the spectroscopy of rare earth ions in crystals. The model of $f \longleftrightarrow f$ transitions presented here gives the opportunity to establish, already at the second order, the importance of relativistic effects in relation to the standard Judd-Ofelt terms. At the same time, the presence of double tensor operators changes the selection rules for the non-vanishing contributions to the transition amplitude.

The extension of the relativistic model by the third-order contributions is rather straightforward. However, the expressions for such new terms are more complex than those of the non-relativistic approach, since the closure procedure has to be performed twice, for the spin and orbital parts of three inter-shell double tensor operators. When the electron correlation effects are taken into account, again at the third-order two particle effective operators are expected as originating from the Coulomb interaction. The third-order relativistic model of the Judd-Ofelt theory is discussed in detail in Chapter 18 of Wybourne and Smentek [13].

10.5 Parameterizations of the f-Spectra

The amplitude of the electric dipole $f \longleftrightarrow f$ transition defined within the standard J-O Theory is determined by one particle effective operators as defined in equation (10.17). It is seen from (10.31) and (10.32) that in order to include the third-order electron correlation contributions to the transition amplitude, only the J-O radial integrals have to be modified by the appropriate radial terms that result from the Coulomb interaction and the u_{HF} potential. The angular terms remain the same as at the second order, thus the Judd-Ofelt one-particle parametrization of the spectra is not changed (remembering that the two-particle effective operators are relatively negligible). This means that when applying the semi-empirical approach, the line strengths of all electric dipole transitions for a given system are determined by at the most three intensity parameters Ω_λ, for $\lambda = 2, 4, 6$. These parameters are modulated by the reduced matrix elements of unit tensor operators with the functions describing the final and the initial states of each transition, namely

$$S_{f\longleftarrow i} = \sum_{\lambda=2,4,6} \Omega_\lambda \mid \langle \Psi_f \mid\mid U^{(\lambda)} \mid\mid \Psi_i \rangle \mid^2.$$

It should be pointed out that the one particle nature of this expression is the only common feature of the semi-empirical version of the J-O Theory and its originally derived, which is the standard *ab initio* version introduced by Judd and Ofelt in their original papers of 1962 [3, 4]. This is why the expression for the $S_{f\longleftarrow i}$ defines the Judd-Ofelt parametrization scheme of the f-spectra rather than the Judd-Ofelt theory, as it is erroneously used in the literature. Indeed, the expression for the line strengths should be read only as a possibility of determining the line strengths of various transitions by three parameters limited to the even values of the unit tensor operators, and without assigning to the parameters any specific physical interpretation or origin. In fact, when analyzing different physical mechanisms and their impact upon the transition amplitude via the *ab initio* calculations, it has been demonstrated that the intensity parameters Ω_λ represent not only the crystal field perturbing influence taken into account at the second order, as the original derivation of the J-O theory might suggests. The values of Ω_λ when determined via the fitting procedure contain the impact of the following physical mechanisms [65]:

1. crystal field potential and its influence taken into account at the second order of the static model; the standard J-O Theory and its original derivation; and in addition:
2. crystal field potential and its influence taken into account at the second order via the dynamic model;
3. electron correlation effects at the third order based on the static and dynamic models;
4. spin-orbit interaction at the third order;
5. mass polarization shift at the third order;
6. hyperfine interactions at second and third orders; and finally
7. the relativistic effects.

It is possible to identify all of these different contributions to the intensity parameters when deriving their tensorial form explicitly at the second, third, or even higher orders of perturbation approach. It is impossible however to go backwards and select parts of fitted parameters and assign to them a particular physical meaning or interpretation. In general, the parameters Ω_λ represent the overall picture of possibly important mechanisms that affect the $f \longleftrightarrow f$ transition amplitudes, without any specification about the order of perturbation, since this particular formalism does not play any role at the point of numerical analysis of their fitted values.

As a consequence, when looking at the expression for $S_{f \longleftarrow i}$ above as a formal introduction of the intensity parameters of one particle scheme there is neither a methodological nor physical objection to increase their number by including those terms that are associated with the unit tensor operators with the odd ranks. Just the opposite! The presence of the odd rank intensity parameters would compensate for an error possibly introduced by assumptions about the energy denominators of the Judd-Ofelt derivation, which limit the rank λ to even values. From a physical point of view it means that these odd-rank terms would also compensate for the errors made when the direct calculations are performed for the average energy of configuration, and with the frozen core orbitals in the case of the excited one electron radial basis sets. The latter approximations, as presented here, are essential for performing the partial closure to derive the effective operators determining the transition amplitude. This is crucial also for application of the perturbed function approach to avoid the problems with the evaluation of the radial terms for the complete basis sets.

Thus, the line strength defined within the one-parametrization Judd-Ofelt scheme has a more general form, namely [66, 67],

$$S_{f \longleftarrow i} = \sum_{\lambda}^{all} \Omega_\lambda \mid \langle \Psi_f \mid\mid W^{(0\lambda)\lambda} \mid\mid \Psi_i \rangle \mid^2. \tag{10.36}$$

The unit tensor operators $U^{(\lambda)}$ of a standard presentation are formally replaced in equation (10.36) by double tensor operators $W^{(0\lambda)\lambda}$ with the zero rank for the spin part of the space. The ranks of these operators λ now have even and also odd values. In this way the selections rules for the allowed electric dipole transitions are changed and expression (10.36), when applied for the fitting procedure, includes in the data set also the $0 \longleftrightarrow 1$ and $0 \longleftrightarrow 3$ transitions observed in the case of the Eu^{3+} ion in various hosts, for example. The latter transitions, forbidden within the standard J-O parametrization scheme of f-spectra, now are allowed and can be theoretically described.

The results of the discussion presented in the previous section indicate however that the semi-empirical approach based on equation (10.36) would neglect the purely relativistic effects represented by the unit tensor operators $W^{(1k)\lambda}$. In order to include these effects in the fitting procedure and to describe the subtle features of f-spectra, the expression from equation (10.36) should be extended by the following term:

$$\sum_{\lambda}\sum_{k}\Omega_{\lambda}^{1k} \mid \langle\Psi_f \mid\mid W^{(1k)\lambda} \mid\mid \Psi_i\rangle \mid^2. \tag{10.37}$$

In the particular case of $k = 1$ in equation (10.37) the effective operator represents the influence of the spin-orbit interaction [68]. It is seen that this very effect is responsible for the unusual transition $0 \longleftrightarrow 0$ observed in Eu^{+3}. Indeed, when $\lambda = 0$, the triangular condition for the non-vanishing reduced matrix element indicates that $k = 1$ and there is a non-zero contribution to the transition amplitude due to the effective operator associated with $W^{(11)0}$.

Combining together the non-relativistic and relativistic parts of the line strength it is possible to write the general expression as follows:

$$\sum_{\lambda}\sum_{\kappa k}\Omega_{\lambda}^{\kappa k} \mid \langle\Psi_f \mid\mid W^{(\kappa k)\lambda} \mid\mid \Psi_i\rangle \mid^2, \tag{10.38}$$

where for $\kappa = 0 \Longrightarrow \delta(k,\lambda)$ and with $\lambda = even \Longrightarrow \Omega_{\lambda}^{0\lambda} \equiv \Omega_{\lambda}$ of the standard Judd-Ofelt parametrization scheme, when $\lambda = odd \Longrightarrow$ extension by the odd terms. For $\kappa = 1$ there are new intensity parameters. The number of these relativistic parameters is limited by the requirement of hermiticity of the operators that is satisfied if $\kappa + k + \lambda$ is even. This means that for $\kappa = 1$ the parity of k is opposite to the parity of λ. For example, the standard Judd-Ofelt parameter $\Omega_2 (\equiv \Omega_2^{02})$ now is accompanied in the fitting procedure by Ω_2^{11} and Ω_2^{13}. This is especially beneficial for the description of the hypersensitive transition $0 \longleftrightarrow 2$ in Eu^{3+}, which requires more sensitive tools for its description. It should be mentioned at this point that in this way not only the number of parameters is increased in comparison to the standard J-O scheme, that might suggest better fitting, although this is not always in practice the case. The most important consequence of this generalized scheme is the possibility to include the impact of all of these physical mechanisms that occur within the spin-orbital space (the spin-orbit interaction, for example).

The possibility for the hypersensitive transition $0 \longleftrightarrow 2$ in Eu^{3+} to be represented in the generalized Judd-Ofelt scheme by more than one intensity parameter provoke an additional conclusion. Namely, the introduction of the intensity parameters presented and understood here does not preclude their negative (fitted) values; very often this is the case of Ω_2 that determines the intensity of the hypersensitive transition $0 \longleftrightarrow 2$ in Eu^{3+}. This very criterion, the negative values of the fitted intensity parameters, is used in the literature to disregard the set of parameters as though they would be directly determined from *ab initio* calculations (when they are evaluated as squared real numbers). This is not the case of semi-empirical calculations, especially when the line strength is determined not by just one but more intensity parameters. This particular interpretation of the intensity parameters of the Judd-Ofelt parametrization scheme of the f-spectra was one of the points discussed with (and approved by) the late Professor Brian Wybourne, one of the founders of the spectroscopy of the lanthanides. And only recently Professor Brian Judd, during the celebration of the Golden Anniversary of the Judd-Ofelt theory in 2012 in Udine gave his support to this very important conclusion, as based on a correct understanding of the Judd-Ofelt theory, separately from the Judd-Ofelt parametrization scheme of f-spectra.

Acknowledgment

The work has been performed under the auspices of the NSF, Award CHE-1209371, entitled *Luminescence as a Structural Probe of Ultra-Small Rare Earth Oxide Nanoparticles*.

References

[1] http://icfe8.uniud.it/?q=node/39 (January 27, 2014).
[2] http://www.aps.org/units/fip/newsletters/201302/juddofelt.cfm (January 27, 2014).
[3] B. R. Judd, Optical absorption intensities of rare-earth ions, *Physical Review*, **127**, 750–761 (1962).
[4] G. S. Ofelt, Intensities of crystal spectra of rare-earth ions, *Journal of Chemical Physics*, **37**, 511–520 (1962).
[5] L. Smentek, Judd-Ofelt theory (in person): concert for three tenors and piano, *J. All. Comp.*, **380**, 2–3 (2004).
[6] L. Smentek, B. G. Wybourne, and B. A. Hess, Jr., Judd-Ofelt theory in a new light on its (almost) 40th anniversary, *J. All. Comp.*, **323–324**, 645–648 (2001).
[7] L. Smentek, B. Andes Hess, Jr., H. C. Manning, and D. J. Bornhop, Density-functional theory structures of 1,4,7,10-tetraazacyclododecane-1,4,7,10-tetraacetic acid complexes for ions across the lanthanide series, *Journal of Chemical Physics*, **123**, 244302 (2005).
[8] B. Andes Hess, Jr., A. Kędziorski, L. Smentek, and D. J. Bornhop, Role of the antenna in tissue selective probes built of lanthanide-organic chelates, *Journal of Physical Chemistry A*, **112**, 2397–2407 (2008).
[9] L. Smentek and B. A. Hess, Jr., Conformational analysis of Eu(III)-(Ethylenedinitrilo)tetrakis-(methylphosphonates) (EDTMP), *Collection of Czechoslovak Chemical Communications*, **73**, 1437–1456 (2008).
[10] L. Smentek, Lanthanides caged by the organic chelates; structural properties, *Journal of Physics C*, **23**, 143202 (2011).
[11] L. Smentek, Theoretical description of the spectroscopic properties of rare earth ions in crystals, *Physics Reports*, **297**, 155–237 (1998).
[12] R. Loudon, *The Quantum Theory of Light*, Oxford University, Oxford (1973).
[13] B. G. Wybourne and L. Smentek, *Optical Spectroscopy of Lanthanides; Magnetic and Hyperfine Interaction*, CRC Press, Taylor & Francis (2007).
[14] G. G. Siu and D. J. Newman, Spin-correlation effects in lanthanide-ion spectroscopy, *Journal of Physics*, **C 16**, 7019–7025 (1983).

[15] Y.Y. Yeung and D.J. Newman, Orbitally correlated crystal field parametrization for lanthanide ions, *Journal of Chemical Physics*, **86**, 6717–6721 (1987).
[16] B. Ng and D. J. Newman, Ab initio calculation of crystal-field correlation effects in $Pr^{3+} - Cl^{-}$, *Journal of Physics C*, **19**, L585–L588 (1986).
[17] B. Ng and D. J. Newman. Many-body crystal field calculations. I. Methods of computation and perturbation expansion, *Journal of Chemical Physics*, **87**, 7096–7109 (1987).
[18] B. Ng and D. J. Newman, Many-body crystal field calculations. II. Results for the system $Pr^{3+} - Cl^{-}$, *Journal of Chemical Physics*, **87**, 7110–7117 (1987).
[19] M. Faucher and D. Garcia, Electrostatic crystal-field contributions in rare-earth compounds with consistent multipolar effects. I. Contribution to k-even parameters, *Physical Review B*, **26**, 5451–5468 (1982).
[20] D. Garcia, M. Faucher, and O. L. Malta, Electrostatic crystal-field contributions in rare-earth compounds with consistent multipolar effects. II. Contribution to k-odd parameters (transition probabilities), *Physical Review B*, **27**, 7386–7392 (1983).
[21] D. Garcia and M. Faucher, Crystal-field parameters in rare-earth compounds: Extended charge contribution, *Physical Review*, **30**, 1703–1707 (1984).
[22] D. Garcia and M. Faucher, Crystal field parameters in rare earth compounds: a mixed "covalo-electrostatic model," *Journal of Chemical Physics*, **82**, 5554–5564 (1985).
[23] S. Edvardsson, M. Wolf, and J. O. Thomas, Sensitivity of optical-absorption intensities for rare-earth ions, *Physical Review B*, **45**, 10918–10923 (1992).
[24] M. Wolf, S. Edvardsson, M. A. Zendejas, and J. O. Thomas, Molecular-dynamics-based analysis of the absorption spectra of Nd^{3+}-doped Na^{+} β''-alumina, *Physical Review B*, **48**, 10129–10136 (1993).
[25] I. Lindgren and J. Morrison, *Atomic Many-Body Theory*, Springer-Verlag Berlin, Heidelberg, New York (1982).
[26] J. C. Morrison, Effect of core polarization upon the f-f interactions of rare-earth and actinide ions, *Physical Review A*, **6**, 643–650 (1972).
[27] J. C. Morrison, Many-body calculations for the heavy atoms. III. Pair correlations, *Journal of Physics B*, **6**, 2205–2211 (1973).
[28] K. Jankowski, L. Smentek-Mielczarek, and A. Sokołowski, Electron-correlation third-order contributions to the electric dipole transition amplitudes of rare earth ions in crystals, *Molecular Physics*, **59**, 1165–1175 (1986).
[29] C. K. Jorgensen and B.R. Judd, Hypersensitive pseudoquadrupole transitions in lanthanides, *Molecular Physics*, **8**, 281–287 (1964).
[30] R. D. Peacock, The intensities of lanthanide f-f transitions, *Structure and Bonding*, **22**, 83–122 (1975).
[31] S. F. Mason, R. D. Peacock, and B. Steward, Ligand-polarization contributions to the intensity of hypersensitive trivalent lanthanide transitions, *Molecular Physics*, **30**, 1829–1841 (1975).
[32] R. D. Peacock, The charge-transfer contribution to the intensity of hypersensitive trivalent lanthanide transitions, *Molecular Physics*, **33**, 1239–1246 (1977).
[33] S. F. Mason, Localized systems mechanisms for f-f transition probabilities in lanthanide coordination compounds, *Journal of the Less Common Metals*, **93**, 45–58 (1983).
[34] S. F. Mason, Independent-systems mechanisms for f-f transition probabilities in lanthanide coordination compounds, *Inorganica Chimica Acta*, **94**, 313–322 (1984).
[35] M. F. Reid , J. J. Dallara and F. S. Richardson, Comparison of calculated and experimental $4f^N - 4f^N$ intensity parameters for lanthanide complexes with isotropic ligands, *Journal of Chemical Physics*, **79**, 5743–5751 (1983).
[36] L. Smentek-Mielczarek, Sign relation between the static and dynamic contributions to the one-photon $f \rightarrow f$ transition amplitude, *Physical Review B*, **48**, 9273–9278 (1993).
[37] O. L. Malta, A. A. S. Da Gama, and S. Canuto, Comment on the average energy denominator method in perturbation theory, *Physics Letters A*, **97**, 333–334 (1983).
[38] M. F. Reid and F. S. Richardson, Electric dipole intensity parameters for lanthanide $4f \rightarrow 4f$ transitions, *Journal of Chemical Physics*, **79**, 5735–5742 (1983).

[39] M. F. Reid and F. S. Richardson, Anisotropic ligand polarizability contributions to lanthanide $4f \longleftrightarrow 4f$ intensity parameters, *Chemical Physics Letters*, **95**, 501–506 (1983).
[40] O. L. Malta and G. F. De Sa, Explanation of the anomalous hypersensitive $^3H_4 \rightarrow {}^3P_2$ transition in Pr^{3+}, *Physical Review Letters*, **45**, 890–893 (1980).
[41] R. Kuroda, S. F. Mason, and C. Rosini, Anisotropic contributions in the ligand polarization model for the f-f transition probabilities of Eu(III) complexes, *Chemical Physics Letters*, **70**, 11–16 (1980).
[42] L. Smentek and B. A. Hess, Jr., Theoretical description of $0 \longleftrightarrow 0$ and $0 \longleftrightarrow 1$ transitions in the Eu^{3+} ion in hosts with C_{2v} symmetry, *Molecular Physics*, **92**, 835-845 (1997).
[43] L. Smentek, A. Kędziorski and B. A. Hess Jr., research in progress: NSF Award CHE-1209371, entitled *Luminescence as a Structural Probe of Ultra-Small Rare Earth Oxide Nanoparticles*.
[44] K. Jankowski and L. Smentek-Mielczarek, Effect of electron correlation on the forced electric dipole transition probabilities in f^N systems. A General Effective Operator Formulation, *Molecular Physics*, **38**, 1445–1457 (1979); Electron correlation effects on transition probabilities of $LaCl_3 : Pr^{3+}$, *ibid* **38**, 1459–1465 (1979); Effect of electron correlation on the forced electric dipole transition probabilities in f^N systems. II. A model study of $Pr : LaCl_3$ and $Eu : LaCl_3$, *ibid* **43**, 371–382 (1981); Effect of electron correlation on the forced electric-dipole transition probabilities in f^N systems. III. Some aspects of the mechanism of hypersensitive transitions, *International Journal of Quantum Chemistry*, **24**(s17), 339–346 (1983).
[45] L. Smentek-Mielczarek, Electron correlation third-order contributions to the electric dipole transition amplitudes of rare earth ions in crystals II. A general effective operator formulation based on static and dynamic models, *Molecular Physics*, **61**, 767–774 (1987).
[46] L. Smentek-Mielczarek and B. A. Hess, Jr., Third-order electron-correlation and crystal-field contributions to the amplitudes of two-photon absorption in f^N systems, *Physical Review B*, **36**, 1811–1817 (1987).
[47] L. Smentek-Mielczarek and B. A. Hess, Jr., Third-order *ab initio* calculations of the intensity parameters for the Pr^{+3} ion in $LaAlO_3$, $NdAlO_3$, and $LaCl_3$ hosts, *Journal of Chemical Physics*, **87**, 3509–3512 (1987).
[48] L. Smentek-Mielczarek and B. A. Hess, Jr., Third-order *ab initio* calculations of the $f \longleftrightarrow f$ transition amplitudes for ions across the lanthanide series, *Journal of Chemical Physics*, **89**, 703–710 (1988).
[49] L. Smentek-Mielczarek, Crystal field, ligand polarization and electron correlation third order contributions to the $f \longleftrightarrow f$ transition probabilities, *Molecular Physics*, **54**, 873–881 (1985).
[50] L. Smentek-Mielczarek and B. A. Hess, Jr., Perturbing influence of doubly excited configurations upon the electric-dipole transition amplitudes of rare-earth ions in crystals, *Physical Review B*, **39**, 517–522 (1989).
[51] L. Smentek-Mielczarek, An alternative formulation of the Judd-Ofelt theory of $f \longleftrightarrow f$ transitions, *Journal of Alloys and Compounds*, **224**, 81–85 (1995).
[52] L. Smentek and B. A. Hess, Jr., Three-particle effective operators of $f \longleftrightarrow f$ transition theory, *Molecular Physics*, **92**, 847–854 (1997).
[53] L. Smentek, Uniform perturbation approach of the theory of one- and two-photon $f \longleftrightarrow f$ transitions, *Spectrochimica Acta A*, **54**, 1545–1556 (1998).
[54] L. Smentek-Mielczarek and B. A. Hess, Jr., Third order *ab initio* calculations of amplitudes of two-photon absorption for ions across the lanthanide series, *Journal of Chemical Physics*, **90**, 1390–1395 (1989).
[55] L. Smentek, Two-center exchange interactions in rare earth doped materials, *International Journal of Quantum Chemistry*, **90**, 1206–1214 (2002).
[56] L. Smentek, Effective operators and spectroscopic properties, *Journal of Alloys and Compounds*, **380**, 89–95 (2004).
[57] B. G. Wybourne, L. Smentek, and A. Kędziorski, Borrowing intensity in rare earth doped materials; magnetic dipole transitions, *Collection of Czechoslovak Chemical Communications*, **70**(7), 905–922 (2005).
[58] L. Smentek and A. Kędziorski, Hyperfine-induced $f \longleftrightarrow f$ transitions: effective operator formulation, *Spectroscopy Letters*, **40**, 293–315 (2007).

[59] P. G. H. Sandars and J. Beck, Relativistic effects in many electron hyperfine structure. I. Theory, *Proceedings of the Royal Society*, **289**, 97–107 (1965).
[60] L. Smentek and B. G. Wybourne, Relativistic $f \longleftrightarrow f$ transitions in crystal fields, *Journal of Physics B*, **33**, 3647–3651 (2000).
[61] L. Smentek and B. G. Wybourne, Relativistic $f \longleftrightarrow f$ transitions in crystal fields: II. Beyond the single-configuration approximation, *Journal of Physics B*, **34**, 625–630 (2001).
[62] L. Smentek, B. G. Wybourne, and J. Kobus, A relativistic crystal field for S-state f electron ions, *Journal of Physics B*, **34**, 1513–1522 (2001).
[63] B. G. Wybourne and L. Smentek, Relativistic effects in lanthanides and actinides, *Journal of Alloys and Compounds*, **341**, 71–75 (2002).
[64] A. Kędziorski, L. Smentek, and B. G. Wybourne, Net-value of the relativistic crystal field effect, *Journal of Alloys and Compounds*, **380**, 151–155 (2004).
[65] L. Smentek, Morphology of the $f \longleftrightarrow f$ intensity parameters, *Molecular Physics*, **98**, 1233–1241 (2000).
[66] A. Kędziorski and L. Smentek, Extended parametrization scheme of f-spectra, *Journal of Luminescence*, **127**, 552–560 (2007).
[67] A. Kędziorski and L. Smentek, New parametrization of spectra of Nd^{3+} and Sm^{3+} in glasses, *Journal of Alloys and Compounds*, **451**, 686–690 (2008).
[68] L. Smentek, Relativistic contributions to the amplitude of electric dipole transitions in rare earth ions in crystals: I. Spin-orbit interaction, *Journal of Physics B*, **32**, 593–606 (1999).

11

Applied Computational Actinide Chemistry

André Severo Pereira Gomes,[1] *Florent Réal,*[1] *Bernd Schimmelpfennig,*[2] *Ulf Wahlgren,*[3] *Valérie Vallet*[1]

[1]*Université Lille CNRS, Laboratoire PhLAM, CNRS UMR 8523*
[2]*Institute for Nuclear Waste Disposal, Karlsruhe Institute of Technology*
[3]*Department of Physics, Stockholm University, AlbaNova University Centre*

11.1 Introduction

The theoretical modeling of lanthanide and actinide complexes in condensed media (in solution or in the solid state) is challenging for theoretical chemistry, due in part to the intrinsic general challenge of devising a reliable model of the species in such environments and the necessity of 'chemical' accuracy (errors of about a kilocalorie per mol). These challenges are exacerbated with lanthanide and actinide elements for two reasons. The first aspect is that in oxidation states II, III, and IV, the lanthanide and actinide elements are present as atomic cations while in higher oxidation states V and VI, which only exist for actinides, the mono-oxo cation PaO^{2+} and trans-dioxo cations $\mathrm{AnO}_2^{+/2+}$ are found. Their high charge induces strong polarization effects on the environment, together with ion-pairing interactions with neighboring counter-ions and neutral ligands, thus making it crucial to design on the one side a proper chemical molecular model to describe their complexes in condensed phase and on the other side a computational protocol to include the environment. The theoretical description of spent nuclear fuel partitioning is an area of application where a balanced treatment of such effects for both lanthanides and actinides is mandatory. The second challenging aspect is the accurate treatment of the electronic structure of the heavy element complex itself, because it has a direct bearing on the accuracy of the properties one is interested in.

Computational Methods in Lanthanide and Actinide Chemistry, First Edition. Edited by Michael Dolg.

In this chapter we aim to discuss the requirements, challenges, and pitfalls associated with attempting to theoretically model molecular properties for such systems that can be directly compared to experimental data, such as valence excitation spectra, core-excitation spectra, thermodynamics of chemical reactions, and redox properties.

A number of methods are available for probing the electronic structure of molecular systems containing heavy elements [1, 2], so before discussing their use for the different properties outlined above we shall briefly outline the most widely used ones at present. These approaches differ essentially in the way they address relativistic effects (of which the perhaps most relevant aspect is the presence of spin-orbit coupling and its consequences to molecular properties) and electron correlation. Although these effects are not simple additive (e.g., spin-orbit coupling can bring different electronic states closer than they would otherwise be, thus affecting electron correlation), there is a substantial body of work that highlights the fact that the description of electron correlation for the different electronic states is the key factor governing the accuracy one may achieve, even though spin-orbit coupling is crucial to get a realistic picture of the electronic spectrum. Because of that, we shall outline the approaches employed to describe both separately in the following subsection, first focusing on (spin-free) correlation methods, followed by a discussion of spin-orbit coupling in Section 11.1.2.

11.1.1 Relativistic Correlated Methods for Ground and Excited States

A thorough review of electronic structure methods is outside the scope of this chapter, and we therefore refer the reader to a non-exhaustive list of recent review papers [3–9] for further reading.

Starting first with methods applicable to single-reference ground states, we have the time-dependent (TD) generalization of density functional theory (DFT), TDDFT [10, 11] where the excitation energies are obtained as the poles of a propagator or, in the language of time-dependent perturbation theory, the poles of the linear response function [12]. A well-known drawback of TDDFT for molecular application is that, within the commonly used adiabatic approximation only single excitations can be described. Different proposals to include higher-order excitations by going beyond the adiabatic approximation by building in the frequency dependence of the exchange-correlation kernel have been put forth in recent years [13–15], but one has yet to see their use in real-life applications. Irrespective of these issues, it is also well-known that the choice of the density functional approximation (DFA) in actual calculations is crucial to achieve a good accuracy [7, 16, 17], provided that a given DFA may be quite accurate for certain properties and not for others.

At the wavefunction level methods based on coupled cluster (CC) theory are among the most reliable ones. For ground-state energetics the CCSD(T) approach is the 'gold standard' of chemistry, whereas for excited states one can use the equation-of-motion (EOM) CC (EOM-CC) method or CC linear response theory (CC-LRT) [4] approaches. Note that the CC-LRT is size-extensive for both energies and properties such as intensities, however for EOM-CC this is true only for energies (unless one uses the closely related similarity transformed (ST)EOM-CC method [18]). As the computational cost of fully iterative (e.g., CCSD, CCSDT, etc.) methods can quickly become prohibitive, perturbative methods [4]

(CC2, CC3, etc.) can be devised offering a good balance between cost and accuracy. Other simple (second-order) perturbative approaches that are comparable to CC2 can be found in the ADC(2) [19] or CIS(D) [20] methods. As is the case for TDDFT, these linear response or propagator formulations are not only interesting from a formal perspective (by allowing one to obtain information from the excited states based on ground-state quantities) but also because of their 'black box' nature, that is, one does not need to construct a complicated, multi-reference wavefunction via the definition of orbital active spaces. However, as we shall see, the restriction to closed-shell references somewhat limits their applicability to the actinides.

General CI methods [9] including higher-order excitations were developed during the 1970s and 1980s, such as (internally or externally) contracted multi-reference single and double excitation methods and can tackle cases where the ground-state wavefunction has an explicit multi-reference character. All excited states can in principle be obtained from MR-CISD. The disadvantage is that the method is computationally cumbersome and the resulting wavefunction is not size consistent. The energies can anyway be corrected by ad hoc size-extensivity corrections, such as Davidson or Pople methods.

As an alternative to diagonalizing large CI matrices, one can apply perturbation theory, as an extension of MP2 perturbation approach to a multi-reference wavefunction. Any number of roots, up to the total number of states in the reference, can be calculated (CASPT2, NEVPT2). The advantage of this method is that it is reasonably fast and reasonably accurate. A disadvantage is that the method only includes double excitations in the perturbative step, and that it to some extent is parameter dependent. Moreover the reference state must be carefully chosen so that the orbitals included in the CASSCF step generate the desired excitation, and that, for high symmetry cases, the proper degenerate partner orbitals (like e_x and e_y) are included. The density Matrix Renormalization Group method (DMRG) offers an alternative, computationally more tractable route to handle large active spaces that are insurmountable to the conventional CASSCF approaches.

For open-shell ground states, the choice of electronic structure methods depends on the character of the electronic ground state. If it is well described by a single-slater determinant, excitation energies can be obtained with TDDFT, CC-LRT, EOM-CCSD, but very few test applications have been reported so far [21]. Instead, multi-reference methods treating ground and excited states in a unified framework are more routine, such as with multi-reference CC calculations with Fock-Space (FSCC) methods, or MRCI calculations, and CASPT2. Because of the computational cost of the first two, CASPT2 is normally the only method that can be readily applied.

At present, implementations of each of these electronic structure methods exist in efficient parallel schemes and approximate tensor factorizations, the most well-known being density fitting (DF) or the resolution of the identity (RI), have become increasingly common in electronic structure theory programs as a means to both accelerate computations and eliminate the storage or generation of the four-index electron repulsion integral (ERI) tensor. Current efforts are devoted to local correlation techniques, which uses localized orbitals to scale down the number of interacting orbitals in wavefunction theory (WFT) correlated methods with the aim of making them routinely applicable to much larger systems (hundreds of atoms) than today.

11.1.2 Spin-Orbit Effects on Heavy Elements

Spin-orbit effects are in general important to account for in excited-states calculations and in chemical reactions involving trans-uranium actinides and in many cases also uranium complexes. In fact, the only uranium state in which spin-orbit effects are unimportant are complexes where only the ground states of U(VI) is involved, the most known of which is the uranyl ion, UO_2^{2+}.

The most rigorous framework to include spin-orbit interaction is to treat it *a priori* in a four- or two-component relativistic framework. For a detailed description of such approaches, we refer the reader to Chapter 3. On the other hand, if one starts from a scalar relativistic framework, in which the wavefunction is expanded on relativistic orbitals, instead of spinors, one can add spin-orbit coupling *a posteriori* by coupling all relevant spin-orbit free states. This was the method of choice, before the opportunity of using two- and four-component methods became possible in the late 1980s. Using this approach, spin-orbit calculations were done already in the 1960s, when for example Walker and Richards calculated spin-orbit coupling constants in some diatomic molecules [22]. During the following decades the theory was developed, and in the early 1990s calculations on systems of chemical interest appeared. See for example the chapters by Hess *et al.* [23, 24].

Spin-orbit calculations were at the time hampered by the calculation of the spin-orbit integrals over the three spin operators, S_x, S_y, and S_z, although a number of applications were done during the 1990s, including studies by Wahlgren *et al.* [25] and Rakowitz *et al.* [26] on the spin-orbit splitting in the $6p$-shell of thallium and by Teichteil *et al.* on iodine [27].

At the time both *LS*-based methods (one component calculations followed by a separate calculation of the spin-orbit contributions), two- and four-component methods were developed. Despite the increase in computer power the high computational cost of four-component methods restricts their applicability to relatively small molecular systems. However, different flavors of two-component Hamiltonian have matured in the past years and are now approaching the computational efficiency of one-component methods (ZORA, X2C, etc.). (See for instance references [1, 28–35]). As a result, for chemical reactions or spectroscopic studies, one-component approaches treating spin-orbit coupling a posteriori are preferred.

In these approaches, the spin-orbit part can be done either at the variation-perturbation level where only a few spin-free states are included in the spin-orbit Hamiltonian, or in a spin-orbit CI, typically including all single excitations from the reference states as in the spin-orbit CI method EPCISO [36]. Fromager *et al.* [37] has shown that the two methods are essentially equivalent provided that the orbitals are relaxed separately in all of the spin-free reference configurations.

For heavy elements the problem with the cumbersome integrals calculations was resolved by realizing that the spin-orbit integrals, which depend on $\frac{1}{r^3}$ are quite local, which allows a one-center approximation and since most of the spin-orbit contributions occur at short distances from the nucleus, a mean-field approximation for a given atom. This method, referred to as the Atomic Mean-Field Integrals (AMFI) method [38, 39], was developed by Schimmelpfennig *et al.* in the later 1990s [40]. The AMFI approximation, which has been included in several quantum chemistry codes including MOLCAS, DALTON, DIRAC, ResPect has been shown to perform well [41–45].

Of course, other approaches to the spin-orbit problem have also been developed. This includes the spin-orbit CI suggested by Yabushita *et al.* [46], and ECPs describing the spin-orbit interaction explicitly, like Dolg *et al.* [47], and Seijo and Barandiarán [48] using the AIMP approach.

11.2 Valence Spectroscopy and Excited States

The UV-Vis spectroscopy of lanthanide and actinide elements directly reflects the electronic structure of the species involved. In oxidation states (III) and (IV), the ground-state electronic configuration is f^n, thus the low-lying spectrum is dominated by f-f transitions that are strictly parity forbidden, and can also be spin-forbidden although spin-orbit coupling attenuates the selection rules. Nevertheless, both restrictions have important consequences, namely that these f-f bands have very low absorption intensities, and the radiative lifetimes of f-f states are often rather large (10^{-3} s) and sensitive to the environment. This is routinely used in Time-Resolved-Laser-Fluoresence-Spectroscopy (TRLFS) of Eu(III) and Cm(III) at approximately 17000 cm^{-1} [49–51]. Metal-centered f-d transitions occur at energies, which are much higher than those of f-f transitions; they generally give rise to intense absorption bands in the UV region. Depending on the relative redox properties of the metal and coordinated ligands, charge-transfer transitions either from the metal to the ligand (MLCT), leading to an oxidation of the metal, or from the ligand to the metal (LMCT), leading to a reduction of the metal can happen. In oxidation states (V) and (VI), actinides formed so-called actinyl ions $\mathrm{AnO}_2^{+/2+}$, in which the actinide is triply bonded to the yl-oxygens by overlapping its 5f and 6d orbitals with the oxygen 2p to form bonding and anti-bonding σ and π orbitals, and some non-bonding orbitals of symmetry $\varphi_u(5f)$, $\delta_u(5f)$, and $\delta_g(6d)$. A special feature of the actinyl ions is the participation of the $6p_\sigma$ orbital in the bond, which leads to the 6p-hole, which is related to the Electric Field Gradient (EFG) as discussed by Larsson and Pyykkö [52]. In uranyl (VI) ground-state, the six bonding orbitals are doubly occupied, and excited states between about 30000 and 50000 cm^{-1} correspond to bonding to non-bonding excitations. In a lower oxidation state [uranyl(V)] and in trans-uranium 'yl' species, the non-bonding orbitals become occupied in the ground-state, thus excitations may either occur within the non-bonding orbital manifold or involve the bonding or anti-bonding orbitals. Actinide valence spectra are thus dense, and when a project involving excited states like calculating spectra or investigating a photochemical process is planned, it is important to anticipate that a large number of states have to be computed, and that the challenge for electronic structure method is to accurately account for differential correlation effects across states of very different electronic character.

11.2.1 Accuracy of Electron Correlation Methods for Actinide Excited States: WFT and DFT Methods

Before initiating a computational study of electronic spectra it is important to decide on the accuracy needed and to choose methods accordingly. As nearly no gas-phase spectra are available for actinide complexes, only comparisons between currently available methods can be used to estimate the expected precision. In this section we will address these issues.

A number of investigations have been done on the accuracy of different methods for calculating spectra of actinide complexes with WFT and DFT methods, focusing on the accuracy of TDDFT because it is the method of choice for larger molecular systems. Réal *et al.* compared calculated spectra of isolated uranyl up to 50000 cm^{-1} in two studies [53, 54] in 2007 and 2009, later complemented by extensive investigations by Bast *et al.* [55] and Tecmer *et al.* [17]. The spectrum of UO_2^{2+} was calculated using LR-CCSD, AQCC, Davidson-corrected MRCI, CASPT2 both all-electron and ECP, TDDFT with the functionals SAOP, BLYP, B3LYP, BHLYP, PBE0, M06, M06-L, M06-2X, CAMB3LYP, and the hybrid DFT-MRCI method.

In the pure WFT framework it was shown that the LR-CCSD, AQCC, and DC-MRCI agree within 1500 cm^{-1}, while CASPT2 systematically underestimate the excitation energies with up to 5000 cm^{-1} for the highest ones. However, in a later investigation Réal *et al.* [54] concluded that all of these methods still lack higher-order corrections to reach FSCC accuracy, but the description of the spectrum is overall similar in term of attribution and spacing among excited states. In an attempt to quantify the effect higher-order excitations, Tecmer *et al.* [21] have applied the active space variant of completely renormalized EOM-CC with singles, doubles, and non-iterative triples [CR-EOM-CCSD(t)]. Although this method yields very good results for detachment energies, this method tends to overshoot all transitions in the bare uranyl ion by up to 0.9 eV with respect to EOM-CCSD. However, EOM-CCSDT reduces the difference down to 0.3 eV. This suggests that triple excitations have indeed a sizable contribution to the transition energies, but that the CR-EOM-CCSD(T) significantly overestimates them.

This illustrates explicitly that the correlation functional plays a key role in the accuracy of the transition energies. This is enhanced by the spectra obtained by TDDFT calculations first for the bared uranyl and second for uranyl coordinated with four chlorides in the equatorial plane. Combining the theoretical spectra obtained at the spin-free and the spin-orbit corrected levels in the energy region 20000–30000 cm^{-1} of the studies of Pierloot [56, 57], Tecmer *et al.* [58], and Gomes *et al.* [59], it appears clearly that a critical factor is the degree of non-locality introduced through the inclusion of HF exchange in hybrids or meta-hybrids. Thus generally hybrid functionals outperform their pure GGA counterparts. Even the ALDA approximation, estimating the frequency-dependent exchange-correlation kernel appeared not satisfactory and shows that it is necessary to go beyond this approximation. In addition, non-local effects should be included to the exchange-correlation kernel, as is done for all (meta)hybrids and CAM-B3LYP or CAM-PBE, if one wishes to approach the accuracy of WFT methods. Tecmer *et al.* [58] compared several properties of the uranyl ion as well as uranyl coordinated with four chlorides as a function of DFT choices in an extended study, and they found a quite reasonable agreement for different versions of CAM-B3LYP, in particular in the lower energy region with high-precision WFT method. However, high precision spectroscopy cannot be reached with TDDFT so far in terms of energy level as well as state ordering.

Turning now to open-shell systems, a number of open-shell triatomic molecules have been studied with multi-reference relativistic methods, using CASSCF/CASPT2 theory and *a posteriori* treatment of spin-orbit effects. (See for instance the works by Kovács *et al.* [60–65] and Bolvin *et al.* [66]). For uranium open-shell cations (U^{5+}, U^{4+}) and actinyl systems, comparisons with four- and two-component FSCC calculations have been conducted to test not only whether the treatment of spin-orbit coupling either *a priori* (in 4C or 2C approaches) or *a posteriori* has an effect on the electronic spectrum, but also how sensitive the treatment of electron correlation might be. The widest comparison of correlation methods was carried out for U^{4+}, for which both coupled-cluster methods of the type of FSCC, but also variational CI and perturbative CASPT2 methods have been tested. CASPT2 method tends to systematically underestimate MR-SDCI transition energies [67–70], but if the virtual $6d$ orbitals are included in the active space, discrepancies diminish, because angular correlations of the $5f$ orbitals through the $6d$ orbitals improves correlation treatments within the $5f^2$ manifold. This has been tested for the U^{4+} ($5f^2$) atom, for which experimental data are available. For this atomic spectrum, both all-electron and relativistic pseudopotential treatments yield as accurate transition energies. Although the treatment of electron correlation was pushed to high limits, using a combination of extended basis sets and large correlation spaces, the transition to the highest $J = 0$ (^1S) state of the $5f^2$ manifold is overestimated by 1500–2000 cm^{-1}, pointing to either the need of higher-order correlation contributions or to an error in the experimental estimate of that transition energy.

11.2.2 Valence Spectra of Larger Molecular Systems

There are fewer computational studies of spectra of larger molecular actinide complexes, which can either be justified from the computational difficulties encountered with WFT-based methods, and the caution to be taken in the choice of the appropriate TDDFT scheme. However, there are few studies of molecular systems involving uranyl in oxidation state (VI) or (V) with a larger first-coordination shell than the elementary 4 halides.

In an attempt to shed light on the origin of the remarkable experimental intensity increase in the lower part of the uranyl-chloride-acetone absorption spectra, upon addition of chloride to uranyl acetone complex, Van Besien and Pierloot [71] applied the CASPT2 method with RASSI spin-orbit coupling to explore the lowest part of the spectra of $[UO_2Cl_2ac_4]$, $[UO_2Cl_2ac_3]$, $[UO_2Cl_3ac_2]^-$, and $[UO_2Cl_3ac]^-$ (ac = acetone) complexes. The active space (uranyl bonding, anti-bonding, and non-bonding orbitals) is identical to that of calculations on the bare uranyl unit or uranyl tetrachloride. If the computed transitions closely correspond to the experimental one, the computed oscillator strengths do not reveal any desired intensity increase. One might infer that the active space needs to be enlarged to include ligand orbitals because the intensity shift may arise from LMCT transitions. This is so far beyond current CASSCF size limits, but one might think of reducing the size of the CAS space following the concepts of restricted active spaces (RASPT2). Alternatively, other complex stoichiometries have to be investigated such as binuclear complexes, consisting of two uranyl units bridged by chloride or acetone ligands, but this again might be beyond the current methods' capabilities.

Wiebke *et al.* [72, 73] have illustrated the strong dependency of the results with respect to the chosen functional by comparing TDDFT excitation spectra of complexes of uranyl(VI) with (bis)salicylhydroxamate, (bis)benzohydroxamate, and benzoate. GGA functionals such as BP86 and PW91 yield very different absorption wavelengths while the hybrid functionals B3LYP and PBE0 give a consistent picture, but transitions are significantly underestimated. This was revealed by calculations with the CIS(D) method (doubles correction to the excitation energy from configuration interaction with single substitution, and tested for the first time on actinide complexes), which match fairly well the experimental spectrum. However, we have to note that this comparison is biased by the lack of spin-orbit interaction. Su *et al.* [74] have applied SO-TDDFT with the SAOP model potential to the calculations of the luminescence spectrum of uranyl with glycine by systematically optimizing the geometries in its luminescent state as done by Tsushima *et al.* [75, 76] and Réal *et al.* [77] in their studies of the photo-reactivity of uranyl. Knowing that the SO-TDDFT (SAOP) transitions tend to be too high compared to the SO-CCSD or SO-CASPT2, the computed transitions were shifted to the experimental origin band, and decorated by the uranyl vibronic lines computed from the Franck-Condon formula. The outcome perfectly matches experiment reflecting the high accuracy of computed vibrational frequencies. The only WFT-based calculation of a relatively large uranyl complex is provided by Tecmer [21] who computed low-lying excited states uranyl(V/VI) saldien complex using, not only TDDFT with the CAM-B3LYP functional, but also the CR-EOM-CCSD(T) method discussed before. (See Section 11.2.1.) Both methods agree within 0.2 eV, a difference that is smaller than that observed for the bare complex. (See Section 11.2.1.) So far, these examples rank heavy-element TDDFT calculations among semi-empirical method and ask for further developments and investigations.

11.2.3 Effects of the Condensed-Phase Environment

The discussion above concerns calculations of isolated species. Most situations of experimental interest, however, involve species in a solid or liquid environment, or at interfaces. In these situations, even though there are developments to enable the use of correlated wavefunctions for very large systems, the approaches discussed so far cannot be employed directly in their current form, and embedding approaches are employed.

In the case of solids, to go beyond the gas-phase approximation, i.e., describe the spectroscopy of actinides in their condensed phase such as in a pure crystal of $Cs_2NpO_2Cl_4$ or $Cs_2UO_2Cl_4$ or doped by Np element, one has to include explicitly the interaction with the surrounding beyond the first sphere of coordination. Benchmarks comparing high-resolution absorption spectroscopy recorded by Denning [78, 79] to ab initio calculations combining different approaches are listed in the following discussion. The first attempt was performed by Matsika and Pitzer [80] in which the environment effect on the AnO_2Cl_4 (U, Np) molecule is included in the Hamiltonian by the use of the six nearest-neighbor cesium atoms depicted as model potential and the rest of the crystal by explicit point charges. In the case of UO_2^{2+} in $Cs_2UO_2Cl_4$ and NpO_2^{2+} in $Cs_2NpO_2Cl_4$, the authors clearly showed the importance of the environment to predict transitions in the vicinity of the experimental data. While in the case of UO_2^{2+} the excitation energy deviations are about a few cm^{-1} up to 4000 cm^{-1}. (See Table 7 of Matsika and Pitzer [80].) The predicted adiabatic transitions are for the most part of the spectrum about a 5000 cm^{-1} overestimated for NpO_2^{2+} in $Cs_2NpO_2Cl_4$. (See Table 14 of Matsika and Pitzer [80]). More recently, these systems have also been investigated with FSCC and use the frozen density embedding method (FDE) to describe the equatorial ligands bound to the actinyls as well as a relatively large portion of the surrounding cesium and uranyl chloride species [59, 81]. By describing a maximum of subsystems by their density in interaction with the central subsystem, FDE reached a good accuracy in terms of state ordering and spacing. Even so a constant IR shift of the total spectrum is observed and calls for further investigations; the overall precision of the approach is an open door for spectroscopy interpretation of more complex systems. They have also concluded that studying the central system in the gas-phase with an explicit inclusion of the ligands is sufficient to capture most of the observables and to make a prediction or an interpretation of spectroscopy experiments.

Let us now turn our attention to species in solution. While approaches such as FDE are promising ways to enable the use of correlated electronic structure methods in the study of solvatochromic shifts [82], the fact remains that solvation effects on electronic absorption have to date been introduced with self-consistent reaction field models, which compute the reaction field, that is the electrostatic field exerted on the solute by the polarization of the solvent due to the presence of the solute, such as polarizable continuum models [83]. Proper use of this model to electronic excitation requires one to take into account properly the time-scales of the various processes involved. Indeed the solute and solvent respond to each other and to the exciting photon much faster than the nuclear motion can respond, making the nuclear motions almost static on the excitation time scale. The static character of the nuclear motion is called the Franck-Condon principle, which is associated to the slowly responding contribution of the reaction field. But in the course of a photochemical process, the solvent and excited-state solute may typically have time to equilibrate not only their electronic distribution, but also their nuclear geometry. Solvent effects on electronic transitions of the isoelectronic U^{4+}, NpO_2^+, PuO_2^{2+} species have been estimated by building a chemical model that includes water molecules in their first hydration sphere and outer-sphere solvent effects through the continuum reaction field [69]. In order to compute the absorption spectra the slow and fast components were equilibrated to the ground-state electronic densities. On these systems where the interesting transitions are within atomic-centered 5*f* manifolds, solvent effects are moderate in magnitude. The addition of the first hydration sphere lifts bare complex degeneracies, thus increasing the number of transitions. Most transitions are up-shifted by a few thousands wave numbers, except the highest U^{4+} f^2 state (J=0), which is more spatially diffuse and thus pushed up by exchange-repulsion effects. The resulting computed transitions are in good agreement with experiment and made it possible to identify the observed transitions.

A step towards more realistic modeling of properties of solvated actinide complexes, requires an exploration of the configurational space of the system at experimental conditions (temperature, pressure). This can be done either by Monte-Carlo sampling or by following molecular dynamics trajectories. Both methods rely on an accurate description of interatomic forces. One can either use ab initio-based MD of the Born-Oppenheimer or Car-Parinello type, which, for computational costs reasons, is so far restricted to using Density Functional Theory [84, 85]. However current functionals have been shown not to overestimate lanthanide/actinide-ligand interactions thus questioning the reliability of such dynamics. Alternatively, interatomic interactions can be described with classical force fields. Over the last years the efforts of several groups in developing appropriate force-field models are progressing but facing difficulties in the complexity of handling many-body effects in these systems. (See for instance the following references [86–90]). Ultimately, representative snapshots can be generated and used to compute a series of quantum calculations on the solvated systems. Because the size of the system is too large for full quantum calculations, embedding methods such as the FDE approach presented in the condensed phase

Section 11.2.3 are promising. No application to heavy element solvated species have been reported, but the experience gathered on lighter molecular systems (acetone [81], uracil [91] in water) can be directly translated to heavier counterparts.

11.2.4 Current Challenges for Electronic Structure Calculations of Heavy Elements

Our discussion of lanthanide and actinide complexes has made it clear that most of the time one has to deal with multi-reference wavefunctions for ground states, excited states, but this is even more true for heavy-element based magnets, mixed-valence systems, etc. Current multi-reference methods reach their limits when the number of unpaired electrons and active orbitals increase. It is thus important to search for efficient approaches to increase their applicability. We have already mentioned the concept of cutting down the size of the active space in terms of exceptions with restricted active spaces methods, in which one can limit excitation levels. Another alternative is the Density Matrix Renormalization Group (DMRG) [92], which can handle active spaces of up to 30 active orbitals, and even larger for longer molecules where locality can be used to optimize the computational cost. DMRG calculations on the CUO molecule surrounded by four rare gas atoms have been recently reported by Tecmer *et al.* [93]. The definition of a large active space was necessary to improve the treatment of electron correlation of the two close-lying states in CUO, the $^{1}\Sigma$ and $^{3}\Phi$ states, the latter being more stabilized than the former by interactions with the rare-gas matrix. We must note, however, that these DMRG calculations may not capture all dynamical correlation effects, and this problem is currently being investigated via coupling of DMRG to other approaches such as perturbation theory, configuration interaction, or cluster operators [92]. Another great step towards the reduction of the computational cost in the calculation of correlation energy is a result of the use of localized orbitals. Since electron correlation is a local phenomenon, by using local orbitals one can neglect long-range interactions so that all orbitals involved in an elementary operation must be in the same region of space to product a nonvanishing effect. In this way, an N-scaling behavior may be obtained. N-scaling multi-reference scaling codes were proposed [94–96] and successfully applied to spectroscopy, magnetism of large molecular systems, suggesting that they will be methods of choice for realistic numerical experiment.

11.3 Core Spectroscopies

Although electronic excitations in the UV-visible or near-IR ranges involving valence electrons have arguably been the main focus of theoretical modeling for actinides and lanthanides, it nevertheless remains the case that processes involving inner electrons can be of particular interest, because they allow for obtaining information on the immediate surroundings of a given atom in a molecular system; while the core level energies tend to be rather characteristic for each atom, these will be shifted from a given reference value (e.g., in a calibration standard or for the species in vacuum) by changes in the atom's oxidation state or its surroundings, be it changes in covalent interactions with nearest neighbors or simply changes in the electrostatic potential due to near neighbors and other species (e.g., in a solution or crystal environment).

11.3.1 X-ray Photoelectron Spectroscopy (XPS)

As its name implies, XPS is a technique that relies on the ionization of the species of interest and that is particularly useful to investigate surface properties, and as such has been widely used to investigate actinide oxides, in particular those of uranium (given its predominant role in nuclear fuels) in different forms such as solids or thin films. There, the core levels associated with the $4f$ electrons are particularly interesting, first due to their distinctive positions and the magnitude of the splitting of the $4f_{5/2}$ and $4f_{7/2}$ due to spin-orbit coupling (about 10 eV in the UO_2 crystal [97]).

The position of these peaks can be obtained, in a first approximation, by considering the orbital energies ($-\epsilon_k$). Here one should note the difference in the meaning of these for Hartree-Fock and DFT approaches (see Bagus *et al.* [98] and references therein) in order to properly correlate them to the underlying physical processes; thus, while the former approach provides a measure of the binding energy of electron in k in the absence of any relaxation effects on the electron density that accompany the ionization, the latter would provide an approximation to the binding energies with some relaxation. In the case of DFT, however, one should carefully consider the functional to use, since deficiencies such as the self-interaction error (SIE) can strongly affect the core orbitals' energies and effectively reduce the electrons' binding energy (see for example [17, 59] for a comparison of some of the first binding energies with different functionals for actinide-containing species). Whatever the case, one can further ameliorate the binding energies obtained from orbital energies by considering the energy difference for the original (N) and ionized (N-1) systems in a ΔSCF procedure, with the caveat that the latter must be optimized under the constraint that the hole is assigned to particular spinors, by employing so-called maximum overlap schemes (see [99] for example).

A second and perhaps more interesting feature of XPS is that it provides information on the oxidation state of a species; in the case of actinides, such information comes from the satellite structure of the two $4f$ peaks [100], which is greatly affected by the occupation of the species' $5f$ shells, due to the coupling of angular momenta for hole created in the $4f$ shell and those of any $5f$ electrons present [101–104]. In order to properly account for such couplings, a rather simple approach such as ΔSCF is in general not sufficient, and one must resort to multi-reference approaches, for which CI-based ones have had a prominent role so far [98]. In those approaches, in order to take the relaxation of the core hole into account, one usually performs calculations on systems containing N and N-1 electrons, something that introduces a certain degree of complexity in the calculation of the peak intensities, since the wavefunctions for neutral and ionized species will not be orthogonal and a cofactor expansion must be used.

There, it is interesting to note that approaches based on more approximate two-component relativistic Hamiltonians, in which spin-orbit coupling is introduced in a mean-field fashion [105], can reproduce quite well the results of more computationally expensive four-component calculations for the $4f$ spectra in U^{5+}, thus opening the perspective of treating systems of relatively larger size.

Due to the solid environments and the importance of long-range electrostatic effects on the states' energies, embedded cluster models [5] are chosen. While the point-charge embedding has been widely used (see Bagus *et al.* [98] and references therein) due to its simplicity, it is not without drawbacks (such as a tendency to spuriously stabilize delocalized states, in particular when large basis sets are used) and more sophisticated approaches (such as, but not restricted to, ab initio model potentials or frozen density embedding) are viable alternatives to it that may be particularly useful in cases where the environment is made up of molecular species (e.g., molecular ions or simply molecular crystals).

11.3.2 X-ray Absorption Spectroscopies

X-ray absorption spectroscopy (XAS) is proving to be a powerful tool for probing actinide speciation in solution [106, 107]. Technically the recorded XAS spectra are very often split in two regions, which are analyzed separately: i) the edge/pre-edge region, also referred to as X-ray absorption near-edge structure (XANES), which correspond to electronic excitations from core to bound states (e.g., low-lying or Rydberg states); and ii) the post-edge region, also referred to as extended X-ray absorption fine structure (EXAFS), corresponding to excitations from core to continuum states.

Unlike XPS, these methods are usually applied to species in solution, which requires that in their computational modeling effects due to species' concentration, temperature, etc., should be taken into account in order to approach the experimental conditions. To that end, the most important factor is arguably the geometrical structure of the species of interest, which can be obtained by using implicit solvent models (such as PCM) or by considering the solvent structure explicitly (e.g., by employing a statistically significant set of snapshots from molecular dynamics simulations), which can be advantageous if specific interactions such as hydrogen bonding are important. From those data, the spectra can be simulated in a subsequent step as described below.

11.3.2.1 XANES

The spectra near the absorbing edge is ideal for determining oxidation states and probing the electronic structure, since the position of the edge and its shape will vary according to changes in these, much like in XPS.

From a methodological perspective, the task of calculating **(pre-)edge spectra** resembles somewhat that of valence spectra, insofar as one is looking for bound states. Because of that there have been a number of efforts to enable the use of the conventional machinery (e.g., TDDFT [108, 109], EOM-CC [110], CIS(D) [111]) for calculating electronic spectra in this domain, notably employing the so-called restricted channel approximation, where excitations are only allowed from one or a few core orbitals of interest. This approximation is motivated by the need to reduce the number of excited states to be calculated, since conventional eigenvalue solvers are devised to obtain a given number of lowest excited states, a great number of which lie below the core excited levels.

To date most applications to actinides have been performed with restricted-channel TDDFT and focused on the *K*-edge lines of the species bound to the actinide, such as the carbons in actinocenes [112], oxygens in uranyl [113], chlorine in metallocenes [114], metal halides [115], or analogues or uranyl chlorides [116] (sulfur K-edge of ligands employed in actinide extraction have also received attention recently [117]). In these investigations one of the main interests was to obtain information on the participation of the 6*d* and 5*f* orbitals in the bonding, by inspecting the intensity and composition of excitations from the chlorine to the 6*d*, 5*f* as well as how these vary along a series of compounds. The only attempt to use another approach than TDDFT is the study of Armbruster *et al.* [118] who used the ADC(2) method to investigate metal-ion complexation effects on the C 1*s*-NEXAFS spectra of carboxylic acids describing the metals with a relativistic effective core potential.

A perhaps fortunate aspect of modeling the *K*-shell spectra is that, unlike the *L*-shells or higher, there is no splitting of core levels due to spin-orbit coupling, merely a shift to higher energies due to scalar relativistic effects. This means that it becomes possible to greatly simplify the calculations and perform non-relativistic or scalar relativistic calculations with pseudopotentials used for the actinide center(s), and correct the calculated origins with an *a posteriori* global shift, in particular since the trends identified from several studies indicate that the relative positions of states and intensities are well-represented by TDDFT (see Atkins *et al.*, for example [119] and references therein).

If, on the other hand, one is interested in investigating *L*-edges or higher, such simplifications would no longer be satisfactory, and spin-orbit coupling should be taken into account, even if only in a perturbative manner (see Roemelt *et al.*, for example [120] and references therein for a discussion of different approaches to transition metals). In addition to that, one should keep in mind that when calculating the intensities at edges at much higher energies than those for first- and second-row elements, for instance the uranium $L_{1,3}$ edges [79], it will become necessary to go beyond the dipole approximation and consider higher-order contributions to the transition moments in a consistent manner [121] in order to avoid a spurious origin dependence on the results.

A second important aspect is that, as was the case for XPS, the standard approaches do not take into account the relaxation of the core hole. A rather pragmatic way of introducing orbital relaxation due to the core hole, which is lacking in the approaches above, is found in the static exchange (STEX) method [122], a Hartree-Fock-based approach in which the orbitals for the system containing N-1 electrons (where an electron having been removed from a core orbital of interest) are used to determine core-excited states energies and wavefunctions via a CI singles (CIS) calculation, and subsequently the transition moments between these core-excited states and the ground-state (calculated with N electrons). Since the two types of wavefunctions will not be orthogonal, a cofactor expansion is employed as well.

Although a generalization to relativistic Hamiltonians is available [123] and has been applied to determine the *K*- and *L*-edges of chemisorbed on a gold surface [124, 125], to our knowledge it has not yet been used to obtain the spectra of actinide-containing species. These applications have nevertheless shown the importance of considering both channels ($p_{1/2}$ and $p_{3/2}$) for the *L*-edges of the light elements, even if this is done in a approximate fashion, e.g., by employing orbitals from average-of-configuration calculations over the $\{p_{1/2}, p_{3/2}\}^5$ spinors), for both peak separation and intensities [123].

An alternative to both standard approaches employing restricted-channel approximation and STEX is the use of the complex polarization propagator (CPP) approach [126, 127], where one directly calculates absorption cross sections for the molecule at different frequencies from the imaginary part of the molecular polarizability, expressed as a sum-over-states with an imaginary damping term modeling the finite lifetime of the excited state(s). In contrast to conventional formulations, where the electronic Hessian is constructed and diagonalized thus requiring the use of the restricted channel approximation, in the CPP approach all channels are implicitly taken into account and, if the underlying electronic structure approach used can effectively describe electron correlation (and relativistic effects), core hole (and relativistic) relaxation effects are also taken into account. A practical downside of this approach is the need to perform several calculations to cover a broad range of excitations. Even though, to the best of our knowledge, the CPP approach has not yet been explored for actinide species, it is available for both relativistic [128] and non-relativistic [126, 127, 129] Hamiltonians for mean-field theories (e.g., DFT) as well as for correlated wavefunctions in the non-relativistic domain [130, 131]. In the latter case, a recent benchmark study on the K-edge spectra of carbon [132] has shown that the CAM-B3LYP functional can yield results of comparable quality to CCSD for relatively modest basis.

A relative weakness of all of the approaches above in the case of actinides, however, is that they are applicable to closed-shell species or relatively simple open-shell cases in the presence of spin-orbit coupling interactions, something that would exclude most chemically interesting species other than uranium (VI) ones. Thus, the development of approaches capable of treating species with strong multi-reference character on their ground and excited states, for instance along the lines of recently described multi-reference coupled-cluster approaches [133], would be a necessary step towards their general applicability.

Beyond the pre-edge, one can also use electronic structure to investigate only the ground state and use this information to aid the interpretation of the experimental spectra [134], or by using it as input for XANES calculations performed with the scattering codes (FFEF [135, 136] and FDMNES [137]), which need as inputs the atomic positions and the electronic configuration (or information such as atomic charges) of the system. Using the latter approach, Fillaux *et al.* [113, 138, 139] have shown that employing charges derived from DFT calculations one improves the agreement between the computed spectra and experiment for the actinide M_5 edge, which is found to be rather sensitive to changes in the electronic structure, in contrast to the L_3-edge.

11.3.2.2 EXAFS

Further from the edge, the extended X-ray absorption fine structure (EXAFS) (see these references, for example [89, 106, 140], for reviews focusing on heavy elements) provides information about the number, identity, and distance of coordinating ligands about the absorbing ion. Experimentally this is generally done from the more energetic L or higher edges of the metal center, which are in general well-separated from signals due to other element's edges; unlike XANES, ligand K-edge EXAFS spectra are often near the strongly absorbing M, N XANES edges of the metal center, and therefore their experimental determination is rather difficult [141].

Because EXAFS spectra are usually recorded in solution they reflect an average structural picture of a system resulting from a thermodynamic averaging across fluctuating metal-ligand configurations. Moreover, these spectra correspond to excitations from core orbitals to the continuum, the inability of the finite basis sets employed in electronic structure calculations to properly describe the continuum wavefunctions leads to a spurious basis set dependences on the calculated spectra unless techniques such as Stieltjes imaging [142, 143] are employed. Such methods are currently available to a number of electronic structure approaches in a non-relativistic framework (e.g., [144–147], but not yet for relativistic ones).

Because of the shortcomings of standard electronic structure methods, the scattering approaches mentioned above are widely used to simulate EXAFS spectra, even if one should keep in mind that certain approximations [148] made in scattering codes might not be very good for a given system; for instance, the muffin-tin potentials used are known to be very poor for hydrogen atoms [149] and, therefore, it is common practice to remove these from the structures prior to the scattering calculations.

As was the case for XANES, here electronic structure calculations including solvent effects are used to draw conclusions on the most probable coordination isomer and from that structure the interatomic bond distances can be directly compared to the computed EXAFS spectra [150–152]. As EXAFS spectroscopy probes the structure of a statistically averaged system, the most appropriate way of comparing theoretical EXAFS data to experimental ones is to use molecular dynamics trajectories to sample the configuration space, select snapshots and finally compute a statistically average spectrum [89, 153] with a direct estimate of the mean square relative disorder (MSRD, also called σ^2 or the EXAFS Debye-Waller term). However, as discussed in Section 11.2.3, the relevance of MD simulations hinges on how accurate intermolecular interactions are, given that these are usually obtained at DFT level (in Car-Parinello MD simulations) or with force-fields (in classical MD simulations).

Significant changes on the computed EXAFS spectra and on the agreement with experimental data have been observed by Atta-Fynn *et al.* [153] when studying the curium(III) aqua ion, and by Spezia *et al.* on the thorium(IV) aqua ion [154]. This supports the key fact that the interaction potential plays a crucial role in the description of the solvated ion. Nevertheless, when comparing simulated EXAFS spectra to experimental data, one must keep in mind that simulations conditions may differ from experimental conditions, in terms of total concentrations and of relative concentrations of the studied cations and counterions. Counterions can indeed induce significant changes in the coordination of heavy element species in solutions as observed experimentally by Wilson *et al.* for instance [155], and theoretically by Spezia *et al.* [154] and Bühl *et al.* [156]. The challenge for theoretical modeling is to improve current models to capture all effects present at the nanoscale, to match experimental conditions and ultimately to make simulated spectra resemble the experimental ones.

11.4 Complex Formation and Ligand-Exchange Reactions

Complex formation reactions in aqueous solution are described by the equilibria of the type:

$$\mathrm{M(aq) + nL(aq) \rightleftharpoons ML(aq).} \tag{11.1}$$

where the charges have been omitted for simplicity. In order to use quantum chemical methods to compute the energy of this reaction 11.1, it is necessary to determine the composition of the first (and sometimes the second) coordination sphere, such as

$$\mathrm{M(H_2O)_x + L(H_2O)_y \rightleftharpoons ML(H_2O)_z + (x + y - z)H_2O.} \tag{11.2}$$

Note that QM calculations refer to zero ionic strength conditions, and in the first step only the reaction energy is obtained; the contribution to get the Gibbs free energy are calculated *a posteriori*. The level of details in the QM model can vary, but the first coordination sphere of the reacting metal ion and the ligand complex, must always be specified, but it is rarely possible to complete a second coordination sphere in such calculations. Thus the long-range solvation are most frequently described using a continuum solvent model. Since the size and the charge of the reactants and products may vary in the model described by equation (11.2), we may expect errors in the estimated solvation energy to have significant influence on the reaction energies. As an alternative, the complex formation reaction can be rewritten as a two-step process, the first one being the formation of an outer-sphere complex between the metal ion and the ligand, and then an exchange between the ligand and water in the first coordination sphere as follows:

$$M(H_2O)_x + L \rightleftharpoons [M(H_2O)_x], L \tag{11.3}$$

$$[M(H_2O)_x], L \rightleftharpoons [ML(H_2O)z], (H_2O)_{x-z} \tag{11.4}$$

One can estimate the equilibrium constant for reaction (11.4) using the Fuoss equation [157], while the Gibbs free energy of the second intramolecular reaction is estimated using QM. In the latter, the reactant and product are very similar as they have the same charge and comparable size, thus we might expect that errors in the solvent model cancel to a large extent. To verify whether this statement holds, and to illustrate the level of theory required for such calculations, we discuss the simplest ligand exchange reaction in hydrated uranyl(VI), that is the water-exchange process between first- and second-sphere water molecules. The reactant being the five-coordinated uranyl aqua ion, the reaction can proceed either via a dissociative pathway forming a dissociative intermediate (D-int) $[UO_2(H_2O)_4]^{2+}, (H_2O)_2$:

$$[UO_2(H_2O)_5]^{2+}, (H_2O) \rightleftharpoons [UO_2(H_2O)_4]^{2+}, (H_2O)_2 \tag{11.5}$$

or via an associative mechanism to form an associative intermediate (A-int) $[UO_2(H_2O)_6]^{2+}$

$$[UO_2(H_2O)_5]^{2+}, (H_2O) \rightleftharpoons [UO_2(H_2O)_6]^{2+} \tag{11.6}$$

To obtain reliable optimized structures, density functional theory using, for instance, the B3LYP functional can be used, as structural parameters are usually in good agreement with CCSD(T) values [158, 159]. However, for energetics, currently available exchange-correlation functionals are not accurate enough to describe metal-ligand binding energies, yielding either too low or too large values depending on the functional and the amount of Hartree-Fock exchange, as compared to MP2 or coupled-cluster results, the latter two being in excellent agreement with each other [160–162]. Our discussion will thus be based on energies computed at the MP2 level, using a relativistic effective core potential for uranium with quadruple-zeta quality basis set and triple-zeta quality sets for oxygen and hydrogen atoms. The objective here is to illustrate the sensitivity of the continuum solvation model to the choice of the cavity shape; the cavity is built from interlocking spheres that are either centered around each atom (individual sphere, IS) or around pseudoatoms referring to an atom and its covalently attached hydrogens (called a united atom model, UA). Both approaches were tested on reactions (11.5) and (11.6) and on reactions (11.7) and (11.8) corresponding to a chemical model with a total number of ten water molecules, five of which partially saturating the second coordination sphere,

$$[UO_2(H_2O)_5]^{2+}, (H_2O)_5 \rightleftharpoons [UO_2(H_2O)_4]^{2+}, (H_2O)_6 \tag{11.7}$$

or via an associative mechanism

$$[UO_2(H_2O)_5]^{2+}, (H_2O)_5 \rightleftharpoons [UO_2(H_2O)_6]^{2+}, (H_2O)_4. \qquad (11.8)$$

The UA and IS models give rather scattered results for differential solvation contributions. This is true not only for the smallest chemical model but also for the model with 10 water molecules. This indicates that the shape of the cavity around the water molecules has a strong impact on the absolute solvation energies and that these effects do not cancel out when taking solvation energy differences. Klamt *et al.* [83] and Cramer *et al.* [165–167] have recently concluded that IS cavities are superior to UA cavities for solvation free energy calculations of neutral and ionic species. However, this conclusion can be questioned from the scattered results obtained by Wåhlin *et al.* [163, 164], see Table 11.1. Looking at the computed ΔH, all models tend to place the A-intermediate lower in energy that the D-intermediate, except for the case of the IS results for the six water model where the two energies are just 3 $\mathrm{kJ \cdot mol^{-1}}$ apart from each other. This demonstrates that the large uncertainties arising from cavities in solvation models may hamper any definitive conclusions with respect to a reaction mechanism, unless the final energies are well separated. This was the case of more complex ligand-exchange reactions, with fluoride, oxalate, carbonates, where we could satisfactorily draw conclusions on the preferred mechanism [168].

One particular and rather complicated exchange reaction, which has been studied theoretically [76, 77, 169] and experimentally [170] is the oxygen exchange between the water solvent and the uranyl ion.

$$U^{17/18}O_2^{2+}(aq) + H_2^{16}O \rightleftharpoons U^{16}O_2^{2+}(aq) + H_2^{17/18}O \qquad (11.9)$$

This reaction is normally very slow in the ground state. Several mechanisms have been explored using quantum chemical methods. At low pH, the reaction pathway involves a binuclear $(UO_2)_2(\mu - OH)_2^{2+}$ complex [76], as proposed by Szabó and Grenthe [170]. At

Table 11.1 *Differential Gibbs free solvation energies $\Delta\Delta G_{solv}$, and reaction enthalpy ΔH in $\mathrm{kJ \cdot mol^{-1}}$ computed at the MP2 level with the UA and IS models at the MP2 level for the water exchange reaction for uranyl(VI) in the electronic ground state, using the six- and ten-water models (See [163, 164])*

Cavity	$[UO_2(H_2O)_5]^{2+}, (H_2O)$ A-int	D-int	$[UO_2(H_2O)_5]^{2+}, (H_2O)_5$ A-int	D-int
		$\Delta\Delta G_{solv}$		
UA	−19	27	−24	−8
IS	−3	2	−5	23
		ΔH		
gas-phase	37	35	36	39
UA	17	63	12	31
IS	34	37	31	62

higher pH, Shamov *et al.* [169] assumed the transfer to be mediated by chains of water molecules, using DFT-based methods, followed by Bühl [171], using Car-Parinello simulations. Interestingly the yl-exchange reaction (11.9) is strongly enhanced by UV radiation. Réal *et al.* [77] followed the reaction path for a number of excited states of the uranyl aqua ion using a TDDFT approach. Although the TDDFT approach is not very accurate in terms of excitation energy values, as discussed in Section 11.2, but it provided a good picture of the excited states involved in the process, and this reaction can now be considered to be understood.

11.5 Calculations of Standard Reduction Potential and Studies of Redox Chemical Processes

Upon reduction a metal complex gains one electron to convert into its reduced form.

$$[\mathrm{ML_n}]^{\mathrm{m+}}(\mathrm{aq}) + \mathrm{e}^- \rightleftharpoons [\mathrm{ML_n}]^{(\mathrm{m-1})+}(\mathrm{aq}) \qquad (11.10)$$

The total change in the Gibbs free energy of reaction (11.10) in solution, $\Delta G_{aq}^{\circ,redox}$, is related to the standard potential E° according to equation (11.11)

$$E^\circ = \frac{-\Delta G_{aq}^{\circ,redox}}{nF} \qquad (11.11)$$

The standard potential with respect to the standard hydrogen electrode ($E^\circ(SHE)$) is

$$E^\circ(SHE) = \frac{-(\Delta G_{aq}^{\circ,redox} - \Delta G^\circ(SHE))}{nF} \qquad (11.12)$$

where n is the number of electron transferred (n=1 in this case) and F is the Faraday constant (96.485 $\mathrm{kJ \cdot mol^{-1} V^{-1}}$). To calculate $\Delta G_{aq}^{\circ,redox}$, we have to use the following thermodynamic cycle (Born-Haber) shown in Figure 11.1,

which results in

$$\Delta G_{aq}^{\circ,redox} = \Delta G_{gas}^{\circ,redox} + \Delta G_{solv}^{\circ}(Red) - \Delta G_{solv}^{\circ}(Ox) \qquad (11.13)$$

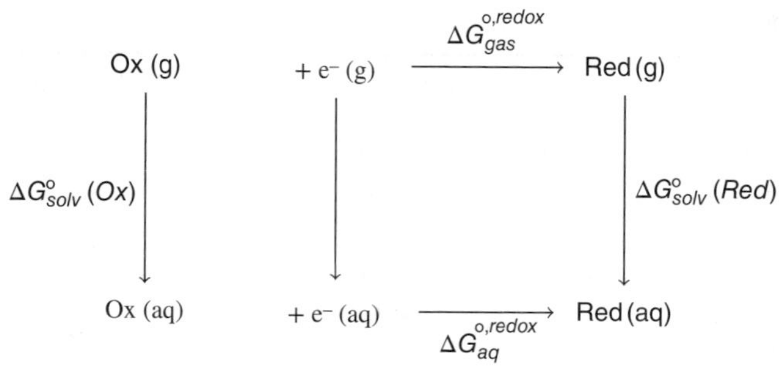

Figure 11.1 *Born-Haber cycle*

where $\Delta G_{gas}^{\circ,redox}$ is the change of standard Gibbs free energy of reaction (11.10) in the gas-phase and $\Delta G_{solv}^{\circ}(Red)$ and $\Delta G_{solv}^{\circ}(Ox)$ are the standard solvation energies of the reduced and oxidized forms. $\Delta G_{gas}^{\circ,redox}$ can be calculated using the adiabatic ionization energy of the reduced form IE, adding the thermal contributions G_{evr} for the oxidized, reduced forms, and the free energy of the electron ($G(e^-) = -3.63 \text{ kJ} \cdot \text{mol}^{-1}$) at 298 K, adding the

$$\Delta G_{gas}^{\circ,redox} = -IE + G_{evr}(red) - G_{evr}(ox) - G(e^-) \tag{11.14}$$

Note that $\Delta G_{aq}^{\circ,redox}$ has a liquid-phase standard state of 1 mol/L, and we can use a gas-phase standard state of either 1 atm or 1 mol/L, as long as we use the same convention for the oxidized and reduced forms. Often the 1 mol/L standard state is used. There are several sources of uncertainties in the calculations of reduction potentials, and we will comment on them by scanning the published literature on actinide elements, for which the most studied redox systems are actinyl aqua ions, with the exception of one study on Pu(VII)/Pu(VIII) [172].

The first comment is that redox potentials are defined with respect to the standard hydrogen electrode corresponding to the following half-equation

$$H_3O^+ + e^- \rightleftharpoons \frac{1}{2}H_2 + H_2O \tag{11.15}$$

In several studies [173–175], the authors computed this reduction potential at the same level of theory as that used for the actinide complexes, with values ranging from −4.92 eV (B3LYP) [173, 174] to −5.20 V (PBE) [174] (the value computed at the CASPT2 level in Tsushima *et al.* [175] is not explicitly given in the paper). Other studies prefer to use the IUPAC recommended value −4.44 V, although Truhlar *et al.* [176] recently revised this value down to −4.28 eV, using a different value of the hydration energy of a proton. The choice of this value will impact the absolute value of the computed reduction potential, however, this error will cancel out when comparing trends.

The second critical ingredient is the estimation of solvation free energies, which, in all studies, have been computed with continuum solvent models. Like in the case of chemical reactions (see Section 11.4), it has been demonstrated for non-actinide systems (such as ruthenium [167] and ferrocene [177]) that the shape of the cavities enclosing the solute seem noticeably influences not only the absolute free energy of solvation but also, and this is more critical, the difference in free energy of solvation between the reduced and oxidized forms (as shown in the paper by Jaque *et al.* [167] on ruthenium complexes) of the order of 100 $\text{kJ} \cdot \text{mol}^{-1}$, which might yield to uncertainties of about 1 eV in the final computed reduction potential. Thus caution must be taken when using continuum solvent models and when comparing results. Among the published data, all use individual spheres, except for the study by Tsushima *et al.* [172, 175], which most likely used the default united atom model set as default in Gaussian 03.

The third critical point is the choice of the appropriate electronic structure method to treat the species in gas phase. The change of electronic configuration from the oxidized to the reduced species implies that electronic correlation effects are different and that spin-orbit coupling is likely to contribute differently for the two species. Most studies rely on DFT using either the PBE functional, a hybrid functional, or more recently a meta-functional of the Minnesota M06 family. Changing the functional can yield changes of the gas-phase ionization potential up to an eV. Unfortunately, we cannot compare these DFT values to the most reliable

Table 11.2 *Reduction potentials in eV for the* $AnO_2(H_2O)_5^{2+/+}$ *(An = U, Np, and Pu) computed with various correlation methods, continuum solvent models, and spin-orbit correction. MUSE (mean unsigned error) values are calculated with respect to standard potential (V)*

	B3LYP	PBE	M06	M06L	BP86	PBE	CASPT2*	std potential
U	−0.024	0.052	0.136	0.136	−0.54	−0.51	0.00	0.088
Np	1.177	1.056	0.989	1.227	1.33	0.87	1.53	1.159
Pu	1.232	1.379	1.374	0.946	0.49	0.43	0.73	0.936
MUSE	0.13	0.19	0.22	0.04	0.42	0.46	0.22	
Np vs U	1.201	1.004	0.853	1.091	1.87	1.38	1.53	1.071
Pu vs U	1.256	1.327	1.238	0.81	1.03	0.94	0.73	0.848

*SO coupling computed with the SO-RASSI-AMFI approach, while other calculations use Hay SO-multiplet correction; Continuum calculations with united atom model.

calculations performed so far using the CASPT2 multi-reference method, since the gas-phase energies are not reported in Tsushima *et al.* study [175]. Concerning spin-orbit, in the latter study it was computed with the SO-RASSI method yielding to spin-orbit lowering values very close to that computed by Hay *et al.* [173] using a minimal spin-orbit CI using an effective spin-orbit operator with effective spin-orbit coupling constants. There is one exception, the case of plutonyl(V) where Tsushima's value (-51.1 kJ $\cdot$ mol^{-1}) is significantly lower (-89.7 kJ $\cdot$ mol^{-1}), the difference being due to a change of the dominating configuration in the spin-orbit ground state, from that of the spin-free one. Horowitz and Martson [178] proposed an alternative treatment of strong correlations and multi-configurational effects in building an Anderson impurity model over the frontier orbitals and defining the parameters from matrix elements of the KS Hamiltonian, and treating spin-orbit coupling in an effective one-electron operator. The outcome of this model yields somewhat scattered agreement with experiment, but it provides some insights into the physics of the 5*f* electrons in actinide complexes.

Table 11.2 reports all computed values gathered by Steele *et al.* [179]. In absolute terms and even relative values with respect to the mean unsigned error (MUSE), the M06-L functional agrees best with experiment, although it is not clear how well it captures multi-configurational effects. The CASPT2 results are in fair agreement with experiment, but there the uncertainty lies in the treatment of solvent effects. We can note that in an earlier study by Vallet *et al.* [180] on the step-wise reduction of uranyl(VI) to uranium(IV), no estimate of absolute solvation energies was made, but relative redox potentials were predicted with a fair accuracy using the U system for calibration.

11.6 General Conclusions

Our review of published studies on computational investigations of actinides complexes clearly indicates that the most studied systems are uranium, neptunium, and plutonium. Yet even for these there are still many unanswered questions and debates persist on topics such as the character of the electronic ground state, magnetism, and the character of bonding.

From the experimental point of view, other species such as americium and curium are of importance for closing the fuel cycle by adapting some Partitioning and Transmutation scheme, and one driving force may arise from subtle differences in the bonding of actinides and lanthanides to the separating ligands. The factors that determine the separation efficiencies are currently poorly understood, partly because of experimental difficulties in obtaining these species (They are produced in rather small amounts.) and later in their manipulation (which requires highly specialized facilities and trained experts for handling due to their short half-life and high activity) and partly because, as they are taken to behave largely as heavy lanthanides, there is not much motivation to overcome the aforementioned experimental difficulties to improve upon our rudimentary understanding of them and, as a consequence, there hasn't been much interest in their theoretical study, either.

Thus, the actinide series as a whole is modestly understood, with the level of comprehension decreasing with atomic number, calling for further joint experimental-theoretical studies to unravel the properties of actinides across the series.

While current computational modeling can offer a great deal of insight, a description of thermodynamics, spectroscopy, and properties with a sufficiently high accuracy requires the use of sophisticated correlated methods as well as more realistic models of the condensed phase, and there are still rather important shortcomings that need to be addressed, particularly for the former. Current developments of new theoretical and computational approaches are promising in this regard, especially for ongoing implementation of a wide range of wavefunction-based methods successful in the one-component domain to the two- and four-component scheme, particularly combined with state-of-the-art resolution-of-the-identity techniques and large-scale parallelization.

Indeed some systems that have an open-shell or multi-configurational ground-state wavefunction in a scalar relativistic framework might turn out to be closed-shell in a relativistic framework (e.g., the molecular astatine monoxide cation AtO^+ [181]) thus making it easy to apply single-reference two- and four-component methods. However, significant efforts on multi-configurational methods applicable to real-size complexes have to be undertaken to tackle more complex electronic configurations. Continued reductions in the cost/performance of computing and improvements in algorithmic details should continue to yield shorter time to solution for increasingly larger systems.

Similarly, improvement can be expected in treatment of the condensed phase environment. Optimization of the accuracy of continuum solvation methods is far from a solved problem. Furthermore, new embedding methods, and QM/MM approaches can be made more accurate, robust, cost-effective, and easy to use. Lastly, sampling algorithms in the condensed phase are crucial for many large-scale applications, and significant advances can be expected as more complex problems are currently being addressed.

Altogether advances in theory and software, and their corresponding applications to systems of increasing size, taken in parallel to advances in experimental techniques will enable explanatory and predictive models to be constructed for the fascinating elements standing at the foot of the periodic table.

References

[1] Saue T. Relativistic Hamiltonians for Chemistry: A Primer. ChemPhysChem. 2011;12:3077–3094.

[2] Fleig T. Invited review: Relativistic wave-function based electron correlation methods. In: Recent Advances and Applications of Relativistic Quantum Chemistry. vol. 395; 2012. p. 2–15.

[3] Capelle K. A bird's-eye view of density-functional theory. Braz J Phys. 2006;36: 1318–1343.

[4] Helgaker T, Coriani S, Jørgensen P, Kristensen K, Olsen J, Ruud K. Recent Advances in Wave Function-Based Methods of Molecular-Property Calculations. Chem Rev. 2012;112:543–631.

[5] Gomes ASP, Jacob CR. Quantum-chemical embedding methods for treating local electronic excitations in complex chemical systems. Annu Rep Prog Chem, Sect C: Phys Chem. 2012;108:222–277.

[6] Lyakh DI, Musiał M, Lotrich VF, Bartlett RJ. Multireference Nature of Chemistry: The Coupled-Cluster View. Chem Rev. 2011;112:182–243.

[7] Adamo C, Jacquemin D. The calculations of excited-state properties with Time-Dependent Density Functional Theory. Chem Soc Rev. 2013;42:845–856.

[8] González L, Escudero D, Serrano-Andrés L. Progress and Challenges in the Calculation of Electronic Excited States. ChemPhysChem. 2012;13:28–51.

[9] Szalay PG, Müller T, Gidofalvi G, Lischka H, Shepard R. Multiconfiguration Self-Consistent Field and Multireference Configuration Interaction Methods and Applications. Chem Rev. 2011;112:108–181.

[10] Casida ME. Time-Dependent Density Functional Response Theory for Molecules. In: Chong DP, editor. Recent Advances in Density Functional Methods, Part 1. Singapore: World Scientific; 1995. p. 155–192.

[11] Ullrich CA. Time-Dependent Density-Functional Theory. Concepts and Applications. New York: Oxford University Press; 2012.

[12] Norman P. A perspective on nonresonant and resonant electronic response theory for time-dependent molecular properties. Phys Chem Chem Phys. 2011;13:20519–20535.

[13] Maitra NT, Zhang F, Cave RJ, Burke K. Double excitations within time-dependent density functional theory linear response. J Chem Phys. 2004;120:5932.

[14] Casida ME. Propagator corrections to adiabatic time-dependent density-functional theory linear response theory. J Chem Phys. 2005;122:054111.

[15] Gritsenko OV, Baerends EJ. Double excitation effect in non-adiabatic time-dependent density functional theory with an analytic construction of the exchange–correlation kernel in the common energy denominator approximation. Phys Chem Chem Phys. 2009;11:4640–4646.

[16] Kümmel S, Kronik L. Orbital-dependent density functionals: Theory and applications. Rev Mod Phys. 2008;80:3–60.

[17] Tecmer P, Gomes ASP, Ekström U, Visscher L. Electronic Spectroscopy of UO_2^{2+}, NUN and NUO^+. Phys Chem Chem Phys. 2011;13:6249–6259.

[18] Sous J, Goel P, Nooijen M. Similarity transformed equation of motion coupled cluster theory revisited: a benchmark study of valence excited states. Mol Phys. 2013; 112:616–638.

[19] Trofimov AB, Krivdina IL, Weller J, Schimer J. Algebraic-diagrammatic construction propagator approach to molecular response properties. Chem Phys. 2006;329:1–10.

[20] Head-Gordon M, Rico RJ, Oumi M, Lee TJ. A doubles correction to electronic excited states from configuration interaction in the space of single substitutions. Chem Phys Lett. 1994;219:21–29.

[21] Tecmer P, Govind N, Kowalski K, de Jong WA, Visscher L. Reliable modeling of the electronic spectra of realistic uranium complexes. J Chem Phys. 2013;139:034301.

[22] Walker TEH, Richards WG. Calculation of Spin-Orbit Coupling Constants in Diatomic Molecules from Hartree-Fock Wave Functions. Phys Rev. 1969;177:100–101.

[23] Hess BA, Marian CM, Peyerimhoff SD. Ab initio calculation of spin-orbit effects in molecules including electron correlation. In: Advanced Series in Physical Chemistry – Vol. 2, Modern Structure Theory Part I. Singapore: World Scientific; 1995. p. 152–278.

[24] Hess BA, Marian CM. Relativistic effects in the calculation of electronic energies. In: Jensen P, Bunker PR, editors. Computational Molecular Spectroscopy. Sussex: Wiley; 2000. p. 169.

[25] Wahlgren U, Sjøvoll M, Fagerli H, Gropen O, Schimmelpfennig B. *Ab initio* calculations of the $^2P_{1/2}$-$^2P_{3/2}$ splitting in the thallium atom. Theor Chem Acc. 1997;97:324–330.

[26] Rakowitz F, Marian CM. An extrapolation scheme for spin-orbit configuration interaction energies applied to the ground state and excited electronic states of thallium hydride. Chem Phys. 1997;225:223–238.

[27] Teichteil C, Pélissier M, Spiegelmann F. Ab initio molecular calculations including spin-orbit coupling. I. Method and atomic tests. Chem Phys. 1983;81:273–282.
[28] Dyall KG, Fægri K. Introduction to relativistic quantum chemistry. New York: Oxford University Press; 2007.
[29] Reiher M, Wolf A. Relativistic Quantum Chemistry: The Fundamental Theory of Molecular Science. Wiley-VCH Verlag; 2009.
[30] Sikkema J, Visscher L, Saue T, Iliaš M. The molecular mean-field approach for correlated relativistic calculations. J Chem Phys. 2009;131:124116.
[31] Liu W, Peng D. Exact two-component Hamiltonians revisited. J Chem Phys. 2009;131:031104.
[32] Nakajima T, Hirao K. The Douglas–Kroll–Hess Approach. Chem Rev. 2011;112:385–402.
[33] Belpassi L, Storchi L, Quiney HM, Tarantelli F. Recent advances and perspectives in four-component Dirac-Kohn-Sham calculations. Phys Chem Chem Phys. 2011;13:12368–12394.
[34] Peng D, Reiher M. Exact decoupling of the relativistic Fock operator. Theor Chem Acc. 2012;131:1081.
[35] Autschbach J. Perspective: Relativistic effects. J Chem Phys. 2013;136:150902.
[36] Vallet V, Maron L, Teichteil C, Flament JP. A two-step uncontracted determinantal effective Hamiltonian-based SO-CI method. J Chem Phys. 2000;113:1391–1402.
[37] Fromager E, Vallet V, Schimmelpfennig B, Macak P, Privalov T, Wahlgren U. Spin-orbit effects in electron transfer in neptunyl(VI)-neptunyl(V) complexes in solution. J Phys Chem A. 2005;109:4957–4960.
[38] Heß BA, Marian CM, Wahlgren U, Gropen O. A mean-field spin-orbit method applicable to correlated wavefunctions. Chem Phys Lett. 1996;251:365–371.
[39] Marian CM, Wahlgren U. A new mean-field and ECP-based spin-orbit method. Applications to Pt and PtH. Chem Phys Lett. 1996;251:357–364.
[40] Schimmelpfennig B. AMFI, an Atomic Mean-Field Integral program; 1996.
[41] Aquilante F, De Vico L, Ferré N, Ghigo G, Malmqvist PÅ, Neogrády P, et al. MOLCAS 7: The Next Generation. J Comput Chem. 2010;31:224–247.
[42] Dalton, a molecular electronic structure program, Release DALTON2013.X (2013); Available from: http://daltonprogram.org.
[43] Aidas K, Angeli C, Bak KL, Bakken V, Bast R, Boman L, et al. "The Dalton quantum chemistry program system." WIREs Comput Mol Sci. 2013.
[44] DIRAC, a relativistic ab initio electronic structure program, Release DIRAC13 written by L. Visscher, H. J. Aa. Jensen, R. Bast, and T. Saue, with contributions from V. Bakken, K. G. Dyall, S. Dubillard, U. Ekström, E. Eliav, T. Enevoldsen, E. Faßhauer, T. Fleig, O. Fossgaard, A. S. P. Gomes, T. Helgaker, J. K. Lærdahl, Y. S. Lee, J. Henriksson, M. Iliaš, Ch. R. Jacob, S. Knecht, S. Komorovský, O. Kullie, C. V. Larsen, H. S. Nataraj, P. Norman, G. Olejniczak, J. Olsen, Y. C. Park, J. K. Pedersen, M. Pernpointner, K. Ruud, P. Sałek, B. Schimmelpfennig, J. Sikkema, A. J. Thorvaldsen, J. Thyssen, J. van Stralen, S. Villaume, O. Visser, T. Winther, and S. Yamamoto. Available from: http://www.diracprogram.org. 2013.
[45] ReSpect, version 3.3.0 (beta), 2013; Relativistic Spectroscopy DFT program of authors Repisky M.; Komorovsky S.; Malkin V. G.; Malkina O. L.; Kaupp M.; Ruud K., with contributions from Bast R.; Ekstrom U.; Knecht S.; Malkin Ondik I.; Malkin E. Available from: http://relqchem.sav.sk.
[46] Yabushita S, Zhang Z, Pitzer RM. Spin-orbit configuration interaction using the graphical unitary group approach and relativistic core potential and spin-orbit operators. J Phys Chem A. 1999;103:5791–5800.
[47] Dolg M, Cao X. Relativistic Pseudopotentials: Their Development and Scope of Applications. Chem Rev. 2012;112:403–480.
[48] Seijo L, Barandiarán Z. The ab initio model potential method: a common strategy for effective core potential and embedded cluster calculations. In: Leszczynski J, editor. Computational Chemistry: Reviews of Current Trends. vol. 4. Singapore: World Scientific; 1999. p. 55–152.
[49] Geipel G. Some aspects of actinide speciation by laser-induced spectroscopy. Coord Chem Rev. 2006;250:844–854.

[50] Billard I, Geipel G. Luminescence Analysis of Actinides: Instrumentation, Applications, Quantification, Future Trends, and Quality Assurance. Springer Ser Fluoresc. 2008;5:465–492.
[51] Panak PJ, Geist A. Complexation and Extraction of Trivalent Actinides and Lanthanides by Triazinylpyridine N-Donor Ligands. Chem Rev. 2013;113:1199–1236.
[52] Larsson S, Pyykkö P. Relativistically parameterized extended Hueckel calculations. IX. An iterative version with applications to some xenon, thorium and uranium compounds. Chem Phys. 1986;101:355–369.
[53] Réal F, Vallet V, Marian C, Wahlgren U. Theoretical investigation of the energies and geometries of photo-excited uranyl(VI) ion: a comparison between wave-function theory and density functional theory. J Chem Phys. 2007;127:214302.
[54] Réal F, Gomes ASP, Visscher L, Vallet V, Eliav E. Benchmarking Electronic Structure Calculations on the Bare UO_2^{2+} ion: How Different are Single and Multireference Electron Correlation Methods? J Phys Chem A. 2009;113:12504–12511.
[55] Bast R, Jensen HJA, Saue T. Relativistic adiabatic time-dependent density functional theory using hybrid functionals and noncollinear spin magnetization. Int J Quantum Chem. 2009;109:2091–2112.
[56] Pierloot K, van Besien E. Electronic structure and spectrum of UO_2^{2+} and $UO_2Cl_4^{2-}$. J Chem Phys. 2005;123:204309.
[57] Pierloot K, van Besien E, van Lenthe E, Baerends EJ. Electronic structure and spectrum of UO_2^{2+} and $UO_2Cl_4^{2-}$ calculated with time-dependent density functional theory. J Chem Phys. 2007;126:194311.
[58] Tecmer P, Bast R, Ruud K, Visscher L. Charge-Transfer Excitations in Uranyl Tetrachloride ($[UO_2Cl_4]^{2-}$): How Reliable are Electronic Spectra from Relativistic Time-Dependent Density Functional Theory? J Phys Chem A. 2012;116:7397–7404.
[59] Gomes ASP, Jacob CR, Réal F, Visscher L, Vallet V. Towards systematically improvable models for actinides in condensed phase: the electronic spectrum of uranyl in $Cs_2UO_2Cl_4$ as a test case. Phys Chem Chem Phys. 2013;15:15153–15162.
[60] Kovács A, Konings RJM, Raab J, Gagliardi L. A theoretical study of AmO_n and CmO_n (n = 1, 2). Phys Chem Chem Phys. 2008;10:1114–1117.
[61] Kovács A, Konings RJM. Molecular structure and thermodynamic properties of the gaseous ThC_2 and ThC_4 species. J Nucl Mat. 2008;372:391–393.
[62] Kovács A, Konings RJM. Computed Vibrational Frequencies of Actinide Oxides $AnO^{0/+/2+}$ and $AnO_2^{0/+/2+}$ (An = Th, Pa, U, Np, Pu, Am, Cm). J Phys Chem A. 2011;115:6646–6656.
[63] Pogány P, Kovács A, Varga Z, Bickelhaupt FM, Konings RJM. Theoretical Study of the Structure and Bonding in ThC2 and UC2. J Phys Chem A. 2012;116:747–755.
[64] Pogány P, Kovács A, Szieberth D, Konings RJM. Theoretical study of the Pu and Am dicarbide molecules. Struct Chem. 2012;23:1281–1289.
[65] Kovács A, Pogány P, Konings RJM. Theoretical Study of Bond Distances and Dissociation Energies of Actinide Oxides AnO and AnO_2. Inorg Chem. 2012;51:4841–4849.
[66] Notter FP, Dubillard S, Bolvin H. A theoretical study of the excited states of AmO_2^{n+}, n = 1,2,3. J Chem Phys. 2008;128:164315.
[67] Infante I, Eliav E, Vilkas MJ, Ishikawa Y, Kaldor U, Visscher L. Fock space coupled cluster study on the electronic structure of the UO_2, UO_2^+, U^{4+}, and U^{5+} species. J Chem Phys. 2007;127:124308.
[68] Danilo C, Vallet V, Flament JP, Wahlgren U. Spin-orbit configuration interaction study of the electronic structure of the $5f^2$ manifold of U^{4+} and $5f$ manifold of U^{5+}. J Chem Phys. 2008;128:154310.
[69] Danilo C, Vallet V, Flament JP, Wahlgren U. Effects of the first hydration sphere and the bulk solvent on the spectra of the f^2 isoelectronic actinide compounds: U^{4+}, NpO_2^+, and PuO_2^{2+}. Phys Chem Chem Phys. 2010;12:1116–1130.
[70] Weigand A, Cao X, Vallet V, Flament JP, Dolg M. Multi-configuration Diract-Hartree-Fock Adjusted Energy-consistent Pseudopotential for Uranium: Spin-orbit Configuration Interaction and Fock-Space Coupled-Cluster Study of U^{4+} and U^{5+}. J Phys Chem A. 2009;113:11509–11516.

[71] van Besien E, Pierloot K, Görller-Walrand. Electronic spectra of uranyl chloride complexes in acetone: a CASSCF/CASPT2 investigation. Phys Chem Chem Phys. 2006;8:4311–4319.
[72] Wiebke J, Moritz A, Glorius M, Moll H, Bernhard G, Dolg M. Complexation of Uranium(VI) with Aromatic Acids in Aqueous Solution: A Combined Computational and Experimental Study. Inorg Chem. 2008;47:3150–3157.
[73] Wiebke J, Weigand A, Weissmann D, Glorius M, Moll H, Bernhard G, et al. Combined Computational and Experimental Study of Uranyl(VI) 1:2 Complexation by Aromatic Acids. Inorg Chem. 2010;49:6428–6435.
[74] Su J, Zhang K, Schwarz WHE, Li J. Uranyl-Glycine-Water Complexes in Solution: Comprehensive Computational Modeling of Coordination Geometries, Stabilization Energies, and Luminescence Properties. Inorg Chem. 2011;50:2082–2093.
[75] Tsushima S. Photochemical Reduction of UO_2^{2+} in the Presence of Alcohol Studied by Density Functional Theory Calculations. Inorg Chem. 2009;48:4856–4862.
[76] Tsushima S. "yl"-Oxygen Exchange in Uranyl(VI) Ion: A Mechanism Involving $(UO_2)_2\ (\mu-OH)_2^{2+}$ via $U--O_{yl}--U$ Bridge Formation. Inorg Chem. 2012;51: 1434–1439.
[77] Réal F, Vallet V, Wahlgren U, Grenthe I. *Ab initio* study of the mechanism for photoinduced yl-oxygen exchange in uranyl(VI) in acidic aqueous solution. J Am Chem Soc. 2008;130: 11742–11751.
[78] Denning RG. Electronic structure and bonding in actinyl ions. Struct Bond. 1992;79: 215–276.
[79] Denning RG. Electronic structure and bonding in actinyl ions and their analogs. J Phys Chem A. 2007;111:4125–4143.
[80] Matsika S, Pitzer RM. Actinyl ions in $Cs_2UO_2Cl_4$. J Phys Chem A. 2001;105:637–645.
[81] Gomes ASP, Jacob CR, Visscher L. Calculation of local excitations in large systems by embedding wave-function theory in density-functional theory. Phys Chem Chem Phys. 2008;10:5353–5362.
[82] Höfener S, Gomes ASP, Visscher L. Molecular properties via a subsystem density functional theory formulation: A common framework for electronic embedding. J Chem Phys. 2012;136:044104.
[83] Klamt A, Mennucci B, Tomasi J, Barone V, Curutchet C, Orozco M, et al. On the Performance of Continuum Solvation Methods. A Comment on "Universal Approaches to Solvation Modeling." Acc Chem Res. 2009;42:489–492.
[84] Bühl M, Sieffert N, Chaumont A, Wipff G. Water versus Acetonitrile Coordination to Uranyl. Effect of Chloride Ligands. Inorg Chem. 2012;51:1943–1952.
[85] Bühl M, Sieffert N, Wipff G. Density functional study of aqueous uranyl(VI) fluoride complexes. Chem Phys Lett. 2008;467:287–293.
[86] Clavaguéra C, Calvo F, Dognon JP. Theoretical study of the hydrated Gd^{3+} ion: Structure, dynamics, and charge transfer. J Chem Phys. 2006;124:074505.
[87] Hagberg D, Bednarz E, Edelstein NM, Gagliardi L. A Quantum Chemical and Molecular Dynamics Study of the Coordination of Cm(III) in Water. J Am Chem Soc. 2007;129: 14136–14137.
[88] Marjolin A, Gourlaouen C, Clavaguéra C, Ren P, Wu J, Gresh N, et al. Toward accurate solvation dynamics of lanthanides and actinides in water using polarizable force fields: from gas-phase energetics to hydration free energies. Theor Chem Acc. 2012;131:1198.
[89] D'Angelo P, Spezia R. Hydration of Lanthanoids(III) and Actinoids(III): An Experimental/Theoretical Saga. Chem Eur J. 2012;18:11162–11178.
[90] Réal F, Trumm M, Schimmelpfennig B, Masella M, Vallet V. Further insights in the ability of classical non-additive potentials to model actinide ions in aqueous solution. J Comput Chem. 2013;34:707–719.
[91] Höfener S, Gomes ASP, Visscher L. Solvatochromic shifts from coupled-cluster theory embedded in density functional theory. J Chem Phys. 2013;139:104106.
[92] Chan GKL, Sharma S. The Density Matrix Renormalization Group in Quantum Chemistry. Ann Rev Phys Chem. 2011;62:465–481.

[93] Tecmer P, Boguslawski K, Legeza O, Reiher M. Unravelling the quantum-entanglement effect of noble gas coordination on the spin ground state of CUO. Phys Chem Chem Phys. 2014;16:719–727.
[94] Chwee TS, Carter EA. Valence Excited States in Large Molecules via Local Multireference Singles and Doubles Configuration Interaction. J Chem Theory Comput. 2010;7:103–111.
[95] Bories B, Maynau D, Bonnet ML. Selected excitation for CAS-SDCI calculations. J Comput Chem. 2007;28:632–643.
[96] Ben Amor N, Bessac F, Hoyau S, Maynau D. Direct selected multireference configuration interaction calculations for large systems using localized orbitals. J Chem Phys. 2011;135:014101.
[97] Baer Y, Schoenes J. Electronic structure and Coulomb correlation energy in UO_2 single crystal. Solid State Commun. 1980;33:885–888.
[98] Bagus PS, Ilton ES, Nelin CJ. The interpretation of XPS spectra: Insights into materials properties. Surf Sci Rep. 2013;68:273–304.
[99] Besley NA, Gilbert ATB, Gill PMW. Self-consistent-field calculations of core excited states. J Chem Phys. 2009;130:124308.
[100] Ilton ES, Bagus PS. XPS determination of uranium oxidation states. Surf Interface Anal. 2011;43:1549–1560.
[101] Bagus PS, Ilton ES. Atomic many-body effects in the 4*f* XPS of the U^{5+} and U^{4+} cations: part II: consequences of orbital relaxation. Theor Chem Acc. 2007;118:495–502.
[102] Bagus PS, Ilton ES, Martin RL, Jensen HJA, Knecht S. Spin-orbit coupling in actinide cations. Chem Phys Lett. 2012;546:58–62.
[103] Bagus PS, Nelin CJ, Ilton ES. Theoretical modeling of the uranium 4*f* XPS for U(VI) and U(IV) oxides. J Chem Phys. 2013;139:244704.
[104] Bagus PS, Ilton ES. Theory for the XPS of Actinides. Top Catal. 2013;56:1121–1128.
[105] Klooster R, Broer R, Filatov M. Calculation of X-ray photoelectron spectra with the use of the normalized elimination of the small component method. Chem Phys. 2012;395: 122–127.
[106] Knope KE, Soderholm L. Solution and Solid-State Structural Chemistry of Actinide Hydrates and Their Hydrolysis and Condensation Products. Chem Rev. 2013;113:944–994.
[107] Antonio MR, Soderholm L. X-ray Absorption Spectroscopy of the Actinides. In: Morss LR, Edelstein NM, Fuger J, editors. The Chemistry of the Actinide and Transactinide Elements. vol. 5. 3rd ed. Dordrecht, The Netherlands: Springer; 2006. p. 3086–3198.
[108] Besley NA, Asmuruf FA. Time-dependent density functional theory calculations of the spectroscopy of core electrons. Phys Chem Chem Phys. 2010;12:12024–12039.
[109] Zhang Y, Biggs JD, Healion D, Govind N, Mukamel S. Core and valence excitations in resonant X-ray spectroscopy using restricted excitation window time-dependent density functional theory. J Chem Phys. 2012;137:194306.
[110] Besley NA. Equation of motion coupled cluster theory calculations of the X-ray emission spectroscopy of water. Chem Phys Lett. 2012;542:42–46.
[111] Asmuruf FA, Besley NA. Calculation of near-edge X-ray absorption fine structure with the CIS(D) method. Chem Phys Lett. 2008;463:267–271.
[112] Minasian SG, Keith JM, Batista ER, Boland KS, Clark DL, Kozimor SA, et al. New evidence for 5f covalency in actinocenes determined from carbon K-edge XAS and electronic structure theory. Chem Sci. 2014;5:351–359.
[113] Fillaux C, Guillaumont D, Berthet JC, Copping R, Shuh DK, Tyliszczak T, et al. Investigating the electronic structure and bonding in uranyl compounds by combining NEXAFS spectroscopy and quantum chemistry. Phys Chem Chem Phys. 2010;12:14253–14262.
[114] Kozimor SA, Yang P, Batista ER, Boland KS, Burns CJ, Clark DL, et al. Trends in Covalency for *d*- and *f*-Element Metallocene Dichlorides Identified Using Chlorine K-Edge X-ray Absorption Spectroscopy and Time-Dependent Density Functional Theory. J Am Chem Soc. 2009;131:12125–12136.
[115] Minasian SG, Keith JM, Batista ER, Boland KS, Clark DL, Conradson SD, et al. Determining Relative *f* and *d* Orbital Contributions to M–Cl Covalency in MCl_6^{2-} (M = Ti, Zr, Hf, U)

and $UOCl_5^-$ Using Cl K-Edge X-ray Absorption Spectroscopy and Time-Dependent Density Functional Theory. J Am Chem Soc. 2012;134:5586–5597.

[116] Spencer LP, Yang P, Minasian SG, Jilek RE, Batista ER, Boland KS, et al. Tetrahalide Complexes of the $[U(NR)_2]^{2+}$ Ion: Synthesis, Theory, and Chlorine K-Edge X-ray Absorption Spectroscopy. J Am Chem Soc. 2013;135:2279–2290.

[117] Daly SR, Keith JM, Batista ER, Boland KS, Clark DL, Kozimor SA, et al. Sulfur K-edge X-ray Absorption Spectroscopy and Time-Dependent Density Functional Theory of Dithiophosphinate Extractants: Minor Actinide Selectivity and Electronic Structure Correlations. J Am Chem Soc. 2012;134:14408–14422.

[118] Armbruster MK, Schimmelpfennig B, Plaschke M, Rothe J, Denecke MA, Klenze R. Metal-Ion Complexation Effects in C 1s-NEXAFS Spectra of Carboxylic Acids — Evidence by Quantum Chemical Calculations. J Electron Spectrosc Relat Phenom. 2009;169:51–56.

[119] Atkins AJ, Bauer M, Jacob CR. The chemical sensitivity of X-ray spectroscopy: high energy resolution XANES versus X-ray emission spectroscopy of substituted ferrocenes. Phys Chem Chem Phys. 2013;15:8095–8105.

[120] Roemelt M, Maganas D, DeBeer S, Neese F. A combined DFT and restricted open-shell configuration interaction method including spin-orbit coupling: Application to transition metal L-edge X-ray absorption spectroscopy. J Chem Phys. 2013;138:204101.

[121] Bernadotte S, Atkins AJ, Jacob CR. Origin-independent calculation of quadrupole intensities in X-ray spectroscopy. J Chem Phys. 2012;137:204106.

[122] Ågren H, Carravetta V, Vahtras O, Pettersson LGM. Direct, atomic orbital, static exchange calculations of photoabsorption spectra of large molecules and clusters. Chem Phys Lett. 1994;222:75–81.

[123] Ekström U, Norman P, Carravetta V. Relativistic four-component static-exchange approximation for core-excitation processes in molecules. Phys Rev A. 2006;73:022501.

[124] Ekström U, Ottosson H, Norman P. Characterization of the Chemisorption of Methylsilane on a Au(1,1,1) Surface from the Silicon K- and L-Edge Spectra: A Theoretical Study Using the Four-Component Static Exchange Approximation. J Phys Chem C. 2007;111: 13846–13850.

[125] Villaume S, Ekström U, Ottosson H, Norman P. Near sulfur L-edge X-ray absorption spectra of methanethiol in isolation and adsorbed on a Au(111) surface: a theoretical study using the four-component static exchange approximation. Phys Chem Chem Phys. 2010;12: 5596–5604.

[126] Ekström U, Norman P, Carravetta V, Ågren H. Polarization Propagator for X-Ray Spectra. Phys Rev Lett. 2006;97:143001.

[127] Ekström U, Norman P. X-ray absorption spectra from the resonant-convergent first-order polarization propagator approach. Phys Rev A. 2006;74:042722.

[128] Villaume S, Saue T, Norman P. Linear complex polarization propagator in a four-component Kohn–Sham framework. J Chem Phys. 2010;133:064105.

[129] Linares M, Stafström S, Rinkevicius Z, Ågren H, Norman P. Complex Polarization Propagator Approach in the Restricted Open-Shell, Self-Consistent Field Approximation: The Near K-Edge X-ray Absorption Fine Structure Spectra of Allyl and Copper Phthalocyanine. J Phys Chem B. 2010;115:5096–5102.

[130] Coriani S, Christiansen O, Fransson T, Norman P. Coupled-cluster response theory for near-edge x-ray-absorption fine structure of atoms and molecules. Phys Rev A. 2012;85:022507.

[131] Coriani S, Fransson T, Christiansen O, Norman P. Asymmetric-Lanczos-Chain-Driven Implementation of Electronic Resonance Convergent Coupled-Cluster Linear Response Theory. J Chem Theory Comput. 2012;8:1616–1628.

[132] Fransson T, Coriani S, Christiansen O, Norman P. Carbon X-ray absorption spectra of fluoroethenes and acetone: A study at the coupled cluster, density functional, and static-exchange levels of theory. J Chem Phys. 2013;138:124311.

[133] Brabec J, Bhaskaran-Nair K, Govind N, Pittner J, Kowalski K. Application of state-specific multireference coupled cluster methods to core-level excitations. J Chem Phys. 2012;137:171101.

[134] Vlaisavljevich B, Diaconescu PL, Lukens WL, Gagliardi L, Cummins CC. Investigations of the Electronic Structure of Arene-Bridged Diuranium Complexes. Organometallics. 2013;32:1341–1352.
[135] Rehr JJ, Kas JJ, Vila FD, Prange MP, Jorissen K. Parameter-free calculations of X-ray spectra with FEFF9. Phys Chem Chem Phys. 2010;12:5503–5513.
[136] Rehr JJ, Kas JJ, Prange MP, Sorini AP, Takimoto Y, Vila F. Ab initio theory and calculations of X-ray spectra. C R Phys. 2009;10:548–559.
[137] Bunău O, Joly Y. Self-consistent aspects of x-ray absorption calculations. J Phys Condens Matter. 2009;21:345501.
[138] Fillaux C, Den Auwer C, Guillaumont D, Shuh DK, Tyliszczak T. Investigation of actinide compounds by coupling X-ray absorption spectroscopy and quantum chemistry. J Alloys Compd. 2007;444–445:443–446.
[139] Fillaux C, Berthet JC, Conradson SD, Guilbaud P, Guillaumont D, Hennig C, et al. Combining theoretical chemistry and XANES multi-edge experiments to probe actinide valence states. C R Chimie. 2007;10:859–871.
[140] Denecke MA. Actinide speciation using X-ray absorption fine structure spectroscopy. Coord Chem Rev. 2006;250:730–754.
[141] Müller A, Wittneben V, Diemann E, Hormes J, Kuetgens U. Electronic structure of thiometalates $[MS_4]^{n-}$ (M=Mo, W, Re). XANES spectra and SCF-Xα-SW calculations. Chem Phys Lett. 1994;225:359–363.
[142] Langhoff PW. Stieltjes imaging of atomic and molecular photoabsorption profiles. Chem Phys Lett. 1973;22:60–64.
[143] Langhoff PW, Corcoran CT, Sims JS, Weinhold F, Glover RM. Moment-theory investigations of photoabsorption and dispersion profiles in atoms and ions. Phys Rev A. 1976;14: 1042–1056.
[144] Ågren H, Carravetta V, Jensen HJA, Jørgensen P, Olsen J. Multiconfiguration linear-response approaches to the calculation of absolute photoionization cross sections: HF, H_2O, and Ne. Phys Rev A. 1993;47:3810–3823.
[145] Carravetta V, Luo Y, Ågren H. Accurate photoionization cross sections of diatomic molecules by multi-configuration linear response theory. Chem Phys. 1993;174:141–153.
[146] Gokhberg K, Vysotskiy V, Cederbaum LS, Storchi L, Tarantelli F, Averbukh V. Molecular photoionization cross sections by Stieltjes–Chebyshev moment theory applied to Lanczos pseudospectra. J Chem Phys. 2009;130:064104.
[147] Cukras J, Coriani S, Decleva P, Christiansen O, Norman P. Photoionization cross section by Stieltjes imaging applied to coupled cluster Lanczos pseudo-spectra. J Chem Phys. 2013;139:094103.
[148] Rehr JJ, Albers RC. Theoretical approaches to x-ray absorption fine structure. Rev Mod Phys. 2000;72:621–654.
[149] Wilson KR, Tobin JG, Ankudinov AL, Rehr JJ, Saykally RJ. Extended X-Ray Absorption Fine Structure from Hydrogen Atoms in Water. Phys Rev Lett. 2000;85:4289–4292.
[150] Vallet V, Szabó Z, Grenthe I. Experimental and quantum chemical studies of structure and reaction mechanisms of dioxouranium(VI) complexes in solution. Dalton Trans. 2004; p. 3799–3807.
[151] Vallet V, Macak P, Wahlgren U, Grenthe I. Actinide chemistry in solution, quantum chemical methods and models. Theor Chem Acc. 2006;115:145–160.
[152] Vallet V, Flament JP, Grenthe I. Quantum chemical calculations of the structure and Debye-Waller factors of actinide complexes; a comparison with experimental EXAFS data. In: Denecke MA, editor. Speciation Techniques and Facilities for Radioactive Materials at Synchrotron Light Sources. Issy-les-Moulineaux, France: OECD Nuclear Energy Agency; 2007. p. 93–107.
[153] Atta-Fynn R, Bylaska EJ, Schenter GK, de Jong WA. Hydration Shell Structure and Dynamics of Curium(III) in Aqueous Solution: First Principles and Empirical Studies. J Phys Chem A. 2011;115:4665–4677.

[154] Spezia R, Beuchat C, Vuilleumier R, D'Angelo P, Gagliardi L. Unravelling the Hydration Structure of ThX_4 (X = Br, Cl) Water Solutions by Molecular Dynamics Simulations and X-ray Absorption Spectroscopy. J Phys Chem B. 2012;116:6465–6475.
[155] Wilson RE, Skanthakumar S, Burns PC, Soderholm L. Structure of the homoleptic thorium(IV) aqua ion $[Th(H_2O)_{10}]Br_4$. Angew Chem Int Ed. 2007;46:8043–8045.
[156] Bühl M, Schreckenbach G, Sieffert N, Wipff G. Effect of Counterions on the Structure and Stability of Aqueous Uranyl(VI) Complexes. A First-Principles Molecular Dynamics Study. Inorg Chem. 2009;48:9977–9979.
[157] Fuoss RM. Ionic Association. III. The Equilibrium between Ion Pairs and Free Ions. J Am Chem Soc. 1958;80:5059–5061.
[158] Vallet V, Schimmelpfennig B, Maron L, Teichteil C, Leininger T, Gropen O, et al. Reduction of uranyl by hydrogen: an *ab initio* study. Chem Phys. 1999;244:185–193.
[159] Vallet V, Wahlgren U, Grenthe I. Probing the Nature of Chemical Bonding in Uranyl(VI) Complexes with Quantum Chemical Methods. J Phys Chem A. 2012;116: 12373–12380.
[160] Gutowski KE, Dixon DA. Predicting the Energy of the Water Exchange Reaction and Free Energy of Solvation for the Uranyl Ion in Aqueous Solution. J Phys Chem A. 2006;110: 8840–8856.
[161] Wåhlin P, Danilo C, Vallet V, Réal F, Flament JP, Wahlgren U. An investigation of the accuracy of different DFT functionals on the water exchange reaction in hydrated uranyl(VI) in the ground state and the first excited state. J Chem Theory Comput. 2008;4:569–577.
[162] Austin JP, Burton NA, Hillier IH, Sundararajan M, Vincent MA. Which density functional should be used to study actinyl complexes? Phys Chem Chem Phys. 2009;11:1143–1145.
[163] Wåhlin P, Schimmelpfennig B, Wahlgren U, Grenthe I, Vallet V. On the combined use of discrete solvent models and continuum descriptions of solvent effects in ligand exchange reactions; a case study of the uranyl(VI) aquo ion. Theor Chem Acc. 2009;124:377–384.
[164] Wåhlin P. Theoretical Actinide Chemistry – Methods and Models [Ph.D. dissertation]. Stockholm University. Stockholm, Sweden; 2011. Available from: http://urn.kb.se/resolve?urn=urn:nbn:se:su:diva-54848.
[165] Cramer CJ, Truhlar DG. A Universal Approach to Solvation Modeling. Acc Chem Res. 2008;41:760–768.
[166] Cramer CJ, Truhlar DG. Reply to Comment on "A Universal Approach to Solvation Modeling." Acc Chem Res. 2009;42:493–497.
[167] Jaque P, Marenich AV, Cramer CJ, Truhlar DG. Computational Electrochemistry: The Aqueous $Ru^{3+}|Ru^{2+}$ Reduction Potential. J Phys Chem C. 2007;111:5783–5799.
[168] Szabó Z, Toraishi T, Vallet V, Grenthe I. Solution coordination chemistry of actinides; thermodynamics, structure and reaction mechanisms. Coord Chem Rev. 2006;250: 784–815.
[169] Shamov GA, Schreckenbach G. The role of peripheral alkyl substituents: a theoretical study of substituted and unsubstituted uranyl isoamethyrin complexes. Inorg Chem. 2008;47: 805–811.
[170] Szabó Z, Grenthe I. On the Mechanism of Oxygen Exchange Between Uranyl(VI) Oxygen and Water in Strongly Alkaline Solution as Studied by ^{17}O-NMR Magnetization Transfer. Inorg Chem. 2010;49:4928–4933.
[171] Bühl M, Schreckenbach G. Oxygen Exchange in Uranyl Hydroxide Via Two "Non Classical" Ions. Inorg Chem. 2010;49:3821–3827.
[172] Tsushima S. Quantum Chemical Calculations of the Redox Potential of the Pu(VII)/Pu(VIII) Couple. J Phys Chem B. 2008;112:13059–13063.
[173] Hay PJ, Martin RL, Schreckenbach G. Theoretical studies of the properties and solution chemistry of AnO_2^{2+} and AnO_2^{+} aquo complexes for An = U, Np, and Pu. J Phys Chem A. 2000;104:6259–6270.
[174] Shamov GA, Schreckenbach G. Relativistic density functional theory study of dioxoactinide(VI) and -(V) complexation with alaskaphyrin and related Schiff-Base macrocyclic ligands. J Phys Chem A. 2006;110:9486–9499.

[175] Tsushima S, Wahlgren U, Grenthe I. Quantum chemical calculations of reduction potentials of AnO_2^{2+}/AnO_2^{+} (An = U, Np, Pu, Am) and Fe^{3+}/Fe^{2+} couples. J Phys Chem A. 2006;110:9175–9182.
[176] Truhlar DG, Cramer DJ, Lewis A, Bumpus JA. Molecular Modeling of Environmentally Important Processes: Reduction Potentials. J Chem Educ. 2004;81:596–604.
[177] Namazian M, Lin CY, Coote ML. Benchmark Calculations of Absolute Reduction Potential of Ferricinium/Ferrocene Couple in Nonaqueous Solutions. J Chem Theory Comput. 2010;6:2721–2725.
[178] Horowitz SE, Martson JB. Strong correlations in actinide redox reactions. J Chem Phys. 2011;134:064510.
[179] Steele H, Guillaumont D, Moisy P. Density Functional Theory Calculations of the Redox Potentials of Actinide(VI)/Actinide(V) Couple in Water. J Phys Chem A. 2013;117:4500–4505.
[180] Vallet V, Maron L, Schimmelpfennig B, Leininger T, Teichteil C, Gropen O, et al. Reduction behavior of the early actinyl ions in aqueous solution. J Phys Chem A. 1999;103:9285–9289.
[181] Gomes ASP, Réal F, Galland N, Angeli C, Cimiraglia R, Vallet V. Electronic structure investigation of the evanescent $AtO+$ ion. Phys Chem Chem Phys. 2014;16:9238–9248.

12

Computational Tools for Predictive Modeling of Properties in Complex Actinide Systems

Jochen Autschbach,[1] *Niranjan Govind,*[2] *Raymond Atta-Fynn,*[3] *Eric J. Bylaska,*[2] *John W. Weare*[4] *and Wibe A. de Jong*[5]

[1]*Department of Chemistry, University at Buffalo, State University of New York*
[2]*Pacific Northwest National Laboratory*
[3]*Department of Physics, University of Texas*
[4]*Department of Chemistry and Biochemistry, University of California San Diego*
[5]*Lawrence Berkeley National Laboratory*

12.1 Introduction

In this chapter we focus on methodological and computational aspects that are key to accurately modeling the spectroscopic and thermodynamic properties of molecular systems containing actinides within the density functional theory (DFT) framework. Our focus is on properties that require either an accurate relativistic all-electron description or an accurate description of the dynamical behavior of actinide species in an environment at finite temperature, or both. The implementation of the methods and the calculations discussed in this chapter were carried out with the NWChem software suite [1]. In the first two sections we discuss two methods that account for relativistic effects, the ZORA and the X2C Hamiltonian. Section 12.2.1 discusses the implementation of the approximate relativistic ZORA Hamiltonian and its extension to magnetic properties. Section 12.3 focuses on the exact X2C Hamiltonian and the application of this methodology to obtain accurate molecular properties. In Section 12.4 we examine the role of a dynamical environment at finite temperature as well as the presence of other ions on the thermodynamics of hydrolysis and exchange reaction mechanisms. Finally, Section 12.5 discusses the modeling of XAS

Computational Methods in Lanthanide and Actinide Chemistry, First Edition. Edited by Michael Dolg.

(EXAFS, XANES) properties in realistic environments accounting for both the dynamics of the system and relativistic effects.

12.2 ZORA Hamiltonian and Magnetic Property Calculations

It is well established that scalar and spin-orbit relativistic effects have to be taken into account for accurate electronic structure calculations of actinides and other heavy elements. Relativistic effects are best described in electronic-structure calculations by solving the Dirac equation, whose solutions are made up of four-component spinor wavefunctions. However, four-component methods are not only fraught with problems such as variational collapse [2], they have in the past been an order of magnitude more expensive. The extra cost stems from the need to properly describe the small component wavefunction. In principle, the cost of the Dirac equation can be reduced by transforming it from a regular four-component eigenvalue equation into a non-regular two-component eigenvalue equation by decoupling the large and small components of the wavefunction. However, this non-regular two-component eigenvalue equation turns out to be difficult to solve, because it has a non-trivial normalization condition and the eigenvalue depends non-linearly upon itself. Over the years, many approximations have been developed to circumvent these difficulties such as the Breit-Pauli Hamiltonian, direct perturbation theory, Dyall's modified Dirac method, and the Douglas-Kroll-Hess (DKH) Hamiltonian [3]. Another approximation, the ZORA method, has become one of the more popular of these approximations. In the following sections we will discuss the ZORA Hamiltonian and the theory and implementation of magnetic properties within the ZORA framework.

12.2.1 ZORA Hamiltonian

The zeroth-order regular approximation (ZORA) was originally developed by Chang, Pelissier, and Durand [4]. This method was rediscovered and developed much further by Baerends, van Lenthe and coworkers [5]. ZORA is a two-component spinor approach for approximately solving the Dirac equation based upon regularizing the wave equation by ignoring the energy dependence of the effective mass of the electron. It has been shown that the solutions of the ZORA equation are reasonable two-component approximations to the fully relativistic Dirac solution for hydrogen-like systems. Since this is a two-component method, calculations can be performed using only the large component.

The ZORA approximation and other regularized two-component methods can be used to treat all of the electrons (core + valence) in the system. However, all-electron calculations are not always necessary, since the most significant relativistic effects on valence shells of heavy elements can be encapsulated using effective core potentials. These approaches yield accurate structures, frequencies, and other properties that depend primarily on the valence electronic structure. However, for properties like XAS, XPS, NMR, EPR, etc., all-electron relativistic approaches are needed.

Atomic units with $\hbar = 1, m_e = 1, 4\pi\varepsilon_0 = 1, c = \alpha^{-1} = 137.035999074(44)$ are used in this section. The two-component ZORA equation is given by

$$\left[\frac{1}{2}(\boldsymbol{\sigma} \cdot \hat{\boldsymbol{p}})\mathcal{K}(\boldsymbol{\sigma} \cdot \hat{\boldsymbol{p}}) + V\right] \psi_n = \epsilon_n^Z \psi_n \tag{12.1}$$

with

$$\mathcal{K} = \frac{2c^2}{2c^2 - V} \tag{12.2}$$

Here, V can be V_{ext}, the external potential generated by the nucleus-electron interaction, or a model potential V_{mp}, similar to the potentials used for the X2C transformation in Section 12.3. The nonrelativistic limit is formally obtained by letting $\mathcal{K} \rightarrow 1$.

Expanding equation (12.1),

$$\left[\frac{1}{2}\hat{\boldsymbol{p}}\mathcal{K} \cdot \hat{\boldsymbol{p}} + \frac{i}{2}\boldsymbol{\sigma} \cdot (\hat{\boldsymbol{p}}\mathcal{K} \times \hat{\boldsymbol{p}}) + V\right]\psi_n = \epsilon_n^Z \psi_n \tag{12.3}$$

As can be seen, the spin-orbit contribution (term 2) in equation (12.3) is present even at this lowest order of the expansion. Equation (12.3) shows the separation of the ZORA operator into a scalar (spin-free) part and the SO operator (the electron spin-dependent term with $\boldsymbol{\sigma}$). With $\mathcal{K} \approx 1 + V_{\text{nuc}}/(2c^2)$, the ZORA SO operator becomes equivalent to the Breit-Pauli one-electron counterpart in order c^{-2}. However, the scalar part of ZORA misses some contributions in order c^{-2}. The ZORA equation can also be written as,

$$\left[\frac{1}{2}\hat{\boldsymbol{p}}^2 + \frac{1}{2}\hat{\boldsymbol{p}}\,(\mathcal{K} - 1) \cdot \hat{\boldsymbol{p}} + \frac{i}{2}\boldsymbol{\sigma} \cdot (\hat{\boldsymbol{p}}\mathcal{K} \times \hat{\boldsymbol{p}}) + V\right]\psi_n = \epsilon_n^Z \psi_n \tag{12.4}$$

where the non-relativistic kinetic energy contribution has been isolated. The advantage of this decomposition is that the ZORA contribution can be treated as a correction that can be added to the kinetic-energy matrix elements in any standard implementation. The spin-free or scalar-relativistic equation can be obtained by eliminating the spin-orbit term, resulting in the following equation:

$$\left[\frac{1}{2}\hat{\boldsymbol{p}}^2 + \frac{1}{2}\hat{\boldsymbol{p}}\,(\mathcal{K} - 1) \cdot \hat{\boldsymbol{p}} + V\right]\psi_n = \epsilon_n^Z \psi_n \tag{12.5}$$

It has been shown [188, 189] that the one-electron energies of the ZORA equation, ϵ_i^Z can be improved by scaling the ZORA energy. Basically, this scaling captures the effects of summing certain higher-order contributions to infinite order. The scaled one-electron energies are given by

$$\epsilon_i^{Scaled} = \frac{\epsilon_i^Z}{1 + \left\langle \phi_i \middle| \boldsymbol{\sigma} \cdot \hat{\boldsymbol{p}} \frac{\mathcal{K}^2}{4c^2} \boldsymbol{\sigma} \cdot \hat{\boldsymbol{p}} \middle| \phi_i \right\rangle} \tag{12.6}$$

Within the DFT framework, the total scaled ZORA energy can be written as

$$\begin{aligned} E^{Scaled} = \sum_{i=1}^{N} \left[\left\langle \phi_i \middle| \boldsymbol{\sigma} \cdot \hat{\boldsymbol{p}} \frac{\mathcal{K}}{2} \boldsymbol{\sigma} \cdot \hat{\boldsymbol{p}} \middle| \phi_i \right\rangle - \epsilon_i^{Scaled} \left\langle \phi_i \middle| \boldsymbol{\sigma} \cdot \hat{\boldsymbol{p}} \frac{\mathcal{K}^2}{4c^2} \boldsymbol{\sigma} \cdot \hat{\boldsymbol{p}} \middle| \phi_i \right\rangle \right] \\ + E_{ne} + E_H + E_{xc} + E_{nn} \end{aligned} \tag{12.7}$$

where the summation runs over the occupied orbitals. The above equation is identical to the non-relativistic total energy functional except for the kinetic energy contribution, which includes the ZORA correction.

In our Gaussian basis set implementation [6], the molecular spinors are expanded in terms of real basis functions as follows:

$$\phi_i = \sum_{\mu} \left\{ c^{\alpha}_{\mu i} \begin{pmatrix} \chi_\mu \\ 0 \end{pmatrix} + c^{\beta}_{\mu i} \begin{pmatrix} 0 \\ \chi_\mu \end{pmatrix} \right\} = \begin{pmatrix} \phi^{\alpha}_i \\ \phi^{\beta}_i \end{pmatrix} \tag{12.8}$$

where χ_μ are Gaussian basis functions and $c^{\alpha}_{\mu i}, c^{\beta}_{\mu i}$ are expansion coefficients, which are, in general, complex. Within the basis set approximation and using the decomposition from equation (12.4), the corrected kinetic energy elements of the Fock matrix can be written as

$$T^{Z}_{\mu\nu} = \left\langle \chi_\mu \left| \frac{(\boldsymbol{\sigma} \cdot \hat{\boldsymbol{p}})(\boldsymbol{\sigma} \cdot \hat{\boldsymbol{p}})}{2} \right| \chi_\nu \right\rangle + \left\langle \chi_\mu \boldsymbol{\sigma} \cdot \hat{\boldsymbol{p}} \left| \frac{\mathcal{K} - 1}{2} \right| \boldsymbol{\sigma} \cdot \hat{\boldsymbol{p}} \chi_\nu \right\rangle \tag{12.9}$$

The non-relativistic part of the kinetic energy (first term in the above equation) is evaluated analytically, while the ZORA correction (second term) is calculated numerically on atom-centered grids. Since the ZORA correction depends on the potential, it is not gauge invariant. There have been a number of attempts to address this issue [7–9]. In our implementation, we address this by using the atomic approximation of van Lenthe and coworkers [10, 11]. Strictly speaking none of these approaches are truly gauge invariant in the general sense, but they help minimize the problem. Within this approximation, the ZORA corrections to the kinetic energy matrix elements are calculated using the superposition of densities of the atoms in the system. As a result only intra-atomic contributions are involved, and no gradient or second derivatives of these corrections need to be calculated. In addition, the corrections only have to be calculated once at the start of the calculation and stored. With this, we can re-express equation (12.9) as

$$T^{Z}_{\mu\nu} = \left\langle \chi_\mu \left| \frac{(\boldsymbol{\sigma} \cdot \hat{\boldsymbol{p}})(\boldsymbol{\sigma} \cdot \hat{\boldsymbol{p}})}{2} \right| \chi_\nu \right\rangle + \left\langle \chi_\mu \boldsymbol{\sigma} \cdot \hat{\boldsymbol{p}} \left| \frac{\tilde{\mathcal{K}} - 1}{2} \right| \boldsymbol{\sigma} \cdot \hat{\boldsymbol{p}} \chi_\nu \right\rangle \tag{12.10}$$

where $\tilde{\mathcal{K}} = \frac{2c^2}{2c^2 - \tilde{V}}$ and $\tilde{V} \approx V^{ne}_{atom} + V^{H}_{atom}$, the sum of the atomic nuclear-electron, Hartree potentials, respectively. The atomic densities are calculated for the neutral atoms at the Hartree-Fock (HF) level and then the ZORA potential corrections are calculated using this density. Note that we ignore the exchange-correlation contribution in the evaluation of the ZORA atomic corrections in the same way as van Lenthe and co-workers [10–12]. This approximation works well, as shown by our results. Our implementation differs from theirs in that they use a resolution of identity (RI) approach with auxiliary basis sets to represent the inverse of the potential to calculate the ZORA corrections, while we use an all-electron grid representation. Several applications using this approach will be discussed in subsequent paragraphs.

12.2.2 Magnetic properties

Over the past 4 years, a set of capabilities has been implemented for relativistic calculations of EFGs [13, 14] (needed to model nuclear quadrupole effects in solid-state NMR and quadrupole line broadening for solution NMR), NMR chemical shifts [15] (in a scalar relativistic version for the time being), as well as electron paramagnetic resonance (EPR) g-factors [15, 16] (or g-shifts = deviations from the free-electron g-value) and electron-nucleus hyperfine coupling (hfc, a) [17, 18]. These developments mostly utilized the approximate two-component relativistic zeroth-order regular approximation (ZORA) [19] framework (Section 12.2.1) due to its proven ability to model the aforementioned properties efficiently and accurately. In the near future, the magnetic property modules will also be interfaced with the X2C functionality discussed in Section 12.3. One advantage of the ZORA approach is that the operators needed for magnetic perturbations are straightforwardly derived at the operator level, and calculations of their AO matrices are easily implemented with the help of numerical integration or a combination of analytic and numerical techniques. We first outline the derivation of perturbation operators for external magnetic fields and nuclear spin magnetic moments, and discuss selected computational studies for actinide compounds subsequently.

The presence of a magnetic field-external, or from a nuclear spin-can be incorporated by minimal substitution for the electron momentum operator,

$$\hat{\boldsymbol{p}} \rightarrow \hat{\boldsymbol{p}} + \boldsymbol{A} \tag{12.11}$$

with the vector potentials for an external field $\boldsymbol{B}_0$ or a collection of nuclear point magnetic moments $\boldsymbol{m}_N$, respectively, being given as

$$\boldsymbol{A}_0 = \frac{1}{2}\boldsymbol{B}_0 \times \boldsymbol{r} \tag{12.12a}$$

$$\boldsymbol{A}_N = \frac{\mu_0}{4\pi} \sum_N^{\text{nuclei}} \frac{\boldsymbol{m}_N \times \boldsymbol{r}_N}{r_N^3} \tag{12.12b}$$

In Hartree atomic units, the conversion of $\mu_0/(4\pi)$ from SI gives a factor of $1/c^2$ that is used below in the hyperfine operators. In equations (12.12a, 12.12b), $\boldsymbol{r}_N$ is an electron-nucleus distance, and $\boldsymbol{r}$ is the distance of an electron to an arbitrarily chosen gauge origin for the external-field vector potential (usually, but not necessarily, coinciding with the laboratory coordinate origin). An unphysical dependence of the results on the chosen gauge origin for the external field vector potential is eliminated by adopting 'gauge-including atomic orbitals' (GIAO) [15, 16, 20]. The vector potentials (12.12a) and (12.12b) satisfy the Coulomb gauge,

$$\boldsymbol{\nabla} \cdot \boldsymbol{A} = 0 \tag{12.13}$$

As it has been shown previously [21], at the ZORA level the magnetic perturbation operators are the same no matter if the minimal substitution is made in the Dirac Hamiltonian before transformation to two-component form or afterwards. Minimal substitution (12.11) gives a set of additional magnetic terms in the ZORA operator:

$$\hat{h}^{Z}(\boldsymbol{B}, \boldsymbol{m}_N) = \hat{h}^{Z} + \hat{h}^{Z}_{\text{mag}} \tag{12.14}$$

with

$$\hat{h}^{Z}_{\text{mag}} = \frac{1}{2}\left([\hat{\boldsymbol{p}} \cdot \mathcal{K}\boldsymbol{A} + \mathcal{K}\boldsymbol{A} \cdot \hat{\boldsymbol{p}}] + i\boldsymbol{\sigma} \cdot [\hat{\boldsymbol{p}} \times (\mathcal{K}\boldsymbol{A}) + (\mathcal{K}\boldsymbol{A}) \times \hat{\boldsymbol{p}}]\right) + \frac{1}{2}\mathcal{K}\boldsymbol{A} \cdot \boldsymbol{A} \tag{12.14a}$$

There are electron spin-independent and spin-dependent paramagnetic terms (linear in the vector potential), and a diamagnetic spin-independent term that (quadratic in the vector potential). By substituting (12.12a) or (12.12b) in the spin-free linear terms, one obtains the ZORA form of the *Orbital Zeeman* (OZ) operator

$$\hat{h}^{\text{OZ}} = \frac{1}{4}\left[\mathcal{K}\boldsymbol{B}_0 \cdot (\boldsymbol{r} \times \hat{\boldsymbol{p}}) + (\boldsymbol{r} \times \hat{\boldsymbol{p}}) \cdot \mathcal{K}\boldsymbol{B}_0\right] \tag{12.14b}$$

and the ZORA form of the *Paramagnetic Orbital* (OP) operator

$$\hat{h}^{\text{OP}} = \frac{1}{2c^2}\sum_N \left[\left(\frac{\boldsymbol{r}_N}{r_N^3} \times \hat{\boldsymbol{p}}\right) \cdot \mathcal{K}\boldsymbol{m}_N + \mathcal{K}\boldsymbol{m}_N \cdot \left(\frac{\boldsymbol{r}_N}{r_N^3} \times \hat{\boldsymbol{p}}\right)\right] \tag{12.14c}$$

A bi-linear perturbation (double perturbation) [22] of the electronic energy by these two operators gives the ZORA analog of the paramagnetic component to the NMR shielding tensor. A bi-linear substitution of (12.12a) and (12.12b) in the $\mathcal{K}\boldsymbol{A} \cdot \boldsymbol{A}$ term in (12.14a) gives the corresponding operator for the *Diamagnetic Shielding*,

$$\hat{h}^{\text{DS}} = \frac{1}{2c^2}\frac{\mathcal{K}}{2}\sum_N \left[(\boldsymbol{m}_N \cdot \boldsymbol{B}_0)\left(\frac{\boldsymbol{r}_N}{r_N^3} \cdot \boldsymbol{r}\right) - (\boldsymbol{m}_N \cdot \boldsymbol{r})\left(\boldsymbol{B}_0 \cdot \frac{\boldsymbol{r}_N}{r_N^3}\right)\right] \tag{12.14d}$$

which is calculated from (12.14d) as an expectation value with the two-component ZORA electron density.

By substituting (12.12a) or (12.12b) in the spin-dependent linear terms in equation (12.14a), one obtains after some manipulations the ZORA version of the *Spin Zeeman* (SZ) operator

$$\hat{h}^{SZ} = \frac{1}{4}\boldsymbol{\sigma} \cdot \{\boldsymbol{B}_0(\boldsymbol{\nabla} \cdot \mathcal{K}\boldsymbol{r}) - (\boldsymbol{B}_0 \cdot \boldsymbol{\nabla})\mathcal{K}\boldsymbol{r}\} \tag{12.14e}$$

and the sum of the *Fermi-contact* and *Spin-Dipole* (FC+SD) hyperfine terms

$$\hat{h}^{FC} + \hat{h}^{SD} = \frac{1}{2c^2}\sum_N \boldsymbol{\sigma} \cdot \left\{\boldsymbol{m}_N\left(\boldsymbol{\nabla} \cdot \frac{\mathcal{K}\boldsymbol{r}_N}{r_N^3}\right) - (\boldsymbol{m}_N \cdot \boldsymbol{\nabla})\frac{\mathcal{K}\boldsymbol{r}_N}{r_N^3}\right\} \tag{12.14f}$$

In the last two equations, curly braces $\{\cdots\}$ indicate that derivatives are only taken in the operator, not of a function to its right. For completeness, we also give the ZORA expressions for the diamagnetic operators for bi-linear perturbations either by a pair of nuclear spins (diamagnetic orbital term [OD] of nuclear spin-spin coupling) or by an external magnetic field (diamagnetic magnetizability [DM]):

$$\hat{h}^{OD} = \frac{1}{2c^4}\mathcal{K}\sum_{M\neq N}\frac{(\boldsymbol{m}_N\cdot\boldsymbol{m}_M)(\boldsymbol{r}_N\cdot\boldsymbol{r}_M)-(\boldsymbol{m}_N\cdot\boldsymbol{r}_M)(\boldsymbol{m}_M\cdot\boldsymbol{r}_M)}{r_N^3 r_M^3} \tag{12.14g}$$

$$\hat{h}^{DM} = \frac{1}{2}\frac{\mathcal{K}}{4}(\boldsymbol{r}\times\boldsymbol{B}_0)\cdot(\boldsymbol{r}\times\boldsymbol{B}_0) \tag{12.14h}$$

It is noted that the well-known nonrelativistic versions of the operators are obtained by formally letting $\mathcal{K}\to 1$ in equations (12.14b) to (12.14h). Afterwards, the derivatives of $\boldsymbol{r}_N/r_N^3$ can be taken in equation (12.14f), which among other terms gives the Fermi contact operator in its usual form written in terms of $\delta(\boldsymbol{r}_N)$ distributions. In the ZORA form, the 'contact' term is actually suppressed, which may be considered as a drastic form of picture change (PC) [21, 23].

EFGs and other electric-field-related properties are dealt with in a somewhat different manner. The EFG and multipole moments are calculated as expectation values with the relevant operators and the electron charge density. To avoid PC errors, if the operators are the four-component versions this charge density has to be the four-component (Dirac) density. The latter differs from the two-component density [24, 25] already in order c^{-2}, which is the same leading order as the relativistic effects on the properties. In the so-called ZORA-4 (Z4) framework, the relevant operators are kept in their four-component form, and an approximate four-component electron charge density is reconstructed from the two-component ZORA density. As was shown by van Lenthe and Baerends [26], the Z4 method eliminates most of the PC errors in order c^{-2}, with relatively small residual errors. In a Kohn-Sham (KS) DFT framework with two-component molecular orbitals φ_i with occupations n_i, the ZORA two-component density is

$$\rho^Z = \sum_i n_i\varphi_i^\dagger\varphi_i \tag{12.15}$$

An approximate small-component density per orbital can be constructed as

$$\rho_i^S(\boldsymbol{r}) = (\boldsymbol{\sigma}\cdot\hat{\boldsymbol{p}}\varphi_i)^\dagger\frac{\mathcal{K}^2}{4c^2}(\boldsymbol{\sigma}\cdot\hat{\boldsymbol{p}}\varphi_i) \quad ; \quad S_i = \int \rho_i^S dV \tag{12.16}$$

The Z4 electron density is then obtained by adding the ZORA and the small-component densities per orbital and renormalizing,

$$\rho^{Z4}(\boldsymbol{r}) = \sum_i n_i\frac{\varphi_i^\dagger\varphi_i+\rho_i^S(\boldsymbol{r})}{1+S_i} \tag{12.17}$$

The Z4 method can alternatively be justified from a 'scaled ZORA' ansatz, as discussed by van Lenthe and Baerends [26] and Aquino *et al.* [13], leading to the same working equations. The relation with scaled ZORA is seen by the appearance of the term $\mathcal{K}^2/(4c^2)$ both of these equations: (12.16) and (12.6). A scalar relativistic variant is obtained by replacing

$(\boldsymbol{\sigma} \cdot \hat{\boldsymbol{p}})\mathcal{K}^2/(4c^2)(\boldsymbol{\sigma} \cdot \hat{\boldsymbol{p}})$ with $\hat{\boldsymbol{p}}\mathcal{K}^2/(4c^2) \cdot \hat{\boldsymbol{p}}$ in the equations. The two-component form (with SO coupling) as well as a scalar relativistic version have been implemented in NWChem [13] for the purpose of EFG calculations. Applications to uranyl EFGs, for the systems discussed in Section 12.3, produced comparable results as the X2C approach. PC effects were not explicitly calculated for uranyl, but for test sets of hydrogen halides (HX, X = F, Cl, Br, I) the PC corrections from Z4 were comparable to those obtained by lower-order DKH [27].

A scalar ZORA module for NMR shielding for DFT calculations was developed subsequently [15]. The contributing operators are the derivatives of equations (12.14b), (12.14c), and (12.14d) with respect to components of the nuclear spin and the external field. The same module framework was also used to set up DFT calculations of EPR g-factors in a linear response (LR) fashion [15]. LR refers here to the treatment of SO coupling as a perturbation, rather than including it variationally in the ground state calculation. The NMR shielding and g-factor calculations both rest on a ground state calculation with the scalar part of the ZORA operator in equation (12.4). The LR g-factor calculations then use a 'spin derivative' [20] of the ZORA SO operator in (12.4), the external magnetic field derivative of the OZ operator (12.14c), and a bi-linear derivative of the SZ operator (12.14e) with respect to 'spin' and the external field. The main difference to the NMR shielding module is that the g-factor starts with a spin-unrestricted open-shell scalar ZORA calculation and works with spin density matrices rather than the total density matrices used in the NMR code. Results from representative calculations are provided below. Related modules were later set up for LR calculations of hfc tensors and paramagnetic NMR chemical shifts [17]. As in the LR g-factor calculations, SO coupling was treated as a linear perturbation. The other operators relevant for hfc are the nuclear spin magnetic moment derivative of the OP term (12.14c), which in conjunction with the SO operator 'spin derivative' gives a spin-orbit OP-SO LR term in the hfc. Finally, an expectation value of the derivative of the FC+SD operator (12.14f) with respect to 'spin' and the nuclear spin magnetic moment gives the FC+SD hfc contribution. The latter is usually the most important one (a notable exception is discussed below), and it is the only term that has a non-vanishing nonrelativistic limit for spin-multiplets. In this limit, the operator probes the electronic spin density at a nucleus of interest [28].

Subsequent developments [16, 18] have focused on g-factor and hfc calculations from two-component DFT with SO coupling included variationally. In this case, it is not necessary to solve LR equations. Instead, the EPR parameters are calculated in an expectation-value fashion with the magnetic-field derivatives of OZ and SZ [as seen in equations (12.14b), (12.14e)] for g-factors, and the nuclear magnetic moment derivatives of OP and FC+SD [see equations (12.14c), (12.14f)] for hfc, respectively. Two forms of SO DFT EPR calculations were developed. One approach allows for spin-polarization and was dubbed the 'magnetic anisotropy' (MA) route, inspired by DFT studies of magnetic anisotropy in van Wüllen [29] and Schmitt *et al.* [30]. Related HF and DFT methods have been put forward by Hrobarik *et al.* [31], Jayatilaka [32], and Malkin *et al.* [33, 34]. The other approach follows works by van Lenthe, Wormer, and van der Avoird (LWA) [35, 36] which does not treat spin polarization but has some advantages due to its simplicity and computational efficiency and tends to perform well for g-factors.

The MA approach [16, 18] rests on a set of generalized-collinear SO calculations in which a spin-quantization direction v is chosen along one of the coordinate axes. In a KS DFT setup with two-spinor KS MOs determined for a given v, components of the EPR g-'tensor' and the hfc 'tensor' a for nucleus N are then calculated via

$$g_{uv} = 1/(\beta_e S) \sum_i n_i \frac{\partial}{\partial B_u} \langle \varphi_i^v(\boldsymbol{B}) | \hat{F}(\boldsymbol{B}) | \varphi_i^v(\boldsymbol{B}) \rangle \quad \text{(with GIAO basis)} \tag{12.18a}$$

$$a_{uv} = g_N \beta_N / S \sum_i n_i \frac{\partial}{\partial m_{N,u}} \langle \varphi_i^v | \hat{F}(\boldsymbol{m}_N) | \varphi_i^v \rangle \tag{12.18b}$$

Here, S is the pseudo-spin of the system, and $\hat{F}$ is the KS Fock operator up to first order in the external field or the nuclear spin magnetic perturbation. LWA is suitable for Kramers doublets. Assuming no spatial degeneracy, a pair of degenerate Kramers orbitals is initially calculated from SO DFT by assigning equal occupations of 0.5 to two frontier orbitals. One then chooses one of them, φ, and constructs from its real and imaginary parts of the spin α and β components the Kramers pair Φ_1, Φ_2 as

$$\Phi_1 = \begin{pmatrix} \varphi_\alpha^R \\ \varphi_\beta^R \end{pmatrix} + i \begin{pmatrix} \varphi_\alpha^I \\ \varphi_\beta^I \end{pmatrix} \tag{12.19a}$$

$$\Phi_2 = \begin{pmatrix} -\varphi_\beta^R \\ \varphi_\alpha^R \end{pmatrix} + i \begin{pmatrix} \varphi_\beta^I \\ -\varphi_\alpha^I \end{pmatrix} \tag{12.19b}$$

The EPR tensors are then calculated via

$$\begin{aligned} g_{ux} \text{ or } a_{ux} &= k \frac{\partial}{\partial K} \operatorname{Re}\langle \Phi_1 | \hat{F} | \Phi_2 \rangle \\ g_{uy} \text{ or } a_{uy} &= k \frac{\partial}{\partial K} \operatorname{Im}\langle \Phi_2 | \hat{F} | \Phi_1 \rangle \\ g_{uz} \text{ or } a_{uz} &= k \frac{\partial}{\partial K} \langle \Phi_1 | \hat{F} | \Phi_1 \rangle = -\frac{\partial}{\partial K} \langle \Phi_2 | \hat{F} | \Phi_2 \rangle \end{aligned} \tag{12.20}$$

where $k = 1/(\beta_e S), K = B_u$ for g-factors, and $k = g_N \beta_N / S, K = m_{N,u}$ for hfc. The Fock operators are assumed to contain terms up to first order in the derivative parameter, and in the case of g-factor calculations the orbital pair is $\boldsymbol{B}$-field dependent because of the GIAO basis set. A set of equations with similar appearance as (12.20) can be used with wavefunctions in place of the orbital pair to calculate EPR tensors, as it was done by Bolvin based on spin-orbit coupled complete active space (CAS) wavefunctions [37].

The NWChem code supports ground state SCF calculations with a spherical Gaussian nuclear model [38, 39] in order to model finite nuclear volume effects. The hyperfine perturbation operators depending on the nuclear spin magnetic moment, in particular, need to be modified accordingly in relativistic calculations of molecular properties such as hfc and NMR J-coupling [40, 187]. A nuclear Gaussian exponent ξ_N is obtained as $\xi_N = 3/(2R_N^2)$ where $R_N = (0.863 M_N^{1/3} + 0.571)$ fm is the RMS nuclear radius and M_N the mass number of the nucleus [39]. The effects from the finite nuclear volume can then be considered in the relevant ZORA hyperfine operators by substituting $r_N^{-3} \boldsymbol{r}_N$ with $P(3/2, \xi_N r_N^2) r_N^{-3} \boldsymbol{r}_N$. Here, $P(q, r) = 1/\Gamma(q) \int_0^r t^{q-1} e^{-t} dt$ is the lower incomplete gamma function ratio. For a derivation and further details see Autschbach [40], Aquino *et al.* [17], and articles cited therein. The hyperfine coupling capability incorporates the Gaussian nuclear model fully.

In the examples discussed below, finite nucleus effects are small. However, for hfc of ^{199}Hg, for instance, such effects may exceed -10% of the result calculated with a point nucleus in magnitude. Other molecular properties, such as Mössbauer isomer shifts, may exhibit even stronger finite nucleus effects [41].

In the following paragraphs a selection of f-element studies that were performed with the methodology described in this section is discussed. We begin with NMR chemical shifts. The trends and magnitudes of the ^{19}F NMR shielding in the series of complexes UF_nCl_{6-n} ($n = 1 - 6$) have long been an unsolved problem. Non-hybrid DFT calculations by Schreckenbach [42] with the Amsterdam Density Functional program gave relatively poor agreement with experiment, and SO effects were shown to be a minor influence on the fluorine shifts. Straka and Kaupp later [43] obtained one of the important trends in the series correctly with the BHLYP 50% global hybrid functional, namely that fluorines trans to other fluorine atoms have more shielded (less deshielded) nuclei than fluorine atoms that are trans to a chlorine. The data set is graphically displayed in Figure 12.1. See color plate section.

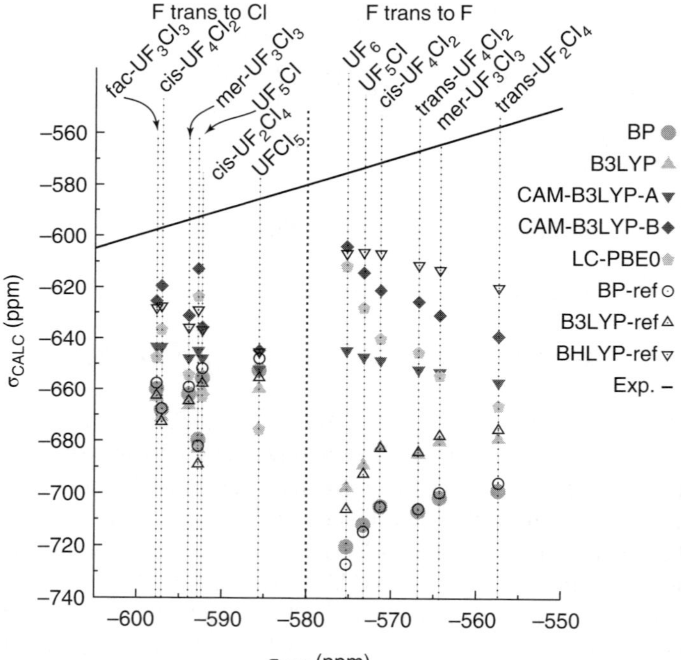

Figure 12.1 *Performance of different functionals in the calculation of ^{19}F shielding for the series of diamagnetic U^{VI} complexes UF_nCl_{6-n} ($n = 1 - 6$). BP-ref, B3LYP-ref, and BHLYP-ref are DFT results employing a scalar relativistic effective core potential for U, from Straka and Kaupp [43]. The other data are from all-electron scalar ZORA calculations with NWChem by Aquino et al. [15]. CAM and LC indicates hybrid functionals with range-separated exchange. The straight black line indicates where calc = expt. Absolute shielding constants converted from experimental shifts taken from Straka and Kaupp [43]. Figure reprinted with permission from Aquino et al. [15]. © 2011 American Chemical Society. See color plate section*

The trends among different compounds for one type of fluorine atom (F-trans-F versus F-trans-Cl) is better reproduced with a non-hybrid functional such as BP, but this class of functionals incorrectly gives an overall larger shielding for F trans to Cl rather than F trans to F. We re-investigated this series recently, and also included several hybrid functionals with range-separated exchange (RSE) in the study (indicated by CAM - Coulomb attenuated method, or LR = long-range corrected). As the plots in Figure 12.1 show, none of the functionals performs well. The RSE functional results are overall not very different from the BHLYP global hybrid data. Two parametrizations of CAM-B3LYP, A with 65% and B with 100% exact exchange in the long range part, are seen to transition, roughly, between B3LYP and BHLYP. A LR variant of PBE produces data in the range between the two CAM-B3LYP versions. In order to arrive at a similar performance as BHLYP it is likely that a larger fraction of exact exchange would be needed in the short-range part of the RSE. The precise reasons for the poor performance of DFT for the set are currently unknown.

Next, consider EPR magnetic data for actinide systems. Table 12.1 collects calculated and, where available, experimental g-factors and hfc constants for selected AnX_6 complexes with $5f^1$ electron configurations. The free-electron g-value is approximately $g_e = 2.0023$. Large deviations from g_e for a molecule are caused by an orbital angular momentum. In the case of the AnX_6 systems, the ground states are orbitally non-degenerate and the enormous g-shifts ($\Delta g = g - g_e$) are caused by the strong SO coupling in the An $5f$ shell. The DFT calculations capture the trend and reproduce most of the g-shift magnitudes. We have previously shown that LR calculations for NpF_6 are also able to produce g-shifts of similar magnitude [15]. This may be surprising as the actinide $5f$ shell SO coupling is evidently very strong. Figure 12.2 shows the dependence of the g-factors of NpF_6 from ZORA/MA calculations where the SO integrals in the ground state DFT calculation have been scaled by a factor λ. See the color plate section. At the non-hybrid DFT level, the behavior is almost linear, indicating why a LR calculation can perform reasonably well for this system. However, LR calculations for UF_6^- and UCl_6^- from Autschbach and Pritchard [20] showed a gross overestimation of g for these systems due to near-degeneracy of frontier orbitals and a concomitant breakdown of the LR approach, indicating the limitations of the approach for

Table 12.1 *AnX_6 EPR data calculated with DFT, LC-PBE0 functional, ZORA MA approach with finite nucleus model. Experimental data[a] in parentheses. Calculated data from Verma and Autschbach [16] and Verma and Autschbach [18]. Isotopes: ^{19}F, ^{35}Cl, ^{235}U, ^{237}Np*

	g	a_{metal}	a_{ligand}
NpF_6	−0.3 (−0.6)	−2200 (−1994)	−51 (−72)
UF_6^-	−0.4 (−0.8)	168	−46
UCl_6^-	−1.0 (−1.1)	192	−7

[a] Experimental data from Butler and Hutchison [44], Clyde A. Hutchison and Weinstock [45], Rigny and Plurien [46], Selbin *et al.* [47]

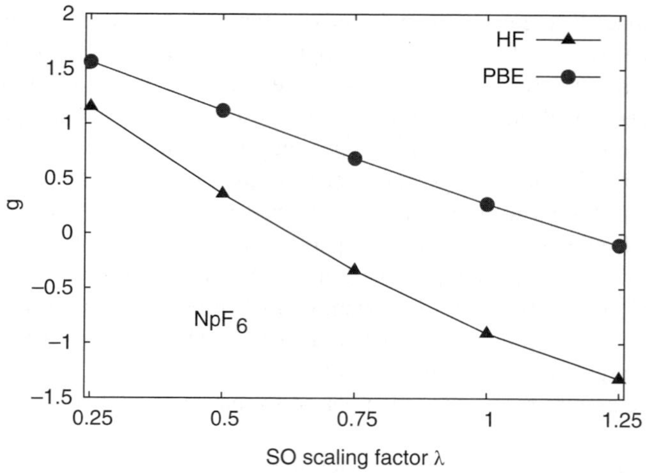

Figure 12.2 *Computed g-factors of NpF_6 using HF theory and a non-hybrid density functional (PBE). ZORA/MA, with SO integrals scaled by a factor λ. Expt. approx. $g = -0.6$. Data to prepare the figure were taken from Verma and Autschbach [16]. See color plate section*

studies of f-element compounds. In comparison, the MA data of Table 12.1 and corresponding LWA data from the same study [16] (not shown) agree much better with experiment. Figure 12.2 indicates a pronounced dependence of the MA results on the presence of absence of exact exchange, with the non-hybrid functional PBE and Hartree-Fock theory being at extreme ends of the spectrum, and a different impact of non-linear SO effects depending on the electronic structure model. For a plot similar to Figure 12.2 but for the Np hfc, reaffirming the functional dependence and different impact of higher order SO effects [18]. The RSE functional LR-PBE0 performed reasonably well overall in MA calculations in the study [16] (Table 12.1), for a set of compounds with elements ranging from light to very heavy. This functional was also among the better performers in a related study of hfc [18].

The g-factors of the AnX_6 series can be rationalized with a crystal-field (CF) model. The high symmetry permits a treatment with a single semi-empirical CF parameter Δ and an SO coupling constant ζ, both of which can be extracted from first-principles calculations [48]. In this case, the model includes both electrostatic (pure CF) effects as well as effects from covalent bonding (ligand-field effects). Figure 12.3 displays the predicted g-factor from such a model in dependence of the ratio $\xi = \Delta/\zeta$. See the color plate section. The free-ion $5f^1$ limit is at $\xi = 0$ (vanishing CF) while the free-electron value of g_e of approximately +2 is approached for vanishing SO coupling. In the latter case, the unpaired orbital is the $5f_{xyz}$ shown in the inset of Figure 12.3, and the orbital angular momentum is completely quenched. The experimental g-factor for NpF_6, at approximately -0.6, is the furthest from the free ion limit among the AnX_6 series [48].

Calculated hfc data for NpF_6 collected in Table 12.2 show a sizable hfc for the ^{19}F ligands, which is reasonably well reproduced by the MA calculations. Verma and Autschbach [18] found a strong dependence on the functional. Moreover, corresponding non spin-polarized

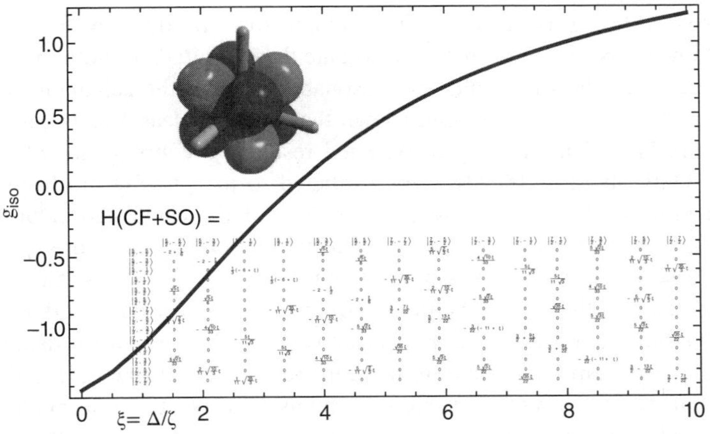

Figure 12.3 *g-factor of NpF_6 as a function of SO splitting ζ and CF splitting Δ for a $\mathcal{O}_h$ crystal field. Symbolic computation using software written by one of the authors (JA). For a description of the CF model see Notter and Bolvin [48]. The inset shows the unpaired Np orbital $5f_{xyz}$ from a scalar ZORA DFT calculation. See color plate section*

Table 12.2 *NpF_6: LMO analysis[a] of $a_{^{237}Np}$ and $a_{^{19}F}$ (MHz). ZORA LR data from Aquino et al. [17], PBE0 functional, point nuclei. SOMO = unpaired ('singly occupied') 5f Np orbital*

LMO type	F	Np
Metal core	−12	−545
F core	−5	0.3
σ(Np-F)	−326	251
F LP_σ	282	
Metal SOMO	−1	−1937
$\sum$ all LP		−137
total calc.	−62	−2368

[a] LP = lone pair. For the F hyperfine coupling, the F core and Np–F bond contribution is for the same atom for which the hfc was calculated. Contributions from the other F cores and Np–F bonds were very small in this case. For the Np hfc, the contributions from all fluorines and Np–F bonds were added.

LWA calculations significantly underestimated the fluorine hfc. The importance of spin-polarization indicates that the Np–F interactions are not purely ionic. SO coupling enables covalent Np – ligand interactions that would not be possible if the unpaired electron were described by a pure spin-free Np $5f_{xyz}$ orbital as shown in the inset of Figure 12.3, as was already pointed out by Case [49].

In order to investigate the nature of the electron-nucleus magnetic interactions for NpF_6 [17] a quantitative breakdown of the calculated isotropic hfc constants was obtained terms of localized molecular orbitals (LMO) representing bonds, lone-pairs (LP), and non-bonding core orbitals. The analysis data are collected in Table 12.2, based on ZORA LR calculations (facilitating the analysis). The comparison with the MA data shows that the LR calculations for NpF_6 somewhat overestimate the magnitude of both the Np and F hfc, but in relative terms the LR results are reasonably close to the calculations that include SO coupling variationally. It has been shown that finite nucleas effects can be neglected for the Np hfc [17]. This initially unexpected result is a consequence of the nature of the Np hfc. Only about -200 MHz are attributed to the FC+SD mechanism, which is sensitive to finite-nucleus effects, while over 90% of the calculated value comes from the OP-SO spin-orbit cross term in the LR calculation. OP-SO is a pure spin-orbit effect that, in the case of NpF_6, involves orbitals other than $s_{1/2}$ and $p_{1/2}$. Consequently, the orbitals implicated in the Np hfc hardly contribute to the electron spin density at the Np nucleus, and therefore finite-nucleus effects are very small. The analysis data in Table 12.2 attribute most of the Np hfc directly to the unpaired $5f$ orbital (the 'SOMO'), with additional contributions from metal outer core orbitals. The positive contributions from the Np–F bonds reduce the magnitude of the Np hfc. The sign is therefore attributed to partial sharing of the unpaired electron with the ligand atoms via covalent interactions. A breakdown of the F hfc showed that about one-third of the total can be attributed to the OP-SO spin-orbit term, while two-thirds of the total coupling is a result of spin-density being transferred from the metal to the ligands via covalent interactions and spin polarization. The analysis data in Table 12.2 show a clear signature of the spin polarization mechanism in the large, opposing, contributions from the Np–F bond and one of the F LP orbitals.

12.3 X2C Hamiltonian and Molecular Properties from X2C Calculations

There has been much excitement in the relativistic quantum chemistry community regarding the possibility of constructing a formally exact two-component Hamiltonian for molecular calculations [50–56], as outlined in several review articles recently [56–59]. To be specific, an exact Hamiltonian can be constructed relatively straightforwardly at the one-electron level. Many-electron effects can be built into the approach in a pragmatic way with the help of model potentials [60–62]. For perspectives on a systematic incorporation of electron correlation into relativistic quantum chemical methods with many-electron wavefunctions, see Kutzelnigg [58]; Liu [59]; Saue [56]; Saue and Visscher [63]. For a perspective on DFT, see van Wüllen [64].

X2C ('eXact 2-Component') is an umbrella acronym [56] for a variety of methods that arrive at an exactly decoupled two-component Hamiltonian, with X2C referring to one-step approaches [65]. Related methods to arrive at formally exact two-component relativistic operators are, for example, infinite-order methods by Barysz and coworkers (BSS = Barysz Sadlej Snijders, IOTC = infinite-order two-component) [66–69] and normalized elimination of the small component (NESC) methods [70–77]. We discuss here an X2C approach as it has been implemented in a full two-component form with spin-orbit (SO) coupling and transformation of electric property operators to account for 'picture-change' (PC) corrections [14].

The two-component operators are constructed directly in matrix form in a basis set representation, assuming that the underlying four-component relativistic equation is expressed in a 'restricted kinetic balance' (RKB) basis set [78, 79]. The RKB small-component basis set is comprised of the set of functions $\{(\boldsymbol{\sigma}\cdot\hat{\boldsymbol{p}})\chi_r\}$ with $\{\chi_r\}$ being the large component basis. In the following paragraphs the reader can find a simplified description that has been adapted from Autschbach *et al.* [14] and Autschbach [80]; the formalism sketched herein was presented in detail by Liu and Peng [55]. A symbol formatted as M indicates an AO matrix in the basis of a set $\{\chi_r\}$ of spin-free basis functions ('atomic orbitals' = AOs). $\boldsymbol{M}$ indicates a 2×2 super-matrix in the space of two-component spinors. Whenever possible, four-component operators are represented in a split notation for the two-spinor components. In places where this is not convenient we use $\mathbb{M}$ to indicate a 4×4 super-matrix representation of a four-component operator or transformation matrix. A one-electron system is considered in the following unless noted otherwise. Atomic units with $\hbar = 1, m_e = 1, 4\pi\varepsilon_0 = 1, c = \alpha^{-1} = 137.035999074(44)$ are used.

The eigenvalue equation for the four-component Dirac operator represented in a basis set reads

$$\mathbb{H}^D\mathbb{C}^D = \mathbb{S}\mathbb{C}^D\varepsilon \tag{12.21}$$

The eigenvectors are collected in a coefficient matrix $\mathbb{C}^D$, containing the large and small wavefunction components, and ε is a diagonal matrix with the eigenvalues. The overlap matrix of the combined large and small basis is $\mathbb{S}$. The four-component Dirac operator couples the 'large' (*L*) and 'small' (*S*) components of the wavefunction (in other works we have used 'upper' and 'lower' component indices for the wavefunction in place of 'large' and 'small'), and it has the structure

$$\mathbb{H}^D = \begin{bmatrix} \boldsymbol{H}^D_{LL} & \boldsymbol{H}^D_{SL} \\ \boldsymbol{H}^D_{SL} & \boldsymbol{H}^D_{SS} \end{bmatrix} \tag{12.22}$$

(using the aforementioned split notation on the right-hand side). The idea underlying the two-component relativistic paradigm is that it ought to be able to de-couple the wavefunction components by block-diagonalizing the Hamiltonian,

$$\mathbb{U}^\dagger\mathbb{H}^D\mathbb{U} = \begin{bmatrix} \boldsymbol{H}^+ & \boldsymbol{0} \\ \boldsymbol{0} & \boldsymbol{H}^- \end{bmatrix} \tag{12.23}$$

and solving an electrons-only equation with the upper-left diagonal block of the Hamiltonian in the large-component basis set (the small-component basis is reflected in the operator matrices). Dropping the superscript '+', the two-component equation reads

$$\boldsymbol{HC} = \boldsymbol{SC}\varepsilon \tag{12.24}$$

The decoupling transformation in equation (12.23) has the block structure

$$\mathbb{U} = \begin{bmatrix} \boldsymbol{U}_{LL} & \boldsymbol{U}_{LS} \\ \boldsymbol{U}_{SL} & \boldsymbol{U}_{SS} \end{bmatrix} \tag{12.25}$$

To construct an electrons-only Hamiltonian only the blocks $\boldsymbol{U}_{LL}$ and $\boldsymbol{U}_{SL}$ need to be known.

Practical approaches to construct the decoupling transformation start from the 'modified Dirac equation' [81–84]

$$\begin{bmatrix} \boldsymbol{V} & \boldsymbol{T} \\ \boldsymbol{T} & (\frac{1}{4c^2}\boldsymbol{W} - \boldsymbol{T}) \end{bmatrix} \begin{bmatrix} \boldsymbol{C}_L \\ \boldsymbol{C}_S \end{bmatrix} = \begin{bmatrix} \boldsymbol{S} & \boldsymbol{0} \\ \boldsymbol{0} & \frac{1}{2c^2}\boldsymbol{T} \end{bmatrix} \begin{bmatrix} \boldsymbol{C}_L \\ \boldsymbol{C}_S \end{bmatrix} \varepsilon, \tag{12.26}$$

which is equivalent to the original four-component equation in the case of a RKB basis [55]. Here, $\boldsymbol{C}_L$ and $\boldsymbol{C}_S$ are separate coefficient matrices for the positive energy solutions for the large and small component two-spinors represented in their respective basis sets. The matrices

$$\boldsymbol{S} = \begin{pmatrix} \mathsf{S} & 0 \\ 0 & \mathsf{S} \end{pmatrix}, \; \boldsymbol{T} = \begin{pmatrix} \mathsf{T} & 0 \\ 0 & \mathsf{T} \end{pmatrix}, \; \boldsymbol{V} = \begin{pmatrix} \mathsf{V} & 0 \\ 0 & \mathsf{V} \end{pmatrix} \tag{12.27}$$

are constructed from the overlap (S), nonrelativistic kinetic energy (T), and one-electron external potential energy (V) AO matrices in the large-component basis. In addition, one needs the potential energy matrix in the small-component RKB basis,

$$\boldsymbol{W} = \begin{pmatrix} \mathsf{W}^0 + i\mathsf{W}^z & \mathsf{W}^y + i\mathsf{W}^x \\ -\mathsf{W}^y + i\mathsf{W}^x & \mathsf{W}^0 - i\mathsf{W}^z \end{pmatrix} \tag{12.28}$$

The matrix elements read

$$W^0_{rs} = \langle r|\hat{p}_x V \hat{p}_x + \hat{p}_y V \hat{p}_y + \hat{p}_z V \hat{p}_z|s\rangle, \tag{12.29a}$$
$$W^x_{rs} = \langle r|\hat{p}_y V \hat{p}_z - \hat{p}_z V \hat{p}_y|s\rangle, \tag{12.29b}$$
$$W^y_{rs} = \langle r|\hat{p}_z V \hat{p}_x - \hat{p}_x V \hat{p}_z|s\rangle, \tag{12.29c}$$
$$W^z_{rs} = \langle r|\hat{p}_x V \hat{p}_y - \hat{p}_y V \hat{p}_x|s\rangle. \tag{12.29d}$$

where $\hat{p}_u = -i\partial/\partial r_u$ is an element of the linear momentum operator, and χ_r, χ_s are basis functions used for the large component. First, the one-electron matrix eigenvalue equation (12.26) is solved. From the positive-energy solutions, one can then construct $\boldsymbol{X}$, which relates the coefficients of the small wavefunction components to those of the large components,

$$\boldsymbol{X}\boldsymbol{C}_L = \boldsymbol{C}_S \tag{12.30}$$

After solving for $\boldsymbol{X}$, the desired components of $\mathbb{U}$ are then obtained as follows [14, 55]:

$$\boldsymbol{U}_{LL} = \boldsymbol{S}^{-1/2}(\boldsymbol{S}^{-1/2}\tilde{\boldsymbol{S}}\boldsymbol{S}^{-1/2})^{-1/2}\boldsymbol{S}^{1/2}; \quad \boldsymbol{U}_{SL} = \boldsymbol{X}\boldsymbol{U}_{LL} \tag{12.31}$$

with

$$\tilde{\boldsymbol{S}} = \boldsymbol{S} + \frac{1}{2c^2}\boldsymbol{X}^\dagger \boldsymbol{T}\boldsymbol{X} \tag{12.32}$$

The X2C two-component Hamiltonian matrix of equation (12.24) is then obtained via

$$\boldsymbol{H} = \boldsymbol{U}_{LL}^{\dagger}\boldsymbol{V}\boldsymbol{U}_{LL} + \boldsymbol{U}_{LL}^{\dagger}\boldsymbol{T}\boldsymbol{U}_{SL} + \boldsymbol{U}_{SL}^{\dagger}\boldsymbol{T}\boldsymbol{U}_{LL} + \boldsymbol{U}_{SL}^{\dagger}\left(\frac{1}{4c^2}\boldsymbol{W} - \boldsymbol{T}\right)\boldsymbol{U}_{SL} \tag{12.33}$$

This is the version of X2C as it has been implemented in NWChem [14]. Due to the RKB condition, for simplicity the code currently requires a fully uncontracted basis set. It has been demonstrated that local decoupling schemes are suitable whereby an atomic and nearest-neighbor partitioning is employed in order to render the construction of the X2C Hamiltonian matrix less CPU and memory intensive [53, 85, 86].

One of the advantages of X2C and related methods is that the decoupling transformation matrices are available explicitly. They can therefore be re-used in order to transform matrix representations of other operators. Among those are the important spin-free operators for electric-field related properties such as dipole and higher multipole moments, and electric-field gradients (EFG) at atomic nuclei. PC errors arise when properties such as EFGs are calculated from a two-component framework but the operators are not, or not correctly, transformed to the two-component framework [24, 25, 27, 87, 88]. Using $\hat{P}$ to indicate a spin-free one-electron operator representing, for instance, a component of the electronic EFG or a dipole operator component, the operator matrix in the large and the small component (RKB) basis sets read

$$P_{rs}^{LL} = \langle r|\hat{P}|s\rangle \tag{12.34a}$$
$$P_{rs}^{SS} = \langle r|(\boldsymbol{\sigma}\cdot\hat{\boldsymbol{p}})\hat{P}(\boldsymbol{\sigma}\cdot\hat{\boldsymbol{p}})|s\rangle \tag{12.34b}$$

One then constructs

$$\boldsymbol{P}_{LL} = \begin{pmatrix} \mathsf{P}^{LL} & 0 \\ 0 & \mathsf{P}^{LL} \end{pmatrix}, \quad \boldsymbol{P}_{SS} = \begin{pmatrix} \mathsf{P}^{SS,0} + i\mathsf{P}^{SS,z} & \mathsf{P}^{SS,y} + i\mathsf{P}^{SS,x} \\ -\mathsf{P}^{SS,y} + i\mathsf{P}^{SS,x} & \mathsf{P}^{SS,0} - i\mathsf{P}^{SS,z} \end{pmatrix} \tag{12.35}$$

The matrix elements $P_{rs}^{SS,\{0,x,y,z\}}$ have the same structure as those of $\boldsymbol{W}$ in equation (12.29), but with the potential replaced by $\hat{P}$. The PC-transformed operator matrix for use in X2C calculations is then

$$\boldsymbol{P} = \boldsymbol{U}_{LL}^{\dagger}\boldsymbol{P}_{LL}\boldsymbol{U}_{LL} + \frac{1}{4c^2}\boldsymbol{U}_{SL}^{\dagger}\boldsymbol{P}_{SS}\boldsymbol{U}_{SL} \tag{12.36}$$

See Cheng and Gauss [89] for further discussion as to whether an electric field dependence of $\mathbb{U}$ also needs to be considered in first-order properties; numerical contributions at a scalar X2C level were essentially negligible in benchmark calculations. Contraction of $\boldsymbol{P}$ with a density matrix from a matching X2C calculation gives the PC-transformed property, whereas contraction of $\boldsymbol{P}_{LL}$ with the density matrix gives the property without PC correction. Examples are provided below. For a general (spin dependent) four-component operator with non-vanishing SL, LS blocks one needs to calculate in addition $\langle\chi_r|\hat{P}(\boldsymbol{\sigma}\cdot\hat{\boldsymbol{p}})|\chi_s\rangle$, $\langle\chi_r|(\boldsymbol{\sigma}\cdot\hat{\boldsymbol{p}})\hat{P}|\chi_s\rangle$, i.e., the operator in the mixed large/small component basis. An example would be the operators representing a perturbation from an external magnetic field or the magnetic field from a nuclear spin magnetic moment to calculate NMR parameters. Because these are second-order properties, the X2C decoupling to first order in the magnetic field is definitely required. An implementation in NWChem is under way. For details regarding the formalism see Sun *et al.* [90].

Table 12.3 lists orbital energies for the Hg atom that were obtained with different four-component and X2C implementations, including the one in the NWChem code. Table 12.4 lists the corresponding SO splittings of the orbital energies. In addition to using the (external) electron-nucleus potential V_{ext} in the de-coupling transformation, a variant using a model potential [60–62] V_{mp} was also implemented. In this case, the transformation to two-component form takes into account, in an approximate sense, the many-electron nature of a system. In NWChem, the model potential is the external potential plus the Hartree potential of the initial nonrelativistic atomic fragment guess used in the computations. This model potential is the same that is used in conjunction with ZORA and ZORA-based molecular property calculations [6, 13] (Section 12.2.1). Improvements are forthcoming. With V_{ext} for the X2C transformation, the results in Tables 12.3 and 12.4 obtained with the X2C implementations in NWChem and in the Dirac code [186] are identical, validating the code. It is noted that a ZORA calculation from Saue [56] gave a 1*s* orbital energy of

Table 12.3 *Orbital energies for the Hg atom (point nucleus) from four-component and X2C Hartree-Fock calculations*

Hg $-\varepsilon_i$/au	Dirac[a] DC	Dirac[a] X2C+AMFI	NWChem[b] X2C V_{mp}	Dirac[b] X2C V_{ext}	NWChem[b] X2C V_{ext}
$1s_{1/2}$	3074.239	3069.951	3076.319	3071.660	3071.660
$2s_{1/2}$	550.250	549.615	550.559	549.866	549.866
$2p_{1/2}$	526.846	526.451	527.412	528.109	528.109
$2p_{3/2}$	455.153	454.824	455.185	454.174	454.174
$3s_{1/2}$	133.110	132.974	133.182	133.030	133.030
$3p_{1/2}$	122.634	122.557	122.760	122.942	122.942
$3p_{3/2}$	106.541	106.477	106.551	106.320	106.320
$3d_{3/2}$	89.432	89.418	89.509	89.737	89.737
$3d_{5/2}$	86.016	85.989	86.033	85.781	85.781
$4s_{1/2}$	30.644	30.613	30.661	30.627	30.627
$4p_{1/2}$	26.119	26.104	26.149	26.203	26.203
$4p_{3/2}$	22.184	22.172	22.185	22.131	22.131
$4d_{3/2}$	14.792	14.792	14.808	14.866	14.866
$4d_{5/2}$	14.048	14.045	14.049	13.997	13.997
$4f_{5/2}$	4.469	4.472	4.474	4.524	4.524
$4f_{7/2}$	4.307	4.310	4.305	4.272	4.272
$5s_{1/2}$	5.099	5.094	5.102	5.097	5.097
$5p_{1/2}$	3.533	3.532	3.538	3.551	3.551
$5p_{3/2}$	2.838	2.837	2.837	2.829	2.829
$5d_{3/2}$	0.646	0.647	0.647	0.656	0.656
$5d_{5/2}$	0.571	0.572	0.571	0.566	0.566
$6s_{1/2}$	0.326	0.326	0.326	0.326	0.326

[a] Results obtained with the Dirac program, taken from Saue [56]. Uncontracted 24s19p12d9f large-component Gaussian-type basis. DC = Dirac-Coulomb. AMFI = two-electron SO atomic mean-field integrals. [b] NWChem and Dirac results with the same basis set as used to generate the data of Saue [56]. V_{mp} = model potential (see text). V_{ext} = 'external' electron-nucleus potential only.

Table 12.4 *Orbital energy SO splittings (ΔSO) for the Hg atom (point nucleus) from four-component and X2C Hartree-Fock calculations*

Hg ΔSO / au	Dirac[a] DC	Dirac[a] X2C+AMFI	NWChem[b] X2C V_{mp}	Dirac[b] X2C V_{ext}	NWChem[b] X2C V_{ext}
2*p*	71.693	71.627	72.227	73.935	73.935
3*p*	16.093	16.080	16.209	16.622	16.622
3*d*	3.416	3.429	3.476	3.956	3.956
4*p*	3.935	3.932	3.964	4.072	4.072
4*d*	0.744	0.747	0.759	0.869	0.869
4*f*	0.162	0.162	0.169	0.252	0.252
5*p*	0.695	0.695	0.701	0.722	0.722
5*d*	0.075	0.075	0.076	0.090	0.090

[a] Results obtained with the Dirac program, taken from Saue [56]. Uncontracted 24s19p12d9f large-component Gaussian-type basis. DC = Dirac-Coulomb. AMFI = two-electron SO atomic mean-field integrals. [b] NWChem and Dirac results with the same basis set as used to generate the data of Saue [56]. V_{mp} = model potential (see text). V_{ext} = 'external' electron-nucleus potential only.

about −3380 au, more than 300 au off, but a 'scaled ZORA' [91] calculation performed better than second-order DKH.

It is well known that two-electron contributions in the potential significantly improve SO splittings. The results of including the model potential in the X2C calculations (Table 12.4) reduces the SO splittings and brings the results close to four-component theory and an X2C calculation with AMFI two-electron SO integrals [92]. Percentage-wise, the improvement is particularly pronounced for the 4*f* and 5*d* shells. It is expected that by using a library of four-component atomic potentials, as proposed by van Wüllen and Michauk [62], the results will be improved further. The AMFI route has the advantage that spin-other-orbit integrals are included explicitly.

As an example for molecular property calculations with the X2C code, and an application to *f*-element chemistry, consider the EFG at the uranium nucleus in UO_2^{2+} and related complexes such as uranyl-carbonate and -nitrate complexes. It has previously been shown that complexation of uranyl by carbonate or nitrate ligands in the equatorial changes the sign of the EFG at the uranium nucleus [13, 93, 94]. Experimental estimates for uranyl coordinated by three nitrate ligands, derived from Mössbauer data, place the uranium EFG in equatorially coordinated UO_2^{2+} complexes at about +8 au [95]. Table 12.5 collects selected results from Autschbach *et al.* [14], showing that there is indeed a sign change of the EFG. An EFG of +8.3 au after PC corrections was obtained for the carbonate complex with a Coulomb-attenuated hybrid functional, which agrees slightly better with the experimental estimate than B3LYP.

Graphical representations of the EFG tensors are provided in Figure 12.4 (See color plate section.); the sign of the EFG represents the sign of its largest-magnitude component, V_{33}. Both for free UO_2^{2+} and the carbonate system, the corresponding principal axis is in the direction of the U–O_{yl} axis, but the EFGs have opposite signs. The positive sign is a signature of a '6*p* hole' [96, 97], i.e., a lack of electron density in the U 6p_σ orbital in UO_2^{2+} due to its involvement in bonding with the axial oxygens. However, the effect is overall more subtle as the 6*p* hole signature is not evident in free uranyl, only in the equatorially coordinated

Table 12.5 *Electric field gradient[a] (V_{33}, in au) at uranium, in uranyl and a uranyl-tris-hydrogen-carbonate complex used as a model for the U environment in $(NH_4)_4UO_2(CO_3)_3$*

UO_2^{2+}	before pcc	after pcc
SO V_{mp}	−5.98	−9.27
SO V_{nuc}	−5.73	−9.12
SC V_{nuc}	−4.65	−8.33
$[UO_2(HCO_3)_3]^-$	before pcc	after pcc
SO V_{mp}	12.8	6.54
SO V_{nuc}	13.2	6.80
SC V_{nuc}	14.1	6.71

[a] Data taken from Autschbach *et al.* [14]. B3LYP functional. pcc = picture-change corrections. SO = spin-orbit X2C calculation. SC = scalar (spin-free) X2C calculation.

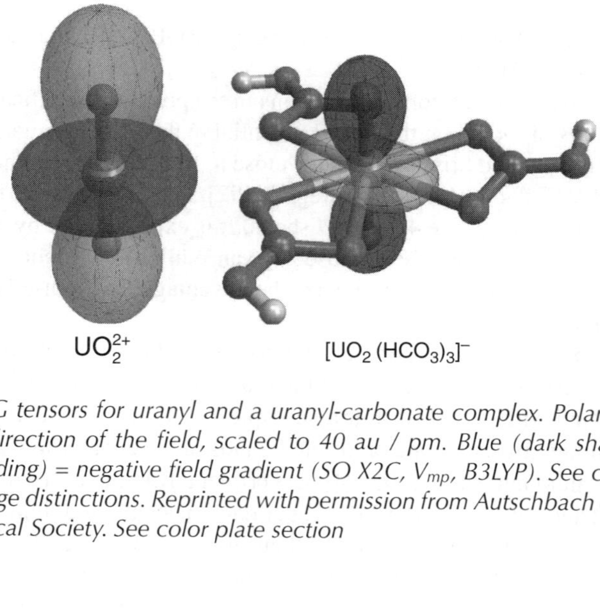

Figure 12.4 *EFG tensors for uranyl and a uranyl-carbonate complex. Polar plots of the field gradient in the direction of the field, scaled to 40 au / pm. Blue (dark shading) = positive, orange (light shading) = negative field gradient (SO X2C, V_{mp}, B3LYP). See color plate section for blue and orange distinctions. Reprinted with permission from Autschbach et al. [14]. © 2012 American Chemical Society. See color plate section*

forms. A detailed analysis of the effects from the equatorial carbonate ligands on the U EFG has been provided in [13] using a two-component ZORA implementation for EFGs that pre-dated the X2C development (see Section 12.2.2).

The data in Table 12.5 show that the PC corrections are of similar magnitude as the final EFG values. Clearly, the use of untransformed operators would be woefully inadequate. With PC effects being relativistic in origin, the differences are less drastic for EFGs of lighter atoms, and smaller overall, compared to EFGs, for valence-shell properties such as dipole and higher multipole moments. However, in the X2C framework where the transformation matrices to two-component form are already available there is little added computational overhead for obtaining PC-corrected electric properties such as EFGs and multipole moments, apart from the need for the AO matrix elements in equation (12.34b), and therefore the correct representation should be used by default to avoid unphysical results. It is noted that numerical integration can be a convenient strategy to obtain such matrix elements in the absence of an analytic integral code, in particular when parallel processing can be used to distribute the numerical grid.

12.4 Role of Dynamics on Thermodynamic Properties

In this section we focus on the thermodynamics of two fundamental reaction mechanisms that determine the ability of metal cations to form complex species: cation hydrolysis and solvent exchange. The thermodynamic modeling strategy is first discussed, followed by the hydrolysis of the U(IV), U(V), and U(VI) aqua ions. Finally, the solvent exchange thermodynamics of Cm(III) in dilute, perchlorate (ClO_4^-), chloride (Cl^-), and bromide (Br^-) solution will be discussed in which the effect counter anions on changes in the Cm(III) primary hydration number.

12.4.1 Sampling Free Energy Space with Metadynamics

The thermodynamic modeling strategy used to probe the free energies and activation barriers associated with solvent reaction and mechanisms was metadynamics with ab initio molecular dynamics (AIMD). The AIMD simulations highlighted here were performed using pseudopotential plane-wave DFT [98] with NVT [99, 100] Car-Parrinello simulations [101]. Details about these type of free energy simulations can be found in the recent work of Atta-Fynn *et al.* [102–104]. Metadynamics is a non-equilibrium molecular dynamics method that accelerates the sampling of the multidimensional free energy surfaces of chemical reactions by adding an *external* time-dependent bias potential, which is a function of a few *collective variables* to the Hamiltonian of the system [105–107]. The collective variable **S** is defined as a continuous and differentiable function of the system coordinates, $\mathbf{R}^N$ (e.g., bond distance, bond angle, etc.) that is capable of describing the chemical reaction or mechanism of interest. Essentially, metadynamics fills the minima of the free energy surface (FES) by periodically depositing repulsive Gaussian kernels centered on the current location of **S**(t) so that the system is forced to explore other regions of the FES. In its most general form, the bias potential, V, at an arbitrary point, **s**, along the trajectory at time t is given by

$$V(\mathbf{s},t) = \sum_{\substack{t'=0,\tau,2\tau,\ldots \\ t'<t}} H \exp\left(\frac{-V(\mathbf{S}(t'))}{\gamma k_B T}\right) \exp\left(-\frac{(\mathbf{s}-\mathbf{S}(t'))^2}{2\omega^2}\right) \tag{12.37}$$

where τ is the time interval between the deposition of successive Gaussians, H and ω are respectively the height and width, T is the simulation temperature, k_B is the Boltzmann constant, and γ is a positive dimensionless tempering (or tuning) parameter. γ is interpreted as follows: $\gamma = \infty$ corresponds to standard metadynamics, while a positive finite value of γ corresponds to well-tempered metadynamics [105]. After a sufficiently large t, $V(\mathbf{s}, t)$ will nearly flatten the FES, $F(\mathbf{s})$, along the collective variable space, and an unbiased estimator of F(s) is given by

$$F(\mathbf{s}) \approx -\lim_{t\to\infty}\left(1+\frac{1}{\gamma}\right)V((s),t) + constant \qquad (12.38)$$

The coordination number collective variable is a good choice for modeling solvent thermodynamics of ions in solution. The coordination number of species A with respect to species B is defined using a Fermi-Dirac function as

$$n_{A-B} = \sum_{B} \frac{1}{1+\exp\left(\kappa\left(r_{AB} - rcut\right)\right)} \qquad (12.39)$$

where r_{AB} is the distance between A and B, r_{cut} is the cut-off distance, and $\kappa > 0$ is the smearing parameter (with dimensions of inverse length). The term under the summation decays rapidly to zero when the bond distance r_{AB} is much larger than the cut-off radius r_{cut}, and approaches unity when r_{AB} is much smaller than r_{cut}.

12.4.2 Hydrolysis constants for U(IV), U(V), and U(VI)

Metadynamics was used to probe thermodynamics of hydrolysis by calculating the first acidity constant, pK_a, for all three oxidation states of aqueous uranium (U^{4+}(aq), UO_2^+(aq), and UO_2^{2+}(aq)). The collective variable employed to describe the deprotonation reactions is the coordination number of an arbitrary first-shell water oxygen atom with respect to all protons, n_{O-H}. The coordination number parameters were chosen to be κ =10Å^{-1}, and r_{cut} = 1.38Å. Starting with well-equilibrated NVT AIMD simulations, standard metadynamics simulations were carried out at 300 K with H = 0.063 kcal/mol and $\omega = \frac{0.1}{\sqrt{2}}$. The time between the addition of Gaussians was $\tau = 100\delta t$ for U^{4+}(aq) and $\tau = 20\delta t$ for UO_2^+(aq) and UO_2^{2+}(aq), where δt is the simulation time step. Once the free energy difference, ΔF, for the reaction was known, the first acid dissociation constant, pK_a, was computed as

$$pK_a = \frac{\Delta F}{k_B T \ln 10} \qquad (12.40)$$

For a metal cation in aqueous solution, hydrolysis occurs when the first-shell water molecule is sufficiently polarized to promote a proton transfer to a second-shell water molecule. The deprotonated species often forms a mixture of new aqua species and precipitates, depending on the pH and other thermodynamic conditions. Therefore, knowledge of the hydrolysis of actinide ions in aqueous solution enhances our ability to accurately predict the chemistry and thermodynamics of nuclear waste remediation. Hydrolysis occurs easily for highly charged actinide ions in solution, since the highly charged ion weakens the O-H bond in the first-shell water molecule, leading to the release of H^+(aq).

The first hydrolysis reactions of U(IV), U(V), and U(VI), which include the first-shell water molecules, are written as

$$U(H_2O)_8^{4+} \Longleftrightarrow U(H_2O)_7^{3+} + H_3O^+ \tag{12.41}$$
$$UO_2(H_2O)_5^{+} \Longleftrightarrow UO_2(H_2O)_4 + H_3O^+ \tag{12.42}$$
$$UO_2(H_2O)_5^{2+} \Longleftrightarrow UO_2(H_2O)_4^{+} + H_3O^+ \tag{12.43}$$

U^{4+}(aq), because of its high charge, is expected to be a much stronger acid in aqueous solution than UO_2^+(aq) and UO_2^{2+}(aq). The measured acid dissociation constants, pK_a, of U^{4+}(aq) is 0.54 ± 0.06 [108], while two different measurements of UO_2^{2+}(aq) are 5.24 ± 0.25 [108] and 5.58 ± 0.24 [109].

In Figure 12.5, the metadynamics free energy profiles are shown, where the reactant-free energy is taken as the zero reference point. From the figure, the free-energy difference between the reactant state and the product state can be estimated by taking the difference in the minima of curve at $n_{O-H} \approx 1$ and $n_{O-H} \approx 2$. The computed free energy difference for U^{4+}(aq) (black curve in Figure 12.5) is 2.52 kcal/mol. However, this free energy needs to be adjusted, since any of the 8 independent first-shell water molecules could have been used for the deprotonation reaction and an entropic energy correction $-TS$, where the entropy (given by $S = k_B \log 8$) must be added to the computed free-energy difference. At T = 300 K, this correction amounts to -1.24 kcal/mol, yielding the final estimate of the free-energy difference of 1.28 kcal/mol, and subsequently a pK_a value of 0.93. The simulated pK_a value is slightly greater than the experimental value by 0.4 pH units (experimental $pK_a = 0.54 \pm 0.06$ [108]). The free-energy profiles for UO_2^+(aq) and UO_2^{2+}(aq) in Figure 12.5 (red and blue curves) yielded values of 12.64 kcal/mol and 7.75 kcal/mol, respectively. Accounting for the entropic correction of $k_B T \log 5$ for each system results in the corrected first-hydrolysis free energies of 11.68 kcal/mol ($pK_a = 8.51$) and 6.79 kcal/mol ($pK_a = 4.95$). The simulated pK_a value for UO_2^{2+}(aq) deviates from one experimental value of 5.24 ± 0.26 [108] by 0.3 pH units and another value of 5.58 ± 0.24 by 0.6 pH units.

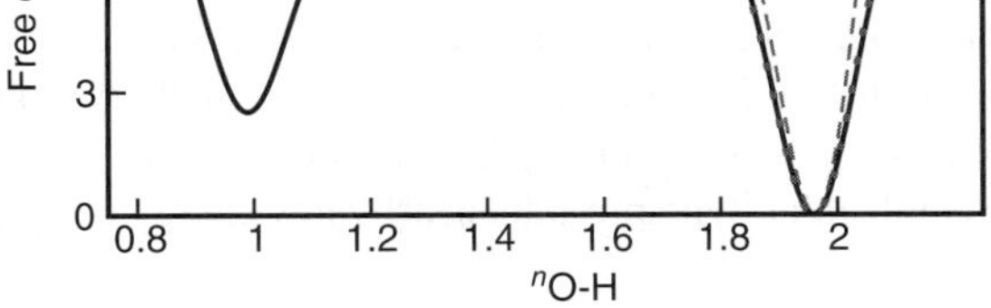

Figure 12.5 *Free energy profiles of the deprotonation of U^{4+} (aq), UO_2^+ (aq), and UO_2^{2+} (aq) as a function of n_{O-H}*

However, we note that if we consider the fact that the experimental value corresponds to the infinite dilution limit, then the simulated values for U^{4+}(aq) and UO_2^{2+}(aq) agree well with the experimental values. The $pK - a$ values of UO_2^{2+}(aq) from previous theoretical studies follow: 6.98 (AIMD simulation) [110], 9.61 [111], and -0.21 [112] (both were gas-phase simulations with continuum treatment of solvent effects). Based on the generally good agreement with experiment for the metadynamics pK_a values for U^{4+}(aq) and UO_2^{2+}(aq), we conservatively estimate the error in the free energy of UO_2^{+}(aq) to be 1 kcal/mol; this translates to a pK_a error of 0.7 pH units. The pK_a value of 8.5 suggests that UO_2^{+}(aq) is a weak acid in solution, and in effect says that UO_2^{+}(aq) may be an important species up to pH 8.5 (excluding the presence of strongly interacting ligands).

12.4.3 Effects of Counter Ions on the Coordination of Cm(III) in Aqueous Solution

The majority of hydrated actinide cation simulations that have been studied to date have been carried out as a single ion immersed in bulk water with electrostatic neutrality achieved by a uniformly distributed background charge of opposite sign; this is the so-called infinite dilution approximation. Alternatively, the charge on the metal center can be explicitly neutralized with counter ions. Because counter ions have spatial gradients, their presence in solution can influence the dynamical processes of nearby solvent molecules [113]. One could infer that several interesting phenomena could occur in simulations with counter ions versus simulations based on the infinite dilution model. Here we focus on the curium [Cm^{3+}(aq)] cation as an example to probe counter ion effects on metal ion hydration because a considerable amount of work has been devoted to understanding the hydration structure of Cm^{3+}(aq). However, despite these efforts its hydration structure and speciation is not fully understood.

Various experimental probes on the hydration-shell structure of Cm^{3+}(aq) reported in literature have yielded a wide range of coordination numbers. To mention a few, EXAFS experiments measured primary hydration numbers of 9 or 7 (based on the truncation of the EXAFS fitting data) in 1 M $HClO_4$ acid [114] and 10 in 0.25 M HCl acid [115]. High energy X-ray scattering (HEXS) experiments yielded a hydration number of 8.8 [114]. Time-resolved laser fluorescence spectroscopy (TRLFS) found coordination numbers between 8 and 9 in glycolic acid [116] and 8.9 in 0.1 M perchloric acid [117].

Theoretical studies also face a somewhat similar conundrum. Classical, ab initio, and QM/MM simulations with no counter ions have predicted coordination numbers of 8 or 9 [118–120]. However the EXAFS of the eightfold and ninefold coordinated structures were nearly indistinguishable [118]. Although this is an inherent limitation of EXAFS in the structural characterization of hydrated ions, it is, nevertheless, of fundamental interest to know the relative stability of the different coordination states. In dynamical simulations of hydrated ions, it is possible (and in fact, this is almost always the case because of exchange barriers) for the system to get locked into a single primary hydration state in dynamical simulations, and in cases where the sampling time is not long enough, which is particularly a problem for ab initio molecular dynamics, the nearby low energy coordination states cannot be accessed.

Given that these are highly concentrated solutions containing a large number of counterions, the coordination number may also be sensitive to the type of counterions. To better understand this effect, metadynamics free energy simulation was used to study the effect that highly concentrated counter ions have on the primary coordination number of Cm^{3+}(aq). These simulations were used to compute the free energy of changes in primary hydration of Cm^{3+}(aq) in the presence and absence of these counter ions. The counter ions were chosen based on their relative sizes and stability (average ligand lifetime) of their respective primary hydration shells. The sizes of the counter ions in descending order are ClO_4^->Br^->Cl^-, while their hydration stabilities in descending order are Cl^->Br^->ClO_4^- [114]. Each simulated system comprised a Cm^{3+}(aq) in a periodic cubic cell containing 64 water molecules (representing a density of water $\approx$1 g/cm^3). The charge on the metal center was neutralized by 3 counter anions; this corresponds to a counter anion concentration of 2.6 M. A uniform background charge was used for the counter anion free simulations. It is important to note that the conventional approach to modeling coordination number changes involves a single solvent molecule associated to or dissociated from the first coordination shell. Metadynamics has a significant advantage over the conventional approach, because the different coordination states are sampled multiple times in a single simulation. The metadynamics simulation is carried out for up to a tenth of a nanosecond to ensure multiple sampling of different coordination states and transition regions, and also to achieve convergence. Metadynamics convergence is particularly important if the reaction energy scales is less than 10 kcal/mol.

The free energy profiles of each system is given in Figure 12.6. See color plate section. Clearly energy profile for the counter anion-free solution (solid black curve) in the 9- and 10-fold coordination regions indicates that the tenfold state was never reached during the metadynamics simulations and that the barrier between 9- and 10-fold coordination is at least 17 kcal/mol. On the other hand, the tenfold state was accessed, albeit with a short coordination shell lifetime, in the counter anion solutions. Thus the analyses will be focused on the sevenfold, eightfold, and ninefold coordination states. In Table 12.6, the equilibrium first-shell coordination number n_{eq}, activation barriers and free energies differences between the equilibrium coordination state and neighboring coordination states are reported. Also reported in Table 12.6 are the relative first-shell solvent concentrations

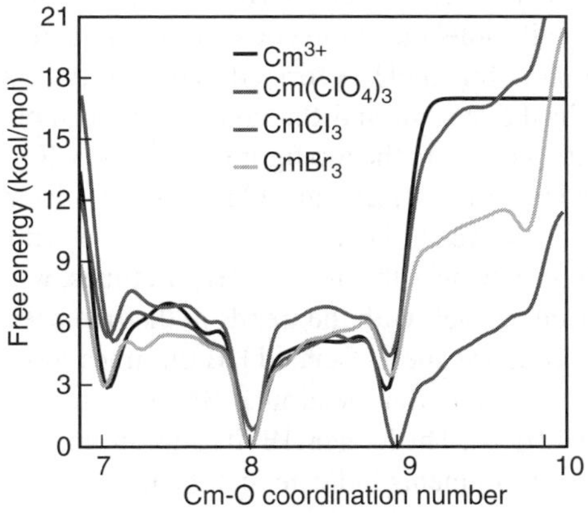

Figure 12.6 *Free energy profiles of Cm coordination with respect to O in Cm^{3+} (aq), Cm^{3+}+3ClO_4^- (aq), Cm^{3+}+3Cl^- (aq), and Cm^{3+}+3Br^- (aq). Reprinted with permission from Atta-Fynn et al. [118]. © 2013 American Chemical Society. See color plate section*

Table 12.6 *Equilibrium coordination numbers (n_{eq}), free energy differences (ΔA), reaction barrier ($A^{\dagger}$) for each system, and estimates of the solvent population from the free energies*

		ΔA (kcal/mol)		$A^{\dagger}$ (kcal/mol)		Coordination state (%)		
	n_{eq}	$8 \rightarrow 9$	$8 \rightarrow 7$	$8 \rightarrow 9$	$8 \rightarrow 7$	[7]	[8]	[9]
Cm^{3+} (aq)	8	2.8	2.9	5.1	7.0	0.76	98.34	0.90
Cm^{3+}+3ClO4$^-$ (aq)	9	−0.8	4.3	4.6	5.7	0.01	19.67	80.31
Cm^{3+}+3Cl^- (aq)	8	4.4	5.4	6.8	7.6	0.01	99.93	0.06
Cm^{3+}+3Br^- (aq)	8	3.5	3.0	6.0	5.5	0.68	99.03	0.29

of the sevenfold, eightfold, and ninefold coordination states. The concentrations were estimated using the chemical equilibrium equations $exp(-\frac{\Delta A_{7\rightarrow 8}}{k_B T}) = \frac{[8]}{[7]}$, $exp(-\frac{\Delta A_{8\rightarrow 9}}{k_B T}) = \frac{[9]}{[8]}$ and $[7] + [8] + [9] = 1$. The predicted [9] : [8] population ratio of 80% : 20% seen in the ClO_4^- simulation, resulting in a weighted coordination number of 8.8, is in excellent agreement with high energy x-ray scattering (HEXS) data for 0.523 M Cm^{3+} in 1.0 M perchloric acid [114]. Our results are also in reasonable agreement with the 90% : 10% measured in time-resolved laser fluorescence spectroscopy (TRLFS) experiment at much lower concentrations of 0.0005 M Cm^{3+}(aq) in 0.1 M perchloric acid,18 showing the subtle effects of perchlorate in the 0.1–1.0 M concentration range. In counter ion-free, Cl^-, and Br^-solutions, the analyses show that the eightfold population is dominant by least 98%.

Solvent exchange around a metal ion, that is, changes in the ion first-shell coordination number, involves the breakage and subsequent reformation of the hydrogen bonds (HB) by the solvent molecules visiting or departing from the shell. The rate at which the HBs are broken and formed is directly related to the characteristic lifetimes of the bonds. The HB lifetimes can be probed from the hydrogen dynamics via HB survival time correlation function $S(t)$ [103, 121–128]. For each fully equilibrated system, we computed $S(t)$ for the HBs formed between (i) the first-shell solvent molecules (those water molecules coordinated with Cm) and all other molecules, (ii) the counter anion molecules and all solvent molecules, and (iii) all molecules (solvent-solvent and anion-solvent) in the system. Figure 12.7 depicts the plots for the HBs formed between the first-shell water molecules to the other molecules in the system and the counter anion molecules to all the solvent molecules. See color plate section. In the top figure, Cm^{3+} + 3ClO_4^- shows the fastest decay followed by Cm^{3+}, with the Cm^{3+} + 3Br^- and Cm^{3+} + 3Cl^- decays being the slowest and nearly degenerate. Thus, relative to the Cm^{3+} (aq) solution with no counter anions, the Cl^- and Br^- counter anions enhances the average HB lifetimes, while ClO_4^- reduces the HB lifetimes. This trend correlates well with the trends in the free energy differences and reaction barriers. In the bottom figure, the associated HB lifetimes show that the lifetime around Cl^-(1.5 ps) is twice as large as the lifetime around Br^- (0.76 ps) and an order of magnitude larger than that of ClO_4^- (0.16 ps). The longer HB lifetime around Cl^- and Br^- is not surprising, since experimental measurements indicate that the mean water residence times around $Cl^-$and $Br^-$are much longer compared with neat liquid water, and that the hydration shells of these ions are sufficiently rigid [129]. ClO_4^- on the other hand, has been characterized in EXAFS experiments as a large, sluggish, weakly coordinating ion in solution [114]. This, again, confirms the trends in the free energy differences and the reaction barriers suggesting

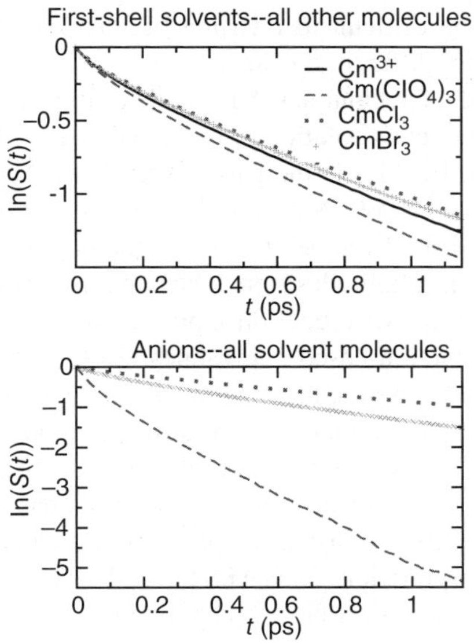

Figure 12.7 *HB survival autocorrelation function of Cm^{3+} (aq), Cm^{3+}+$3ClO_4^-$ (aq), Cm^{3+}+$3Cl^-$ (aq), and Cm^{3+}+$3Br^-$ (aq). The vertical axis is on a logarithmic scale. Reprinted with permission from Atta-Fynn et al. [118]. © 2013 American Chemical Society. See color plate section*

that large reaction barriers imply that solvent exchange rate is slower and hence the mean HB lifetime is relatively longer.

Overall, ab initio molecular dynamics studies of aqueous Cm(III) in the absence and presence of counter anions has revealed that the Cl^- and Br^- solutions makes the first coordination shell more stable relative to the counter ion-free solution, and consequently, the eightfold coordination state is the preferred primary hydration number with a 98% probability. The ClO_4^- counter ion, on the other hand, makes the first coordination shell less stable relative to the counter ion-free solution, with the ninefold and eightfold states existing in an 80% to 20% ratio. The stability of the coordination shells were rationalized by the dynamics of the mean hydrogen bond lifetimes. We found that the more stable coordination shells (i.e., Cl^- and Br^- solutions) have relatively long mean hydrogen bond lifetimes, while less stable coordination shells imply otherwise. In essence, our work indicates that the stability of the coordination shell of Cm(III) in different solution environments can be characterized through free energy landscape of the ion coordination. This approach is useful for differentiating between the effects of different ligands on hydration properties of other metal ions.

12.5 Modeling of XAS (EXAFS, XANES) Properties

Many import chemical and separation processes in actinide systems are dictated by the solute-solvent structure and dynamics in water and other fluids that are under extremely non-ideal conditions such as high concentrations, and high temperatures and pressures. However, many of these processes are difficult to interpret because the structure and dynamics of solute species in solution are difficult to observe.

One of the few experimental techniques that can determine the local water structure surrounding actinide ions in aqueous solution is Extended X-Ray Absorption Fine Structure spectra or EXAFS spectra. EXAFS is the part of the X-ray Absorption Spectrum (XAS) that is approximately in the 50–1000 eV region above the absorption edge of an atom. Since EXAFS involves backscattering from nearby atoms, the physics is essentially captured by the phase shifts and scattering amplitudes from which the spectrum can be calculated using a damped spherical photoelectron wavefunction approximation. Nowadays these are routinely computed using Green's function-based approaches [130, 131]. It is a particularly useful technique because it is able to provide information about an atom's local environment. It is most recognized as a technique for analyzing the structure of amorphous crystalline solids; however, in recent years it has been extended to the analysis of aqueous species and other non-crystalline materials. A key advance in these analyses has been the use of molecular dynamics (MD-EXAFS) [132] and first-principles-based molecular dynamics simulations (AIMD-EXAFS) in the interpretation of EXAFS [134]. Even though MD-EXAFS has been around for over a decade, it has never been exceptionally accurate. The origin of the errors has always been suspected to be from the imprecision of classical molecular dynamics potentials used. This proposition has recently been shown to be most likely correct. Recent studies by Fulton *et al.* have shown that highly accurate EXAFS experiments on a number of strongly interacting aqueous divalent and trivalent metal ions can be reliably interpreted when highly accurate first principles molecular simulations are used in the MD-EXAFS analysis, aka AIMD-EXAFS [133, 134].

Another closely related technique that can be used for obtaining local structure in complex condensed phases is X-ray absorption near edge spectroscopy or XANES. XANES focuses on the XAS absorptions $\approx$50 eV from the rising edge and involves excitations from the relevant core state to increasingly high-energy virtual states. These excitations give insight into the chemical state of the atom (for example, coordination, bonding, oxidation state). The analysis of XANES is considerably more difficult than EXAFS, because it requires a full electronic structure treatment of the absorbing center and neighboring atoms constituting the chemical environment. Over the years various approaches have been pioneered like the static exchange approximation (STEX) and static approaches [135–137]. All of these approaches incorporate the relaxation of the core hole and valence orbitals indirectly by manual preparation of the core hole state and/or use of modified core potentials that mimic the core relaxation when a core electron is excited. The absorption spectra are computed by calculating the transition dipole elements of the initial and final states of interest. In a nutshell, these methods are very sensitive to the choice of initial and final states. Recently, linear-response time-dependent density functional theory (LR-TDDFT) has been extended to treat core level excitations [138–142]. The advantage of this approach is that no manual preparation of the core hole state is needed and peak separations and trends are properly captured. However, the quality of the results is dependent on the exchange-correlation functional choice [143]. This approach has been applied successfully to several studies over the last few years. Other approaches have also been reported [144–146]. Since this is not a complete overview of all theoretical approaches that have been used to simulate XANES, we refer the reader to comprehensive reviews [130, 131, 147–149] on the subject.

We also highlight recent use of MD-XAFS with ab initio molecular dynamics (AIMD) simulations of U(IV) and U(V) actinide ions. This type of analysis has also been done by us for Cm(III) [118] and U(VI) [150]. The goals of the simulations were to shed light on the solvent-shell geometries and electronic structure of these ions and test the reliability of the aforementioned simulation schemes by comparing our data to available experimental data.

Finally we conclude this section by discussing a recent application of our LR-TDDFT approach for core excitations, which has been developed in the NWChem program [1, 140] to study the Cl K-edge spectra of the closed-shell $[UOCl_5]^-$ complex and the open-shell tetravalent actinide hexahalides series $[UCl_6]^{2-}$, $[NpCl_6]^{2-}$, and $[PuCl_6]^{2-}$ containing two, three, and four 5f-electrons, respectively. Cl K-edge spectroscopy has been used to study the electronic structure of actinide complexes, and to probe the role of 5f and 6d orbitals in the formation of covalent bonds [151, 152]. Across the actinides series the 5f-orbitals become more contracted, and their role in bonding changes, which can be captured by the Cl K-edge spectra. We also probe the role of the environment of the host crystal and dynamical effects due to atomic motion on the shape of the spectra.

12.5.1 EXAFS of U(IV) and U(V) Species

It is well established experimentally that the coordination shell structure of U(VI), which exists in solution as the uranium di-cation, UO_2^{2+}(aq), is characterized by two oxygens covalently bound axially to U and five solvent molecules in an equatorial plane [158, 163–165]. We must mention however that recent high energy x-ray scattering studies have revealed the possibility of a fourfold coordinated UO_2^{2+}(aq) [165]. U(V), which exists in solution as the UO_2^+(aq), is generally thought to be short-lived in aqueous solution and rapidly disproportionates into UO_2^{2+}(aq) and U^{4+}(aq). However, U(V) can be stabilized by various organic ligands or in concentrated aqueous carbonate solutions and nonaqueous solutions [166–169]. Unfortunately, experiments have not searched for, nor identified, U(V) under environmentally relevant conditions, since the disproportionation reaction is assumed to render U(V) insignificant.

Table 12.7 lists the structural properties of the hydration shell of UO_2^+(aq) and UO_2^{2+}(aq). For the purposes of comparison, past experimental and theoretical data for the first shell of AnO_2^+ (An = U, Np, Pu) are also reported in Table 12.7. The hydration shell structure of UO_2^{2+}(aq) has been described in detail elsewhere [150] so we will focus on the shell structure of UO_2^+(aq). As shown in Figure 12.8, the AIMD simulations indicate that the first shell of UO_2^+(aq) has five water molecules in the equatorial plane, in contrast to the QM/MM prediction of 4.51. The predicted U(V)=O_{ax} distance is very close to previous measurements of other actinyl(V) ions (Np(V) and Pu(V)) and are greater than the previous predicted value by 0.07Å. Also, our average first-shell U-O_{eq} bond distance is slightly longer than the previous simulated value, which is expected since the first shell of the AIMD simulations contains more water ligands. Previous gas-phase structures exhibit slightly longer UO_{eq} bonds as expected. Relative to UO_2^{2+}(aq), UO_2^+(aq) shows a lengthening of 0.08Å and 0.1Å for the U=O_{ax} and UO_{eq} bonds, respectively, because of reduced electrostatic attraction. Other first-shell properties of UO_2^+(aq) and UO_2^{2+}(aq), such as the intramolecular water geometry and tilt angles, compare closely.

Table 12.7 *Average first hydration shell parameters of AnO_2^+ (An = U, Np, Pu)*

System	$N(H_2O)$	$R(An{=}O_{ax})^b$	$R(An\text{-}O_{eq})^c$	ψ^d_{tilt}
UO_2^+ (aq)	5	1.85	2.54	34.9
UO_2^+ (aq) (QM/MM)[e]	4	1.78	2.51	
$UO_2(H_2O)_5^+$ (gas-phase)[f]	5	1.81–1.83	2.56–2.62	
$UO_2(H_2O)_5^+$ (H_2O) (gas-phase)[g]	5	1.78	2.62	
NpO_2^+ (aq) (EXAFS)[h]	5	1.78	2.62	
PuO_2^+ (aq) (XANES)[i]	4	1.84	2.45	
UO_2^{2+} (aq)	5	1.77	2.44	32.9
U^{4+} (aq)	8.7		2.45	24.6
U^{4+} (aq) (QM/MM)[j]	9		2.45	
U^{4+} (aq) (EXAFS)	9–11[k], 9–10[l]		2.42, 2.40	

[a]Number of water molecules in the first coordination shell. [b]An=O_{ax} bond distance in Å, where O_{ax} denotes an axial O atom bonded to An. [c]An-O bond distance in Å, O belongs to a first-shell water. [d]Average water tilt angle. [e] [153], [f] [154–156]. [g] [157]. [h] [158]. [i] [159]. [j] [160]. [k] [161]. [l] [162].

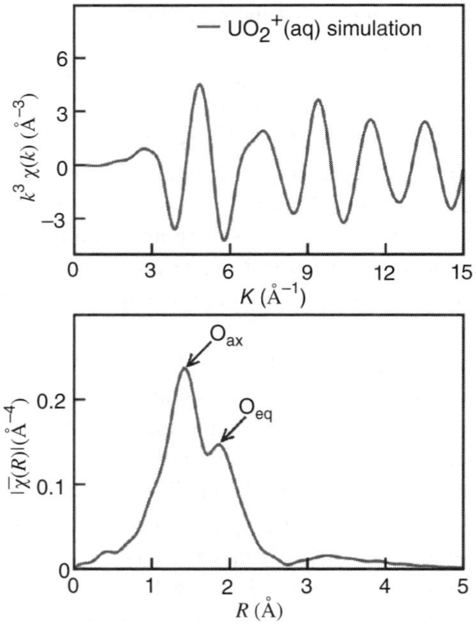

Figure 12.8 *Simulated EXAFS spectra of UO_2^+ (aq). The upper panel shows a $k^3\chi(k)$ EXAFS spectrum; the lower panel shows the magnitude of the Fourier transform, $|\bar{\chi}(R)|$, of $k^3\chi(k)$. Reprinted with permission from Atta-Fynn et al. [104]. © 2012 American Chemical Society*

To the best of our knowledge, there is no experimental EXAFS data for $UO_2^+(aq)$; therefore, the simulated EXAFS is a prediction that could be of some use in probing the hydration shell structure in experimental measurements. It has been shown, using the MD-EXAFS method, that the spectra of UO_2^{2+} from AIMD agrees well with experimental data. In Figure 12.8, the $k^3\chi(k)$ of UO_2^+ (top panel), and the magnitude of Fourier transform of $k^3\chi(k)$, $|\bar{\chi}(R)|$ (bottom panel) are depicted (phase-shift corrections were not included in the radial distances). The axial and equatorial oxygen locations in the spectrum are shown. There is a strong similarity between the simulated UO_2^+ EXAFS and the reported theoretical and experimental spectra for /usix. This is obviously due to the similarities in coordination shell geometries and radial distance distributions.

Looking at the $U^{4+}(aq)$ data in Table 12.7, it is evident that the average primary hydration number of 8.7 (mixture of eightfold and ninefold coordinated geometries) agrees with previous simulations and the lower limit of EXAFS measurements. We note that the underestimation of the first-shell coordination number of actinide ions appears to be a generic problem with DFT-GGA. The analyses of the first-shell geometry indicated that the dominant eightfold geometry, $U^{4+}(aq)$, was square antiprismatic (SAP), while the dominant eightfold geometry, $[U(H_2O)_9]^{4+}$, was tricapped trigonal prism (TTP).

In Figure 12.9, plots of the simulated $k^3\chi(k)$ EXAFS (top panel) and $|\bar{\chi}(R)|$ (bottom panel) are depicted and compared with recent experimental data by Ikeda-Ohno *et al.* [162]

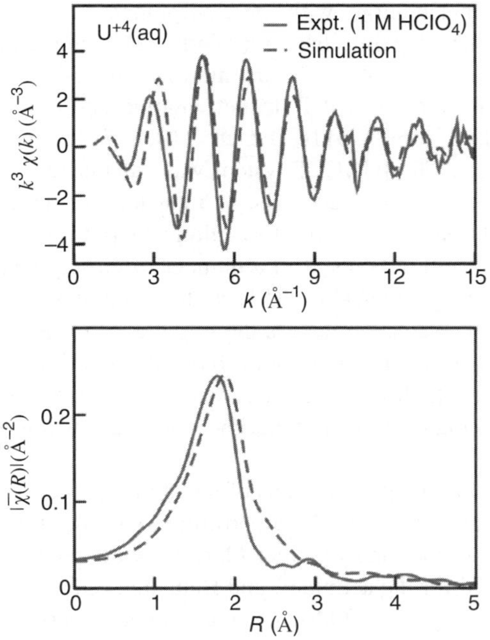

Figure 12.9 *Comparison of the simulated and experimental EXAFS spectra of U_4^+ (aq). The upper panel shows the $k^3\chi(k)$ EXAFS spectra; the lower panel shows the magnitude of the Fourier transform, $|\bar{\chi}(R)|$, of $k^3\chi(k)$. Reprinted with permission from Atta-Fynn et al. [104]. © 2012 American Chemical Society*

for 0.05 M U^{4+} in a 1 M $HClO_4$ solution. As can be seen in the top panel, the simulated spectrum almost matches the frequency and amplitude oscillations of the experimental curve. The agreement can also be seen in the Fourier-transformed data in the bottom panel. The central peak mismatch in the bottom panel is due to the fact that the simulated average UO distance is longer than the corresponding experimental value.

12.5.2 XANES Spectra of Actinide Complexes

In our Gaussian basis set-based molecular calculations, the actinide atoms (U, Np, Pu) were represented with small core Stuttgart relativistic ECPs which replace 60 core electrons and corresponding basis sets containing g-functions [170], while the ligand atoms (Cl and O) were represented with the aug-cc-pvtz basis [171]. The actinide basis sets are of triple zeta quality. In general and as expected, the absolute positions of the core spectra are dependent on the choice of exchange-correlation functional as well as the basis set [143]. We have tested the B3LYP (20% HF) [172–175] PBE0 (25% HF) [176] and Becke Half-and-Half (50% HF) [172] functionals, respectively. The PBE0 functional gave the best agreement with experiment for all of the actinide complexes considered in this study [151]. The spectra were shifted by ≈58.4 eV to match experiment and uniformly Lorentzian broadened (1.0 eV). Unlike valence and charge-transfer (CT) excitations where range-separated exchange-correlation functionals offer significant improvement, they offer little or no improvement for the strongly localized core spectra. For the molecular calculations with the environment, the actinide complex in each case was placed in a set of point charges (determined using a Mulliken analysis) representing the host crystalline solid. The actinide hexahalides $[XCl_6]^{2-}$(X= U, Np, Pu) were performed with unrestricted DFT (triplet, quartet, and quintet), and the calculated ground state $\langle S^2 \rangle$ values were 2.0116 (exact: 2.0), 3.7725 (exact: 3.75), and 6.0833 (exact: 6.0), respectively. The $[UOCl_5]^-$system was calculated with restricted DFT. The observed small spin-contamination could be an indication of the presence of higher spin state character. In addition, the spin-orbit coupling for $[UCl_6]^{2-}$and $[NpCl_6]^{2-}$is ≈0.20–0.25 eV [177, 178] and will also result in a number of closely spaced excited states that may not be experimentally resolvable [151] as the natural width of Cl K-edge transitions is ≈0.64 eV [179]. We cannot assess the importance of these contributions because our implementation does not include spin-orbital and multiple excited-state electronic configuration effects [148]. However, the key features and the overall shape of the spectra are in qualitative agreement with experiment as we will discuss in the subsequent paragraphs.

To capture the role of dynamical motions of the atoms, planewave-based Car-Parrinello AIMD simulations discussed earlier were performed with the PBE exchange-correlation functional [179]. Norm-conserving Troullier-Martins pseudopotentials [180] with relativistic corrections were used for the actinides, and Hamann pseudopotentials [181] were used for the other atoms in the system. Other calculation parameters include an energy cutoff of 100 Ry, density cutoff of 200 Ry, a simulation time step of 0.121 fs, and a fictitious electron mass of 600 au, respectively. The unit cell dimensions, symmetry, and the atomic coordinates were taken from experimental data [182, 183]. Replacing all hydrogen atoms with deuterium allows for larger integration time steps. The simulation temperature was set at 300 K and controlled using a Nose-Hoover thermostat [184]. Approximately 10 ps of dynamics were collected for each of the structures. In our sampling, the first 3 ps were considered the equilibration phase and ignored.

Cl K-edge core absorption spectra calculations were performed on all of the complexes using 1) experimental molecular geometries without the environment, 2) experimental molecular geometries with the environment, and 3) 100 molecular geometry snapshots extracted from periodic AIMD runs, which were averaged to form the final spectra. A total of 300 excited states from the Cl 1s states were calculated for each system. The geometries for all of the complexes including the crystal structures were taken from experimental data [182, 183]. The optimized geometries do not change the overall shape and nature of the spectrum.

The ground state of $[UCl_6]^{2-}$ is an open-shell triplet ($5f^2$) where the HOMO/LUMO character of the alpha electrons are dominated by Cl 3p (HOMO-2), U 5f (HOMO-1), U 5f (HOMO), U 5f (LUMO), U 5f (LUMO+(1-4)), while orbitals containing U 6d are found at higher energy. For the beta electrons we observe the Cl 3p (HOMO-2), Cl 3p (HOMO-1), Cl 3p (HOMO), U 5f + Cl 3s (LUMO), and U 5f and U 5f + Cl 3s/3p (LUMO +(1-4)), respectively. All of the higher states are a mixture of U and Cl states. The shape of the spectrum (Figure 12.10) remains largely unaffected with and without the environment. The shoulder between 2821.75 eV and 2823 eV, which consists of two close peaks separated by $\approx$0.75 eV and composed of excitations from the Cl 1s $\rightarrow$ U 5f states, is mostly smoothened out when the averaged spectrum is calculated using AIMD snapshots. The prominent peak at $\approx$2824 eV arises from excitations from the Cl 1s $\rightarrow$ U 6d + Cl 3s/3p states. On the whole, our calculated spectrum is in better agreement with experiment, in terms of the peak intensities and locations, compared with published gas phase computational results [151].

The ground state of $[NpCl_6]^{2-}$ is an open-shell quartet ($5f^3$) where the HOMO/LUMO character of the alpha electrons are dominated by Cl 3p (HOMO-3), Np 5f (HOMO-2), Np 5f + Cl 3p (HOMO-1), Np 5f + Cl 3p (HOMO), Np 5f (LUMO), Np 5f (LUMO+(1-3)).

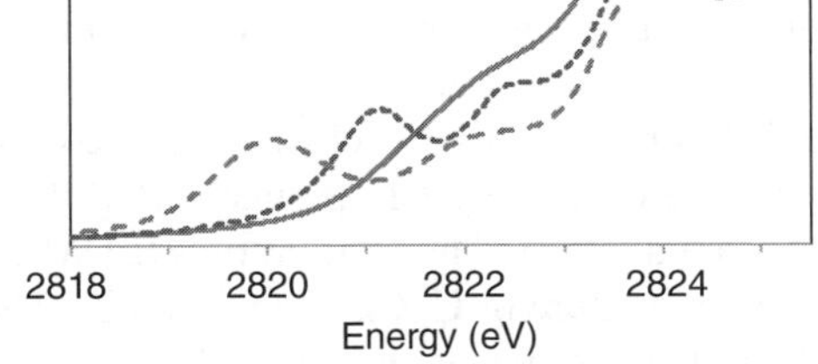

Figure 12.10 *Comparative Cl K-edge spectra of $[UCl_6]^{2-}$, $[NpCl_6]^{2-}$, $[PuCl_6]^{2-}$ in the presence of the host environment and averaged over 100 configurations extracted from AIMD simulations. The spectra have been normalized relative to each other so that the peaks $\approx$2826 have the same intensity. Reprinted with permission from Govind and de Jong [185]. © 2014 Springer*

Although the 5f orbitals in uranium were found to be atomic-like, a much larger mixing of 5f with the Cl 3p is found for Np. For the beta electrons we observe the Cl 3p (HOMO-3), Cl 3p (HOMO-2), Cl 3p (HOMO-1), Cl 3p (HOMO), and Np 5f (LUMO), and Np 5f and Np 5f + Cl 3s/3p (LUMO+(1-3)), respectively. The Np 6d orbitals are found above the Np 5f orbitals, and all of the higher states are a mixture of Np and Cl orbitals. The shoulder (Figure 12.10), which is composed of two closely spaced features in $[UCl_6]^{2-}$, are clearly separated by ≈1.5 eV and red-shifted in $[NpCl_6]^{2-}$. The first feature ≈ 2821 eV arises from transitions from Cl 1s → Np 5f states, while the second peak ≈2822.5 eV is dominated by transitions from Cl 1s → Np 5f + Cl 3s/3p states, respectively. The third peak ≈2824 eV is composed of excitations from Cl 1s → Np 6d + Cl 3s/3p states. Although there are differences with and without the environment and including dynamics, the overall shape and the number of features are preserved.

The ground state of $[PuCl_6]^{2-}$ is an open-shell quintet ($5f^4$) where the HOMO/LUMO character of the alpha electrons are dominated by Cl 3p (HOMO-3), Pu 5f + Cl 3p (HOMO-2), Pu 5f + Cl 3p (HOMO-1), Pu 5f + Cl 3p (HOMO), Pu 5f (LUMO), and Pu 5f (LUMO + (1–2)). A further increase in mixing of the 5f with the Cl 3p is observed for Pu compared to both U and Np. For the beta electrons we observe the Cl 3p (HOMO-3), Cl 3p (HOMO-2), Cl 3p (HOMO-1), Cl 3p (HOMO), Pu 5f + Cl 3s/3p (LUMO), and Pu 5f and Pu 5f + Cl 3s/3p (LUMO+(1–2)). The Pu 6d orbitals are found above the Pu 5f orbitals, and higher states are a mixture of Pu and Cl orbitals. There are two clear early features (Figure 12.10) separated by ≈2.0 eV and further red-shifted compared with $[NpCl_6]^{2-}$. The first feature ≈2820 eV arises from transitions from Cl 1s → Pu 5f states, while the second peak ≈2822 eV is dominated by transitions from Cl 1s → Pu 5f + Cl 3s/3p states, respectively. The third peak 2824 eV is composed of excitations from Cl 1s → Pu 6d + Cl 3s/3p states. As in the $[NpCl_6]^{2-}$spectra, there are differences in the spectra, however, the overall nature of the spectrum remains largely unaffected as in $[PuCl_6]^{2-}$even with inclusion of dynamics.

Figure 12.10 also shows a clear trend. The first and second features are a direct consequence of the behavior of the 5f electrons in the U, Np, Pu series. In addition to not being shielded by the filled 6s and 6p shells, the 5f electrons also shield each other poorly from the nucleus resulting in a lowering of their orbital energies with increasing atomic number. This is reflected in the increased spacing and red-shifting of the first and second features as one goes from $[UCl_6]^{2-}$to $[PuCl_6]^{2-}$. The second peak, as discussed above, is a result of the mixing of the actinide 5f orbitals with the Cl 3s/3p states, which increases as one moves from U → Pu. The prominent feature at ≈2824 eV is mostly unaffected in the three spectra, reflecting the much smaller shift in orbital energies compared to the 5f in the U, Np, Pu series.

The ground state of $[UOCl_5]^-$ has a closed-shell configuration ($5f^0$) where HOMO/LUMO character is dominated by Cl 3p states (HOMO-2, HOMO-1, HOMO), while the unoccupied states are dominated by U 5f orbitals (LUMO, LUMO+1, LUMO+2, LUMO+3). Higher states are a mixture of U, Cl, and O orbitals. The computed Cl K-edge spectra in Figure 12.11 shows two prominent features ≈2821 eV (Cl 1s → U 5f) and 2822.5 eV (Cl 1s → U 5f + Cl 3s + O 2p) that are distinct from the $[UCl_6]^{2-}$complex. The Cl 1s → U 6d + Cl 3s/3p transitions are split into two peaks: one at ≈2824 eV and a second (shoulder) at ≈2826 eV. These are similar in nature to the analogs in the actinide hexahalide complexes and are also located in the same position in the spectrum. On the whole, the spectra

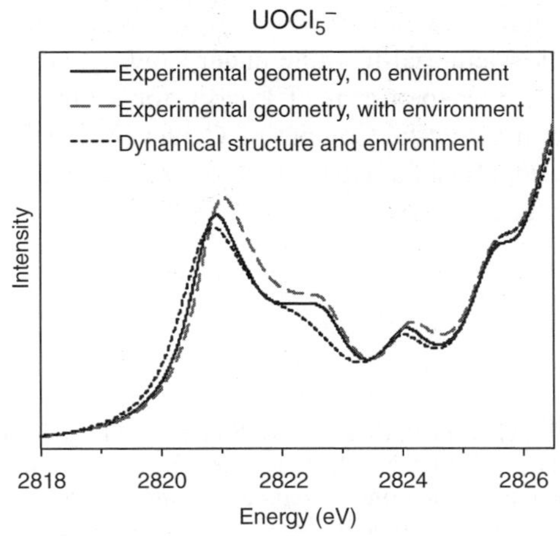

Figure 12.11 *Cl K-edge spectra of $[UOCl_5]^-$ in the gas phase, in the presence of the host crystal and including dynamical effects using AIMD simulations. Reprinted with permission from Govind and de Jong [185]. © 2014 Springer*

with and without the host environment and including dynamical effects are very similar. The agreement with experiment is reasonable. However, we do not observe the shoulder at ≈2820.5 eV that is clearly visible in the experimental data [151] by either introducing the crystal field via the host environment or dynamics. We analyzed this further by analyzing the structural changes needed to reproduce the experimental spectrum and specifically the shoulder feature.

Experimentally, the axial U-O and U-Cl bonds are 1.78 Å and 2.43 Å, and the equatorial U-Cl bonds are all 2.54 Å, respectively. If the axial U-Cl bond is elongated by ≈0.25 Å and two of the equatorial U-Cl bonds (opposite to each other) are elongated by ≈0.25 Å, all the experimental features are reasonably captured including the shoulder. However, these bond length elongations (≈10%) are unphysical and cannot be accounted for by a weak host crystal field, dynamics and different oxidation states (which only accounts for ≈0.08 Å per charge of charge). This feature may be related to spin-flip effects, which cannot be assessed because our implementation does not include these effects.

Acknowledgments

The research was funded by the BES Heavy Element Chemistry program in the Division of Chemical Sciences, Geosciences, and Biosciences, Office of Basic Energy Sciences, U.S. Department of Energy. J.A. acknowledges support by the U.S. Department of Energy, Office of Basic Energy Sciences, Heavy Element Chemistry program, grant DE-FG02-09ER16066, which supported the developments for magnetic properties and the X2C method and applications thereof. Calculations were performed using the Molecular Science Computing Capability at EMSL, a national scientific user facility sponsored by the U.S. Department of Energy, Office of Biological and Environmental Research and located at Pacific Northwest National Laboratory (PNNL). PNNL is operated for the Department of Energy by the Battelle Memorial Institute under Contract DE-AC06-76RLO-1830. We also wish to thank the U.S. Department of Energy for a grant of computer time at the National Energy Research Scientific Computing Center (NERSC), Berkeley, CA. Discussions with and access to unpublished $[UOCl_5]^-$ structural data from Stosh Kozimor (LANL) are gratefully acknowledged.

References

[1] Valiev M, Bylaska EJ, Govind N, Kowalski K, Straatsma TP, Van Dam HJ, Wang D, Nieplocha J, Apra E, Windus TL *et al.* 2010 Nwchem: a comprehensive and scalable open-source solution for large scale molecular simulations. *Computer Physics Communications* **181**(9), 1477–1489.

[2] Barysz M 2000 The relativistic scheme for eliminating small components hamiltonian: Analysis of approximations. *The Journal of Chemical Physics* **113**(10), 4003–4007.

[3] Faegri Jr K *et al.* 2007 *Introduction to relativistic quantum chemistry*. Oxford University Press.

[4] Chang C, Pelissier M and Durand P 1986 Regular two-component pauli-like effective hamiltonians in dirac theory. *Physica Scripta* **34**(5), 394.

[5] van Lenthe E, Baerends EJ and Snijders JG 1993a Relativistic regular two-component hamiltonians. *The Journal of chemical physics* **99**(6), 4597–4610.

[6] Nichols P, Govind N, Bylaska EJ and de Jong WA 2009 Gaussian basis set and planewave relativistic spin-orbit methods in nwchem. *J. Chem. Theory Comput.* **5**(3), 491–499.

[7] Filatov M and Cremer D 2005b A gauge-independent zeroth-order regular approximation to the exact relativistic hamiltonian—formulation and applications. *The Journal of chemical physics* **122**, 044104.

[8] van Lenthe E, Ehlers A and Baerends EJ 1999 Geometry optimizations in the zero order regular approximation for relativistic effects. *The Journal of chemical physics* **110**(18), 8943–8953.

[9] van Wüllen C 1998 Molecular density functional calculations in the regular relativistic approximation: Method, application to coinage metal diatomics, hydrides, fluorides and chlorides, and comparison with first-order relativistic calculations. *The Journal of chemical physics* **109**(2), 392–399.

[10] van Lenthe J and Van Lingen J 2006 Note on the calculation of analytical hessians in the zeroth-order regular approximation (zora). *International journal of quantum chemistry* **106**(12), 2525–2528.

[11] van Lenthe J, Faas S and Snijders J 2000 Gradients in the ab initio scalar zeroth-order regular approximation (zora) approach. *Chemical Physics Letters* **328**(1), 107–112.

[12] Faas S, van Lenthe J, Hennum A and Snijders J 2000 An ab initio two-component relativistic method including spin–orbit coupling using the regular approximation. *The Journal of Chemical Physics* **113**(10), 4052–4059.

[13] Aquino F, Govind N and Autschbach J 2010 Electric field gradients calculated from two–component relativistic density functional theory including spin–orbit coupling. *J. Chem. Theory Comput.* **6**, 2669–2686.

[14] Autschbach J, Peng D and Reiher M 2012 Two-component relativistic calculations of electric-field gradients using exact decoupling methods: Spin-orbit and picture-change effects. *J. Chem. Theory Comput.* **8**, 4239–4248.

[15] Aquino F, Govind N and Autschbach J 2011 Scalar relativistic computations of nuclear magnetic shielding and *g*-shifts with the zeroth-order regular approximation and range-separated hybrid density functionals. *J. Chem. Theory Comput.* **7**, 3278–3292.

[16] Verma P and Autschbach J 2013b Variational versus perturbational treatment of spin-orbit coupling in relativistic density functional calculations of electronic g-factors: Effects from spin-polarization and exact exchange. *J. Chem. Theory Comput.* **9**, 1052–1067.

[17] Aquino F, Pritchard B and Autschbach J 2012 Scalar relativistic computations and localized orbital analysis of nuclear hyperfine coupling and paramagnetic nmr chemical shifts. *J. Chem. Theory Comput.* **8**, 598–609.
[18] Verma P and Autschbach J 2013a Relativistic density functional calculations of hyperfine coupling with variational versus perturbational treatment of spin-orbit coupling. *J. Chem. Theory Comput.* **9**, 1932–1948.
[19] van Lenthe E, Baerends EJ and Snijders JG 1993b Relativistic regular two-component Hamiltonians. *J. Chem. Phys.* **99**, 4597–4610.
[20] Autschbach J and Pritchard B 2011 Calculation of molecular g-tensors using the zeroth-order regular approximation and density functional theory: Expectation value versus linear response approaches. *Theor. Chem. Acc.* **129**, 453–466.
[21] Autschbach J and Ziegler T 2000a Nuclear spin-spin coupling constants from regular approximate relativistic density functional calculations. I. Formalism and scalar relativistic results for heavy metal compounds. *J. Chem. Phys.* **113**(3), 936–947.
[22] Autschbach J and Ziegler T 2003 Double perturbation theory: A powerful tool in computational coordination chemistry. *Coord. Chem. Rev.* **238/239**, 83–126.
[23] Autschbach J and Ziegler T 2000b Nuclear spin-spin coupling constants from regular approximate relativistic density functional calculations. II. Spin-orbit coupling effects and anisotropies. *J. Chem. Phys.* **113**(3), 9410–9418.
[24] Autschbach J and Schwarz WHE 2000 Relativistic electron densities in the four-component Dirac representation and in the two-component picture. *Theor. Chem. Acc.* **104**, 82–88.
[25] Baerends EJ, Schwarz WHE, Schwerdtfeger P and Snijders JG 1990 Relativistic atomic orbital contractions and expansions: magnitudes and explanations. *J. Phys. B.* **23**(19), 3225–3240.
[26] van Lenthe E and Baerends EJ 2000 Density functional calculations of nuclear quadrupole coupling constants in the zero-order regular approximation for relativistic effects. *J. Chem. Phys.* **112**(19), 8279–8292.
[27] Mastalerz R, Barone G, Lindh R and Reiher M 2007 Analytic high-order douglas-kroll-hess electric field gradients. *J. Chem. Phys.* **127**(7), 074105.
[28] Eriksson LA 1998 Esr hyperfine calculations In *Encyclopedia of Computational Chemistry* (ed. von Ragué Schleyer P) Wiley Chichester, UK pp. 952–958.
[29] van Wüllen C 2009 Magnetic anisotropy from density functional calculations. comparison of different approaches: Mn12O12 acetate as a test case. *J. Chem. Phys.* **130**(19), 194109–14.
[30] Schmitt S, Jost P and van Wüllen C 2011 Zero-field splittings from density functional calculations: Analysis and improvement of known methods. *J. Chem. Phys.* **134**(19), 194113–11.
[31] Hrobarik P, Repisky M, Komorovsky S, Hrobarikova V and Kaupp M 2011 Assessment of higher-order spin-orbit effects on electronic g-tensors of d 1 transition-metal complexes by relativistic two- and four-component methods. *Theor. Chem. Acc.* **129**(3-5), 715–725.
[32] Jayatilaka D 1998 Electron spin resonance g tensors from general hartree-fock calculations. *J. Chem. Phys.* **108**, 7587–7594.
[33] Malkin E, Repisky M, Komorovsky S, Mach P, Malkina OL and Malkin VG 2011 Effects of finite size nuclei in relativistic four-component calculations of hyperfine structure. *J. Chem. Phys.* **134**(4), 044111–8.
[34] Malkin I, Malkina OL, Malkin VG and Kaupp M 2005 Relativistic two-component calculations of electronic g-tensors that include spin polarization. *J. Chem. Phys.* **123**(24), 244103–16.
[35] van Lenthe E, van der Avoird A and Wormer PES 1998 Density functional calculations of molecular hyperfine interactions in the zero order regular approximation for relativistic effects. *J. Chem. Phys.* **108**, 4783–4796.
[36] van Lenthe E, Wormer PES and van der Avoird A 1997 Density functional calculations of molecular g-tensors in the zero order regular approximation for relativistic effects. *J. Chem. Phys.* **107**, 2488–2498.
[37] Bolvin H 2006 An alternative approach to the g-matrix: Theory and applications. *ChemPhysChem* **7**, 1575–1589.
[38] Andrae D 2000 Finite nuclear charge density distributions in electronic structure calculations for atoms and molecules. *Phys. Rep.* **336**, 413–527.

[39] Visscher L and Dyall K 1997 Dirac–Fock atomic electronic structure calculations using different nuclear charge distributions. *At. Data Nucl. Data Tables* **67**(2), 207–224.
[40] Autschbach J 2009 Magnitude of finite nucleus size effects in relativistic density functional computations of indirect nmr nuclear spin–spin coupling tensors. *ChemPhysChem* **10**, 2274–2283.
[41] Knecht S, Fux S, Meer R, Visscher L, Reiher M and Saue T 2011 Mössbauer spectroscopy for heavy elements: a relativistic benchmark study of mercury. *Theor. Chem. Acc.* **129**(3-5), 631–650.
[42] Schreckenbach G 2005 Density functional calculations of f-19 and u-235 nmr chemical shifts in uranium (vi) chloride fluorides uf6-ncln: Influence of the relativistic approximation and role of the exchange-correlation functional. *Int. J. Quantum Chem.* **101**(4), 372–380.
[43] Straka M and Kaupp M 2005 Calculation of f-19 nmr chemical shifts in uranium complexes using density functional theory and pseudopotentials. *Chem. Phys.* **311**(1–2), 45–56.
[44] Butler JE and Hutchison, Jr. CA 1981 Electron paramagnetic resonance and electron nuclear double resonance of 237-neptunium hexafluoride in uranium hexafluoride single crystals. *J. Chem. Phys.* **74**(6), 3102–3119.
[45] Clyde A. Hutchison J and Weinstock B 1960 Paramagnetic resonance absorption in neptunium hexafluoride. *J. Chem. Phys.* **32**(1), 56–61.
[46] Rigny P and Plurien P 1967 Resonance paramagnetique dans les fluorures complexes d'uranium (v) de type uf6m. *J. Phys. Chem. Solids* **28**(12), 2589–2595.
[47] Selbin J, Ortego JD and Gritzner G 1968 Electronic spectral studies of some uranium(v) compounds. *Inorg. Chem.* **7**(5), 976–982.
[48] Notter FP and Bolvin H 2009 Optical and magnetic properties of the $5f^1$ anx_6^{q-} series: A theoretical study. *J. Chem. Phys.* **130**(18), 184310–11.
[49] Case DA 1985 Spin-orbit and spin-polarization effects in neptunium hexafluoride. *J. Chem. Phys.* **83**(11), 5792–5796.
[50] Ilias M, Aa. Jensen HJ, Kello V, Roos BO and Urban M 2005 Theoretical study of PbO and the PbO anion. *Chem. Phys. Lett.* **408**(4–6), 210–215.
[51] Ilias M and Saue T 2007 An infinite-order two-component relativistic hamiltonian by a simple one-step transformation. *J. Chem. Phys.* **126**(6), 064102.
[52] Kutzelnigg W and Liu W 2005 Quasirelativistic theory equivalent to fully relativistic theory. *J. Chem. Phys.* **123**(24), 241102–4.
[53] Liu W 2010 Ideas of relativistic quantum chemistry. *Mol. Phys.* **108**, 1679–1706.
[54] Liu W and Kutzelnigg W 2007a Quasirelativistic theory. II. theory at matrix level. *J. Chem. Phys.* **126**(11), 114107–14.
[55] Liu W and Peng D 2009 Exact two-component hamiltonians revisited. *J. Chem. Phys.* **131**(3), 031104–4.
[56] Saue T 2011 Relativistic hamiltonians for chemistry: A primer. *ChemPhysChem* **12**(17), 3077–3094.
[57] Autschbach J 2012 Perspective: Relativistic effects. *J. Chem. Phys.* **136**, 150902–15.
[58] Kutzelnigg W 2012 Solved and unsolved problems in relativistic quantum chemistry. *Chem. Phys.* **395**, 16–34.
[59] Liu W 2011 Perspectives of relativistic quantum chemistry: The negative energy cat smiles. *Phys. Chem. Chem. Phys.* **14**(1), 35–48.
[60] Liu W and Kutzelnigg W 2007b Quasirelativistic theory. II. Theory at matrix level. *J. Chem. Phys.* **126**, 114107.
[61] Peng D, Liu W, Xiao Y and Cheng L 2007 Making four- and two-component relativistic density functional methods fully equivalent based on the idea of "from atoms to molecule". *J. Chem. Phys.* **127**, 104106.
[62] van Wüllen C and Michauk C 2005 Accurate and efficient treatment of two-electron contributions in quasirelativistic high-order Douglas–Kroll density-functional calculations. *J. Chem. Phys.* **123**, 204113.
[63] Saue T and Visscher L 2003 Four-component electronic structure methods for molecules *Theoretical Chemistry and Physics of Heavy and Superheavy Elements* vol. 11 of *Progress in Theoretical Chemistry and Physics* Kluwer Dordrecht pp. 211–267.

[64] van Wüllen C 2010 Relativistic density functional theory In *Relativistic Methods for Chemists* (ed. Barysz M and Ishikawa Y) vol. 10 of *Challenges and Advances in Computational Chemsitry and Physics* Springer Dordrecht pp. 191–214.

[65] Peng D and Reiher M 2012a Exact decoupling of the relativistic Fock operator. *Theor. Chem. Acc.* **131**, 1081.

[66] Barysz M and Sadlej AJ 2001 Two-component methods of relativistic quantum chemistry: from the Douglas-Kroll approximation to the exact two-component formalism. *J. Mol. Struct. (THEOCHEM)* **573**, 181–200.

[67] Barysz M and Sadlej AJ 2002 Infinite-order two-component theory for relativistic quantum chemistry. *J. Chem. Phys.* **116**(7), 2696–2704.

[68] Kedziera D and Barysz M 2004 Two-component relativistic methods for the heaviest elements. *J. Chem. Phys.* **121**(14), 6719–6727.

[69] Kedziera D and Barysz M 2007 Non-iterative approach to the infinite-order two-component (iotc) relativistic theory and the non-symmetric algebraic riccati equation. *Chem. Phys. Lett.* **446**, 176–181.

[70] Dyall KG 1997 Interfacing relativistic and nonrelativistic methods. I. Normalized elimination of the small component in the modified Dirac equation. *J. Chem. Phys.* **106**, 9618–9626.

[71] Dyall KG 1998 Interfacing relativistic and nonrelativistic methods. II. Investigation of a low-order approximation. *J. Chem. Phys.* **109**, 4201–4208.

[72] Dyall KG 2001 Interfacing relativistic and nonrelativistic methods. IV. One- and two-electron scalar approximations. *J. Chem. Phys.* **115**, 9136–9143.

[73] Dyall KG 2002 A Systematic Sequence of Relativistic Approximations. *J. Comput. Chem.* **23**, 786–793.

[74] Dyall KG and Enevoldsen T 1999 Interfacing relativistic and nonrelativistic methods. III. Atomic 4-spinor expansions and integral approximations. *J. Chem. Phys.* **111**, 10000–10007.

[75] Filatov M and Cremer D 2003 Representation of the exact relativistic electronic Hamiltonian within the regular approximation. *J. Chem. Phys.* **119**, 11526–11540.

[76] Filatov M and Cremer D 2005a Connection between the regular approximation and the normalized elimination of the small component in relativistic quantum chemistry. *J. Chem. Phys.* **122**, 064104–8.

[77] Zou W, Filatov M and Cremer D 2011 An improved algorithm for the normalized elimination of the small-component method. *Theor. Chem. Acc.* **130**(4), 633–644.

[78] Stanton RE and Havriliak S 1984 Kinetic balance: A partial solution to the problem of variational safety in dirac calculations. *J. Chem. Phys.* **81**(4), 1910–1918.

[79] Sun Q, Liu W and Kutzelnigg W 2011 Comparison of restricted, unrestricted, inverse, and dual kinetic balances for four-component relativistic calculations. *Theor. Chem. Acc.* **129**(3-5), 423–436.

[80] Autschbach J n.d. Relativistic calculations of magnetic resonance parameters: Background and some recent developments. *J. Phil. Trans. A* pp. accepted for publication (DOI not yet accessible on–line).

[81] Dyall KG 1994 An exact separation of the spin-free and spin-dependent terms of the Dirac–Coulomb–Breit Hamiltonian. *J. Chem. Phys.* **100**, 2118–2127.

[82] Dyall KG, Grant IP and Wilson S 1984 Matrix representation of operator products. *J. Phys. B* **17**, 493–503.

[83] Kutzelnigg W 1984 Basis Set Expansion of the Dirac Operator without Variational Collapse. *Int. J. Quantum Chem.* **25**, 107–129.

[84] Kutzelnigg W 1997 Relativistic one-electron Hamiltonians 'for electrons only' and the variational treatment of the Dirac equation. *Chem. Phys.* **225**, 203–222.

[85] Peng D and Reiher M 2012b Local relativistic exact decoupling. *J. Chem. Phys.* **136**, 244108–11.

[86] Seino J and Nakai H 2012 Local unitary transformation method for large-scale two-component relativistic calculations. II. extension to two-electron coulomb interaction. *J. Chem. Phys.* **137**, 144101–15.

[87] Kellö V and Sadlej AJ 1998 Picture change and calculations of expectation values in approximate relativistic theories. *Int. J. Quantum Chem.* **68**, 159–174.

[88] Pernpointner M, Schwerdtfeger P and Hess BA 2000 Accurate electric field gradients for the coinage metal chlorides using the PCNQM method. *Int. J. Quantum Chem.* **76**(3), 371–384.
[89] Cheng L and Gauss J 2011 Analytic energy gradients for the spin-free exact two-component theory using an exact block diagonalization for the one-electron dirac hamiltonian. *J. Chem. Phys.* **135**(8), 084114–084114–7.
[90] Sun Q, Xiao Y and Liu W 2012 Exact two-component relativistic theory for NMR parameters: General formulation and pilot application. *J. Chem. Phys.* **137**(17), 174105–20.
[91] van Lenthe E 1996 *The ZORA Equation* PhD thesis Vrije Universiteit Amsterdam, Netherlands.
[92] Heß BA, Marian CM, Wahlgren U and Gropen O 1996 A mean–field spin–orbit method applicable to correlated wavefunctions. *Chem. Phys. Lett.* **251**, 365–371.
[93] Belanzoni P, Baerends E and van Lenthe E 2005 The uranyl ion revisited: the electric field gradient at u as a probe of environmental effects. *Mol. Phys.* **103**(6), 775–787.
[94] de Jong WA, Visscher L and Nieuwpoort WC 1998 On the bonding and the electric field gradient of the uranyl ion. *J. Mol. Struct.* **458**(1–2), 41–52.
[95] Monard JA, Huray PG and Thomson JO 1974 Mössbauer studies of electric hyperfine interactions in u234, u236, u238. *Phys. Rev. B* **9**(7), 2838–2845.
[96] Larsson S and Pyykkö P 1986 Relativistically parameterized extended hückel calculations. ix. an iterative version with applications to some xenon, thorium and uranium compounds. *Chem. Phys.* **101**, 355–369.
[97] Pyykkö P and Seth M 1997 Relativistic effects in nuclear quadrupole coupling. *Theor. Chem. Acc.* **96**, 92–104.
[98] Bylaska E, Tsemekhman K, Govind N and Valiev M 2011 Large-scale plane-wave-based density-functional theory: formalism, parallelization, and applications. *Computational methods for large systems: electronic structure approaches for biotechnology and nanotechnology. Wiley, Hoboken* pp. 77–116.
[99] Hoover WG 1985 Canonical dynamics: equilibrium phase-space distributions. *Physical Review A* **31**(3), 1695.
[100] Nosé S 1984 A unified formulation of the constant temperature molecular dynamics methods. *The Journal of Chemical Physics* **81**(1), 511–519.
[101] Car R and Parrinello M 1985 Unified approach for molecular dynamics and density-functional theory. *Physical review letters* **55**(22), 2471.
[102] Atta-Fynn R, Bylaska EJ and De Jong WA 2012a Free energies and mechanisms of water exchange around uranyl from first principles molecular dynamics *MRS Proceedings*, vol. 1383, pp. mrsf11–1383 Cambridge Univ Press.
[103] Atta-Fynn R, Bylaska EJ and de Jong WA 2013 Importance of counteranions on the hydration structure of the curium ion. *The Journal of Physical Chemistry Letters* **4**(13), 2166–2170.
[104] Atta-Fynn R, Johnson DF, Bylaska EJ, Ilton ES, Schenter GK and De Jong WA 2012b Structure and hydrolysis of the u (iv), u (v), and u (vi) aqua ions from ab initio molecular simulations. *Inorganic chemistry* **51**(5), 3016–3024.
[105] Barducci A, Bussi G and Parrinello M 2008 Well-tempered metadynamics: A smoothly converging and tunable free-energy method. *Phys. Rev. Lett.* **100**, 020603.
[106] Laio A and Gervasio FL 2008 Metadynamics: a method to simulate rare events and reconstruct the free energy in biophysics, chemistry and material science. *Reports on Progress in Physics* **71**(12), 126601.
[107] Laio A and Parrinello M 2002 Escaping free-energy minima. *Proceedings of the National Academy of Sciences* **99**(20), 12562–12566.
[108] Guillaumont R, Fanghanel T, Fuger J, Grenthe I, Neck V, Palmer DA and Rand MH 2003 *Chemical Thermodynamics Vol. 5: Update on the Chemical Thermodynamics of Uranium, Neptunium, Plutonium, Americium, and Technetium. F. J. Mompean and M. Illemassene and C. Domenech-Orti and K. Ben Said,Editors:*. Elsevier, Amsterdam.
[109] Zanonato P, Di Bernardo P, Bismondo A, Liu G, Chen X and Rao L 2004 Hydrolysis of uranium(vi) at variable temperatures (10–85 c). *Journal of the American Chemical Society* **126**(17), 5515–5522.
[110] Bühl M and Kabrede H 2006 Acidity of uranyl(vi) hydrate studied with first-principles molecular dynamics simulations. *ChemPhysChem* **7**(11), 2290–2293.

[111] Hay PJ, Martin RL and Schreckenbach G 2000a Theoretical studies of the properties and solution chemistry of ano22+ and ano2+ aquo complexes for an = u, np, and pu. *The Journal of Physical Chemistry A* **104**(26), 6259–6270.

[112] Tsushima S, Yang T and Suzuki A 2001 Theoretical gibbs free energy study on uo2(h2o)n2+ and its hydrolysis products. *Chemical Physics Letters* **334**(46), 365–373.

[113] Bühl M, Schreckenbach G, Sieffert N and Wipff G 2009 Effect of counterions on the structure and stability of aqueous uranyl(vi) complexes. a first-principles molecular dynamics study. *Inorganic Chemistry* **48**(21), 9977–9979. PMID: 19780567.

[114] Skanthakumar S, Antonio MR, Wilson RE and Soderholm L 2007 The curium aqua ion. *Inorganic Chemistry* **46**(9), 3485–3491.

[115] Allen PG, Bucher JJ, Shuh DK, Edelstein NM and Craig I 2000 Coordination chemistry of trivalent lanthanide and actinide ions in dilute and concentrated chloride solutions. *Inorganic Chemistry* **39**(3), 595–601.

[116] Stumpf T, Fanghanel T and Grenthe I 2002 Complexation of trivalent actinide and lanthanide ions by glycolic acid: a trlfs study. *J. Chem. Soc., Dalton Trans.* pp. 3799–3804.

[117] Lindqvist-Reis P, Klenze R, Schubert G and Fanghl T 2005 Hydration of cm3+ in aqueous solution from 20 to 200 c. a time-resolved laser fluorescence spectroscopy study. *The Journal of Physical Chemistry B* **109**(7), 3077–3083. PMID: 16851323.

[118] Atta-Fynn R, Bylaska EJ, Schenter GK and de Jong WA 2011 Hydration shell structure and dynamics of curium(iii) in aqueous solution: First principles and empirical studies. *The Journal of Physical Chemistry A* **115**(18), 4665–4677.

[119] Hagberg D, Bednarz E, Edelstein NM and Gagliardi L 2007 A quantum chemical and molecular dynamics study of the coordination of cm(iii) in water. *Journal of the American Chemical Society* **129**(46), 14136–14137.

[120] Yang and Bursten BE 2006 Speciation of the curium(iii) ion in aqueous solution: A combined study by quantum chemistry and molecular dynamics simulation. *Inorganic Chemistry* **45**(14), 5291–5301. PMID: 16813391.

[121] Chandra A 2000 Effects of ion atmosphere on hydrogen-bond dynamics in aqueous electrolyte solutions. *Phys. Rev. Lett.* **85**, 768–771.

[122] Luzar A 2000 Resolving the hydrogen bond dynamics conundrum. *The Journal of Chemical Physics* **113**(23), 10663–10675.

[123] Luzar A and Chandler D 1993 Structure and hydrogen bond dynamics of waterdimethyl sulfoxide mixtures by computer simulations. *The Journal of Chemical Physics* **98**(10), 8160–8173.

[124] Luzar A and Chandler D 1996a Effect of environment on hydrogen bond dynamics in liquid water. *Phys. Rev. Lett.* **76**, 928–931.

[125] Luzar A and Chandler D 1996b Hydrogen-bond kinetics in liquid water. *Nature* **379**, 55–57.

[126] Rapaport D 1983 Hydrogen bonds in water. *Molecular Physics* **50**(5), 1151–1162.

[127] Starr FW, Nielsen JK and Stanley HE 1999 Fast and slow dynamics of hydrogen bonds in liquid water. *Phys. Rev. Lett.* **82**, 2294–2297.

[128] Stillinger FH 2007 *Theory and Molecular Models for Water*. John Wiley & Sons, Inc.

[129] Kropman MF and Bakker HJ 2001 Dynamics of water molecules in aqueous solvation shells. *Science* **291**(5511), 2118–2120.

[130] Rehr J and Ankudinov A 2005 Progress in the theory and interpretation of xanes. *Coordination Chemistry Reviews* **249**(1), 131–140.

[131] Rehr JJ and Albers R 2000 Theoretical approaches to x-ray absorption fine structure. *Reviews of Modern Physics* **72**(3), 621.

[132] Palmer BJ, Pfund DM and Fulton JL 1996 Direct modeling of exafs spectra from molecular dynamics simulations. *The Journal of Physical Chemistry* **100**(32), 13393–13398.

[133] Cauët E, Bogatko S, Weare JH, Fulton JL, Schenter GK and Bylaska EJ 2010 Structure and dynamics of the hydration shells of the zn2+ ion from ab initio molecular dynamics and combined ab initio and classical molecular dynamics simulations. *The Journal of chemical physics* **132**(19), 194502.

[134] Fulton JL, Bylaska EJ, Bogatko S, Balasubramanian M, Cauët E, Schenter GK and Weare JH 2012 Near-quantitative agreement of model-free dft-md predictions with xafs observations

of the hydration structure of highly charged transition-metal ions. *The Journal of Physical Chemistry Letters* **3**(18), 2588–2593.
[135] Cavalleri M, Ogasawara H, Pettersson L and Nilsson A 2002 The interpretation of x-ray absorption spectra of water and ice. *Chemical Physics Letters* **364**, 363–370.
[136] Triguero L, Pettersson LGM and Ågren H 1998 Calculations of near-edge x-ray-absorption spectra of gas-phase and chemisorbed molecules by means of density-functional and transition-potential theory. *Phys. Rev. B* **58**, 8097–8110.
[137] Villaume S, Ekström U, Ottosson H and Norman P 2010 Near sulfur l-edge x-ray absorption spectra of methanethiol in isolation and adsorbed on a au (111) surface: a theoretical study using the four-component static exchange approximation. *Physical Chemistry Chemical Physics* **12**(21), 5596–5604.
[138] Besley NA and Asmuruf FA 2010 Time-dependent density functional theory calculations of the spectroscopy of core electrons. *Physical Chemistry Chemical Physics* **12**(38), 12024–12039.
[139] Liang W, Fischer SA, Frisch MJ and Li X 2011 Energy-specific linear response tdhf/tddft for calculating high-energy excited states. *Journal of Chemical Theory and Computation* **7**(11), 3540–3547.
[140] Lopata K, Van Kuiken BE, Khalil M and Govind N 2012 Linear-response and real-time time-dependent density functional theory studies of core-level near-edge x-ray absorption. *Journal of Chemical Theory and Computation* **8**(9), 3284–3292.
[141] Ray K, DeBeer George S, Solomon EI, Wieghardt K and Neese F 2007 Description of the ground-state covalencies of the bis (dithiolato) transition-metal complexes from x-ray absorption spectroscopy and time-dependent density-functional calculations. *Chemistry-A European Journal* **13**(10), 2783–2797.
[142] Stener M, Fronzoni G and De Simone M 2003 Time dependent density functional theory of core electrons excitations. *Chemical physics letters* **373**(1), 115–123.
[143] Van Kuiken BE, Valiev M, Daifuku SL, Bannan C, Strader ML, Cho H, Huse N, Schoenlein RW, Govind N and Khalil M 2013 Simulating ru l3-edge x-ray absorption spectroscopy with time-dependent density functional theory: Model complexes and electron localization in mixed-valence metal dimers. *The Journal of Physical Chemistry A* **117**(21), 4444–4454.
[144] Ikeno H, de Groot FM, Stavitski E and Tanaka I 2009 Multiplet calculations of l2, 3 x-ray absorption near-edge structures for 3d transition-metal compounds. *Journal of Physics: Condensed Matter* **21**(10), 104208.
[145] Nelin CJ, Bagus PS and Ilton ES 2014 Theoretical analysis of the ul 3-edge nexafs in u oxides. *RSC Advances* **4**(14), 7148–7153.
[146] Roemelt M, Maganas D, DeBeer S and Neese F 2013 A combined dft and restricted open-shell configuration interaction method including spin-orbit coupling: Application to transition metal l-edge x-ray absorption spectroscopy. *The Journal of chemical physics* **138**(20), 204101.
[147] De Groot F and Kotani A 2008 *Core level spectroscopy of solids*. CRC press.
[148] Moore KT and van der Laan G 2009 Nature of the 5 f states in actinide metals. *Reviews of Modern Physics* **81**(1), 235.
[149] Stöhr J 1992 *NEXAFS Spectroscopy*. Springer-Verlag: New York.
[150] Nichols P, Bylaska EJ, Schenter GK and de Jong W 2008 Equatorial and apical solvent shells of the uo22+ ion. *The Journal of chemical physics* **128**(12), 124507.
[151] Minasian SG, Keith JM, Batista ER, Boland KS, Clark DL, Conradson SD, Kozimor SA, Martin RL, Schwarz DE, Shuh DK *et al.* 2012b Determining relative f and d orbital contributions to m–cl covalency in mcl62–(m= ti, zr, hf, u) and uocl5–using cl k-edge x-ray absorption spectroscopy and time-dependent density functional theory. *Journal of the American Chemical Society* **134**(12), 5586–5597.
[152] Spencer LP, Yang P, Minasian SG, Jilek RE, Batista ER, Boland KS, Boncella JM, Conradson SD, Clark DL, Hayton TW *et al.* 2013 Tetrahalide complexes of the [u (nr) 2] 2+ ion: Synthesis, theory, and chlorine k-edge x-ray absorption spectroscopy. *Journal of the American Chemical Society* **135**(6), 2279–2290.
[153] Frick RJ, Hofer TS, Pribil AB, Randolf BR and Rode BM 2010 Structure and dynamics of the uo+ 2 ion in aqueous solution: an ab initio qmcf-md study. *Physical Chemistry Chemical Physics* **12**(37), 11736–11743.

[154] Austin JP, Sundararajan M, Vincent MA and Hillier IH 2009 The geometric structures, vibrational frequencies and redox properties of the actinyl coordination complexes ([ano2 (l) n] m; an= u, pu, np; l= h2o, cl-, co32-, ch3co2-, oh-) in aqueous solution, studied by density functional theory methods. *Dalton Transactions* (30), 5902–5909.
[155] Hay PJ, Martin RL and Schreckenbach G 2000b Theoretical studies of the properties and solution chemistry of ano22+ and ano2+ aquo complexes for an= u, np, and pu. *The Journal of Physical Chemistry A* **104**(26), 6259–6270.
[156] Shamov GA and Schreckenbach G 2005 Density functional studies of actinyl aquo complexes studied using small-core effective core potentials and a scalar four-component relativistic method. *The Journal of Physical Chemistry A* **109**(48), 10961–10974.
[157] Vallet V, Privalov T, Wahlgren U and Grenthe I 2004 The mechanism of water exchange in amo2 (h2o) 52+ and in the isoelectronic uo2 (h2o) 5+ and npo2 (h2o) 52+ complexes as studied by quantum chemical methods. *Journal of the American Chemical Society* **126**(25), 7766–7767.
[158] Allen P, Bucher J, Shuh D, Edelstein N and Reich T 1997 Investigation of aquo and chloro complexes of uo22+, npo2+, np4+, and pu3+ by x-ray absorption fine structure spectroscopy. *Inorganic chemistry* **36**(21), 4676–4683.
[159] Conradson SD 1998 Application of x-ray absorption fine structure spectroscopy to materials and environmental science. *Applied Spectroscopy* **52**, 252A–252A.
[160] Frick RJ, Pribil AB, Hofer TS, Randolf BR, Bhattacharjee A and Rode BM 2009 Structure and dynamics of the u4+ ion in aqueous solution: An ab initio quantum mechanical charge field molecular dynamics study. *Inorganic chemistry* **48**(9), 3993–4002.
[161] Moll H, Denecke M, Jalilehvand F, Sandström M and Grenthe I 1999 Structure of the aqua ions and fluoride complexes of uranium (iv) and thorium (iv) in aqueous solution an exafs study. *Inorganic chemistry* **38**(8), 1795–1799.
[162] Ikeda-Ohno A, Hennig C, Tsushima S, Scheinost AC, Bernhard G and Yaita T 2009 Speciation and structural study of u (iv) and-(vi) in perchloric and nitric acid solutions. *Inorganic chemistry* **48**(15), 7201–7210.
[163] Hennig C, Tutschku J, Rossberg A, Bernhard G and Scheinost A 2005 Comparative exafs investigation of uranium (vi) and-(iv) aquo chloro complexes in solution using a newly developed spectroelectrochemical cell. *Inorganic chemistry* **44**(19), 6655–6661.
[164] Neuefeind J, Soderholm L and Skanthakumar S 2004 Experimental coordination environment of uranyl (vi) in aqueous solution. *The Journal of Physical Chemistry A* **108**(14), 2733–2739.
[165] Soderholm L, Skanthakumar S and Neuefeind J 2005 Determination of actinide speciation in solution using high-energy x-ray scattering. *Analytical and bioanalytical chemistry* **383**(1), 48–55.
[166] Ferri D, Grenthe I and Salvatore F 1983 Studies on metal carbonate equilibriums. 7. reduction of the tris (carbonato) dioxouranate (vi) ion, uo2 (co3) 34-, in carbonate solutions. *Inorganic Chemistry* **22**(21), 3162–3165.
[167] Horeglad P, Nocton G, Filinchuk Y, Pécaut J and Mazzanti M 2009 Pentavalent uranyl stabilized by a dianionic bulky tetradentate ligand. *Chemical Communications* (14), 1843–1845.
[168] Ikeda A, Hennig C, Tsushima S, Takao K, Ikeda Y, Scheinost AC and Bernhard G 2007 Comparative study of uranyl (vi) and-(v) carbonato complexes in an aqueous solution. *Inorganic chemistry* **46**(10), 4212–4219.
[169] Madic C, Hobart D and Begun G 1983 Raman spectrometric studies of actinide (v) and-(vi) complexes in aqueous sodium carbonate solution and of solid sodium actinide (v) carbonate compounds. *Inorganic Chemistry* **22**(10), 1494–1503.
[170] Cao X and Dolg M 2004 Segmented contraction scheme for small-core actinide pseudopotential basis sets. *Journal of Molecular Structure: THEOCHEM* **673**(1), 203–209.
[171] Woon DE and Dunning Jr TH 1993 Gaussian basis sets for use in correlated molecular calculations. iii. the atoms aluminum through argon. *The Journal of chemical physics* **98**(2), 1358–1371.
[172] Becke AD 1993 Density-functional thermochemistry. iii. the role of exact exchange. *The Journal of Chemical Physics* **98**(7), 5648–5652.
[173] Lee C, Yang W and Parr R 1994 Phys. rev. b 1988, 37, 785–789; c) pj stephens, fj devlin, cf chabalowski, mj frisch. *J. Phys. Chem* **98**, 11623–11627.

[174] Stephens P, Devlin F, Chabalowski C and Frisch MJ 1994 Ab initio calculation of vibrational absorption and circular dichroism spectra using density functional force fields. *The Journal of Physical Chemistry* **98**(45), 11623–11627.

[175] Vosko SH, Wilk L and Nusair M 1980 Accurate spin-dependent electron liquid correlation energies for local spin density calculations: a critical analysis. *Canadian Journal of Physics* **58**(8), 1200–1211.

[176] Adamo C and Barone V 1999 Toward reliable density functional methods without adjustable parameters: The pbe0 model. *The Journal of chemical physics* **110**(13), 6158–6170.

[177] Menzel ER and Gruber JB 1971 Absorption spectrum of cs2npcl6. *The Journal of Chemical Physics* **54**(9), 3857–3864.

[178] Wagner W, Edelstein N, Whittaker B and Brown D 1977 Spectral properties of bis (tetraethylammonium) uranium hexaiodide bis (tetraethylammonium) uranium hexafluoride. *Inorganic Chemistry* **16**(5), 1021–1026.

[179] Perdew JP, Burke K and Ernzerhof M 1996 Generalized gradient approximation made simple. *Phys. Rev. Lett.* **77**, 3865–3868.

[180] Troullier N and Martins JL 1991 Efficient pseudopotentials for plane-wave calculations. *Physical Review B* **43**(3), 1993.

[181] Hamann D 1989 Generalized norm-conserving pseudopotentials. *Physical Review B* **40**(5), 2980.

[182] Kozimor SA 2013 private communication.

[183] Minasian SG, Boland KS, Feller RK, Gaunt AJ, Kozimor SA, May I, Reilly SD, Scott BL and Shuh DK 2012a Synthesis and structure of (ph4p) 2mcl6 (m= ti, zr, hf, th, u, np, pu). *Inorganic chemistry* **51**(10), 5728–5736.

[184] Frenkel D and Smit B 2001 *Understanding molecular simulation: from algorithms to applications* vol. 1. Academic press.

[185] Govind N and de Jong WA 2014 Simulating cl k-edge x-ray absorption spectroscopy in mcl6 2-(m= u, np, pu) complexes and uocl5- using time-dependent density functional theory. *Theoretical Chemistry Accounts* **133**(4), 1–7.

[186] DIRAC 2012. DIRAC, a relativistic ab initio electronic structure program, Release DIRAC12, written by H. J. Aa. Jensen, R. Bast, T. Saue, and L. Visscher, with contributions from V. Bakken, K. G. Dyall, S. Dubillard, U. Ekström, E. Eliav, T. Enevoldsen, T. Fleig, O. Fossgaard, A. S. P. Gomes, T. Helgaker, J. K. Lærdahl, Y. S. Lee, J. Henriksson, M. Iliaš, Ch. R. Jacob, S. Knecht, S. Komorovský, O. Kullie, C. V. Larsen, H. S. Nataraj, P. Norman, G. Olejniczak, J. Olsen, Y. C. Park, J. K. Pedersen, M. Pernpointner, K. Ruud, P. Sałek, B. Schimmelpfennig, J. Sikkema, A. J. Thorvaldsen, J. Thyssen, J. van Stralen, S. Villaume, O. Visser, T. Winther, and S. Yamamoto (see http://www.diracprogram.org).

[187] Malkin E, Malkin I, Malkina OL, Malkin VG and Kaupp M 2006 Scalar relativistic calculations of hyperfine coupling tensors using the douglas-kroll-hess method with a finite-size nucleus model. *Phys. Chem. Chem. Phys.* **8**, 4079–4085.

[188] van Lenthe E, Baerends EJ and Snijders JG 1994 Relativistic total energy using regular approximations. *The Journal of chemical physics* **101**(11), 9783–9792.

[189] van Lenthe E, Snijders J and Baerends E 1996 The zero-order regular approximation for relativistic effects: The effect of spin–orbit coupling in closed shell molecules. *The Journal of chemical physics* **105**(15), 6505–6516.

[190] Krause MO and Oliver J 1979 Natural widths of atomic k and l levels, kα x-ray lines and several kll auger lines. *Journal of Physical and Chemical Reference Data* **8**(2), 329–338.

13

Theoretical Treatment of the Redox Chemistry of Low Valent Lanthanide and Actinide Complexes

*Christos E. Kefalidis, Ludovic Castro, Ahmed Yahia, Lionel Perrin, and Laurent Maron**

Laboratory of Physics and Chemistry of Nano-Objects, National Institute of Applied Sciences
**Corresponding Author*

13.1 Introduction

Despite the fact that calculations on divalent lanthanides are still rather scarce in the literature, the peculiar properties of these atoms have been experimentally known for several decades. For instance, traces of organolanthanide complexes begin with the early report of Wilkinson in 1954 of a series of Cp_3Ln complexes (where Cp is cyclopentadienyl) synthesized from rare earth halides and sodium cyclopentadienyl in THF. [1] In a related paper published two years later, some of the physical properties including magnetic susceptibility were reported. [2] Wilkinson described the Cp rare earth complexes as ionic complexes, as shown by their easy exchange with iron chloride, and attributed the magnetic moment to the only presence of the lanthanide atom. The reported magnetic moment of Cp_3Yb appeared however low compared to that observed for other molecular complexes of Yb(III). Instead of the expected value of 4.3–4.5 μ_B, Wilkinson measured a magnetic moment of 4.0 μ_B. Since this collection of Cp_3Ln complexes, the synthesis of the first reactive divalent organolanthanides Cp_2Ln (Ln = Eu and Yb) appeared. [3] At about the same time, the first discussions came in about the degree of "covalency" in them, [4] which is a rather strange conjecture since f-orbitals are known to be buried in the core of the atom and therefore do not allow overlap with ligand-based orbitals in any great extent.

Computational Methods in Lanthanide and Actinide Chemistry, First Edition. Edited by Michael Dolg.

A decade later, Andersen reported the synthesis of another divalent complex of ytterbium, $Cp_2^*YbOEt_2$ (where Cp* is pentamethylcyclopentadienyl), and became interested in looking at the monoelectronic transfer from the metal center to N-aromatic heterocycles as bipyridine for example. This new type of reactivity, named single electron transfer (SET) reactivity, started to play an important role in organolanthanide chemistry [5] since many interesting reactions were reported [6] but, still, did not address the intriguing question of the bonding. Witnesses of the paradigm that occupy the organolanthanides compounds are the very simple molecules $Cp_2^*Yb(bipy)$, Cp_3Yb, and Cp_3Eu.

At the beginning of the 1990s, Amberger embarked in a systematic parametric crystal field analysis of organolanthanides of high symmetry. [7] Among other, Cp_3Eu has intrigued and notably the measurement of its Mössbauer spectrum opened more questions than it answered. [8] Indeed, if isomeric shift in Mössbauer spectroscopy is associated with oxidation state, the reported one of Cp_3Eu is neither in agreement with a +II oxidation nor a +III oxidation state. This observation was to relate with nephelauxetic effects observed in optical spectra in europium complexes and again here associated by the authors with some "covalency" in molecules of this nature. [9] Cp_3Yb falls in the same hole than its europium congener: its photoelectron spectrum and magnetic susceptibility (3.5 μ_B at room temperature, *i.e.*, even lower that the one measured by Wilkinson in 1956) are again neither typical for Yb(II) nor for Yb(III). [10] Denning and Green gathered their later results on Cp_3Yb and calculations in a recent paper to conclude that "there is Cp $\rightarrow$ Yb^{3+} charge-transfer configuration in the ground state wave function" and this may happen because of possible hybridization of the 5d shell with ligand orbitals. [11] From the two precedent examples, the charge transfer is occurring from the ligand to the metal but it may come from the metal center to the ligand. In this light, $Cp^*{}_2Yb(bipy)$ was synthesized in the late 1970s [12] but was not published until 2002.[6c] The reason for this rather long period of time finds its answer in the odd magnetic behavior of these molecules that makes its electronic structure not obvious. Two limit structures are possible: a neutral bipyridine and a Yb(II) (f^{14} configuration) or a bipyridine radical anion and an ytterbium (III) (f^{13} configuration). Again, both configurations cannot be explained by the observed magnetic data (reduced magnetic moment at room temperature). Moreover, the ytterbium L_{III}-edge X-ray absorption near-edge (XANES) spectra show the presence of f^{13} and f^{14} features that are independent of temperature from 30 K to 400 K, with the f^{13} configuration dominant. A physical model derived from CASSCF calculations, consistent with all experimental observations, is that the ground state is an open-shell singlet consisting of $f^{13}(\pi^*)^1$ and $f^{14}(\pi^*)^0$ configurations that are lower in energy than the triplet configuration. [13] Again, the question of energetic match between the ligand-based orbitals and the 4f-shell is crucial. [14] As it appears, organolanthanide molecules have the ability to form unusual electronic strfuctures, and ligands are to play a significant role in it due to their redox "non-innocence." If, from a fundamental perspective, those concepts remain fascinating, they may also be extremely useful in reactivity. As organolanthanides are known to provide multiple different reactivities [15], that may be enhanced by the precise knowledge of what triggers forming and breaking chemical bonds. Important issues include the understanding of magnetic [16] and optical properties [17] in the objective of development of new molecular-based materials. [18]

From an experimental point of view, there have been impressive improvements on the synthesis procedure and more importantly on the chemical stabilization of the synthesized complexes. Indeed, for a long time, the synthesis of divalent organolanthanides has been limited to the Yb, Eu, and Sm metal as they are the easiest to reduce, [19] and it was the common belief that they would be the only ones accessible in a laboratory. It is therefore not surprising if those are the most studied divalent metal complexes. The groups of Evans and Andersen in California have shown over the last three decades an enormous amount of reactivity associated with them, namely reductive chemistry or electron transfer reactivity.[15b] Among them, the most popular one was the reductive coplanar coordination of N_2 in sandwich fashion between two Cp^*_2Sm fragments. [20] But as it is related by Evans in a fairly recent review,[15b] "*the assumption that this was the limit of reductive chemistry was completely wrong*." Nowadays, most of the lanthanides have been obtained in their divalent form, although some questions remain on their electronic structure. [21] In typical procedures, the classical divalent lanthanides halides Yb and Eu may be synthesized from the metal and NH_4I in liquid ammonia. This very straightforward reaction needs special care but allows the synthesis of such complexes in high purity and in good yield. [22] The Kagan's synthesis of SmI_2 [23] remains a typical synthesis that works well and yields pure material. Tm, Dy, and Nd halides are a little more complicated to prepare in their pure form because they are extremely sensitive toward oxidation. Tm, Nd, and Dy solution chemistry in their divalent state failed until 1997 when Evans proposed a reaction of finely powdered Tm in a DME solution of iodide to yield the first source of divalent Tm, $TmI_2(DME)_3$. [24] Since then convenient methods of synthesis of NdI_2, DyI_2, and TmI_2 were reported by the same investigator by simply reacting metal powder and iodine at high temperature in a quartz Schlenk. [25] Little reactivity is reported directly with the halides of Tm, Nd, and Dy because of their low stability in THF solutions, and the need of ancillary ligands appeared quickly as a solution to this problem. [26] It is however worth noting that it is found that TmI_2 is reacting with pyridine in THF to form the $[(C_5H_5N)_4TmI_2]_2(\mu\text{-}C_{10}H_{10}N_2)$, that contains the 1,1'-dihydro-4,4'-bipyridyl bridging ligand. [27] After this observation, Evans started to work on molecular complexes of Tm, Dy, and Nd. In the attempt to obtain Cp^*_2Tm in THF, he observed the doubly reduction of $N_2$to form $[Cp^*_2Tm]_2(N_2)$ [28], whereas in Et_2O under Argon atmosphere, to avoid N_2 reduction, only a product of Et_2O cleavage was obtained. [28] By increasing the steric hindrance of the cyclopentadienyl ligand using $Me_3Si_2C_5H_3$ instead of Me_5C_5 for example, the first molecular complex of Tm was synthesized. [29] Therefore, it seemed that by increasing the ligand's steric congestion was indeed a way to obtain stable molecular complexes of those non-classical lanthanides. Nief and his collaborators have worked out using several diverse ways to obtain stable molecular complexes of divalent Tm, Dy, and Nd using different bulky versions of the Cp ligands, or phospholes, or even of Tp type one (Figure 13.1). [26] As predicted, the molecular complexes of those metals are highly reactive and resulting in different types of reactivity.

As aforementioned, the magnetism of divalent lanthanide complexes can be tricky to interpret. In a typical description of the lanthanides magnetic behavior, it is assumed that the good quantum number is J since the spin orbit coupling constant is big compared to the crystal field splitting. Therefore, a Russell-Saunders coupling scheme may be used to determine the spectroscopic ground state. [30] Then, the crystal field (although usually a minor perturbation) splits the spin-orbit states along with the symmetry of the molecule in several crystals field states ($m_J$states also called ligand field or Stark states) that may be rather close in energy (within several hundred of wavenumbers). This means that at room

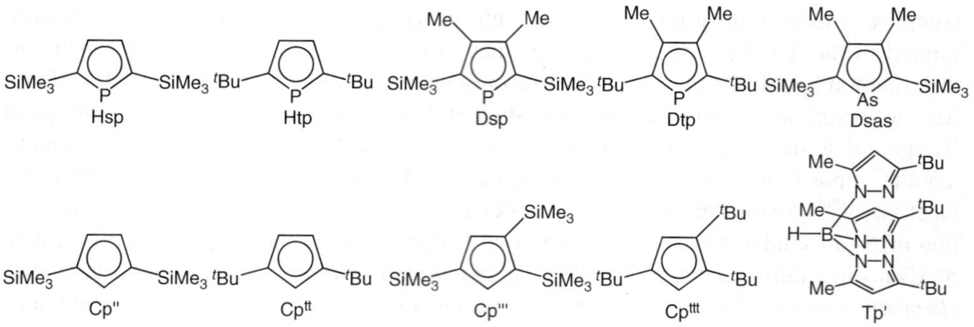

Figure 13.1 *Bulky ligands used for the stabilization of divalent organolanthanides [26]*

temperature most crystal field states are populated and the expected theoretical magnetic moment is the Russell-Saunders one. For example, the magnetism of an ytterbium complex in its +III oxidation state is reported to be between 4.3–4.5μ_B at room temperature that is very close to the theoretical one (4.54 μ_B). However at lower temperatures, the observed magnetic moment is slightly decreases, being around 4.0 μ_B at 5K for instance. This is due to higher crystal field states that are depopulated at this temperature, and therefore their contribution in the global magnetism is vanished. [31] This is even more pronounced when the lanthanides complexes are non-Kramers systems and may have a non-magnetic single crystal field ground state. [32] From a magnetic perspective, it is then possible to have insight into the oxidation state as well as the crystal field energy ($m_J$separation and distribution). [33] Of course, the electron transfer complicates the situation because one unpaired spin is transferred on the ligand side and accounts for the new electronic structure. Recent articles have shown that the exchange coupling (2J) between radical and lanthanides ions may be relatively important in the case of the $\{[(Me_3Si)_2N]_2Ln\}_2(\mu - \eta^y{:}\eta^y\text{-}N_2)$ (-54 cm^{-1}for Gd and -102 cm^{-1} for Dy case) [16] and may be even more pronounced in multiconfigurational systems such as the Cp*$_2$Yb(bipy) complex (-920 cm^{-1}). [14]

As magnetic measurement on its own is hard to be fully interpreted and is insufficient at the same time to provide a clear description of the low-lying crystal field (m_J), optical spectroscopy and in particular luminescence can shed complementary information. Thanks to the unique photophysical properties of f-block elements (sharp line-like emission bands, long lifetime etc.) and with the help of the phenomenological Judd-Ofelt theory, which is well established for the case of inorganic solids and coordination complexes, the fine splitting of the f-f emission spectra can serve as a signature of the local symmetry of the metal, [34] a property widely used in biological sensing. [35] In particular, the number of Stark sublevels directly depends on the molecular symmetry, and a well resolved spectrum allows determining the energy level of each of the m_J states but not their identification. In this respect, very recently some experimental groups have illustrated the complementarity between magnetic and luminescence information in the determination of $m_J$states for various Yb and Dy complexes. [36] It is worth noting that contrarily to ground state magnetic measurements, luminescence comes from the first excited state and the m_J state population is not subjected to a temperature-dependent Boltzmann distribution. Consequently, it is possible to observe m_J states even at low temperatures. Interestingly, whereas luminescence of coordination complexes or inorganic solids containing f-block elements have been extensively studied, the emission properties of organometallic lanthanide complexes is a quasi-open field of research. One can mention the pioneer works of Ellis and co-workers in the mid 1980s on Cp^*_3Yb and related divalent $Cp^*_2Ln(OEt_2)$ complexes (Ln = Eu, Yb). [37] In the latter case, Eu(II) presents the classical broad emission centered in the blue region (around 450 nm) as already observed in many phosphors. [38] and assigned to an interconfigurational $4f^65d{\rightarrow}4f^7$ transition. On the other hand, $Cp^*_2Yb(OEt_2)$ presents a very uncommon broad band emission centered at 900 nm that remains to date unexplained. Nevertheless, the huge difference between divalent and trivalent lanthanide emission gives a unique opportunity to investigate electron transfer processes *via* luminescence spectroscopy.

It is nowadays well established that quantum chemistry calculations on f-element compounds must take into account relativity to obtain a reliable description of the electronic structure. Pyykkö presented in a very popular review the decisive role of relativistic effects in the predictions of chemical, structural, and spectroscopic properties of heavy element systems. [39] Furthermore, the presence of open shells, due to the partial occupancy of the d and f orbitals of the metal, may require the use of accurate theoretical methods but that are expensive in terms of computer time. Among all possible theoretical treatments, the most "accurate" framework consists in the use of a full four-component Dirac-Coulomb Hamiltonian, but it is limited to the study of small systems of very few atoms. In order to be able to carry out calculations on bigger systems, approximations have to be applied. One of them is the quasi-relativistic two-component framework that includes the spin-orbit coupling using, for example, the zeroth-order regular approximation (ZORA) Hamiltonian. Nevertheless, the most affordable method is the one-component approach based on relativistic effective core potentials (RECP) to replace the inner core electrons. In the case of lanthanides, the use of RECP is defined by the size of the core region: if 4f orbitals are defined in the core, the RECP is called "large-core," while if the 4f orbitals are in the valence region, the RECP is then considered to be "small-core." The choice of the RECP for the lanthanide atom is then based on its oxidation state, with the most popular RECPs used in the literature being the one developed by Dolg and his collaborators. [40]

Many theoretical groups attempted to constitute proof of a possible existence of a partial covalency in the bonding of various lanthanide complexes. However, this issue still continues to be debated between theoreticians, as well as between theoreticians and experimentalists. [41] For instance, the geometry of trihalogenated lanthanide complexes, LnX_3, was the "apple of discord" of a long debate between theoreticians and experimentalists. [42] The fact that some of such complexes prefer to adopt a pyramidal geometry is difficult to interpret using a model of a pure ionic Ln-X bond. DFT and *ab initio* studies concluded that the 5d orbitals are responsible of the pyramidalization but without a participation of the 4f shell. [43, 44] In this regard, a recent study by Denning *et al.* describes clearly for the first time the complementarities between DFT calculations and spectroscopic measurements. These have been interpreted as a 4f-shell perturbation due to the presence of the ligand, suggesting their participation in the metal-ligand bonding. [45]

Similarly to divalent lanthanides, low-valent actinides, mainly in the oxidation state +III, have attracted much interest experimentally. Unlike the divalent lanthanide, the "covalency" in early actinide complexes is well established and thus more experimental work is devoted to understanding their reactivity. Indeed, theoretical insights into the bonding situation of some uranyl compounds reveal that 5f orbitals can hybridize with 6d and 7s in order to be available for participating in the bonding. [46] Using post Hartree-Fock calculations, that rigorously include the relativistic effects and the electronic correlations, it was shown the participation of actinides 5f orbitals in chemical bonding, especially for the actinide set Pa, U, Np, Pu. [46] Unlike in the case of lanthanides for the 4f orbitals, it is difficult to include 5f orbitals in the effective core potential (ECP). Quite unexpectedly, in the literature, DFT calculations lead to outstanding results regarding the geometrical features and the vibrational frequencies, even for systems with a high multireference character. [47] Since DFT approach is found to describe correctly the actinide-ligand interactions, it may apply in tackling reactivity problems involving actinide complexes. Among others, the activation of small molecules has been of great interest for many decades and is still an important chemical target, particularly in the context of environmentally friendly "green" chemistry. [48] Current research projects mainly focus on the transformation of the greenhouse gas CO_2 into useful chemicals, [49] the C-H activation of alkanes, [50] the conversion of N_2 into nitrogen-containing organic products, [51] the degradation of chlorofluorocarbons *via* C-F activations, [52] and the use of renewable feedstock like CO and CH_4 for the production of carbon-based molecules [53] (Figure 13.2).

This kind of activation can be readily achieved by electron-rich metal complexes. Since 1985 and the pioneering work of Andersen, [54] uranium (III) complexes have increasingly attracted much attention for their impressive ability to reduce or activate (through binding) small inert molecules. Indeed the oxidation state +IV of uranium is known to be more thermodynamically stable than the oxidation state +III, being the driving force for complexes of uranium (III) to be oxidized by substrates in order to form very often bimetallic uranium (IV) complexes. The substrate is usually doubly reduced and ends up sandwiched between two U(IV) centers (see Figure 13.2). This consecutive double reduction can lead to a domino of unexpected chemical reactivity. Andersen's contributions in that direction included the activation of heteroallenes by tris-cyclopentadienyl U(III) complexes. [54] The group of Cloke in Sussex has been interested in the rich reactivity offered by a variety of mixed-sandwiched U(III) complexes, which are composed of a cyclopentadienyl (Cp) ligand and an eight-membered carbocycle cyclooctatetraene (COT). [55] These complexes readily react with CO, CO_2, NO, and N_2. In the first case, cyclic aromatic oxocarbons of the generic structure $[C_nO_n]^{2-}$ are formed *via* consecutive couplings of CO molecules, which is an unprecedented kind of chemistry.[55d] The group of Meyer in Erlangen works mainly on the reactivity of bulky tris-aryloxide complexes with heteroallenes of the type CE_2 (E = O, S, Se), since the seminal work of isolation of $[((^{Ad}ArO)_3tacn)U^{IV}(CO_2)]$ in 2004. [56] Arnold has recently worked on tris-aryloxide and tris-amide U(III) complexes and their reactivities towards CO, CO_2, and N_2. [57] Finally, Mazzanti and colleagues are interested in the use of siloxide ligands in U(III) complexes, for the facile reduction of small molecules. [58] Some of the most studied U(III) complexes in that context are presented in Figure 13.3.

Alongside such an abundant amount of experimental studies, theoretical chemistry can be of great help to understand the mechanisms of the observed reactions. Density functional

$$2[U^{III}] + S \longrightarrow [U^{IV}]\text{–}(S^{2-})\text{–}[U^{IV}] \longrightarrow \text{Further reactivity}$$

Figure 13.2 *Typical reaction between a $[U^{III}]$ complex and a substrate S*

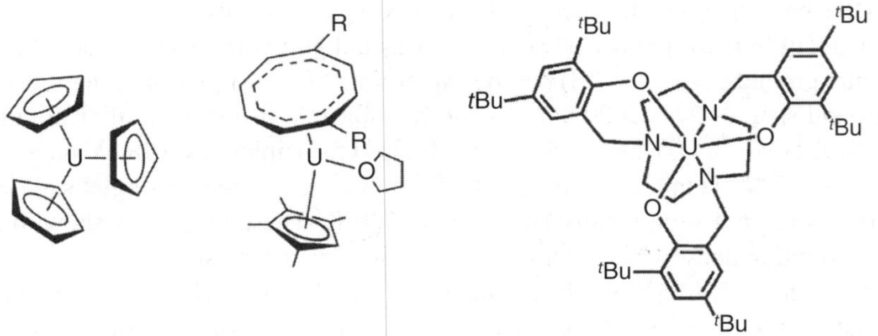

Figure 13.3 Examples of U(III) complexes able to reduce and activate substrates like CO_2

theory in particular is an important tool for the prediction of electronic structures, plausible mechanisms, and spectroscopic data involving complexes of d-transition, lanthanide, and actinide metals. In this regard, the innovative research of Eisenstein and co-workers on the reactivity of organolanthanide complexes by using DFT methodology, has paved the way for the modeling of organoactinide reactivity as well. [59]

13.2 Divalent Lanthanides

13.2.1 Computing the Nature of the Ground State

The concept of a ligand in a metal compound acting as a single-electron acceptor is a topic of much recent interest [e.g., see "Forum" in volume 50 (issue 20) of *Inorganic Chemistry*]. [60] The accessibility of an empty orbital on a ligand in a coordination complex was originally referred to as a "non-innocent" ligand, but this terminology does not clearly distinguish between a metal-to-ligand charge transfer (MLCT) process, in which a pair of electrons is transferred to an empty orbital and a single electron is transferred to an empty ligand orbital that is referred to as a "redox non-innocent" or "redox active ligand." The latter process generates an electron-transfer complex in which an electron resides in the ligand's LUMO with an electron hole remaining on a metal-based molecular orbital. The ground state electronic structure is then determined by how the biradical correlates the two electrons, either forming a triplet state ($M_S = 1$) or an open-shell singlet state ($M_S = 0$). Complexes of d-transition metals with redox active ligands have been extensively and intensively studied. [60] In contrast, complexes of the f-block elements, although they are known, are not as well studied, with most work only appearing recently. [61]

As mentioned in the introduction, the 2,2'-bipyridine adducts of Cp^*_2Yb are of interest because its synthesis and magnetic properties were reported in the 1970s, but an interpretation was only provided more than thirty years after. These complexes has been shown, both experimentally and computationally, to have multiconfigurational open-shell singlet electronic ground states in which ytterbium is "intermediate" valent.[61c,61d] In this context, Scarborough and Wieghardt [62] systematized and classified the often confusing and/or contradictory literature of the 2,2'-bipyridine and related adducts with d-transition metals using a DFT broken-symmetry methodology. A comparison between the electronic ground state of $Cp_2Ti(bipy)$ [63] and $Cp*_2Yb(bipy)$ was enlightening. Both adducts have open-shell singlet ground state[61d,63] (S) but the triplet state (T) in $Cp_2Ti(bipy)$ lies close enough to the ground state ($-2J = 600$ cm^{-1}) that the molecule is a spin equilibrium molecule, S($M_S = 0$) ΔT($M_S = 1$), whereas for $Cp*_2Yb(bipy)$ the triplet lies 0.28 eV (calculated) or $-2J = 0.11$ eV (920 cm^{-1}, experimental)[61k] above the open-shell singlet state, with the triplet state being not significantly populated at 300 K. These properties show that strong exchange coupling does indeed occur in these ytterbium complexes.

Studies of the 2,2'-Me_2bipyridine adducts of $Cp*_2Yb$ have shown that the electronic ground state is multiconfigurational, that is, the ground state wavefunction $\Psi = c_1|Yb(III, f^{13})(bipy^-) > + c_2|Yb(II, f^{14})(bipy)^0 >$ is an admixture of two configurations, where c_1 and c_2 are their respective coefficients.[61d] The $f^{13}(bipy^-)$ configuration is an open-shell singlet and $f^{14}(bipy)^0$is a close-shell singlet and configuration interaction is possible since they have the same symmetry. The outcome of the configuration interaction is that the valence of the ytterbium in $Cp*_2Yb(bipy)$ is neither +II, where $c_1 = 0$, nor +III, where $c_2 = 0$, but it is in between these two extreme values and therefore intermediate valent. Experimentally, the valence of the ytterbium obtained from the L_{III}-edge XANES and expressed as n_f, the number of f-holes, that is, the value of c_1^2 and is independent of temperature. For example, in $Cp*_2Yb(bipy)$ is c_1^2= 0.83(2) and the $f^{13}(bipy^-)$ is the dominant configuration. The concept that the ground state is multiconfigurational was developed from CASSCF calculations, initially using the model $Cp_2Yb(bipy)$,[61d] then extended to $Cp*_2Yb(bipy)$ and including four empty bipy orbitals, $\pi*_1$, $\pi*_2$, $\pi*_3$, and $\pi*_4$, in the active space. The calculated value of n_f was 0.86, in agreement with experiment.[61c]

When the 2,2'-bipyridine ligand in $Cp*_2Yb(bipy)$ was replaced by methyl substituted 2,2'-bipyridine ligands, symbolized as x-Mebipy or x,x'-Me_2bipy (where x indicates the position of the methyl group[s] in the 2-pyridyl rings), the ytterbium atoms are intermediate valent, but the value of n_f depends on temperature.[61c] For example when x is 5, the value of n_f in $Cp*_2Yb(5\text{-}Mebipy)$ changes from 0.42 to 0.75 at 300 K and 30 K, respectively. The temperature dependence is not due to a closed shell singlet–open-shell triplet equilibrium, a valence tautomerism, but an equilibrium between two open-shell singlet states in which the thermodynamic constants, ΔH and ΔS, are obtained using the Boltzmann equation. The computational model showed that two open-shell singlet states, SS1 and SS2, lie below the triplet by 0.58 eV and 0.11 eV, respectively, in agreement with the experimental findings. The evolving model of the electronic structure of the bipyridine adducts of $Cp*_2Yb$ follows: (i) the ground state is an open-shell singlet, (ii) the configurations of both fragments, $Cp*_2Yb$ and the bipyridines, are multiconfigurational, (iii) the number and position of the methyl groups on the bipyridine ligand, and (iv) the n_f values qualitatively track the reduction potential of the free bipyridine ligand, at least for those bipyridines that are in the literature. The latter also provides a strategy for manipulating the value of n_f and therefore the magnetic properties of the adducts.

The reduction potential of a ligand is a measure of the HOMO-LUMO energy gap, which can be affected by substituents that are electron-donors or electron-acceptors, as well as by the torsion angle between the 2-pyridyl rings, usually reported as the NCCN angle obtained from solid-state crystal structures. Electron-donating groups, such as a methyl group, increase the reduction potential relative to hydrogen (Table 13.1), while a phenyl group decreases it. Increasing the torsion angle should raise the reduction potential, since

Table 13.1 *CASSCF calculation results*

	S-T gap (eV)	n_f (calc)	n_f (exp)
$Cp_2Yb(bipy)$	−0.6	0.32	0.30
$Cp^*_2Yb(bipy)$	−0.06	0.86	0.83[1]
$Cp^*_2Yb(3,3'\text{-}Me_2bipy)$	−0.28	0.27	0.17
$Cp^*_2Yb(4,4'\text{-}(MeO)_2bipy)$	−0.007	0.11	0.13

twisting the 2-pyridyl groups out of coplanarity will move the energy of π^*_1 closer to the energy of π^*_2 and π^*_3 antibonding orbitals. Although no experimental reduction potentials are reported for bipyridine ligands with large torsion angles, the energy of the empty orbital should behave much like those in twisted ethylene. [63] The CASSCF methodology gives the calculated singlet-triplet energy separations, with the corresponding n_f values listed in Table 13.1.

The calculational results are consistent with those obtained on related bipy adducts in that the ground state are singlet states that are multiconfigurational. Only one singlet is found below the triplet. The singlet state is composed of two configurations $f^{13}(\pi_1{}^*)^1$ and $f^{14}(\pi_1{}^*)^0$. The calculated values of n_f agree with the experimental values very well for $Cp_2Yb(bipy)$ and $Cp^*_2Yb(4,4'\text{-}(OMe)_2bipy)$, but the agreement is in less agreement for the 3,3'-Me_2 bipyadduct. In each case, the dominant configuration is f^{14}, in contrast to the results for Cp^*_2 Yb(bipy).

Although bipyridine and related ligands, such as diazadienes, attached to d- and f-block metallocenes have attracted the most attention, adducts with 1,10-phenanthroline have been largely ignored. It is shown that the ground state of $Cp^*_2Yb(phen)$ is a triplet, in contrast to the open-shell singlet ground state of $Cp^*_2Yb(bipy)$. One chemical ramification of the triplet electronic configuration is that the phenanthroline ligands in the individual monomers units are coupled by a C-C bond at 4,4'-positions resulting in a dimer. The related adduct, $Cp^*_2Yb(3,8\text{-}Me_2\text{-}phen)$ exists in solution in a dimer Δ monomer equilibria, with variable temperature ^1H NMR spectra giving $\Delta H = -8$ kcal · mol^{-1} and $\Delta S = -30$ cal · mol^{-1}· K^{-1}. The CASSCF methodology is extented to the monomeric phenanthroline adducts, $Cp^*_2Yb(phen)$ (1), $Cp^*_2Yb(3,8\text{-}Me_2phen)$ (3) and $Cp^*_2Yb(5,6\text{-}Me_2phen)$ (7). The ground state of $Cp^*_2Yb(phen)$ is calculated to be comprised of two nearly degenerate triplet states, T_1 and T_2, which are 2.12 eV lower in energy than an open-shell singlet state. The state configuration for the f-orbitals are therefore pure (100 %) f^{13}, and the T_1 and T_2 configuration for the π^*-orbitals are $0.72\ \pi^*_1 + 0.28\ \pi^*_2$ and $0.28\ \pi^*_1 + 0.72\ \pi^*_2$, respectively. The calculated charge-transfer ground state is in accordance with the observation of two LMCT bands in the Vis-NIR spectrum in toluene solution near 500 cm^{-1}.[61e] The calculated ground states for the 3,8-Me_2phen and 5,6-Me_2phen adducts are similar to each other but somewhat different than that of the unsubstituted phenanthroline adduct. Thus, the calculated ground states are spin triplets (pure f^{13}) and the open-shell singlet states are only 0.08 eV and 0.09 eV higher in energy, respectively. The excited-state open-shell singlets are multiconfigurational, in which the dominant configuration is f^{13}; in 3,8-Me_2phen, the f^{13}:f^{14} contributions are 0.75:0.25 and the π^*_1 is the only configuration that contributes to the ligand. In 5,6-Me_2phen adduct, the f^{13}:f^{14} contributions in the excited open-shell singlet state are 0.85:0.15 and the π^*_1 and π^*_2contributions are 0.95 and 0.05, respetcively. The calculated spin triplet ground states in these three phenanthroline adducts are in contrast to the open-shell singlet ground states obtained in all of the bipyridine adducts of Cp*Yb. The triplet ground states in these phenantholine adducts mean that the exchange coupling is weak and is the origin of the different physical and chemical properties displayed by the phenantholine and bipyridine adducts. Interestingly, experimentally, the phentantroline complex of ytterbium is found to be dimerized in solution. The dimerization reaction was studied using DFT methods as the two complexes were Yb(III). A transition state was calculated 15.4 $kcal \cdot mol^{-1}$ (in Gibbs energy) above the monomeric form of the phenanthroline adduct and the dimerization reaction is exothermic by 3.1 $kcal \cdot mol^{-1}$ (Figure 13.1), in good agreement with the rather low enthalpy obtained from experimetal data in solution. The calculated distance of the C-C bond formed in 4,4'-positions (1.596 Å) is in good agreement with the elongated C-C bond found in the solid state structure (1.619 Å). The transition state also involves two molecules of $Cp^*_2Yb(phen)$ with similar bond distances as the final dimer but in which the C-C bond in the in 4,4'-positions is 1.800 Å.

Theoretical approaches have been extensively used to predict the nature of the ground state of several Yb(II) complexes, using the multireference CASSCF method. However, this computational approach is hardly tractable for bimetallic complexes and reactivity studies involving divalent lanthanide complexes. There was thus a need for a simplified approach but without, if possible, too much loss of precision.

13.2.2 Single Electron Transfer Energy Determination in Divalent Lanthanide Chemistry

13.2.2.1 Indirect Method

A first attempt in order to estimate theoretically the single electron transfer from the metal to the substrate in lanthanide(II) chemistry has been recently proposed. [64] In this study, among others, two model experimental systems were considered: the reductive dimerizations of pyridine promoted by $(Cp^*)_2Tm^{II}$ model complex, and the reduction of CO_2 mediated by $(Cp^*)_2Sm^{II}$. The choice of these model complexes was based on their different *f*-orbital occupations that results also in different spin multiplicities. For instance, the electronic configuration (EC) of the Tm(II)-pyridine complex is $4f^{13}$, S =1/2, while after one-electron promotion to pyridine its EC is $4f^{12}$ S = 3/2. The two different spin multiplicities observed when passing from Tm^{II} to Tm^{III}, have easily allowed the estimation of the energetic cost of the redox step by directly computing the energies of the corresponding species by using small-core RECPs at B3PW91 level (Figure 13.4). In particular, the energy that is required for this monoelectronic transfer was estimated to be 19.9 $kcal \cdot mol^{-1}$, in terms of Gibbs free energy. In addition, the validity of this direct measurement of the energy of the SET step was further assessed by the use of CASSCF calculations. In these calculations the active space, for both spin states, consisted of 13 electrons in eight active orbitals, with the seven 4*f* orbitals of Tm and the π^* of the pyridine. The outcome of the latter further supports the single determinant nature of the corresponding reduced complex of thulium (C_I of 0.92), verifying the validity of the computational protocol that has been used further (B3PW91/SC-SDDALL(Tm)/6-31+G*(others)).

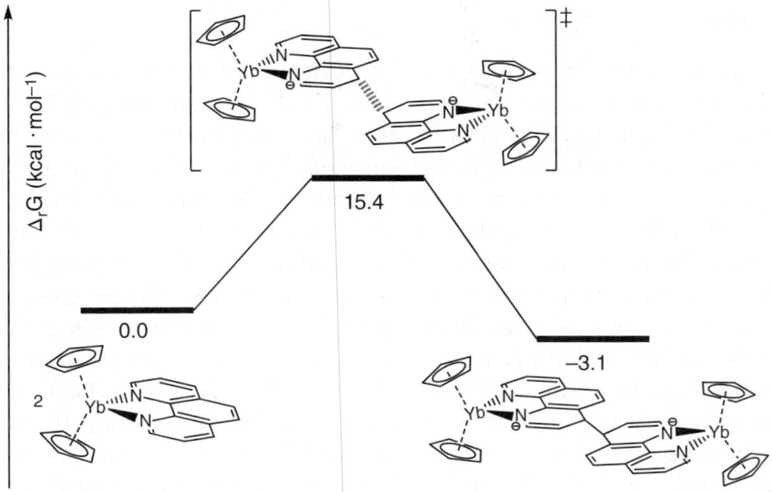

__Figure 13.4__ Energy profile for the dimerization reaction observed with phenanthroline adducts. The calculations were performed with Cp instead of Cp ligands*

Nevertheless, dealing with the single electron transfer in samarium(II), reactivity is both computationally complicated and challenging at the same time. This is due to its less than half filled f subshell. In fact, the $(Cp^*)_2Sm^{II}$-CO_2 adduct has an EC in the ground state $4f^6$, S =3, which after the redox step becomes $4f^5$, S =3, implying no change in the overall spin multiplicity. However, the case of the reduction of CO_2 is peculiar. Indeed, by populating with one electron the lowest unoccupied molecular orbital of the linear CO_2 molecule results in a first-order Jahn-Teller effect, leading to the bending of the molecule. The latter geometrical change allows the direct estimation of the energetics of the SET process. It should be noted, that the energy estimation of a consecutive second electron transfer from a newly introduced fragment $Sm(Cp^*)_2$ to an already monoreduced CO_2 adduct was studied. The latter process has been found to correspond to a barrierless one, and the formation of the final binuclear species corresponds to an exothermic process.

But what about the direct estimation of the SET when precursors with a high spin multiplicity in their ground state as Sm(II) react with unsaturated substrates as pyridines, alkynes, alkenes, ketones, etc. [65], in which a geometrical change is not as clearly marked as in the case of CO_2? Someone can think that the use of the large core ECPs adapted in a fixed oxidation states II and III of the samarium can help in this way. However, the straight comparison of the corresponding total energies is not possible, due to the different construction of each basis set. To address this issue an indirect estimation of the thermodynamics of the redox step was considered by using a hypothetical redox thermodynamic cycle involving CO_2 for the oxidation step followed by an isodesmic substrate exchange from the oxidized species, as is shown in Figure 13.4. This strategy was applied to the indirect estimation of the pyridine promoted by $(Cp^*)_2Tm^{II}$ complex and proved to give almost identical results compared to the direct estimation (19.5 vs. 19.9 kcal · mol^{-1}).

13.2.2.2 Other Indirect Method

In a recent contribution, the estimation of the energetics of the redox step of the reduction of benzophenone ($Ph_2C{=}O$) by SmI_2 in neat THF was studied. [66] This step was experimentally estimated to be slightly exoergic. However, based on the available experimental measured redox potentials, the single electron transfer from SmI_2 to benzophenone is slightly endoergic. This discrepancy was attributed by the experimentalists to the binding energy of Sm(III) to the bound $Ph_2C{=}O$ with radical anion character. Hence, the latter computational scheme, referring to the indirect method using CO_2, was applied in order to reproduce the experimental value for this reductive process. Nevertheless, using this methodology, it was not feasible to obtain any value due to the failure of locating a "hypothetical" CO_2 adduct of $SmI_2(THF)_4$. From the other side, as was expected, the direct calculation of the balance $[SmI_2(THF)_5] + Ph_2CO = [SmI_2(THF)_4(OCPh_2)] + THF$ was also found to be unsuccessful, because it was described previously. In particular, careful inspection of the geometry of the $SmI_2(THF)_4(OCPh_2)$ in the septet spin state, reveals that the $O{=}CPh_2$ bond is not reduced as its C=O bond distance remains unchanged with respect to the free molecule of the benzophenone. Yet, by imposing a quintet spin state multiplicity on adduct's geometry, someone can enforce the reduction of the ketone to be realized, by populating the π^* orbital of the C=O bond, but in an anti-parallel spin fashion. The comparison of the geometries of the small core calculation in quintet and the large core in fixed oxidation state gave identical geometries. In this case, the reduction of the $OCPh_2$ that leads to a possible formation of the reduced complex, in which the samarium is in the quintet spin state, was not able to reproduce the experimental results. A possible reason for that is probably a result of the energy that is needed for the spin-flip process. The failure of the above described methods to reproduce the energetics that underlie the redox step, led us to examine the possibility of using different thermodynamic cycle. The newly modified cycle includes three steps (Figure 13.5), with the first referring to the complete disruption of $SmI_2(THF)_4$ into Sm^{2+}, free iodide and THF, $\Delta_r H(1)$. A second, in which the Sm^{2+} is oxidized to Sm^{3+} and benzophenone is reduced to $Ph_2^{\bullet}C{-}O$, $\Delta_r H(2)$, and a third, by which, the formation of $[SmI_2(THF)_4]^+$ from separated fragments is taking place $\Delta_r H(3)$.

It should be noted that since the energy difference between the free Sm^{2+} and Sm^{3+} cations was required to be computed, the spin-orbit coupling contribution was taken into account by using the Landé coupling scheme. The computed enthalpy for the redox step was found to be in good agreement with the experimental value, $\Delta_r H_{redox}$. [67] However, the $\Delta_r H_{activ}$, which is the sum of the reduction energy of the substrate and the binding energy, was found to be higher than the experimentally estimated. The latter inconsistency further highlights the totally different behavior of the SmI_2 systems with respect to other lanthanide congeners,

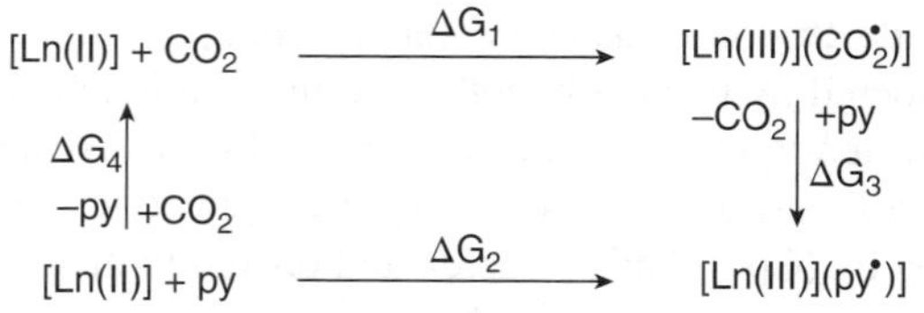

Figure 13.5 *A Hess cycle to determine the SET energy using DFT methods*

and makes more imperative than ever the need for finding a general method or strategy that will be simple and effective.

13.2.2.3 The SOMO-LUMO Gap

Finally, a quantitative way to estimate the SET energy in various of lanthanide(II) complexes having alkynes (phenylacetylene, hex-1-yne) or hetero-aromatic compounds was found by calculating the SOMO-LUMO gap energy of the orbitals of these complexes. [68] In particular, multi-configurational (CASSCF) calculations were performed on the DFT geometry of $(Cp^*)_2Sm^{II}$(HCCPh) adduct, septet spin state, by distributing six electrons in eight orbitals (the seven *f*'s and the π^* of the substrate) in order to determine the SET energy. It worth noting that the aforementioned system corresponds to an experimental one, from which reactivity can be accessed *via* two consecutive SET steps to give as final products the trienediyl complexes with a simultaneous release of H_2. The CASSCF have shown that the lowest root corresponds to a f^5-π^{*1} occupation, and in a smaller extent, to a f^6 one. The decryption of these interesting results reveals that the substrate in the septet spin state is already reduced upon coordination, and the SET energy was estimated to be $-78.6\ \text{kcal}\cdot\text{mol}^{-1}$. From the other side, by applying the indirect method by using the Born-Haber thermodynamic cycle, the energy was found to be endergonic by $7.0\ \text{kcal}\cdot\text{mol}^{-1}$. Thus, the sign of the determined SET energy is found to be very different as well as the magnitude of the energy. However, carefully inspecting the restricted open-shell Hartree-Fock orbitals that have been used as a guess function for the multi-reference calculation revealed that the π^* of the substrate was already occupied, and that the LUMO orbital is corresponding to a pure *f* one, in line with the picture of the orbitals of a reduced substrate. The same however was true by checking the DFT(B3PW91) calculated molecular orbitals. More interestingly, it was observed that the energy of the "SOMO-LUMO gap" in CASSCF is $-88.7\ \text{kcal}\cdot\text{mol}^{-1}$ being very close to the estimated SET energy at the same level of theory. Another interesting outcome was derived by the CASSCF calculation of the quintet spin state. The calculated energy for this electronic configuration was found to be higher than the corresponding in septet, stating hence that the quintet state is a high energy excited state. This coordination induced SET was further verified by checking the SOMO and LUMO orbitals of different samarium complexes with substrates as bipyridine or terpyridine and found to be reminiscent to the phenylacetylene case. The same also holds true, for the case of the $SmI_2(THF)_4(Ph_2CO)$ system, that all of the previous methodologies failed to give coherent results. It is also worth noting that the SOMO-LUMO gap energy is computed and was found to be in very close agreement with that coming from the electrochemical results, and having as well the same sign. Finally, we extended to other lanthanide systems, as europium and ytterbium cyclopentadienyls having different alkynes as substrates, but leading to different reactivity. In the case of the europium it was found that europium keeps the same oxidation state (II) throughout the reaction with phenylacetylene. This different behavior serves as a perfect candidate to verify further the validity of our method. In particular, in the europium adduct (octet spin state multiplicity) inspection of the MO spectra has revealed that the SOMO corresponds to an *f* orbital, when the LUMO is the π^* orbital of the alkyne bond, assessing hence further the validity of this approach. From the other side, ytterbocene(II) complexes can react with dimethylacetylene to afford the adduct in which the Yb oxidation state remains unchanged. Like in the case of europium the corresponding

SOMO and LUMO orbitals are corresponding to an f and π^*, respectively, pointing out again the validity of this approach. Hence, by using this very simple method someone can define very quickly if the SET step corresponds to an endoergic or to exothermic. In the first case, the use of computationally heavy methods as multi-reference one is essential, as it influences the whole reaction process as much as the subsequent bimetallic reactivity. For the second case, where the SET step is just favorable (coordination induced SET and exothermic SET) so that it is the subsequent bimetallic reactivity that is crucial there is no need for extra theoretical investigation. Finally, this method, which is simple to handle, appears to be powerful to predict the reduction ability of lanthanide(II) complexes allowing someone to proceed safely on the reactivity computational studies.

13.3 Low-Valent Actinides

13.3.1 Actinide(III) Reactivity

13.3.1.1 Methodology for the Modeling of the Redox Reactivity of U(III) Compounds

In modern computational methods, it is common to model organolanthanide complexes by using large core effective core potentials (ECP) including the 4f electrons since 4f orbitals are buried and do not participate in their reactivity. We have additionally shown that 5f electrons show no substantial activity in the organometallic chemistry of actinides by comparing energy profiles calculated with both small core and 5f-in-core ECPs. [69] Energetic values corresponding to the ortho C–H activation of pyridine N-oxide and other derivatives by $Cp_2U^{IV}(CH_3)_2$ are reported in Table 13.2 and demonstrate this clearly. It is noteworthy that the energies of the adducts and products of the reaction are nearly equivalent with a small core and a 5f-in-core ECP. On the other hand, the calculated activation barriers are slightly larger when a 5f-in-core ECP is used. This is due to a lack of core-valence correlation, which can be fixed by the addition of a core polarization potential, and not because of the implicit treatment of the 5f electrons. Globally, the overestimation of the transition state energies lies around 3-5 kcal · mol^{-1}.

Table 13.2 *Energies of the adducts, the transition states and the products with respect to the reactants for the reaction $Cp_2U^{IV}(CH_3)_2 + C_5H_5NO \rightarrow Cp_2U^{IV}(CH_3)(C_5H_4NO) + CH_4$ (in kcal · mol^{-1})*

	Adduct	TS	Products
Small core ECP			
Pyridine N-oxide (ortho C–H)	+4.5	+15.2	−31.6
2-Picoline N-oxide (ortho C–H)	+3.0	+15.2	−31.7
2-Picoline N-oxide (methyl C–H)	+3.2	+21.1	−27.7
2,6-Lutidine N-oxide (methyl C–H)	+10.4	+21.6	−27.1
5f-in-core ECP			
Pyridine N-oxide (ortho C–H)	+6.3	+18.4	−31.0
2-Picoline N-oxide (ortho C–H)	+6.4	+18.6	−30.5
2-Picoline N-oxide (methyl C–H)	+5.7	+26.2	−25.2
2,6-Lutidine N-oxide (methyl C–H)	+9.3	+26.9	−24.5

These results show that it is often safe to use a 5f-in-core ECP adapted to the oxidation state +IV to model the reactivity of U(IV) complexes. This is particularly interesting for the redox chemistry of U(III) complexes because a large part of the reactivity involves bimetallic U(IV) complexes (Figure 13.1). These species are usually quite large (e.g., with the bulky ligands used by the group of Meyer) and modeling the uranium centers with a small core ECPs becomes too expensive computationally, suffering as well by convergence problems of their wavefunction. On the other side, the main problem that arises from the use of 5f-in-core ECPs is that they are adapted to only one oxidation state. Thus, computing the redox step for systems that an oxidation change is involving, like for instance for U(III)/U(IV), this becomes problematic, and a method to achieve it has been explained above. [64] However, let us give a more analytic description of that. The proposal is to separate the equation of formation of the bimetallic complex into two equations (see Figure 13.6). Equation (2) in Figure 13.7 can be computed with a small core ECP since it only involves monometallic complexes whilst Equation (3) in Figure 13.7 can be computed with the 5f-in-core ECP as it only involves the oxidation state +IV of uranium. Concomitantly, the energy of equation (1) in Figure 13.7 is simply the sum of the energies of equations (2) and (3). Such a process is easily achieved with, for example, S = CO_2 as $[U^{IV}]$-(CO_2^-) has a very specific geometry. When CO_2 is reduced, the O-C-O angle is no longer linear and computed bond lengths increase. By forcing a bent CO_2 geometry in interaction with uranium, structural optimization leads directly to the oxidized form of the uranium complex in interaction with (CO_2^-). Despite success with CO_2, extension of this methodology to CO is more challenging as optimizations with a small core ECP lead more readily to a neutral CO molecule interacting with a U(III) center. A method to avoid that problem is described in Figure 13.7; the oxidative binding of CO_2 can be computed with the small core ECP, while the ligand exchange

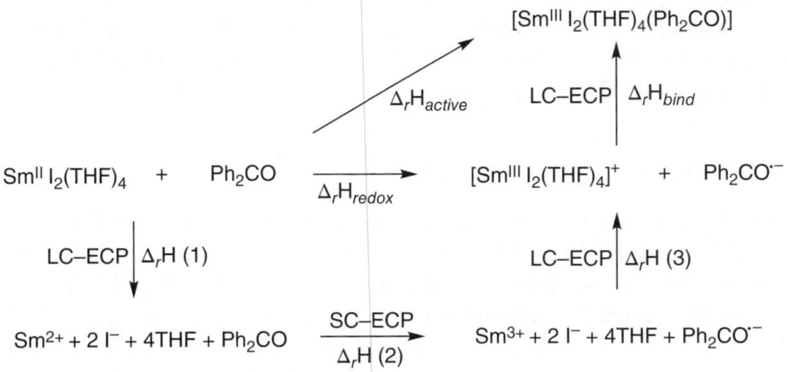

Figure 13.6 *A schematic representation of the SET energy determination using DFT*

$2[U^{III}] + S$	$\longrightarrow$	$[U^{IV}]–(S^{2-})–[U^{IV}]$	(1)
$2[U^{III}] + 2S$	$\longrightarrow$	$2[U^{IV}]–(S^-)$	(2) Small core ECP
$2[U^{IV}]–(S^-)$	$\longrightarrow$	$[U^{IV}]–(S^{2-})–[U^{IV}] + S$	(3) 5f-in-core ECP

Figure 13.7 *Separation of the global equation of the double reduction of S into two other equations. (1) = (2) + (3)*

from CO_2 to CO does not require any change in the oxidation state at U and can therefore be computed with the 5f-in-core ECP.

13.3.1.2 *Reactivity of U(III) Complexes with CO_2*

Several theoretical investigations regarding the reactivity of U(III) complexes with CO_2 have been carried out so far. [58, 70] These studies showed that three different mechanisms are in competition. In all schemes, the bimetallic species $[U^{IV}]$-(CO_2^{2-})-$[U^{IV}]$ is initially formed. This complex can then become bereft of CO to form the μ-oxo complex $[U^{IV}]$-(O^{2-})-$[U^{IV}]$, which may further react with another CO_2 molecule to yield the μ-carbonate product $[U^{IV}]$-(CO_3^{2-})-$[U^{IV}]$. The carbonate complex can also be formed by an alternative mechanism where the attack of CO_2 and release of CO occur in concerted manner. The third mechanism is a direct C-C coupling between free CO_2 and the CO_2^{2-} moiety of $[U^{IV}]$-(CO_2^{2-})-$[U^{IV}]$ to yield the μ-oxalate product $[U^{IV}]$-$(C_2O_4^{2-})$-$[U^{IV}]$. Regardless of the U(III) complex studied, the μ-oxalate is always found as the most stable product. However, the steric bulk from the ligands of the U(III) reactant have a marked effect on the height of the activation barriers of the different pathways. In the case of $[((^{Me}ArO)_3mes)U^{III}]$, the transition state of C-C coupling is 10 $kcal \cdot mol^{-1}$ higher than the transition state of release of CO, which is in agreement with experimental evidence since the μ-carbonate product is observed. To illustrate the effect of the steric bulk, three different pathways are shown in Figure 13.2. The low-lying pathway corresponds to the reaction between $[((^{Me}ArO)_3mes)U^{III}]$ and CO_2 to yield the μ-carbonate product *via* the formation of the μ-oxo intermediate. It is an accessible reaction since the ligands of this complex are quite flexible. The high-lying profile is the same reaction with the bulkier $[U^{III}(C_8H_6(1,4\text{-}Si^iPr_3)_2)(Cp^*)]$ complex. Steric hindrance dramatically increases the height of the barriers as well as the energy of the oxo intermediate. For this second system, the only way to form the carbonate product is to keep a long U-U distance maintained throughout the pathway. This is the last profile reported in Figure 13.2 – a concerted mechanism where the electrophilic attack of CO_2 and the release of CO are concomitant.

13.3.1.3 *Reactivity of U(III) Complexes with CO*

An additional theoretical study of importance concerning U(III) redox reactivity is the reaction of $[U^{III}(C_8H_6\{Si^iPr_3\text{-}1,4\}_2)(Cp^*)]$ with CO. [71] Experimentally, the reaction leads to the formation of the deltate product $[U^{IV}(C_8H_6\{Si^iPr_3\text{-}1,4\}_2)(Cp^*)]_2$-$(C_3O_3^{2-})$. If the stoichiometry is carefully controlled, the linear ynediolate-bridged $[U^{IV}(C_8H_6\{Si^iPr_3\text{-}1,4\}_2)(Cp^*)]_2$-$(C_2O_2^{2-})$ is obtained. McKay *et al.* reported two different mechanisms leading to the respective formations of both products. [71] The ynediolate-bridged complex is found to result from a very facile C-C coupling (+2.0 $kcal \cdot mol^{-1}$) of two monometallic adducts $[U^{IV}(C_8H_6\{Si^iPr_3\text{-}1,4\}_2)(Cp^*)(CO^-)]$, where CO is coordinated to U by the carbon atom. The coupling leads to a bimetallic complex where a zig-zag $C_2O_2^{2-}$ unit is sandwiched between both U^{IV} centers. By a succession of kinetically accessible isomerisation steps (the highest activation barrier being +6.9 $kcal \cdot mol^{-1}$), the zig-zag structure becomes linear to yield the experimental complex, thanks to a strong thermodynamic driving force.

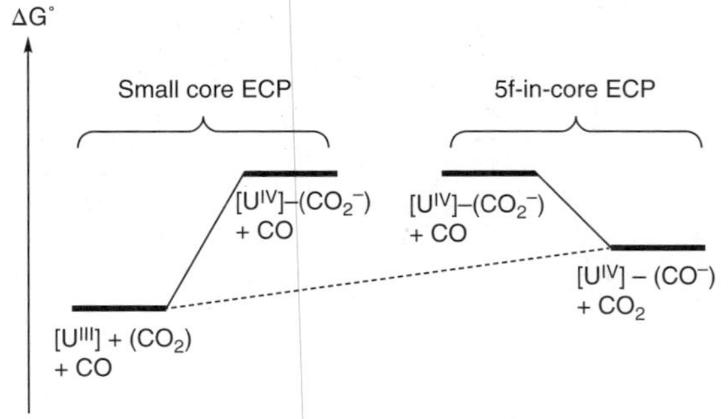

Figure 13.8 *Indirect method to calculate the oxidation step [U^{III}] + CO → [U^{IV}]–(CO)*

The formation of the deltate complex has also been investigated. The key intermediate of this reaction is found to be a bimetallic ketene complex where $C_2O_2^{2-}$ lies between both U^{IV} centers. This intermediate is obtained *via* a C-C coupling between a side-on bimetallic CO complex and a free CO molecule. The side-on complex has never been identified experimentally, despite being isoelectronic with known dinitrogen-bridged diuranium complexes. A summary of the two pathways is reported in Figure 13.8.

It is noteworthy that the experimental linear ynediolate complex does not react with CO to form the deltate complex. Theoretical studies show that this complex has to isomerize into the ketene intermediate of the second pathway before being able to react with CO, which is kinetically inaccessible.

13.3.1.4 Other Substrates or Complexes

It was experimentally observed that a dinuclear homoleptic siloxide complex of uranium(III) can react with CS_2 to afford the corresponding two electron reduced bimetallic product, while reaction with CO_2 is leading to the carbonate dinuclear product upon release of one CO molecule. [58] Based on these experimental observations, DFT calculations were performed to elucidate the mechanism that underlies behind this remarkable reactivity. In this respect, we have shown that in the case of CS_2, subsequent one electron oxidation of two uranium(III) centers to uranium(IV) is induced by the coordination of CS_2, resulting in the formation of a highly stable bimetallic $(CS_2)^{-2}$ bridged complex, the so-called "key intermediate." Further reactivity from these bimetallic species was considered in order to gain a deeper insight in the inactiveness of this complex. All of the possible reactions with an additional CS_2 molecule to yield a possible sulfide, trithiocarbonate, or tetrathiooxalate products were found to be kinetically or thermodynamically unfavoured, in line with the experimental results. In contrast with CS_2, carbon dioxide showed a reductive disproportionation type of reactivity. To address this diverse behavior we have proposed two possible mechanisms; one that is passing from a classical oxo intermediate and the other from a concerted one-electron reduction of two CO_2 molecules transient intermediate. It should be noted that the latter mechanism was proposed for the first time in the uranium chemistry, and found as the operating one (Figure 13.3).

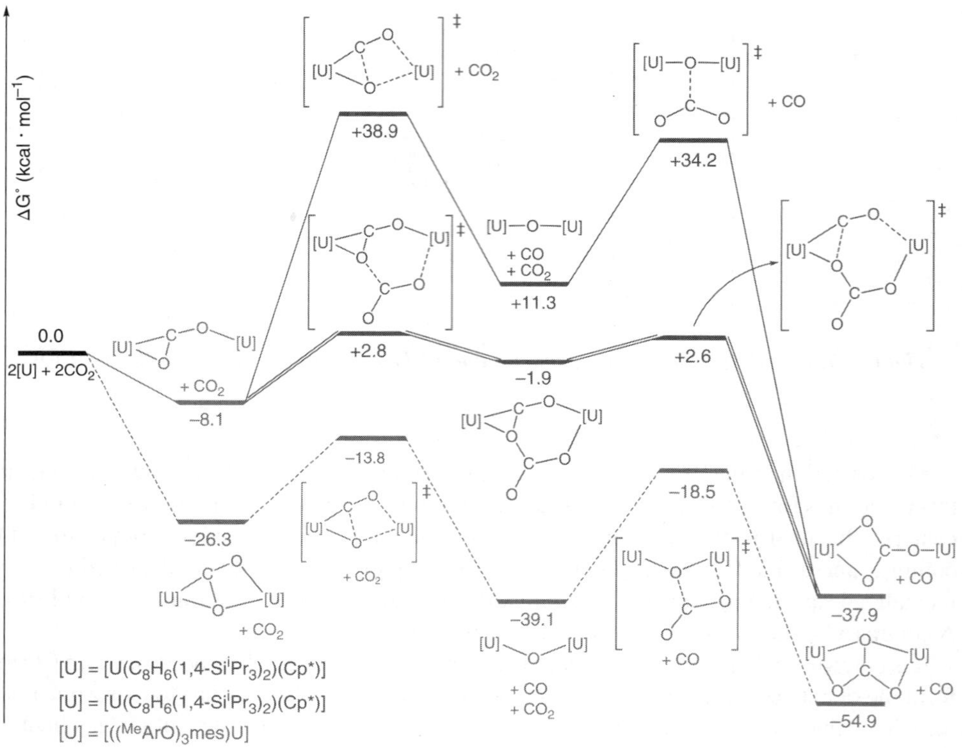

Figure 13.9 *Gibbs free energy profiles of the reactions of U(III) complexes and CO_2. The profile on the bottom concerns $[((^{Me}ArO)_3mes)U^{III}]$ while both other pathways concern $[U^{III}(C_8H_6(1,4\text{-}Si^iPr_3)_2)(Cp^*)]$. Both complexes are simplified as [U]. See color plate section*

Although the reactivity has concerned more the allene-like reduction chemistry, the concept was extended to the reaction mechanism of the formation of the first bis-vinyl bridged dinuclear complexes of uranium(IV) by addition of terminal mono- or bis-alkynes to $^{N}U^{III}$(DME) complexes (N stands for N-anchored tris-aryloxides ligands. [72] DFT calculations revealed that the most thermodynamically favored and kinetically accessed mechanism for the C-C coupling of terminal alkynes consists of two steps: first, oxidation and subsequent formation of the activated binuclear alkynyl complex, followed by C-C coupling after the addition of a second terminal alkyne, to afford the corresponding vinyl bridged binuclear uranium complex (IV) (Figure 13.4). The same mechanism was found to operate for terminal bis-alkynes, resulting in the *cyclo*-products through C-C coupling cyclization step. A possible C-C coupling between two terminal alkynes in the monomer was examined as well. The latter was found to be kinetically not easily accessible and it was ruled out. The results derived from calculations highlighted the generality of this particular mechanism as well as the importance of the formation of the thermodynamically very stable "key intermediate," which corresponds to the binuclear alkynyl complex.

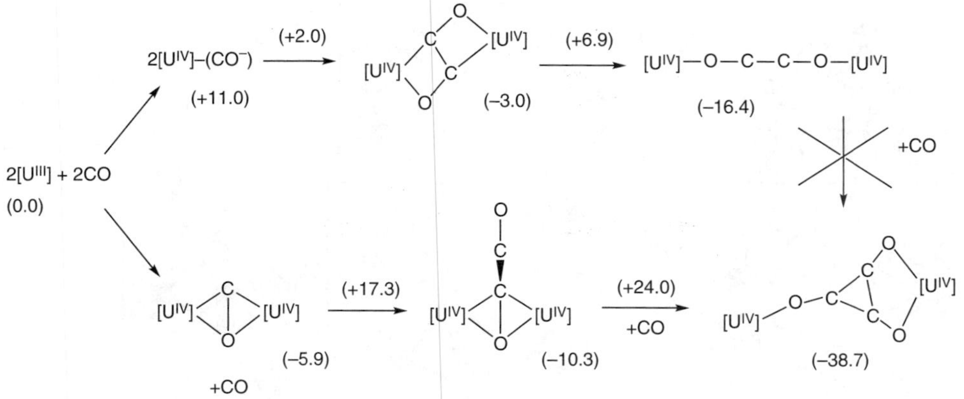

Figure 13.10 Summary of the theoretical investigation of the reactivity of $[U^{III}(C_8H_6\{Si^iPr_3\text{-}1,4\}_2)(Cp^*)]$ with CO. The complex is simplified as $[U^{III}]$. The numbers in parentheses are the calculated Gibbs free energies in kcal · mol^{-1}. The values near the arrows are relative activation barriers; the values near the compounds are their free energies of formation

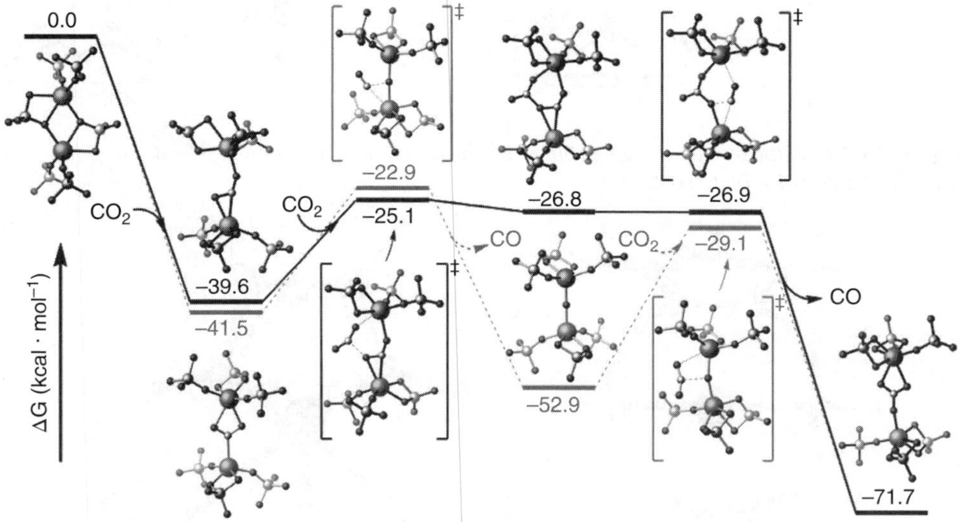

Figure 13.11 Gibbs free energy profiles of the reactions between $[U^{III}(OSi(OtBu)_3)_3]$ and CO_2. See color plate section

13.3.2 Other Oxidation State (Uranyl…)

As already shown, DFT methods can be used to describe and explain the reactivity of actinide complexes, even when SET is involved. Interestingly, it is worth noting that DFT allowed theoretical treatment of the experimental systems. In particular, the geometry and

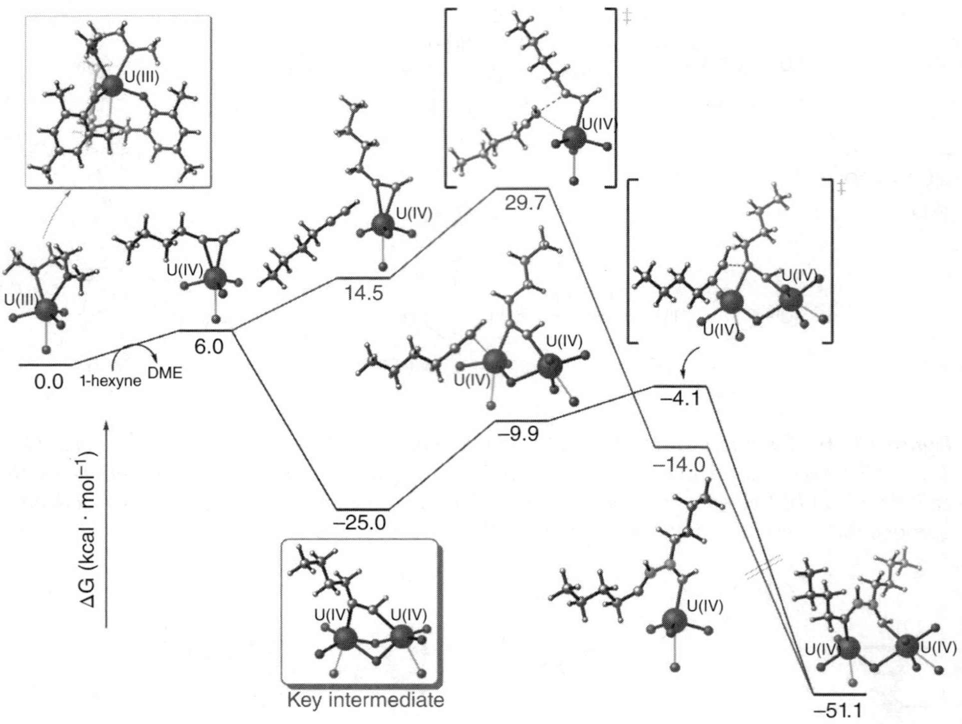

Figure 13.12 Gibbs free energy profiles of the reactions between $[((^{Me}ArO)_3N)U^{III}]$ and n-BuCCH. See color plate section

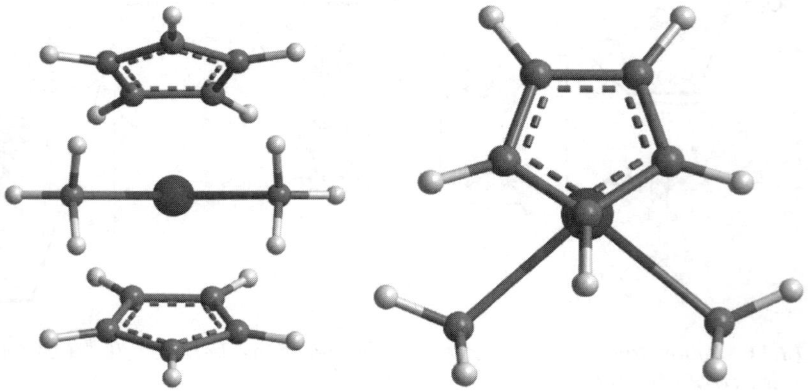

Figure 13.13 Cyclopentadienyl complex $Cp_2An(IV)R_2$

the electronic structure of cyclopentadienyl complex (Figure 13.5) which are very important in organoactinide chemistry [73] are correctly reproduced.

In these cases, interaction between actinides and the ligands were analyzed [74] and confirmed the participation of actinides 5f orbitals for chemical bonding, leading to new modes of actinide-ligand interactions. DFT was also used in order to study the reactivity

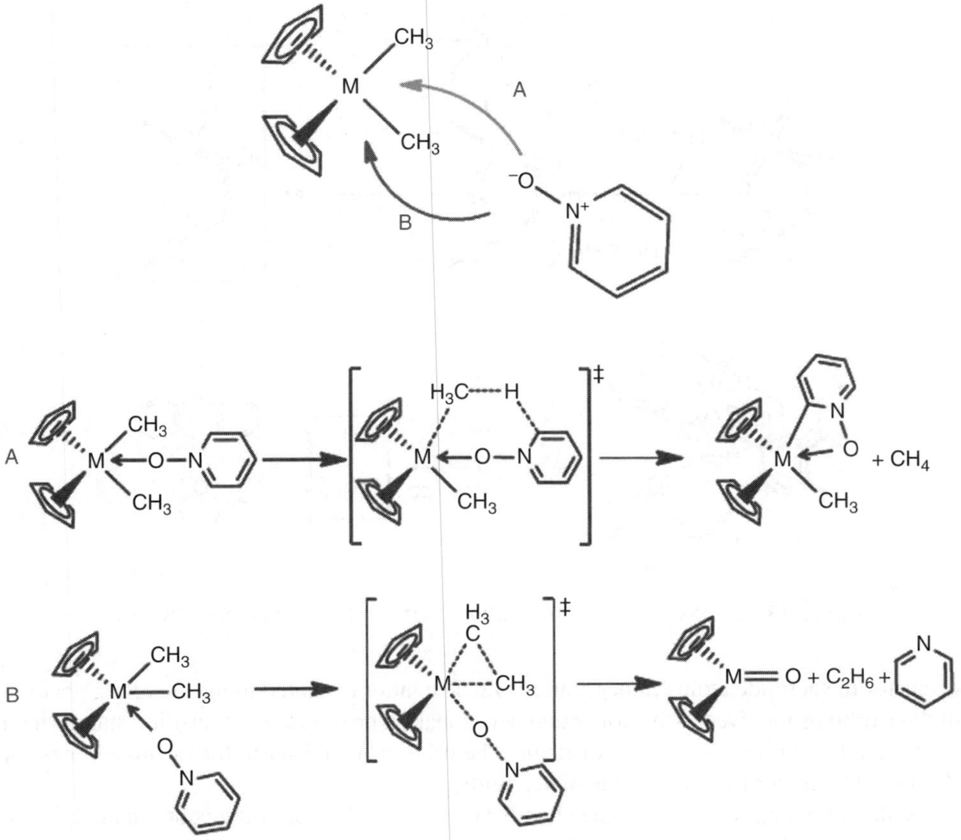

Figure 13.14 *Possible reactions between the complexes $Cp_2[M]Me_2$ and pyridine N-oxide in the case of U*

of organoactinides compounds. For example, Kiplinger [75] reported experimentally a difference of reactivity between actinides (Th, U) and transition metals (Zr) in the case of the reaction between biscyclopendienyl-bismethyl ($Cp_2[M]Me_2$) complexes and pyridine N-oxide (Figure 13.6).

Experimentally, Kiplinger *et al.* reported *ortho* C-H activation (Figure 13.6, channel A) rather than oxygen transfer (Figure 13.6, channel B) in the case of actinides, and no reaction with transition metals analogs. Calculations were performed [76] in order to study these reactions and revealed an excellent agreement with the experimental observations, giving some explanations for this difference in reactivity. For instance, the putative oxygen transfer reaction is the thermodynamically most favorable reaction; kinetically is forbidden. The lack of reactivity of the zirconium complex reported experimentally was also proved theoretically. By comparing the results for Zr, Th, and U, it was shown that despite its electronic configuration, the thorium atom behaves more as an actinide than as a transition metal. The ionic nature of the M-Me bonds, the use of the 5f orbitals at the transition state, and the energetic data obtained for both the energy profiles led us to the conclusion that thorium clearly behaves as an actinide. They also reveal that it is possible to explain why the reaction should

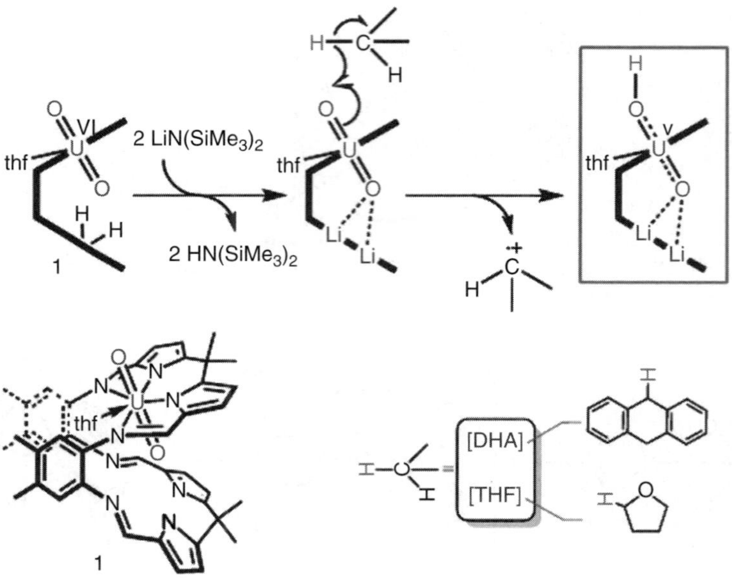

Figure 13.15 Reactions of oxofunctionalisation of UO_2(pacman) by DHA

occur for the actinide atoms rather than for Zr. The latter is related to the energetic position of the d orbitals involved in the donation/back-donation process at the transition state, which makes the bond breaking easy or difficult. The orbitals are closer to the occupied ones for Zr but much higher for the actinide under study.

As already mentioned earlier, the chemistry of high-valent uranium is dominated by the dioxo or uranyl dication, $[UO_2]^{2+}$, which is found both in aqueous solutions and in the solid state. It is chemically robust and shows little propensity to participate in the myriad reactions that are characteristic of its Group 6 transition metal analogues, $[MO_2]^{2+}$. Furthermore, this stability coupled with its mobility in the aqueous phase means that it is a problematic environmental contaminant. There is in the literature a long range of experimental and theoretical study on actinyl, and particularly uranyl. [46, 77] All of these theoretical and computational studies are focused on structure and electronic properties of actinyl using DFT and post Hartree-Fock calculations methods. But often, the theoretical studied systems are not corresponding to the real experimental one. With the increase of computational power, it is now possible to treat full experimental systems, and even more to study the reactivity of these species at the DFT level adequately.

Experimental work by the group of Arnold proves that it is possible to functionalize the oxo group of uranyl in a macrocyclic environment with a silane $SiMe_4$ or make a C-H activation of dihydroanthracene (DHA), even though it is well known that the interaction between uranium and oxygen is very strong and leads to a chemically inert oxo group (Figure 13.7).

A theoretical investigation of the reductive oxo-group silylation reaction of the uranyl dication held in a Pacman macrocylic environment has been carried out by us. [78] The effect of the modeling of the Pacman ligand on the reaction profiles is found to be important, with the dipotassiation of a single oxo group identified as a key component in promoting the reaction between the Si X and uranium-oxo bonds. This reductive silylation reaction is also proposed to occur in an aqueous environment but was found not to operate on bare ions; in this latter case, substitution of a ligand in the equatorial plane was the most likely reaction. These results demonstrate the importance of the presence but not the identity of the equatorial ligands upon the silylation of the uranyl U-O bond. The sequential addition of a lithium metal base to the uranyl ion constrained in a 'Pacman' environment results in lithium coordination to the U=O bonds and single-electron reduction. This reaction depends on the nature and stoichiometry of the lithium reagent and suggests that competing reduction and C-H bond activation reactions are occurring at DHA. Calculations demonstrate that the reactions are kinetically and thermodynamically favorable only for DHA compared to a set of compounds that are kinetically and/or thermodynamically disfavored.

13.4 Conclusions

Current computational tools allow a good treatment of a range of divalent lanthanide and actinides species in different oxidation state. They lead to an excellent agreement with the experimental observations as structure, spectroscopy, electronic properties, and even more on reactivity aspects. Although multireference calculations are mandatory to properly account for magnetism of divalent lanthanide complexes, DFT-based computational strategies have proven their ability to explain the structure and reactivity of both divalent lanthanide and actinide complexes. In the case of the reactivity of divalent lanthanides, DFT methods allow to reproduce well the SET energy that appears to be the crucial step in the reactivity. For the subsequent reactivity, that does not involve any change in oxidation state, f-in-core RECPs together with DFT allow to describe the mechanisms that involve even bimetallic complexes. Challenging projects, such as the rich reactivity offered by SmI_2 reagents, remain almost untouched theoretically but with the improvement of computers this might not be true in the next decade.

Uranium (III) redox reactivity with small molecules always consists in the preliminary double reduction of the latter by two U(III) complexes. Thus, the reactivity concerns only U(IV) bimetallic complexes. DFT calculations can be used for this problem thanks to the recent development of 5f-in-core ECPs by Moritz *et al.* and a methodology to compute the preliminary redox step *via* the combination of small core and 5f-in-core calculations. Recent theoretical studies have mainly concerned the reduction of CO_2 and CO, but attention has also to be paid to the reduction of CS_2, COS, PhNCO, and PhN_3, the C-C coupling of terminal bis-alkynes to form U(IV) vinyl complexes and the reduction of arenes. However, in order to have more elements to discuss about actinide properties, one must perform calculations at a multireference post Hartree-Fock level to take into account the important effect of electronic correlation in these systems; however, there is the obstacle of the computational power to perform calculations of real systems at this level that stands still.

References

[1] Cyclopentadienyl Compounds of Sc, Y, La, Ce and Some Lanthanide Elements, G. Wilkinson, J. M. Birminham, Journal of the American Chemical Society 1954, 76, 6210.
[2] The Cyclopentadienides of Scandium, Yttrium and Some Rare Earth Elements, J. M. Birmingham, G. Wilkinson, Journal of the American Chemical Society 1956, 78, 42.

[3] Uber dicyclopentadienyleuropium und dicyclopentadienylytterbium und tricyclopentadienyle des terbiums, holmiums, thuliums und lutetiums, E. O. Fischer, H. Fischer, Journal of Organometallic Chemistry 1965, 3, 181.
[4] (a) On the behaviour of f-electrons influenced by usual ligands field II. The absorption spectrum of tricyclopentadienylthulium, R. D. Fischer, H. Fischer, Journal of Organometallic Chemistry 1967, 8, 155. (b) Innere Elektronenanregungen von Tricyclopentadienyl-Komplexen des Ytterbiums, R. D. Fischer, H. Fischer, Journal of Organometallic Chemistry 1965, 4, 412.
[5] Reductive coupling of pyridazine and benzaldehyde azine and reduction of bipyridine by samarium complex (C5Me5)2Sm(THF)2, W. J. Evans, D. K. Drummond, Journal of the American Chemical Society 1989, 111, 3329.
[6] (a) Weak paramagnetism in compounds of the type $Cp\prime_2Yb$(bipy), M. D. Walter, M. Schultz, R. A. Andersen, New Journal of Chemistry 2006, 30, 238. (b) Coordination Complexes of Decamethylytterbocene with 4,4‘-Disubstituted Bipyridines?: An Experimental Study of Spin Coupling in Lanthanide Complexes, M. D. Walter, D. J. Berg, R. A. Andersen, Organometallics 2006, 25, 3228. (c) Coordination of 2,2‘-Bipyridyl and 1,10-Phenanthroline to Substituted Ytterbocenes?: An Experimental Investigation of Spin Coupling in Lanthanide Complexes, M. Schultz, J. M. Boncella, D. J. Berg, T. D. Tilley, R. A. Andersen, Organometallics 2002, 21, 460.
[7] (a) Electronic structures of organometallic complexes of f elements LXXIX: Truncated crystal field splitting patterns of monomeric pseudo trigonal planar $Yb(\eta^5\text{-}C_5H_4R)_3$ (R = Me, Et, tBu, $SiMe_3$) complexes and polymeric $[(\eta^5\text{-}C_5H_5)_2Yb(\mu - \eta^5{:}\eta^1\text{-}C_5H_5)]$ as well as selected mono adducts, H.-D. Amberger, H. Reddmann, G. Liu, Journal of Organometallic Chemistry 2012, 716, 138. (b) Electronic Structures of Highly Symmetrical Compounds of f Elements. 42 Derivation and Simulation of the Crystal Field Splitting Pattern of Tris(bis(trimethylsilyl)amido)ytterbium(III), S. Jank, H. Reddmann, C. Apostolidis, H.-D. Amberger, Zeitschrift Fur Anorganische Und Allgemeine Chemie 2007, 633, 398. (c) Electronic Structures of Highly Symmetrical Compounds of f Elements. 34 [1] Synthesis and Spectroscopic Characterization of Biscyclohexylisocyanide Adducts derived from the Tris(bis(trimethylsilyl)amido)lanthanide(III) Moiety as well as Crystal, Molecular, and Electronic Structure of the Corresponding Neodymium Compound, S. Jank, J. Hanss, H. Reddmann, H. D. Amberger, N. M. Edelstein, Zeitschrift Fur Anorganische Und Allgemeine Chemie 2002, 628, 1355. (d) The Electronic-Structure of Highly Symmetrical Compounds of the F-Elements .25. Growth of Ultrapure Tris[Bis(Trimethylsilyl)amido]lanthanide(III) Single-crystals for Chemical Syntheses and Physical Investigations, C. Hagen, H. D. Amberger, Zeitschrift Fur Naturforschung Section B-a Journal of Chemical Sciences 1993, 48, 1365.
[8] 4f Orbital Covalence in (.eta.5-C5H5)3Eu(THF) as Revealed by Europium-151 Mossbauer Spectroscopy, G. Depaoli, U. Russo, G. Valle, F. Grandjean, A. F. Williams, G. J. Long, Journal of the American Chemical Society 1994, 116, 5999.
[9] Electronic structures of highly symmetrical compounds of f elements: XXXIX. Do more covalent europium(III) compounds exhibit an anti-nephelauxetic behavior?, H. D. Amberger, H. Reddmann, C. Hagen, Inorganica Chimica Acta 2005, 358, 3745.
[10] Variable Photon Energy Photoelectron Spectroscopy and Magnetism of $YbCp_3$ and $LuCp_3$, M. Coreno, M. de Simone, R. Coates, M. S. Denning, R. G. Denning, J. C. Green, C. Hunston, N. Kaltsoyannis, A. Sella, Organometallics 2010, 29, 4752.
[11] Covalency in the 4f Shell of tris-Cyclopentadienyl Ytterbium ($YbCp_3$)—A Spectroscopic Evaluation, R. G. Denning, J. Harmer, J. C. Green, M. Irwin, Journal of the American Chemical Society 2011, 133, 20644.
[12] R. A. Andersen, T. D. Tilley. unpublished results, phD thesis of T. Don Tilley.
[13] (a) Decamethylytterbocene Complexes of Bipyridines and Diazabutadienes: Multiconfigurational Ground States and Open-Shell Singlet Formation, C. H. Booth, M. D. Walter, D. Kazhdan, Y.-J. Hu, W. W. Lukens, E. D. Bauer, L. Maron, O. Eisenstein, R. A. Andersen, Journal of the American Chemical Society 2009, 131, 6480. (b) Intermediate-Valence Tautomerism in Decamethylytterbocene Complexes of Methyl-Substituted Bipyridines, C. H. Booth, D. Kazhdan, E. L. Werkema, M. D. Walter, W. W. Lukens, E. D. Bauer, Y.-J. Hu, L. Maron, O. Eisenstein, M. Head-Gordon, R. A. Andersen, Journal of the American Chemical Society 2010, 132, 17537.

[14] Application of the Hubbard Model to Cp*$_2$Yb(bipy), a Model System for Strong Exchange Coupling in Lanthanide Systems, W. W. Lukens, N. Magnani, C. H. Booth, Inorganic Chemistry 2012, 51, 10105.
[15] (a) Perspectives in reductive lanthanide chemistry, W. J. Evans, Coordination Chemistry Reviews 2000, 206, 263. (b) The Importance of Questioning Scientific Assumptions: Some Lessons from f Element Chemistry,W. J. Evans, Inorganic Chemistry 2007, 46, 3435.
[16] (a) A N_2^{3-} Radical-Bridged Terbium Complex Exhibiting Magnetic Hysteresis at 14 K, J. D. Rinehart, M. Fang, W. J. Evans, J. R. Long, Journal of the American Chemical Society 2011, 133, 14236. (b) Strong exchange and magnetic blocking in N_2^{3-}-radical-bridged lanthanide complexes, J. D. Rinehart, M. Fang, W. J. Evans, J. R. Long, Nature Chemistry 2011, 3, 538.
[17] (a) d-f Heterobimetallic Association between Ytterbium and Ruthenium Carbon-Rich Complexes: Redox Commutation of Near-IR Luminescence, E. Di Piazza, L. Norel, K. Costuas, A. Bourdolle, O. Maury, S. Rigaut, Journal of the American Chemical Society 2011, 133, 6174. (b) Synthesis, Structural Studies, Theoretical Calculations, and Linear and Nonlinear Optical Properties of Terpyridyl Lanthanide Complexes:? New Evidence for the Contribution of f Electrons to the NLO Activity, K. Senechal-David, A. Hemeryck, N. Tancrez, L. Toupet, J. A. G. Williams, I. Ledoux, J. Zyss, A. Boucekkine, J.-P. Guegan, H. Le Bozec, O. Maury, Journal of the American Chemical Society 2006, 128, 12243.
[18] Ytterbocene Charge-Transfer Molecular Wire Complexes, C. N. Carlson, C. J. Kuehl, R. E. Da Re, J. M. Veauthier, E. J. Schelter, A. E. Milligan, B. L. Scott, E. D. Bauer, J. D. Thompson, D. E. Morris, K. D. John, Journal of the American Chemical Society 2006, 128, 7230.
[19] Advances in f element reductive reactivity as a paradigm for expanding lanthanide and actinide science and technology, W. J. Evans, Journal of Alloys and Compounds 2009, 488, 493.
[20] Isolation and x-ray crystal structure of the first dinitrogen complex of an f-element metal, [(C5Me5)2Sm]2N2, W. J. Evans, T. A. Ulibarri, J. W. Ziller, Journal of the American Chemical Society 1988, 110, 6877.
[21] (a) Lanthanum Does Form Stable Molecular Compounds in the +2 Oxidation State, P. B. Hitchcock, M. F. Lappert, L. Maron, A. V. Protchenko, Angewandte Chemie-International Edition 2008, 47, 1488. (b) Reactivity of $(C_5Me_5)_3LaL_x$ Complexes: Synthesis of a Tris(pentamethylcyclopentadienyl) Complex with Two Additional Ligands, $(C_5Me_5)_3La(NCCMe_3)_2$, W. J. Evans, T. J. Mueller, J. W. Ziller, Journal of the American Chemical Society 2009, 131, 2678. (c) Isolation of Dysprosium and Yttrium Complexes of a Three-Electron Reduction Product in the Activation of Dinitrogen, the $(N_2)^{3-}$ Radical, W. J. Evans, M. Fang, G. Zucchi, F. Furche, J. W. Ziller, R. M. Hoekstra, J. I. Zink, Journal of the American Chemical Society 2009, 131, 11195. (d) Expanding Rare-Earth Oxidation State Chemistry to Molecular Complexes of Holmium(II) and Erbium(II), M. R. MacDonald, J. E. Bates, M. E. Fieser, J. W. Ziller, F. Furche, W. J. Evans, Journal of the American Chemical Society 2012, 134, 8420.
[22] T. D. Tilley, J. M. Boncella, D. J. Berg, C. J. Burns, R. A. Andersen, G. A. Lawless, M. A. Edelman, M. F. Lappert, in *Inorganic Syntheses*, John Wiley & Sons, Inc., 2007, pp. 146.
[23] Divalent lanthanide derivatives in organic synthesis. 1. Mild preparation of samarium iodide and ytterbium iodide and their use as reducing or coupling agents, P. Girard, J. L. Namy, H. B. Kagan, Journal of the American Chemical Society 1980, 102, 2693.
[24] Synthesis and Structure of the First Molecular Thulium(II) Complex: $[TmI_2(MeOCH_2CH_2OMe)_3]$, M. N. Bochkarev, I. L. Fedushkin, A. A. Fagin, T. V. Petrovskaya, J. W. Ziller, R. N. R. Broomhall, Dillard, W. J. Evans, Angewandte Chemie-International Edition 1997, 36, 133.
[25] (a) $[NdI_2(thf)_5]$, the First Crystallographically Authenticated Neodymium(II) Complex, M. N. Bochkarev, I. L. Fedushkin, S. Dechert, A. A. Fagin, H. Schumann, Angewandte Chemie-International Edition 2001, 40, 3176. (b) A New Route to Neodymium(II) and Dysprosium(II) Iodides, M. N. Bochkarev, A. A. Fagin, Chemistry, A European Journal 1999, 5, 2990. (c) Large Scale Synthesis of Dysprosium and Neodymium Diiodides, W. J. Evans, N. T. Allen, P. S. Workman, J. C. Meyer, Inorganic Chemistry 2003, 42, 3097.
[26] Non-classical divalent lanthanide complexes, F. Nief, Dalton Transactions 2010, 39, 6589.

[27] Reduction of 2,5-di-tert-butylcyclopentadienone and pyridine with thulium diiodide. Structures of the complexes TmI2(THF)2[η5-But2C5H2O]TmI2(THF)3 and [TmI2(C5H5N)4]2(μ2-N2C10H10), I. L. Fedushkin, V. I. Nevodchikov, M. N. Bochkarev, S. Dechert, H. Schumann, Russian Chemical Bulletin 2003, 52, 154.
[28] Facile Dinitrogen Reduction via Organometallic Tm(II) Chemistry, W. J. Evans, N. T. Allen, J. W. Ziller, Journal of the American Chemical Society 2001, 123, 7927.
[29] Expanding Divalent Organolanthanide Chemistry: The First Organothulium(II) Complex and the In Situ Organodysprosium(II) Reduction of Dinitrogen, W. J. Evans, N. T. Allen, J. W. Ziller, Angewandte Chemie-International Edition 2002, 41, 359.
[30] Abragam A, Bleaney B, *Electron Paramagnetic Resonance of Transition Ions*, Clarendon Press, Oxford, 1970.
[31] O. Kahn, *Molecular Magnetism*, John Wiley & Sons, 1993.
[32] Static Magnetic-Field-Induced Phase Lag in the Magnetization Response of Tris(dipicolinato) lanthanides, M. Sugita, N. Ishikawa, T. Ishikawa, S.-y. Koshihara, Y. Kaizu, Inorganic Chemistry 2006, 45, 1299.
[33] Exploiting single-ion anisotropy in the design of f-element single-molecule magnets, J. D. Rinehart, J. R. Long, Chemical Science 2011, 2, 2078.
[34] S. V. Eliseeva, J.-C. G. Bunzli, *Vol. 7* (Eds.: P. Hänninen, H. Härmä), Springer Series on Fluorescence, 2010.
[35] Engineering emissive europium and terbium complexes for molecular imaging and sensing, S. Pandya, J. Yu, D. Parker, Dalton Transactions 2006, 2757.
[36] (a) Magnetic Anisotropy in a Dysprosium/DOTA Single-Molecule Magnet: Beyond Simple Magneto-Structural Correlations, G. Cucinotta, M. Perfetti, J. Luzon, M. Etienne, P.-E. Car, A. Caneschi, G. Calvez, K. Bernot, R. Sessoli, Angewandte Chemie-International Edition 2012, 51, 1606. (b) A bifunctional luminescent single-ion magnet: towards correlation between luminescence studies and magnetic slow relaxation processes, J. Long, R. Vallat, R. A. S. Ferreira, L. D. Carlos, F. A. Almeida Paz, Y. Guari, J. Larionova, Chemical Communications 2012, 48, 9974. (c) A redox-active luminescent ytterbium based single molecule magnet, F. Pointillart, B. Le Guennic, S. Golhen, O. Cador, O. Maury, L. Ouahab, Chemical Communications 2013, 49, 615.
[37] Diethyl ether adducts of bis(pentamethylcyclopentadienyl)europium(II) and -ytterbium(II). Excited-state energy transfer with organolanthanoid complexes, A. C. Thomas, A. B. Ellis, Organometallics 1985, 4, 2223.
[38] New developments in the field of luminescent materials for lighting and displays, T. Jüstel, H. Nikol, C. Ronda, *Angewandte Chemie-International Edition* 1998, 37, 3084.
[39] Relativistic effects in structural chemistry, P. Pyykkö, Chemical Reviews 1988, 88, 563.
[40] Relativistic Pseudopotentials: Their Development and Scope of Applications, M. Dolg, X. Cao, Chemical Reviews 2012, 112, 403.
[41] (a) Existence and Stability of Lanthanide-Main Group Element Multiple Bonds. New Paradigms in the Bonding of the 4f Elements. A DFT Study of Cp_2CeZ ($Z = F^+$, O, NH, CH^-, CH_2) and the Ligand Adduct $Cp_2Ce(CH_2)(NH_3)$, D. L. Clark, J. C. Gordon, P. J. Hay, R. Poli, Organometallics 2005, 24, 5747. (b) Structural and Electronic Analysis of Lanthanide Complexes: Reactivity May Not Necessarily Be Independent of the Identity of the Lanthanide Atom — A DFT Study, S. Schinzel, M. Bindl, M. Visseaux, H. Chermette, Journal of Physical Chemistry A 2006, 110, 11324. (c) Covalent Bonding and the Trans Influence in Lanthanide Compounds, K. Krogh-Jespersen, M. D. Romanelli, J. H. Melman, T. J. Emge, J. G. Brennan, Inorganic Chemistry 2010, 49, 552. (d) Evaluating f-Element Bonding from Structure and Thermodynamics, S. G. Minasian, J. L. Krinsky, J. Arnold, Chemistry, A European Journal 2011, 17, 12234.
[42] (a) Prediction of the Molecular Shape of Lanthanide Trihalides, J. Molnar, M. Hargittai, Journal of Physical Chemistry 1995, 99, 10780. (b) Molecular Structure of Metal Halides, M. Hargittai, Chemical Reviews 2000, 100, 2233. (c) On the effect of 4f electrons on the structural characteristics of lanthanide trihalides: computational and electron diffraction study of dysprosium trichloride, G. Lanza, Z. Varga, M. Kolonits, M. Hargittai, Journal of Chemical

Physics 2008, 128, 74301. (d) Topological approach in the structural and bonding characterization of lanthanide trihalide molecules, L. Joubert, B. Silvi, G. Picard, Theoretical Chemistry Accounts 2000, 104, 109.

[43] (a) A Theoretical Study of Bonding in Lanthanide Trihalides by Density Functional Methods, C. Adamo, P. Maldivi, Journal of Physical Chemistry A 1998, 102, 6812. (b) Ionic versus covalent character in lanthanide complexes. A hybrid density functional study, C. Adamo, P. Maldivi, Chemical Physics Letters 1997, 268, 61.

[44] (a) New Relativistic Atomic Natural Orbital Basis Sets for Lanthanide Atoms with Applications to the Ce Diatom and LuF, B. O. Roos, R. Lindh, P-Å Malmqvist, V. Veryazov, P-O Widmark, A. C. Borin, Journal of Physical Chemistry A 2008, 112, 11431. (b) Quasirelativistic f-in-core pseudopotentials and core-polarization potentials for trivalent actinides and lanthanides: molecular test for trifluorides, A. Weigand, X. Cao, J. Yang, M. Dolg, Theoretical Chemical Accounts 2010, 126, 117.

[45] Covalency in the 4f Shell of tris-Cyclopentadienyl Ytterbium ($YbCp_3$)—A Spectroscopic Evaluation, R. G. Denning, J. Harmer, J. C. Green, M. Irwin, Journal of the American Chemical Society 2011, 133, 20644.

[46] (a) Theoretical studies of the actinides: method calibration for the UO_2^{2+} and PuO_2^{2+} ions, N. Ismail, J. L. Heully, T. Saue, J. P. Daudey, C. J. Marsden, Chemical Physics Letters 1999, 300, 296. (b) Reduction of uranyl by hydrogen: an ab initio study, V. Vallet, B. Schimmelpfennig, L. Maron, C. Teichteil, T. Leininger, O. Gropen, I. Grenthe, U. Wahlgren, Chemical Physics 1999, 244, 185.

[47] (a) Density functional calculations on actinide compounds: Survey of recent progress and application to $[UO_2X_4]^{2-}$ (X=F, Cl, OH) and AnF_6 (An=U, Np, Pu), G. Schreckenbach, P. J. Hay, R. L. Martin, Journal of Computational Chemistry1999, 20, 70. (b) Theoretical studies of organometallic complexes of uranium involving nitrogen ligands using density functional approaches, P. J. Hay, Faraday Discussions 2003, 124, 69. (c) Can density functional methods be used for open-shell actinide molecules? Comparison with multiconfigurational spin-orbit studies, C. Clavaguera-Sarrio, V. Vallet, D. Maynau, C. J. Marsden, Journal of Chemical Physics 2004, 121, 5312. (d) Relativistic energy-consistent ab initio pseudopotentials as tools for quantum chemical investigations of actinide systems, X. Cao, M. Dolg, Coordination Chemistry Reviews 2006, 250, 900.

[48] G. Wilkinson, F. G. A. Stone, F. W. Abel, *Comprehensive organometallic chemistry*, vol 1. Pergamon, Oxford, 1982.

[49] Catalysis Research of Relevance to Carbon Management: Progress, Challenges, and Opportunities, H. Arakawa, M. Aresta, J. N. Armor, M. A. Barteau, E. J. Beckman, A. T. Bell, J. E. Bercaw, C. Creutz, E. Dinjus, D. A. Dixon, K. Domen, D. L. DuBois, J. Eckert, E. Fujita, D. H. Gibson, W. A. Goddard, D. W. Goodman, J. Keller, G. J. Kubas, H. H. Kung, J. E. Lyons, L. E. Manzer, T. J. Marks, K. Morokuma, K. M. Nicholas, R. Periana, L. Que, J. Rostrup-Nielson, W. M. H. Sachtler, L. D. Schmidt, A. Sen, G. A. Somorjai, P. C. Stair, B. R. Stults, W. Tumas, Chemical Reviews 2001, 101, 953.

[50] Understanding and exploiting C–H bond activation, J. A. Labinger, J. E. Bercaw, Nature 2002, 417, 507.

[51] (a) Dinitrogen Coordination Chemistry: On the Biomimetic Borderlands, B. A. MacKay, M. D. Fryzuk, Chemical Reviews 2004, 104, 385. (b) Protonation of coordinated dinitrogen, G. J. Leigh, Accounts of Chemical Research 1992, 25, 177.

[52] Recent Advances in C–F Bond Activation, J. Burdeniuc, B. Jedlicka, R. H. Crabtree, Chemische Berichte Recueil 1997, 130, 145.

[53] Building Molecules with Carbon Monoxide Reductive Coupling, B. Wayland, X. Fu, Science 2006, 311, 790.

[54] (a) Electron-transfer reactions of trivalent uranium. Preparation and structure of the uranium metallocene compounds (MeC5H4)3U:NPh and [(MeC5H4)3U]2[.mu.-.eta.1,.eta.2-PhNCO], J. G. Brennan, R. A. Andersen, Journal of the American Chemical Society 1985, 107, 514. (b) Chemistry of trivalent uranium metallocenes: electron-transfer reactions with carbon disulfide. Formation of [(RC5H4)3U]2[.mu.-.eta.1.eta.2-CS2], J. G. Brennan, R. A.

Andersen, Inorganic Chemistry 1986, 25, 1756. (c) Chemistry of trivalent uranium metallocenes: electron-transfer reactions. Synthesis and characterization of [(MeC5H4)3U]2E (E = S, Se, Te) and the crystal structures of hexakis(methylcyclopentadienyl)sulfidodiuranium and tris(methylcyclopentadienyl)(triphenylphosphine oxide)uranium, J. G. Brennan, R. A. Andersen, Inorganic Chemistry 1986, 25, 1761.

[55] (a) Reductive disproportionation of carbon dioxide to carbonate and squarate products using a mixed-sandwich U(III) complex, O. T. Summerscales, A. S. P. Frey, F. G. N. Cloke, P. B. Hitchcock, Chemical Communications 2009, 198. (b) U^{III}-Induced Reductive Co-Coupling of NO and CO to Form U^{IV} Cyanate and Oxo Derivates, A. S. P. Frey, F. G. N. Cloke, M. P. Coles, P. B. Hitchcock, Chemistry, A European Journal 2010, 16, 9446. (c) Steric Effects in the Reductive Coupling of CO by Mixed-Sandwich Uranium(III) Complexes, N. Tsoureas, O. T. Summerscales, F. G. N. Cloke, S. M. Roe, Organometallics 2013, 32, 1353. (d) Reductive Cyclotrimerization of Carbon Monoxide to the Deltate Dianion by an Organometallic Uranium Complex, O. T. Summerscales, F. G. N. Cloke, P. B. Hitchcock, J. C. Green, N. Hazari, Science 2006, 311, 829.

[56] (a) A Linear, O-Coordinated η^1-CO_2 Bound to Uranium, I. Castro-Rodriguez, H. Nakai, L. N. Zakharov, A. L. Rheingold, K. Meyer, Science 2004, 305, 1757. (b) Carbon Dioxide Reduction and Carbon Monoxide Activation Employing a Reactive Uranium(III) Complex, I. Castro-Rodriguez, K. Meyer, Journal of the American Chemical Society 2005, 127, 11242. (c) Insights into the mechanism of carbonate formation through reductive cleavage of carbon dioxide with low-valent uranium centers, O. P. Lam, S. C. Bart, H. Kameo, F. W. Heinemann, K. Meyer, Chemical Communications 2010, 46, 3137. (d) C-C Bond Formation through Reductive Coupling of CS_2 to Yield Uranium Tetrathiooxalate and Ethylenetetrathiolate Complexes, O. P. Lam, F. W. Heinemann, K. Meyer, Angewandte Chemie International Edition 2011, 50, 5965.

[57] (a) Uranium-mediated activation of small molecules, P. L. Arnold, Chemical Communications 2011, 47, 9005. (b) Carbon monoxide coupling and functionalisation at a simple uranium coordination complex, P. L. Arnold, Z. R. Turner, R. M. Bellabarda, R. P. Tooze, Chemical Science 2011, 2, 77. (c) Small Molecule Activation by Uranium Tris(aryloxides): Experimental and Computational Studies of Binding of N_2, Coupling of CO, and Deoxygenation Insertion of CO_2 under Ambient Conditions, S. M. Mansell, N. Kaltsoyannis, P. L. Arnold, Journal of the American Chemical Society 2011, 133, 9036. (d) Thermally Stable Uranium Dinitrogen Complex with Siloxide Supporting Ligands, S. M. Mansell, J. H. Farnaby, A. I. Germeroth, P. L. Arnold, Organometallics 2013, 32, 4214.

[58] Siloxides as Supporting Ligands in Uranium(III)-Mediated Small-Molecule Activation, V. Mougel, C. Camp, J. Pécaut, C. Coperet, L. Maron, C. Kefalidis, M. Mazzanti, Angewandte Chemie International Edition 2012, 51, 12280.

[59] (a) Do f Electrons Play a Role in the Lanthanide-Ligand Bonds? A DFT Study of $Ln(NR_2)_3$; R = H, SiH_3, L. Maron, O. Eisenstein, Journal of Physical Chemistry A 2000, 104, 7140. (b) Are the Carbon Monoxide Complexes of Cp_2M (M = Ca, Eu, or Yb) Carbon or Oxygen Bonded? An Answer from DFT Calculations, L. Perrin, L. Maron, O. Eisenstein, R. A. Andersen, Journal of American Chemical Society 2002, 124, 5614. (c) Hydrogen for Fluorine Exchange in $CH_{4-x}F_x$ by Monomeric $[1,2,4\text{-}(Me_3C)_3C_5H_2]_2CeH$: Experimental and Computational Studies, E. L. Werkema, E. Messines, L. Perrin, L. Maron, O. Eisenstein, R. A. Andersen, Journal of the American Chemical Society 2005, 127, 7781.

[60] See the entire volume of *Inorganic Chemistry* 2012, 50.

[61] (a) Self-Contained Kondo Effect in Single Molecules, C. H. Booth, M. D. Walter, M. Daniel, W. W. Lukens, R. A. Andersen, Physical Review Letters 2005, 95, 267202. (b) Cerocene Revisited: The Electronic Structure of and Interconversion Between $Ce_2(C_8H_8)_3$ and $Ce(C_8H_8)_2$, M. D. Walter, C. H. Booth, W. W. Lukens, W. W., R. A. Andersen, Organometallics 2009, 28, 698. (c) Intermediate-Valence Tautomerism in Decamethylytterbocene Complexes of Methyl-Substituted Bipyridines, C. H. Booth, D. Kazhdan, E. L. Werkema, M. D. Walter, W. W. Lukens, E. D. Bauer, Y.-J. Hu, L. Maron, O. Eisenstein, M. Head-Gordon, R. A. Andersen, Journal of the American Chemical Society 2010, 132, 17537. (d) Decamethylytterbocene Complexes of Bipyridines and Diazabutadienes: Multiconfigurational Ground States and Open-Shell Singlet

Formation, C. H. Booth, M. D. Walter, D. Kazhdan, Y.-J. Hu, W. W. Lukens, E. D. Bauer, L. Maron, O. Eisenstein, R. A. Andersen, Journal of the American Chemical Society 2009, 131, 6480. (e) Electrochemical and Spectroscopic Characterization of the Novel Charge-Transfer Ground State in Diimine Complexes of Ytterbocene, R. E. Da Re, C. J. Kuehl, M. G. Brown, R. C. Rocha, E. D. Bauer, K. D. John, D. E. Morris, A. P. Shreve, J. L. Sarrao, Inorganic Chemistry 2003, 42, 5551. (g) Direct Comparison of the Magnetic and Electronic Properties of Samarocene and Ytterbocene Terpyridine Complexes, J. M. Veauthier, E. J. Schelter, C. N. Carlson, B. L. Scott, R. E. Da Re, J. D. Thompson, J. L. Kiplinger, D. E. Morris, K. D. John, Inorganic Chemistry 2008, 47, 5841. (h) Reactions of Ytterbocenes with Diimines: Steric Manipulation of Reductive Reactivity, A. A. Trifonov, European Journal of Inorganic Chemistry 2007, 3151. (i) C-C Coupling and C-H Bond Activation—Unexpected Pathways in the Reactions of $[Yb(\eta^5\text{-}C_{13}H_9)_2(thf)_2]$ with Diazadienes, A. A. Trifonov, E. A. Fedorova, G. K. Fukin, N. O. Druzhkov, M. N. Bochkarev, Angewandte Chemie International Edition 2004, 43, 5045. (j) Solvent-Mediated Redox Transformations of Ytterbium Bis(indenyl)diazabutadiene Complexes, A. A. Trifonov, E. A. Fedorova, V. N. Ikorskii, S. Dechert, H. Schumann, M. N. Bochkarev, European Journal of Inorganic Chemistry 2005, 2812. (k) Application of the Hubbard Model to $Cp^*_2Yb(bipy)$, a Model System for Strong Exchange Coupling in Lanthanide Systems, W. W. Lukens, N. Magnani, C. H. Booth, Inorganic Chemistry 2012, 51, 10105. (l) Is there a molecular analogue of a Kondo singlet state?,C. S. Neumann, P. Fulde, Zeitschrift Fur Physik B - Condensed Matter 1989, 74, 277. (m) Formally tetravalent cerium and thorium compounds: a configuration interaction study of cerocene $Ce(C_8H_8)_2$ and thorocene $Th(C_8H_8)_2$ using energy-adjusted quasirelativistic ab initio pseudopotentials, M. Dolg, P. Fulde, H. Stoll, H. Preuss, A. Chang, R. M. Pitzer, Chemical Physics 1995, 195, 71.

[62] Electronic Structure of 2,2/-Bipyridine Organotransition-Metal Complexes. Establishing the Ligand Oxidation Level by Density Functional Theoretical Calculations, C. C. Scarborough, K. Wieghardt, K. Inorganic Chemistry 2011, 50, 9773.

[63] W. T. Bordon, *Effects of Electron Repulsion in Diradicals,* New York, 1982.

[64] Theoretical Treatment of Redox Processes Involving Lanthanide(II) Compounds: Reactivity of Organosamarium(II) and Organothulium(II) Complexes with CO_2 and Pyridine, S. Labouille, F. Nief, L. Maron, Journal of Physical Chemistry A 2011, 115, 8295.

[65] (a) Divalent lanthanide derivatives in organic synthesis—II: Mechanism of SmI_2 reactions in presence of ketones and organic halides, H. B. Kagan, J. L. Namy, P. Girard, Tetrahedron 1981, 37, 175. (b) Synthesis and crystallographic characterization of an unsolvated, monomeric samarium bis(pentamethylcyclopentadienyl) organolanthanide complex, (C5Me5)2Sm, W. J. Evans, L. A. Hughes, T. P. Hanusa, Journal of the American Chemical Society 1984, 106, 4270. (c) Synthesis and x-ray crystal structure of bis(pentamethylcyclopentadienyl) complexes of samarium and europium: (C5Me5)2Sm and (C5Me5)2Eu, W. J. Evans, L. A. Hughes, T. P. Hanusa, Organometallics 1986, 5, 1285. (d) A mild and convenient method for the reduction of organic halides by using a SmI2-THF solution in the presence of hexamethylphosphoric triamide (HMPA), J. Inanaga, M. Ishikawa, M. Yamaguchi, Chemical Letters 1987, 1485. (e) Carbon-carbon bond formation by coupling of two phenylethynyl ligands in an organolanthanide system, W. J. Evans, R. A. Keyer, J. W. Ziller, Organometallics 1990, 9, 2628. (f) Investigation of organolanthanide-based carbon-carbon bond formation: synthesis, structure, and coupling reactivity of organolanthanide alkynide complexes, including the unusual structures of the trienediyl complex [(C5Me5)2Sm]2[.mu.-.eta.2:.eta.2-Ph(CH2)2C:C:C:C-(CH2)2Ph] and the unsolvated alkynide [(C5Me5)2Sm(C.tplbond.CCMe3)]2, W. J. Evans, R. A. Keyer, J. W. Ziller, Organometallics 1993, 12, 2618. (g) Synthesis and Reactivity of Organometallic Complexes of Divalent Thulium with Cyclopentadienyl and Phospholyl Ligands, F. Jaroschik, F. Nief, X.–F. Le Goff, L. Ricard, Organometallics 2007, 26, 3552. (h) Decamethylytterbocene Complexes of Bipyridines and Diazabutadienes: Multiconfigurational Ground States and Open-Shell Singlet Formation, C. H. Booth, M. D. Walter, D. Kazhdan, Y.-J. Hu, W. W. Lukens, E. D. Bauer, L. Maron, O. Eisenstein, R. A. Andersen, Journal of the American Chemical Society 2009, 131, 6480. (i) Intermediate-Valence Tautomerism in Decamethylytterbocene Complexes of Methyl-Substituted Bipyridines, C. H. Booth, D. Kazhdan, E. Werkema, M. D.

Walter, E. D. Bauer, Y.-J. Hu, L. Maron, O. Eisenstein, M. Head-Gordon, R. A. Andersen, Journal of the American Chemical Society 2010, 132, 17537. (k) Thermal Dihydrogen Elimination from Cp^{*}_2Yb(4,5-diazafluorene), G. Nocton, C. H. Booth, L. Maron, R. A. Andersen, Organometallics 2013, 32, 1150.

[66] Preliminary Theoretical Insights into SmI_2-Mediated Reactions: Activation of Ketones in THF, C. E. Kefalidis, L. Perrin, L. Maron, European Journal of Inorganic Chemistry 2013, 4042.

[67] Quantifying the Electrostatic Driving Force behind SmI_2 Reductions, H. Farran, S. Hoz, Organic Letters 2008, 10, 4875–4877.

[68] Qualitative Estimation of the Single Electron Transfer Step Energetics mediated by Sm(II) complexes: a SOMO-LUMO Gap Approach, C. E. Kefalidis, S. Essafi, L. Perrin, L. Maron, *Inorganic Chemistry* 2014, 53, 3427–3433.

[69] Are 5f electrons really active in organoactinide reactivity? Some insights from DFT studies, L. Castro, A. Yahia, L. Maron, *ChemPhysChem* 2010, 11, 990.

[70] (a) Carbonate Formation from CO_2 via Oxo versus Oxalate Pathway: Theoretical Investigations into the Mechanism of Uranium-Mediated Carbonate Formation,L. Castro, O. P. Lam, S. C. Bart, K. Meyer, L. Maron, Organometallics 2010, 29, 5504. (b) Formation of a Uranium Trithiocarbonate Complex via the Nucleophilic Addition of a Sulfide-Bridged Uranium Complex to CS_2, O.P. Lam, L. Castro, B. Kosog, F.W. Heinemann, L. Maron, K. Meyer, Inorganic Chemistry 2012, 51, 781. (c) Insights into the Mechanism of Reaction of $[(C_5Me_5)_2Sm^{II}(thf)_2]$ with CO_2 and COS by DFT Studies, L. Castro, S. Labouille, D.R. Kindra, J.W. Ziller, F. Nief, W.J. Evans, L. Maron, Chemistry, A European Journal 2012, 18, 7886. (d) Insight into the Reaction Mechanisms of $(MeC_5H_4)_3U$ with Isoelectronic Heteroallenes CS_2, COS, PhN_3, and PhNCO by DFT Studies: A Unique Pathway that Involves Bimetallic Complexes, L. Castro, L. Maron, Chemistry, A European Journal 2012, 18, 6610.

[71] Computational insight into the reductive oligomerisation of CO at uranium(III) mixed-sandwich complexes, D. McKay, A. S. P. Frey, J. C. Green, F. G. N. Cloke, L. Maron, Chemical Communications 2012, 48, 4118.

[72] Uranium(III)-Mediated C–C-Coupling of Terminal Alkynes: Formation of Dinuclear Uranium(IV) Vinyl Complexes, B. Kosog, C. E. Kefalidis, F. W. Heinemann, L. Maron, K. Meyer, Journal of the American Chemical Society 2012, 134, 12792.

[73] (a) Electronic Structure, Excited States, and Photoelectron Spectra of Uranium, Thorium, and Zirconium Bis(Ketimido) Complexes $(C_5R_5)_2M[-NCPh_2]_2$ (M = Th, U, Zr; R = H, CH_3), A. E. Clark, R. L. Martin, J. H. Hay, J. C. Green, K. C. Jantunen, J. C. Kiplinger, Journal of Physical Chemistry A 2005, 109, 5481. (b) The electronic structure of actinide-containing molecules: a challenge to applied quantum chemistry, M. Pepper, B. Bursten, Chemical Reviews 1991, 91, 719. (c) Quantum Chemistry Study of Actinide(III) and Lanthanide(III) Complexes with Tridentate Nitrogen Ligands, D. Guillaumont, Journal of Physical Chemistry A 2004, 108, 6893.

[74] A comparative theoretical study of bonding in UO_2^{++},UO_2^{+},UO_2,UO_2^{-}, OUCO, $O_2U(CO)_2$ and UO_2CO_3,D. Majumdar, K. Balasubramanian, H. Nitsche, Chemical Physics Letters 2002, 361, 143.

[75] A New Mode of Reactivity for Pyridine N-Oxide: C-H Activation with Uranium(IV) and Thorium(IV) Bis(alkyl) Complexes, J. A. Pool, B. L. Scott, J. L. Kiplinger, Journal of the American Chemical Society 2005, 127, 1338.

[76] Is Thorium a d Transition Metal or an Actinide? An Answer from a DFT Study of the Reaction between Pyridine N-Oxide and $Cp_2M(CH_3)_2$ with M = Zr, Th, and U, A. Yahia, L. Maron, Organometallics 2009, 28, 672.

[77] (a) Theoretical Studies of the Properties and Solution Chemistry of AnO_2^{2+} and AnO_2^{+} Aquo Complexes for An = U, Np, and Pu, P. J. Hay, R. L. Martin, G. Schreckenbach, Journal of Physical Chemistry A 2000,104, 6259. (b) Linear Uranium Complexes X_2UL_5 with L=Cyanide, Isocyanate: DFT Evidence for Similarities between Uranyl (X=O) and Uranocene (X=Cp) Derivatives, N. Iché-Tarrat, N. Barros, C.J. Marsden, L. Maron, Chemistry, A European Journal 2008, 14, 2093. (c) On the inverse trans influence. Density functional studies of $[MOX_5]^{n-}$ (M = Pa, n= 2; M = U, n= 1; M = Np, n= 0; X = F, Cl or Br), E. O'Grady, N. Kaltsoyannis, Dalton Transactions 2002, 1233. (e) A DFT study of the single electron reduction and

silylation of the U–O bond of the uranyl dication in a macrocyclic environment, A. Yahia, P. L. Arnold, J. B. Love, L. Maron, Chemical Communications 2009, 2402. (f) Synthesis of the Imido Analogs of the Uranyl Ion, T. W. Hayton, J. M. Boncella, B. L. Scott, P. D. Palmer, E. R. Batista, P. J. Hay, Science 2005, 310,1941. (g) Recent developments in computational actinide chemistry, N. Kaltsoyannis, Chemical Society Reviews, 2003, 32, 9. (h) The Mechanism for Water Exchange in $[UO_2(H_2O)_5]^{2+}$ and $[UO_2(oxalate)_2(H_2O)]^{2-}$, as Studied by Quantum Chemical Methods, V. Vallet, U. Wahlgren, B. Schimmelpfennig, Z. Sbazo, I. Grenthe, Journal of the American Chemical Society 2001, 123,11999. (i) Solvent Effects on Uranium(VI) Fluoride and Hydroxide Complexes Studied by EXAFS and Quantum Chemistry, V. Vallet, U. Wahlgren, B. Schimmelpfennig, H. Moll, Z. Sbazo, I. Grenthe, Inorganic Chemistry 2001, 40, 3516. (j) Modeling Complexes of the Uranyl Ion UO2L2n+: Binding Energy, Geometries and Bonding Analysis, C. Clavaguera-Sarrio, S. Hoyau, N. Ismail, C.J. Marsden, Journal of Physical Chemistry A, 2003, 107, 4515. (k) Theoretical Investigations of Uranyl-Ligand Bonding: Four- and Five-Coordinate Uranyl Cyanide, Isocyanide, Carbonyl, and Hydroxide Complexes, J. L. Sonnenberg, P. J. Hay, R. L. Martin, B. E. Bursten, Inorganic Chemistry 2005, 44, 2255. (l) Noble Gas-Actinide Compounds: Complexation of the CUO Molecule by Ar, Kr, and Xe Atoms in Noble Gas Matrices, J. Li, B. E. Bursten, B. Y. Liang, L. Andrews, Science 2002, 295, 2242. (m) The Coordination of Uranyl in Water: A Combined Quantum Chemical and Molecular Simulation Study, D. Hagberg, G. Karlstrom, B. O. Roos, L. Gagliardi, Journal of the American Chemical Society 2005, 127, 14250. (n) Quantum Chemical Calculations Show that the Uranium Molecule U_2 has a Quintuple Bond, L. Gagliardi, B. O. Roos, Nature 2005, 433, 848. (o) Covalency in the uranyl ion: A polarized x-ray spectroscopic study, R. G. Denning, J. C. Green, T. E. Hutchings, C. Dallera, A. Tagliaferri, K. Giarda, N. B. Brookes, L. Braicovich, Journal of Chemical Physics 2002, 117, 8008. (p) The importance of spin-orbit coupling and electron correlation in the rationalization of the ground state of the CUO molecule, I. Infante, L. J. Visscher, Journal of Chemical Physics, 2004, 121, 5783. (q) Vibrational Spectroscopy of Mass-Selected $[UO_2(ligand)_n]^{2+}$ Complexes in the Gas Phase: Comparison with Theory, G. S. Groenewold, A. K. Gianotto, K. C. Cossel, M. J. Van Stipdonk, D. T. Moore, N. Polfer, J. Oomens, W. A. De Jong, L. Visscher, Journal of the American Chemical Society 2006, 128, 4802.

[78] The Effect of the Equatorial Environment on Oxo-Group Silylation of the Uranyl Dication: A Computational Study, A. Yahia, P. L. Arnold, J. Love, L. Maron, Chemistry, A European Journal 2010, 16, 4881.

14

Computational Studies of Bonding and Reactivity in Actinide Molecular Complexes

Enrique R. Batista,[1] *Richard L. Martin*[1] *and Ping Yang*[2]

[1]*Theoretical Division, Los Alamos National Laboratory (LANL)*
[2]*Environmental Molecular Sciences Laboratory, Pacific Northwest National Laboratory (PNNL)*

14.1 Introduction

One of the biggest achievements of theoretical chemistry in the last 50 years is the development of effective computational methodologies for studying the electronic structure of molecules and solids. Even though the exact solution of the Schrödinger equation for a molecular system still eludes us, approximations have been developed that are capable of sufficient accuracy to complement the interpretation of experimental data and sometimes to lead experiment into new insights. These advances in methods, coupled with the equally impressive growth in the speed and power of available computers, have made computational electronic structure a very fertile field.

The challenges associated with theoretical studies of heavy elements are twofold: the need to include relativistic effects, and the electron correlation problem. In many senses the former can now be considered largely under control, although the routine inclusion of the spin-orbit interaction is still uncommon and deserves more attention. The correlation problem is an issue for the lighter elements as well but is particularly exacerbated in molecules and solids of the lanthanides and actinides. This book is testimony to the development

Computational Methods in Lanthanide and Actinide Chemistry, First Edition. Edited by Michael Dolg.

of computational methodologies to address these two challenges, with chapters dedicated to relativistic treatments using the full Dirac equation or via the effective core potential approach, and to the inclusion of correlation energy using sophisticated wavefunction methods for small molecules, or via density functional theory (DFT) for larger systems. The combination of these tools allows us to study problems ranging from detailed analysis of the optical spectra of small molecules and/or cluster models of solids to the nature of the high energy excited states involved in x-ray absorption by the core electrons of the molecule; from the dissociation energy of UF_6 to the chemistry of UO_2^{2+} ions in solution; from studies of the properties and diffusion of defects in nuclear fuels to the strong correlations and magnetic properties of many actinide and lanthanide solids.

In this chapter we concentrate on computational studies that relate directly to the fundamental issue in chemistry: making and breaking chemical bonds. We emphasize that this contribution is not meant to be a review of such studies in f-element complexes. For that we refer the reader to recent review articles. [1–4] Rather, we simply touch on some important considerations particularly important for lanthanides and actinides. Specifically, we are interested in using DFT to characterize reaction pathways. Section 14.2 begins with a definition of some terms and a discussion of some of the issues associated with relativity and how they are typically treated within a DFT framework. It also includes an overview of the density functional approximations available for use and describes strengths, weaknesses, and typical errors expected from DFT computations of thermochemistries. Section 14.3 focuses on the study of actinide-ligand bonds and the competition between the 6d and 5f orbitals in the bonding. In Section 14.4, we describe in some detail two examples of reaction mechanism studies: sp^2 vs. sp^3 CH bond activation in 2-picoline by $Cp^*{}_2U(IV)$ and Cp*2Th(IV) complexes, and exchange reactions in bis(imido) uranium complexes.

14.2 Basic Considerations

14.2.1 Bond Energies

In order to define a bond energy one has to first define the chemical bond between two atoms. There is a vast amount of literature discussing this, the most fundamental concept in chemistry. [5–10] In this chapter we will avoid discussing what a bond *is* and concentrate instead on the fundamental property of a bond, which is to produce a combined system that is more stable than the fragments, that is, two fragments are bonded if the enthalpy of the fragments in their thermodynamic stable state, separated an infinite distance, is higher than the enthalpy of the complex formed by bringing the two fragments together. With that concept in mind, the bond dissociation energy (BDE) for a ligand bound by a single bond can be defined as the difference between the enthalpy of the fragments minus the enthalpy of the complex. Hence, for a complex AB, the bond dissociation energy can be defined as

$$\mathrm{BDE_{AB}} = [\mathrm{H(A)} + \mathrm{H(B)}] - \mathrm{H(AB)} \tag{14.1}$$

where H is the total electronic energy plus the zero-point energy of the fragments, plus the vibrational, rotational and translational energies.

$$\mathrm{H} = \mathrm{E_0} + \mathrm{ZPE} + \mathrm{E_{vib}} + \mathrm{E_{rot}} + \mathrm{E_{trans}} + \mathrm{RT} \tag{14.2}$$

This definition correlates with the heat measured in a calorimeter upon the reaction of reactants A and B to form complex AB. If the fragments are not connected by a single bond, defining a bond energy is much more difficult and open to arbitrariness. It is in this case perhaps even impossible because one needs to determine the actual electronic behavior that leads to the formation of a given bond, i.e., the stabilization energy, and proceed to subtract it from the complex. The reason a definition of this type of bond energy can be very arbitrary relates to the quantum nature of the electrons and the non-local interactions that rule their behavior. Perhaps one intuitive definition of this type of energy is found in the fragmentation analysis developed by Ziegler and Rauk, although even those definitions are based on *ad hoc* arguments needed to recover the macroscopic picture. [11]

In the rest of this section we review the corrections due to relativistic effects in the electronic structure. We go into some detail because these effects are usually of negligible size in studies of main-row elements, which are the most commonly found in the literature. We finish with examples of application of several computational methods to calculations of bond dissociation energies in actinide-containing molecules.

14.2.2 Scalar Relativistic Corrections

In the case of atoms with small atomic number Z, the kinetic energy of the electrons does not approach the relativistic limit, allowing for the dismissal of corrections that come, for example, from the dependence of the mass of the electrons on their velocity. In those cases, to a good approximation one can use the rest mass of the electron. For heavier elements this is not the case. The large charge of the nucleus tightly binds the inner shell electrons and causes the kinetic energy term to grow to relativistic scales, i.e., to the point where $p_e^2c^2$ is of similar magnitude as m^2c^4, p_e being the momentum of the electron, m the rest mass, and c the speed of light. A simple correction for the mass-velocity correction [12] takes the form:

$$W_{mv} = \frac{\mathbf{p}^4}{8m_ec^2} \tag{14.3}$$

To illustrate the magnitude of this term, recall that for a hydrogen-like model this correction is of order $(Z\alpha)^2$. Hence, in the heavy elements, where Z is comparable to the inverse of the fine structure constant, $1/\alpha = 137$, this term cannot be ignored. Because this term in the Hamiltonian is an operator on the spatial wavefunction of the electron, it is referred as a scalar relativistic effect, to distinguish it from the vector terms that depend on more than one component of the spin states, as we discuss below.

The second relativistic contribution of scalar nature is the one-electron Darwin term, W_D. This term derives from the non-relativistic expansion of Dirac's equation, in powers of (v/c), and results in a non-local interaction between the electron and the nucleus. The interaction extends over a region in space of size roughly that of the Compton wavelength of the electron. The order of magnitude of this term in the Hamiltonian is also $(Z\alpha)^2$ making it non-negligible for heavy elements. [12] These scalar relativistic terms have significant effects on the radial extent of the inner core orbitals.

We should point out that these magnitudes are estimated for the most extreme relativistic situations, those associated with the 1s orbital. For higher shells the magnitude of these terms are much smaller. After all, the valence electrons are bound by energies of the order

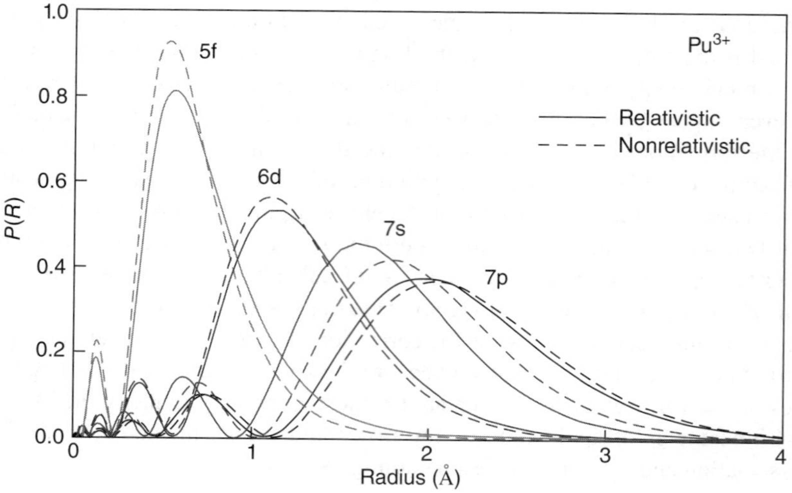

Figure 14.1 *Relativistic effect on the radial distribution function of* Pu^{3+}. *P. Jeffrey Hay Los Alamos Science No. 26, vol II, p. 371. See color plate section*

of eV, not the keV associated with deep core states. However, there are significant relativistic effects in the spatial extent of the outer orbitals that stem from rather complicated indirect effects associated with requirements of orthogonality to the lower core orbitals and from mixing with orbitals of higher energy. [13] As an example, Figure 14.1 shows the effect of relativity on the radial wavefunctions for the outer-shells of Pu^{3+}. See color plate section. The charge distribution is significantly affected leading to orbital expansions for the 5f and 6d orbitals and contractions for the 7s and 7p orbitals.

14.2.3 Spin-Orbit Corrections

From the reference frame of the electron, the spin-orbit term is generated by the interaction between its intrinsic magnetic moment (spin) and the magnetic field produced by the current loop generated as the nucleus revolves about it. The general form of this term is

$$H_{SO} = \frac{1}{2}\left(\frac{Ze^2}{4\pi\varepsilon_0}\right)\left(\frac{g_s}{2m_e^2c^2}\right)\frac{L*S}{r^3} \tag{14.4}$$

and its order of magnitude, for inner shells, is also $(Z\alpha)^2$, increasing in prominence as the mass of the nucleus becomes comparable to 137.

The most important effect of the spin-orbit interaction is a splitting of the electronic states that is manifest as an increase in spectroscopic complexity in molecules or band broadening in solid-state materials. The effect on bond dissociation energies, at least in the early actinides, is usually minor, compared to the BDE accuracy provided by DFT. In rough terms, this contribution can be on the order of 1 to 2 kcal/mol. When energies of that order are being considered, for example in studies of molecular reorganization energies, this term cannot be ignored. It is also important to note that we are not saying that the spin-orbit correction to the total energy is small; it is simply that it often cancels out in the computation of the thermochemistry and BDE. As mentioned earlier, in studies of reactivity this term is often ignored, but as improved methods for computing it in a DFT context emerge it should become more routine.

One example in which the spin-orbit interaction can play an important role involves cases where the spin state and formal f-orbital occupancy of the fragments differs from that of the complex. In those cases the spin-orbit contributions to the complex and fragment energies will not tend to cancel. The dissociation of UF_6 provides an example where the complex state is $5f^0$ and the fragment $5f^1$ and is discussed in some detail below. More critically, however, are situations where the spin-orbit splitting in the outer shells is so large that one approaches the *jj* coupling limit. Clearly the strongest effect will show up in the heaviest atoms. Studies in the hypothetical dihalogenide molecules involving elements 116 and 117, $(116)(117)_2$, [14] show that the large spin-orbit splitting in the 7p manifold, leads to bonding orbitals that involve only the $7p_{3/2}$ atomic orbitals of each element, while the $7p_{1/2}$, being much higher in energy, remain empty. The 7p electrons are in the $7p_{3/2}$ shell that is severely stabilized due to the spin-orbit splitting. Note once again that while the qualitative description of the bonding is dramatically changed in this limit, if the number of electrons occupying the $7p_{3/2}$ manifold differs little between the complex and the fragments, the influence of the spin-orbit coupling on the BDE may be smaller than one might think.

14.2.4 Relativistic Effective Core Potentials (RECPs)

Although early implementations of RECPs were motivated in large part by the reduction of computational time associated with removing the chemically inert core electrons from the problem, their major attraction now is to provide a relatively simple way to include the effects of scalar relativity on the valence orbitals without requiring an explicit treatment of relativistic wavefunctions. One method for developing these functional forms is by computing the all-electron wavefunction for the atomic species in question and then solving the inverse problem for the valence electrons, i.e., find a potential that yields the same orbitals (outside a chosen "core radius") as the full solution. A more comprehensive description of the development of effective core potentials can be found in Chapter 3. This is the approach taken by Wadt and Hay, Christiansen, Ermler and Pitzer, and other pioneers in this area. [15–18] Note that if this is done using a relativistic atomic wavefunction, the one-electron relativistic effects are implicitly incorporated into this effective potential. Standard *non-relativistic* quantum chemistry techniques can then be used to treat the valence electron problem, which is the effects of the relativistic core being propagated to the valence region via the RECPs. Another important technique for determining RECPs is to fit matrix elements of the valence Hamiltonian, the approach taken by the Stuttgart group. These effective potentials, which implicitly contain the mass-velocity and one-electron Darwin terms, are referred to as scalar relativistic potentials. The caveats mentioned above about the spin-orbit term still hold and that term may still need to be included explicitly depending on the type of study of interest. Both approaches can and have been used to generate effective spin-orbit potentials. [19]

An important consideration in designing a calculation for actinide containing species relates to the size of the valence space, or alternatively, the size of the core excluded from the calculations. The actinide elements have a core consisting of closed shells for $n = 1 - 4$, plus a set of shallow core orbitals associated with the filled 5s, 5p, and 5d subshells. At first glance, electrons in these shallow core orbitals are not expected to have a significant effect on the chemistry and could be folded into the core of the RECPs, leading to a core containing 78 electrons for the actinides. In fact, many of the early RECPs for actinides utilized these "large-core" potentials. [15, 20, 21] However, it was quickly realized that it is imperative that the $n = 5$ shell be included explicitly in the valence space, leading to RECPs with 60 electrons in the core, exemplified for example by the "small-core" SDD potentials. [22–24] The origin of the problem with the 78 electron "large-core" potentials is similar in spirit to that associated with the LANL1(small-core) and LANL2(small-core) RECPs for the main-group and transition metal systems. It is ultimately associated with the fact that although the shallow core orbitals are not involved directly in the interaction, they are necessary in order to enforce the proper nodal behavior of the pseudo-orbitals in the valence region. In the case of the actinide series, the shallow core 5d orbital must be treated explicitly in order to enforce a radial node in the 6d orbital, which is node-less when using a large-core potential. This is critical because the "real" 6d orbital happens to have a node in the region of space where the 5f orbital is large. If this node is not accounted for properly the density in this region is significantly in error. One dramatic example is provided by calculations on the uranyl ion, UO_2^{2+}. This unit is ubiquitous in actinide chemistry, aligning in a linear fashion with a heat of formation rivaling that of CO_2. It is predicted by a wide variety of approximations to be bent when using large-core RECPs. The inclusion of the shallow core in the small-core RECPs restores linearity, provides reliable bond distances, vibrational frequencies, and dissociation energies.

14.2.5 Basis Sets

All quantum chemistry calculations rely on an appropriate choice of finite basis set. Many codes are based on expansions using Gaussian-type functions centered at each atomic center. Others rely on a basis of Slater-type orbitals to expand the wavefunction/density. It is our experience that for simulations based on DFT, basis sets for the metal at the double-zeta level of accuracy can yield good predictions at moderate cost. Indeed, for heavy elements Gaussian basis functions with exponents smaller than 0.05 a.u. can sometimes lead to difficulties with self-consistent-field convergence. We therefore often omit these functions with small exponent (large radial extent) from the basis sets provided for use in wavefunction calculations For ligand atoms there is a huge family of basis sets [25] for different levels of expansion. As a rule of thumb we have found the Pople basis sets, such as 6–31g*, [26] to give a good representation of the density and to be appropriate for most geometry optimizations, vibrational frequencies, and bonding interactions. The '*' is used to represent the polarization sets, which includes basis functions of one angular momentum higher than the shell being represented. These extra functions are found to be important due to the high oxidation states and strong electric fields in these molecules.

Finally we would like to point out that, whereas for wavefunction methodologies one should use larger expansions than the ones mentioned. DFT is a functional of the density, not the wavefunction, and it is the square of the wavefunction that must be well represented by the basis. The reader is made aware that the comments above represent rules of thumb based on our experience in this area. The reader is referred to Chapter 7 by F. Weigend, and Chapter 8 by K.A. Peterson and K.G. Dyall, for a thorough discussion on basis sets.

14.2.6 Density Functional Approximations for Use with f-Element Complexes

The development of Density Functional Theory (DFT) has been one of the most impactful achievements in electronic structure computations. Indeed, this development was recognized with the 1998 Nobel Prize in Chemistry awarded to Walter Kohn. [27] In our experience, DFT provides an excellent compromise among accuracy, computational cost, and ease of interpretation. Full descriptions of DFT have been extensively published in other sources in the 50 years since the first publication of the theory and it does not need to be revisited here. [28–30] In the context of lanthanide and actinide chemistry, there are still problems to be worked out associated with the reliance of DFT on a single-determinant description, particularly with the description of spin and angular momentum multiplets generated by open-shell molecules, but for many of the cases of interest to us the ground states are high-spin and described adequately by a single determinant.

The original formulation of DFT proves the existence of a functional of the density for the Hamiltonian, but it does not provide the actual form of this functional. The standard approach for developing the Hamiltonian functional has been to include a kinetic energy term and a Columbic term, leaving the additional interactions due to the quantum nature of the electrons for the "exchange-correlation" term. The exchange-correlation functional then distinguishes among the flavors of DFT: the local density approximation (LDA), [31] where the electron density is assumed to be uniform; the generalized gradient approximation (GGA), [32] where first order corrections due to inhomogeneity of the density are included; and the meta-GGA, [33, 34] which includes a term involving the Laplacian of the density. The last approximation actually corrects the kinetic energy density and should be a reminder that the proper "exchange-correlation" functional actually contains corrections to the kinetic energy as well as exchange and correlation. A family of "hybrid" functionals, where the exact form of the exchange interaction is included in the functional has lead to hybrid DFT, a mixture of density functional and wavefunction formulations. [35–38] In all, hundreds of functionals have been developed (for example, the Gaussian 09 code includes over 800 functionals) with different ranges of success. (This number is a ballpark estimation based on the possible combinations that the user can make of exchange functional and correlation functionals, plus the possibility of adding a fraction of exact exchange to the mix and optimize it. Other possible functionals include long-range corrections and empirical corrections to dispersion interaction.) Our experience is that within a given class of functionals (LDA, GGA, meta-GGA) the differences are rather subtle. The results of two GGA calculations, one using the PBE functional and the other PW91, should not lead to qualitatively different results. There are more important and well-known differences among the classes, however. Our working rule of thumb, based on past experience, is that the hybrid functionals give a better representation of actinide complexes than the local or semi-local approximations, yielding typical errors in interatomic distance errors of about 1% and bond energies to the level of a few ($<\sim$5) kcal/mol.

14.2.7 Example of application: Performance in Sample Situation ($UF_6 \rightarrow UF_5 + F$) [39, 40]

To illustrate the various effects described above we discuss here the calculation of the bond dissociation energy to detach a fluorine atom from the UF_6 molecule. This very simple case exemplifies many of the more general features found in calculations in organoactinide chemistry. UF_6 is a closed-shell molecule; the U site is formally U(VI), $5f^0$. One of the fragments, UF_5, formally contains U(V), $5f^1$, a spin doublet with unpaired spin density on the metal. Note that this is an example in which the formal charge on the metal site changes during the reaction and therefore the formal f-electron count; as such it qualifies as a case where we should be aware that the inclusion of spin-orbit coupling may be a particular concern in computing accurate energies.

Table 14.1 gives a comparative analysis of the performance of various levels of theory, from simple Hartree-Fock theory, two classes of DFT (LDA and GGA), and two variants of hybrid DFT. Besides comparing the different functionals, this table also compares the effect of using large and small core RECPs. Comparing with the experimental data for BDE in this system one finds that the small core RECPs in combination with the hybrid DFT functionals performs satisfactorily. Introducing zero-point energy corrections (ZPE) the computed enthalpies are within a few kcal/mol of the experimental numbers. Also note that the hybrid DFT bond energies appear to be slightly higher than experiment. The spin-orbit interaction will stabilize the $5f_{5/2}$ subshell and destabilize the $5f_{7/2}$. UF_5 contains an electron in this $5f_{5/2}$ subshell, which is unoccupied in UF_6, consequently stabilizing UF_5 and reducing the computed BDE. The calculated effect of including the spin-orbit term in the BDE is on the order of 3.5kcal/mol improving the agreement with experimental numbers. [39, 40]

Table 14.1 *Fluorine-uranium bond dissociation energy for the reaction $UF_6 \rightarrow UF_5 + F$. The table illustrates the critical effect of including the closed shells 5d, 5p, and 5s in the valence space of the DFT calculation. Reprinted with permission from Batista et al. [39] Copyright 2004, American Institute of Physics.*

	ΔE		ZPE		ΔE+ZPE	
	RECP Core		RECP Core		RECP Core	
Method	$78e^-$	$60e^-$	$78e^-$	$60e^-$	$78e^-$	$60e^-$
HF	−28.11	−5.70	−2.02	−2.05	−30.15	−7.72
LSDA	86.68	124.41	−1.59	−1.39	85.29	122.82
PBE (GGA)	48.45	98.56	−1.59	−0.99	47.46	96.97
PBE0	32.06	73.79	−1.73	−1.93	30.13	72.05
B3LYP	19.07	75.01	−1.69	−1.46	17.61	73.32
Expt [41]					70 ± 2	
Expt [42]					69 ± 5	
Expt [43]					73	

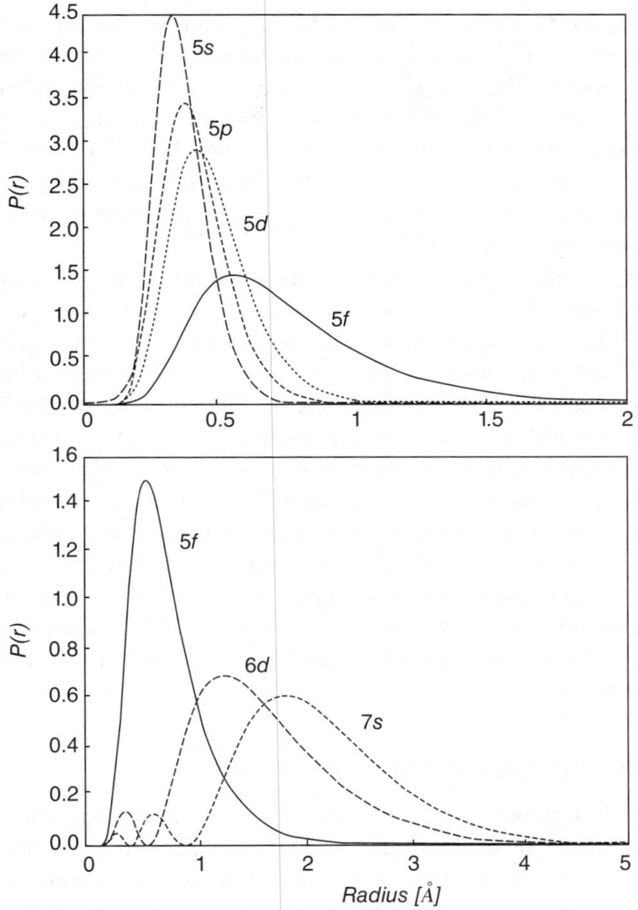

Figure 14.2 *Radial distribution function [$r^2 R_{nl}^2$] of the atomic pseudo-orbitals of uranium for the 60e− RECP. Note the node in the radial distribution for the 6d orbital (lower panel) in the region where the 5f has its maximum. The node is determined by the requirement that the 6d orbital be orthogonal to the 5d orbital, and effect present in all-electron approaches, but missing in the large-core RECPs, which fold the 5d orbital into the core. Reprinted with permission from Batista et al. [39] Copyright 2004, American Institute of Physics*

Another striking conclusion from Table 14.1 is that, when using the small core RECPs, DFT improves in accuracy as one progresses from LDA to GGA to hybrid. However, this trend is not seen when the large-core RECP is employed. As mentioned earlier, this dramatic improvement with the small-core potential is due to the enforcement of a node in the 6d orbital imposed by inclusion of the 5d in the valence space. Figure 14.2 presents the radial pseudo-orbitals using the $60e^-$ small core RECPs. Note in the top panel that 5s, 5p, and 5d pseudo-orbitals have negligible amplitude in the region where bonding interactions might occur. Only the 5f orbital has appreciable amplitude at distances of the order of an Angstrom. The lower panel compares the radial distribution of the 5f with the virtual 6d and 7s orbitals. Most importantly, note that the 6d orbital has a radial node around 0.6Å, which coincides almost exactly with the maximum in the 5f orbital. The node also redistributes the 6d orbital amplitude, although it is not apparent in Figure 14.2. Because this node is missing in the large-core potential, the density in this region is described incorrectly, resulting in an incorrect description of the energy associated with the interplay between the 5f and 6d orbitals contributions to the bonding.

At the same time, both large and small core RECPs were found to give good predictions of molecular structures, both within 1% of the experimental bond distance, and reasonable agreement with vibrational spectroscopy. [39] As generally observed, there is a tendency for Hartree-Fock theory to underestimate bond lengths leading to an overestimate of the vibrational frequencies. In the DFT family the LDA functional is typically found to predict bond distances in reasonable agreement with experiment, or slightly longer. However, upon inclusion of the gradient corrections, the error in bond distances does not correct itself and the GGA family overestimates bond lengths and underestimates vibrational frequencies. For that reason many researchers tend to use the LDA functional to estimate molecular structure but other levels of theory for electronic properties. The hybrid functionals are found to predict structures in good agreement with experiment and good agreement in frequencies as well. The reason some researchers continue to use the LDA functional to predict structures is that, in some codes, it is faster to compute as the non-local Hartree-Fock exchange calculation is not needed.

14.2.8 Molecular Systems with Unpaired Electrons

We touch finally on a problem inherent to DFT but particularly exacerbated in lanthanide and actinide complexes. In systems with many unpaired electrons occupying essentially degenerate orbitals, such as the open-shell configurations that characterize f^n manifolds, there are often a large number of determinants needed to span the Hilbert space necessary to describe spin and orbital angular momentum multiplets. For example, an f^3 atomic configuration can give rise to a quartet and two independent doublet spin states. An unrestricted single determinant method can account for the quartet, but only one doublet state, an average of the two independent doublet eigenstates. Similarly, a single determinant can span the space associated with the highest angular momentum multiplet, but not those of lower L. In general, most of the situations we have encountered in actinide complexes are governed by high-spin f-orbital couplings, as suggested by Hund's first rule, but the accuracy of DFT becomes more suspect as the number of unpaired electrons in the f-orbitals becomes larger. Figure 14.3 shows the calculation of the BDE for UF_n and UCl_n as function of the number of halide ligands. [44] The hybrid functionals that were shown to perform well in the BDE calculation for $n = 6$ above are used here for the whole series. Although the agreement with experimental data is not too bad, the error grows as the number of ligands decreases and the number of unpaired electrons increases. BDE predictions begin to extend outside of the experimental error bars beginning with $n = 4$, corresponding to an f^2 ground state dissociating to an f^3 fragment. More sophisticated wavefunction-based computational methodologies are highly recommended for accurate calculations of the energies in these types of situations.

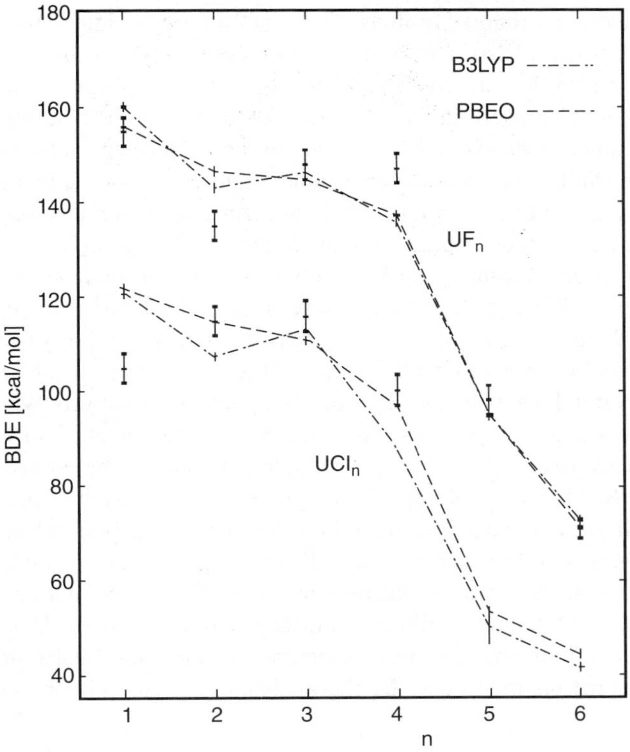

Figure 14.3 *Bond dissociation energy for UF_n and UCl_n as function of the number of fluorine and chlorine ligands. The experimental data points (with error bars) are compared to two hybrid functionals. Reprinted with permission from Batista et al. [44] Copyright 2004, American Institute of Physics*

14.3 Nature of Bonding Interactions

The studies of BDE demonstrate the importance of careful consideration of scalar and spin-orbit relativistic effects, suitable basis sets, and the choice of effective-core potentials and density functional. The appropriate inclusion of these factors allows not only for the evaluation of strength of the chemical bonds, but also provides understanding of the nature of actinide-ligand bonding interactions, such as the nature of the orbitals participating in bonding and the relative importance of f- and d-orbital involvement. In this section we will demonstrate how can we picture the bonding interactions and how the theoretical studies of bonding can be connected to spectroscopic measurements that can be used to validate the theoretical predictions. Understanding the nature of bonding interactions can lead to a better comprehension of bond formation and breakage, thus chemical reactivity.

The electronic structure resulting from a calculation allows us to visualize both bonding orbitals and their corresponding unoccupied antibonding orbitals. Optical spectroscopy techniques provide information about bonding interactions by promoting electrons from occupied orbitals to unoccupied orbitals. The resulting spectra (signature transition peaks and spectra profiles) are sensitive to the bonding nature and can be used to work backwards to understand the bonding orbitals. Therefore, coupling spectroscopic techniques with theoretical studies is a powerful tool to obtain a full picture of the bonding orbitals.

The spatial extent of the 5f and 6d orbitals in the valence space (see Figure 14.1 and Figure 14.2) are similar, and at first glance either might be important in the actinide-ligand bond. The relative role of 5f and 6d orbitals in actinide bonding has long been a subject of interest. The classic and simplest example is the linear uranyl (UO_2^{2+}) unit, which has been known for over a century. The U≡O bond is now well understood to be described by a bond order of three. [1,45] The actinide 5f and 6d orbitals form six bonds, two σ, and four π bonds to oxygen atoms, σ_g, σ_u, two π_g, and two π_u, as displayed in the left column in Figure 14.4. See color plate section. The populations of the atomic orbitals in each molecular orbital were obtained by Mulliken population analysis, which for this modest basis set gives a good representation of the relative participation of each atom in the bonds. After many years of effort in the actinide community, the family of isoelectronic imido ligand (NR^{2-}) analogs of uranyl was synthesized recently by Boncella and Hayton [$U(NR)_2^{2+}$]. [46] The computed bonding orbitals of fragment $U(NMe)_2^{2+}$ are depicted to the right side in Figure 14.4. Much as in uranyl, the bonding consists of six orbitals, two σ, and four π bonds from the interaction of the nitrogen 2p with the uranium the $5f_\pi$, $6d_\pi$, $5f_\sigma$, and $6d_\sigma$. Overall, the U≡N bonding follows a similar pattern to that of the UO_2^{2+} unit, although with minor changes in the participation of the uranium atomic orbitals. [46] The main difference is that in the bis-imido unit the π orbitals are pushed up in energy in $U(NMe)_2^{2+}$,

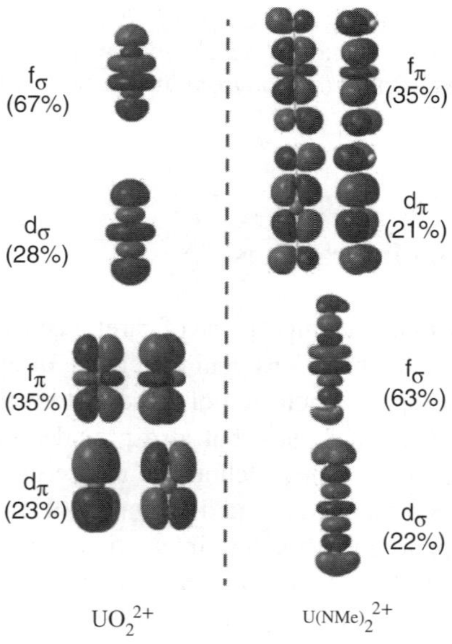

Figure 14.4 *Bonding orbitals of UO_2^{2+} and $U(NMe)_2^{2+}$. The column on the sides shows the type of uranium atomic orbital participating in each MO, along with its percentage contribution to the total MO. See color plate section*

while the σ bonds are stabilized through the interactions with the N-C bonds. This change makes the imido ligand more reactive than the oxo one, and in spite of the two being isoelectronic, that opens the door to new reactivity and a new area in synthetic chemistry (see Section 14.4.2).

In order to explore potentially different chemistries of the isoelectronic units $U(NR)_2^{2+}$ and UO_2^{2+}, a comparison of the ion $[UO_2Cl_4]^{2+}$ to $[U(NR)_2Cl_4]^{2+}$ was carried out, focusing on the effect of the different axial ligands on the interaction between the equatorial chlorine and the metal center. [47–49] Experimentally, chlorine K-edge x-ray absorption spectroscopy bring in a powerful approach for probing the nature of the actinide-chlorine bond. [4] The Cl 3p symmetry-adapted linear combinations form U-Cl σ-bonding and π-bonding interactions with a nonbonding set of chlorine lone pairs, reflected in the measured spectra, and DFT calculations identifies these signatures as hybridization of the Cl 3p with the uranium 5f orbitals. [50] The uranium participation in these orbitals is about 7%, indicating a mostly ionic interaction for both axial ligands, oxo and imido. The agreement between experimental spectra and the computationally predicted excited states mutually validate both approaches to probe-bonding interactions and reinforces the conclusions resulting from them.

14.4 Chemistry Application: Reactivity

As evidenced in the earlier sections, DFT methods have been employed to explore actinide-ligand interactions in a variety of complexes. With appropriate treatment of relativistic effects and electron correlation, the geometrical structures, thermochemistry, vibrational, and optical spectroscopic properties are in good agreement with the experimental structural and spectroscopic results. By comparison, rather little mechanistic information is available for organoactinide reactions, even though such studies are central to homogeneous catalysis. The complexity of actinide systems, especially in reactions where one expects unpaired f-electrons, makes identification of the saddle points and reaction pathway a significant theoretical challenge.

Here we will discuss the first few studies of reactivity of actinide-containing complexes for early actinides elements, such as bis(pentamethylcyclopentadienyl) uranium(IV) and thorium(IV) complexes [5, 7], and bis(imido) uranium complexes. [47] Related studies were later published in the literature, [51–53] and the reader can refer to them for further illustration. These studies further demonstrated that density functional theory is a useful tool to probe the electronic structures and reactivity of actinide complexes.

Conceptually, there are three fundamental steps to investigate reactivity: a) bringing reactant into the proximity of actinide center; b) cleavage of bond in reactant; c) thermodynamic stability of the newly formed An-L bond. In this section, we will illustrate the capability of characterizing these three aspects for two distinctive reaction mechanisms using hybrid density functional theory: a) σ-bond metathesis C–H activation by actinide complexes and b) [2+2] cycloaddition of imido reactivity.

14.4.1 First Example: C–H Bond Activation Reaction

The bis(pentamethylcyclopentadienyl thorium(IV) and uranium(IV) alkyl complexes $Cp^*_2AnR_2$ (where $Cp^* = \eta\text{-}C_5Me_5$, An = Th, U; R = CH_3, CH_2Ph, Ph) have proven to be

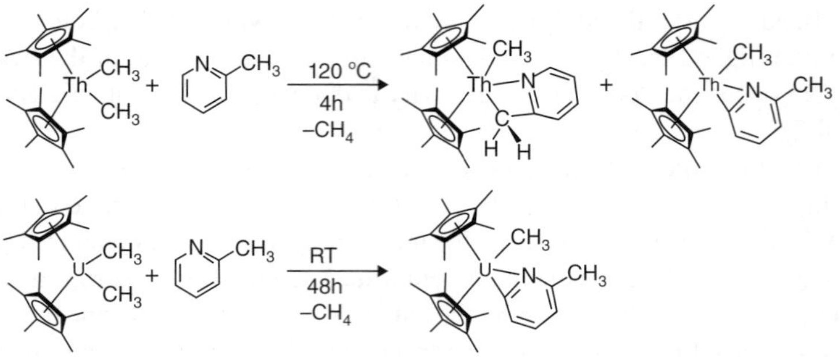

Scheme 14.1 *Reaction of Cp_2AnMe_2 with 2-picoline. The thorium alkyl complex $Cp^*{}_2Th(CH_3)_2$ and 2-picoline react to give preferential sp^3 C–H bond activation in the presence of a reactive sp^2 C–H bond, while the analogous uranium $Cp^*{}_2U(CH_3)_2$ readily reacts with only the ortho 2-picoline sp^2 C–H bond. Reprinted with permission from Yang et al. [5] Copyright 2008, American Chemical Society*

versatile starting materials for the synthesis of a diverse array of actinide organometallic systems containing An-N bonds such as imido, hydrazonato, and ketimido complexes It has been observed that complexes of lanthanide, actinide, and transition metals activate hydrocarbon substrates by different mechanisms. Recently, Kiplinger and co-workers reported that these actinide alkyl complexes undergo interesting C–H and C–N bond cleavage chemistry with N-heterocycles. [54, 55] For example, the Th(IV) complexes $Cp^*{}_2Th(CH_3)_2$ and $Cp^*{}_2Th(CH_2Ph)_2$ readily react with the sp^2 C–H bonds in pyridine *N-oxide* and the sp^3 C–H bonds in 2,6-lutidine *N-oxide,* whereas the corresponding U(IV) complexes activate only the sp^2 C–H bonds in pyridine *N-oxide*. [54] Later studies were carried out on 2-picoline (2-methylpyridine), which possesses both sp^2 and sp^3 hybridized C–H bonds. Deuterium labeling studies demonstrated that the thorium and uranium $Cp^*{}_2An(CH_3)_2$ complexes react with 2-picoline by different mechanistic reaction pathways. [55] The thorium alkyl complex $Cp^*{}_2Th(CH_3)_2$ selectively activates a sp^3 C–H bond on the 2-picoline methyl group to give the kinetic R-picolyl product, $Cp^*{}_2Th(CH_3)[\eta^2\text{-}(N,C)\text{-}2\text{-}CH_2\text{-}NC_5H_3]$, which reacts with additional 2-picoline to afford the thermodynamic η^2-pyridyl product, $Cp^*{}_2Th(CH_3)$ $[\eta^2\text{-}(N,C)\text{-}6\text{-}CH_3\text{-}NC_5H_3]$. This is in marked contrast with the uranium system that only reacts with a sp^2 C–H bond on the 2-picoline aromatic ring to give the η^2-pyridyl product $Cp^*{}_2U(CH_3)[\eta^2\text{-}(N,C)\text{-}6\text{-}CH_3\text{-}NC_5H_3]$, see Scheme 14.1.

We performed a computational study of the competitive sp^2 versus sp^3 C–H bond activation chemistry with 2-picoline and $Cp^*{}_2An(CH_3)_2$ for An = Th and U. [5] The energy profiles for reaction precursors, adducts, transition sates, and products are identified.

Hybrid B3LYP functional was employed with the relativistic small-core ECP. [23, 24] and the all-electron basis set 6–31G* for the light elements. Harmonic vibrational analyses were performed to confirm that structures were minima or saddle points and to obtain the thermochemical corrections at the standard state (298.15 K and 1 atm) to the energy, entropy, and Gibbs free energy. Solvation effects were included using polarizable continuum models (PCM). [56, 57] in the solvent toluene. Inclusion of these effects results in changes of ~1 kcal/mol in reaction energies and does not alter the relative ordering. The intrinsic reaction

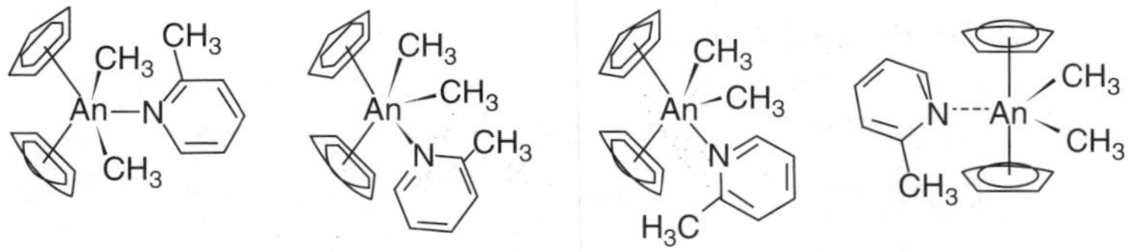

An = Th	5A (0.0)	5B (2.05)	5C (2.20)	5D (6.52)
An = U	6A (0.0)	6B (1.02)	6C (1.56)	6D (11.00)

Figure 14.5 *The possible adducts involved along the reaction pathways. The numbers shown in parentheses are the energy difference relative to the most stable configuration (all values in kcal/mol). Reprinted with permission from Yang et al. [5] Copyright 2008, American Chemical Society*

coordinate (IRC) method was used to follow the reaction path in both directions from the transition state.

The structures and thermochemistry of reactants and products arising from sp^2 and sp^3 C–H bond activation are examined for both Cp* and simplified model Cp ligand (cyclopentadienyl, Cp = η-C_5H_5). The model Cp complexes reproduce well the experimental geometric parameters and accurately reflect the relative thermochemical relationships. Therefore, the model Cp systems will be employed to search for transition states and reaction pathways. Th(IV) complexes possess a closed shell electronic ground state (5f), while U(IV) systems represent a high-spin ($5f^2$) state with two unpaired electrons in f-orbitals.

14.4.1.1 Formation of Adducts

Although the existence of stable adducts is not required along reaction paths leading to C–H bond activation, the C–H bond needs to be brought to the proximity to the metal center. As illustrated in Figure 14.5, the 2-picoline ligand can approach the metal center in one of the three ways, leading to possible adduct structures: 1) from the front between two methyl groups; 2) from a side of the metallocene, where the methyl group can either point toward or away from the actinide methyl groups; 3) from the back forming a linear metallocene complexes with two Cp rings parallel. These are all local minima. For both thorium and uranium systems, the most stable adducts are **5A** and **6A**, which are the only exothermically stable adducts with energy of −1.80 and −0.16 kcal/mol, respectively, causing the least conformational change of Cp_2AnMe_2 thus the most likely approach to happen.

14.4.1.2 Transition Sates

The essential aspect of the σ-bond metathesis mechanism is the concerted exchange of a metal-ligand σ-bond with one of an incoming substrate where the reaction proceeds via a $[2_\sigma + 2_\sigma]$ cycloaddition, as illustrated in the transitions state in Figure 14.6. The key concept is that σ-bond metathesis is a one-step reaction with two σ-bonds breaking (M–C and C'–H) and two σ-bonds forming (M–C' and C–H). Therefore, there is no change involved in the oxidation state of the metal center. This reaction mechanism has been proposed for the d^0 and d^0f^n metal compounds, and has been recently reviewed for transition metals. [58, 59]

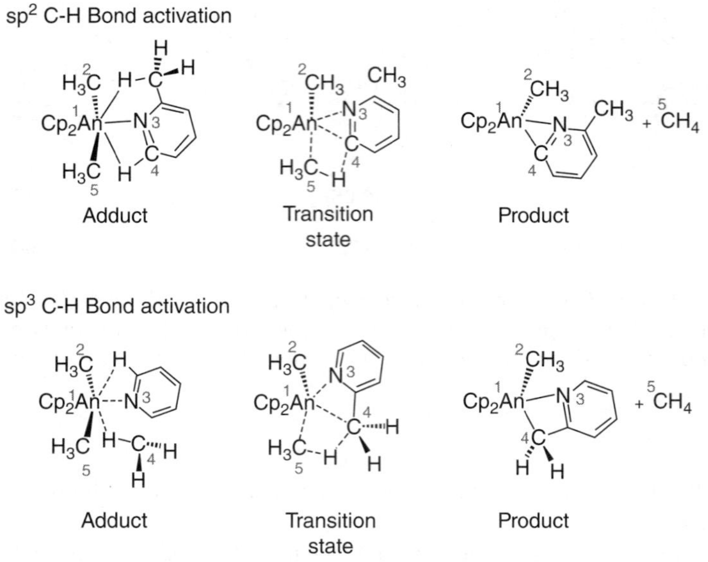

Figure 14.6 *The labeling scheme for adducts, transition states, and products. Reprinted with permission from Yang et al. [5] Copyright 2008, American Chemical Society*

As the deuterium labeling studies suggested that the 2-picoline C–H activation chemistry proceeded by σ-bond metathesis for both the thorium and uranium $Cp^*_2AnMe_2$ complexes, we examined the role of the actinide metal center in hydrogen atom migration. We identified the transition states for the sp^2 and sp^3C–H bond activation pathways if the proposed σ-bond metathesis mechanism was operative.

Figure 14.6 illustrates the numbering and labeling scheme employed for the following discussions regarding adducts, transition state, and product structures. The summarized geometric parameters of optimized stable adduct, transition states, and products for thorium and uranium systems are tabulated in Table 14.2 and Table 14.3, respectively. TS stands for transition state and the subscript refers to the reactant and product. For example, $\mathbf{TS}_{5A-1A}$represents the transition state that occurs along the sp^2 C–H bond activation pathway starting from adduct **5A** and giving product **1A**. Figure 14.7 shows the transition state structure for the sp^2 and sp^3C–H bond of 2-picoline for thorium. See color plate section. The analogous uranium system presents high similarity with the thorium systems, as shown in Table 14.2 and Table 14.3.

The transition state confirms that it is a one-step concerted reaction mechanism. However, the most important and distinct characteristic for these transition states with respect to the conventional σ-bond metathesis is that the agostic structures involved the actinide center directly mediates the migration of the activated H atom. For the thorium sp^2 and sp^3 C–H activation chemistry, the distance between thorium and the activated hydrogen atom is 2.328 and 2.532 Å, respectively. The activated hydrogen atom is bound to the 2-picoline and methyl group simultaneously as the activated C–H bond distances are of in the range of 1.34 to1.51 Å, as shown in Figure 14.7. The Th-C(Me) bond are significantly activated

Table 14.2 *Geometric data for adduct, transition states, and products for the C–H bond activation chemistry between $Cp_2Th(CH_3)_2$ and 2-Picoline. Reprinted with permission from Yang et al. [5] Copyright 2008, American Chemical Society*

	sp^2 C–H activation				sp^3 C–H activation		
	Adduct **5A**	Transition state $TS_{5A\text{-}1A}$	Product **1A**	Expt	Transition state $TS_{5A\text{-}2A}$	Product **2A**	Expt
$R(Th^1\text{-}C^2)$	2.538	2.516	2.507	2.530	2.459	2.508	2.902
$R(Th^1\text{-}N^3)$	2.842	2.264	2.497	2.444	2.654	2.599	2.574
$R(Th^1\text{-}C^4)$		2.719	2.498	2.471	2.880	2.660	2.642
$R(Th^1\text{-}C^5)$	2.538	2.796			2.776		
$R(Th^1\text{-}H(sp^2))$	3.736	2.328					
$R(Th^1\text{-}H(sp^3))$	3.299				2.352		

Table 14.3 *Geometric data for adduct, transition states, and products for the C–H bond activation chemistry between $Cp_2U(CH_3)_2$ and 2-Picoline. Reprinted with permission from Yang et al. [5] Copyright 2008, American Chemical Society*

	sp^2 C–H activation				sp^3 C–H activation	
	Adduct **6A**	Transition state $TS_{6A\text{-}3A}$	Product **3A**	Expt	Transition state $TS_{6A\text{-}4A}$	Product **4A**
$R(U^1\text{-}C^2)$	2.476	2.459	2.439	2.467	2.456	2.508
$R(U^1\text{-}N^3)$	2.778	2.553	2.428	2.394	2.678	2.599
$R(U^1\text{-}C^4)$		2.636	2.423	2.396	2.792	2.660
$R(U^1\text{-}C^5)$	2.476	2.774			2.706	
$R(U^1\text{-}H(sp^2))$	3.455	2.247				
$R(U^1\text{-}H(sp^3))$	3.401				2.258	

with the Th–C distance greater than 2.77 Å. Intrinsic reaction coordinate calculations verified both directions from these transition states leading to the anticipated protectants and reactants. The agostic Th· · ·H bond can be compared with the thorium hydride distance of 2.116 determined for the model complex $Cp_2Th(H)(Me)$. This indicates there is a significant bonding interaction that exists between the thorium metal center and the migrating H atom when the H atom is transferring from one carbon atom to another.

High structural similarity is observed for the uranium system. The lowest transition states for both sp^2 and sp^3 C–H bond activation of 2-picoline are the five-member agostic σ-bond metathesis structure. The agostic U· · ·H distances are 2.247 and 2.258 Å, respectively in the sp^2 and sp^3 transition states, which are longer than the uranium hydride bond in $Cp_2U(H)(Me)$. As a result of highly activated U-C(Me) bond, the leaving methyl group binds weakly with a long coordination distance greater than 2.71 Å.

For such a synergistic nonadditive H-migration process, we refer to this reaction mechanism as agostic interactions assisted σ-bond metathesis. Given its concerted nature, this

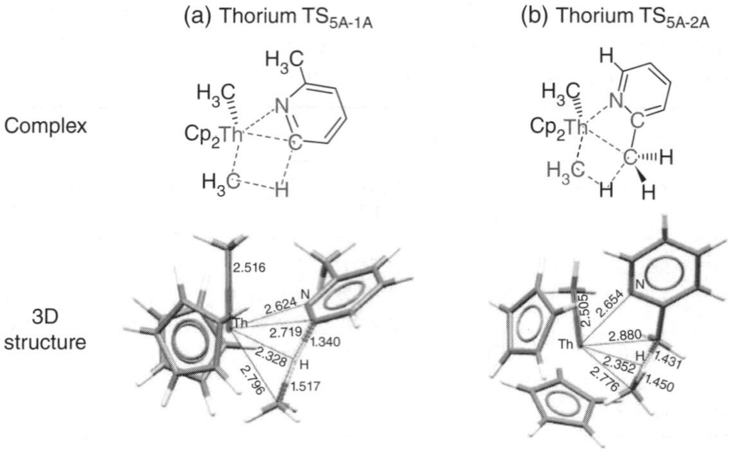

Figure 14.7 *Illustrated and 3D structures of transition states for the sp² and sp³ C–H bond activation. Notice the close resemblance between the geometries of the thorium and uranium complexes. Reprinted with permission from Yang et al. [5] Copyright 2008, American Chemical Society. See color plate section*

reaction mechanism is very stereospecific. A similar reaction mechanism has been recently reported for cyclometalation of palladium and iridium. [60, 61] These data suggest that the operative mechanism for the C–H activation in Th(IV) and U(IV) centers involving additional agostic interactions between the metal center and the activated H atom. This differs from the conventional σ-bond metathesis where the activated H atom does not coordinate with its diagonal metal center. It has been hypothesized for more than 30 years that agostic structures play an important role in transition metal complexes and catalytic reaction pathways. [59, 62–64] Now the early actinides are added to the list.

14.4.1.3 Reaction Pathways

The reaction diagram for the sp^2 and sp^3 C–H activation pathways starting from the most stable adduct **5A** and **6A** for thorium and uranium complex is shown in Figure 14.8 and Figure 14.9, respectively. For the thorium system, the sp^3 insertion pathway leads to the formation of the transition state $\mathbf{TS}_{5A-2A}$ with activation energy of 21.46 kcal/mol and **2A** as a kinetic product. Along the sp^2 pathway the transition state $\mathbf{TS}_{5A-1A}$ surmounts at higher activation energy of 22.00 kcal/mol forming the thermodynamic product **1A**. These results are consistent with experimental observation that the C–H bond activation of 2-picoline occurs to give mixture of **1A** and **2A**. The product ratio is dependent on temperature, time, and concentration of 2-picoline; greater concentrations of 2-picoline, longer reaction times, or higher temperatures afforded greater yields of the sp^2 C–H bond activation product **1A**. [55] Therefore the small energy difference between the transition state of sp^2 and sp^3 C–H bond activation (0.56 kcal/mol) allows the formation of kinetic product **2A** and the conversion to the thermodynamic product **1A**.

In marked contrast to the thorium reaction, the sp^3 C–H bond activation ($\mathbf{TS}_{6A-4A}$) of uranium system is associated with higher activation energy than the sp^2 activation ($\mathbf{TS}_{6A-3A}$)

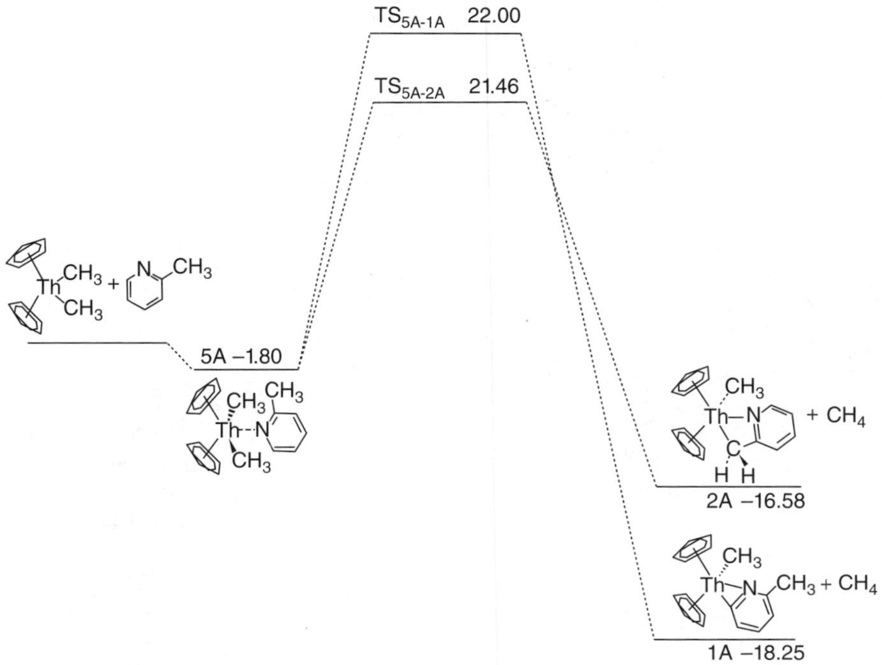

Figure 14.8 *Reaction energy diagram for* Cp_2ThMe_2 *and 2-picoline. Reprinted with permission from Yang et al. [5] Copyright 2008, American Chemical Society*

by 1.24 kcal/mol. As a result, the thermodynamically favored product is also kinetically favored for the uranium complex, leading to the only observed product through sp^2 activation. [55] The overall results regarding the trends in sp^2 versus sp^3 activation pathways are encouraging in that they delineate the experimental observations to date, despite the small energy differences. Note that these activation energies are calculated for the model Cp systems. Larger differences usually are anticipated for Cp* systems when the five methyl groups on the Cp are fully included. Indeed, we observed larger energy differences between product isomers using Cp* as ligand instead of the model Cp ligand. However, in order to correctly identify the transition states, a reduced degree of freedom is usually very helpful. The chemical nature reviewed using the Cp model system will be transferable to the Cp* systems.

A natural question to ask is the cause of the reactivity difference between the thorium and uranium complexes. We will examine the electronic properties of the reactants, products, and transitions in order to understand the similarities and differences. The natural populations of 6d and 5f for all species are depicted in Figure 14.10; the 5f population at the metal center roughly keeps consistent along the reaction pathways. See color plate section. The higher 5f populations for U reflect the extra two unpaired electrons at the uranium center, which does not participate in the reaction directly. Thus, there is no oxidation state of the metal center consistent with the proposed σ-bond metathesis and experimental observations.

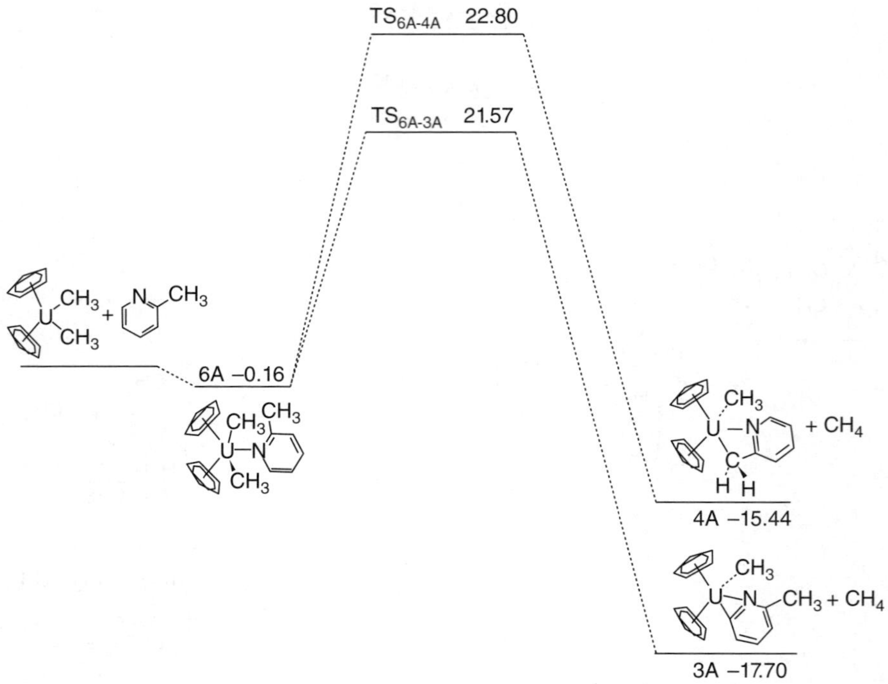

Figure 14.9 *Reaction energy diagram for the* Cp_2UMe_2 *and 2-picoline. Reprinted with permission from Yang et al. [5] Copyright 2008, American Chemical Society*

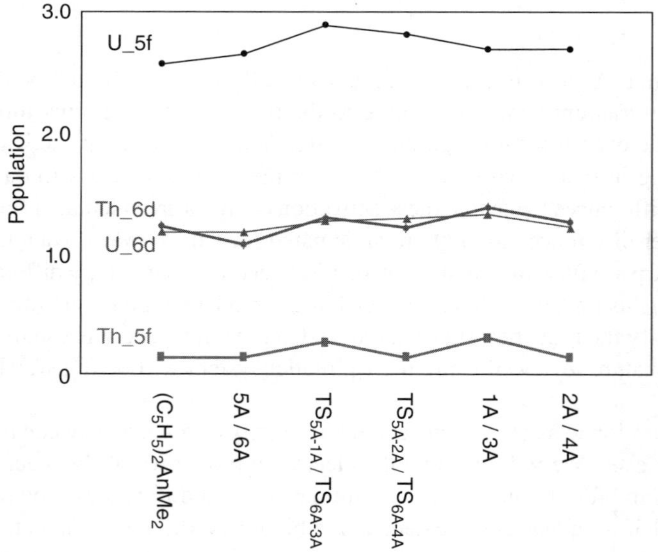

Figure 14.10 *Orbital population on the actinide metal center along reaction pathways. Reprinted with permission from Yang et al. [5] Copyright 2008, American Chemical Society. See color plate section*

The influence of solvation effects on the reaction mechanism was also evaluated using the polarizable continuum models (PCM). Because the solvent used in the experiments is toluene with a small dielectric constant of 2.38, one does not anticipate large changes in reaction energies. Indeed, the solvation effects are of the same degree because the complexes along the reaction pathway have similar dipole moments. The solvent-induced dipole moment are about 15% uniformly larger for reactants, adducts, transitions states, and products for both Th and U systems. Overall the role of solvent effects is negligible giving the same results as those gas-phase conclusions regarding the thermodynamic and kinetic pathways for the C–H activation.

In summary, theoretical study provides mechanistic insights into the aspects of C–H activation chemistry involving actinide metal center. The results of density functional theoretical exploration are consistent with the experimental outcome. A weakly bound adduct is formed to bring the C–H to the proximity to the metal center. This is followed by the activation of adjacent C–H bonds through an agostic interactions-assisted σ-bond metathesis mechanism. The origin of the regioselectivity rests in the highly ordered nature of the concerted transition states. Despite many common features found between thorium and uranium systems, including similar geometries of the products, adducts, and agostic interactions in the transitions state, the relative activation energies between sp^2 and sp^3 activation switch the order with small energy difference. These results are in good agreement with the experimental observation that the sp^2 insertion product is the thermodynamic product in both cases, but the sp^3 insertion product is the kinetic product for thorium complex. The differences in the competition between sp^2 and sp^3 pathways for the two actinide centers are subtle. Electronic property analyses conform that 5f electrons are not involved in the C–H activation. A non-conventional σ-bond metathesis mechanism involved in agostic interactions is proposed for the cyclometalation mechanism of C–H activation of N-heterocycles by actinocene complexes.

14.4.2 Study of Imido-Exchange Reaction Mechanism

In the second example we present a study where theoretical helped unravel an unexpected result in the laboratory: a reaction expected to lead to the thermodynamically most stable species (uranium oxo), instead generated a uranium imido species. As mentioned earlier, the uranyl ion possesses U-O bonds with high thermodynamic stability and extreme kinetic inertness. As a result, the majority of uranyl ion reaction chemistry involves substitution of equatorially coordinated ligands while leaving the U-O bond unaffected. Its isoelectronic bis(imido) $[U(NR)_2]^{2+}$ analog, while possessing many of the bonding features found in UO_2^{2+} [46] exhibits reactivity quite distinct from UO_2^{2+}.

Organic isocyanates have often been used in transition metal chemistry to effect transformations of M=N imido functional groups. [65] The reaction of R*NCO with M=NR imido comounds have been reported to produce either (1) carbodiimides (R*NCNR) and metal oxo complexes or (2) a new isocyanate (RNCO) and M=NR* metal imido compound. It is suggested that the reaction proceeds by the formal [2 + 2] cycloaddition of either the C = O or C = N bond of the isocyanate across the M=N imido functional group. Given the high strength of U = O bond, it was anticipated that the reaction of bis(imido) complex U(N^{t}Bu)2- (I)2(OPPh3)2 (**7**) with aryl isocyanates would yield the oxo-imido

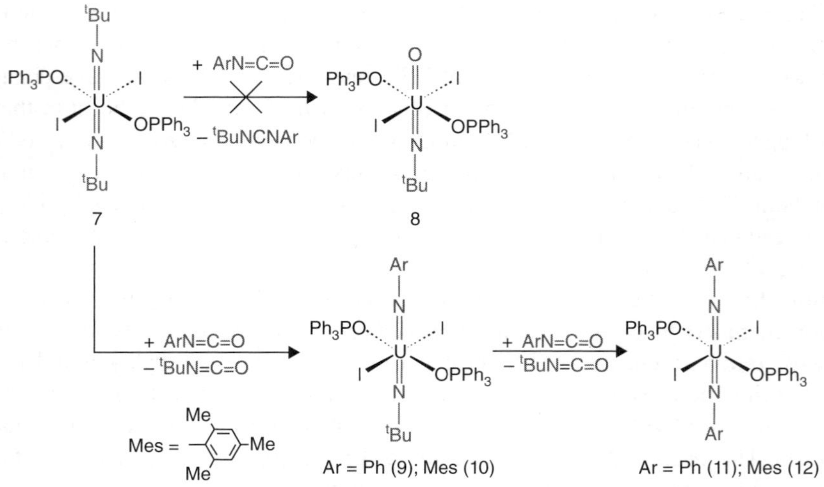

Scheme 14.2 *Potential reaction pathways bis(imido) uranium complex with aryl isocyanates. Reprinted with permission from Spencer et al. [47] Copyright 2008, American Chemical Society. See color plate section*

complex **8** (Scheme 14.2). See color plate section. Instead, a mixed bis(imido) complex U(NPh)(N^t- Bu)(I)2(OPPh3)2 (**9**) was isolated. [47]

Density functional theory (DFT) calculations were performed to elucidate the relative energies of the intermediates and products of this reaction. There are two reaction pathways that could generate the bis(imido) product **9** and the oxo-imido species **8** from U(N^{t}Bu)2(I)2(OPPh3)2 (Paths 1 and 2, Scheme 14.3). See color plate section. The first pathway involves the [2 + 2] cycloaddition of the C = N bond of the aryl isocyanate to form an *NN*-bound ureato intermediate (**13**), which can isomerize to form species **14** with the -NPh group *trans* to the *tert*-butyl imido moiety. Compound **14** can then eliminate ^{t}BuNCO to generate the unsymmetrical bis(imido) complex **9** (top line in Scheme 14.2). Alternatively, *NO*-bound carbamate intermediates **15** and **16** could form that result from the [2 + 2] cycloaddition of the C = O bond of the aryl isocyanate across the U = N imido bond. Elimination of a substituted carbodiimide would generate the oxo-imido complex **8** (bottom line in Scheme 14.2).

The computational results suggest the lowest energy pathway involves the [2 + 2] cycloaddition of the C = N bond of phenyl isocyanate to form the *NN*-bound ureato intermediate **13** (Path 1, Scheme 14.3). In the calculations performed, it is assumed that OPPh3 dissociation occurs in order to generate intermediates **13** and **14**. Experimentally, the reactions between **7** and PhNCO proceed much more slowly in the presence of excess OPPh3 as is consistent with this assumption. Overall the transformation of U(N^{t}Bu)2(I)2(OPPh3)2 (**7**) to U(NPh)(N^{t}Bu)(I)2(OPPh3)2 (**9**) is exergonic, with the free energy of the bis(*tert*-butyl)imido uranium complex (**7**) + PhNCO higher than the mixed imido species **9** + ^{t}BuNCO by 6.2 kcal $\cdot$ mol^{-1}.

In contrast to this mechanism, the calculated [2 + 2] C = O cycloaddition bond pathway (Path 2) involves the formation of the higher energy *NO*-bound ureato intermediates **15**

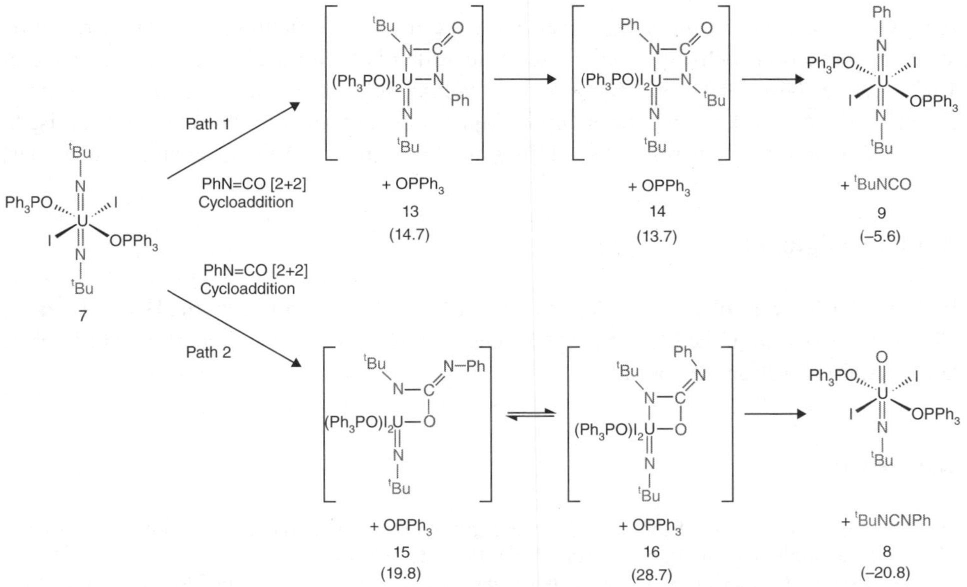

Scheme 14.3 *Relative free energies of the products and intermediates in the potential pathways for the formation of 8 and 9. The energies are provided in parentheses in kcal/mol. Reprinted with permission from Spencer et al. [47] Copyright 2008, American Chemical Society. See color plate section*

(19.8 kcal · mol^{-1}) and **16** (28.7 kcal · mol^{-1}). Complex **15** can then eliminate the mixed carbodiimide ^{t}BuN = C = NPh and form the oxo-imido species **8**. As was anticipated, the relative energy of **8** is substantially lower in energy than the unsymmetrical imido species **9** (15.2 kcal · mol^{-1}). This energy difference between U = O and U = N bond formation has also been observed in cyclopentadienyl-substituted U^{IV} complexes. [7]

Given the evidence from DFT calculations and the results from $^{[15]}$N-labeling studies, it appears the mechanism for the formation of **9** involves the [2 + 2] cycloaddition of the aryl isocyanate C = N bond across the U = N imido moiety. These results are quite surprising given the thermodynamic and kinetic stability of U = O bonds. The further reaction at elevated temperature gives a mixed oxo-imido complex **8** which is consistent with the relative energies determined by DFT calculations.

14.5 Final Remarks

In this chapter we tried to point out certain general observations and our personal preferences for choosing computational tools that represent a good compromise between accuracy and computational cost. When dealing with the heavy elements, relativistic corrections are imperative and, to a good approximation one can include them via small-core relativistic effective core potentials. All electron calculations with relativistic Hamiltonians are becoming more "doable" for large molecular systems, but with the exception of certain problems they do not necessarily afford an increase in accuracy worth the extra cost. Density functional theory has been proven to be a useful tool to probe the electronic structures and reactivity of actinide complexes. Given the experimental difficulties accessing transition states and reaction pathways, even a semi-quantitative theoretical investigation provides value. While these DFT approaches are not yet capable of chemical accuracy, computational chemistry working in partnership with judicious experiments shows great promise to unravel reaction mechanisms and shed light on the chemistry of the fascinating molecules.

Acknowledgment

ERB and RLM gratefully acknowledge their funding for this work from the Heavy Element Chemistry Program of the Basic Energy Sciences Division of the US Department of Energy, at Los Alamos National Laboratory.

References

[1] Pepper, M. & Bursten, B. E. The electronic structure of actinide-containing molecules: A challenge to applied quantum chemistry. *Chem. Rev.* **91**, 719–741 (1991).
[2] Kaltsoyannis, N., Hay, P. J., Li, J., Blaudeau, J.-P. & Bursten, B. E. in *The Chemistry of the Actinide and Transactinide elements* **3**, (Springer).
[3] Dognon, J.-P. Theoretical insights into the chemcal bonding in actinide complexes. *Coordination Chemistry Reviews* **266–267**, 110–122 (2014).
[4] Neidig, M. L., Clark, D. L. & Martin, R. L. Covalency in f-element complexes. *Coordination Chemistry Reviews* **257**, 394–406 (2013).
[5] Yang, P., Warnke, I., Martin, R. L. & Hay, P. J. Theoretical Studies of the sp^2 versus sp^3 C–H Bond Activation Chemistry of 2-Picoline by $(C_5Me_5)_2An(CH_3)_2$ Complexes (An = Th, U). *Organometallics* **27**, 1384–1392 (2008).
[6] Lewis, G. N. Valence and Tautomerism. *J Am. Chem. Soc.* **35**, 1448–1455 (1913).
[7] Barros, N. *et al.* Single but Stronger UO, Double but Weaker UNMe Bonds: The Tale Told by Cp_2UO and Cp_2UNR. *Organometallics* **26**, 5059–5065 (2007).
[8] Heitler, W. & London, F. Wechselwirkung neutraler Atome und homöopolare Bindung nach der Quantenmechanik. *Zeitschrift für Physik A Hadrons and Nuclei* **44**, 455–472 (1927).
[9] Pauling, L. *Nature of the Chemical Bond.* (Cornell University Press, 1960).
[10] Perkins, P. G. *Elementary Molecular Bonding Theory.* (Chapman and Hall, 1972).
[11] Ziegler, T. & Rauk, A. Calculation of Bonding Energies by Hartree-Fock Slater Method. 1. Transition-state method. *Theoretica chimica acta* **46**, 1–10 (1977).
[12] Cohen-Tannoudji, C., Diu, B. & Laloë, F. *Quantum Mechanics, Vol. 2.* **2**, (Wiley-VCH, 1991).
[13] Baerends, E. J., Schwarz, W. H. E., Schwerdtfeger, P. & Snijders, J. G. Relativistic atomic orbital contractions and expansions: magnitudes and explanations. *J. Phys. B: At. Mol. Opt. Phys.* **23**, 3225–3240 (1990).
[14] van Wüllen, C. & Langermann, N. Gradients for two-component quasirelativistic methods. Application to dihalogenides of element 116. *J Chem. Phys.* **126**, 114106 (2007).
[15] Hay, P. J. & Wadt, W. R. Ab initio effective core potentials for molecular calculations. Potentials for K to Au including the outermost core orbitals. *J Chem. Phys.* **82**, 299 (1985).
[16] Fernandez Pacios, L. & Christiansen, P. A. Ab initio relativistic effective potentials with spin-orbit operators. I. Li through Ar. *J Chem. Phys.* **82**, 2664 (1985).
[17] Ermler, W. C., Lee, Y. S., Christiansen, P. A. & Pitzer, K. S. AB initio effective core potentials including relativistic effects. A procedure for the inclusion of spin-orbit coupling in molecular wavefunctions. *Chem. Phys. Lett.* **81**, 70–74 (1981).
[18] Kahn, L. R. Ab initio effective core potentials: Reduction of all-electron molecular structure calculations to calculations involving only valence electrons. *J Chem. Phys.* **65**, 3826–3853 (1976).

[19] Dolg, M. & Cao, X. Relativistic Pseudopotentials: Their Development and Scope of Applications. *Chem. Rev.* **112**, 403–480 (2012).
[20] Wadt, W. R. & Hay, P. J. Abinitio Effective Core Potentials for Molecular Calculations — Potentials for Main Group Elements Na to Bi. *J Chem. Phys.* **82**, 284–298 (1985).
[21] Hay, P. J. & Martin, R. L. Theoretical studies of the structures and vibrational frequencies of actinide compounds using relativistic effective core potentials with Hartree–Fock and density functional methods: UF_6, NpF_6, and PuF_6. *J Chem. Phys.* **109**, 3875 (1998).
[22] Dolg, M., Stoll, H., Savin, A. & Preuss, H. Energy-adjusted pseudopotentials for the rare earth elements. *Theoretica chimica acta* **75**, 173–194 (1989).
[23] Küchle, W., Dolg, M., Stoll, H. & Preuss, H. Energy-adjusted pseudopotentials for the actinides. Parameter sets and test calculations for thorium and thorium monoxide. *J Chem. Phys.* **100**, 7535–7542 (1994).
[24] Cao, X., Dolg, M. & Stoll, H. Valence basis sets for relativistic energy-consistent small-core actinide pseudopotentials. *J Chem. Phys.* **118**, 487–496 (2003).
[25] *Basis Set Exchange*. (David Feller) at <http://bse.pnl.gov/>.
[26] Hehre, W. J., Ditchfield, R. & Pople, J. A. Self—Consistent Molecular Orbital Methods. XII. Further Extensions of Gaussian—Type Basis Sets for Use in Molecular Orbital Studies of Organic Molecules. *J Chem. Phys.* **56**, 2257–2261 (1972).
[27] Kohn, W. Nobel Lecture: Electronic structure of matter-wave functions and density functionals. *Rev. Mod. Phys.* **71**, 1253–1266 (1999).
[28] Hohenberg, P. & Kohn, W. Inhomogeneous Electron Gas. *Phys. Rev. B* **136**, 864–871 (1964).
[29] Kohn, W. & Sham, L. J. Self-Consistent Equations Including Exchange and Correlation Effects. *Phys. Rev.* **140**, 1133–1138 (1965).
[30] Payne, M. C., Teter, M. P., Allan, D. C., Arias, T. A. & Joannopoulos, J. D. Iterative minimization techniques for ab initio total-energy calculations: molecular dynamics and conjugate gradients. *Rev. Mod. Phys.* **64**, 1045–1097 (1992).
[31] Vosko, S. H., Wilk, L. & Nusair, M. Accurate spin-dependent electron liquid correlation energies for local spin density calculations: a critical analysis. *Canadian Journal of Physics* **58**, 1200–1211 (1980).
[32] Perdew, J. P., Burke, K. & Ernzerhof, M. Generalized Gradient Approximation Made Simple. *Phys. Rev. Lett.* **77**, 3865–3868 (1996).
[33] Perdew, J. P., Tao, J., Staroverov, V. N. & Scuseria, G. E. Meta-generalized gradient approximation: Explanation of a realistic nonempirical density functional. *J Chem. Phys.* **120**, 6898–6911 (2004).
[34] Tao, J., Perdew, J., Staroverov, V. & Scuseria, G. Climbing the Density Functional Ladder: Nonempirical Meta–Generalized Gradient Approximation Designed for Molecules and Solids. *Phys. Rev. Lett.* **91**, 146401 (2003).
[35] Becke, A. D. Density-Functional Thermochemistry . 3. The Role of Exact Exchange. *J Chem. Phys.* **98**, 5648–5652 (1993).
[36] Ernzerhof, M. & Scuseria, G. E. Assessment of the Perdew–Burke–Ernzerhof exchange-correlation functional. *J Chem. Phys.* **110**, 5029 (1999).
[37] Adamo, C. & Barone, V. Toward reliable density functional methods without adjustable parameters: The PBE0 model. *J Chem. Phys.* **110**, 6158–6170 (1999).
[38] Staroverov, V. N., Scuseria, G. E., Tao, J. & Perdew, J. P. Comparative assessment of a new nonempirical density functional: Molecules and hydrogen-bonded complexes. *J Chem. Phys.* **119**, 12129–12137 (2003).
[39] Batista, E. R., Martin, R. L., Hay, P. J., Peralta, J. E. & Scuseria, G. E. Density functional investigations of the properties and thermochemistry of UF_6 and UF_5 using valence-electron and all-electron approaches. *J Chem. Phys.* **121**, 2144–2150 (2004).
[40] Peralta, J. E., Batista, E. R., Scuseria, G. E. & Martin, R. L. All-electron hybrid density functional calculations on UFn and UCln ($n = 1 - 6$). *J Chem Theory Comput* **1**, 612–616 (2005).
[41] Hildenbrand, D. L. & Lau, K. H. Redetermination of the thermochemistry of gaseous UF_5, UF_2, and UF. *J Chem. Phys.* **94**, 1420–1425 (1991).
[42] Compton, R. N. On the formation of positive and negative ions in gaseous UF_6. *J Chem. Phys.* **66**, 4478 (1977).

[43] McDowell, R., Asprey, L. B. & Paine, R. T. Vibrational-Spectrum and Force-Field of Uranium Hexafluoride. *J Chem. Phys.* **61**, 3571–3580 (1974).
[44] Batista, E. R., Martin, R. L. & Hay, P. J. Density functional investigations of the properties and thermochemistry of UF_n and UCl_n (n = 1, . . . , 6). *J Chem. Phys.* **121**, 11104–11111 (2004).
[45] Denning, R. G. Electronic Structure and Bonding in Actinyl Ions and their Analogs. *J Phys. Chem. A* **111**, 4125–4143 (2007).
[46] Hayton, T. W. *et al.* Synthesis of imido analogs of the uranyl ion. *Science* **310**, 1941–1943 (2005).
[47] Spencer, L. P., Yang, P., Scott, B. L., Batista, E. R. & Boncella, J. M. Imido exchange in bis (imido) uranium (vi) complexes with aryl isocyanates. *J Am. Chem. Soc.* **130**, 2930–2931 (2008).
[48] Spencer, L. P., Yang, P., Scott, B. L., Batista, E. R. & Boncella, J. M. Uranium(VI) bis(imido) chalcogenate complexes: synthesis and density functional theory analysis. *Inorg. Chem.* **48**, 2693–2700 (2009).
[49] Spencer, L. P. *et al.* Cation-cation interactions, magnetic communication, and reactivity of the pentavalent uranium ion $[U(N^tBu)_2]$. *Angew. Chem. Int. Ed. Engl.* **48**, 3795–3798 (2009).
[50] Spencer, L. P. *et al.* Tetrahalide Complexes of the $[U\ (NR)_2]^{2+}$ Ion: Synthesis, Theory, and Chlorine K-Edge X-ray Absorption Spectroscopy. *J Am. Chem. Soc.* **135**, 2279 (2013).
[51] Sharma, M., Andrea, T., Brookes, N. J., Yates, B. F. & Eisen, M. S. Organoactinides Promote the Dimerization of Aldehydes: Scope, Kinetics, Thermodynamics, and Calculation Studies. *J Am. Chem. Soc.* **133**, 1341–1356 (2011).
[52] Yahia, A. & Maron, L. Is Thorium a d Transition Metal or an Actinide? An Answer from a DFT Study of the Reaction between Pyridine N-Oxide and $Cp_2M(CH_3)_2$with M = Zr, Th, and U. *Organometallics* **28**, 672–679 (2009).
[53] Castro, L., Yahia, A. & Maron, L. A DFT study of the reactivity of actinidocenes (U, Np and Pu) with pyridine and pyridine N-oxide derivatives. *Dalton Transactions* **39**, 6682–6692 (2010).
[54] Pool, J. A., Scott, B. L. & Kiplinger, J. L. A New Mode of Reactivity for Pyridine N-Oxide: C–H Activation with Uranium(IV) and Thorium(IV) Bis(alkyl) Complexes. *J Am. Chem. Soc.* **127**, 1338–1339 (2005).
[55] Kiplinger, J. L., Scott, B. L., Schelter, E. J. & Pool-Davis-Tournear, J. A. sp^3 versus sp^2 C–H bond activation chemistry of 2-picoline by Th(IV) and U(IV) metallocene complexes. *Journal of Alloys and Compounds* **444–445**, 477–482 (2007).
[56] Miertuš, S., Scrocco, E. & Tomasi, J. Electrostatic interaction of a solute with a continuum. A direct utilizaion of AB initio molecular potentials for the prevision of solvent effects. *Chem. Phys.* **55**, 117–129 (1981).
[57] Cammi, R., Mennucci, B. & Tomasi, J. Fast Evaluation of Geometries and Properties of Excited Molecules in Solution: A Tamm-Dancoff Model with Application to 4-Dimethylaminobenzonitrile. *J Phys. Chem. A* **104**, 5631–5637 (2000).
[58] Waterman, R. σ-Bond Metathesis: A 30-Year Retrospective. *Organometallics* **32**, 7249–7263 (2013).
[59] Balcells, D., Clot, E. & Eisenstein, O. C–H Bond Activation in Transition Metal Species from a Computational Perspective. *Chem. Rev.* **110**, 749–823 (2010).
[60] Davies, D. L., Donald, S. M. A. & Macgregor, S. A. Computational Study of the Mechanism of Cyclometalation by Palladium Acetate. *J Am. Chem. Soc.* **127**, 13754–13755 (2005).
[61] Davies, D. L. *et al.* N-H versus C–H Activation of a Pyrrole Imine at {Cp*Ir}: A Computational and Experimental Study. *Organometallics* **25**, 5976–5978 (2006).
[62] Grubbs, R. H. & Coates, G. W. α-Agostic Interactions and Olefin Insertion in Metallocene Polymerization Catalysts. *Accounts of Chemical Research* **29**, 85–93 (1996).
[63] Brookhart, M., Green, M. L. H. & Parkin, G. Agostic interactions in transition metal compounds. *P Natl Acad Sci Usa* **104**, 6908–6914 (2007).
[64] Omae, I. Agostic bonds in cyclometalation. *Journal of Organometallic Chemistry* **696**, 1128–1145 (2011).
[65] Hazari, N. & Mountford, P. Reactions and Applications of Titanium Imido Complexes. *Accounts of Chemical Research* **38**, 839–849 (2005).

15

The 32-Electron Principle: A New Magic Number

Pekka Pyykkö,[1] *Carine Clavaguéra*[2] *and Jean-Pierre Dognon*[3]

[1]*Department of Chemistry, University of Helsinki*
[2]*Laboratoire de chimie moléculaire, Département de chimie, École Polytechnique, CNRS*
[3]*CEA/Saclay, UMR 3299 CEA/CNRS SIS2M, Laboratoire de chimie moléculaire et de catalyse pour l'énergie*

15.1 Introduction

15.1.1 Mononuclear, ML_n systems

The Lewis octets and the 18-electron principle are well-known, and are thought to correspond to filling the (s,p)-like and (s,p,d)-like shells, respectively, for an ML_n complex. For the heavier main-group elements the s-like shell can form a ('stereochemically active' or 'stereochemically inactive') lone pair, but the level order and electron count remain the same, $s < p$. The 18-electron (18e) principle was first suggested by Langmuir [1]. Its interpretation was recently discussed by Pyykkö [2] (and references therein). The point was that the s- and d-like orbitals indeed have considerable central-atom character while the p-like population on M may be small. Nevertheless, the latter molecular orbital lies energetically in between, imposed by its intermediate number of angular nodes and, consequently, by the intermediate angular kinetic energy. Therefore, the p-like level must be filled, even when it is a 'free passenger.' The level order presumed here, s < p < d < f, is valid for a screened Coulomb potential with a single dominant centre.

Computational Methods in Lanthanide and Actinide Chemistry, First Edition. Edited by Michael Dolg.

15.1.1.1 Beyond 18e with a central actinide atom

As a precursor to the present topic we mention the closed-shell $Th(COT)_2$ [2] and the linear triatomic LThL' [3], where formal electron counts of 20 and 24 were predicted at Th, respectively. At the time, the step to 32-electron (32e) in an ML_n system seemed insurmountable.

15.1.1.2 'Thin-shell molecules'

Approximating the ligand shell by a sphere at radius R [2], the kinetic energies of the states with an orbital angular momentum l become

$$E_l = \frac{l(l+1)}{2R^2} \tag{15.1}$$

in atomic units.

A similar order of symmetries seems to occur in molecules consisting of a single shell of the L_n on the surface. Examples are the hypothetical 32e system Au_{32} of symmetry I_h [4], or the equally hypothetical 72e system Au_{72} of the chiral symmetry I [5]. A list of these single-shell magic numbers is given in Table 15.1. Note the absence of radial nodes in this case.

15.1.2 Metal Clusters as 'Superatoms'

A richer zoo of magic numbers can be obtained by filling the sphere. Then the qualitative types (a)–(d) in Figure 15.1 can be imagined. Case (a) is that of the thin attractive shell or particle *on* a sphere. Case (b) is a 'wine bottle' potential, with a repulsive excess at the centre. Case (c) is the classic problem of a particle *in* a sphere. Case (d) is a parabolic one, possibly flattened at the bottom.

An example on a radial potential is the 'jellium' one, see for example the review by de Heer [6]. The radial potential of the jellium model rather is of the 'wine-bottle' type. The level order of the particle in a sphere is qualitatively similar to the jellium one. See Figure 2 of de Heer [6]. It gives the level order

$$1S < 1P < 1D \leq 2S < 1F < 2P < 1G \ldots \tag{15.2}$$

Table 15.1 *The magic numbers, N, for occupied shells $l = 0 - L_{max}$*

L_{max}	Shells filled	N
0	S	2
1	SP	8
2	SPD	18
3	SPDF	32
4	SPDFG	50
5	SPDFGH	72

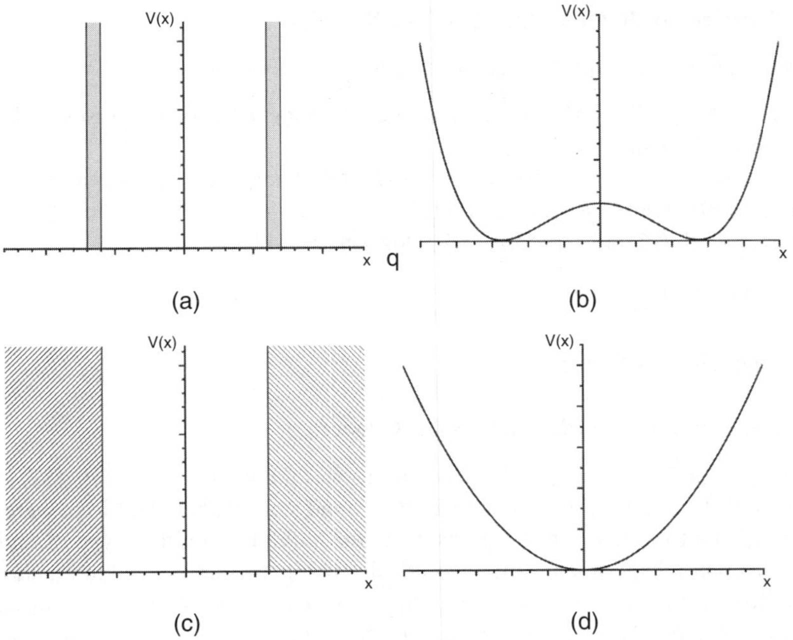

Figure 15.1 *Some types of potential wells for quasispherical systems. Case (a) corresponds to a thin attractive shell, case (b) is the 'wine bottle' potential, case (c) is the particle in a box, and case (d) is a spherical harmonic oscillator potential*

Concerning the notation, $S, P, D, \ldots$ have their usual meaning. Concerning the principal quantum number, n, it starts from 1 to every L. The corresponding magic numbers are 2, 8, 18, 20, 34, 40, 58, ... for sodium clusters, as seen from Figure 1 in Knight *et al.* [7].

In an atomic nucleus, the order of proton levels or neutron levels is qualitatively of the same type. Spin-orbit splitting is so far neglected.

An example on introducing heteroatoms is the magic number $N = 20$ occurring in Si@Au_{16} (see Walter and Häkkinen [8]). A wide application of such magic numbers was found for the compact inner part of clad gold clusters (see Häkkinen [9]). Some occurring magic numbers for the inner gold cluster are 8, 18, 20, 34, and 58. He also discusses the magic numbers for planar, quantum-dot models.

A further example on a filled 18e system is the predicted [10] and experimentally prepared [11] icosahedral WAu_{12}. Isoelectronic species with M = Mo, V^-, Nb^-, and Ta^- were both predicted and prepared. Incidentally, uranium as the central atom was not found to work.

A sufficient amount of wine-bottle character will raise 2S above 1F, and give a magic number of 32e, instead of 34e. For the spherical harmonic oscillator one obtains the equidistant energy levels [12]:

$$1S < 1P < 2S + 1D < 2P + 1F < 3S + 2D + 1G < 3P + 2F + 1H < \ldots \qquad (15.3)$$

Filling all levels with independent electrons would give the magic numbers 2, 8, 20, 40, 70, 112, ...

15.1.3 The Present Review: An@L_n-Type Systems

Our purpose here is to cover the recently discussed 32e systems:

(a) Pu@Pb_{12} [13] with Pu@Sn_{12} and further thermochemical data discussed later by Dognon *et al.* [14],
(b) The case of An@C_{28} was already experimentally known, but not previously identified as a 32e system until Dognon *et al.* [15],
(c) The anionic $[U@Si_{20}]^{6-}$ predicted by Dognon *et al.* [16].

Other recent cases are also mentioned.

15.2 Cases So Far Studied

15.2.1 The Early Years: Pb_{12}^{2-} and Sn_{12}^{2-} Clusters

The starting point for this case as described in (a) was the work of Cui *et al.* [17, 18] when a new class of Zintl ions, plumbaspherene, Pb_{12}^{2-}, and stannaspherene, Sn_{12}^{2-}, were characterized by photoelectron spectroscopy as hosts for building up a new type of endohedral clusters. Gas-phase experiments and theoretical investigations predicted a remarkable stability for the icosahedral plumbaspherene Pb_{12}^{2-} and the stannaspherene Sn_{12}^{2-} cluster anions of I_h point symmetry. These empty polyanions have not yet been synthesized in crystalline form.

The filled icosahedral M@Pb_{12}^{x} (x= +1,0,−1,−2 and M=Al, Pt, Ni, Pd, Co...) were studied both experimentally using mass spectrometry, ^{207}Pb NMR, and X-ray analysis by Esenturk *et al.* [19, 20], Neukermans *et al.* [21], Zhang *et al.* [22], and theoretically using density functional theory (DFT) calculations by Chen *et al.* [23, 24] and Neukermans *et al.* [21]. In all of these studies, the clusters were found to be very stable, opening a branch of novel, chemically inert cluster species. The Pb_{12}^{2-} and Sn_{12}^{2-} clusters are 12-vertex polyhedra (see Figure 15.2) with highly regular icosahedral structure.

The cage electronic structure is characterized by the remarkable availability of "free" 26 electrons at the top of the occupied 6p or 5p bands of Pb_{12}^{2-} and Sn_{12}^{2-}, respectively. This is

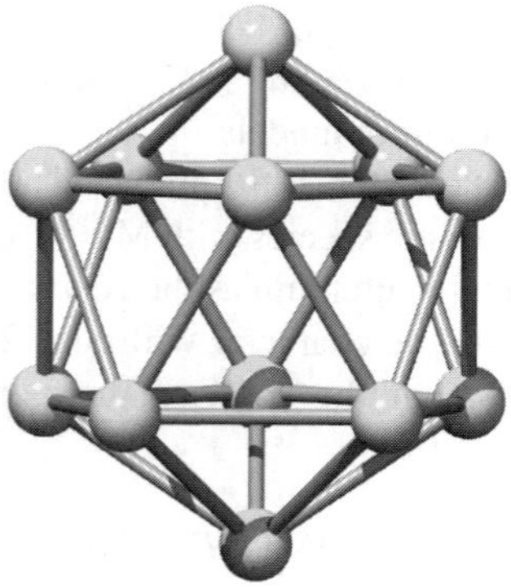

Figure 15.2 Pb_{12}^{2-} (I_h) optimized geometry. From DFT/B3LYP (see Dognon et al. [13] for details)

allowed by the large energy separation between the 6s and 6p (5s and 5p) shells as observed in the total density of state (DOS) (Figure 15.3). This separation is a consequence of the stabilization of the 6s-based MOs due to the "inert pair effect" arising from the relativistic effects [25]. This is clearly demonstrated from the comparison of the scalar relativistic and non-relativistic Pb_{12}^{2-} DOS (Figure 15.3). Because of this large energy separation, the $6s^2(5s^2)$ electron pairs of Pb_{12}^{2-} (Sn_{12}^{2-}) are localized on Pb or Sn atoms leaving the two 6p (5p) electrons per atom to participate in the bonding inside the cage. Among the 13 scalar-relativistic valence MOs of Pb_{12}^{2-} shown in Figure 15.4 and Figure 15.5, there are four radial π orbitals (a_g and t_{1u}) and nine in-sphere σ MOs (g_u and h_g) as previously calculated by Cui *et al.* [17]. See color plate section. The bonding pattern in the stannaspherene Sn_{12}^{2-} is very similar. Filling the cage in icosahedral symmetry, the s orbitals transform as a_g, p orbitals as t_{1u}, and d orbitals as h_g, while f orbitals are split into the g_u and t_{2u} irreps, and similarly g orbitals into h_g and g_g irreps. Thus, the s, p, and f orbitals of a central atom can only mix with cage s, or p or f orbitals. The d orbitals of a central metal, however, are not only allowed to interact with cage d orbitals, but also with the h_g set of the cage g orbitals.

These considerations were at the origin of our first predicted 32-electron system: $Pu@Pb_{12}$ and related systems [13]. Our specific idea was, that with 6e more, on top of the 26e plumbaspherene ion 6p band $a_g \oplus g_u \oplus h_g \oplus t_{1u}$ (HOMO), we could obtain a perfect $a_g \oplus g_u \oplus h_g \oplus t_{1u} \oplus t_{2u}$ 32e configuration. The t_{2u} from the actinide 5f levels is then added to the count. Indeed, if any f-contributions are desired, an actinide is required, the 4f shells of the lanthanides being far too compact. We also have to note that the 5f shell of the heavier actinides might not be chemically accessible.

In order to establish the "proof of concept," the closed-shell $Yb@Pb_{12}$, $Th@Pb_{12}^{4-}$, $U@Pb_{12}^{2-}$, $Pu@Pb_{12}$, $Am@Pb_{12}^{+}$, $Cm@Pb_{12}^{2+}$ endohedral clusters were explored using DFT, including relativistic effects. The $M@Pb_{12}^{n}$ have different structures for the most stable

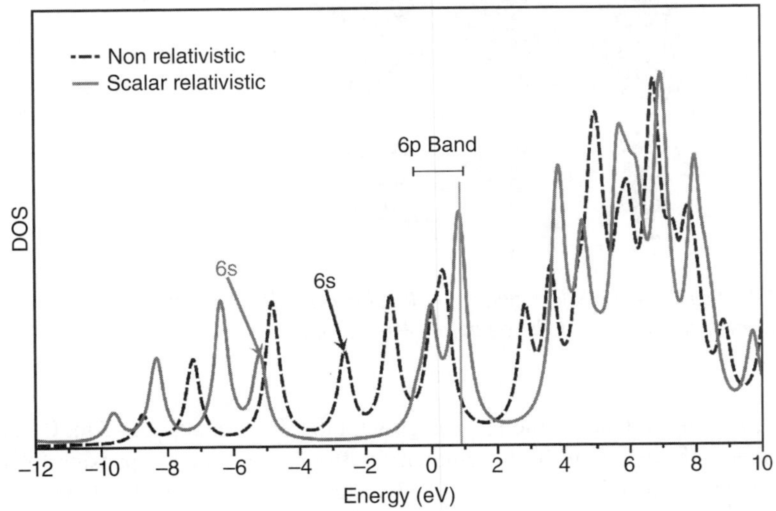

Figure 15.3 *Pb_{12}^{2-} (I_h) total density of state. From DFT/B3LYP (see Dognon et al. [13, 14] for details). The Fermi level is shown by a vertical gray line (scalar-relativistic)*

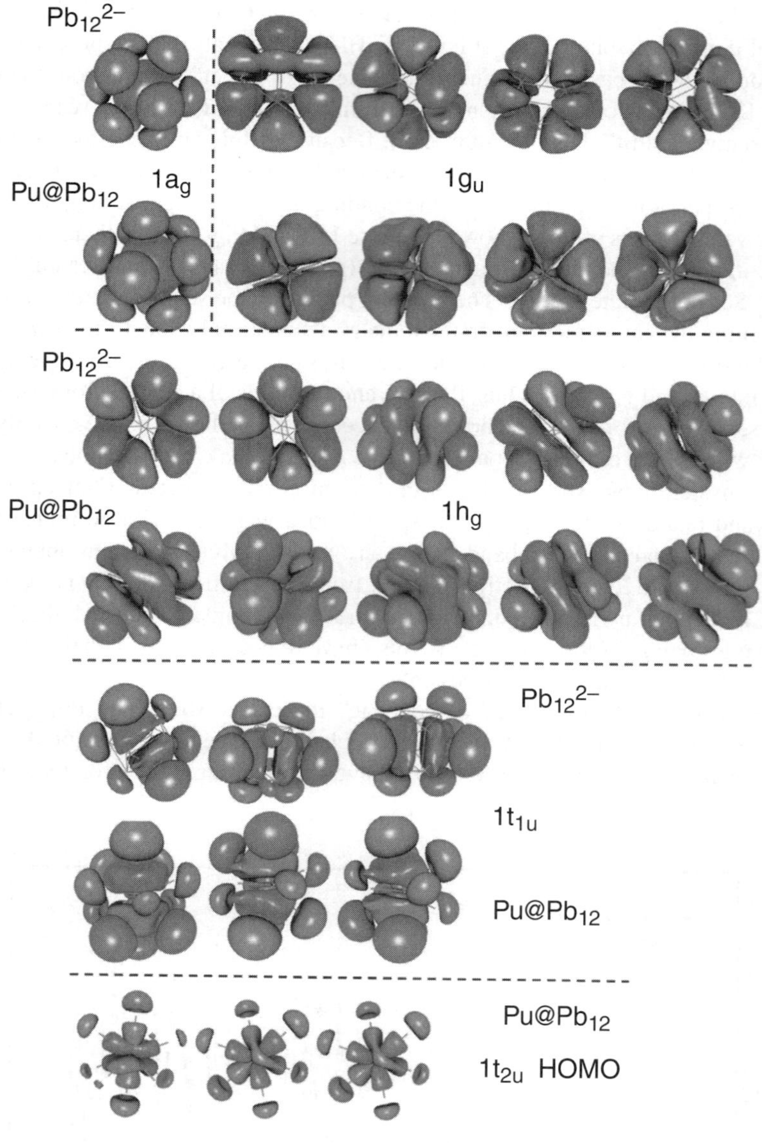

Figure 15.4 *Scalar relativistic valence molecular orbitals of Pu@Pb_{12}^{2-} (I_h) and Pb_{12}^{2-} (I_h). Adapted from Dognon et al. [13]. See color plate section*

isomer as M changes. Doping the Pb_{12}^{2-} cage with the 5f elements, Pu, Am, Cm, and with the 4f element Yb, maintain the I_h symmetry with a slight radial expansion of the initial cage (with a maximum of 0.23 Å for the M-Pb distance). In contrast, the Th, U, Np have C_5, D_{5h}, and D_{5h} point groups, respectively. The HOMO-LUMO gap is more than 1 eV with binding energies, with respect to the $[Pb_{12}]^{2-}$ and M^x fragments, larger than -15 eV (see Table 15.2). This makes the An@Pb_{12}^{2-} endohedral clusters very stable species.

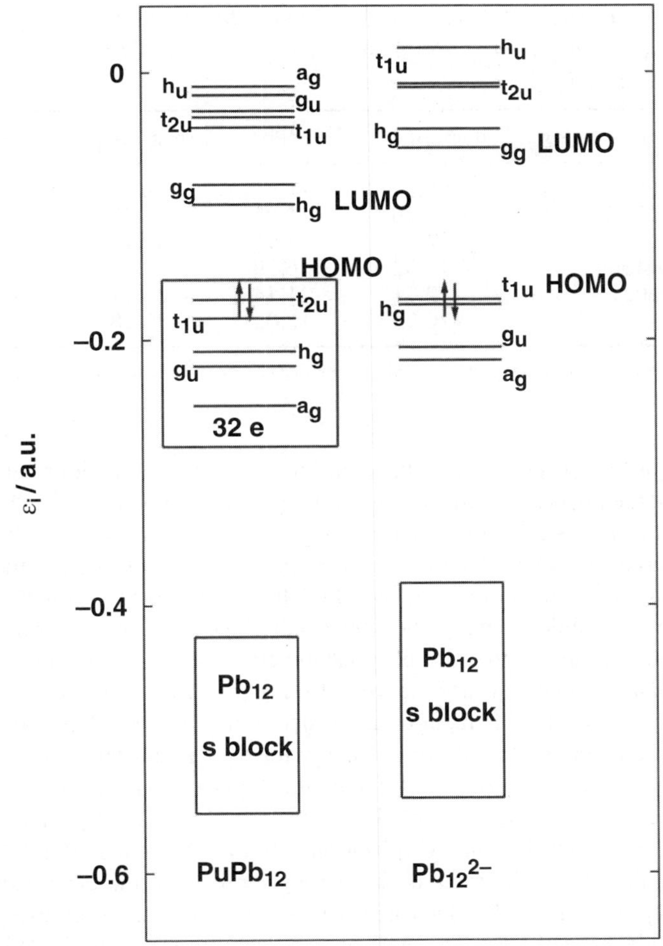

Figure 15.5 *Scalar relativistic energy levels of Pu@Pb_{12}^{2-} (I_h) and Pb_{12}^{2-} (I_h). The latter have been shifted to make the HOMOs equal. Adapted from Dognon et al. [13]*

For comparison, binding energies in the range of -2.47/-5.60 eV were reported for neutral M@Pb_{12} (M = B, Al, Ga, In and Tl) endohedral clusters by Chen *et al.* [23, 24].

The electronic structure is usually described by inspecting shape and occupation of the orbitals or by calculating charges or bond orders. These traditional quantum-chemistry measures can lead to counterintuitive results when applied to bonding across the 5f series (see Prodan *et al.* [26]) and are insufficient for an in-depth interpretation of the bonding. The joint and complementary charge, orbital, quantum chemical topology, and energy decomposition analyses (EDA) are a more powerful approach for a detailed electronic structure analysis [27]. Intuitively, the compounds seem to be similar to endohedral clusters like $[Pt@Pb_{12}]^{2-}$, which was earlier synthesized by Esenturk and coworkers [20]. The bonding in the latter species is characterized by strong peripheral interactions among the lead

Table 15.2 *HOMO-LUMO gap, binding energy (BE) analysis with respect to the $[Pb_{12}]^{2-}$ and M^x fragments (in eV). From scalar relativistic DFT/PBE (see Dognon et al. [13] for details)*

	Pb_{12}^{2-}	$[U@Pb_{12}]^{2-}$	$Pu@Pb_{12}$	$[Am@Pb_{12}]^{+}$	$[Cm@Pb_{12}]^{2+}$
Symmetry	I_h	D_{5h}	I_h	I_h	I_h
HOMO-LUMO	3.08	1.26	1.93	2.45	0.85
BE		−17.59	−22.17	−39.18	−69.33
Pauli repulsion		15.52	16.39	15.91	15.27
Electrostatic		−12.11	−21.54	−21.11	−33.25
Orbital		−20.93	−17.02	−27.98	−51.36

The BE is the sum of Pauli, electrostatic, and orbital contributions. Negative values are binding.

atoms forming a Pb_{12}^{2-} cage, whereas the contribution of the interactions between the central Pt atom and the Pb_{12} cage to the overall stability of the compound is much smaller. Each Pb atom donates two electrons to cluster bonding and the Pt atom does not donate electrons (it is qualified as a zerovalent Pt atom by Cui *et al.* [17]). The insertion of the Pt atom into the cage only further stabilizes the cage by filling the empty center. In contrast, the insertion of a lanthanide or actinide atom highly stabilizes the cage because of a strong Pb-An interaction that originates in an attractive electrostatic effect and in a large orbital interaction. The Pauli repulsion remains nearly the same from U to Cm. The orbital term contributes ~45% to the total attractive interactions and is determined by the 6d orbital energy for the Pu and Am clusters and by the 5f orbital energy for the Cm cluster [13]. Further analysis from the valence molecular orbital (MO) diagram (Figure 15.4) revealed the formation of 16 MOs between the cage and the central atom (a_g, g_u, h_g, t_{1u}, and t_{2u}). The Pu 7s, 7p, 6d, and 5f orbitals are involved in the hybridization with the Pb_{12}^{2-} cage orbitals. In $Yb@Pb_{12}$, $[Cm@Pb_{12}]^{2+}$, the 4f and 5f levels do not strongly mix with the cage because they are already too deep in energy. The electronic structure was also analyzed using the electron localizability indicator (ELI-D). The electron localizability indicator (ELI) is a functional introduced by Kohout [28, 29] that describes the correlation of electronic motion. In fact ELI designates a family of indicators. Among them, one of the most pertinent for the study of the bonding is the ELI-D that can be seen as being proportional to the charge that is needed to form an electron pair. In the $[An@Pb_{12}]^n$ clusters, the ELI-D map exhibits localization domains between the central atom and the cage as shown in diagram (a) in Figure 15.6: See color plate section. They result from the interaction between Pb and Pu for example, and are a consequence of the formation of the a_g, g_u, h_g, t_{1u}, and t_{2u} orbitals. For example, diagram (b) in Figure 15.6 depicts the pELI-D contributions of the $1h_g$ orbitals in the real space. This accounts for the existence of a strong interaction between the central atom and the cage. The combination charge, orbital, quantum chemical topology and energy decomposition analyses provides an unambiguous picture of the bonding in the $[An@Pb_{12}]^n$ clusters with notable covalency between the central actinide atom and the surrounding cages. These results have enabled us to conclude that $Pu@Pb_{12}$ and $[Am@Pb_{12}]^+$ compounds are the first examples of centered 32e systems.

As a matter of fact, the formation of a covalent bond between the constituent monomers is required to achieve the present endohedral clusters in the gas phase (none of them have

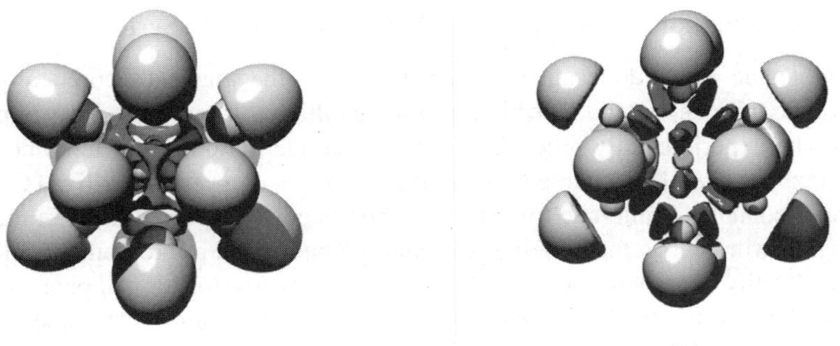

Figure 15.6 *ELI-D isosurfaces of the [Pu@Pb$_{12}$] (I_h) : magenta domains (a) visualize the interaction between the Pb cage and the Pu atom. Blue domains (b) are pELI-D isosurfaces (0.25) of summation over 5 MOs (1h_g) visualizing the 6p(Pb)-6d(Pu) orbital mixing. From scalar relativistic all-electron DFT/PBE. See color plate section*

yet been prepared). The reaction of encapsulation of the plutonium ion in the plumbaspherene and stannaspherene cage is exothermic. For example, the $Pu^{2+} + Pb_{12}^{2-} \rightarrow Pu@Pb_{12}$ reaction Gibbs free energy was calculated at −2100 kJ/mol (T = 298.15 K, P = 1 atm.) from DFT/B3LYP harmonic vibrational frequencies and standard statistical thermodynamics (assuming ideal gas). In order to locate thermodynamically the $Pu@Sn_{12}$ and $Pu@Pb_{12}$ gas-phase clusters with respect to the solid phase, we have used the experimental standard enthalpy of formation and entropy at 298.15 K from Morss *et al.* [30] and Cox *et al.* [31] (see Dognon *et al.* [14] for details). $Pu@Sn_{12}$ and $Pu@Pb_{12}$ are evaluated at about 1950 kJ/mol and 650 kJ/mol for the enthalpy and at about 1550 kJ/mol and 250 kJ/mol for the Gibbs free energy above Pu(s) + 12 Sn(s) or Pu(s) + 12 Pb(s), respectively. The existence of the present cluster compounds is made more likely by the experimental observation of the mixed metallic phases Pu_3Pb, Pu_5Pb_3, Pu_5Pb_4, Pu_4Pb_5, $PuPb_2$, $PuPb_3$, in the bulk [30].

The covalent bonding between the cage and the central atom is likely to affect the optical properties (UV/Visible/IR) of these clusters. Due to I_h symmetry, only the T_{1u} modes are IR-active and only the T_{1u} symmetry electronic transitions are electric-dipole-allowed. The DFT/B3LYP calculated infrared spectra show that the cage has vibration frequencies at 90 cm^{-1} and 126 cm^{-1} for Pb_{12}^{2-} and Sn_{12}^{2-}, respectively. Due to the rigidity of the cage, the inclusion of the actinide ion does not change significantly the cage deformation modes (e.g., an increase of 17 cm^{-1} for $Pu@Pb_{12}$). The intra-sphere motion of the actinide atom can be viewed as a particle-in-a box translation with a very small vibrational frequency (e.g., 30 cm^{-1} for $Pu@Pb_{12}$).

For the Pb_{12}^{2-} and Sn_{12}^{2-} cages, the calculated electronic absorption spectra at the adiabatic time-dependent density functional theory (TDDFT) level are similar. The absorptions are mainly in the ultraviolet with a strong absorption at about 220–230 nm and a weak absorption between 290–480 nm. The modifications in the absorption spectra induced by the encapsulation of Pu^{2+} ion are mainly the introduction of an absorption in the red (700–750 nm) with a blueshift from Sn_{12} to Pb_{12}. The major MOs involved in this excitation are t_{1u} [6p(Pu)/5p(Sn) or 6p(Pb)] to h_g [(6d(Pu)/5p(Sn) or 6p(Pb)].

15.2.2 The Validation: An@C_{28} (An = Th, Pa^+, U^{2+}, Pu^{4+}) Series

The C_{20} molecule with a dodecahedral cage structure is the smallest theoretically possible member of the fullerene family. The highly strained fullerene C_{20} consists only of pentagons and cannot be experimentally produced by carbon condensation or cluster annealing processes. Prinzbach *et al.* [32] claimed a gas-phase detection of a very short-lived C_{20} species. Therefore, the smallest fullerene to form in carbon vapor is expected to be in the C_{24}-C_{30} cluster region with a C_{28} cluster being particularly abundant. From empirical rules, which relates the stability of carbon cages mainly to the existence and location of pentagonal rings, Kroto [33] showed that the T_d C_{28} had a particular stability compared to other clusters of approximately the same size with an expected open-shell electronic structure. Further evidence in support of the C_{28} fullerene in the gas phase was provided by Guo *et al.* [34] who suggested that C_{28} may be remarkably stabilized by uranium encapsulation. This was the starting point for several theoretical papers on the electronic structure of C_{28} cage and U@C_{28} endohedral cluster. From ab initio HF calculations, the empty C_{28} fullerene is found by Guo *et al.* [35] to have a $^5A_2$ open-shell ground state and behaves as a hollow superatom with an effective valence of 4, both toward the outside and inside of the carbon cage. It was noted that the ground state is not subject to Jahn-Teller distortion and that C_{28} cage is just about the right size to accommodate a U atom. Their calculations show that the binding energy of the M@C_{28} species (M=Mg, Al, Si, S, Ca, Sc, Ti, Ge, Zr, and Sn) is a good indicator of the success in experimentally trapping the metal atoms (M) inside the fullerene cage.

As pointed out by Rösch *et al.* [36] in a study of Ce@C_{28}, the C_{28} molecule can be viewed as a spherical shell on which a single π orbital per carbon atom can delocalize to form molecular orbitals that can be filled by a central atom. Furthermore, they found that there is a significant Ce(4f) covalency identifiable in photoelectron spectra. The first insights on the stability of the U@C_{28} complex were provided by Jackson *et al.* [37] from DFT/LDA calculations. They suggest that its origin can be found in the 5f(U) and 6d (U) and 2p(C) orbital mixing including a 7s(U) state that produces very strong bonding between U and the C_{28} cage. In addition, their analysis points to the following elements as likely candidates for producing stable M@C_{28} complexes: Mo, W, Ru, Os, Ce, Th, and Pu (+4 oxidation state). Electronic structure calculations, including relativistic core potentials and the spin-orbit interaction, have been carried out on the C_{28}, Pa@C_{28}, and U@C_{28} species by Zhao and Pitzer [38]. They conclude that the U@C_{28} complex has a $(\pi^*)^1(5f)^1$ diamagnetic ground state as opposed to $(\pi^*)^2$ or $(5f)^2$, the latter being common in uranium organometallic complexes. The lowest $(5f)^2$ state is found to be 0.553 eV higher. As with Pa@C_{28}, the C_{28} π MOs mix with the metal orbitals, suggesting large binding, consistent with the experimentally observed high yields. All of these works mention the An-cage covalent bonding but did not analyze it in detail. As previously stated, the joint and complementary charge, orbital, quantum chemical topology, and energy decomposition analyses are mandatory to go further in the understanding of chemical bonding.

Stabilization of C_{28} by an endohedral species is a fascinating topic especially as the C_{28} cage provides itself 28 electrons, and has the ability to accept four electrons in the LUMO allowing a route to a 32 electrons closed-shell M@C_{28} system. Because of the quasi monoreference (94.3%) nature of the ground state of C_{28} [38] one can expect that the DFT is an appropriate theoretical framework for the study of the An@C_{28} closed-shell complexes.

Our theoretical study of the $M@C_{28}$ (M=Ce, Th, Pa^{+}, U^{2+}, Pu^{4+} confirms the remarkable chemical stabilization caused by the large HOMO-LUMO gap and the high bonding energy between −11 and −38 eV (Table 15.3).

Figure 15.7 shows the geometry of the $Pu^{4+}@C_{28}$. Encapsulation of the central M atom preserves the T_d symmetry and does not result in a significant distortion of the cage since a small radial expansion of up to 11pm in the initial empty C_{28} radius is noticed.

Following the energy decomposition analysis with respect to the C_{28} and M^x fragments (Table 15.3), the cluster is stabilized by an attractive electrostatic interaction ($Pu^{4+}@C_{28}$ excepted) and a large orbital interaction. The latter contributes 68–100% of the total attractive energy reporting bonds that involve both cage and central atom orbitals. The overlap between the actinide atomic orbitals and the C_{28} cage orbitals in the $6a_1$, $10t_2$, $5t_1$, $9t_2$, $8t_2$, 4e, $5a_1$ valence molecular orbitals leads to a 32-electron system, i.e., 4 electrons provided from the An atom and 28 electrons from the $2p_{\pi}$(C). As an illustration, the valence molecular orbitals of $Pu^{4+}@C_{28}$ are given in Figure 15.8. See color plate section.

This is a 60-electron diagram with the 32-electron system ($6a_1$, $10t_2$, $5t_1$, $9t_2$, $8t_2$, 4e, $5a_1$) plus the 'pure cage' orbitals (5e, $4t_1$, $1a_2$, $7t_2$, 3e, $6t_2$) that do not interact with the metal center. The 7s,7p, 6d, and 5f orbitals of the actinide mix with the C_{28} cage orbitals. The $6a_1$, $10t_2$, and $5t_1$ MOs mix the 2p(C) with the 5f(An), the $9t_2$, $8t_2$ MOs mix the 2p(C) with the 6d(An) and 7p(An), and the 4e and $5a_1$ MOs mix the 2p(C) with the 6d(An) and 7s(An). From Th to Pu, the participation of the 5f orbital in the $6a_1$, $10t_2$, $5t_1$ MOs increases from 9 to 46% (symmetrized fragment orbital analysis, see te Velde *et al.* [39]). In the lanthanide

Table 15.3 *HOMO-LUMO gap and binding energy (BE) analysis with respect to the C_{28} and M^x fragments (in eV)*

	$Ce@C_{28}$	$Th@C_{28}$	$Pa^{+}@C_{28}$	$U^{2+}@C_{28}$	$Pu^{4+}@C_{28}$
HOMO-LUMO	3.5	3.5	3.7	4.0	3.9
BE	−15.13	−11.03	−14.66	−34.59	−38.13
Pauli repulsion	141.10	171.68	103.49	121.50	30.38
Electrostatic	−47.99	−54.58	−37.22	−29.29	4.06
Orbital	−108.25	−128.13	−80.93	−126.80	−72.56

From scalar relativistic DFT/B3LYP (see Dognon *et al.* [15] for details).

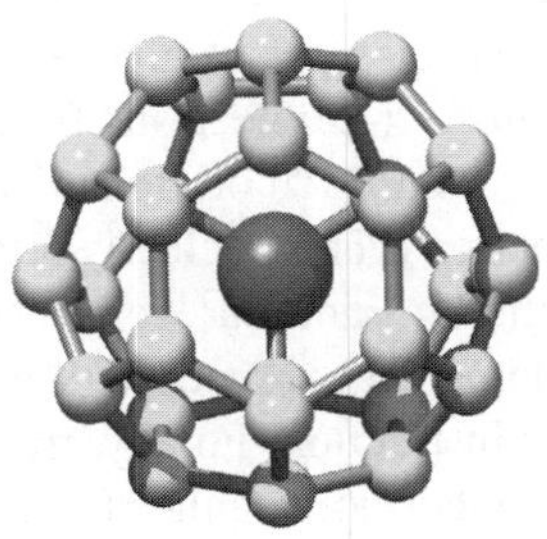

Figure 15.7 *$Pu^{4+}@C_{28}$ (T_d) optimized geometry. From scalar relativistic DFT/PBE (see Dognon et al. [15] for details)*

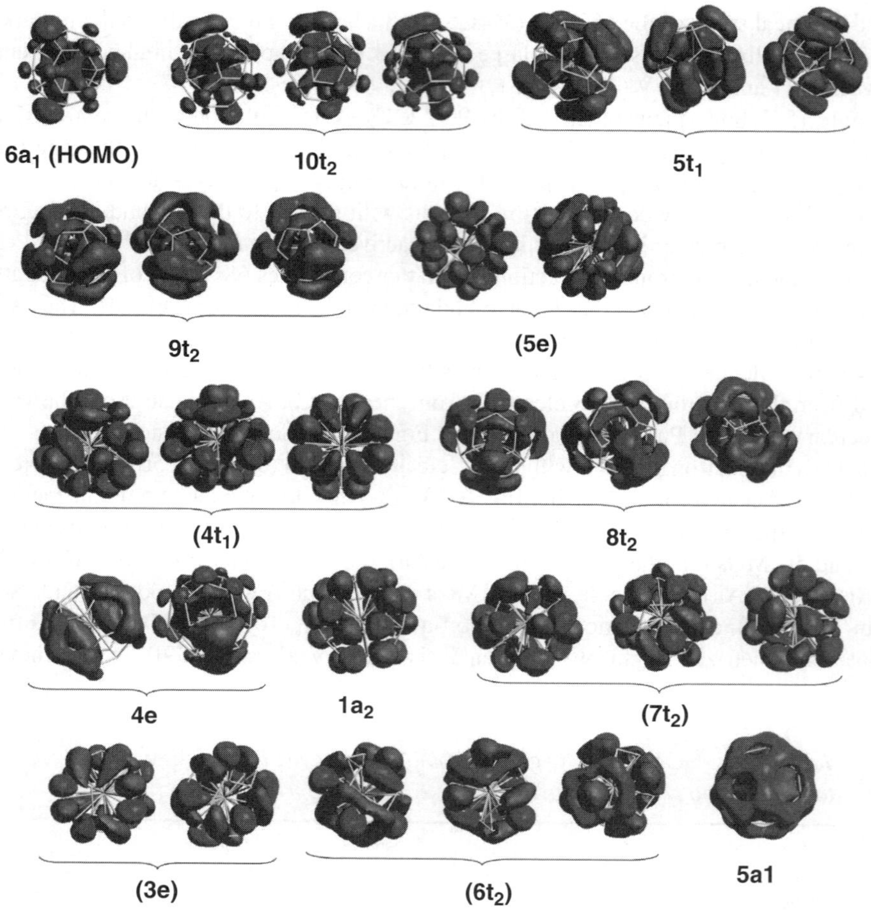

Figure 15.8 *Scalar-relativistic valence molecular orbitals of $Pu^{4+}@C_{28}$ (T_d). Adapted from Dognon et al. [15]. The 'pure cage' orbitals ($5e$, $4t_1$, $1a_2$, $7t_2$, $3e$, $6t_2$) are given in parentheses. See color plate section*

case, a slight hybridization of the 4f(Ce) orbitals (~6–8%) was found, but this system cannot be qualified as a 32-electron species. As might be expected, this 'covalence' is also shown by a significant charge transfer (1.4 electrons from a Voronoi deformation density analysis) from the cage to the U^{2+} and Pu^{4+} orbitals. For the other clusters of our series, the value ranges from 0 to 0.3 electrons. This is clearly reflected by a larger binding energy and orbital interaction term, for the U^{2+} and Pu^{4+} cases (see Table 15.3). The C-An interaction was also probed using topological approaches (see Dognon *et al.* [15]) that support the existence of a strong bonding interaction. Finally, from the DFT harmonic vibrational frequencies and standard statistical thermodynamics (assuming ideal gas), an estimation of the thermodynamic stabilities of $M@C_{28}$ can be done (see Table 15.4). The encapsulation is exothermic and the calculated Gibbs free energies account for a great stability of the clusters and particularly for the $U^{2+}@C_{28}$ and $Pu^{4+}@C_{28}$ compounds.

Table 15.4 *Reaction enthalpies and Gibbs free energies (kJ/mol) for an actinide encapsulation in a C_{28} cage (T=298.15 K, P=1 atm.)*

Reaction	$\Delta_r H$	$\Delta_r G$
$Ce + C_{28} \rightarrow Ce@C_{28}$	−1181	−1123
$Th + C_{28} \rightarrow Th@C_{28}$	−1440	−1385
$Pa^{+} + C_{28} \rightarrow Pa^{+}@C_{28}$	−1431	−1373
$U^{2+} + C_{28} \rightarrow U^{2+}@C_{28}$	−3636	−3577
$Pu^{4+} + C_{28} \rightarrow Pu^{4+}@C_{28}$	−3641	−3582

From scalar relativistic DFT/PBE (see Dognon *et al.* [15] for details).

The UV/Visible/IR optical properties are sensitive to the C-M interaction and accordingly, to the nature of the central atom. In the T_d point group symmetry, only the T_2 modes are IR-active and T_2 symmetry electronic transitions are electric-dipole-allowed. Calculated harmonic vibrational frequencies reflect the ion cage-bonding. For the $Pu^{4+}@C_{28}$ ($U^{2+}@C_{28}$) clusters, the IR absorption spectra are characterized by a strong (weak) band with a maximum near 345 cm^{-1} (326 cm^{-1}). The corresponding vibrational mode can be assigned to the coupled motions of the cage and of the central atom in the cage. From Th to Pu^{4+} the displacement of the band related to this mode is consistent with the increase in binding energy. UV/Visible absorption spectra of $An^{n+}@C_{28}$ were computed at the adiabatic TDDFT level and compared with that of the C_{28} cage. A strong absorption in the ultraviolet at ~355 nm is calculated for the C_{28} cage. Without spin-orbit coupling, as a result of the encapsulation of the central atom, the absorption band is shifted toward the visible (380 nm). The introduction of spin-orbit coupling will be discussed in Section 15.3.

$Th@C_{28}$, $Pa^{+}@C_{28}$, $U^{2+}@C_{28}$, and $Pu^{4+}@C_{28}$ qualify as new examples of 32-electron species, in addition to our previous $An@Pb_{12}$ clusters (quoted by Kemsley [40]). This raises an isolated curiosity to a principle. Moreover, the experimentally already known, the neutral $U@C_{28}$ is shown to have a similar 32e bonding system.

15.2.3 The Confirmation: $[U@Si_{20}]^{6-}$-like Isoelectronic Series

Silicon is an important material in semiconductor and microelectronic industries. Therefore silicon clusters attracted wide attention these last years, e.g., as building blocks for nanomaterials. Although fullerene-shaped carbon cages exhibit unusual stability due to sp^2 hybridization of the carbon atoms and surface conjugation, a hollow Si cage is unstable because sp^2 hybridization is unfavorable in silicon where sp^3 bonding is preferred. Doping silicon cages with a metal can stabilize these clusters with the possibility to fine-tune the chemical, electronic, optical, or magnetic properties. Extensive theoretical and experimental studies have been carried out on metal-doped Si clusters to understand their structures and properties [41]. A large number of polyhedral cages $M@Si_n$ of different sizes have been stabilized by endohedral doping of a metal atom. Among them, a prominent structure of high symmetry (tetrahedral) is $Ti@Si_{16}$ with a Frank-Kasper polyhedron structure (e.g., Figure 15.9) [42]. The largest Si cage that can be stabilized by metal encapsulation has been predicted (from projected augmented plane wave method within DFT/GGA) to be $Th@Si_{20}$

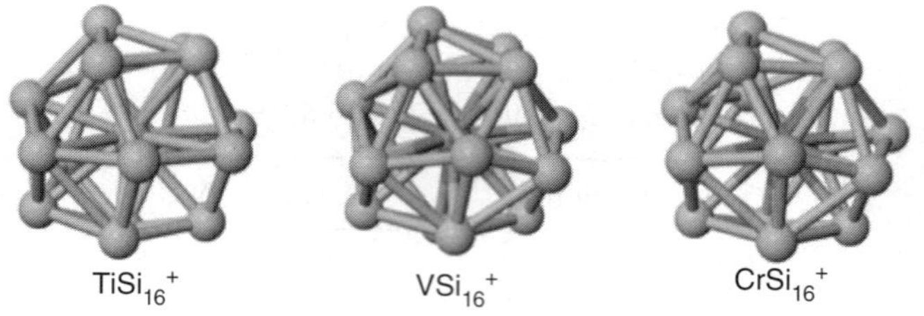

Figure 15.9 *Ground state Frank-Kasper geometries of Si_{16} endohedral clusters. Reprinted with permission from D. Palagin, M. Gramzow, and K. Reuter., J. Chem. Phys. 134, 244705 (2011). Copyright 2011, AIP Publishing LLC*

by Singh *et al.* [43] with a dodecahedral structure of Si_{20} and icosahedral cluster symmetry similar to that of C_{20}. Thorium was found to be the only element in the periodic table that stabilizes Si_{20} with icosahedral symmetry in the neutral state. With a similar theoretical approach, Singh *et al.* [43] explored doping with Ce, Pr, Tb in the lanthanide series and Pa, U, and Pu atoms in the actinide series (+4 oxidation states), but, in each case, the cage distorts or lowers its symmetry. Kumar *et al.* [44] reported stabilization of the Si_{20} cage by encapsulation of a variety of lanthanide and actinide atoms forming neutral $M@Si_{20}$ (M = Ce, Pa, Sm, Pu, and Tm) endohedral clusters with T_h symmetry.

Following Singh *et al.* [43], under I_h symmetry, the Si_{20} dodecahedral cage has 80 valence electrons (VE) and 30 Si-Si bonds. These sigma-bonds take 60 VE, leaving 20 lone π-electrons. The stability of the icosahedral $Th@Si_{20}$ structure can be understood starting from a spherical potential model according to which the highest occupied level of the Si cage has d-character with four holes. Adding four electrons from the metal, these occupy the following global orbitals of the $Th@Si_{20}$ cluster:

$$\begin{aligned}&\sigma\text{: } 1S(a_g)^2 + 1P(t_{1u})^6 + 1D(h_g)^{10} + 1F(g_u + t_{2u})^{14} + 1G(g_g + h_g)^{18} + 1H(h_u)^{10},\ \Sigma = 60\ (1)\\&\pi\text{: } 2S(a_g)^2 + 2P(t_{1u})^6 + 2D(h_g)^{10} + 2F(t_{2u}\text{ only})^6(\text{HOMO}),\ \Sigma = 24\ (2)\end{aligned} \tag{15.4}$$

Here 'σ' and 'π' refer to the surface of the cage. The lowest empty orbital (LUMO) then is the g_u component of the 2F shell. A reasonable HOMO-LUMO gap was found, but the gap from LUMO to LUMO+1 was even larger. Moreover, the LUMO had rather more Th 5f character than the occupied orbitals. This suggests to us that by giving the system eight further electrons, filling the previous $2F(g_u)$ LUMO, we would have a fair chance of obtaining a new 32-electron system. Both 6d and 5f orbitals may contribute to the bonding of the early actinides. For a single, neutral atom the d-f crossing has been placed at Pa [4] (see Figure 1 in [44]). For finding further 32-electron compounds, the choice of a later actinide than thorium would both give some of the required electrons, and more appropriate 5f orbitals, until the 5f electrons strongly localize at Am, or so. We first demonstrated that the idea in principle works for the model system $[U@Si_{20}]^{6-}$, and we then considered the actinide series $[An@Si_{20}]^{n-}$ (An=Np, Pu, Am, Cm). With reference to $[U@Si_{20}]^{6-}$, adding the 20 π-electrons of the Si_{20} cage, 6 electrons from the actinide and 6 electrons from the charge we could obtain a 32e configuration.

Starting from an initial icosahedral structure, the symmetry is reduced for An=Np, Pu, Am to T_d (Figure 15.10a) while it is preserved for U and Cm (see Figure 15.10b). All endohedral $[An@Si_{20}]^{n-}$ clusters are very stable species with a large HOMO-LUMO gap (Table 15.5). The energetics of the encapsulation of the actinide in the cage can be gauged by the evaluation of the binding energies starting with Si_{20}^{n-} and An fragments that mimic:

$$[Si_{20}]^{n-} + An \rightarrow [An@Si_{20}]^{n-} \tag{15.5}$$

Along the series, all of the reactions are exoenergetic, with large BE values (Table 15.5), reflecting energy stabilization during encapsulation of the metal. The variations in the BE, Pauli repulsion, electrostatic, and orbital terms are small along the Np, Pu, Am, Cm series, even if the values are slightly lower for Cm. Nevertheless, uranium is a special case. The $[U@Si_{20}]^{6-}$ is strongly stabilized by electrostatic and orbital interactions with respect to the other actinides. Once more, the origin of the stabilization is in the creation of covalent bonds between the cage and the central atom. In the I_h symmetry, the 2F energy levels of the cage split into a threefold degenerate HOMO (t_{2u}) that accommodates two electrons and an empty fourfold degenerate g_u level (LUMO). After U encapsulation, the 12 valence electrons (6 from the U atom and 6 from charging of the cage) fill the t_{2u} and g_u orbitals. The fourfold-degenerate LUMO of the Si_{20} cage mixes strongly with the 5f states of U due to their energetic proximity. The central atom orbitals participate in the a_g, t_{1u}, h_g, t_{2u},

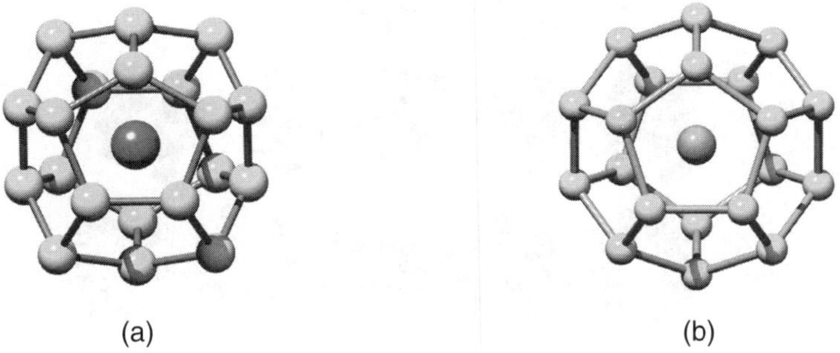

Figure 15.10 *$[Am@Si_{20}]^{3-}$ T_d (a) and $[U@Si_{20}]^{6-}$ I_h (b) optimized geometries. From DFT/B3LYP (see Dognon et al. [16] for details)*

Table 15.5 *HOMO-LUMO gap and binding energy (BE) analysis with respect to the Si_{20}^{n-} and An fragments (in eV)*

Symmetry	$[U@Si_{20}]^{6-}$ I_h	$[Np@Si_{20}]^{5-}$ T_h	$[Pu@Si_{20}]^{4-}$ T_h	$[Am@Si_{20}]^{3-}$ T_h	$[Cm@Si_{20}]^{2-}$ I_h
HOMO-LUMO	2.01	1.62	1.53	1.55	1.54
BE	−42.03	−15.46	−14.61	−14.08	−14.83
Pauli repulsion	36.44	34.69	34.96	35.04	29.34
Electrostatic	−31.35	−21.24	−21.58	−21.47	−18.54
Orbital	−47.12	−28.9	−27.99	−27.65	−25.63

From scalar relativistic DFT/OPBE (see Dognon *et al.* [16] for details).

and g_u valence MOs. The 7s, 7p, 6d, and 5f atomic orbitals are involved in hybridization with the Si_{20} cage (Figure 15.11). See color plate section. In Figure 15.11, the other MOs are pure Si_{20} cage orbitals that do not interact with the metal center. In the $[U@Si_{20}]^{6-}$ cluster, the HOMO (g_u) and HOMO-1 (t_{2u}) mix the 5f(U) and 3p(Si) orbitals. The h_g, t_{1u}, and a_g MOs correspond to 6d(U), 7p(U), and 7s(U), respectively, each one mixing with the cage. The density of states of the endohedral clusters are plotted in Figure 15.12. A band structure is observed for the cage corresponding to the pure 3s(Si) band, and a 32-electron band involving the hybridized cage and actinide orbitals. The width of the 32-electron band reflects the stability of the clusters. From U to Cm, the DOS is shifted deeper in energy due to the 5f(An)/3p(Si) mixing according to our previous work on the $Pu@Pb_{12}$ and related systems. The nature of the interaction is confirmed with an ELI-D topological analysis. For example, one can see U-Si basins in Figure 15.13 reflecting a significant electron sharing between actinide and silicon atoms. See color plate section.

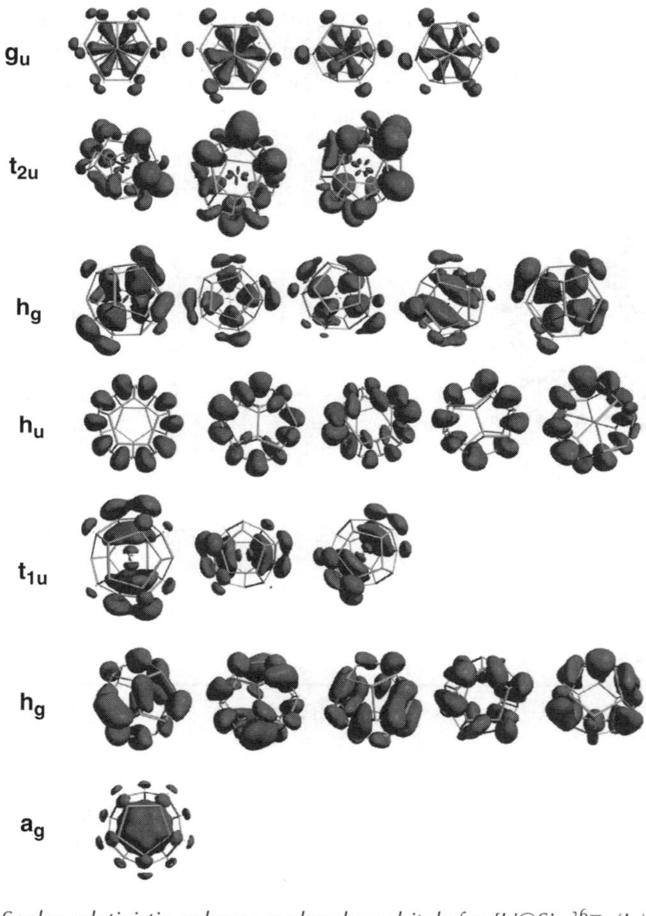

Figure 15.11 *Scalar relativistic valence molecular orbitals for $[U@Si_{20}]^{6-}$ (I_h). Adapted from Dognon et al. [16]. See color plate section*

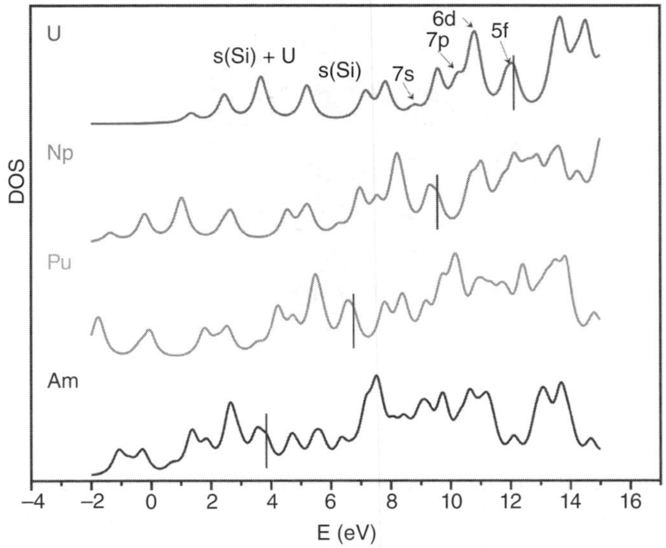

Figure 15.12 *Densities of states for* $[U@Si_{20}]^{6-}$*,* $[Np@Si_{20}]^{5-}$*,* $[Pu@Si_{20}]^{4-}$ *and* $[Am@Si_{20}]^{3-}$ *as a function of the orbital energy. The Fermi levels are shown by vertical black lines. From scalar relativistic DFT/OPBE (see Dognon et al. [16] for details)*

Figure 15.13 $[U@Si_{20}]^{6-}$ *(I_h) ELI-D isosurface (0.75). A cutting plane is used to show the U-Si basins (in magenta). U atom in blue. See color plate section*

This analysis is valid along the proposed actinide series. Accordingly, combining the 20 electrons from the Si_{20} cage, the 7s, 7p, 6d, and 5f electrons from the actinide plus the charge of the cluster, give the required electrons to form the 16 MOs of a centered 32-electron system (quoted by Pichon [46]).

In order to check the possibility to build neutral systems, we use lanthanide atoms to form neutral molecular compounds such as $La_2[U@Si_{20}]$ (Figure 15.14). A lanthanide atom is placed on a pentagonal face of the Si_{20} cage at two opposite sites. Each lanthanide transfers

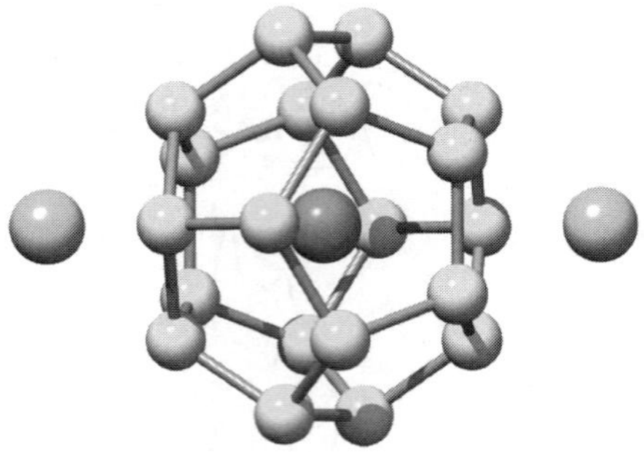

Figure 15.14 *$La_2[U@Si_{20}]$ (C_{5v}) optimized geometry. From DFT/B3LYP (see Dognon et al. [16] for details)*

three electrons to the silicon cage. The geometry is then stabilized in a C_{5v} symmetry. The addition of La atoms outside the cage does not lead to a significant modification in the neighboring Si-Si bond lengths for the corresponding pentagon. The La-Si bond length is 2.96 Å. The HOMO-LUMO gap is reduced but still large with the corresponding value of 1.42 eV. The binding energy with respect to the atomic dissociation limit is large with a value of −124 eV. By adding outside electron donor ligands, the stabilization of the endohedral cluster is higher than for the $[An@Si_{20}]^{n-}$ (An=Np, Pu, Am, Cm) series. These results point out the possibility to build new very stable silicon-based neutral compounds by an appropriate mixing of lanthanide, actinide, and silicon atoms in the $Ln_n[An@Si_{20}]$ composition. Concerning the electronic structure (see Dognon *et al.* [16] for details), with respect to the different symmetries, similar molecular orbital arrangements were found with a similar mixing scheme between the actinide and the cage orbitals leading to a fully 32-electron system.

15.3 Influence of Relativity

The previous results and discussions were based on the zero-order regular approximation (ZORA) to the Dirac equation in the scalar relativistic (SR) framework, i.e., in which spin-orbit (SO) interaction was neglected. Of course, due to the nature of the elements involved, one can expect that spin-orbit coupling affects some properties although relatively small effects may be anticipated in these closed shell molecules compared to open shell compounds, for example. As the SO effects scale as Z^2, the relativistic effects will be significant for Pb or Sn cages, but the effect due to the presence of the heavy central atom alone can significantly alter the properties of the endohedral cluster. Among the 32e compounds, the $An@C_{28}$ doped fullerenes case is interesting as no direct relativistic effects are expected on the electronic properties localized on the cage. We considered the ZORA implementation in the ADF program package [39, 47]. The spin-orbit operator, in regularized form, is already present in this zero-order Hamiltonian. When the spin-orbit coupling is included, double group symmetry adapted functions have to be used. In $Pu^{4+}@C_{28}$, the spin-orbit effect on the Pu-C distance never exceeds an expansion of ~3 pm, and a decrease on the

HOMO-LUMO gap stays less than 7% with respect to the SR values. In Figure 15.15, scalar relativistic and first-order spin-orbit split valence molecular orbitals levels are shown.

The spin-orbit effect is small, and the most significant splittings are obtained for the $9t_2$ and $8t_2$ MOs with values of 0.3 and 0.5 eV, respectively. They correspond to a mixing of the 6d and 7p orbitals with the 2p(C) of the cage. The binding energy is increased by 2.9 eV. The spin-orbit coupling effects are 'diluted' by the presence of the fullerene cage, and these effects do not invalidate the previous conclusions about the 32-electron principle.

As already stated, due to the nature of the involved chemical elements, a relativistic quantum chemistry approach is required. Indeed, these effects play the main role for orbital energy positioning leading to the observed orbital mixing in the clusters. The relativistic effects induce a contraction and an energy stabilization of the s and p shells and an expansion and destabilization of d and f shells. As a consequence, in a non-relativistic framework, e.g., the 5f(Pu), 6d(Pu) orbitals, are too low in energy, and the 7s(Pu) orbital is too high

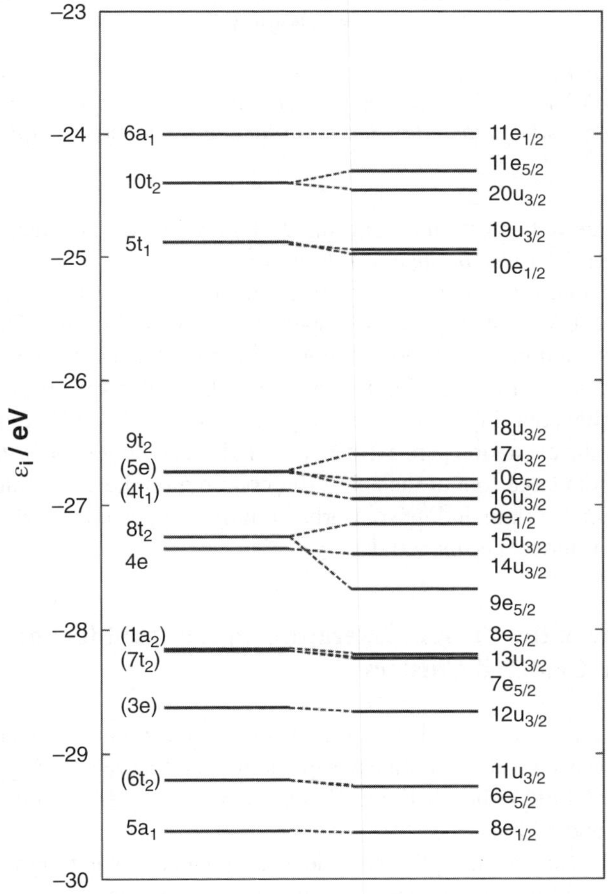

Figure 15.15 *Scalar relativistic (left-hand side) and first-order spin-orbit split (right-hand side) valence molecular orbital levels for Pu^{4+}@C_{28} (T_d)*

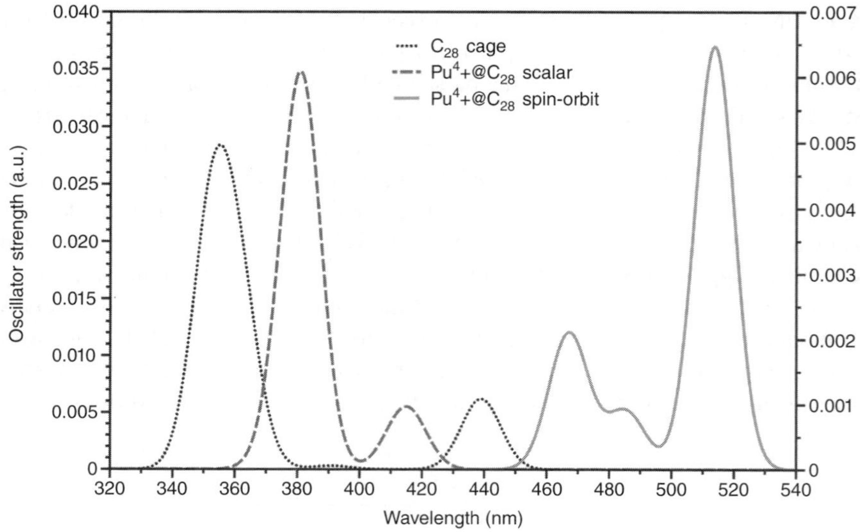

Figure 15.16 *TDDFT UV-visible spectra for $Pu^{4+}@C_{28}$ without (dashed line) and with spin-orbit coupling (solid line, right y axis) compared with the C_{28} cage spectrum (dotted line). The band shapes are approximated by Gaussian functions with a full width at half-maximum of 15 nm*

in energy. They are not able to mix with the 2p(C) orbitals of the cage. Accordingly, the 32-electron system *is broken in the non-relativistic case.*

Obviously, the spin-orbit coupling can affect the optical properties of these clusters and more strongly the UV/Visible absorption spectra as observed on the TDDFT absorption spectra. The introduction of spin-orbit coupling in the calculation gives an absorption in the blue (470–475 nm) and green (525 nm) region of the visible spectrum with an oscillator strength strongly lowered (Figure 15.16, right y axis). For the $Pu^{4+}@C_{28}$ clusters, the major MOs involved in the excitations are 5f(Pu)/2p(C) hybrids without and with spin-orbit coupling. So, due to the large role of the spin-orbit coupling on the electronic spectra, further extended theoretical work including spin-orbit coupling will be necessary for a quantitative comparison with future experimental data.

15.4 A Survey of the Current Literature on Lanthanide- and Actinide-Centered Clusters

After our paper, Dognon *et al.* [15] appeared, there have been later studies on $An@C_n$. The case $n = 28$ was considered both experimentally and theoretically by Dunk *et al.* [48] and Ryzhkov *et al.* [49]. Other theoretically considered n values include 20 [50], 24 [51], 26 [52], 36 [53], and 40 [54].

The results of Dunk *et al.* [48] provide experimental insight into how endohedral fullerenes are formed in carbon vapor and allow for a complete description of fullerene formation to be developed. Larger $U@C_{2n}$ endohedral metallofullerenes $U@C_{36}$ and $U@C_{44}$ are shown to form with a small probability based on $U@C_{28}$ as their precursor.

Uranium appears to be very efficient in catalyzing initial fullerene formation. In addition, if an encapsulating species is small enough to form M@C_{28}, it must also stabilize C_{28} by charge transfer sufficiently to be observed, or all of the species may grow to larger fullerenes after initial formation in carbon vapor. It begins to be possible to decipher why certain elements more efficiently form endohedral fullerenes than those that are not able to form, or at least not readily form [48].

Following our ideas and computational methodologies, Manna *et al.* [50–53] investigated the structures, bonding, stability, and spectroscopic properties of the endohedral metallofullerenes, M@C_{20}, Pu@C_{24}, M@C_{26} (M = Pr^-, Pa^-, Nd, U, Pm^+, Np^+, Sm^{2+}, Pu^{2+}, Eu^{3+}, Am^{3+}, Gd^{4+}, and Cm^{4+}), and U@C_{36}. Among them, only the closed-shell M@C_{26} (D_{3h}) fulfills the 32e principle with 26 electrons provided by the C_{26} cage and 6 electrons by the metal central atom/ion. Note that the 32e systems are not explicitly identified in the series. For example, some lanthanides or actinides cannot be 32e systems due to too deep 4f or 5f shells. The M@C_{20} (I_h) series are found to be highly stable clusters as compared to the empty C_{20} cage with a large HOMO-LUMO gap. The orbital mixing between the cage and the central atom/ion provides a 26e system. The M@C_{26} endohedral cluster is calculated as a metastable system with respect to U@C_{28} system. The U@C_{36} is stabilized in C_{6v} symmetry with a closed-shell electronic structure. The large HOMO-LUMO gap and binding energy confirm its stability and the experimental observation of Dunk *et al.* [48].

Ryzhkov *et al.* [49] carried out a study of the electronic structure of neutral endohedral An@C_{28} (An = Th-Md) confirming our results from fully relativistic discrete variational method. The 6d and 5f contributions to the bonding were found to be comparable for the earlier actinides. In addition, the actinides (Th-Md) series stabilize a C_{40} cage with a noticeable overlap between the 5f, 6d, 7s, 7p orbitals of the central actinide atom and the 2p(C) of the cage [54]. The most stable complex was found to be Pa@C_{40}.

More recently, a novel, potential 32e system $[Th(BH_4)_6]^{2-}$ was synthesized by Girolami *et al.* (as quoted by Ritter [55]). It has 16 hydrogens around the central thorium.

15.5 Concluding Remarks

We have presented a handful of examples on putative 32-electron systems, one or two of which already are experimentally made. These form an interesting possible extension of the previously discussed 8- or 18-electron magic numbers. In particular, we have demonstrated that there are contributions from the (s, p, d, f) shells of the central actinide atom. In other words, this 32e magic number is not a coincidence for electrons in a particular quasispherical potential well. The experiments by Cui *et al.* [56, 57] indicate that all transition metal or f-block atoms can be trapped inside stannaspherene (see for example Gd@Sn_{12}^-). Consequently, 32e compounds theoretically predicted as very stable species should be accessible experimentally. The future will show how many 32e systems will turn up.

References

[1] Langmuir I. Types of valence. Science. 1921;54:59–67. This paper mentions the 8, 18 and 32-electron closed shells and uses on pp. 65–66 $Fe(CO)_5$, $Ni(CO)_4$ and $Mo(CO)_6$ as examples on 18e.

[2] Pyykkö P. Understanding the eighteen-electron rule. J Organomet Chem. 2006;691:4336–4340.

[3] Hrobárik P, Straka M, Pyykkö P. Computational study of bonding trends in the metalloactinyl series EThM and MThM' (E = N^-, O, F^+; M, M' = Ir^-, Pt, Au^+). Chem Phys Lett. 2006;431: 6–12.

[4] Johansson MP, Sundholm D, Vaara J. Au_{32}: A 24-carat golden fullerene. Angew Chem Int Ed. 2004;43:2678–2681. Version in German: *Angew. Chem.* 2004, 116, 2732-2735.

[5] Karttunen AJ, Linnolahti M, Pakkanen TA, Pyykkö P. Icosahedral Au_{72}: a predicted chiral and spherically aromatic golden fullerene. Chem Comm. 2008;p. 465–467.

[6] de Heer WA. The physics of simple metal clusters: experimental aspects and simple models. Rev Mod Phys. 1993;65(3):611–676.

[7] Knight WD, Clemenger K, de Heer WA, Saunders WA, Chou MY, Cohen ML. Electronic Shell Structure and Abundances of Sodium Clusters. Phys Rev Lett. 1984;52(24):2141–2143.

[8] Walter M, Häkkinen H. A hollow tetrahedral cage of hexadecagold dianion provides a robust backbone for a tuneable sub-nanometer oxidation and reduction agent via endohedral doping. Phys Chem Chem Phys. 2006;8:5407–5411.

[9] Häkkinen H. Atomic and electronic structure of gold clusters: understanding flakes, cages and superatoms from simple concepts. Chem Soc Rev. 2008;37:1847–1859.

[10] Pyykkö P, Runeberg N. Icosahedral WAu_{12}: A predicted closed-shell species, stabilized by aurophilic attraction and relativity and in accord with the 18-electron rule. Angew Chem Int Ed. 2002;41:2174–2176.

[11] Li X, Kiran B, Li J, Zhai HJ, Wang LS. Experimental observation and confirmation of icosahedral $W@Au_{12}$ and $Mo@Au_{12}$ molecules. Angew Chem Int Ed. 2002;41:4786–4789. 'German version' Vol. 114, pp. 4980–4983.

[12] Flügge S. 65. Spherical oscillator, 66. Degeneracy of the spherical oscillator. In: Practical Quantum Mechanics. Berlin: Springer; 1974. p. 166–174.

[13] Dognon JP, Clavaguéra C, Pyykkö P. Towards a 32-electron principle: $Pu@Pb_{12}$ and related systems. Angew Chem Int Ed. 2007;46:1427–1430.

[14] Dognon JP, Clavaguéra C, Pyykkö P. Chemical properties of the predicted 32-electron systems $Pu@Sn_{12}$ and $Pu@Pb_{12}$. Compt Rend Chimie. 2010;13:884–888.

[15] Dognon JP, Clavaguéra C, Pyykkö P. A predicted organometallic series following a 32-electron principle: $An@C_{28}$ (An = Th, Pa^+, U^{2+}, Pu^{4+}). J Am Chem Soc. 2009;131:238–243.

[16] Dognon JP, Clavaguéra C, Pyykkö P. A new, centered 32-electron system: the predicted $[U@Si_{20}]^{6-}$-like isoelectronic series. Chem Sci. 2012;3:2843–2848.

[17] Cui LF, Huang X, Wang LM, Li J, Wang LS. Pb_{12}^{2-}: Plumbaspherene. J Phys Chem A. 2006;110(34):10169–10172.

[18] Cui LF, Huang X, Wang LM, Zubarev DY, Boldyrev AI, Li J, et al. Sn_{12}^{2-}: Stannaspherene. J Am Chem Soc. 2006;128(26):8390–8391.

[19] Esenturk EN, Fettinger J, Lam YF, Eichhorn B. $[Pt@Pb_{12}]^{2-}$. Angew Chem Int Ed. 2004;43(16):2132–2134.

[20] Esenturk EN, Fettinger J, Eichhorn B. The Pb_{12}^{2-} and Pb_{10}^{2-} Zintl Ions and the $M@Pb_{12}^{2-}$ and $M@Pb_{10}^{2-}$ cluster series where M = Ni, Pd, Pt. J Am Chem Soc. 2006;128(28): 9178–9186.

[21] Neukermans S, Janssens E, Chen ZF, Silverans RE, v R Schleyer P, Lievens P. Extremely Stable Metal-Encapsulated $AlPb_{10}^{+}$ and $AlPb_{12}^{+}$ Clusters: Mass-Spectrometric Discovery and Density Functional Theory Study. Phys Rev Lett. 2004;92(16):163401.

[22] Zhang X, Li G, Xing X, Zhao X, Tang Z, Gao Z. Formation of binary alloy cluster ions from group-14 elements and cobalt and comparison with solid-state alloys. Rapid Commun Mass Spectrom. 2001;15(24):2399–2403.

[23] Chen DL, Tian WQ, Lu WC, Sun CC. Special stability of cationic MPb_{12}^{+} clusters and superalkali character of neutral MPb_{12} clusters (M = B, Al, Ga, In, and Tl). J Chem Phys. 2006;124(15):154313. Erratum: J. Chem. Phys., 2006, 125, 049901.

[24] Chen DL, Tian WQ, Lu WC, Sun CC. Erratum: "Special stability of cationic MPb_{12}^{+} clusters and superalkali character of neutral MPb_{12} clusters (M = B,Al,Ga,In, and Tl)" [J. Chem. Phys.124, 154313 (2006)]. J Chem Phys. 2006;125:049901.

[25] Pyykkö P. Relativistic effects in structural chemistry. Chem Rev. 1988;88:563–594.

[26] Prodan ID, Scuseria GE, Martin RL. Covalency in the actinide dioxides: Systematic study of the electronic properties using screened hybrid density functional theory. Phys Rev B. 2007 07;76(3):033101.
[27] Dognon JP. Theoretical insights into the chemical bonding in actinide complexes. Coord Chem Rev. 2014;266–267(0):110–122.
[28] Kohout M. Bonding indicators from electron pair density functionals. Faraday Discuss. 2007;135(0):43–54.
[29] Kohout M. A measure of electron localizability. Int J Quantum Chem. 2004;97(1):651–658.
[30] Morss LR, Edelstein NM, Fuger J, Katz JJ. The Chemistry of the Actinide and Transactinide Elements. Third ed. Springer; 2008.
[31] Cox JD, Wagman DD, Medvedev VA. CODATA Key Values for Thermodynamics. New York: Hemisphere Publishing Corp.; 1989.
[32] Prinzbach H, Weiler A, Landenberger P, Wahl F, Worth J, Scott LT, et al. Gas-phase production and photoelectron spectroscopy of the smallest fullerene, C_{20}. Nature. 2000;407:60–63.
[33] Kroto HW. The stability of the fullerenes C_n, with n = 24, 28, 32, 36, 50, 60 and 70. Nature. 1987;329:529–531.
[34] Guo T, Diener MD, Chai Y, Alford MJ, Haufler RE, McClure SM, et al. Uranium Stabilization of C_{28}: A Tetravalent Fullerene. Science. 1992;257:1661–1664.
[35] Guo T, Smalley RE, Scuseria GE. Ab initio theoretical predictions of C_{28}, $C_{28}H_4$, $C_{28}F_4$, $(Ti@C_{28})H_4$, and $M@C_{28}$ (M=Mg, Al, Si, S, Ca, Sc, Ti, Ge, Zr, and Sn). J Chem Phys. 1993;99(1):352–359.
[36] Rösch N, Häberlen OD, Dunlap BI. Bonding in Endohedral Metal–Fullerene Complexes: f-Orbital Covalency in $Ce@C_{28}$. Angew Chem, Int Ed Engl. 1993;32(1):108–110.
[37] Jackson K, Kaxiras E, Pederson MR. Bonding of Endohedral Atoms in Small Carbon Fullerenes. J Phys Chem. 1994;98(32):7805–7810.
[38] Zhao K, Pitzer RM. Electronic Structure of C_{28}, $Pa@C_{28}$, and $U@C_{28}$. J Phys Chem. 1996;100(12):4798–4802.
[39] te Velde G, Bickelhaupt FM, Baerends EJ, Fonseca Guerra C, van Gisbergen SJA, Snijders JG, et al. Chemistry with ADF. J Comput Chem. 2001;22(9):931–967.
[40] Kemsley JN. Stable Caged Actinides Proposed. Chem & Eng News. 2008 December 15;86(50):29.
[41] Kumar V. In: Design of Nanomaterials by Computer Simulations. CRC Press; 2010. p. 1–23.
[42] Palagin, D, Gramzow M, Reuter K. On the stability of "non-magic" endohedrally doped Si clusters: A first-principles sampling study of MSi16+ (M = Ti,V,Cr)." The Journal of Chemical Physics, 2011;134, 244705.
[43] Singh AK, Kumar V, Kawazoe Y. Stabilizing the silicon fullerene Si_{20} by thorium encapsulation. Phys Rev B. 2005;71(11):115429.
[44] Kumar V, Singh AK, Kawazoe Y. Charged and magnetic fullerenes of silicon by metal encapsulation: Predictions from ab initio calculations. Phys Rev B. 2006;74(12):125411.
[45] Pyykkö P, Laakkonen LJ, Tatsumi K. REX calculations. 12. Iteration parameters for the 5f-element organometallics of thorium-neptunium. Geometries of thorium dioxide and uranyl ion revisited. Inorg Chem. 1989;28(10):1801–1805.
[46] Pichon A. Cluster chemistry: Filling in the blanks. Nat Chem. 2012 09;4(9):690–690.
[47] van Lenthe E, Snijders JG, Baerends EJ. The zero-order regular approximation for relativistic effects: The effect of spin-orbit coupling in closed shell molecules. J Chem Phys. 1996;105(15):6505–6516.
[48] Dunk PW, Kaiser NK, Mulet-Gas M, Rodríguez-Fortea A, Poblet JM, Shinohara H, et al. The Smallest Stable Fullerene, $M@C_{28}$ (M = Ti, Zr, U): Stabilization and Growth from Carbon Vapor. J Am Chem Soc. 2012;134(22):9380–9389.
[49] Ryzhkov MV, Ivanovskii AL, Delley B. Electronic structure of endohedral fullerenes $An@C_{28}$ (An=Th-Md). Comp Theor Chem. 2012;985:46–52.
[50] Manna D, Ghanty TK. Theoretical Prediction of Icosahedral $U@C_{20}$ and Analogous Systems with High HOMO-LUMO Gap. J Phys Chem C. 2012;116(31):16716–16725.
[51] Manna D, Sirohiwal A, Ghanty TK. $Pu@C_{24}$: A New Example Satisfying 32-Electron Principle. J. Phys. Chem. C. 2014; 118:7211–7221.

[52] Manna D, Ghanty TK. Prediction of a New Series of Thermodynamically Stable Actinide Encapsulated Fullerene Systems Fulfilling the 32-Electron Principle. J Phys Chem C. 2012;116(48):25630–25641.
[53] Manna D, Ghanty TK. Enhancement in the Stability of 36-Atom Fullerene through Encapsulation of a Uranium Atom. J Phys Chem C. 2013;117(34):17859–17869.
[54] Ryzhkov MV, Delley B. Electronic structure of predicted endohedral fullerenes An@C_{40} (An=Th-Md). Comp Theor Chem. 2013;1013:70–77.
[55] Ritter KS. Chemical connections. Chem & Eng News. 2013;91(36):28–33.
[56] Cui LF, Huang X, Wang LM, Li J, Wang LS. Endohedral Stannaspherenes M@Sn_{12}^{-}: A Rich Class of Stable Molecular Cage Clusters. Angew Chem Int Ed Engl. 2007;46(5):742–745.
[57] Cui LF, Wang LS. Stable icosahedral hollow cage clusters: stannaspherene and plumbaspherene. Int Rev Phys Chem. 2008;27(1):139–166.

16

Shell Structure, Relativistic and Electron Correlation Effects in f Elements and Their Importance for Cerium(III)-based Molecular Kondo Systems

Michael Dolg

Institute for Theoretical Chemistry, University of Cologne, Germany

16.1 Introduction

Lanthanides (Ln: ^{57}La - ^{71}Lu) and actinides (An: ^{89}Ac - ^{103}Lr), as shown in Table 16.1, comprise about one-fourth of the known elements collected in the periodic table, however, despite many interesting chemical and technical applications, they are so far certainly to a much smaller fraction represented in quantum chemical studies [1]. The underlying reasons are manifold and mostly quite obvious, i.e., the complexity of their electronic structure as well as large electron correlation and relativistic effects.

Lanthanide and actinide atoms possess a large number of electrons in shells with angular quantum numbers l up to 3, which already requires for the electronic ground states the usage of quite large and thus also costly basis sets for an accurate description at the independent-particle level. Especially the ground and low-lying excited configurations of the elements in the middle of the two series are typically characterized by many unpaired electrons in the valence f shell, resulting in a much higher number of possible electronic states than it is the case for open d shells of transition metals or open p shells of main group elements. Figure16.1 reveals that the possible number of LS states, J levels, and determinants

Computational Methods in Lanthanide and Actinide Chemistry, First Edition. Edited by Michael Dolg.

***Table 16.1** Electronic ground state J levels, assigned LS terms and (leading) configurations of the lanthanides and actinides*

La	Ce	Pr	Nd	Pm	Sm	Eu	
$^2D_{3/2}$	$^1G_4$	$^4I_{9/2}$	$^5I_4$	$^6H_{5/2}$	$^7F_0$	$^8S_{7/2}$	
d^1s^2	$f^1d^1s^2$	f^3s^2	f^4s^2	f^5s^2	f^6s^2	f^7s^2	
Gd	Tb	Dy	Ho	Er	Tm	Yb	Lu
$^9D_2$	$^6H_{15/2}$	$^5I_8$	$^4I_{15/2}$	$^3H_6$	$^2F_{7/2}$	$^1S_0$	$^2D_{3/2}$
$f^7d^1s^2$	f^9s^2	$f^{10}s^2$	$f^{11}s^2$	$f^{12}s^2$	$f^{13}s^2$	$f^{14}s^2$	$f^{14}d^1s^2$
Ac	Th	Pa	U	Np	Pu	Am	
$^2D_{3/2}$	$^3F_2$	$^4K_{11/2}$	$^5K_6$	$^6L_{11/2}$	$^7F_0$	$^8S_{7/2}$	
d^1s^2	d^2s^2	$f^2d^1s^2$	$f^3d^1s^2$	$f^4d^1s^2$	f^6s^2	f^7s^2	
Cm	Bk	Cf	Es	Fm	Md	No	Lr
$^9D_2$	$^6H_{15/2}$	$^5I_8$	$^4I_{15/2}$	$^3H_6$	$^2F_{7/2}$	$^1S_0$	$^2P_{1/2}$
$f^7d^1s^2$	f^9s^2	$f^{10}s^2$	$f^{11}s^2$	$f^{12}s^2$	$f^{13}s^2$	$f^{14}s^2$	$f^{14}p^1s^2$

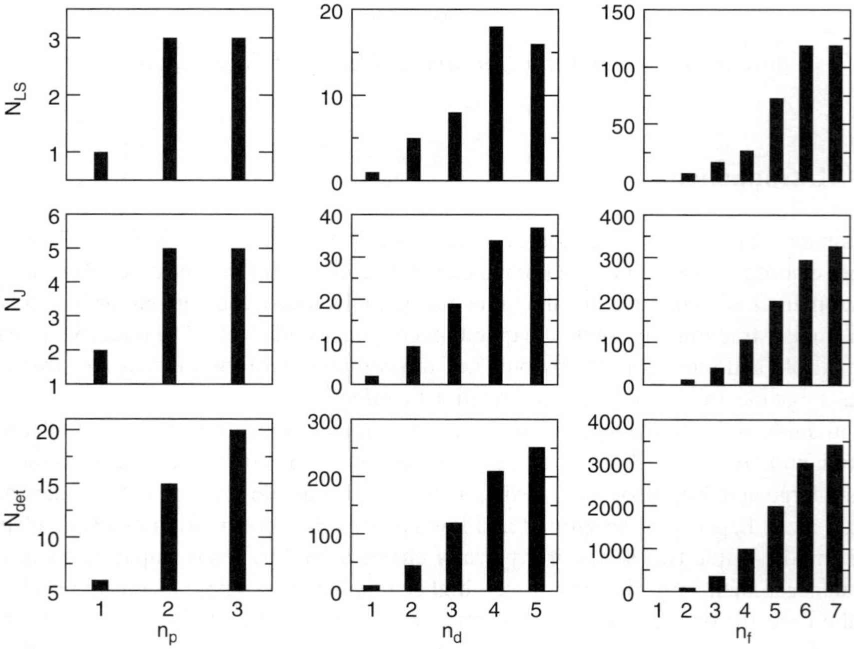

***Figure 16.1** Number of LS states, J levels, and determinants arising from atomic p^n, d^n, and f^n configurations*

for an open shell increases by roughly one order of magnitude when going from p to d, respectively, from d to f elements. Often the open shells in lanthanides and actinides are of different main quantum number, e.g., 4f (5f), 5d (6d), and possibly 6s (7s) for Ln (An). They are thus spatially somehow separated and as a consequence relatively weakly coupled, which together with the larger number of possible states leads to a high density of electronic states. Figure 16.2 compares the spread of the J level energies as well as the mean energy gap between adjacent J levels for the elements in the middle of the 3d (Cr), 4d (Mo), and 5d (W) transition metal series with those in the middle of the 4f (Gd) and 5f (Cm) series [2]. The chosen configurations, i.e., d^5s^1 for the d elements and $f^7d^1s^2$ for the f elements, correspond to the configurations assigned for the experimental ground state of Cr, Mo and Gd, Cm, respectively, but to the low-lying first excited configuration for W. It is obvious that, despite the larger energy range over which the levels are spread, the average energy difference between adjacent J levels is for Gd ($\approx$80 cm^{-1}) and Cm ($\approx$62 cm^{-1}) much lower than for the transition metals. Fortunately, near the ground J level the separations between adjacent levels are typically larger than the average values.

However, the situation might become even more complex, since often other low-lying excited configurations are present, e.g., for Gd one finds besides the $4f^75d^16s^2$ (term energies of arising J levels $\geq$ 0 cm^{-1}) atomic ground state configuration also $4f^75d^26s^1$ ($\geq$6378 cm^{-1}), $4f^86s^2$ ($\geq$10947 cm^{-1}), $4f^76s^26p^1$ ($\geq$13434 cm^{-1}), and $4f^75d^16s^16p^1$ ($\geq$14036 cm^{-1}) [3], which are all low enough in energy to play a role for the formation of molecules. If more than one element with an open f shell is present, very high spin multiplicities may arise, e.g., a $^{19}\Sigma_g^-$ ground state was postulated for Gd_2 [4–6] in agreement with an assignment to $^{19}\Sigma$ from later ESR experiments [7]. The high number of unpaired electrons results from nearly atomic-like $4f^7$ ^8S high-spin substates on each Ce center, leading to a $\sigma_g^1\,\pi_u^2\,\delta_g^2\,\phi_u^2\,\phi_g^2\,\delta_u^2\,\pi_g^2\,\sigma_u^1$ $^{15}\Sigma_u^+$ molecular substate contributed by the core-like 4f shells, and

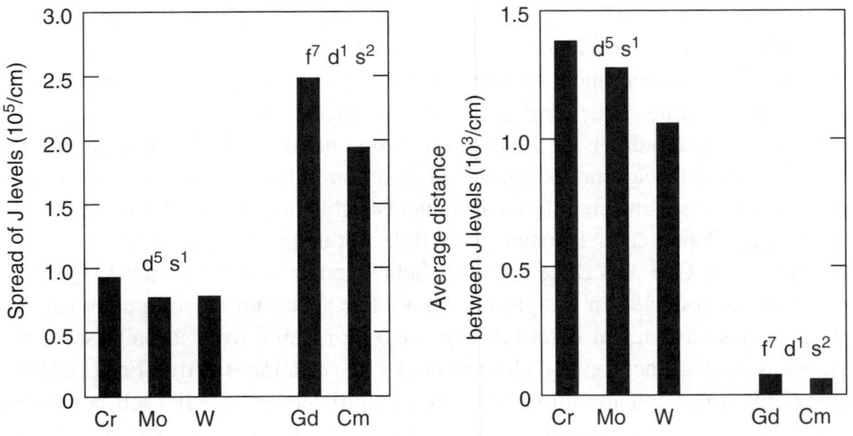

Figure 16.2 *Spread of energies of J levels and average energy gap between adjacent J levels for the Cr, Mo, W d^5s^1 (74 J levels) and Gd, Cm $f^7d^1s^2$ (3106 J levels) configurations from MCDHF/DC calculations [2]. Except for W the chosen configurations are the ones assigned for the experimental ground state*

a $\sigma_g^2\ \sigma_u^1\ \sigma_g^1\ \pi_u^2\ {}^5\Sigma_u^-$ molecular substate arising from the 5d and 6s valence orbitals, which are coupled to form the ${}^{19}\Sigma_g^-$ ground state with the possibly highest spin-multiplicity in a diatomic molecule.

Dynamic electron correlation contributions are non-negligible for accurate calculations, especially when cases are investigated where a change of the f occupation number on Ln (An) is important. Due to the slow convergence of the dynamic correlation energy with the highest angular quantum number present in the basis set, it is unavoidable to include basis functions of very high angular momentum, e.g., up to at least h or i functions, for accurate work. Usually the correlation treatment cannot be restricted to the valence orbitals, i.e., Ln (An) 4f (5f), 5d (6d), and 6s (7s), but has to include also semicore shells, i.e., at least Ln (An) 5s (6s), 5p (6p), which are energetically much lower in energy than the valence orbitals, but are spatially more diffuse than Ln (An) 4f (5f). In addition core-valence correlation contributions resulting from the Ln (An) 4d (5d) shells, which radially overlap significantly with the 4f (5f) shells, are often non-negligible. Finally, it is usually also necessary for quantitative work to account for near-degeneracy effects such as Ln $6s^2 \to 6p^2$, $5d^2$ (An $7s^2 \to 7p^2$, $6d^2$) [8].

The large nuclear charges of the f elements make a relativistic quantum chemical description, including spin-orbit contributions for quantitative work, indispensable for lanthanide and actinide systems. As will be described below it is not only the leading Dirac one-particle relativity that makes important contributions, but also corrections due to the two-particle Breit interaction as well as to a lesser extent even corrections arising from quantum electrodynamics and finite nucleus effects may play a role. As a result of the frequently encountered energetic proximity of levels with the same J value, arising in easy to handle cases from different LS terms due to the same electronic configuration, but in more difficult ones from LS terms due to different electronic configurations, a multi-configurational treatment, i.e., an intermediate coupling scheme, is often necessary even for a correct qualitative description. Thus, molecules containing lanthanides and/or actinides with open f shells pose very high challenges to quantum chemistry.

Relativistic quantum chemistry underwent a spectacular development since the early pioneering work of Pyykkö and Desclaux [9], Pitzer [10], and others dating back more than three decades. The development of new relativistic methods as well as the concomitant increased understanding of the origins and consequences of relativistic effects [11–13] was also especially beneficial for the field of quantum chemistry for lanthanide and actinide systems. The rapid development in this field of research becomes obvious from a number of review articles focusing mainly on f elements published during the last three decades by Pyykkö [1], Pepper and Bursten [14], Balasubramanian [15], Dolg and Stoll [16], Kaltsoyannis [17], Cao and Dolg [18–20], Schreckenbach and Shamov [21], as well as the contributions collected in the present book. The spectrum of computational methods described in these articles also reflects the ongoing change from initially semiempirical approaches as well as uncorrelated Hartree-Fock (HF) or Dirac-Hartree-Fock (DHF) calculations towards correlated ab initio studies using multi-reference (MR) self-consistent field (SCF) methods with a subsequent multi-reference configuration interaction (CI) treatment and related approaches, or at least first-principles calculations in the framework of modern density functional theory (DFT). Despite the tremendous progress made in relativistic quantum chemistry, when it comes to lanthanides or actinides only a small fraction of the problems one would like to address can currently be successfully dealt with, especially when quantitative answers are sought from ab initio approaches, i.e., without resorting to approximate computational schemes based on empirical parameters.

This chapter intends to remind the reader briefly of some important aspects of the shell structure, relativistic effects, and electron correlation effects for lanthanide and actinide atoms and molecules, mainly using the example of Ce and Th. It then turns to a discussion of the electronic structure of cerium-bis(η^8-cyclooctatetraene), cerocene, which fascinated this author for more than two decades, as well as a related cerium(III)-based molecular Kondo system, i.e., bis(η^8-pentalene)cerium. These systems feature many problems of an accurate relativistic correlated description of their electronic structure and moreover leave plenty of room for alternative interpretations of their electronic ground state as well as the involvement of 4f orbitals in chemical bonding.

16.2 Shell Structure, Relativistic and Electron Correlation Effects

In order to describe accurately the electronic structure of many-electron systems, highly correlated approximate solutions of the Schrödinger equation built from the best possible relativistic Hamiltonian are needed. Relativistic and electron correlation effects, respectively, arise from the wish to know how much of the problem could still be understood with a nonrelativistic and uncorrelated approach. Relativity changes the nonrelativistic electron density, e.g., by contractions and expansions of individual shells due to dominating direct and in direct relativistic effects, respectively. Changes in the electron density of course lead to changes in the total energy, including changes in the electron correlation energy, because it is obvious from the local density approximation of density functional theory applied to the electron correlation energy. Thus, it is expected and also well known that relativistic and electron correlation contributions to results of quantum chemical calculations are not simply additive for heavier elements, e.g., as demonstrated by Visscher and Dyall for halogen dimers [22]. They are also somewhat dependent on the details of their evaluation [23]. An example for the non-additivity of relativistic and correlation contributions is given in Figure 16.3, i.e., the fourth ionization potential IP_4 of Ce and Th. The non-additive contributions of relativitiy and correlation are of the order 0.10–0.15 eV [24].

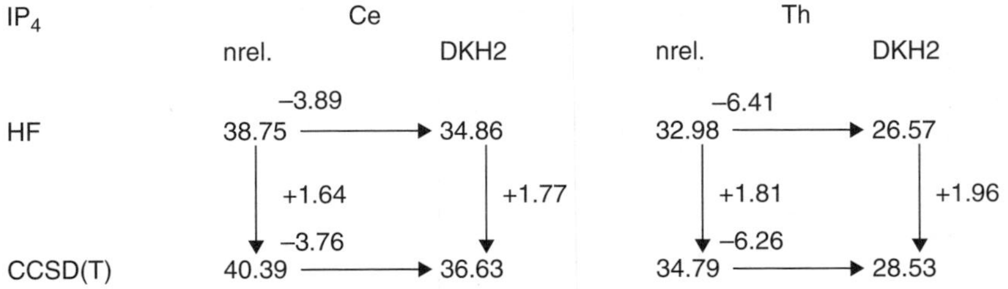

Figure 16.3 *Fourth ionization potentials IP_4 of Ce and Th (eV) as well as corresponding relativistic and electron correlation contributions (eV) from basis set extrapolated all-electron HF and CCSD(T) calculations without and with the DKH2 Hamiltonian [24]. Arrows from left to right denote relativistic, those from top to bottom correlation contributions. Spin-orbit coupling contributions (as shown in Figure 16.4) were not considered*

Shell structure effects are more dubious and less well defined than relativistic and electron correlation effects. They are related to the specific filling of the one-particle levels with electrons, and thus can only be discussed at the nonrelativistic and/or relativistic uncorrelated level. Hypothetical questions such as how the chemistry of the heavier elements would look like, if, e.g., a filled 3d shell or filled 4f shell would not be present in the core and for compensation the nuclear charge would be reduced by 10 or 14 units, respectively, are related to shell structure effects. It is established that such shell structure effects and relativistic effects are also not really independent from each other [25–27]. Clearly, since electron correlation goes beyond the independent-particle model, it makes no sense to ask for couplings between shell structure and electron correlation effects.

In order to keep the picture simple shell structure, relativistic and electron correlation contributions will be discussed separately in the following sections. The emphasis hereby is put on Ce and its heavier homologue Th, which will be considered as representatives of the lanthanide and actinide series, respectively.

16.2.1 Shell Structure

Shell structure effects arise from the filling of the atomic shells with electrons according to the Aufbau principle and Hund's rules. Since the one-particle energies of the shells as well as their spatial extent might be modified in heavy atoms due to relativistic effects, as shown in Figure 16.4 for the one-particle energies, it is obvious that shell structure and relativistic effects are coupled. Interesting features in the shell structure of lanthanides and actinides arise from the fact that the ordering of the shells based on energetic and spatial criteria is partially different. For the Ce $4f^1$ $5d^1$ $6s^2$ configuration at the AE DHF/DC level, the ordering of the outermost shells based on their $< r >$ expectation values is $4f < 5s < 5p < 5d < 6s$, whereas for the one-particle energies ϵ the order $5s < 5p < 4f < 5d < 6s$ is obtained [2]. In case of the Th $5f^1$ $6d^1$ $7s^2$ configuration, the ordering according to the

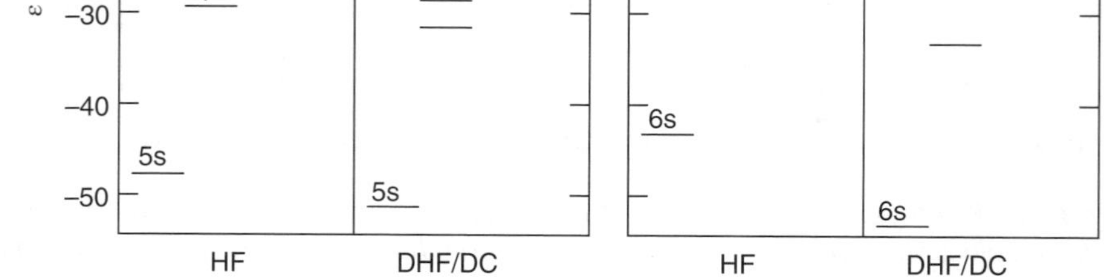

Figure 16.4 Orbital energies of Ce and Th from nonrelativistic (HF) and relativistic (DHF/DC) calculations for the f^1 d^1 s^2 configuration [2]

one-particle energies ϵ is analogous, i.e., 6s < 6p < 5f < 6d < 7s, however, partly due to the strong indirect relativistic effects 5f is less compact, so that an order 6s < 5f < 6p < 6d < 7s is obtained for the $< r >$ expectation values [2].

Due to their lower main quantum number compared to the other valence shells, i.e., 4f vs. 5d and 6s for Ln and 5f vs. 6d and 7s for An, the 4f shells of lanthanides and the 5f shells of (heavier) actinides are spatially quite compact and exhibit a core-like character. This is apparent from the Ce and Th radial functions plotted in Figure 16.5. Note that due to the incomplete shielding of the nuclear charge, the Ln 4f and An 5f shells become more core-like for the heavier members of the series, i.e., the one-particle energies decrease and the radial function become more compact with increasing atomic number, thus becoming also less accessible for chemical bonding.

The especially core-like nature of the lanthanide 4f shell is also related to the missing orthogonality constraints to inner shells bearing the same angular momentum quantum number. A similar situation was observed for the 3d and 4d shells of first- and second-row transition metals by Pyykkö, who coined the expression *missing primogenic repulsion* for the absent orthogonality constraints [28]. Figure 16.5 illustrates the situation for the radial densities of the 4f, 5d, and 6s shells of Ce as well as those of the 5f, 6d, and 7s shells of Th. As a reference the radial densities of the 2s and 2p orbitals of a C atom located at the experimental metal-carbon distance in the sandwich complexes cerocene and thorocene are superimposed. It can be expected that the Th-ring interaction has significantly larger Th 5f contributions (besides 6d contributions) than the Ce-ring interaction has Ce 4f contributions (besides 5d contributions).

Most lanthanides and actinides possess in their atomic ground state a partially occupied f shell, in some cases also an additional singly occupied d shell, as shown in Table 16.1.

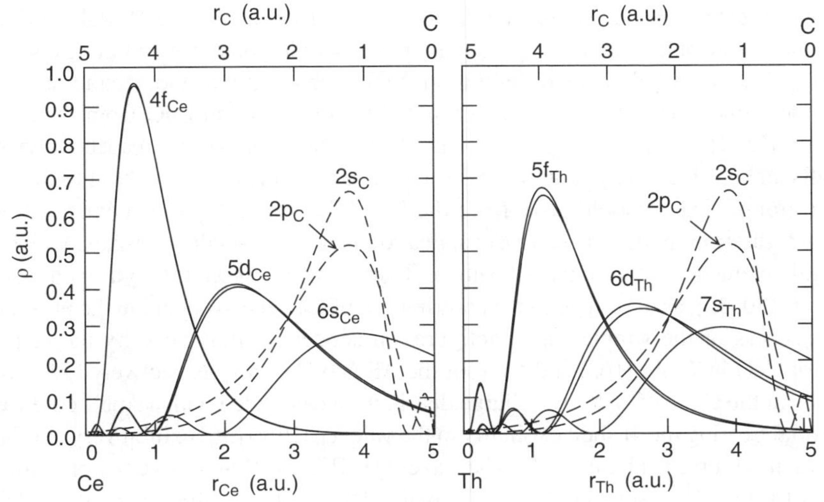

Figure 16.5 *Radial densities of Ce and Th (solid lines, lower axis) as well as C (dashed lines, upper axis) from relativistic DHF/DC calculations [2]. The distances between the origins of the r-axes for Ce (left, bottom) and C (top) as well as Th (right, bottom) and the C (top) correspond to the experimental Ce-C and Th-C distances in cerocene and thorocene, respectively*

This usually gives rise to a large number of energetically low-lying electronic states, as demonstrated already for single open shells in Figure 16.1. The energy gaps between adjacent states in f elements are for single configurations already significantly smaller than for d transition metals or even main group elements, as shown in Figure 16.2. In some cases more than a single configuration gives rise to low-lying electronic states. In a molecular environment the atomic energy levels are further split due to the lower symmetry. Since the f shells of the lanthanides and heavier actinides are well shielded by more diffuse filled semi-core shells, e.g., Ln 5s, 5p, and An 6s, 6p, from the spatial valence region, the resulting splittings are often relatively small. As a result one often encounters even for simple diatomic molecules containing a lanthanide or an actinide element a very high density of low-lying electronic states and very complex electronic spectra [29].

A real challenge for *ab initio* quantum chemistry is the description of such systems with open f shells. In relatively easy cases where only a single f^n subshell is present, e.g., for the trifold positive atomic ions or their hydration complexes, the application of LSJ-coupling is usually sufficient. However, there are numerous examples where only an intermediate coupling scheme adequately describes the electronic structure. Note for example, that the Ce $4f^1$ $5d^1$ $6s^2$ J=4 atomic ground state is decomposed to 55% of ^1G and 29% of ^3H. The lowest excited J levels J=2,3 and 4 arise to 66%, 85%, and 34%, respectively, from a leading ^3F term, and further do not only have contributions of other LS terms stemming from the $4f^1$ $5d^1$ $6s^2$ configuration, but also from LS terms resulting from $4f^1$ $5d^2$ $6s^1$ [3].

The core-like character of the Ln 4f shells was exploited by Field in a simple electronic structure model for lanthanide diatomics, which turned out to be extremely useful to rationalize the complex electronic spectra of these systems [29]. According to Field the Ln $4f^n$ subconfiguration retains an atomic character also in molecules and does not participate directly in bonding, whereas the other valence orbitals such as 5d, 6s, and 6p are involved in (polar) covalent bonding in a MO-LCAO-type fashion and are coupled to form a valence substate. All electronic states that can be constructed from a specific $4f^n$ subconfiguration and a valence substate exhibit very similar spectroscopic constants and are thus assigned to belong to a so-called superconfiguration. Field proposed that the ground states of the lanthanide monoxides arise for LaO to SmO, TbO to TmO and LuO from the superconfiguration [Ln $4f^n$] $\sigma^2_{O2p}\pi^4_{O2p}\sigma^1_{Ln6s,6p}$. For EuO and YbO a half-filled respectively filled 4f subshell can be achieved by promoting one electron from the $\sigma_{Ln6s,6p}$ orbital to the 4f shell, and the ground states should arise from the [Ln $4f^{n+1}$]$\sigma^2_{O2p}\pi^4_{O2p}$ superconfiguration. The superconfiguration model was later translated to 4f-in-core pseudopotentials, which model lanthanide elements with a fixed $4f^n$ subconfiguration corresponding to a fixed valency in molecules [30, 31], as described in the contribution of Cao and Weigand in the present book.

The success of the simple superconfiguration concept is illustrated by the results collected for CeO in Tables 16.2 and 16.3 for the ΛS and Ω states, respectively [32–35]. All 8 ΛS states of the Ce^{2+} $4f^1$ $6s^1$ O^{2-} charge distribution calculated with a small-core Ce pseudopotential treating the 4f shell explicitly in the valence (PPv) [36] exhibit very similar bond distances and vibrational constants. Their average (PPv avg.) agrees very well with results obtained for the $4f^1\sigma^1$ superconfiguration with a large-core Ce pseudopotential attributing the $4f^1$ subshell into the core (PPc a). Both results are in reasonable agreement with the reference data obtained by averaging over the experimentally observed Ω levels (exp. avg.). In order to obtain these with the f-in-core pseudopotential approach, a ligand field treatment

Table 16.2 *Spectroscopic constants for the ΛS states of the $4f^1\sigma^2\pi^4\sigma^1$ superconfiguration of CeO and corresponding averaged values for the superconfiguration*

Method	$^{2S+1}\Lambda$	R_e	D_e	T_e	ω_e	Method	SC	R_e	D_e	ω_e
PPv	$^3\Phi$	1.827	7.28	0.00	838					
exp.	$^3\Phi$	1.820	8.22	0.00	824					
PPv	$^1\Phi$	1.827		0.04	837					
PPv	$^3\Delta$	1.819		0.10	840					
PPv	$^1\Delta$	1.818		0.15	836					
PPv	$^3\Pi$	1.819		0.17	833					
PPv	$^1\Pi$	1.826		0.21	829					
PPv	$^3\Sigma$	1.816		0.18	835					
PPv	$^1\Sigma$	1.817		0.31	834	PPv	avg.	1.821	7.16	836
						PPc a	$4f^1\sigma^1$	1.819	7.25	834
						PPc b	$4f^1\sigma^1$	1.926	6.34	822
						exp.	avg.	1.811	7.94	823

Results from scalar-relativistic 4f-in-valence (PPv) and $4f^1$-in-core (PPc) pseudopotential configuration interaction (CISD) calculations [32, 33] in comparison to experimental data [34, 35]. Bond lengths (R_e) in Å, binding energies (D_e) and term energies (T_e) in eV, vibrational constants (ω_e) in cm^{-1}. PPv: Ce PP with 30 valence electrons, (12s11p9d8f)/[9s8p6d5f]; PPc: Ce PP with 11 valence electrons, a: 4f occupation number n(4f) $\geq$ 1, b: 4f occupation number n(4f) = 1, (7s6p5d4f)/[5s4p3d4f]; O (9s6p1d)/[4s3p1d]. avg. denotes an average over results for individual ΛS states, whereas $4f^1\sigma^1$ denotes a calculation for the superconfiguration.

has been added. Recently a similar 5f-in-core pseudopotential approach was also set up for the actinides, which is also outlined in the Chapter 6 by Cao and Weigand.

Calculations for CeO with f-in-core pseudopotentials allowing (PPc a) and not allowing (PPc b) charge transfer from the ligand to the Ce 4f shell allow to estimate roughly the bonding contributions of the the 4f shell to be about 1 eV of the total binding energy of 8.18 eV and about −0.1 Å of the bond distance of 1.80 Å, as shown in Table 16.2. In the heavier homologue ThO, which has formally a [Th $5f^0$] $\sigma^2_{O2p}\pi^4_{O2p}\sigma^2_{Th7s,7p}$ ground state configuration, the not explicitly occupied 5f shell still contributes with about 1.7 eV to the total binding energy of 8.87±0.15 eV and reduces the bond distance by about 0.15 Å [37].

16.2.2 Relativistic Effects

Relativistic effects can be defined as the difference between results of relativistic and nonrelativistic quantum chemical calculations, i.e., calculations with the speed of light at its correct value ($c \approx 137.036$ a.u.) and the nonrelativistic limit ($c \to \infty$), respectively. Since there are various choices of Hamiltonians, which may be applied in various quantum chemical approaches for the approximate solution of the Schrödinger equation, the magnitude of relativistic contributions somehow depends on the way they are evaluated [23].

There is a wealth of information in the form of review articles [11–13, 38], edited books [39–43], monographs [44–46] as well as a database in printed [47–49] and electronic

***Table 16.3** Spectroscopic constants for the Ω states of the $4f^1\sigma^2\pi^4\sigma^1$ superconfiguration of CeO*

Ω	T_0	T_e		$\Delta G_{1/2}$	ω_e		B_0	B_e	
	exp.	PPv	PPc	exp.	PPv	PPc	exp.	PPv	PPc
2	0	0	0	824	837	828	3545	3519	3496
3	82	119	101	824	836	828	3570	3520	3496
1	813	913	923		838	823	3573	3546	3543
							3568		
2	912	1045	968	823	838	832	3622	3548	3539
0^-	1679	1396	1589		831	819	3579	3549	3614
1	1875	1476	1679		828	838	3183	3542	3620
							3614		
0^+	1925	1715	1769		835	820	4003	3546	3607
4	2042	2139	2302	822	838	828	3533	3518	3492
3	2143	2286	2487	824	837	829	3566	3519	3495
3	2618	2872	3086	828	841	833	3577	3550	3532
2	2771	3039	3165	823	839	833	3602	3551	3532
2	3462	3386	3771	821	836	834	3538	3554	3562
1	3635	3391	3766	820	828	834	3656	3552	3596
							3421		
0^-	3819	3476	4120		836	843	3815	3559	3597
1	4134	3605	4249		843	847	3524	3541	3566
							3621		
0^+	4458	4234	4314		837	843	3677	3558	3595

Results from 4f-in-valence (PPv) and $4f^1$-in-core (PPc) pseudopotential calculations [32, 33] in comparison to experimental data [34, 35]. Term energies ($T_{0,e}$) and vibrational constants ($\Delta G_{1/2}$, ω_e) in cm^{-1}, rotational constants ($B_{0,e}$) in 10^{-4} cm^{-1}.

[50, 51] form available for relativistic electronic structure theory and only a few aspects will be summarized here.

For hydrogen and hydrogen-like ions, the nonrelativistic energy increases with the nuclear charge Z to the second power (Z^2), whereas the leading relativistic energy contributions increase with the nuclear charge Z to the fourth power (Z^4/c^2, with c denoting the velocity of light in vacuum). It is thus obvious that relativity cannot be neglected in calculations for heavy atoms such as lanthanides and actinides. For accurate calculations of many-electron systems it is necessary to go beyond the Dirac one-particle relativity terms in the Hamiltonian and to include also relativistic two-particle terms such as the Breit interaction. Aside from this it seems that also a finite nuclear model has noticeable effects, at least on the atomic and possibly also on the molecular spectra.

For the shells of many-electron atoms one can observe the following major relativistic effects. Replacing the nonrelativistic Hamiltonian by a relativistic one leads for all shells to an energy lowering, i.e., a stabilization, and a concomitant contraction. The magnitude of the direct effects decreases with increasing angular quantum number l, i.e., in the order s > p > d > f. It also decreases with increasing main quantum number n, i.e., the outer valence shells are less affected than the inner core shells. In addition to the scalar (spin-orbit averaged) relativistic effects, one also observes a spin-orbit splitting of shells with angular quantum number $l > 0$ into two subshells with total angular quantum numbers $j = l - 1/2$ and $j = l + 1/2$, at lower and higher energies, respectively.

Since the inner, mainly s and p shells of a heavy atom are contracted by direct relativistic effects, they shield the nuclear charge more efficiently for the outer shells. Thus the outer, mainly d and f shells experience an indirect destabilization due to a reduced effective nuclear charge and undergo a concomitant expansion. The magnitude of the indirect effects usually increases with increasing angular quantum number, i.e., in the order f > d > p > s. It also increases with increasing main quantum number. Destabilized and expanded shells on the other hand less efficiently shield the nuclear charge and thus lead to a higher effective nuclear charge for outer shells, resulting in an indirect stabilization and contraction.

Both direct and indirect effects act on all shells of an atom, however to a different extent. For the valence shells of a many-electron atom one typically observes a contraction and stabilization of the s and p shells and an expansion and destabilization of the d and f shells. The consequences for the ground state configurations of the lanthanide and actinide atoms are depicted in Figure 16.6. Note for example, that at the nonrelativistic Hartree-Fock (HF) level of theory the ground state configurations of Ce and Th are $4f^2\ 6s^2$ and $5f^2\ 7s^2$,

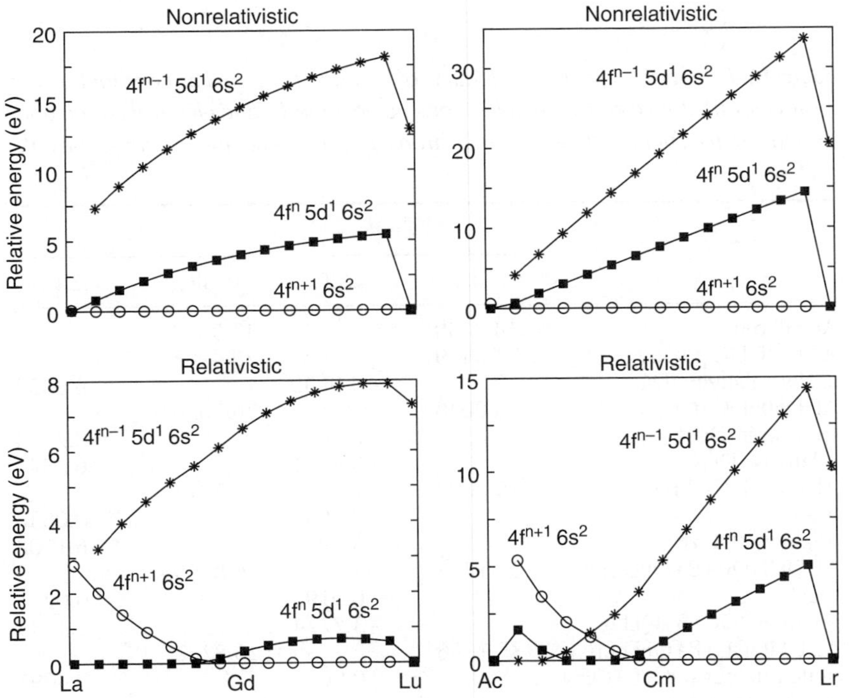

Figure 16.6 *Relative energies of low-lying configurations of lanthanides and actinides from nonrelativistic Hartree-Fock and relativistic Dirac-Hartree-Fock all-electron calculations [2]. The energy of the lowest configuration is set to zero. Note that the data depicted here corresponds to average energies of all J levels belonging to the configurations and that electron correlation effects are not included*

respectively, whereas at the relativistic Dirac-Hartree-Fock level they are $4f^1\ 5d^1\ 6s^2$ and $6d^2\ 7s^2$, respectively, in agreement with the interpretation of the experimental spectra [3, 52].

Taking IP_4 of Ce ($Ce^{3+}\ 4f^1\ {}^2F_{5/2} \rightarrow Ce^{4+}\ {}^1S_0$) and Th ($Th^{3+}\ 5f^1\ {}^2F_{5/2} \rightarrow Th^{4+}\ {}^1S_0$) as an example, various contributions arising from relativity, as well as from electron correlation and the nuclear model are summarized in Table 16.4. Accounting accurately for these contributions is important for calculations of systems where Ce might be tri- and/or tetravalent, e.g., the cerium complexes discussed at the end of this contribution. It is seen that the major relativistic contributions to IP_4 arise from substituting the non-relativistic kinetic energy operator by the relativistic one, i.e., at the Dirac-Coulomb (DC) Hamiltonian level. The reductions at the scalar-relativistic level by 4.06 and 6.53 eV correspond to about 10 and 20% of the nonrelativistic IP_4 values for Ce and Th, respectively, and are due to the dominating strong indirect relativistic destabilization of the f shells of Ce^{3+} and Th^{3+}. An order of magnitude smaller and non-negligible contributions arise at the DC Hamiltonian level from spin-orbit coupling (0.16 and 0.28 eV), as well as at the Dirac-Coulomb-Breit (DCB) Hamiltonian level from supplementing the non-relativistic Coulomb interaction between the electrons with an approximate relativistic term, i.e., the Breit interaction (0.10 and 0.08 eV). An other order of magnitude smaller, but still noticeable are corrections from quantum electrodynamics (QED) (0.02 and 0.03 eV), as well as due to a finite nuclear model (<0.01 and 0.01 eV). Much larger and more difficult to account for accurately are differential

Table 16.4 *Fourth ionization potentials of Ce and Th, as well as contributions of relativity, quantum electrodynamics, finite nuclear model and electron correlation, as derived from finite-difference all-electron calculations [2] and experimental data [3, 52]*

	Ce		Th	
	IP_4	ΔIP_4	IP_4	ΔIP_4
AE HF p.n.	38.9059		32.9620	
AE DHF/DC p.n. avg.	34.8419		26.4282	
scalar relativity (DC)		−4.0640		−6.5338
AE DHF/DC p.n.	35.0048		26.7076	
spin-orbit (DC)		0.1629		0.2794
relativity (DC)		−3.9011		−6.2544
AE DHF/DC+B p.n.	35.1087		26.7912	
Breit		0.1039		0.0836
relativity (DC+B)		−3.7972		−6.1708
AE DHF/DC+B+QED p.n.	35.1285		26.8185	
QED		0.0198		0.0273
relativity (DC+B+QED)		−3.7774		−6.1435
AE DHF/DC+B+QED f.n.	35.1296		26.8282	
finite nucleus (DC+B+QED)		0.0011		0.0097
exp.	36.906(9)		28.648(25)	
correlation		1.776		1.820

AE: all-electron; HF: Hartree-Fock; DHF: Dirac-Hartree-Fock; DC: Dirac-Coulomb-Hamiltonian; +B: Breit interaction in quasi-degenerate perturbation theory; +QED: quantum electrodynamic corrections (vacuum polarisation, self-energy); p.n.: point nucleus; f.n.: finite (Fermi) nucleus; exp.: experimental data.

electron correlation effects (1.78 and 1.82 eV), which were here evaluated as the difference between results of the best uncorrelated calculations (AE MCDHF with DC+B+QED f.n. Hamiltonian [2]) and experimental data [3, 52] and are further discussed in the next section.

16.2.3 Electron Correlation Effects

Electron correlation effects can be defined as the difference between results obtained from the exact solution of a Schrödinger equation with a specific Hamiltonian, and the results obtained at the uncorrelated level, e.g., at the Hartree-Fock or Dirac-Hartree-Fock level. Since for all but the simplest problems the exact solution of the Schrödinger equation is not accessible and usually approximate correlated wavefunctions are used instead. Sometimes experimental values are used rather than the results for the exact solution, which is reasonable as long as the Hamiltonian used for the uncorrelated solution includes all important terms, e.g., with regard to relativistic contributions, influence of the environment of the studied system, etc. As for relativistic effects, the magnitude of electron correlation effects depends to some extent on the details of their evaluation [23].

Static electron correlation plays an important role, e.g., in atomic configurations, where a Ln (An) $6s^2$ ($7s^2$) occupation is present. It is usually necessary for quantitative work to account for the near-degeneracy effects such as Ln $6s^2 \rightarrow 6p^2, 5d^2$ (An $7s^2 \rightarrow 7p^2, 6d^2$), e.g., for the evaluation of the first ionization potentials of Ac to U including An $7s^2 \rightarrow 7p^2$ and $7s^2 \rightarrow 6d^2$ excitations in the complete active space self-consistent field (CASSCF) reference wavefunction of scalar-relativistic all-electron averaged coupled-pair functional (ACPF) calculations increases the values by about 0.10–0.15 and 0.04–0.06 eV, respectively [8]. Dynamic electron correlation effects become especially big, e.g., for energy differences between atomic configurations with different Ln 4f (An 5f) occupation. Often these differential electron correlation effects act in opposite direction to the differential relativistic effects, e.g., whereas electron correlation favors configurations with high f occupation, relativity favors those with low f occupation. Figure 16.7 compares, as a simple example where only dynamical correlation matters, differential relativistic and electron correlation effects in the fourth ionization potentials of Ce (Ce^{3+} $4f^1$ $^2F \rightarrow Ce^{4+}$ ^1S) and Th (Th^{3+} $5f^1$ $^2F \rightarrow Th^{4+}$ ^1S). The results were obtained from scalar-relativistic calculations without and with the second-order Douglas-Kroll-Hess (DKH2) Hamiltonian at the Hartree-Fock (HF) and (basis set extrapolated) coupled cluster level with single and double substitutions and perturbative triple substitutions [CCSD(T)]. Whereas the strong indirect relativistic effects lower the ionization potentials by disfavoring the occupation of the f shells, the differential correlation effects tend to increase them by favoring the occupation of the f shells. The relativistic contributions amount to about -10 and -20% of the experimental ionization potential of Ce and Th, respectively, and the correlation contributions are about +5 to +6% in both cases. As mentioned before the Ce 4f and Th 5f shells have a quite large radial overlap with the core orbitals of the same main quantum number, i.e., 4s, 4p, and 4d for Ce and 5s, 5p and 5d for Th. Correlation of these orbitals in addition to the outer semi-core (Ce 5s, 5p; Th 5s, 5p) and valence orbitals (Ce 4f; Th 5f) amounts to 38.0 and 24.7% of the total differential correlation contributions to IP_4 of Ce and Th, respectively, at the basis set extrapolated DKH2 level, as shown in Figure 16.3. The larger percentage in case of Ce can be explained by the more compact character of the Ce 4f shell compared to the Th 5f shell.

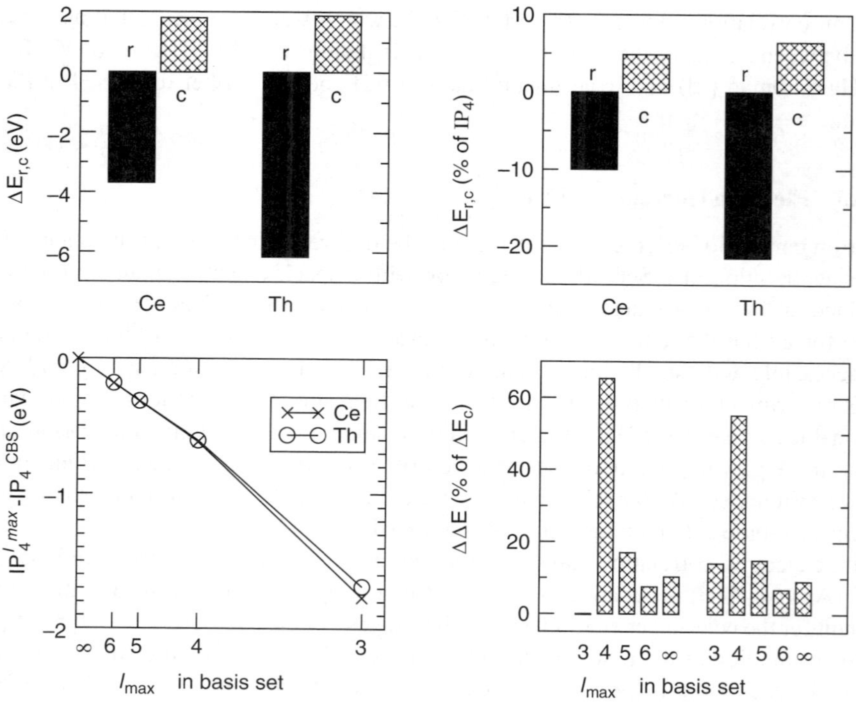

Figure 16.7 *a) Relativistic (r) and electron correlation (c) contributions to IP_4 of Ce and Th (eV). b) as a) but in % of IP_4. c) deviations from the complete basis set limit for IP_4 when basis functions with angular quantum number up to l_{max} are included in the basis set at the DKH2 CCSD(T) level. d) correlation contributions to IP_4 of Ce and Th for basis functions with angular quantum numbers $l_{max} \leq 3$, $l_{max} = 4, 5$, and 6, as well as $l_{max} > 6$. All results were obtained with MOLPRO [24]*

Relativity can nowadays quite easily and accurately be accounted for by a proper choice of the Hamiltonian, e.g., the DKH2 Hamiltonian in the current example, whereas it turns out to be by far more difficult to include properly electron correlation. It is seen that at the CCSD(T) level the correlation corrections behave linearly when plotted vs. $1/l_{max}^3$, where l_{max} denotes the highest angular quantum number included in a radially nearly saturated basis set. The fact that higher angular momenta than $l_{max} = 6$ (i functions) still contribute with about 10% to the correlation corrections, corresponding to about 0.2 eV, leaves only little hope to establish accurately by means of similar quantum chemical calculations energy differences between, e.g., molecular Ce^{III} and Ce^{IV} systems. Contributions of basis functions beyond i functions are thus of similar magnitude as spin-orbit corrections and contributions of the Breit term, as shown in Table 16.4. Since all have the same sign, there is no possibility for an error compensation when neglecting these contributions.

Intermediate Hamiltonian Fock-space coupled cluster (IHFSCC) calculations using the DCB Hamiltonian, a finite nuclear model and a (35s30p25d20f11g9h9i7k7l) basis set obtained for IP_4 of Th a value of 28.65 eV [53], in excellent agreement with the experimental result of 28.65±0.03 eV [52]. A similar result of 28.66 eV was obtained at the IHFSCC level with a MCDHF/DC+B-adjusted pseudopotential and a (16s15p12d10f8g7h7i) basis set [8], however, when using a (14s13p10d8f6g)/[6s6p5d4f3g] standard basis set, which is also suitable for molecular calculations, the IHFSCC result of 28.36 eV already deviates by 0.3 eV from the experimental value. Even worse results of 28.02 and 27.92 eV are obtained with standard basis sets for the more common methods CCSD(T) and MRACPF, both including spin-orbit corrections [8]. It still has to be seen if and how r_{12}-dependent wavefunction approaches can be of any help here. An accurate and balanced description of both relativistic effects and electron correlation effects, if possible at a moderate computational cost allowing also molecular applications, is thus a valid goal of the development of quantum chemical methods suitable for lanthanide and actinide systems.

16.3 Molecular Kondo-type Systems

Cerium is the lightest element in the periodic table that has an occupied 4f orbital in its atomic ground state configuration (Ce $4f^1$ $5d^1$ $6s^2$ $^1G_4$). In the most frequently encountered trivalent oxidation state in the chemistry of cerium, the energetically low-lying 4f shell keeps a single electron (e.g., Ce^{3+} $4f^1$ $^2F_{5/2}$), whereas in the tetravalent state a closed-shell core remains (e.g., Ce^{4+} $^1S_0$). The interplay between the tri- and tetravalent states plays a role, e.g., in the CeO_2/Ce_2O_3 cycle for hydrogen production or in cerimetry used in chemical analysis. As pointed out above, an exact *ab initio* determination of energy differences between Ce^{III} and Ce^{IV} compounds is a substantial challenge for quantum chemical approaches.

In the solid phase Ce keeps a trivalent state characterized by a singly occupied 4f shell on each center, however, when subjected to high pressure a change to a tetravalent state with an empty 4f shell occurs. The so-called Kondo effect well-known in solid state physics denotes an unusual scattering mechanism of conduction electrons in a metal due to local magnetic impurities, i.e., centers with unpaired electrons in atomic-like d or f shells, causing a minimum in the resistivity-temperature curve of dilute magnetic alloys at low temperatures [54]. Figure 16.8 provides a schematic representation of the corresponding wavefunction

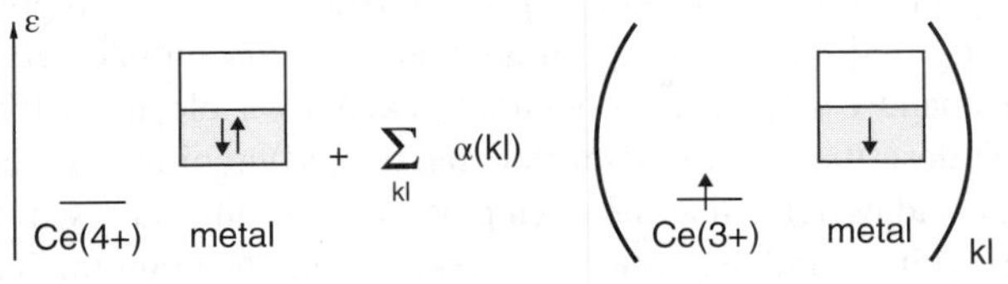

$$\epsilon_{4f} < \epsilon_{\text{conduction band}}$$

$$|\Psi_{S=0}\rangle = A(1 + \sum_{kl} \alpha(kl) f_1^{\dagger} c_k)|\Psi_0\rangle$$

Figure 16.8 *Schematic representation of a Kondo-type wavefunction for a metal with a cerium impurity. The part given after the sum in parentheses stands for a corresponding (antisymmetric) determinant. In case of a molecular analogue the conduction band of the metal has to be replaced by ligand orbitals*

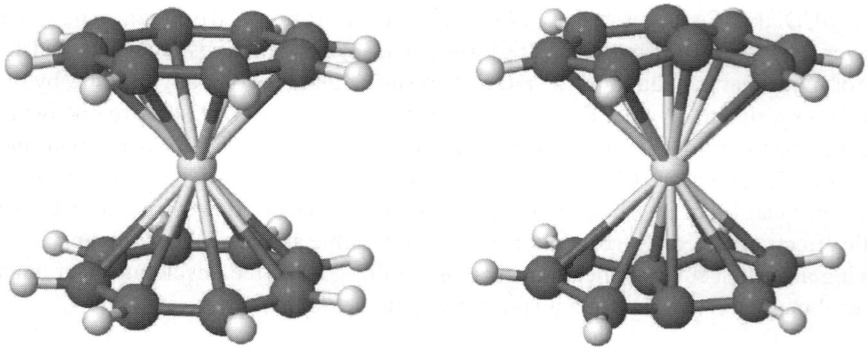

Figure 16.9 *Structures of $Ce(C_8H_8)_2$ and $Ce(C_8H_6)_2$*

for a Ce impurity in a metal, which is a mixture of a configuration with a Ce^{4+} ion and the metal valence electrons in the conduction band, and configurations where a single electron is localized on a Ce^{3+} center and coupled in a singlet fashion to the left unpaired electron in the conduction band. In a molecular analogue to a Kondo system one has to replace the conduction band of the metal by one or more suitable valence orbitals of the ligand(s). In the following section two such systems will be discussed, i.e., bis(η^8-cyclooctatetraenyl)cerium also known as cerocene $Ce(C_8H_8)_2$, and bis(η^8-pentalene)cerium $Ce(C_8H_6)_2$, as shown in Figure 16.9. The former molecule was the first for which a molecular cerium(III)-based Kondo ground state was postulated and later also experimentally confirmed, whereas the latter, due to its lower symmetry (D_{2h} or D_{2d} instead of D_{8h}), can already be understood with a simple two-electron-two-orbital model.

16.3.1 Bis(η^8-cyclooctatetraenyl)cerium

Uranocene $U(C_8H_8)_2$, probably the most famous organouranium compound, was discovered by Streitwieser and Müller-Westerhoff in 1968 [55]. The compound has D_{8h} symmetry in crystalline form and in an ionic limit can be considered to be composed of a U^{4+} ion with a $5f^2$ subconfiguration, leading to an angular quantum number of 3 for the ground state, and two aromatic $C_8H_8^{2-}$ rings containing each 10 π electrons [56, 57]. The compound thus can be described as bis(η^8-cyclooctatetraenyl)uranium(IV). Experimental and quantum chemical studies agree that metal-ring bonding results mainly from the interaction of the uranium 6d orbitals, and to a somewhat lesser extent the uranium 5f orbitals, with the rings π orbitals [58, 59]. The direct contribution of the U 5f orbitals in covalent metal-ring bonding is both due to shell structure effects (e.g., orthogonality constraint of the 5f shell with respect to the core-like 4f shell) and a dominating indirect relativistic destabilization and expansion.

In 1976 cerocene $Ce(C_8H_8)_2$ was synthesized by Greco *et al.* as a lanthanide analogue to uranocene [60]. In their 1983 quasirelativistic self-consistent field-scattered wave study, Rösch and Streitwieser found an unexpected similarity of the electronic structure of cerocene to the one of the heavier homologue thorocene, bis(η^8-cyclooctatetraenyl)thorium(IV), as well as to uranocene, i.e., an unprecedented involvement of the Ce 4f orbitals in metal-ring bonding and a f electron count as high as 1.62 [59]. However, the photoelectron spectra (PES) of cerocene and thorocene both do not exhibit a peak at a lower energy than those arising from the highest ligand π orbitals in e_{2u} (Ce: $\approx$6.75 eV, Th: $\approx$6.75 eV) and e_{2g} (Ce: $\approx$7.68 eV Th: $\approx$7.85 eV) symmetry, whereas uranocene has such a low-energy PES peak ($\approx$6.15 eV) arising from the partially occupied 5f shell below those of the π_{e2u} ($\approx$6.80 eV) and π_{e2g} ($\approx$7.75 eV) ligand orbitals [61]. Cerocene was therefore assigned as bis(η^8-cyclooctatetraenyl)cerium(IV), although Streitwieser and coworkers also noted that already the existence of cerocene should be considered to be remarkable, since in the complex a strong oxidizing agent Ce^{4+} coexists with two strong reducing agents $C_8H_8^{2-}$ [62]. When comparing the orbital energies and the radial spatial extent of the Ce 4f and Th 5f shells, as shown in Figures 16.4 and 16.5, it is also somewhat surprising that the Ce 4f shell should be involved similar to the Th 5f shell in covalent metal-ring bonding in bis(η^8-cyclooctatetraenyl) complexes. It is also noteworthy that Ce keeps a singly occupied 4f orbital in its atomic $4f^1\ 5d^1\ 6s^2\ {}^1G_4$ ground state, whereas for Th the corresponding $5f^1\ 6d^1\ 7s^2\ {}^3H_4$ state lies 7795 cm^{-1} above the $6d^2\ 7s^2\ {}^3F_2$ ground state [3,52]. A similar difference in the f occupation is observed, as discussed above, for the ground states of CeO and ThO.

A quite different view of the electronic structure of the cerocene ground state was presented by Neumann and Fulde in 1989 [63]. The authors proposed that cerocene might be a molecular analogue of a cerium(III)-based Kondo lattice system. In particular it was speculated that a Ce^{3+} ion is complexed by two $C_8H_8^{1.5-}$ rings and that the unpaired electrons in the metal 4f and the ligand π orbitals are coupled to yield an open-shell singlet ground state. Based on semiempirical estimates a very small singlet-triplet splitting of 0.006 eV and a related temperature-dependent magnetic moment was predicted. Large-scale multi-configuration self-consistent field (MCSCF) calculations by Dolg *et al.* using a relativisic pseudopotential for Ce confirmed an open-shell ${}^1A_{1g}$ ground state, with contributions of 82.8% $4f^1_{e2u}\pi^3_{e2u}$ and 17.2% $4f^0_{e2u}\pi^4_{e2u}$, however, the singlet-triplet splitting turned out to be about two orders of magnitude larger than previously estimated [64]. Subsequent MCSCF, multi-reference configuration interaction (MRCI), and multireference averaged coupled-pair functional (MRACPF) calculations of the same group, including also corrections for spin-orbit effects, yielded *ab initio* results for the electronic spectrum, the metal-ring distance and the symmetric metal-ring vibration of cerocene and thorocene [65]. All results were in agreement with leading configurations of $4f^1\ \pi^3_{e2u}$ and $5f^0\ \pi^4_{e2u}$ for the electronic ground states of cerocene and thorocene, respectively. In particular an f electron count of 1.08 was calculated for the cerocene ground state using a Mulliken population analysis, however, it was argued that due to the core-like character of the Ce 4f shell, a corresponding peak in the photoelectron spectrum of cerocene should occur at clearly higher energies than those for the ligand π_{e2u} and π_{e2g} orbitals, whereas due to the strong indirect relativistic destabilization of the U 5f shell the corresponding peak occurs at lower energies (see also Figure 16.4). Thus, the absence of a low-energy f peak in the photoelectron spectrum supports the assignment of thorocene to be a bis(η^8-cyclooctatetraenyl)thorium(IV) complex, but it does not contradict an assignment of cerocene as a bis(η^8-cyclooctatetraenyl)cerium(III) compound [65].

Since orbitals and electron configurations built from them are just objects used for the quantum chemical modelling of the electronic structure, rather than quantum mechanical observables, and both suggestions for the cerocene electronic structure lead to the same $^1A_{1g}$ ground state, it is experimentally difficult, if not impossible, to decide which of the two pictures is more appropriate. Evidence supporting the suggestion of Neumann and Fulde of a bis(η^8-cyclooctatetraenyl)cerium(III) compound was presented in 1996 Edelstein and coworkers by X-ray absorption near-edge structure (XANES) studies, using the Ce L- and K-edges to determine the Ce oxidation state [66]. Two substituted cerocene systems exhibited spectra closer to typical Ce^{III} compounds than to reference Ce^{IV} compounds. In addition, low-temperature absorption and luminiscence spectra as well as magnetic circular dicroism (MCD) measurements lead Amberger and coworkers to conclude that cerocene is more likely a cerium(III) compound [67]. Booth *et al.* presented magnetic measurements and XANES studies, leading to a f electron count of 0.89±0.03 and providing further evidence for a molecular cerium(III)-based Kondo system [68]. Recently Walter *et al.* described new synthetic procedures for the preparation of cerocene together with extensive experimental studies such as magnetic measurements, extended X-ray absorption fine structure (EXAFS) and XANES spectroscopy, confirming a configurational mixture of $4f^1_{e2u}\pi^3_{e2u}$ and $4f^0_{e2u}\pi^4_{e2u}$, where the first configuration is most likely dominating [69].

On the other hand additional evidence was also collected in favor of the originally assumed bis(η^8-cyclooctatetraenyl)cerium(IV) complex. Streitwieser *et al.* found in extensive chemical studies published in 2004 that cerocene behaves remarkably like bis(η^8-cyclooctatetraenyl)actinide(IV) compounds [70]. Kaltsoyannis and coworkers performed all-electron (AE) complete active space second-order perturbation theory (CASPT2) calculations taking scalar-relativistic contributions into account with the second-order Douglas-Kroll-Hess Hamiltonian (DKH2) [71]. They found an f electron count of 0.90±0.04, which only weakly depends on the details of the calculation. A two-state (ground and first excited state) averaged calculation yielded 58.1% of the ground state to have $4f^0$ character, whereas only 23.4% $4f^1$ and 8.7% $4f^2$ character were observed. The authors conclude that their analysis is in agreement with the experimental findings of Streitwieser *et al.* and that cerocene is best described as a Ce^{IV} system. A transfer of electron density from the ligands to the metal leads to an effective oxidation state lower than the formal value of +4, which might in certain cases even be closer to +3. A recent article of Kerridge reports the analysis of the complete active space self-consistent field (CASSCF) wavefunction and arrives at the same conclusion [72]. The authors point out that the concept of a formal oxidation state is not well defined for a multiconfigurational ground state as the one of cerocene. They rather recommend the analysis of the wavefunction in terms of natural orbital occupation numbers. They further argue that a Ce^{III} compound is characterized by a metal-localized singly occupied f orbital, whereas this is absent for a Ce^{IV} system. Despite the significant f density, cerocene is found to fit better into the latter category.

In the most recent work Mooßen and Dolg analyzed the cerocene CASSCF ground state wavefunction with respect to mixings of Ce 4f and ligand π orbitals in the e_{2u} irreducible representation of D_{8h} symmetry as well as of the configurations that can be built from the resulting mixed orbitals to yield a $^1A_{1g}$ ground state [73]. It was concluded that cerocene can be described similarly well in two ways. The molecule may be viewed as a bis(η^8-cyclooctatetraenyl)cerium(IV) complex with a significant metal-ring covalency involving besides the Ce 5d and ligand π orbitals also the Ce 4f orbitals. In this case the fourfold occupied ligand π in the e_{2u} irredicible representation of D_{8h} symmetry have about 25% Ce 4f character, explaining roughly the 4f electron count close to one. Alternatively, it can be thought of as a Kondo-type open-shell singlet bis(η^8-cyclooctatetraenyl)cerium(III) system, characterized by an atomic-like singly occupied Ce 4f shell. In the opinion of the authors the latter description is preferable, since it is based on nearly pure Ce 4f and ring π orbitals and results in a more compact wavefunction with only two significantly contributing configurations, i.e., about 70% $4f^1\pi^3$ and 30% $4f^0\pi^4$. The MCSCF approach applied in the early studies of cerocene [64, 65] was restricted to these two configurations and resulted in almost pure Ce 4f and ligand π orbitals, leading quite straightforwardly to the interpretation as a predominantly Ce(III) system, whereas the natural orbitals of a CASSCF wavefunction show a strong mixing of these orbitals making the analysis more difficult.

16.3.2 Bis(η^8-pentalene)cerium

Bis(η^8-pentalene)cerium is closely related to cerocene, as shown in Figure 16.9. Considering only the ligands π electron systems and the Ce center both compounds are isoelectronic. Its lower symmetry compared to cerocene permits to simplify the wavefunction to fit into an effective two-electron-two-orbital model, which still allows to discuss the role of the Ce 4f orbital(s) in metal-ring bonding. In 2007 two combined experimental and quantum chemical investigations came to different conclusions concerning its electronic ground state, i.e., the formal oxidation state of the central atom in $Ce(C_8H_4(Si\text{-}^iPr_3\text{-}1,4)_2)_2$ was assigned to be Ce^{IV} [74], whereas the one of $Ce(C_8Me_6)_2$ was proposed to be Ce^{III} [75]. From L_{III}-edge XANES measurements, an f electron count of 0.87±0.05 was derived for the latter compound, very similar to the value of 0.89±0.03 obtained for cerocene [68].

Geometry optimizations at the complete active space multiconfiguration self-consistent field (CASSCF) level followed by contracted second-order Rayleigh-Schrödinger perturbation theory (RS2C) using the DKH2 Hamiltonian yielded for the unsubstituted bis(η^8-pentalene)cerium system the staggered (D_{2d}) and the eclipsed (D_{2h}) configuration to be very close in energy, with the former being slightly lower. The mixing of Ce 4f and ligand π orbitals is largest for orbitals that have a shape corresponding to a δ-type symmetry when viewed along the principal axis of rotation. These are located in the irreducible representations a_1 and a_2 of D_{2d} and b_{1u} and a_u of D_{2h}. The leading configurations of the CASSCF wavefunctions of both cases look very similar, when the irreducible representations a_1 and b_{1u} as well as a_2 and a_u of D_{2d} and D_{2h}, respectively, are related to each other. However, D_{2h} can be fully exploited at the correlated level by quantum chemistry codes such as MOLPRO [24], whereas due to the presence of a doubly degenerate irreducible representation, the D_{2d} case has to be treated in one of its subgroups, e.g., in C_{2v}, leading to higher computational costs. Therefore, in the following only the D_{2h} structure will be discussed, for which the most important active orbitals in the CASSCF wavefunction are the 5th and 6th orbital in the a_u irreducible representation, and to a slightly lesser extent the 14th and 15th orbital in the b_{1u} irreducible representation. Using the simplest possible CASSCF wavefunction for the $^1A_{1g}$ ground state with two electrons in two orbitals, i.e., the Ce 4f and ring π orbitals in the a_u irreducible representation, an analysis similar to the one published recently for cerocene [73] can be carried out. For this we consider the orbitals a and b, which stand for every possible orthonormalized pair of orbitals composed of the Ce 4f and ligand π orbitals in the a_u irreducible representation of D_{2h}. The main results are summarized in Figure16.10 on the right side. The natural orbitals (rotation angle $\Phi = 0^o$) are essentially 50%:50% mixtures of the Ce 4f and ring π orbitals. The wavefunction for these orbitals has

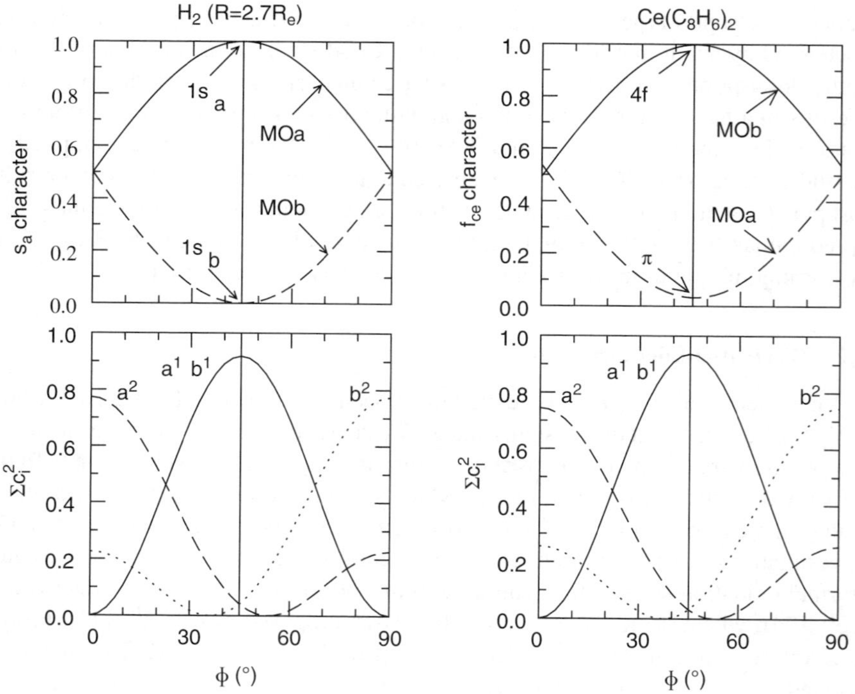

Figure 16.10 Active orbital composition (upper half) and corresponding configuration contributions (lower half) for the simplest possible complete active space wavefunctions (two electrons in two orbitals) describing the H_2 molecule at an intermediate distance of 2 Å (left half) and the $Ce(C_8H_6)_2$ complex at equilibrium geometry (right half). A value of the rotation angle Φ of 0° corresponds to the natural orbitals, whereas one of 45° corresponds to pure H 1s as well as almost pure Ce 4f and ligand π orbitals

only contributions of the two closed shell configurations a^2 (75%) and (with opposite sign of the expansion coefficient) b^2 (25%). Rotation of these orbitals by about $\Phi = 45^o$ leads to orbitals with maximum Ce f and ring π character (denoted by the vertical line in the figure). For these localized orbitals the leading configuration has both orbitals singly occupied, i.e., $4f^1\ \pi^1$ (a^1b^1), with an admixture of the closed shell configurations π^2 (a^2) and $4f^2$ (b^2) of less than 5%. Thus, in contrast to cerocene, the description as an open shell-singlet is here clearly superior to the one of two admixed closed-shell configurations.

It is interesting to compare the $Ce(C_8H_6)_2$ ground state wavefunction with a corresponding one of H_2 in its $^1\Sigma_g^+$ ground state at a stretched bond distance of 2 Å (2.7 R_e), as shown in Figure16.10 on the left side. Here the two electrons are treated in an active space built by the σ_g and σ_u linear combinations of the 1s orbitals on each of the two atoms. The orbitals *a* and *b* now represent any orthonormalized pair of orbitals built from these 1s orbitals. The natural orbitals (rotation angle $\Phi = 0^o$) correspond to the σ_g and σ_u orbitals, delocalized between the two atoms. A rotation by $\Phi = 45^o$ leads to the 1s orbitals localized on the two atoms. In the first case the wavefunction is composed of the closed-shell configurations σ_g^2 (a^2, 77%) and (again with opposite sign of the expansion coefficient) σ_u^2 (b^2, 23%), whereas in the second case it is dominated by the covalent configuration with both orbitals singly occupied $1s_a^1\ 1s_b^1$ (a^1b^1), and only small admixtures of the ionic configurations $1s_a^2$ (a^2) and $1s_b^2$ (b^2). It is well known that near the equilibrium distance the dominating configuration is σ_g^2 with only small contributions of σ_u^2, whereas for the separated atoms at infinite distance both configurations contribute equally, but with opposite sign of the coefficients, in order to quench the ionic terms in the wavefunction leaving only $1s_a^1\ 1s_b^1$. In this case the figure for the configuration contributions becomes symmetric to the line at $\Phi = 45^o$ and the dashed and dotted lines for the contributions of the a^2 and b^2 configurations coincide. The first wavefunction describes a typical covalent bond, whereas the second results for the separated neutral atoms. This comparison leads to the conclusion that the interaction of the Ce 4f orbital and the ring π orbital in $Ce(C_8H_6)_2$ is similar to a significantly stretched (and thus quite weak) covalent bond.

When using localized orbitals covalent interactions can be analyzed by orbital occupation number fluctuations and the local spin, calculated with respect to one or more orbitals localized on a specific atom [77, 78]. The orbital occupation number fluctuations are defined as $\sqrt{\langle \hat{N}_a^{\,2} - \langle \hat{N}_a \rangle^2 \rangle}$, where $\hat{N}_a$ is the occupation number operator and a denotes a localized orbital or a group of localized orbitals associated to the same atom. For an ideal covalent single bond the orbital occupation number fluctuation is $\sqrt{2}/2$, whereas for separated atoms it is 0. Values near zero are thus obtained for spatially well separated and non-interacting atoms but also for van der Waals molecules like He_2, which virtually do not have covalent bonding contributions. The local spin defined as $\langle \hat{S}_a^{\,2} \rangle$ is 0.375 for an ideal covalent bond, 0.75 for a dissociated single covalent bond, and 0 for van der Waals interactions. Occupation number fluctuations and local spin thus can be used to detect covalent bonding contributions in interatomic interactions and also possibly to estimate their strength.

Figure 16.11 compares the orbital occupation number fluctuations and local spin of covalently bonded system such as H_2 and Au_2, a van der Waals bonded system such as He_2 and formally van der Waals bonded systems that exhibit significant covalent bonding contributions such as Hg_2 and Be_2 to the values obtained for the 4f-π interaction in $Ce(C_8H_6)_2$. It is obvious also from these results that the latter case corresponds to a stretched covalent bond, i.e., values similar to H_2 at a bond distance of 2 Å are observed. The bonding is entirely different from covalent contributions arising in the interaction between closed-shell bonding partners in unpolar systems such as Hg_2 or Be_2, as well as polar/ionic systems such as LiF.

The results of the simple analysis presented here still hold when Ce 4f and ligand π orbitals in other irreducible representations are included in the active space, however, the situation becomes more complex, since now more than one orbital rotation angle has to be considered to obtain the wavefunction build from possibly pure Ce 4f and ligand π orbitals [76]. It is also noteworthy to point out that neglected contributions tend to favor a $4f^1$ occupation of cerium, e.g., spin-orbit coupling (which is largely quenched in the singlet ground state) and dynamic correlation effects, as shown in Table 16.4.

Summing up, similar to cerocene the most compact description of the ground state electronic structure of $Ce(C_8H_6)_2$ is by the leading $4f^1\ \pi^1$ configuration, built from nearly pure Ce 4f and ring π orbitals. Thus, the assignment to bis(η^8-pentalene)cerium(III) seems to be the most appropriate.

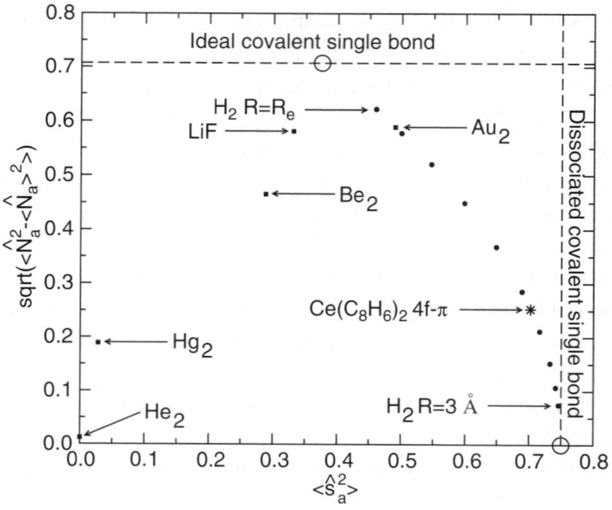

Figure 16.11 Occupation number fluctuations $\sqrt{\langle \hat{N}_a^2 \rangle - \langle \hat{N}_a \rangle^2}$ and local spin $\langle \hat{S}_a^2 \rangle$ on one hydrogen atom of H_2 depending on the internuclear distance R from CASSCF calculations with two electrons in an active space of two orbitals (filled dots at R_e=0.74Å, and $R_i = (1 + i \times 0.25)$Å). The limiting values for an ideal covalent single bond as well as a dissociated covalent single bond are indicated by empty circles. The result for the polar covalently bonded LiF at the equilibrium distance was derived for comparison also for the two electrons in two orbitals active space (filled square). The 4f-π interaction in $Ce(C_8H_6)_2$ at the equilibrium geometry (star) evaluated also at the two electrons in two orbitals active space CASSCF level corresponds to a stretched covalent bond. Corresponding CASSCF results using an active space built from the s and p valence orbitals are given for comparison for the van der Waals bonded He_2, the covalently bonded Au_2, and the intermediate cases Hg_2 and Be_2 exhibiting significant covalent contributions at their equilibrium distances (filled squares) [78]

16.4 Conclusions

Shell structure, relativistic and electron correlation effects play an important role for the electronic structure of lanthanide and actinide systems. The importance and the magnitude of these effects have been examplified for the atoms Ce and Th, and the contributions of the Ce 4f and Th 5f shell to chemical bonding have been reviewed for the monoxides CeO and ThO, respectively. Currently quantitatively correct results for lanthanides and actinides can only be obtained from *ab initio* calculations for the easiest cases, e.g., small systems (atoms, diatomics containing one f element) with a possibly small number of unpaired f electrons and/or problems related only to configurations with the same f occupation number. In other cases *ab initio* quantum chemistry can at least help to interprete experimental findings. As an example organometallic cerium sandwich complexes such as cerocene were discussed, which may be considered to be molecular analogues of cerium(III)-based Kondo lattice systems.

References

[1] P. Pyykkö, Recent developments in the theory of f-element molecules, *Inorg. Chim. Acta*, **139**, 243–245 (1987).

[2] K. G. Dyall, I. P. Grant, C. T. Johnson, F. A. Parpia, E. P. Plummer, GRASP—a general-purpose relativistic atomic-structure program. *Comput. Phys. Commun.*, **55**, 425–456 (1989).

[3] http://physics.nist.gov/PhysRefData/ASD/levels_form.html (accessed February 19, 2014).

[4] M. Dolg, H. Stoll, H. Preuss, Homonuclear diatomic lanthanoid compounds: a pseudopotential configuration interaction and correlation energy density functional study, *J. Molec. Struct. (THEOCHEM)*, **277**, 239–249 (1992).

[5] M. Dolg, W. Liu, S. Kalvoda, Performance of relativistic density functional and ab initio pseudopotential approaches for systems with high-spin multiplicities: Gadolinium diatomics GdX (X = H, N, O, F, P, S, Cl, Gd), *Int. J. Quant. Chem.*, **76**, 359–370 (2000).

[6] X. Cao, M. Dolg, Electronic structure of lanthanide dimers, *Mol. Phys.*, **101**, 1967–1976 (2003).

[7] R. J. van Zee, S. Li, W. Weltner, Gd_2 — the highest spin diatomic molecule?, *J. Chem. Phys.*, **100**, 4010–4012 (1994).

[8] A. Weigand, X. Cao. T. Hangele, M. Dolg, Relativistic small-core pseudopotentials for actinium, thorium and protactinium, *J. Phys. Chem. A*, **118**, 2519–2530 (2014).

[9] P. Pyykkö, J. P. Desclaux, Relativity and the periodic system of elements, *Acc. Chem. Res.*, **12**, 276–281 (1979).

[10] K. S. Pitzer, Relativistic effects in chemical properties, *Acc. Chem. Res.*, **12**, 272–276 (1979).

[11] P. Pyykkö, Relativistic quantum chemistry, *Adv. Quant. Chem.*, **11**, 353–409 (1978).

[12] P. Pyykkö, Relativistic effects in structural chemistry, *Chem. Rev.*, **88**, 563–594 (1988).

[13] W. H. E. Schwarz, An introduction to relativistic quantum chemistry, in M. Barysz, Y. Ishikawa (Eds), *Relativistic Methods for Chemists. Challenges and Advances in Computational Physics*, Springer, London, vol. 10, pp. 1–63 (2010).

[14] M. Pepper, B. Bursten, The electronic structure of actinide-containing molecules — a challenge to applied quantum chemistry, *Chem. Rev.*, **91**, 719–741 (1991).

[15] K. Balasubramanian, Relativistic effects and electronic structure of lanthanide and actinide molecules, in K. A. Gschneidner Jr., L. Eyring, G. R. Choppin, G. H. Lander (Eds), *Handbook on the physics and chemistry of rare earths*, vol. 18, Elsevier, Amsterdam, pp. 29–158 (1993).

[16] M. Dolg, H. Stoll, Electronic structure calculations for molecules containing lanthanide atoms, in K. A. Gschneidner Jr. and L. Eyring (Eds), *Handbook on the physics and chemistry of rare earths*, vol. 22, Elsevier, Amsterdam, pp. 607–729 (1996).

[17] N. Kaltsoyannis, Recent developments in computational actinide chemistry, *Chem. Soc. Rev.*, **32**, 9–16, (2003).

[18] M. Dolg, X. Cao, The relativistic energy-consistent ab initio pseudopotential approach and its application to lanthanide and actinide compounds, in K. Hirao, Y. Ishikawa (Eds), *Recent Advances in Computational Chemistry*, World Scientific, New Jersey, pp. 1–35 (2004).

[19] X. Cao and M. Dolg, Relativistic energy-consistent ab initio pseudopotentials as tools for quantum chemical investigations of actinide systems, *Coord. Chem. Rev.*, **250**, 900–912, (2006).

[20] M. Dolg, X. Cao, Computational methods: lanthanides and actinides, in E. I. Solomon, R. A. Scott, and R. B. King (Eds), *Computational inorganic and bioinorganic chemistry*, Wiley, Chichester, pp. 503–515 (2009).

[21] G. Schreckenbach, G. A. Shamov, Theoretical actinide molecular science, *Acc. Chem. Res.*, **43**, 19–29 (2010).

[22] L. Visscher, K. G. Dyall, Relativistic and electron correlation effects on molecular properties. I. The dihalogens F_2, Cl_2, Br_2, I_2, and At_2, *J. Chem. Phys.*, **104**, 9040–9046 (1996).

[23] M. Dolg, X. Cao, Relativistic pseudopotentials: their development and scope of applications, *Chem. Rev.*, **112**, 403–480 (2012).

[24] H.-J. Werner, P. J. Knowles, G. Knizia, F. R. Manby, M. Schütz, MOLPRO: a general-purpose quantum chemistry program package, *WIREs Comput. Mol. Sci.*, **2**, 242–253 (2012).

[25] P. S. Bagus, Y. S. Lee, K. S. Pitzer, Effects of relativity and lanthanide contraction on atoms from hafnium to bismuth, *Chem. Phys. Lett.*, **33**, 408–411 (1975).

[26] P. Pyykkö, Interpretation of the secondary periodicity in the periodic system, *J. Chem. Res.*, **11**, 380–381 (1979).
[27] M. Seth, M. Dolg, P. Fulde, P. Schwerdtfeger, Lanthanide and actinide contractions: relativistic and shell structure effects, *J. Am. Chem. Soc.*, **117**, 6597–6598 (1995).
[28] P. Pyykkö, Dirac-Fock one-centre calculations, part 8. The $^1\Sigma$ states of ScH, YH, LaH, AcH, TmH, LuH and LrH, *Phys. Scr.*, **20**, 647–651 (1979).
[29] R. W. Field, Diatomic molecule electronic structure beyond simple molecular constants, *Ber. Bunsenges. Phys. Chem.*, **86**, 771–779 (1982).
[30] M. Dolg, H. Stoll, A. Savin, H. Preuß, Energy-adjusted pseudopotentials for the rare earth elements, *Theor. Chim. Acta*, **75**, 173–194 (1989).
[31] M. Dolg, H. Stoll, H. Preuß, A combination of quasirelativistic pseudopotential and ligand field calculations for lanthanoid compounds, *Theor. Chim. Acta*, **85**, 441–450 (1993).
[32] M. Dolg, H. Stoll, H. Preuss, The low-lying electronic states of cerium monoxide CeO: ab initio calculations using energy-adjusted pseudopotentials and spin-orbit operators, *J. Mol. Struct. (Theochem)*, **231**, 243–255 (1991).
[33] M. Dolg, H. Stoll, H. Preuss, A combination of quasirelativistic pseudopotential and ligand field calculations for lanthanoid compounds, *Theor. Chim. Acta*, **85**, 441–450 (1993).
[34] C. Linton, M. Dulick, R. W. Field, P. Carette, P. C. Leyland, R. F. Barrow, Electronic states of the CeO molecule: absorption, emission and laser spectroscopy, *J. Mol. Spectrosc.*, **102**, 441–497 (1983).
[35] L. A. Kaledin, J. E. McCord, M. C. Heaven, Laser spectroscopy of CeO: characterization and assignment of states in the 0-3 eV range, *J. Mol. Spectrosc.*, **158**, 40–61 (1993).
[36] M. Dolg, H. Stoll, H. Preuß, Energy-adjusted ab initio pseudopotentials for the rare earth elements, *J. Chem. Phys.*, **90**, 1730–1734 (1989).
[37] W. Küchle, M. Dolg, H. Stoll, H. Preuss, Energy-adjusted pseudopotentials for the actinides. Parameter sets and test calculations for thorium and thorium monoxide, *J. Chem. Phys.*, **100**, 7535–7542 (1994).
[38] P. Pyykkö, The physics behind chemistry and the periodic table, *Chem. Rev.*, **112**, 371–384 (2011).
[39] B. A. Hess (Ed.), *Relativistic effects in heavy-element chemistry and physics. Wiley series in theoretical chemistry*; Wiley, Chichester, vol. 12 (2002).
[40] P. Schwerdtfeger (Ed.), *Relativistic electronic structure theory. Part 1, Fundamentals. Theoretical and computational chemistry*; Elsevier, Amsterdam, vol. 11 (2002).
[41] P. Schwerdtfeger (Ed.), *Relativistic electronic structure theory. Part 2, Applications. Theoretical and computational chemistry*, Elsevier, Amsterdam, vol. 14 (2004).
[42] K. Hirao, Y. Ishikawa (Eds), *Recent advances in relativistic molecular theory. Recent advances in computational chemistry*, World Scientific, London, vol. 6 (2004).
[43] M. Barysz, Y. Ishikawa (Eds), *Relativistic methods for chemists. Challenges and advances in computational physics*, Springer, London, vol. 10, (2010).
[44] K. G. Dyall, K. Faegri Jr., *Introduction to relativistic quantum chemistry*, Oxford University Press, Oxford (2007).
[45] I. P. Grant, *Relativistic quantum theory of atoms and molecules. Theory and computations*, Springer, New York (2007).
[46] M. Reiher, A. Wolf, *Relativistic quantum chemistry: the fundamental theory of molecular science*, Wiley-VCH, Weinheim (2009).
[47] Pyykkö, P. In *Lecture Notes in Chemistry*, vol. 41, Spinger, Berlin, (1986).
[48] Pyykkö, P. In *Lecture Notes in Chemistry*, vol. 60, Spinger, Berlin, (1993).
[49] Pyykkö, P. In *Lecture Notes in Chemistry*, vol. 76, Spinger, Berlin, (2000).
[50] P. Pyykkö, *Relativistic Quantum theory of atoms and molecules database*, http://rtam.csc.fi/, (accessed February 19, 2014).
[51] P. Pyykkö, The RTAM electronic bibliography, Version 17.0, on relativistic theory of atoms and molecules, *J. Comp. Chem.*, **34**, 2667 (2013).
[52] http://web2.lac.u-psud.fr/lac/Database/Contents.html (accessed February 19, 2014).

[53] E. Eliav, U. Kaldor, Transition energies of Rn- and Fr-like actinide ions by relativistic intermediate Hamiltonian Fock-space coupled-cluster methods, *Chem. Phys.*, **392**, 78–82 (2012).
[54] J. Kondo, Resistance minimum in dilute magnetic alloys, *Prog. Theor. Phys.*, **32**, 37–39 (1964).
[55] A. Streitwieser, U. Müller-Westerhoff, Bis(cyclooctatetraenyl)uranium (uranocene). A new class of sandwich complexes that utilize atomic f orbitals, *J. Am. Chem. Soc.*, **90**, 7364 (1968).
[56] A. H. H. Chang, R. M. Pitzer, Electronic structure and spectra of uranocene, *J. Am. Chem. Soc.*, **111**, 2500–2507 (1989).
[57] W. Liu, M. Dolg, P. Fulde, Low-lying electronic states of lanthanocenes and actinocenes $M(C_8H_8)_2$ (M = Nd, Tb, Yb, U), *J. Chem. Phys.*, **107**, 3584–3591 (1997).
[58] J. P. Clark, J. C. Green, An investigation of the electronic structure of bis(eta-cyclo-octatetraene)-actinoids by helium-(I) and -(II) photoelectron spectroscopy, *J. Chem. Soc. Dalton Trans.*, 505–508 (1977).
[59] N. Rösch, A. Streitwieser, Quasirelativistic SCF-X_α scattered-wave study of uranocene, thorocene, and cerocene, *J. Am. Chem. Soc.*, **105**, 7237–7240 (1983).
[60] A. Greco, S. Cesca, G. Bertolini, New π-cyclooctatetraeny and π-cyclopentadienyl complexes of cerium, *J. Org. Chem.*, **113**, 321–330 (1976).
[61] I. Fragala, G. Condorelli, P. Zanella, E. Tondello, Photoelectron spectroscopy of actinide organometallic compounds I. Bis(cyclooctatetraene)actinide(IV) complexes, *J. Organomet. Chem.*, **122**, 357–363 (1976).
[62] A. Streitwieser, S. A. Kinsley, J. T. Rigsbee, I. L. Fragala, E. Ciliberto, Photoelectron spectra and bonding in cerocene, bis-(π-[8]annulene) cerium(IV), *J. Am. Chem. Soc.*, **107**, 7786–7788 (1985).
[63] C.-S. Neumann, P. Fulde, Is there a molecular analogue of a Kondo singlet state? *Z. Phys. B: Cond. Mat.*, **74**, 277–278 (1989).
[64] M. Dolg, P. Fulde, W. Küchle, C.-S. Neumann, H. Stoll, Ground state calculations of di-π-cyclooctatetraene cerium, *J. Chem. Phys.*, **94**, 3011–3017 (1991).
[65] M. Dolg, P. Fulde, H. Stoll, H. Preuss, A. Chang, R. M. Pitzer, Formally tetravalent cerium and thorium compounds: A configuration interaction study of cerocene $Ce(C_8H_8)_2$ and thorocene $Th(C_8H_8)_2$ using energy-adjusted quasirelativistic ab initio pseudopotentials, *Chem. Phys.*, **195**, 71–82 (1995).
[66] N. Edelstein, P. G. Allen, J. J. Bucher, D. K. Shuh, C. D. Sofield, N. Kaltsoyannis, G. H. Maunder, M. R. Russo, A. Sella, The oxidation state of Ce in the sandwich molecule cerocene, *J. Am. Chem. Soc.*, **118**, 13115–13116 (1996).
[67] H.-D. Amberger, H. Reddmann, F. T. Edelmann, Welche Oxidationszahl hat Cer im tiefvioletten 1,1',4,4'-Tetrakis(trimethylsilyl)cerocen? *J. Org. Chem.*, **690**, 2238–2242 (2005).
[68] C. H. Booth, M. D. Walter, M. Daniel, W. W. Lukens, R. A. Andersen, Self-Contained Kondo effect in single molecules, *Phys. Rev. Lett.*, **95**, 267202, (2006).
[69] M. D. Walter, C. H. Booth, W. W. Lukens, R. A. Anderson, Cerocene revisited: The electronic structure of and interconversion between $Ce_2(C_8H_8)_3$ and $Ce(C_8H_8)_2$, *Organomet.*, **28**, 698–707 (2009).
[70] A. Streitwieser, S. A. Kinsley, C. H. Jenson, J. T. Rigsbee, Synthesis and properties of di-π-[8]annulenecerium(IV), cerocene, *Organomet.*, **23**, 5169–5175 (2004).
[71] A. Kerridge, R. Coates, N. Kaltsoyannis, Is cerocene really a Ce(III) compound? All-electron spin-orbit coupled CASPT2 calculations on $M(\eta^8\text{-}C_8H_8)_2$ (M = Th, Pa, Ce), *J. Phys. Chem. A*, **113**, 2896–2905 (2009).
[72] A. Kerridge, Oxidation state and covalency in f-element metallocenes (M = Ce, Th, Pu): a combined CASSCF and topological study, *Dalton Trans.*, **42**, 16428–16436 (2013).
[73] O. Mooßen, M. Dolg, Two interpretations of the cerocene electronic ground state. *Chem. Phys. Lett.*, **594**, 47–50 (2014).
[74] G. Balazs, G. O. Cloke, J. C. Green, R. M. Harker, A. Harrison, P. B. Hitchcock, C. N. Jardine, R. Walton, Cerium(III) and cerium(IV) bis(η^8-pentalene) sandwich complexes: synthetic, structural, spectroscopic, and theoretical studies, *Organomet.*, **26**, 3111–3119 (2007).

[75] A. Ashley, G. Balazs, A. Cowley, J. Green, C. H. Booth, D. O'Hare, Bis(permethylpentalene)cerium — another ambiguity in lanthanide oxidation state, *Chem. Commun.*, 1515–1517 (2007).
[76] M. Dolg, unpublished results.
[77] M. Mödl, M. Dolg, P. Fulde, H. Stoll, Analysis of large-scale multi-configuration self-consistent field wavefunctions by expectation values of local operators, *J. Chem. Phys.*, **105**, 2353–2363 (1996).
[78] M. Yu, M. Dolg, Covalent contributions to bonding in group 12 dimers M_2 (M = Zn, Cd, Hg), *Chem. Phys. Lett.*, **273**, 329–336 (1997).

Index

Note: Page numbers in *italics* refer to Figures; those in **bold** to Tables

Computational Methods in Lanthanide and Actinide Chemistry, First Edition. Edited by Michael Dolg.